DATE DUE

DEMCO 38-296

THE
ELECTRICAL
NATURE
of STORMS

THE
ELECTRICAL
NATURE
of STORMS

DONALD R. MacGORMAN
W. DAVID RUST

New York Oxford • Oxford University Press 1998

Oxford University Press

Oxford   New York
Athens   Auckland   Bangkok   Bogota   Bombay   Buenos Aires
Calcutta   Cape Town   Dar es Salaam   Delhi   Florence   Hong Kong
Istanbul   Karachi   Kuala Lumpur   Madras   Madrid   Melbourne
Mexico City   Nairobi   Paris   Singapore   Taipei   Tokyo   Toronto   Warsaw

and associated companies in
Berlin   Ibadan

Library of Congress Cataloging-in-Publication Data
MacGorman, D. R.
The electrical nature of storms / Donald R. MacGorman,
W. David Rust.
p. cm.
Includes bibliographical references and index.
ISBN 0-19-507337-1
1. Atmospheric electricity.   2. Storms.   I. Rust, W. D.
II. Title.
QC961.M18   1997
551.5'6—dc20   96-27498

1 3 5 7 9 8 6 4 2

Printed in the United States of America
on acid-free paper.

Preface

Our motivation for writing this book has grown out of two facets of our experience: our research at the National Severe Storms Laboratory (NSSL), one of the Environmental Research Laboratories of the National Oceanic and Atmospheric Administration (NOAA), and our teaching as adjunct faculty in both the School of Meteorology and the Department of Physics and Astronomy at the University of Oklahoma. Class notes for a graduate course we have taught several times over the last decade formed the core of several chapters of this book. We appreciate the support the NSSL and its staff have provided for our research, the opportunities the University has given us to teach, and the curiosity our many students have brought to what has been for us a stimulating exploration of the electrical nature of storms.

Our goal has been to produce a modern, graduate level textbook on the electrical processes of storms. The level of our treatment is designed for those with a basic understanding of meteorology or physics and provides the tools needed to explore this cross-disciplinary topic. Where background material is needed to understand the primary topics of this book, we present basic concepts from the broader context of atmospheric electricity, but these are not treated in as much detail as electrical processes of storms. In each subject area, we include enough references to the scientific literature to make this book useful to professionals, as well as to students.

Some topics in storm electricity progressed rapidly during the 5½ years we were writing this book. During the last months of writing, we added brief references to newly available work in many areas. In some sections, we try to give a summary of "where we are" in overall understanding. Unfortunately, many areas of storm electricity lack sufficient observations, theory, and models for us to attempt such overviews without our being presumptuous or scientifically unsound.

We received help from many and are very grateful for it. There are several we wish to mention specifically: Dr. Robert A. Maddox, who as Director of the NSSL made doing the book possible; Joan O'Bannon who accomplished the major task of drafting the figures for this book, including those previously published, and who strove for uniformity in format, clarity, and ease in interpretation; and those who reviewed significant portions or entire chapters of this book (listed alphabetically)—Dr. John H. Helsdon, Dr. John Latham, Dr. Thomas C. Marshall, Dr. Bradley F. Smull, Dr. Maribeth Stolzenburg, and Dr. Conrad L. Ziegler. Many others offered helpful comments, information, and encouragement. We thank the NOAA/ University of Oklahoma Cooperative Institute for Mesoscale Meteorological Studies for providing facilities for work on this book. To all our readers we state emphatically that the book is much improved for all the comments we received and that its shortcomings are entirely our responsibility.

SYMBOLS AND CONVENTIONS

We have tried to make our symbols agree with common usage. There is a list of selected symbols in Appendix A. Our notation convention in equations is the following: (1) A variable is italicized, as in E; (2) A vector is written as $\vec{E}$; and (3) A function is written as $E(t)$. When we specifically mean the vector, we will write it as in (2) whether it is in a numbered equation or in the text. When an important word or phrase is defined the first time, it is italicized. Definitions of how terms are used, polarity conventions, and so on are given at the relevant point in the text. When a symbol can replace an often used and obvious word or phrase in the text, we use the symbol liberally. References are listed by chapter at the end of the book.

TUTORIAL

The first section of the book is a brief tutorial on basic electricity and magnetism. It is included for those not routinely using and thus conversant with the basics of these subjects. The material can be omitted or included as best fits the needs of a particular class or individual.

COMMENTS, ERRORS, AND SO FORTH

We welcome your comments, suggestions, and reports of any errors, which we will collect for a possible second edition. If you feel we have presented your work incorrectly or omitted relevant citations, we would appreciate hearing from you to consider changes if there is a second edition. Please send us your comments via any method convenient for you. Our mailing address is National Severe Storms Laboratory, 1313 Halley Circle, Norman, Oklahoma 73069, U.S.A. Our current Internet addresses are available on the WorldWide Web under the homepage for the National Severe Storms Laboratory and under the Atmospheric Electricity Homepage.

Norman, Oklahoma   D.R.M.
November 1996   W.D.R.

Contents

THE
ELECTRICAL
NATURE
of STORMS

Tutorial

Basic Electricity and Magnetism

T.1. INTRODUCTION

Atmospheric electricity is a multidisciplinary topic, and one of its fundamental disciplines is electromagnetics. Since meteorology curricula sometimes do not emphasize a working familiarity with electromagnetics, we include this optional section of selected electromagnetic theory. Our goal is to provide the fundamental framework of electromagnetic theory needed by students to understand the material in our book. What we present is intended only as a tutorial covering the essentials needed by this text, not a detailed treatment of electromagnetics. This section can be omitted easily from the course if students already have working knowledge of the material. In this chapter and throughout this text, we assume that students are familiar with the basics of differential and integral calculus of vector fields.

Most of what we cover in this tutorial is *electrostatics*, defined as the study of interactions between stationary charges. Many of the electrical properties of the atmosphere can be treated by electrostatics. Moving particles often can be treated electrostatically when their relative positions and not their motions cause the interactions. Even lightning, which is obviously not static, can be elucidated in part by using electrostatics. Electrostatics rules out interactions, such as magnetic and thermal interactions, that are due to the motion of charge and particles. These other interactions can be treated as additional effects added to electrostatic effects.

We present nothing original in this tutorial; it can all be found in numerous textbooks and study aids.

Because the information is available so widely, specific references are not provided. We appreciate and acknowledge that some material in this tutorial is based on *Electrostatics and Its Applications* (A.D. Moore ed., John Wiley and Sons, 1973), *The Feynman Lectures of Physics, Vol. I and II* (R.P. Feynman, R.B. Leighton, and M. Sands, Addison-Wesley Publishing, 1977), and *Introduction to Electrodynamics* (D.J. Griffiths, Prentice Hall, 1989).

We use the rationalized *MKS* system of units throughout this textbook, so the fundamental units are meter, kilogram, and second. The unit of electrical charge is the coulomb. Other units are derived from these, and in this tutorial we spell out and give the symbols for the units of many physical properties as we introduce them. In the text, we sometimes use units, such as the gram, from another system of units, to remain consistent with common usage.

T.2. CHARGE, ELEMENTARY PARTICLES, AND IONS

Electric *charge* is a fundamental property of "elementary" particles, which in this book we consider to be electrons, protons, and neutrons. The observational evidence for charge may predate the historical record, but it is linked with the change in matter such as occurs when a rubber rod is rubbed with wool. The rod can then influence some types of material around it (e.g., a small bit of paper). The term charged is used to describe the rod's condition. Our understanding of physics tells us that electrons were transferred from the wool to the rod. By virtue of there being two kinds of charge, this makes the rod negatively charged and the wool positively charged. In many processes, the electron is the only important particle, because it is small and mobile and because some electrons are relatively loosely tied to atoms and molecules. The mass of an electron is 9.1 $\times$ 10^{-31} kilogram (kg), and its electric charge, q_e = -1.6×10^{-19} coulomb (C). Here the electron is considered to be the smallest elementary particle, and the only one that has a negative charge. The proton is another elementary particle of which materials are composed. At 1.7×10^{-27} kg, it is much more massive than the electron. The proton has a charge of $+1.6 \times 10^{-19}$ C. Neutrons are the third elementary particle. The neutron is essentially identical to the proton in mass, but it is electrically neutral.

A critical property of charge shown consistently by observations is that we can neither create nor destroy it. So if we observe charge, it had to come from somewhere; likewise, if that charge disappears, it went somewhere, but was not destroyed. This property leads to a basic law of physics: *conservation of charge.* The terms "charge generation" and "charge separation" are often used interchangeably in cloud electrification. What we will mean in this text by these terms is that the preexisting charge will move between particles or will be systematically redistributed in a way that forms regions of net charge in the atmosphere or the cloud. Although particles or regions gain charge, the whole system does not, so conservation of the total charge has not been violated.

A molecule is a combination of two or more atoms that are bound together by strong interatomic forces. For example, a water molecule is made up of two atoms of hydrogen and one of oxygen. If we remove one or more electrons from any single neutral molecule or atom, we produce a positive ion. If we add one or more extra electrons, we produce a negative ion. The term *ion* comes to us from electrochemistry, where it denotes the carriers of *electric current* (moving charged particles) in conductive solutions. We make a distinction between ions and larger charged particles (e.g., charged cloud particles). An *ion* is defined as a charged particle, either positive or negative, consisting of an atom, a molecule, or a cluster of molecules (molecules that are bound together loosely). Small, fast atmospheric ions play an important role in atmospheric and storm electricity.

T.3. COULOMB'S LAW AND ELECTRIC FIELD

Coulomb's law is based on experimental observations; it is a mathematical statement of an observed relationship. Stated in words, Coulomb's law is

The force exerted on one point charge by another is proportional to the magnitude of each of the charges, in-

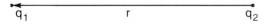

Fig. T.1. Two point charges, q_1 and q_2, separated by a distance r. The arrow denotes the direction of the force on charge q_1 due to charge q_2.

versely proportional to the square of the distance between the charges, and has a direction along the line joining the two charges. It is an attractive force if the charges are opposite in sign and repulsive if the charges have the same sign.

Mathematically, *Coulomb's law* is written

$$\vec{F}_{21} = \frac{1}{4\pi\varepsilon} \frac{q_1 q_2}{r^2} \hat{a}_{r_{21}}, \qquad (T.1)$$

where $\vec{F}_{21}$ is the force on charge 1 due to charge 2, q_1 and q_2 are the two charges (the polarity and magnitude of each are indicated as a signed number), r is the distance between the charges (Fig. T.1), and $\hat{a}_{r}$ is a unit vector pointing along r from q_2 to q_1. The unit of force is the newton (N). The term $1/4\pi\varepsilon$ is the constant of proportionality for the rationalized MKS system of units, and for free space is $10^{-7} c^2$, where c is the speed of light.

If we divide both sides of Eq. T.1 by q_1, we get a ratio, the force exerted by q_2 per unit charge experiencing the force at a particular point, i.e.,

$$\frac{\vec{F}_{21}}{q_1} = \frac{1}{4\pi\varepsilon} \frac{q_2}{r^2} \hat{a}_{r_{21}}. \qquad (T.2)$$

This ratio is called the *electric field,* symbolized by $\vec{E}$. It has been found to be a useful abstract concept, because it can be used to determine the force that any charge (often called a test charge) would experience when placed at a particular location relative to q_2. Because only one charge is involved on the right side of Eq. T.2, the subscripts can be dropped. Thus, $\vec{E}$ at a radius r from a point charge q is

$$\vec{E} = \frac{1}{4\pi\varepsilon} \frac{q}{r^2} \hat{a}_r, \qquad (T.3)$$

where $\hat{a}_r$ is the unit vector along a radial from q to the point being considered. The unit of E is newton per coulomb, which is equivalent to a volt (V) per meter (V m^{-1}). The unit volt is tied to the concept of potential described in Section T.4. We define the *polarity* of the electric field by the direction in which a positive test charge moves under the force $\vec{F}$. Thus, $\vec{E}$ produced by a positive point charge q is directed outward along radials from the charge (Fig. T.2), and that produced by a negative point charge is directed inward. Regardless of the polarity or magnitude of a point charge, the electric field that it produces is spherically symmetric about the charge. From these equations, we see that the force

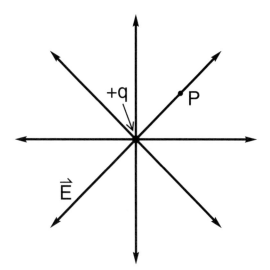

Fig. T.2. A positive point charge produces an outward directed electric field, $\vec{E}$.

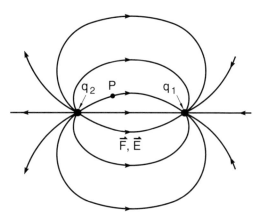

Fig. T.3. Lines of force and electric field lines around a positive and a negative point charge that are close enough to influence each other. The point, P, is an arbitrary point at which E (or F) might be calculated.

on a test charge q_1 in an electric field caused by other charges is simply

$$\vec{F} = q_1\vec{E}. \tag{T.4}$$

This remains true even when $\vec{E}$ is produced by a complex charge distribution instead of a single point charge.

T.3.1. Electric field lines

To explain the influence of one charged object on another, the concept of *electric field lines* has been developed. The electric field is tangent to the line everywhere along an electric field line, which is directed such that it originates on positive charge and terminates on negative charge. Electric field lines can be represented pictorially, as in Fig. T.3. They can be thought of as elastic bands stretched between the positive and negative charges. Furthermore, the elastic lines repel each other in a direction perpendicular to their line at every point. The visualization of field lines in this way can be a powerful tool to provide physical insight into electrostatics, but it quickly becomes misleading if used on anything but extremely simple charge configurations.

T.3.2. Example calculation of E at a point

The following is a simple example of calculating $\vec{E}$. Referring to Fig. T.4, we calculate the magnitude of $\vec{E}$ at the point, P. From Eq. T.3, the magnitude of the electric field is

$$|\vec{E}| = \left| \frac{1}{4\pi\varepsilon} \frac{q}{r^2} \hat{a}_r \right| . \tag{T.5}$$

We separate $\vec{E}$ into Cartesian components as follows:

$$|E| = \frac{1}{4\pi\varepsilon} \frac{q}{x^2 + z^2}$$

$$E_x = \frac{1}{4\pi\varepsilon} \frac{q}{x^2 + z^2} \sin\theta = \frac{1}{4\pi\varepsilon} \frac{qx}{\sqrt{(x^2 + z^2)^3}} \tag{T.6}$$

$$E_z = \frac{1}{4\pi\varepsilon} \frac{q}{x^2 + z^2} \cos\theta = \frac{1}{4\pi\varepsilon} \frac{qz}{\sqrt{(x^2 + z^2)^3}} .$$

T.3.3. Superposition

If there are N point charges instead of one, each point charge will contribute to $\vec{E}$ at a given point, P, in space. To determine the resultant $\vec{E}$, the $\vec{E}$ from each charge is added vectorially to those from each of the other charges. We must include direction as well as magni-

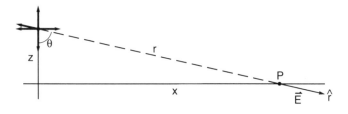

Fig. T.4. Geometry of two-dimensional calculation of $\vec{E}$ and its Cartesian components at point, P. Examples of E lines are shown around the charge. The unit vector for the particular line along which the calculation is made points away from the charge because it is positive.

tude in combining the field intensities. This is called *superposition* and can be written as

$$\vec{E}_{tot} = \sum_{1}^{N} \vec{E}_i$$

$$= \frac{1}{4\pi\varepsilon} \sum_{1}^{N} \frac{q_i}{|\vec{r} - \vec{r}_i|^2} \hat{a}_{ri}, \tag{T.7}$$

where $\vec{E}_i$ is the electric field produced at the point P by the charge q_i, $\vec{r}$ is the radial vector from the origin to P, $\vec{r}_i$ is the radial vector from the origin to q_i, and $\hat{a}_{ri}$ is the unit vector along the difference vector $\vec{r}-\vec{r}_i$. The $\vec{E}_i$ are vectors, and the summation must be done vectorially. From Eq. T.4, we see that superposition also applies to the force at a point from a collection of charges.

The following demonstrates superposition for the simple charge distribution in the x-y plane in Fig. T.5. An easy way to solve such a problem is to solve for the components (x and y) and then calculate a magnitude and direction for $\vec{E}_{tot}$. The x-component is determined as follows:

$$E_{totx} = E_{1x} + E_{2x} + E_{3x}$$

$$E_{1x} = \frac{1}{4\pi\varepsilon} \frac{q_1\cos\alpha}{r_1^2}$$

$$E_{2x} = \frac{1}{4\pi\varepsilon} \frac{-q_2\cos\beta}{r_1^2} = 0 \tag{T.8}$$

$$E_{3x} = \frac{1}{4\pi\varepsilon} \frac{q_3\cos\gamma}{r_3^2}.$$

Note that q_i in Eq. T.8 refers to the magnitude of charge only, and the polarity of charge is shown explicitly by the sign of the expression. Negative charge causes $\vec{E}_2$ to be directed toward q_2. The cosine factors make E_{2x} zero and E_{3x} negative. Similarly, the y-component is given by

$$E_{toty} = E_{1y} + E_{2y} + E_{3y}$$

$$E_{1y} = \frac{1}{4\pi\varepsilon} \frac{q_1\sin\alpha}{r_1^2}$$

$$E_{2y} = \frac{1}{4\pi\varepsilon} \frac{-q_2\sin\beta}{r_1^2} \tag{T.9}$$

$$E_{3y} = \frac{1}{4\pi\varepsilon} \frac{q_3\sin\gamma}{r_3^2}.$$

From the two components, we can then determine that $\vec{E}_{tot}$ has a magnitude given by

$$E_{tot} = \sqrt{E_{totx}^2 + E_{toty}^2} \tag{T.10}$$

along a direction, θ, of

$$\theta = \arctan \frac{E_{toty}}{E_{totx}}. \tag{T.11}$$

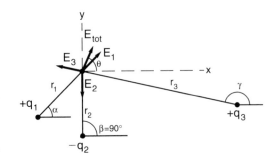

Fig. T.5. Example of superposition of Es from multiple charges at various locations from a point which, for simplicity, is at the origin of an x–y coordinate system. See equations in text for calculation of E_{tot} and its direction angle, θ.

T.3.4. Image charges

A useful concept can be illustrated by a point charge above a conducting plane, which is similar to a charged cloud above Earth (Fig. T.6a). A positive charge and its field lines are shown in the figure. A conductor is a substance in which charge moves freely in response to an electric field. One consequence is that the electric field inside a conductor is zero, and electric field lines from any charge outside a conductor must be perpendicular to the surface of the conductor over its entire surface. If there is a horizontal component of $\vec{E}$ at the surface, it causes charge to move along the surface of the conductor until the horizontal component of $\vec{E}$ from the charge on the surface exactly counters the horizontal component of the external $\vec{E}$ at the surface.

Let us assume we are interested in $\vec{E}$ at some point on the conductor (Earth). A technique that makes answering this question easy is called the method of images. This technique consists of replacing the conducting surface with a charge that produces the same field lines outside the conductor as would be produced by the charge distribution on the surface of the conductor. For our example, it turns out that the image charge has the same magnitude as the real charge, but is of opposite polarity and is located directly underneath the real charge, the same distance below the conductor as the real charge is above it. Hence, it is like a mirror image (Fig. 6b). $\vec{E}$ can then be calculated by the superposition of the real and image point charges. You can show that $\vec{E}$ at the surface in this example has only a z component, given by

$$E_z = -\frac{1}{2\pi\varepsilon} \frac{qz}{\sqrt{(x^2 + z^2)^3}}, \tag{T.12}$$

where x is the horizontal distance from the line (assumed to be the z-axis) joining the charges.

T.4. ELECTROSTATIC POTENTIAL

We discuss the concept of electrostatic potential (usually called potential) initially using $\vec{E}$ from a positive

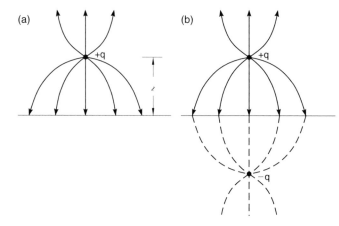

Fig. T.6. (a) Electric field lines from positive point charge a distance, z, above a conducting plane; (b) electric field lines from a positive charge and its negative image charge. The E lines are the same as in (a), showing the equivalence of replacing a conductor with an image charge.

point charge. Electric field lines point radially outward. A two-dimensional diagram of the magnitude of $\vec{E}$ in the vicinity of a point charge would look like a 'hill' (Fig. T.7). The hill is symmetrical, and the magnitude of E varies as $1/r^2$. From Eq. T.4, we know that a small positive charge q' placed at a distance r from the positive charge q would experience a force, $\vec{F} = q'\vec{E}$, where $\vec{E}$ is from the charge q. The direction of the force on the positive test charge q' is away from the positive charge q. If we want to keep q' at its position, we must exert a force to hold it there. Thus, it requires an applied force on q' such that the applied force exactly balances the electric force: $\vec{F}_{applied} = -q'\vec{E}$.

Now if we let q' move a small distance, dl, in the presence of q, an amount of work $\vec{F} \cdot d\vec{l}$ is done by electric force in our system: This is work being done by the energy in the internal system. As work is defined as a force acting over a distance, an applied force does work on q' when it acts against a resisting electric force to move q'. We define energy leaving our electrical system (i.e., work done by the electric field) as a negative change and that coming in (i.e., work done by an external force) as a positive change. In terms of the energy of our electrical system, we write the change in energy, dW, as

$$dW = -\vec{F} \cdot d\vec{l}$$
$$= -q'\vec{E} \cdot d\vec{l}. \qquad \text{(T.13)}$$

The unit of energy (and also of work) is the joule (J).

Notice that only charge motion parallel to the direction of the electric force (and consequently also to $\vec{E}$) can change the energy of an electrical system, and because the direction of the electric force on q' at a given point is constant regardless of the direction of motion of q', the system is *conservative*. That is, the total energy change from a movement of q' along any path whatsoever is zero if q' is returned to its original position. This is equivalent to saying that the right-hand term in Eq. T.13 is an exact differential. If we divide the incremental work by the charge q', we have an expression for an energy change per unit charge. We define this as a change in the *electrostatic (or electric) potential, Φ,*

$$\frac{dW}{q'} = d\Phi, \qquad \text{(T.14)}$$

where the units are joules per coulomb (J C^{-1}). Thus,

$$d\Phi = -\vec{E} \cdot d\vec{l} \qquad \text{(T.15)}$$

also gives the change in potential, and joules per coulomb is equivalent to a volt (V).

Since $d\Phi$ is an exact differential, we can write the difference in electric potential at two points, A and B, as

$$\Phi_B - \Phi_A = -\int_A^B \vec{E} \cdot d\vec{l}. \qquad \text{(T.16)}$$

The potential difference depends only on the locations of the points A and B and not on the path taken by the

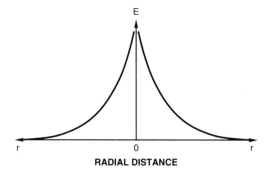

Fig. T.7. Amplitude of E versus radial distance, r, from a positive point charge. (From Hendricks, C.D., In *Electrostatics and Its Applications*, A.D. Moore, ed., copyright © 1973, John Wiley & Sons. Reprinted by permission of John Wiley & Sons Inc.)

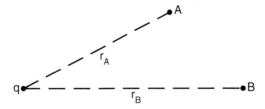

Fig. T.8. Geometry for calculation of potential difference between two points, A and B, at different distances from a point charge.

charge q' to get from A to B. This is a property of the conservative nature of the electric field. Let us examine the potential difference between two points at different distances from a point charge q as shown in Fig. T.8. Then we have

$$\Phi_B - \Phi_A = \frac{q}{4\pi\varepsilon} \left(\frac{1}{r_B} - \frac{1}{r_A} \right). \quad (T.17)$$

Because we can define only potential difference and never an absolute potential, it is convenient to have a reference potential that we designate as zero potential. This is analogous to assigning zero altitude to sea level. In the case of a point charge configuration, we let point A be at infinity and arbitrarily assign to Φ_A the value of zero. Then $\Phi = \Phi_B - 0$, is the potential at point B with respect to the potential at infinity, so we write

$$\Phi = \frac{q}{4\pi\varepsilon r}, \quad (T.18)$$

where $r = r_B$. The physical significance of potential is that it is the potential energy of a unit test charge if brought from a reference point (usually infinity) through a non zero electric field caused by other charges to a specific point in space.

The potential is a scalar and has no direction; thus, the potentials of multiple charges may be combined without regard to vector addition. The potential at a given point due to a number of point charges is just the sum of the potentials of the individual charges and is expressed by

$$\Phi = \sum_{j=1}^{N} \frac{q_j}{4\pi\varepsilon r_j}. \quad (T.19)$$

If we have a continuous distribution of charge, the summation becomes an integral. As it is easier to perform one scalar integration to find the potential due to a charge distribution than it is to perform the three-dimensional vector integration, it is often advantageous to look for a method of obtaining $\vec{E}$ from the potential.

To find a suitable method relating $\vec{E}$ to the potential, we can begin with Eq. T.15. If ds is the magnitude of

the displacement $d\vec{l}$ and α is the angle between the direction of $\vec{E}$ and the direction of $d\vec{l}$, Eq. T.15 can be written

$$d\Phi = -E \cos \alpha \, ds. \quad (T.20)$$

The potential changes most when the displacement is parallel to $\vec{E}$. Then $\cos \alpha = 1$, and dividing both sides of Eq. T.20 by ds, we find that the electric field magnitude is simply the negative derivative of potential with respect to this displacement. Thus, the magnitude of $\vec{E}$ is the negative spatial derivative of potential taken along the direction of the maximum rate of change of potential with distance, that is,

$$|\vec{E}| = - \left(\frac{d\Phi}{ds} \right)_{max}. \quad (T.21)$$

The electric field is given by the maximum rate of change in the potential, and its direction is opposite to the direction of the maximum rate of change. Since the gradient of the potential (usually called the *potential gradient*) provides the direction and magnitude of the maximum rate of change in Φ, this statement is equivalent to saying that the electric field is the negative of the potential gradient:

$$\vec{E} = -\vec{\nabla}\Phi. \quad (T.22)$$

Eq. T.22 is the expression for the electric field in terms of the potential, a basic and very important relationship to keep in mind when studying atmospheric electricity. From this relationship, it follows that the electric field line passing through any point is always perpendicular to the equipotential surface passing through the same point. Note that the principle of superposition applies to potential.

From our definition of Φ and the property that $|\vec{E}| = 0$ throughout the interior of a conductor, it follows that the potential at every point on the surface of a conductor must be the same. That is, the surface of a conductor is an *equipotential surface*.

We alert you that authors of much of the literature in atmospheric electricity have used "field" or even "electric field" when they meant potential gradient (i.e., the right-hand term of Eq. T.22 without the minus sign). This so-called atmospheric electricity sign convention was used extensively in the past, and to a lesser extent is used still. The reader of atmospheric electricity papers needs to be vigilant to establish which sign convention is being used in a given article. In this course, we will try to adhere to the classical physics sign convention, expressed in Eq. T.22, unless we specifically state otherwise.

For those unfamiliar with electric potential, we alert you that the symbol, V, is used often for potential in place of Φ. It can be a bit confusing in that the unit of Φ is volt, whose symbol is also V. Furthermore, the potential difference between two points is often referred to as the voltage between two points. In this book, we tend to use V as the symbol for potential difference and

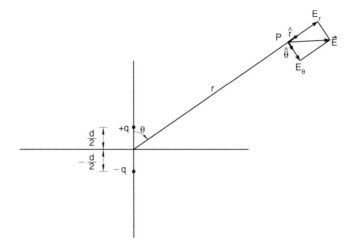

Fig. T.9. Example of simple dipole, which is two closely spaced charges of opposite polarities. The point, P, denotes the location for calculation of potential or electric field.

to call V the voltage between two points (or objects), rather than the potential difference.

T.4.1. Potential distribution of a dipole

As an example, we calculate the potential from a dipole charge configuration. A *dipole* is a pair of charges of opposite polarity that are closely spaced, separated by a distance d (Fig. T.9). On the atomic level, a dipole can be caused by an electric field: The protons and electrons sense opposite forces from $\vec{E}$ and thus are displaced relative to each other, along the direction of $\vec{E}$. The same dipole formation can occur on the molecular scale. From Eq. T.19, the potential produced at a given point by a vertical dipole centered on the origin can be expressed as

$$\Phi(x, y, z) = \frac{1}{4\pi\varepsilon}$$

$$\left[\frac{q}{\sqrt{x^2 + y^2 + \left(z - \dfrac{d}{2}\right)^2}} \right. \tag{T.23}$$

$$\left. + \frac{-q}{\sqrt{x^2 + y^2 + \left(z + \dfrac{d}{2}\right)^2}} \right] .$$

This expression for Φ of two charges is exact and frequently useful. However, in situations in which d becomes very small compared with the distance from the dipole at which Φ is measured, another expression is useful.

To begin deriving this expression, consider that Φ is axisymmetric about the dipole at long ranges. Therefore, we replace $x^2 + y^2 + z^2$ with r^2 and express the vertical displacement of each charge from the origin along the z axis by using the binomial expansion

and keeping only terms that are first-order in d, e.g.,

$$\left(z - \frac{d}{2}\right)^2 \approx z^2 - zd. \tag{T.24}$$

The left-hand term on the right side of Eq. T.23 can then be approximated by

$$\frac{q}{\left[x^2 + y^2 + \left(z - \dfrac{d}{2}\right)^2\right]^{\frac{1}{2}}} \approx \frac{q}{r\left(1 - \dfrac{zd}{r^2}\right)^{\frac{1}{2}}}. \tag{T.25}$$

The process can be repeated for the right-hand term in Eq. T.23, and then both can be expanded again to get

$$\Phi \approx \frac{1}{4\pi\varepsilon} \left(\frac{qzd}{2r^3} + \frac{qzd}{2r^3} \right) = \frac{qzd}{4\pi\varepsilon r^3}. \tag{T.26}$$

Since $z/r = \cos\theta$, this can be written as

$$\Phi \approx \frac{qd\cos\theta}{4\pi\varepsilon r^2}. \tag{T.27}$$

This expression is written often in terms of the *dipole moment*, $p = qd$, as

$$\Phi \approx \frac{1}{4\pi\varepsilon} \frac{p\cos\theta}{r^2}. \tag{T.28}$$

As d approaches zero, this approximation becomes increasingly accurate, so it usually is written as an equality in terms of a point dipole $\vec{p}$, which is an idealized vector of magnitude qd pointing from the negative to the positive charge in the dipole:

$$\Phi \approx \frac{1}{4\pi\varepsilon r^2} \vec{p} \cdot \hat{r}. \tag{T.29}$$

The electric field of a dipole can be calculated from the potential. It is often expressed as having a component in the z-direction and then a transverse one in the

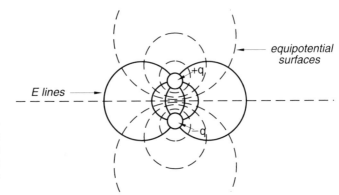

Fig. T.10. Electric field lines and equipotential surfaces in an electric dipole. (From Hendricks, C.D., In *Electrostatics and Its Applications*, A.D. Moore, ed., copyright © 1973, John Wiley & Sons. Reprinted by permission of John Wiley & Sons Inc.)

x-y plane. The $\vec{E}$ for a dipole can also be expressed as

$$\vec{E} = \frac{p}{4\pi\varepsilon r^3}(2\cos\theta\hat{a}_r + \sin\theta\hat{a}_\theta), \qquad (T.30)$$

where the parameters are as shown in Fig. T.9. Shown in Fig. T.10 are equipotential lines (surfaces when rotated about the axis) and electric field lines for a dipole. These approximations for the potential and $\vec{E}$ of a dipole are often called the *point dipole approximation.*

T.5. ELECTRIC FLUX AND GUASS'S LAW

A different form of Coulomb's law is called Gauss's law. To understand it, we begin with the concept of the electric flux through a surface. In free space, the *electric flux*, $\varPsi$, through an infinitesimal surface is the component of $\vec{E}$ normal to the surface at that location times the infinitesimal area. Electric flux is a scalar, and the electric flux through an arbitrary surface is obtained by integrating over the entire surface. Inside substances, however, the electric field can rearrange charge on a molecular scale, and this rearrangement contributes to the net electric field and, hence, to the flux. To account for this effect on the flux, we define a vector $\vec{D}$ that is the density of flux across an infinitesimal surface at a point, which is sometimes called the electric displacement. It is written as

$$\vec{D} = \frac{d\varPsi}{dS}\hat{a}, \qquad (T.31)$$

where $\hat{a}$ is the unit vector normal to the closed surface, S, at a given point on the surface. The surface can be any imaginary one we construct; it does not have to be a physical surface. Then we can express the flux through the infinitesimal surface as

$$d\varPsi = \vec{D} \cdot d\vec{S}, \qquad (T.32)$$

where the surface element has a magnitude, dS, and a direction perpendicular to the surface. Flux has units of charge (coulombs in the MKS system). $\vec{D}$ is measured in units of coulomb per meter squared (C m^{-2}).

Flux often provides an easy means for determining

E. By definition, 1 C of charge causes an electric flux of 1 C through a surface that surrounds the charge. This means that flux can be linked to the concept of electric field lines. *Gauss's law* states:

> The total electric flux through a closed surface is equal to the total net charge within the surface. In other words, if a closed surface of any shape is constructed in a region in which an electric field is present, the surface integral of the normal component of the Maxwell displacement over the surface is equal to the net free charge enclosed by the surface.

In mathematical form, Gauss's law is written in terms of the surface integral of flux around an enclosed surface:

$$\oint d\varPsi = q_{enclosed}, \qquad (T.33)$$

which leads to the more common form of Gauss's law:

$$\oint \vec{D} \cdot d\vec{S} = q_{enclosed}. \qquad (T.34)$$

It can be shown that the electric field and flux density are related by

$$\vec{D} = \varepsilon\vec{E}. \qquad (T.35)$$

Let us examine a rather unsophisticated development of Gauss's law from the expression for electric field obtained from Coulomb's law. Recall that for a point charge

$$\vec{E} = \frac{q}{4\pi\varepsilon r^2}\hat{a}_r, \qquad (T.36)$$

where the electric field spreads radially outward from the charge q. So for a point charge, the electric displacement is

$$\vec{D} = \frac{q}{4\pi r^2}\hat{a}_r. \qquad (T.37)$$

We can choose an arbitrary surface for Gauss's law, so here we choose a spherical one and use spherical coordinates, r, θ, and ϕ. As $\vec{D}$ is spherically symmetric and is directed radially, it is parallel to the surface element

$d\vec{A} = (rd\theta)(r\sin\theta d\phi)$ $\hat{a}_r$ everywhere on the sphere. It is simple then to demonstrate that

$$\oint \vec{D} \cdot d\vec{A} = \int_{\theta=0}^{\pi} \int_{\phi=0}^{2\pi} \frac{q}{4\pi r^2} (r^2 \sin\theta d\theta d\phi) = q. \tag{T.38}$$

Although this example with a point charge allowed a simple demonstration, Guass's law applies to extended charge distributions and irregular shapes. As will be shown in later examples, it can be a powerful tool to calculate the electric field when a more complex charge distribution has some useful symmetry.

Gauss's law can be written in differential form as

$$\vec{\nabla} \cdot \vec{E} = \frac{\rho}{\varepsilon}. \tag{T.39}$$

In our use of this form of Gauss's law in this text, we sometimes will replace ρ with ρ_{tot} to emphasize that all charge carriers contribute to the variation of $\vec{E}$ in space. The differential form of Gauss's law has been used extensively in one-dimensional calculations to infer the vertical structure of thunderstorm charge from measurements of $\partial E_z/\partial z$ from balloon-borne soundings.

The combination of Gauss's law with $\vec{E} = -\vec{\nabla}\Phi$ describes all electrostatic phenomena. Although this statement is true mathematically, we need to be clever when setting up each problem to use these two relationships successfully to solve the problem. We now proceed with some examples of calculations.

T.5.1. Gauss's law for spherical charge distribution

We wish to calculate the electric field from a sphere of radius, R, having a uniform charge density, ρ (Fig. T.11). We write Gauss's law as

$$\varepsilon \oint \vec{E} \cdot d\vec{S} = q, \tag{T.40}$$

and we pick a spherical surface of radius r < R over which to integrate. The resulting magnitude of $\vec{E}$ is

$$E = \frac{1}{4\pi\varepsilon} \frac{q}{r^2} \quad r < R, \tag{T.41}$$

where q is the charge contained within any radius and is given by

$$q = \frac{4}{3}\pi r^3 \rho \quad r < R. \tag{T.42}$$

In terms of the charge density, we can write E as

$$E = \frac{\rho r}{3\varepsilon} \quad r < R$$

$$= \frac{\rho R^3}{3\varepsilon r^2} \quad r \geq R. \tag{T.43}$$

Can you satisfy yourself that once outside the sphere, the radius of the charge distribution does not affect the result? Thus, outside a spherical charge distribution, the charge can be treated as a point charge of magnitude equal to the total charge in the sphere. This is useful to remember.

T.5.2. Gauss's law for a line charge

Consider a line of charge with density, λ_q (units of C m^{-1}). The Gaussian surface we use is a cylinder (Fig. T.12). Gauss's law can be written as

$$\oint_{S_1} \varepsilon\vec{E} \cdot d\vec{S} + \oint_{S_2} \varepsilon\vec{E} \cdot d\vec{S} + \oint_{S_3} \varepsilon\vec{E} \cdot d\vec{S} = q_{enclosed}, \tag{T.44}$$

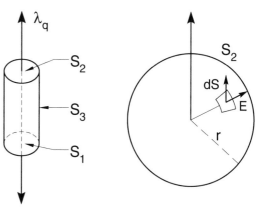

Fig. T.12. Cylindrical Gaussian surface enclosing a portion of a very long (or infinitely long) line charge with line charge density, λ_q. On the right is an enlargement of the top of the cylinder, S_2, showing that the directions of an incremental surface normal and E are orthogonal. Thus they contribute nothing to the integration to obtain the total inclosed charge.

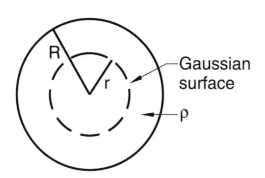

Fig. T.11. Example of a spherical Gaussian surface within a sphere of radius, R, and uniform charge density, ρ.

where we are assuming that all the charge within the surface is on the line. From the symmetry, we see that $\vec{E}$ must point radially outward from the line if it is positively charged. Because the top and bottom surfaces (circular) are orthogonal to $\vec{E}$, the integrals over those two surfaces are zero. The charge enclosed is $\lambda_q L$, where L is the length of the cylindrical Gaussian surface. Thus,

$$E = \frac{\lambda_q}{2\pi r}. \tag{T.45}$$

This field has cylindrical symmetry and is inversely proportional to the distance from the line of charge (measured perpendicular to the line).

T.5.3. Gauss's law for a sheet of charge

We consider a sheet of charge with uniform surface charge density, σ_q. We assume that the linear dimensions of the sheet are much larger than the distance from the sheet at which $\vec{E}$ is measured, so that symmetry requires $\vec{E}$ to be directed perpendicular to the sheet, pointing outward for positive charge or inward for negative charge. From the Gaussian surface in Fig. T.13, we see that there are two contributors to the total $\vec{E}$. In this problem, there is no flux through the curved part of the Gaussian surface, where $\vec{E}$ is everywhere parallel to the curved surface. Gauss's law for this case gives

$$\varepsilon A E_1 + \varepsilon A E_2 = q_{enclosed}. \tag{T.46}$$

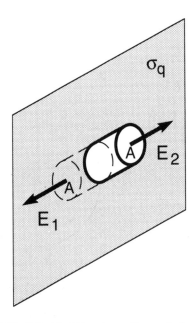

Fig. T.13. Cylindrical Gaussian surface enclosing a portion of a very large (or infinite) sheet of charge with uniform positive charge density, σ_q. The circular ends have area, A.

The enclosed charge is just $\sigma_q A$. As $E_1 = E_2$, we find that E is

$$E = \frac{\sigma_q}{2\varepsilon}. \tag{T.47}$$

Note that the magnitude of $\vec{E}$ is independent of distance from the sheet of charge, as long as our assumption that distances are small compared with the linear dimensions of the sheet is valid.

T.6. ELECTRIC FIELD ON AND INSIDE CONDUCTORS

We now use Gauss's law to investigate E on and inside a conductor. Conductors have the property that, if an electric field E is applied, many electrons can move freely in response to E, and thereby rearrange charge. They move only until they transport enough charge that E, and thus the force on the electrons, becomes zero everywhere inside the conductor. This process of nullifying E inside conductors proceeds at speeds near that of light. Once E is nullified, the electrons no longer are forced to move in the conductor and so become stationary (in the macro sense, ignoring the orbital or vibrational motions in atoms and molecules). Because any Gaussian surface completely within the conductor must have zero flux through it, the charge density must also be zero everywhere inside a conductor.

At the surface of the conductor, on the other hand, the electrons that moved under the influence of a component of E normal to the surface were forced to stop, as atomic forces kept them from leaving the conductor. Thus, a normal component of E can exist at the surface. Components of E tangent to the surface cannot exist, however, because any non zero tangential component causes electrons to move until they nullify the tangential component. For a positively charged conductor, as in Fig. T.14, the electric field will be zero inside the conductor and normal to the surface at the conductor's surface. If there is a Gaussian surface like that shown in Fig. T.14, the only flux will be through the end of the cylinder outside the conductor, and this implies that charge can reside on a conductor's surface.

We change the situation now by making a hollow volume inside the conductor (Fig. T.15). We just learned that inside the conductor $E = 0$, but what about inside the interior cavity? We construct our Gaussian surface so that it does not cut through the cavity, but is in the conductor all around the cavity. Because $E = 0$ in the conductor, there is no flux through the Gaussian surface, and we conclude that the net charge enclosed must be zero. If there is no charge inside the cavity, then no Gaussian surface within the conductor or cavity can enclose charge, so $E = 0$ inside the cavity. If we somehow bring external charge into the cavity, E will be nonzero inside the cavity, according to Gauss's law, but an equal charge of opposite polarity must form on the boundary of the cavity to keep $E = 0$ inside the conductor. Thus, while charge inside the cavity creates an

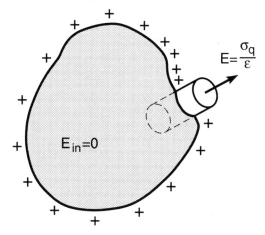

$$E = \frac{\sigma_q}{\varepsilon}$$

$E_{in} = 0$

Fig. T.14. A positively charged solid conductor with surface charge density given by σ_q at any point. No tangential component of E can exist at the surface, as electrons would move to make it zero. The E is everywhere perpendicular.

electric field inside the cavity, the electric field does not extend beyond the cavity. We are left with the result that no external E can create an electric field inside a closed cavity in a conductor, and no interior E can create an electric field outside the conductor. For a conductor, we say that E inside is shielded from E outside, and vice-versa. The following is an interesting experimental conformation of this.

T.6.1. Faraday cage

We are indebted to Michael Faraday for performing an experiment demonstrating a principle that has wide practical application. In 1837, he insulated from the

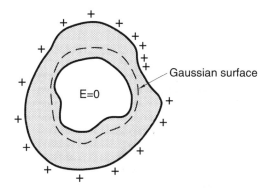

E=0

Gaussian surface

Fig. T.15. A hollow cavity inside a conductor that is positively charged. Inside the cavity, E = 0, regardless of the shape of the conductor or the cavity, as can be shown by Gauss's law.

ground a large conducting metal box, and he shut himself up in it. Inside, he had instruments for measuring E. The outside of the conducting box was then charged, but Faraday detected no evidence of charge inside the box. Faraday described it thus: "I went into the cube and lived in it, and using lighted candles, electrometers, and all other tests of electrical states, I could not find the least influence upon them . . . though all the time the outside of the cube was very powerfully charged, and large sparks and brushes were darting off every part of its outer surface."

This experiment demonstrated that E is zero inside a charged conducting box and that the net charge on a conducting body is found only on the outside surfaces, not on the inner surfaces. It is interesting that Ben Franklin had observed this phenomenon in 1755, but apparently could not explain it. Faraday's result led to the concept of using a metal enclosure (often called a Faraday cage) for protection against lightning. It works, but any openings in the conductor (e.g., windows or even much smaller openings) can allow E—especially the high-frequency components of E produced by rapid transient processes of lightning—to get inside. Thus, while a motor vehicle with a metallic body offers its occupants much protection from lightning, lightning or the effects of lightning can occasionally penetrate a vehicle and cause injury or death.

T.7. CHARGE TRANSPORT AND MOBILITY

Up to now, we have been concerned with electrostatics, but charges obviously move from one place to another under the influence of an electric field. Such movement constitutes an electric current. *Electric current, I,* through some surface is defined as the net charge passing through that surface per second, as in Fig. T.16. Current is measured in amperes (A), which is equivalent to coulombs per second. Often, we need the current passing through a unit area of surface. This quantity, defined as the *current density, $\bar{J}$,* has both magnitude and direction and is measured in units of amperes per meter squared (A m^{-2}).

The medium through which charge moves determines much about the ease with which charge is moved by E. Let us first consider the motion of charge in a metallic solid. The atoms of the material are rigidly

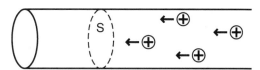

S

Fig. T.16. Positive charges moving through a surface, S, of arbitrary shape. The amount of charge in coulombs per second across the surface determine the current flow across the surface.

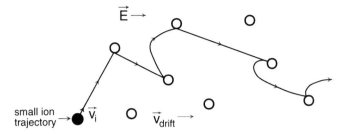

Fig. T.17. Depiction of the drift velocity of a population of particles with a net average velocity and embedded in an E. The solid lines with arrows are the small ions depicted colliding with and rebounding randomly from larger particles (open circles).

held in place, but some of the outer electrons of the atoms may be almost totally free to move in the solid. When this is the case, we class the material as a *conductor.*

Many materials, including metals, obey *Ohm's law,* which is written in terms of electric field intensity and current density, J, as

$$\vec{J} = \sigma\vec{E}, \tag{T.48}$$

where σ is called the *conductivity* of the material, with units of siemen per meter (S m^{-1}). (In atmospheric electricity, λ is often used for conductivity, but in most physics and engineering texts σ is used.) Conductivity is a measure of the ease with which electrons move through the medium, and it is large for metals. This equation is called *Ohm's law at a point,* as it is valid in the limit of a point within the substance through which charge moves. A form of Ohm's law used for electronic circuits is $V = I\,R$, where V is the voltage; I, the current; and R, the resistance.

Note that Ohm's law is a linear relationship. When an electric current does not obey Ohm's law, it is said to be nonohmic. In such a case, other expressions describe the current (e.g., the displacement current of magnitude dD/dt). Although some currents in thunderstorms obey Ohm's law, nonohmic currents also occur in storm environments.

In the atmosphere, ohmic currents often are caused by the movement of larger particles, such as a small ion (sometimes called a fast ion), besides electrons. Atmospheric small ions, which are essentially clusters of molecules, carry a net single charge. Because they are charged, they experience a force in an electric field.

We can calculate the velocity of ions under the influence of E by considering the concept of drift velocity. If the density of the gas in which ions were embedded was small enough that ions never collided, all ions would move in the same direction. However, in a relatively dense gas such as the atmosphere, an ion cannot move far before it collides with another particle. When it collides, its acceleration by E is interrupted, and its momentum changes. Acceleration by E then resumes until the next collision. Without an electric field, an ion in a gas in thermal equilibrium has random motions between collisions, so there is no systematic motion of charge. These thermal motions still are present in an electric field, but the acceleration by the electric field between collisions gives ions a net average velocity,

which is also known as the *drift velocity.* This process in depicted in Fig. T.17.

The drift or average velocity can be expressed as

$$\vec{v} = k\vec{E}, \tag{T.49}$$

where k is called the *mobility* (in our example it would be the small-ion mobility). The unit of mobility is meter per second per volt per meter (m s^{-1} V m^{-1}), which is sometimes written in the equivalent, but not so intuitively meaningful, form: V m^2 s^{-1}. As current density can be written

$$\vec{J} = nq\vec{v}, \tag{T.50}$$

where n is the number of charges per unit volume traveling with a velocity $\vec{v}$, the current density is dependent on the number of carriers per unit volume and their velocity and charge. The combination of Eqs. T.49 and T.50 gives

$$\vec{J} = nqk\vec{E}. \tag{T.51}$$

If more than one carrier type is present, the current density is the sum of the individual current densities from each carrier type. For the two types in our example, positive and negative small ions, we write

$$\vec{J} = n_+ q_+ k_+ \vec{E} + n_- q_- k_- \vec{E} \tag{T.52}$$

for the current density. Thus, the small-ion conductivity in our example is

$$\sigma_{tot} = n_+ q_+ k_+ + n_- q_- k_-. \tag{T.53}$$

In addition to the motion of charges under the influence of E, charge can be transported on moving material to which the charge is attached in a process called convective charging; a good example is found in a Van de Graaff high-voltage generator (Fig. T.18). Charge is sprayed on a moving belt that transports the charge mechanically to a collector where the charge is removed from the belt. Because large quantities of charge can be moved to an insulted, large, and smoothly rounded conductor, very high voltages (i.e., differences in potential) can be produced. If we carry a surface charge density σ_q on the best at a velocity v, and if the belt is W meters wide, a current

$$I = \sigma_q v W \tag{T.54}$$

is carried into the dome. In a time, t, a charge, q, such that

$$q = It, \tag{T.55}$$

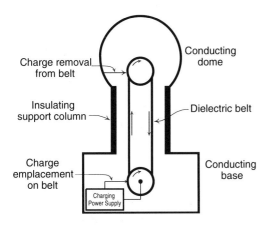

Fig. T.18. Sketch of Van de Graaff high-voltage generator. Charge produced with a high-voltage supply is placed on the dielectric belt. At the top of the insulating support, the charge is removed to the conducting dome. In this way, large potential differences (voltages) between the dome and the conducting base are created. (From Hendricks, C.D., In *Electrostatics and Its Applications,* A.D. Moore, ed., copyright © 1973, John Wiley & Sons. Reprinted by permission of John Wiley & Sons Inc.)

is placed on the dome. The potential difference or voltage, *V*, between the dome and the base will be

$$V = \frac{q}{C}, \qquad (T.56)$$

where *C* is a constant of proportionality called the capacitance (see Sec. T.12). Note that *V* is equivalent to *Φ*; when we talk about voltages applied, voltage difference between two plates, etc., we use the symbol *V* instead of *Φ*, as is commonly done in many contexts. Notice that the symbol and the symbol for its units are the same (e.g., *V* = 200 V). We see that if the current carried up by the belt is greater than the downward current from all leakage mechanisms (e.g., currents through the air in which the belt runs), the charge on the dome will increase. The voltage from the dome to the base will increase until the insulating property of the air breaks down and the dome produces a spark.

T.8. CONDUCTION IN GASES AND BREAKDOWN STRENGTH

A few aspects of electric conduction are unique to gases. In the atmosphere, electrons and ions are produced mainly by cosmic rays and naturally occurring radioactive decay. If an electric field is applied to air between two electrodes, a current will flow and will tend to separate the electrons and ions by polarity. If *E* is steadily increased, the current increases until eventually all naturally occurring carriers are removed as

rapidly as they are produced. Further increase of *E* will produce almost no increase in current until other effects begin to occur. If *E* becomes large enough, electrons will be accelerated to velocities sufficient to knock other electrons from neutral atoms or molecules when they collide, so the number of charge carriers in the gaseous medium (atmosphere) increases. This process is called an *electron avalanche.* The result is a phenomenon called *corona* (described in Ch. 4). Further increase of *E* significantly increases the current flowing between the electrodes, but in a noncatastrophic manner. Still further increase of *E* results in current densities great enough to heat local regions of the electrodes to the point at which thermionic emission of electrons occurs. The current density rises catastrophically, and an electric arc develops. For the atmosphere at sea level under standard conditions, breakdown occurs at about 3×10^6 V m^{-1}. The electric field at which breakdown occurs decreases linearly with pressure.

The process just described is similar, but not identical, to the process of a lightning discharge. Another form of discharge that is important for storm electricity is corona in the air near grounded conducting points, such as trees and grass. This process, sometimes called point discharge, is described in Chapter 4.

T.9. DIELECTRICS

We now turn to *insulators,* a class of materials that do not conduct electricity well. These materials are also called *dielectrics.* One property of dielectrics is their ability to store energy when placed in an electric field. This ability is indicated by the *permittivity* of a dielectric, and the permittivity of all materials is larger than the permittivity of a vacuum. The unit of permittivity is coulomb squared per newton-meter squared (C^2 N^{-1} m^{-2}), which is written as a farad per meter (F m^{-1}). It is convenient to write the permittivity of a material in terms of the permittivity of empty space (often referred to as *permittivity of free space*), ε_0:

$$\varepsilon = \varepsilon_{rel}\varepsilon_0, \qquad (T.57)$$

where ε_{rel} is the *relative permittivity* or, as it is also called, the *dielectric constant,* which we will denote with the symbol κ. Common dielectric constants range from 1.0 for a vacuum to about 80.0 for water in a constant electric field. (They also can depend on the frequency of a varying electric field.) Many common insulators have a dielectric constant of <10. The behavior of dielectrics can be quite complicated. For example, both electrical and mechanical changes can occur in dielectric solids in the presence of E.

The caption for Fig. T.19 describes the effect of *E* on a dielectric. The result is that the net field, $\vec{E}_{di}$, inside the dielectric is less than the field, E_o, outside the dielectric:

$$\vec{E}_{di} = \vec{E}_o - \vec{E}'. \qquad (T.58)$$

Fig. T.19. How a dielectric affects the electric field: (a) The dielectric block has random distribution of charges if the field outside it, $\vec{E}_0 = 0$. (b) In an external $\vec{E}_0 > 0$, the internal charges separate slightly. (c) A net charge occurs at the surfaces perpendicular to the $\vec{E}_0$ and creates a field in the dielectric, $\vec{E}'$, that opposes the $\vec{E}_0$. The resultant field, in the dielectric $\vec{E}_{di}$, is thus less than the electric field outside the dielectric.

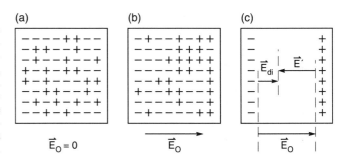

It can be shown that the dielectric constant is related to the electric field in this example by

$$\kappa = \frac{\vec{E}_O}{\vec{E}_{di}}, \tag{T.59}$$

so the larger the dielectric constant, the lower the electric field inside the dielectric.

Insulating materials are important in various instruments used for measurements in atmospheric electricity. We define an insulator as a material having a very low conductivity; an idealized, perfect insulator would have zero conductivity. Quartz and plastics such as polystyrene, Teflon, and KEL-F are among the best insulators. An important issue when very good quality insulators are needed is that of surface cleanliness. Regardless of how good an insulator's bulk properties are, if the surface of an insulator is wet or dirty, current can leak across the surface when the insulator is embedded in an electric field. A common construction technique for insulators is to have slots, valleys, etc., in the insulator. They are effective in reducing leakage, in part because they increase the path length on the surface of the insulator.

Now consider how to quantify the insulating ability of a substance. Assume that we have two blocks of different material that we somehow charge. After being charged, these two blocks are set on a conducting surface, which is connected to ground as in Fig. T.20 (b).

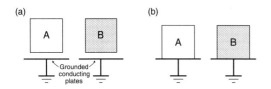

Fig. T.20. Determining if a material is an insulator: (a) Two blocks of different electrical conductivity are charged uniformly and with the same charge while insulated from ground. (b) The blocks are placed on grounded plates, and their charge is measured at specific times. The text describes the decision process in determining whether each block is a conductor or insulator. (From Hendricks, C.D., In *Electrostatics and Its Applications,* A.D. Moore, ed., copyright © 1973, John Wiley & Sons. Reprinted by permission of John Wiley & Sons Inc.)

After a long time, t_1, we measure the charge left on the two blocks and find that it is below the minimum sensitivity of our charge measuring device. If we never measure the charge before t_1, we will always find that the charge has flowed away and will conclude that both materials are conductors.

Now let us perform the experiment again, this time making our measurements in a shorter time $t_2 < t_1$. We find that the charge on block A (Fig. T.20 (b)) is relatively unchanged from the initial value. The charge on block B is again zero. This time we conclude that block A is an insulator and block B is a conductor.

Now we perform the experiment in a still shorter time, $t_3 < t_2 < t_1$. When we make our charge measurements, we find that neither block A nor block B has lost any charge. We thus conclude that both materials are insulators. So what parameter is of true importance in the characterization of an insulator? From this thought experiment, we must conclude that the quantity is the time for charge transfer. The relevant time constant, τ, is

$$\tau = \frac{\varepsilon}{\sigma}. \tag{T.60}$$

Where ε is the permittivity of the material. This time constant is called the *relaxation time constant* of the substance, and its unit is seconds. If a charge density, ρ, is somehow put into a material and released at $t = 0$, the charge density at any later time is given by

$$\rho = \rho_0 e^{-\frac{\sigma t}{\varepsilon}}. \tag{T.61}$$

The time constant, τ, in Eq. T.60 is the time required for the charge density to decrease to $1/e$ of its value, where $e \approx 2.718$ is the base of natural logarithms. Thus, the value falls *exponentially* to about 37% ($=1/e$) of its original value in one time constant. Note a critical assumption of this thought experiment: The charge on the material flows to ground, not into the surrounding environment, so this experiment is best visualized in a charge-free vacuum.

T.10. POLARIZATION AND DIPOLE MOMENT

Polarization can be thought of as the static electrification that occurs internally within a dielectric that is embedded in an electric field. The notion of polarization is also associated with the definition of a dipole. Let us visualize this by considering a single atom in the dielectric as having discrete, but overlapping, spherical positive and negative charges. If we apply E to the atom, the spherical charges move apart slightly. Separating equal and opposite charges a small distance, d, produces a *dipole moment, p,* that can be given in vector form as

$$\vec{p} = q\vec{d}. \qquad (T.62)$$

In this example, $\vec{E}$ and $\vec{d}$ are aligned in the same direction, and $\vec{d}$ points from the negative to the positive charge. If we turn off E, the atoms of most materials will go back to their original state, in which there was no separation of charge. This is analogous to the spring returning to its unstressed position after being released.

However some materials, called *polar dielectrics,* have permanent dipole moments. In the absence of an electric field, the dipoles in the substance are aligned randomly. When an electric field is applied to the material, the electric force tends to align the dipoles.

Regardless of the type of dielectric, the alignment of dipoles in a volume of dielectric material creates net positive and negative charge on opposite surfaces of the substance. Inside the substance, the positive charge of one atomic or molecular dipole is offset by the negative charge of an adjoining dipole. However, this offsetting cannot occur at the surface, so the surface toward which positive charge is aligned has a net positive charge and the surface toward which negative charge is aligned has a net negative charge. For a piece of dielectric material, we call the dipole moment per unit volume the *polarization,* defined by

$$\vec{P} \equiv \lim_{\Delta V \to 0} \frac{Nq\vec{d}}{\Delta V}, \qquad (T.63)$$

where V is the volume and N is number of atoms per unit volume in the dielectric.

There are three types of polarization: *induced* or *electric* (the center of negative charge is displaced from the center of positive charge, as in the above example), *ionic* (ions of one sign in a crystal are moved by E relative to ions of the opposite sign), and *orientation* (permanent dipole moments of polar molecules are aligned by E). Regardless of type, the individual dipoles tend to align with the direction of an applied electric field. For liquids that are polar dielectrics, both the induced and orientation polarizations can make significant contributions to the liquid's behavior as dielectrics and can cause its permittivity to be large.

To try to understand polarization better, consider a dielectric slab embedded in a constant electric field, such as would be approximately attained between the plates of a parallel-plate capacitor (capacitors are explained in Sec. T.12). If the dielectric has uniform $\vec{P}$ throughout, then only at the edge is net charge created by polarization, as noted below. Just as in the case of the atom, there is a systematic shift of the negative charge toward the positively charged plate and of the

positive charge toward the negative plate on the opposite side (see Fig. T.19). The polarization, $\vec{P}$, is equal to the surface charge density at the edges of the dielectric slab. Because polarization is proportional to E, it is written as

$$\vec{P} = \chi \varepsilon \vec{E}, \qquad (T.64)$$

where χ is the proportionality constant called *electric susceptibility,* which is defined by

$$\chi \equiv \kappa - 1 \qquad (T.65)$$

and is dimensionless. Then,

$$\vec{P} = \vec{D} - \varepsilon_0 \vec{E}, \qquad (T.66)$$

so the polarization is just the additional flux density that is produced by the alignment of dipoles in the dielectric.

The water molecule is a polar dielectric of particular interest to us in storm electrification. Remember that polar dielectrics consist of molecules that have a permanent dipole moment. Electrons are shared between atoms of the molecule in such a way that the location of the oxygen atom has a net negative charge and the locations of the hydrogen atoms have a net positive charge. Because the molecular structure is not linear, but forms an obtuse angle with oxygen at its vertex, an electric field can exert torque to rotate the molecule toward alignment of the dipole moment. (A linear triatomic molecule does not do this.) So, while the net charge on a water molecule is zero, it has more positive charge on one 'side' and more negative charge on the other. When a polar dielectric is subjected to an electric field, it not only has the same type of polarization induced in a nonpolar dielectric, but also has orientation polarization from E, aligning the molecules to produce a net dipole moment per unit volume. For this reason, polar molecules, such as water, with high dielectric constants may not be good insulators even though a large κ indicates a long relaxation time constant.

T.11. INDUCTION CHARGING

When an electric field is created in a region containing a conductor insulated from other conductors or when a conductor is moved into a region where a different E exists, charges are moved to new locations on the surface of the conductor. A classic experiment in electrostatics to find an exact law for induction was performed by Michael Faraday. The principle that he demonstrated is used in modern instruments to measure charge on precipitation particles (Ch. 6). He lowered a charged object suspended from an insulating thread into a closed conducting (metal) ice pail (Fig. T.21). He connected an electroscope to the outside of the pail to determine the presence and relative amounts of charge. An electroscope is made of a glass Erlenmeyer flask, an insulating stopper, and a conducting rod. The top of the rod has a conducting sphere on it; the rod inside the flask has two gold foil leaves. When a charge is placed

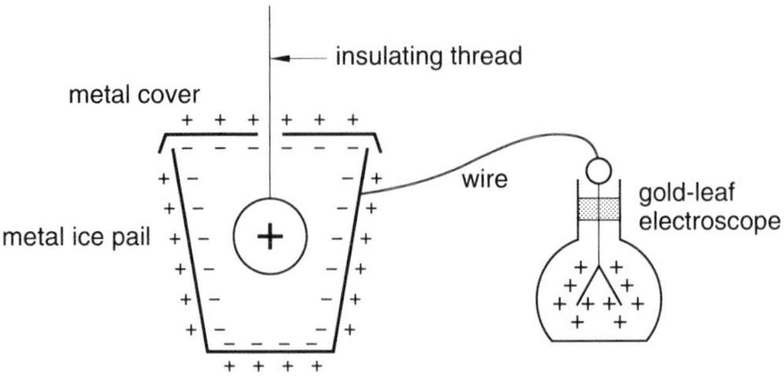

Fig. T.21. Schematic of the Faraday ice pail experiment. (From Hendricks, C.D., In *Electrostatics and Its Applications*, A.D. Moore, ed., copyright © 1973, John Wiley & Sons. Reprinted by permission of John Wiley & Sons Inc.)

on the knob, the gold leaves are identically charged and thus repel each other. If the charge is removed, the leaves hang straight down. Electroscopes can be calibrated by determining the angular deflection of the leaves as a function of charge. When Faraday lowered the charged object into the pail, the leaves on the electroscope diverged, and when the charged object was removed, the leaves collapsed. The maximum divergence occurred when the object was in the container with a metal cover in place.

The analysis of this experiment is basic to electrostatics. We first note that the metal container, the lid, and the electroscope had a total net charge of zero. With the charged object in place, but not touching the container, the net charge on the container and the electroscope was still zero. However, the positively charged object caused an E that acted on electrons in the pail, which in turn redistributed the charge on the container and the electroscope. Electrons were removed from the outside of the container and from the electroscope until there was a total negative charge on the inside container wall equal to the positive charge on the suspended object. (This assumes that no field lines escape through the small hole for the thread in the lid.) An equal positive charge was left on the outside surface of the container and the electroscope. The magnitude is the same as the charge on the object lowered into the pail. In this configuration, the external manifestations of E are not such that the position of the object inside the container can be determined. Gauss's law tells us that the normal component of the electric displacement $\vec{D}$ for any closed surface must be equal to the charge enclosed by the surface. If we enclosed the entire container and electroscope with a Gaussian surface, we get the charge inside, which is equal to that on the object. If the Gaussian surface is just around the object, we get the same result. If the charged object never touched the pail and was withdrawn, the leaves relaxed back because the electroscope and the pail again had zero charge.

Faraday repeated the experiment, but allowed the charged object to rest against the wall or bottom of the container before being removed. He found that the leaves of the electroscope did not collapse when the object was removed. The leaves remained deflected by the same amount, showing that the charge on the object had

been transferred to the pail. If the pail was grounded while the positively-charged object was inside the container, the electroscope leaves collapsed, which indicated no net charge within the system. If the ground was then disconnected and the charged object removed, the electroscope leaves diverged to the original maximum, indicating that the container had been left with a negative charge whose magnitude was equal to that of the charged object.

T.12. CAPACITANCE AND CAPACITORS

Consider two uncharged, isolated conductors of any shape or size. The potential difference between the two conductors is initially zero. If we place charge on one conductor and measure the potential difference between the two conductors each time charge is added, we find that the charge and potential difference are related linearly to one another. This may be expressed by

$$q = C\,\Phi, \tag{T.67}$$

where C is the constant of proportionality. This constant is called the *capacitance* of the system. If the size or relative position of the conductors changes, or the medium in which they are embedded changes, the capacitance will be different. Capacitance has units of coulomb per volt ($C\ V^{-1}$), which has been named the farad (F).

The permittivity of the medium in which a capacitor is embedded affects the properties of the capacitor in a way that has practical applications. As an example, consider a capacitor made with two parallel conducting plates of area, A, separated a distance, d; the two plates constitute what is called a parallel-plate capacitor. As noted above, the magnitude of q on the plates is proportional to the potential difference between them. If a dielectric fills the space between the two plates, the constant of proportionality, C, is given by

$$C = \frac{\varepsilon A}{d}. \tag{T.68}$$

For a given potential difference, Φ, a dielectric with a larger permittivity produces a larger magnitude of charge on the plates. It is always true that if we fill the space between the component conductors of a capacitor with a dielectric, the capacitance is increased by a factor of κ, because we alter ε by that amount.

As an example, we compute the capacitance of a pair of concentric, hollow, conducting spherical shells of radii a and b, where a $<$ b. We place positive charge on one conductor and negative charge on the other. Then we find E between the spheres and the potential difference between them. If we place a charge, q, on the inner sphere, we can use Gauss's law to compute that E is

$$E = \frac{q}{4\pi\varepsilon r^2} \ (a < r < b). \tag{T.69}$$

Then the potential difference between the spheres is

$$\Phi_a - \Phi_b = \frac{q}{4\pi\varepsilon} \left(\frac{1}{a} - \frac{1}{b} \right). \tag{T.70}$$

From Eq. T.67, we can show that the capacitance of the spheres is

$$C = \frac{4\pi\varepsilon ab}{b - a}. \tag{T.71}$$

The limiting case, in which the radius of the outer shell gets very large (i.e., $b \to \infty$ in Eq. T.70), gives an isolated sphere, so the capacitance of an isolated sphere is

$$C = 4\pi\varepsilon a. \tag{T.72}$$

T.13. ENERGY STORAGE

If we consider the force on a positive charge in a vertically pointing electric field, we can visualize attaching a string to the charge and suspending a weight from it. If the electric force is large enough, it would move the charge upward and raise the weight. Energy would be extracted from the electric field. Thus, we can say that this energy was initially stored in the electric field.

As an example, let us examine the energy that is stored in a parallel-plate capacitor. Assume that, initially, the capacitor is uncharged and hence has zero potential difference between the plates. To move the charge from one plate to the other, an external system must put energy into our system. The required work can be calculated from Eq. T.14 and is

$$W = \int_0^q \Phi dq. \tag{T.73}$$

Substituting for Φ from Eq. T.67 shows that

$$W = \int_0^q \frac{qdq}{C} \tag{T.74}$$

$$= \frac{1}{2} \frac{q^2}{C}$$

is the energy stored in the capacitor. When the energy is expressed as a function of q and C, it is called the

stored energy in the field. Again substituting from Eq. T.67, we can write an equivalent expression,

$$W = \frac{1}{2} C\Phi^2, \tag{T.75}$$

which is called the *stored coenergy.*

T.14. MAXWELL'S EQUATIONS

As all electromagnetics are described by Maxwell's equations, we list their free-space, differential versions here:

$$\vec{\nabla} \cdot \vec{E} = \frac{\rho}{\varepsilon_0}, \tag{T.76}$$

$$\vec{\nabla} \times \vec{E} = -\frac{\partial \vec{B}}{\partial t}, \tag{T.77}$$

$$\vec{\nabla} \times \vec{B} = \mu_0\varepsilon_0 \frac{\partial \vec{E}}{\partial t} + \mu_0\vec{J}, \tag{T.78}$$

$$\vec{\nabla} \cdot \vec{B} = 0. \tag{T.79}$$

The term, μ_0, is the magnetic permeability of free space, with units of newton per ampere squared, which usually are given as henry per meter (H m^{-1}). The speed of light in a vacuum, c, is related to the other constants by

$$c = \frac{1}{\sqrt{\varepsilon_0\mu_0}}. \tag{T.80}$$

The units of *magnetic field*, $\vec{B}$, are volt second per meter squared, which are called webers per meter squared (Wb m^{-2}). We alert you to an inconsistent use of terminology in texts: Many texts call H the magnetic field intensity and use another term for B, such as magnetic flux density. H is convenient to use when dealing with a medium that has an intrinsic magnetic dipole moment, $\vec{M}$. If $\vec{M}$ is defined as the magnetic dipole moment per unit volume, then $\vec{H}$ is given by

$$\vec{H} = \frac{1}{\mu_0}\vec{B} - \vec{M}. \tag{T.81}$$

If $|\vec{M}| = 0$, as in free space, this simplifies to $\vec{B} = \mu_0\vec{H}$. The fundamental quantity is $\vec{B}$, which is called by many others (and us) the *magnetic field*. The other term, H, is just named "H" in this convention.

To determine the polarity of a magnetic field, we consider a positive test charge moving at a velocity, $\vec{v}$, past a point in space. If the charge experiences a force that is not caused by an electric field and is perpendicular to the particle's direction of motion, we infer that a magnetic field is present at that point. The force from the magnetic field is given by

$$\vec{F} = q(\vec{v} \times \vec{B}). \tag{T.82}$$

If both electric and magnetic fields are present; the total electromagnetic force on our test particle is

$$F = q[\vec{E} + (\vec{v} \times \vec{B})]. \qquad \text{(T.83)}$$

Finally, we note that Maxwell's equations are used in an endless variety of forms to account for all sorts of differences in propagation, such as propagation in a conductor versus that in a dielectric. Radiation equations can be written using either the electric and magnetic fields or vector potentials. There is a lot of material on these subjects. For more information, we refer you to any good textbook on the subject, such as the books referenced at the beginning of this tutorial.

T.14.1. Wave characteristics

An important ramification of Maxwell's equations is that the electric and magnetic fields propagate with wave characteristics. Some of what we will discuss later regarding lightning will involve the wave characteristics of its radiation. The subject of electromagnetic waves has many aspects; we will only touch upon the topic. We will consider waves in free space, which is the foundation for electromagnetic waves in the atmosphere.

One relevant issue is what happens to the waves when they encounter a more conducting boundary, such as occurs when lightning radiation hits the ionosphere or Earth. Let us consider the conditions at the boundary of two different materials. The electric and magnetic fields can be shown to obey the following conditions for their perpendicular, $\perp$, and parallel, $\parallel$, components at the boundary between two different media, 1 and 2:

$$\begin{aligned}
\varepsilon_1 E_{1\perp} - \varepsilon_2 E_{2\perp} &= \sigma_q \\
E_{1\parallel} - E_{2\parallel} &= 0 \\
B_{1\perp} - B_{2\perp} &= 0 \qquad \text{(T.84)} \\
\frac{1}{\mu_1} B_{1\parallel} - \frac{1}{\mu} B_{2\parallel} &= 0,
\end{aligned}$$

where σ_q is the surface charge density on the boundary between the two materials.

When an electromagnetic wave reaches the interface, it can have both reflected and transmitted (crossing the boundary into the other medium) components. Consider an oft-used, simple example of a horizontally polarized wave (i.e., its amplitude oscillates in the y-plane) that impinges upon the interface, a plane boundary at x = 0. The total intensities of the reflected and transmitted fields must equal the total field intensity that impinged upon the interface. When the incident wave strikes the interface at oblique incidence, the reflected and transmitted waves are a function of incident angle, and the angle of reflection must equal the angle of incidence (i.e., $\theta_I = \theta_R$). As the angle of the incident radiation increases, measured from the normal to the surface, the reflected field intensity increases until eventually all the incident radiation is reflected at an angle called the critical angle and at all angles greater than the critical angle. If the material on which the wave impinges is a metal, the normal polarity of the field reverses upon reflection from incident angles smaller than the critical angle and has a more complicated relationship at angles greater than the critical angle.

We take as our example Eqs. T.77 and T.78 for free space (i.e., the current density, J, is zero). We take the curl of Eq. T.77 and get

$$\vec{\nabla} \times \vec{\nabla} \times \vec{E} = \vec{\nabla} \times \left(-\frac{\partial \vec{B}}{\partial t} \right) \qquad \text{(T.85)}$$

$$= -\frac{\partial}{\partial t} (\vec{\nabla} \times \vec{B}).$$

Substituting from Eq. T.78 yields

$$\nabla^2 \vec{E} = -\mu_0 \varepsilon_0 \frac{\partial^2 \vec{E}}{\partial t^2}. \qquad \text{(T.86)}$$

The equation for the magnetic field has exactly the same form. Each equation is what is called a general wave equation. In free space, the product of the two constants is just $1/c^2$. In other nonconducting media, μ_0 and ε_0 are replaced by their values for the medium $\mu\varepsilon$, which equals $1/v^2$, where v is the velocity of waves in those media. Since there are no source terms in the equations, the equations are called the homogeneous wave equations.

T.15. ELECTROMAGNETIC RADIATION

The description of lightning will require us to understand the basic concepts of what is known as electromagnetic radiation, which contains both electric and magnetic field components. *Radiation* is a process by which energy is transferred from a source (e.g., as happens when a lightning channel radiates out into the atmosphere). Electromagnetic radiation is different from acoustical radiation, as the former can travel through a vacuum. Radiation occurs neither in static situations (i.e., when charge does not move) nor in cases where the charge moves at constant velocity; radiation occurs only when there is acceleration of charge. The acceleration can be from the charge changing speed or direction, or both. What is sometimes called the radiation component of electromagnetic waves applies to terms in Maxwell's equations that are dependent upon the inverse distance from the source, not any higher order dependence. Expanded forms for Maxwell's equations reveal that the reason for this is that higher order terms will be vanishingly small compared to the 1/R terms, where R is the range from the source to the observation point. Radiation fields transport energy from the source all the way to infinity as their magnitudes drop by 1/R. Thus, we generally call the 1/R terms in an equation the radiation component even though under the definition of radiation above, all terms contribute to the energy transfer. The term with $1/R^3$ is the electrostatic term and that with $1/R^2$ the inductive term. We note, however, that if the source is close enough to the observation point, the higher order terms are significant.

The electric and magnetic radiation fields are perpendicular to each other, and their magnitudes are related to each other by $|\vec{E}| = c|\vec{B}|$. The rate and direction at which radiation transfers energy is determined from the *Poynting vector,* $\vec{S}$, which is described by

$$\vec{S} = \frac{1}{\mu_0} \, \vec{E} \times \vec{B}. \qquad (\text{T.87})$$

The units are watts per meter squared (W m^{-2}).

When calculating or analyzing electromagnetic radiation, it is important to remember that it takes time for the radiation to propagate from its source to an observer. We can detect only what the source radiated at an earlier time. You probably are familiar with this phenomenon regarding the light from stars, which takes many years to reach us. The past time at which a source emits radiation that we receive at a given time, t, is called *retarded time,* τ. It is calculated by subtracting propagation time from the time radiation is received at a range, R:

$$\tau = t - \frac{R}{c}. \qquad (\text{T.88})$$

There are various ways of expressing and denoting retarded time; we will use the symbol, $'$, as a superscript on variables to denote they are evaluated at the retarded time.

As $\vec{\nabla} \cdot \vec{B} = 0$, we can express the magnetic field as the curl of a vector, $\vec{A}$:

$$\vec{B}(\vec{r}, t) = \vec{\nabla} \times \vec{A}. \qquad (\text{T.89})$$

Substituting this expression for B in Eq. T.77 gives

$$\nabla \times \left(\vec{E} + \frac{\partial \vec{A}}{\partial t} \right) = 0. \qquad (\text{T.90})$$

The term within the parentheses must equal a quantity whose curl is zero, and it turns out to be $-\vec{\nabla}V$. Setting the term equal to $-\vec{\nabla}V$ gives

$$\vec{E}(\vec{r}, t) = -\vec{\nabla}V - \frac{\partial \vec{A}}{\partial t}. \qquad (\text{T.91})$$

Eqs. T.89 and T.91 are often used to describe the magnetic and electric fields involved in electromagnetic radiation. Developing the radiation equations fully would require an extensive course on electromagnetics. However, we give the portions of the development that use a transformation of *scalar potential,* V, and the *vector potential,* $\vec{A}$, called the *Lorentz gauge* or *condition,* which is defined by

$$\vec{\nabla} \cdot \vec{A} = -\frac{1}{c^2} \frac{\partial V}{\partial t}. \qquad (\text{T.92})$$

The potentials can be expressed as

$$\nabla^2 V - \frac{1}{c^2} \frac{\partial^2 V}{\partial t^2} = \frac{\rho}{\varepsilon_0} \qquad (\text{T.93})$$

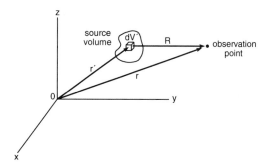

Fig. T.22. Geometry in an orthogonal system for Maxwell's equations at an arbitrary field point (observation location). The radiation source is in the elemental volume, which is at a different arbitrary location relative to the coordinate system origin.

and

$$\nabla^2 \vec{A} - \frac{1}{c^2} \frac{\partial^2 \vec{A}}{\partial^2 t} = -\mu_0 \vec{J}. \qquad (\text{T.94})$$

Note their similarity to the wave equation, Eq. T.86. These two solutions to Maxwell's equations separate the relationship between space charge density and the scalar potential (Eq. T.93) from that between the current density and the vector potential (Eq. T.94). The terms on the right of each equation are the source terms for the radiation. As sources are present, these are called inhomogeneous wave equations.

We consider the source of the radiation in an elemental volume in three-dimensional space (Fig. T.22), and we calculate the electric and magnetic radiation at some other arbitrary observation (often called "field") point. Notice that the range from the source to the observation point is $R = |\vec{r} - \vec{r}'|$. We can write the scalar and vector potentials, respectively, as

$$V(\vec{r}, t) = \frac{1}{4\pi\varepsilon_0} \int \frac{\rho(\vec{r}', \tau)}{R} \, dV' \qquad (\text{T.95})$$

and

$$\vec{A}(\vec{r}, t) = \frac{\mu_0}{4\pi} \int \frac{\vec{J}(\vec{r}', \tau)}{R} dV'. \qquad (\text{T.96})$$

Note that dV' is volume, not potential. These potentials often are referred to as *retarded potentials,* as the sources that created them occurred at the retarded time.

T.15.1. Radiation from an electric dipole

Recall the electric dipole and its dipole moment from Fig. T.9 and related text. The dipole has a net charge of zero, but the charge can flow back and forth between the two charge locations in the dipole. If it does so at a frequency, ω (in units of radian per second, s^{-1}), we can express the time varying charge of the dipole source to be

$$q(t) = q_0\cos(\omega t) \qquad \text{(T.97)}$$

In terms of its moment, the dipole can be expressed as

$$\vec{p}(t) = q(t)\vec{d}$$
$$= p_0 \cos(\omega t)\hat{k}, \qquad \text{(T.98)}$$

where $p_0 = q_0 d$. Using Eqs. T.95 and T.96, we can write the differential form of the electric and magnetic fields radiated by a dipole. It is advantageous to use simplifications that come from the nature of the dipole; the current has only a z-component if the dipole is aligned with the z-axis. Approximations that are used in the development are that (1) the dipole separation is much less than the range to the field point (i.e., $d \ll R$), (2) the dipole separation is much less than the wavelength of the oscillations (i.e., $d \ll \lambda$), and (3) the range is much greater than the wavelength (i.e., $R \gg \lambda$). The last approximation allows us to consider only the radiation term. As we have mentioned, this means that expansion terms involving $1/R$ to a power ≥ 2 are too small to be significant, because the observation point is far from the source dipole. The solutions of Eqs. T.95 and T.96 for the potentials lead to the following expressions for the electric and magnetic fields:

$$\vec{E} = -\frac{\mu_0 p_0 \omega^2}{4\pi R} \sin\theta \cos\left[\omega\left(t - \frac{R}{c}\right)\right]\hat{\theta} \quad \text{(T.99)}$$

and

$$\vec{B} = -\frac{\mu_0 p_0 \omega^2}{4\pi c R} \sin\theta \cos\left[\omega\left(t - \frac{R}{c}\right)\right]\hat{\phi}, \quad \text{(T.100)}$$

where ϕ is the azimuthal angle, and θ is the angle from the z-axis. The waves described by these equations are spherical about the dipole. At long ranges, however, they can be closely approximated as a plane wave.

T.15.2. Radiation bands

Because they are used so much in storm research, we list the names and frequencies of several commonly used frequency bands in Table T.1. Note that the demarcation frequencies are sometimes defined with slightly different values.

Table T.1 Selected frequency band names and frequencies that are commonly used in atmospheric electricity.

Symbolic Designator	Name	Frequency
ELF	Extremely low frequency	30–300 Hz
VF	Voice frequency	300–3000 Hz
VLF	Very low frequency	3–30 kHz
LF	Low frequency	30–300 kHz
HF	High frequency	3–30 MHz
VHF	Very high frequency	30–300 MHz
UHF	Ultra high frequency	300–3000 MHz
	L band	0.39–1.55 GHz
	S band	1.55–3.90 GHz
	C band	3.90–6.20 GHz
	X band	6.20–10.90 GHz
	K band	10.90–36.00 GHz

T.16. MAGNETIC INDUCTION

While there is no such thing as a magnetic monopole, it turns out that the magnetic field at a distance from a conducting loop is analogous to the electric field from a static dipole. Furthermore, the simple closed loop is an important observational apparatus for detecting and locating lightning ground strikes (Sec. 6.10). For that reason, we give its basics here. The equations for a loop of wire moving in a magnetic field and for a changing magnetic field impinging upon a stationary loop are the same. However, the physical explanations are much different. When the magnetic field changes, it induces an electric field that causes a current to flow, which is called an electromotive force (emf). The applicable Maxwell's equation is Eq. T.77; it can be shown that

$$\frac{dB}{dt} = \frac{\mu_0}{2\pi} \frac{NA}{R^3} \frac{dI}{dt}, \qquad \text{(T.101)}$$

where N is the number of loops of wire, A is the loop area, R is the range of the source to our detecting loop (antenna), and I is the current that flows in the antenna.

1

Overview of the Electrical Nature of the Earth's Atmosphere

1.1. BRIEF HISTORY

Before electricity was discovered in the early 1700s, it was obviously impossible to know that Earth's atmosphere is filled with electric currents, that thunderstorms are electrified, and that lightning is a form of electricity. However, thunder and lightning are awesome phenomena, which have always spurred people to seek to explain them. Some ancient civilizations attributed them to gods: Thunder and lightning were struck from Thor's hammer in Nordic mythology, hurled by Zeus in Greek mythology, and governed by a collection of gods in Chinese mythology. When people began studying natural philosophy, they attempted rational explanations. Aristotle suggested that thunder and lightning were caused by interactions of moist and dry exhalations: As clouds condensed and cooled, the dry exhalation was expelled forcefully and struck other clouds. Thunder was the sound of clouds being struck, and lightning was a burning wind produced by the impact of the dry exhalation on clouds. In the 1600s, Descartes suggested that thunder was caused by a resonance of the air between clouds when one cloud descended on another.

As the electrical properties of matter were first being discovered, several scientists noticed that the sound and appearance of sparks resembled thunder and lightning, although the naturally occurring phenomena were much more powerful. The scientists suggested, therefore, that thunderstorms somehow generated electricity, which was discharged by lightning. Benjamin Franklin became famous for designing experiments to

test this hypothesis, but he was not the first to perform them. A few years before his famous kite experiment, which he conducted in June 1752 before learning of the success of others, Franklin had suggested in a letter that electricity could be drawn from a cloud by a tall metal pole. If the pole were insulated from ground and an observer brought a grounded wire near the pole, then a spark would jump from the pole to the wire when an electrified cloud was overhead (Fig. 1.1).

After Franklin's letters on electricity were collected and published in 1751, the French royal court was intrigued and tried a number of the experiments he described. D'Alibard decided to perform the experiment to test whether clouds were electrified, and so he constructed a pole on an insulated platform. On May 10, 1752, sparks were drawn from the pole when a thunderstorm was near. Other scientists continued these experiments with insulated metal poles, kites, and balloons. The danger involved in performing these experiments was manifested a little more than a year after D'Alibard's experiment, when Richman, a scientist in Russia, was killed by lightning that struck the pole he was observing.

Franklin continued to examine the nature of thunderstorm electrification. He had been the first to suggest that electricity was caused by a surplus of either positive or negative substance, and he was curious to learn the polarity of thunderstorms. In September 1752, he erected a 3-m long lightning rod above his house and ran the grounding wire through the well of his inside staircase. He broke the grounding wire and horizontally separated the ends by 15 cm with a bell attached

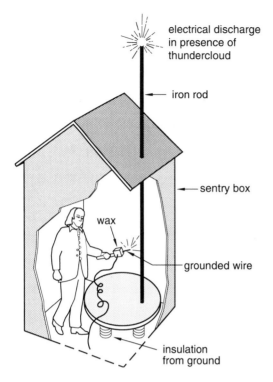

Fig. 1.1. Franklin's experiment that showed thunderclouds are electrified.

velopment of new instrumentation and to ingenuity in using and interpreting measurements to overcome the considerable obstacles posed by the phenomena. These obstacles include difficulty in accessing regions of interest; an environment hostile to in situ instruments; and a range of volume, distance, and temporal scales that in many cases is still impossible to sample adequately. Although considerable progress has been made, significant questions remain about such fundamental issues as the mechanisms by which thunderstorms produce lightning.

The focus of this textbook will be the electrical nature of storms, particularly of thunderstorms. Thunderstorms, by definition, are clouds that produce thunder and, hence, lightning. As information about cloud electrifi-

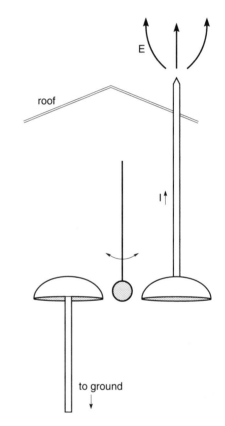

Fig. 1.2. Franklin's experiment to determine the polarity of thunderstorms. When an electrified cloud induced charge on the right bell, connected to a metal rod through the roof, it in turn induced an opposite charge on the suspended ball. The ball was then attracted to the right bell, hit the bell, and picked up charge from the bell. Since the ball had the same charge as the right bell, it was repelled by that bell, and so swung back toward the left bell. When it struck the left bell, it lost its charge, swung back toward the center, and repeated the cycle. The ringing alerted Franklin that an electrified cloud was nearby.

to each end (Fig. 1.2). An insulated metal ball suspended between the bells would swing back and forth and strike them when an electrified cloud passed overhead. By comparing the charge on the wire from the lightning rod with a known positive charge, Franklin determined that the bottom of storm clouds usually was negative.

Also in 1752, the French scientist Lemonnier detected weak electrification in the atmosphere when there were no clouds and determined that the intensity of fair weather electricity varied from night to day. This was confirmed in 1775 by the more sensitive experiments of the Italian scientist Beccaria, who determined further that the polarity of charge in the atmosphere was positive in fair weather and that it reversed to negative when thunderstorms were overhead, consistent with Franklin's observations.

There was relatively little progress in learning about these phenomena during the next hundred years, although there was considerable progress in learning about electricity and magnetism and in making machines to harness the newly discovered power of electricity. Comparable new discoveries about thunderstorms, lightning, and fair weather electrification had to await the development of photographic and electrical instruments to make the necessary measurements. Progress since then has been linked closely to the decation and lightning is presented, we will try also to point out gaps in the present knowledge. Before we deal with clouds and lightning in subsequent chapters, we will summarize in the present chapter what is known about the electrical properties of the atmosphere in which clouds occur. This will serve the dual purposes of placing clouds in their context and of introducing some concepts needed in later chapters.

1.2. EARTH'S MAGNETIC FIELD

Earth has a magnetic field that permeates the atmosphere and extends above the atmosphere into space. Its source is described to first order as a magnetic dipole whose axis is tilted approximately 11° with respect to Earth's spin axis (Fig. 1.3). The primary source of the magnetic field is electric currents flowing deep in Earth's interior. This source is poorly understood, but is thought to involve the interaction of rotation and convection in the electrically conducting fluid core of Earth. The magnetic field varies slowly, changing in magnitude by roughly 0.05% per year and precessing slowly around Earth's spin axis. Over time scales of hundreds of thousands of years, Earth's magnetic field has decreased to zero and recovered, sometimes reversing polarity. The magnetic field now points downward near the North Pole and upward near the South Pole. It is given approximately by

$$\vec{B} = \frac{-0.6R_E^3}{r^3}\sin\phi\,\hat{a}_r + \frac{0.3R_E^3}{r^3}\cos\phi\,\hat{a}_\phi, \quad (1.1)$$

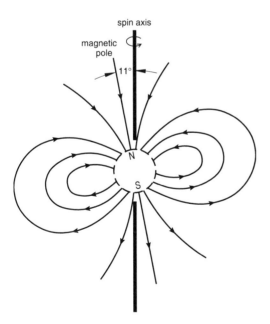

Fig. 1.3. Dipole magnetic field lines extending from Earth's surface into a vacuum.

where $\vec{B}$ is the magnetic field measured in 10^{-4} T (1 tesla $= 10^4$ gauss), R_E is the radius of Earth (mean value of about 6371 km), r is distance from the center of Earth, ϕ is latitude from the geomagnetic equator, and the unit vectors point radially outward and meridionally in the direction of increasing latitude, respectively. At the surface of Earth, the magnitude ranges from about 25,000 nT at the equator to about 60,000 nT near the poles; above the surface, it decreases as the cube of the radial distance. It is sometimes convenient to consider individual magnetic field lines, each of which is defined by starting at the point of interest and continuing parallel to the magnetic field at all points along its path. A particular magnetic field line is denoted by its L value, given by

$$r = L\cos^2\phi, \quad (1.2)$$

where r is measured in multiples of the Earth's radius.

1.3. THE IONOSPHERE

As we will see later, the ionosphere is important to electricity in Earth's lower atmosphere. The ionosphere is the region in the upper atmosphere where there are enough electrons and ions (see Sec. 1.4) to make the atmosphere a reasonably good conductor. Charged particles are created when solar radiation at wavelengths shorter than 102.7 nm is absorbed by atmospheric molecules and atoms. The energy from this extreme ultraviolet radiation is transferred to an electron in the molecule, which then escapes to become a free electron, leaving the molecule with an excess of positive charge. This process is called *photoionization*. Essentially, all solar radiation at extreme ultraviolet frequencies is absorbed in the ionosphere. Below the ionosphere, the atmosphere is weakly conducting. Cosmic rays from space and radioactive decay from the surface of Earth ionize a minute fraction of atmospheric molecules, enabling them to move in response to an electric field. However, the sparsity of ions below the ionosphere keeps conductivity below the ionosphere relatively low.

In the ionosphere overall, neutral atoms and molecules greatly outnumber electrons and ions, but there are still enough charged particles to create a discontinuity in conductivity between the ionosphere and lower atmosphere. Even in regions of the ionosphere where there is a surplus of positive or negative charge, the excess is such a small fraction of the total charge of each polarity that the number of positive ions can be considered to be equal to the sum of negative ions plus free electrons. In much of the ionosphere, the number density of negative ions is negligible. In any case, the conductivity of the ionosphere is due primarily to free electrons, because the mobility of free electrons is much greater than the mobility of ions. As a matter of convention, therefore, ionization of the ionosphere is often described in terms of the number density of electrons, N_e.

The N_e varies considerably with time of day, altitude, latitude, convergence or divergence of high alti-

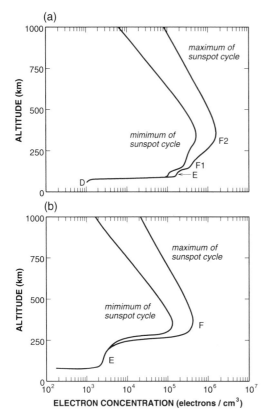

Fig. 1.4. Typical midlatitude distributions of electron number density at the extremes of the sunspot cycle for (a) daytime and (b) nighttime conditions. (From Hanson, 1961, with permission.)

tude winds, variations in the solar output of extreme ultraviolet radiation, and other local effects. However, the average pattern is for N_e to increase with height to a maximum at roughly 300 km. Figure 1.4 shows typical vertical profiles of N_e for both quiet and active periods in the 11-year solar sunspot cycle. Secondary maxima or changes in the slope of the profile below 300 km delineate the different layers or regions of the ionosphere. The main peak is in the F region. A secondary peak below the main peak is often referred to as the F1 region, with the main peak then being referred to as the F2 region. The E region is immediately below the F region, and the D region is still lower.

The structure of the vertical profile is the result of many complex interactions of sources and sinks. Much about the sources can be understood from a relatively simple model developed by Chapman (1931a, b). The production of ions in a given volume is proportional to the flux of radiation at the appropriate frequencies and to the number of molecules in the volume that can absorb and be ionized by the radiation. As the radiation penetrates deeper into the atmosphere, some of it is ab-

sorbed by atmospheric molecules and atoms, so the flux decreases and less is available to ionize a given volume. However, the number density of molecules increases with depth, so the percentage of remaining flux that will be absorbed in a given volume increases. As shown in Fig. 1.5, the result of these opposite tendencies with height is that there is some altitude at which ion production is maximized. Above that altitude, the number density of molecules is small enough that the decrease in density with height dominates the increasing flux, and ionization decreases. Below that altitude, the flux of radiation is small enough that the decrease in flux with depth dominates the increasing number density.

Chapman treated these tendencies quantitatively under a number of simplifying assumptions. If F is the flux of radiation at a particular wavelength, then the change in flux caused by absorption across a path length $dl = dz/\mu$ ($\mu = \cos\psi$ and ψ is the angle of the sun from local zenith) is given by

$$dF = \frac{\sigma_{ab}n(z)F}{\mu} \, dz, \quad (1.3)$$

where $n(z)$ is the number density of species that absorb that wavelength, and σ_{ab} is the absorption cross section by the species for that wavelength. Integrating from z to the top of the atmosphere (i.e., z goes to ∞) gives the flux in terms of F_0, the flux at the top of the atmosphere:

$$F(z) = F_0 \exp\left[-\frac{\sigma_{ab}}{\mu} \int_z^\infty n(z)dz \right]. \quad (1.4)$$

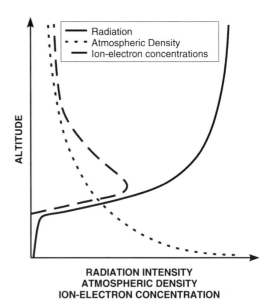

Fig. 1.5. Trends in radiation flux and particle number density that result in a Chapman layer. (From Whitten and Poppoff, 1971, with permission.)

Although we will not derive an expression for n(z), one can be found by considering that the equilibrium pressure of a given parcel of air at height, z, must be equal to the weight of the overlying atmosphere. For an ideal gas, this implies that

$$n(z) = n_0 e^{-\frac{(z-z_0)}{H}}, \qquad (1.5)$$

where n_0 is the number density at a reference height z_0, H is the scale height given by $H = k_B T/mg$, and k_B is Boltzmann's constant; T is temperature in K; m is molecular weight, and g is acceleration of gravity. For an isothermal atmosphere, we can substitute this expression for n(z) in Eq. 1.4 and integrate to get

$$F(z) = F_0 \exp\left[-\frac{\sigma_{ab} n_0 H}{\mu} e^{-\frac{(z-z_0)}{H}} \right]. \quad (1.6)$$

The ionization rate, q(z), from this flux is given simply by $n(z)\beta F(z)$, where β is the molecular cross section for ionization. Therefore,

$$q(z) = \beta n_0 F_0 \exp\left[-\frac{z-z_0}{H} - \frac{\sigma_{ab} n_0 H}{\mu} e^{-\frac{(z-z_0)}{H}} \right]. \tag{1.7}$$

To determine the number density of ions and electrons, we must consider also their destruction by recombination. In much of the ionosphere, the recombination rate is $q_r(z) = \alpha N_e^2$, where α is the recombination coefficient. Because the condition for equilibrium is the production rate equal to the recombination rate, then at equilibrium

$$N_e(z) = \sqrt{\frac{q(z)}{\alpha}}$$

$$= \left(\frac{\beta n_0 F_0}{\alpha} \right)^{\frac{1}{2}} \exp\left[-\frac{z-z_0}{2H} - \frac{\sigma_{ab} n_0 H}{2\mu} e^{-\frac{(z-z_0)}{H}} \right]. \tag{1.8}$$

To determine z_{max}, the height of maximum N_e, we take the derivative of the right side of Eq. 1.8, set it equal to zero, and solve for z. The result is that

$$z_{max} - z_0 = H \ln\left(\frac{\sigma_{ab} n_0 H}{\mu} \right). \qquad (1.9)$$

Substituting this expression in Eqs. 1.7 and 1.8 gives

$$q(z_{max}) = \frac{\beta F_0 \mu}{\sigma_{ab} H} e^{-1}$$

$$N_e(z_{max}) = \left(\frac{\beta F_0 \mu}{\sigma_{ab} \alpha H} \right)^{\frac{1}{2}} e^{-\frac{1}{2}}. \qquad (1.10)$$

The dependence of these quantities on μ and T (through their dependence on H) explains much about the diurnal and seasonal variations in the ionosphere, which we will discuss later.

The above treatment assumes an isothermal atmosphere, monochromatic radiation, plane geometry, thermodynamic equilibrium, no diffusion of ions and electrons, and recombination proportional to N_e^2. These assumptions are not strictly valid, but they help give an understanding of the formation of layers of ionization. The next order of approximation is to consider all ionizing species and wavelengths in a given layer. With these improvements, the Chapman theory describes the E and F2 regions reasonably well. Such a layer is called a *Chapman layer*.

1.3.1. Regions of the ionosphere

The formation of each region of the ionosphere depends largely on the wavelengths and species that dominate photoionization and recombination in that layer. In the D region, for example, the flux of radiation at many wavelengths has been drastically reduced by absorption in higher layers. Longer wavelengths of extreme ultraviolet radiation ($\lambda < 111.8$ nm), x-rays, and solar cosmic rays are primarily responsible for ionization. These ionize NO, O_2, and N_2, but O_2^+ and N_2^+ readily transfer charge to NO. Electrons quickly attach to O_2 ions in the D region to form negative ions, so N_e is not equal to the number density of ions, although the region is neutral. This ready capture of electrons also quickly depletes the free electron population at night, when the ionization rate is essentially zero, so conductivity in the D region becomes almost negligible at night. Recombination of charge occurs primarily by mutual neutralization of pairs of positive and negative molecular ions ($X^+ + Y^- \rightarrow X + Y$). The N_e in the D region cannot be described as a Chapman layer, because in the D region the cosmic ray flux is only weakly dependent on altitude, and much of the absorption of ionizing wavelengths is by molecular species that do not become ionized.

The next layer up, the E region, is described reasonably well by a Chapman layer. Lyman β, extreme ultraviolet ($\lambda < 100$ nm), and soft x-rays are the primary ionizing radiation. O_2 and N_2 are the primary species ionized, although NO and atomic oxygen O also contribute. In the E region at 110–20 km MSL (mean sea level), a transition occurs from well mixed molecular species in the lower atmosphere to each species attaining its own balance between buoyancy and pressure from the overlying atmosphere. Therefore, in upper parts of the E region, number densities of heavier molecules fall off more rapidly than for lighter molecules and atoms, and photoionization of atomic oxygen becomes increasingly important. Because the density of molecules is much less than in the D region, the main loss mechanism is by dissociative recombination of a molecule that reacts with an electron (e.g., $O_2^+ + e \rightarrow O + O$), and the fraction of electrons attaching themselves to neutral O_2 to create negative ions is negligible.

The F1 layer also is described reasonably well as a Chapman layer. The principal source is ionization of atomic oxygen by radiation at even shorter wavelengths ($\lambda < 91.1$ nm), with some contributions from ionization of N_2. Although the loss mechanism still involves dissociative recombination of molecules, it must proceed as a two-step process in the F1 region, because the dominant source of ionization is photoionization of an atom, not a molecule. In the first step, O^+ reacts with O_2 or N_2 by atom-ion interchange (e.g., $O^+ + NO \rightarrow NO^+ + O$) to create ionized molecules. The ionized molecules then recombine with electrons by dissociative recombination. In the upper part of the F1 region, the reaction rate for atom-ion interchange is slower and so controls the recombination rate. In the lower part, the reaction rate for dissociative recombination is the slower of the two, because N_e is smaller, while concentrations of neutral O_2 and N_2 are much larger.

The F2 region is a modified Chapman layer. Like the F1 region, the principal source is photoionization of atomic oxygen, with contributions from N_2, although the reactions are dominated by shorter wavelengths of radiation ($\lambda < 80$ nm). Also, the loss of electrons still occurs in a two-step process, and like the upper part of the F1 region, the reaction rate for atom-ion interchange controls recombination rates. This reaction rate is proportional to the first power of N_e, not the square. The F2 region also differs from lower regions in that diffusion eventually becomes fast enough at high altitudes to dominate chemical reaction rates in establishing equilibrium concentrations of N_e. When this happens, the tendency (in the absence of electric coupling) is for N_e to decrease less rapidly with height than atomic ion concentrations, because electrons have much less mass than atoms. However, the resulting

separation of positive and negative charge (called ambipolar diffusion) creates an electric field that pulls ions upward and electrons downward, thereby coupling them as an ion-electron gas. It can be shown that ambipolar diffusion causes the concentration of positive ions and electrons to decrease with altitude about half as rapidly as the concentration of neutral atoms (e.g., see Chamberlain and Hunten 1987).

1.3.2. Variations in the ionosphere

Variations in the vertical profile of N_e are to be expected from the properties of the regions discussed above. The primary diurnal variation of N_e is a result of the diurnal variation in solar radiation. After the sun sets below the local horizon, N_e decreases as the photoionization rate $q(z)$ becomes essentially zero (Fig. 1.6). (Slight ion production continues from galactic cosmic rays and from light scattered into the night sky.) Because the onset of dusk is later at higher altitudes, N_e begins to decrease later in higher regions. Furthermore, recombination of electrons is slower in higher regions, as discussed above, so N_e persists longer at night in higher regions.

One result of the unique characteristics of the F2 layer is that the layer stabilizes at significant values of N_e a few hours after sunset (demonstrated, for example, by Chamberlain and Hunten 1987). Initially, the altitude of maximum N_e increases, as in other layers. After the region stabilizes, however, the altitude at which N_e peaks remains about the same, and the subsequent fractional decrease of N_e is about the same at all heights in the F2 region, with an exponential time

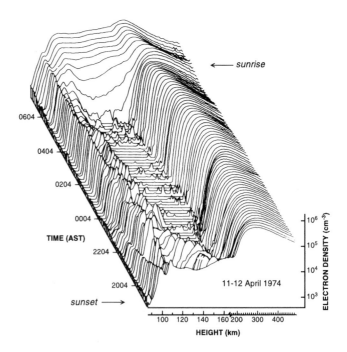

Fig. 1.6. Electron number density from sunset until after sunrise during a typical night over Arecibo, Puerto Rico. (From Shen et al. 1976, with permission.)

constant of roughly 10 h. This is consistent with the long-established observation that the F layer is the only layer to survive the whole night, although sporadic regions of ionization occur in the E region.

Seasonal variations of N_e are a consequence of seasonal variations of (1) the minimum zenith angle that the sun attains during the day and (2) changes in temperature, which affect global wind patterns and concentrations of molecular species in the upper atmosphere. Recall from Eq. 1.10 that the maximum values of both N_e and the ionization rate $q(z)$ increase as μ increases toward 1 (i.e., the minimum zenith angle approaches closer to zero), but decrease with increasing scale height. Since the minimum zenith angle of the sun varies with the season at a given latitude, the maximum daily value of N_e at a given latitude should have a corresponding seasonal variation that is a function of latitude. However, the effects of changes in temperature, circulation, and molecular concentrations can dominate the effect of changes in minimum zenith angle. For example, warming in the summer ionosphere increases concentrations of molecular species at high altitudes, which increases recombination rates and results in generally lower values of N_e in the summer F region.

The ionosphere also varies with the 11-year solar cycle. The flux of extreme ultraviolet radiation from the sun during active periods can increase by more than a factor of two over values during the quiet phase of the solar cycle. Also, during active periods, solar x-ray fluxes can increase by an order of magnitude, and there is a localized increase in fluxes of electrons into polar regions. These changes affect the ionosphere in several ways. First, the increased flux increases ionization rates and rates of photodissociation of molecules into atoms, thereby increasing N_e. Second, the upper atmosphere is heated, sometimes by hundreds of degrees, so it expands, and concentrations of heavier species such as O_2 and N_2 increase at higher altitudes. Third, the greater flux of particles into polar regions provides localized heating that for periods of hours to days can equal total heating of the upper atmosphere from solar radiation. This localized heating modifies global wind circulation patterns above 100 km, and the resulting changes in chemical concentrations affect N_e. The average effect of all changes during the active phase of the solar cycle is that N_e and the height of maximum N_e increase. However, the degree of increase fluctuates considerably with the rapidly varying flux of radiation and particles from the sun during its active phase.

1.4. GLOBAL ELECTRICAL PROPERTIES BELOW THE IONOSPHERE

In *fair weather*, the electrical state of the lower to middle atmosphere is in quasi-static equilibrium (i.e., the charge moving into a region equals the charge leaving the region). The definition of fair weather can be as simple as "no thunderstorms around," or it can be more specific (no hydrometeors, no significant dust blowing,

etc.). In quasi-static equilibrium, the vertical distribution of charge would be essentially the same at different times, and the laws of electrostatics apply. Relative to storm electricity, we consider fair weather as clear air or clouds that are not electrified enough to reverse the polarity of the electric field at the ground.

The overall electrical structure of the atmosphere below the ionosphere is usually described as a spherical capacitor filled with a slightly conductive medium—the atmosphere. Coulomb (1795) discovered that air is conductive, but the idea did not gain acceptance until much later. The outer shell of the capacitor is the highly conductive region of the upper atmosphere; Earth's surface, also highly conductive compared with the lower atmosphere, is the inner shell. The capacitor in this description is charged, with roughly 5×10^5 C of negative charge on Earth and an equal positive charge in the atmosphere. Because the atmosphere is weakly conducting, there is a leakage current that would neutralize the charge on Earth and in the atmosphere on a time scale of roughly 10 min. That Earth is negatively charged was stated by Peltier (1842), but he did not address how to maintain the charge on Earth in the presence of a leakage current. Wilson (1920) completed the circuit concept by proposing that "a thunder-cloud or shower-cloud is the seat of an electromotive force which must cause a current to flow through the cloud between Earth's surface and the upper atmosphere." That there is an ever present positive conduction current to the ground in fair weather is not disputed; its source is generally attributed to thunderstorm generators. What follows is a simplified version of the air-earth circuit (i.e., the global circuit of the lower atmosphere). Extensive reviews are included in the papers referenced in this chapter.

Some caution is needed when reading this literature, because for decades many who investigated electrical properties of the atmosphere used "electric field" when they really meant "potential gradient." The physical property measured by these terms is the same, but the *electric field* is the negative of the *potential gradient*, $\vec{E} = -\vec{\nabla}V$. In this book, we mostly follow the conventions often used in physics. Specifically, the positive direction is up, and the direction of the electric field is the direction in which a positive test charge would move under the influence of the field. Therefore, for example, a positive electric field means that a positive test charge would move upward. We are specific with the terms; field always mean the electric field, never the potential gradient.

In fair weather, the electric field at Earth's surface is caused by net positive charge overhead, which drives a positive charge downward, and so is negative. The electric field at the ground in nonmountainous terrain is about -100 V m^{-1}, and thus the often used potential gradient is $+100$ V m^{-1}. The magnitude of the electric field decreases with altitude. For example, Gish (1944) gave the relationship with height as

$$E(z) = -[81.8\ e^{(-4.52z)} + 38.6\ e^{(-0.375z)} \\ + 10.27\ e^{(-0.121z)}], \qquad (1.11)$$

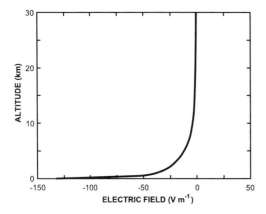

Fig. 1.7. Profile of the electric field in fair weather using a relationship derived by Gish (1944).

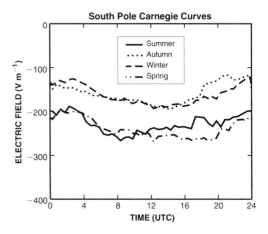

Fig. 1.9. Southern hemisphere seasonal variation in fair weather electric field. (After Cleary et al. 1997, with permission.)

where E is electric field in volts per meter and z is altitude in kilometers. A plot of this relationship up to a height of 30 km is shown in Fig. 1.7. Subsequently, other functional forms have been suggested for E(z), but the profiles are nearly identical.

The diurnal and seasonal variations of electrical properties of the lower atmosphere were first clearly identified by measurements of the electric field and other parameters on the research vessel Carnegie (Torreson et al. 1946). This unique sailing vessel covered over 200,000 km on its seventh and final voyage during 1928–29 before fire destroyed it. The research done with the Carnegie during several voyages provided much of our understanding of Earth's fair weather atmospheric electrical environment. A curve of hourly values of the electric field averaged over many days, the so-called *Carnegie curve* in Fig. 1.8,

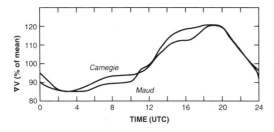

Fig. 1.8. Diurnal variation of fair weather potential gradient at the ocean surface from measurements aboard the sailing research ships, Carnegie and Maud, versus universal time (UTC). The vertical axis shows percentage of the mean potential gradient for each. This variation in potential gradient is often referred to as the Carnegie curve. (From Whipple and Scrase 1936, © Crown copyright. Reproduced with the permission of the Controller of Her Majesty's Stationery Office.)

shows a repeatable diurnal variation as a function of universal time. The diurnal pattern is attributed to total thunderstorm activity around the world. A seasonal variation is also seen in Fig. 1.9. The curves for the four seasons have about the same shape, but two distinct seasonal amplitude modes exist (e.g., Cleary et al. 1997).

Historically, the question of whether the global circuit is really driven by thunderstorm activity has been addressed by comparing total thunderstorm activity with the potential gradient measured at the surface. Appleton (1925) discussed the diurnal variation and assumed that the pattern was linked to storm occurrence, with most storms occurring in a "wide equatorial belt." He also concluded that South Africa and South America are the "centers of gravity" of thunderstorm regions. The best known early attempt to show this was begun by Whipple (1929), who used thunderstorm data compiled by Brooks (1925). Whipple then estimated the thunderstorm activity in 2-h blocks to get diurnal activity. He converted thunderstorm data to land mass area affected, which put the greater diurnal variation in summer in the southern hemisphere. If the electric field and thunderstorm frequency curves are normalized, the amplitude variations are not the same (the thunderstorm curve has greater variations). The electrical activity of individual generators was not considered.

Whipple's analysis showed that the number of thunderstorms varied by 15% more during summer in the southern hemisphere than during summer in the northern hemisphere. Whipple and Scrase (1936) plotted these two parameters and noted the similarity (Fig. 1.10). Their work has continued to be the standard against which the hypothesis of the thunderstorm role has been compared. Israël (1973) calculated a correlation coefficient between Whipple and Scrase's two curves of 0.94. The concept of the thunderstorm generator in the fair weather electric circuit is shown in

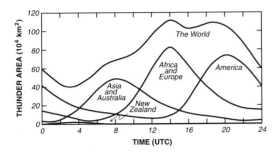

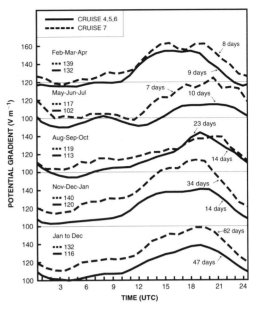

Fig. 1.10. Diurnal variation of worldwide areas of thunder over land. Individual curves are labeled to indicate the continental regions whose thunderstorm activity is plotted. The world curve is a composite of all the others. (From Whipple and Scrase 1936, © Crown copyright. Reproduced with the permission of the Controller of Her Majesty's Stationery Office.)

Fig. 1.12. Mean diurnal variations of the potential gradient for 3-month periods measured during fair weather on four cruises of the Carnegie during the years 1915–21 (cruises 4–6) and 1928–29 (cruise 7). The numbers by the line codes in each are the mean potential gradient for the curve. (From Torreson et al. 1946, with permission.)

Fig. 1.11. Discussion of the currents below the storm from lightning, precipitation, and point discharge will be discussed in later chapters, but are shown in the figure for completeness.

Attempts to reproduce the results of the Carnegie measurements at land stations have often been disappointing, a result of local effects such as convection currents, increased aerosols, etc. Land measurements have always required a lot of averaging to reveal a convincing, repeatable diurnal pattern. For example, in his review article, Ogawa (1985) noted that the diurnal variation showed a "local" effect due to cities, and this effect extended up to 100 km away. In addition, as Fig. 1.12 shows, the diurnal pattern determined from four Carnegie cruises varied as a function of location and season. The overall pattern, however, appeared much the same, supporting the contention that thunderstorms are the global generator.

Other measurements have been used to look for the diurnal effect. These include measurement of the conduction current, which according to current continuity and conservation of charge should be constant with

height, and the potential of the upper plate of the spherical capacitor. Both are affected by local contamination, especially if the measurements are made at the ground or with vertical soundings to only a few kilometers. The upper plate in the model has been called the *electrosphere*. The term ionosphere is also used widely in the literature, especially when discussing the potential of the electrosphere, but it is technically incorrect, at least from the perspective of atmospheric electricity. We will use the term electrosphere because of its unambiguous meaning for atmospheric electric-

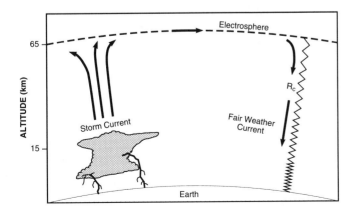

Fig. 1.11. Simple conceptual model of the main global circuit. Thunderstorm 'generators' drive current to the highly conductive electrosphere and back to ground as fair weather current. R_c is the columnar resistance, which is greater in the lower atmosphere. The altitudes are approximate, typical ones; not to scale.

ity. The *electrospheric potential*, $\Phi_E(z)$, can be found by integrating a vertical profile (i.e., *sounding*) of the electric field using

$$\Phi_E(z) = \int_0^{H_E} E(z)dz, \qquad (1.12)$$

where H_E is the height of the electrosphere. Mühleisen (1977) and Markson (1986) have carried out many measurements of electrospheric potential. Mühleisen conducted and reported on radiosonde measurements of the electric field for more than 15 years. As seen in Fig. 1.13, the average of 125 flights at different times of the day closely matches the Carnegie curve. Mühleisen pointed out that although Φ_E is the best global parameter, individual values vary significantly, with his measurements from a larger data set of over 300 soundings having a range of $\Phi_E = 145$–608 kV with a mean of 278 kV.

Markson (1986) summarized his 14 years of airborne measurements of the electric field and conductivity to a height of about 7 km. Although there are some exceptions, he found that a single day's data often fit the Carnegie curve; no longer term averaging was needed to show the classic diurnal pattern. He calculated a yearly average $\Phi_E = 240$–260 kV. Soundings to only 7 km give an estimate of Φ_E, but profiles to greater heights are inherently more accurate, prompting Kasemir (1977) to suggest that we should report the potential to the height measured and that profiles to 20 km are adequate for good estimates of Φ_E, as about 99% of the potential drop occurs in that interval.

There have been attempts to monitor the global circuit by monitoring the conduction current with soundings. The pros and cons of this method compared with Φ_E were discussed by Markson (1987) and Few and Weinheimer (1987).

1.4.1. Atmospheric ions

As mentioned previously, the lower atmosphere is a weak conductor because it contains trace concentrations of ions. Ions are created initially by cosmic rays from galactic space and radioactive decay at Earth's surface, which produces alpha and beta particles and gamma rays. Radiation from both sources interacts with neutral molecules, generally nitrogen and oxygen in the lower atmosphere, to create free electrons and positive ions by ionization. In simple terms, ionization means that an electron is stripped from a molecule or atom. The molecule then has a net positive charge (i.e., it is a positive ion). The electron quickly attaches to one of the neutral molecules (which are much more numerous than ions) to make a negative ion.

Positive and negative ions move in opposite directions under the influence of an electric field, so current flows in the atmosphere whenever an electric field is present. In the fair weather atmosphere, the relationship between currents and electric fields is given by Ohm's law, $J = \sigma E$, where σ is conductivity. Currents can also be created by winds advecting charge and by non-ohmic breakdown leading to corona or sparks, but these types of current normally are ignored when dealing with fair weather.

The conductivity of the lower atmosphere is due primarily to small ions. Larger ions are present as well, especially near the ground and, in fact, a variety of ion categories have been defined. However, large ions contribute only a negligible part of the current driven by an electric field, because their larger mass makes them accelerate much more slowly under a given force. Therefore, small ions dominate the motions of charges that create fair weather currents. Diameters of small ions are 10^{-10}–10^{-9} m. Their mass is sufficiently small that the ions move significantly under the force of the electric field, even the relatively weak fair weather electric field. The *small-ion charge* is a single elementary charge (i.e., equivalent to one electron's charge of 1.6×10^{-19} C).

Small-ion concentrations depend on the balance between production and loss mechanisms. Near Earth's surface, small ions are produced at about 10 ion-pairs cm^{-3} s^{-1} (in MKS units it is 10^7 ion-pairs m^{-3} s^{-1}). Ion-pair production over land is about twice that over oceans; large water surfaces have no significant ra-

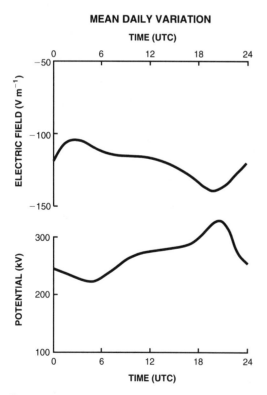

MEAN DAILY VARIATION

Fig. 1.13. Mean diurnal variations of electric field from the Carnegie (top) compared with the potential of the electrosphere (bottom) from balloon soundings of the potential gradient. (After Mühleisen 1977, with permission.)

dioactive emanation. Cosmic radiation produces about half the ions at the surface. In contrast, at altitudes beginning at roughly 1 km, cosmic rays produce most ions in the fair weather atmosphere, regardless of the presence of land below.

Atmospheric ions are destroyed by recombination of small ions and scavenging by aerosols. In equilibrium, ion-pair production as a function of altitude, p(z), must balance destruction, so

$$p(z) = \alpha(z)n(z)^2 + \beta(z)\, n(z)\, N(z), \qquad (1.13)$$

where z is height, α is the recombination coefficient between small ions, n is the small ion concentration, β is the attachment coefficient between small ions and aerosols, and N is the aerosol concentration. The second term is negligible in clean air environments and at high altitudes, but dominates the first when there are significant concentrations of large particulates from pollution or other sources. Typical lifetimes of small ions range from about 10 s at low heights in cities, where N is large, to 300 s over the oceans well away from land. As a result, conductivity is much smaller near the ground in cities than in remote areas, but E is larger.

A detailed description of the atmospheric ions causing atmospheric conductivity is complicated, with many chemical species involved. We can, however, offer a few generalities. First, even the most basic study of ion formation reveals a variety of possible formation mechanisms, including simple charge transfer, third body reactions, switching mechanisms, proton transfer reactions, and clustering (Mohnen 1977). The ion population at any particular place is influenced by the trace gases that are present, their concentrations, and other ion formation parameters. According to Keese and Castleman (1985), N_2^+, O_2^+, N^+, and O^+ are the primary positive ions below 50 km. From these fundamental ions, others such as $H^+(H_2O)_n$, where n is the number of water molecules in the cluster, are rapidly created. Negative ions begin with oxygen capturing a free electron. From there, a whole sequence of ions can occur, with the family of ions of $NO_3^-(HNO_3)_n$ dominating below 30 km. However, a comprehensive understanding of which ions contribute to conductivity in the lower atmosphere is lacking, in part because of the complexity of chemical reactions and the variety of chemical species involved.

1.4.2. Mobility of ions

The characteristic of ions that relates directly to their ability to take part in the flow of electric current is their mobility. *Mobility*, k, is defined as the average velocity acquired by a small ion as it moves under the force exerted on it by the electric field. Mobility is expressed by $k = v/E$. The unit of mobility is meter per second per volt per meter, but it is often written as the less physically obvious meters squared per volt second ($m^2\, V^{-1}\, s^{-1}$). A representative value of small-ion mobility at sea level is about $10^{-4}\, m^2\, V^{-1}\, s^{-1}$. The mobility of

large ions is three orders of magnitude lower, which is why they do not contribute to conductivity. In addition, k decreases by about 18% as relative humidity ranges from 0–100%. The electric field accelerates the ion, but a "terminal" velocity is reached because of the collisions of the ion with air molecules. Stokes's law for particle motion in a viscous fluid (such as the atmosphere) governs mobility. Therefore, mobility is inversely proportional to pressure, and we expect it to increase at higher altitudes. Temperature and pressure affect mobility by

$$k(P,\, T) = k_s\, \frac{P_s}{P}\, \frac{T}{T_s}, \qquad (1.14)$$

where the subscript s refers to standard conditions. Since mobility is a function of collisions, it and the mean free path of an ion increase as air density decreases. For this reason, mobility at altitude is often calculated as reduced mobility,

$$k_r = \frac{\rho_A}{\rho_s}\, k_s, \qquad (1.15)$$

where ρ_A is the ambient air density, and ρ_s is air density at standard temperature and pressure.

1.4.3. Conductivity

The ability of the atmosphere to conduct an electric current is expressed in terms of its conductivity. It is an important parameter in both fair weather and stormy conditions. The study of the conductivity caused by small ions covers a broad spectrum of laboratory and theoretical work. For our purposes, the following summary will suffice. The increase of total conductivity, σ_{tot}, with height can be expressed in several basic forms, including

$$\sigma_{tot}(z) = q_e \sum_i n_i(z)k_i(z), \qquad (1.16)$$

where the summation covers all varieties of small ions, and q_e is the elementary charge on each ion. The total conductivity is the sum of that due to negative and positive ions,

$$\sigma_{tot}(z) = \sigma_+(z) + \sigma_-(z). \qquad (1.17)$$

Combining the above two equations and assuming equality of mobility and number density for each polarity gives this relationship for conductivity:

$$\sigma_{tot}(z) \approx 2q_e n(z)k(z). \qquad (1.18)$$

Figure 1.14 shows measurements of the variation of positive ion concentration n_+ with altitude; it varies from about 1000 cm^{-3} near ground to about 6000 cm^{-3} at 15 km. For greater heights, Mitchell and Hale (1973) calculated from several rocket flights that n^+ ranges from 3000–7000 cm^{-3} within 30–75 km, with a minimum of <1000 cm^{-3} at about 65 km (Fig. 1.15). Owing to the increase of mobility with altitude, conductivity continues to increase.

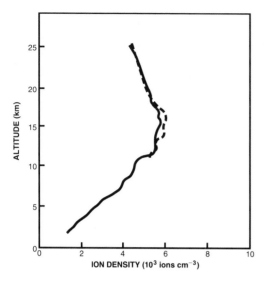

Fig. 1.14. Smoothed profiles of balloon-borne measurements of the positive ion density over Wyoming on 24 February 1978. The solid line is ascent; the dashed line is descent (From Rosen and Hofman 1981: American Geophysical Union, with permission.)

Early measurements of conductivity were made by Gerdien (1905) who developed an instrument, which still carries his name, to measure the conductivity directly. An important pioneering study of the vertical distribution of conductivity was carried out by Gish and Sherman (1936). They mounted an aspirated Gerdien cylinder on a gondola beneath a large balloon, named Explorer II, and measured conductivity to 22 km. They provided a detailed analysis of their experimental procedure, comparison with laboratory measurements, and hypotheses to explain their measurements. They found the ratio of positive to negative

conductivity averaged about 0.8. Owing to the higher mobility of small negative ions compared to positive ones, we could expect the negative conductivity to be 1.4 times the positive. What is found from measurements, however, is that this is only sometimes true. This is in part explained by the number density of positive ions being about 1.2 times that for negative (Wilkening 1985).

Following Gish and Sherman (1936), Woessner et al. (1958) made simultaneous measurements of both polarities of conductivity from the ground to 26 km. They found the following expressions (converted here to MKS units) to describe their conductivity measurements and the ratio of conductivity with height:

$$\sigma_+(z) = 3.33 \times 10^{-14}\, e^{(0.254z - 0.00309z^2)}, \quad (1.19)$$

$$\sigma_-(z) = 5.34 \times 10^{-14} e^{(0.222z - 0.00255z^2)}, \text{ and} (1.20)$$

$$\frac{\sigma_+(z)}{\sigma_-(z)} = 0.624\, e^{(3.2 \times 10^{-2} z - 5.4 \times 10^{-4} z^2)}, \quad (1.21)$$

where z is in kilometers. Graphs of these relationships are shown in Figs. 1.16 and 1.17.

The conductivity of the atmosphere gives it a property called *relaxation time,* which is the time for an isolated charged object to become $1/e$ of its original charge. The relaxation time is given by $\tau = \varepsilon/\sigma_{tot}$, where ε is the *permittivity of air* ($\approx 8.86 \times 10^{-12}\, \text{F m}^{-1}$). Nearly all the charge will be gone from the object within about 5τ. Just above the ground in fair weather, $\tau \approx 7$ min. As conductivity increases with height in the atmosphere (Fig. 1.18), τ decreases, eventually becoming four orders of magnitude smaller at the electrosphere.

1.4.4. Conduction current

Conduction current is a fundamental parameter in the global circuit and is defined by $\vec{J} = \sigma \vec{E}$. Note that J is a current density, but it is usually referred to simply as conduction current. This is a general form of Ohm's

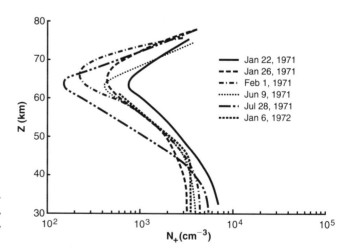

Fig. 1.15. Positive ion density from several rocket flights from Wallops Island, Virginia. (After Mitchell and Hale 1973, with permission.)

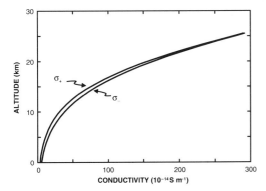

Fig. 1.16. Positive and negative conductivity for fair weather from the relationship derived by Woessner et al. (1958) from their balloon-borne Gerdien cylinders.

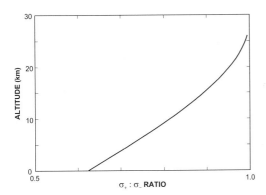

Fig. 1.17. Average ratio of positive to negative conductivity from measurements by Woessner et al. (1958).

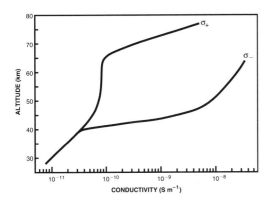

Fig. 1.18. Measured variations in conductivity above 30 km. (After Mitchell and Hale 1973, with permission.)

law. It has been put forth as a measurement suitable for use in monitoring the diurnal change in the global circuit, but it has practical deficiencies for this task. In particular, measurements made at Earth's surface suffer from the noise of convection current, radioactive decay, the electrode layer, etc. Conduction current can be measured better aloft, either directly with balloon-borne current sondes or calculated from electric field and conductivity measurements made with balloons or airplanes. A discussion of measurement problems with sondes having long current sensing antennas was given by Few and Weinheimer (1986).

Conservation of charge and current continuity indicate that equilibrium values of the vertical current density should be approximately independent of height in the lower atmosphere. Otherwise, net charge would collect at some height and modify the electric field and current density until the modified vertical current density was uniform. Measurements of electric field and conductivity with airplanes have shown that conduction current is approximately constant above the layer of mixing near the surface. Over the ocean, Kraakevik (1961) found that J varied only 10% from above 15 m to 6 km. (A low-level anomaly was attributed to positive charge emanating from the sea surface and also could have been a manifestation of the electrode effect; see Sec. 1.4.6).

Kraakevik also measured current density to the ground, which is sometimes called the *air-earth current*. He found that values of the air-earth current ranged from about 3.9×10^{-12} A m^{-2} in clean Arctic air to about 1.3×10^{-12} A m^{-2} in continental air. At heights greater than 100 m, he found an average atmospheric conduction current density of 2.7×10^{-12} A m^{-2}. He estimated that the total fair weather current to Earth is $+1400$ A (in this work the convention was that positive charge to the surface of Earth was a positive air-earth current).

1.4.5. Space charge

Fair weather *space charge density* (often called *space charge*), ρ, and the vertical gradient of the electric field are related by Gauss's law, $\rho = \varepsilon \vec{\nabla} \cdot \vec{E}$. If we assume that the horizontal gradients of the electric field can be ignored, then Gauss's law in one dimension becomes

$$\frac{\partial E_z(z)}{\partial z} = \frac{\rho(z)}{\varepsilon} , \qquad (1.22)$$

where $\rho(z)$ is the altitude-dependent space charge density at height z. Using Eq. 1.11 for $E_z(z)$ gives

$$\rho_{\text{tot}}(z) = \varepsilon[0.370 \, e^{-4.52z} + 0.0145 \, e^{-0.375z} + 0.0012 \, e^{-0.121z}]. \qquad (1.23)$$

The fair weather charge density decreases rapidly with height. To examine the height distribution of charge further, we can estimate the fraction of the total charge below a height z_0. Integrating Eq. 1.22 and evaluating

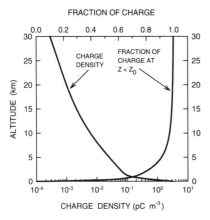

Fig. 1.19. Variation of charge with height in fair weather.

it for the total atmosphere and for the layer up to z_0 gives

$$\frac{Q(z < z_0)}{Q_{tot}} = 1 - \frac{E_z(z_0)}{E_z(0)} \qquad (1.24)$$

$$= 1 - [0.626\,e^{-4.52z} + 0.295\,e^{-0.375z}$$

$$+ 0.079\,e^{-0.121z}].$$

Plots of Eqs. 1.23 and 1.24 up to a height of 30 km are shown in Fig. 1.19. The plots indicate that when E_z is given by Eq. 1.11, more than 70% of the charge in the fair weather atmosphere is below 1 km, and 90% is below 4.5 km. The total charge below 30 km in the entire fair weather atmosphere, obtained by integrating the charge density of Eq. 1.23, is approximately 1.9×10^5 C, of which 1.4×10^5C is below 1 km.

The variation of both E and ρ_{tot} with height can be viewed as a response to the variation of conductivity with height. As we have discussed, the variation of conductivity in the lower to middle atmosphere is determined primarily by that region's ionizing radiation, hydrostatics, and chemistry. If the fair weather current has a particular value and is independent of height, Ohm's law indicates that the height dependence of conductivity will determine the height dependence of E, and Eq. 1.22 shows that this is equivalent to determining how ρ varies with height. A similar treatment of the fair weather space charge was given by Mason (1971, p. 483 ff.).

Note that our estimate of the amount of positive charge residing in the atmosphere below 30 km is approximately 40% of the magnitude of charge commonly estimated to be on the capacitor formed by the Earth and the electrosphere in the electrostatic global circuit. However, both of these estimates are very rough, so 40% may be erroneous. The fact that the magnitude of the vertical electric field at 30 km is much weaker than that near ground suggests that the amount of net charge in the atmosphere above 30 km is much smaller than the amount in the atmosphere below 30

km. In fact, this analysis suggests that most of the net positive charge in the atmospheric component of the electrostatic circuit resides in the lowest 1 km of the atmosphere.

1.4.6. Electrode layer

One manifestation of the conductive atmosphere and the electric field is the occurrence of an electrode layer. The negative electric field in fair weather drives positive ions downward and negative ions upward. Both polarities of ions contribute to the total current. If we consider an infinitesimal layer of air in equilibrium well above the ground (Fig. 1.20), the flux of ions of each polarity moving across the layer's upper boundary equals the flux across the lower boundary, so there is no further build-up of ions. When the lower boundary of the layer is the ground, the situation changes: No ions enter from below because of the solid boundary. Because negative ions are the ones that move upward (in fair weather), there is no flux of negative ions from below the layer to replace those moving out from the top boundary. A net positive charge accumulates in the layer, retarding the subsequent flux of both polarities through the upper boundary. The upper boundary of the electrode layer occurs at the height where the concentration of negative ions at lower heights is sufficient to supply the retarded upward flux required by the modified electric field.

Space charge densities in fair weather have often been reported in numbers of elementary charges per volume (often written as el cm^{-3}). Since a small ion has a single unit of charge, the magnitude of space charge density expressed in elementary particles per

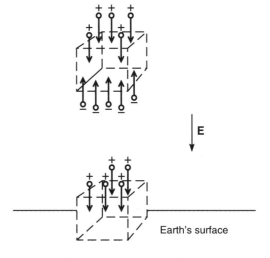

Fig. 1.20. Sketch depicting the flux of ions in fair weather (i.e., negative electric field). The absence of negative ions moving upward in the volume at Earth's surface causes the electrode effect.

volume is the same as small-ion number density for the same unit volume. Crozier (1965) made measurements in the first 3 m above barren ground in New Mexico and found typical values of several hundred elementary charges per cubic centimeter at 0.25 m and about 100 el cm^{-3} at 1.0 m. Crozier was measuring the normal electrode effect (i.e., the one predicted by electrostatic theory). At night under very low winds, the normal electrode layer is replaced by one having negative space charge above positive space charge. Sometimes Crozier observed a net of several thousand positive elementary charges per cubic centimeter at 0.25 m and only a few hundred negative ones at 1.0 m. This occurs both in day and night, although other effects from convection can occur with daytime heating of the ground. Crozier's careful work showed that the normal electrode effect exists during night with moderate winds, which he defined as <7 m s^{-1}, and during daytime with zero to moderate winds. The electrode layer can be dramatically influenced by turbulence and the concentrations of large ions, aerosols, condensation nuclei, etc. The electrode layer is often limited to about 1 m above the ground, except when convection currents deepen it to a few meters.

A quantitative description of the electrode layer is complicated in all but the simplest cases. Realistic depiction of the electrode layer is possible only with numerical models. Willett (1985) used a model to show that radon daughter products may be a more significant source of ionization than trapped radon gas where the ground is covered with vegetation; he also showed deviations from the simple concept of electrode layer over bare ground. He found that this phenomenon can even reverse in polarity from the electrode layer.

1.4.7. The upper electrode

Part of the global circuit theory mentioned previously is that the atmosphere has such high conductivity at some height that it essentially becomes a conductor (i.e., a layer having constant electric potential and serving as the upper terminal of fair weather current). Lord Kelvin (i.e., William Thomson 1860) noted this as early as 1860. Heaviside (1902) suggested that the upper atmosphere was adequately conductive to explain long-range radio wave propagation, and, for this reason, the E-region of the ionosphere is sometimes referred to as the *Heaviside Layer*. However, reconciling the height and properties of the ionosphere with requirements of the spherical capacitor model of the global circuit has created considerable confusion, although the ionosphere provides the only pseudo-discontinuity to large enough conductivity to serve as the upper conducting sphere in this global circuit model.

It turns out that the atmosphere becomes conductive enough to be the upper electrode of the electrostatic global circuit at an altitude lower than the 100-km high Heaviside layer. Several scientists have searched for this conductive region, sometimes called the *electrosphere* (Chalmers 1967) or *equalization*

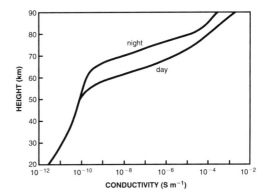

Fig. 1.21. Model calculations of the daytime and nighttime conductivity of the middle atmosphere during quiet solar conditions (From Reid 1986, with permission.)

layer (Dolezalek 1972). A primary question in this concept of an electrosphere above Earth is: Does the layer really exist as hypothesized and, if so, at what height. *Israël* (1970) summarized and updated his earlier work and obtained a height of about 60 km. Reid (1986), in a review, used results of Mitchell and Hale (1973) to corroborate the earlier findings. Their rocket-borne measurements of conductivity, which began at 30 km, showed that $\sigma_+ \approx \sigma_- \approx 1\text{--}3 \times 10^{-11}$ S m^{-1}, σ_- increased rapidly from 5×10^{-11} to 1×10^{-8} S m^{-1} between 40 and 45 km, and σ_+ increased rapidly from 6×10^{-11} to 2×10^{-9} S m^{-1} between 65 and 75 km. In Fig. 1.21 we see that rapid increase to $\sigma > 10^{-10}$ S m^{-1}, and presumably to the equalization layer, is about 10 km lower during the day.

The spherical capacitor paradigm has severe limitations of which we need to be aware: Atmospheric conductivity increases continuously, not suddenly at some upper boundary. The extent to which the electrosphere is a good conductor depends on the time scale of the electromagnetic field change being considered. One consequence is that the capacitor is leaky, particularly to fast electric field changes and the currents they drive. The faster the change in the electric field, the farther up it penetrates, until a threshold is reached where all field changes that are at least that fast penetrate into Earth's magnetosphere. Another implication is that the charge that resides on the capacitor is not really all at an upper and lower boundary; as discussed previously, $\gtrsim 90\%$ of the net positive space charge in the atmosphere is below 5 km, and this charge shields the atmosphere above it from some of the negative charge on the ground. Another consequence is that as the atmospheric resistance is larger near the ground, the work of keeping the global circuit charged is done primarily in the lower atmosphere, consistent with thunderstorms being important in the global circuit.

There is a latitudinal effect on conductivity; the atmosphere is more conductive at higher geomagnetic latitudes. It has been thought that this is due to cosmic rays producing ion pairs at a greater rate at higher latitudes (ion-pair production varies as $\sin^4\phi$, where ϕ is geomagnetic latitude) (Heaps 1978). However, from balloon soundings of conductivity into the stratosphere, Byrne et al. (1988) found that this effect was too small to explain their observed variation in conductivity. They also concluded that when stratospheric aerosols were near their background level (e.g., no recent volcanic production), ion-aerosol attachment, while improving agreement between observation and theory, could not explain the latitudinal difference either. They proposed instead that the latitudinal effect on the atmospheric temperature profile must be included to explain the observed conductivity variations of about 25% between geomagnetic latitudes of 42° N and 63° S. (Note: A combination of Eqs. 1.14 and 1.18 shows that $\sigma \propto \mathrm{k}(P,T)$.)

An alternate view and theoretical treatment of a different global circuit model was put forth by Kasemir (1977). Major differences are that the potential of Earth, Φ_E, is not zero, but is -300 kV, and $\Phi = 0$ at infinity. Even though Φ decreases by 99% within 20 km of Earth's surface, the circuit extends to infinity. Earth is a negative current source driving a conductive medium, whose conductivity increases exponentially to infinity. There is no equipotential at the electrosphere. Kasemir included all generators, but the negatively charged Earth is the fair weather generator in his theory. Thunderstorms are the stormy weather contribution to his current flow model. The potential of his model can be converted to that of the spherical capacitor model by subtracting -300 kV from Φ.

1.4.8. Columnar resistance

The resistance of the atmosphere in a column of unit area from Earth's surface to the electrosphere was named the *columnar resistance* by Gish (1944). It can be expressed as

$$R_c = \int_0^H \frac{1}{\sigma_+(z) + \sigma_-(z)}\, dz, \qquad (1.25)$$

where H is the height of the electrosphere. An example from conductivity measurements to 26 km is shown in Fig. 1.22. Any portion of the total columnar resistance can be found by integrating to any height $z(<H)$. Gish stated that this concept of columnar resistance assumes that Ohm's law is valid and that J is constant with height. It also assumes that all the current is conduction with no convection currents. All it takes to violate this assumption is for the horizontal distribution of atmospheric space charge not to be uniform but to be in globs. Space charge usually is not horizontally uniform: (1) around thunderstorms and (2) in the lower atmosphere under clear skies when solar heating causes convection currents. Examples of nonconstant J are shown in Fig. 1.23. The concept of columnar resistance is useful only for fair weather. Values of the columnar resistance de-

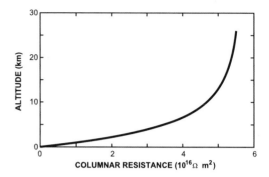

Fig. 1.22. Variation of the columnar resistance, R_C, with altitude calculated using conductivity measurements of Woessner et al. (1958). Ninety-two percent of the total R_C is below the 26-km height of their measurements (assuming the electrosphere at 65 km).

pend mostly on the conductivity profile in the lowest 20 km, with typical values of $R_c \approx 1 \times 10^{17}\ \Omega$ m². Variations in the columnar resistance occur with other local effects, especially the changes in ions in the lower atmosphere. These changes in turn will affect the total current flowing at a given location.

Since the columnar resistance is the resistance of a square meter of atmosphere for the entire column from Earth's surface to the electrosphere, the total resistance of the atmosphere can be obtained simply by dividing by the surface area of Earth. This yields about 230 Ω. Mühleisen (1977) corrected this to about 200 Ω by a 15% reduction for the columnar resistance above mountainous areas. Note that both the columnar and to-

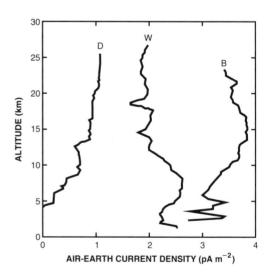

Fig. 1.23. Three examples (D, W, and B) of profiles of fair weather conduction current measured with balloons. (From Kasemir 1977, with permission.)

tal resistance will fluctuate significantly over land surfaces.

1.4.9. Currents flowing in storm environments

A test of whether thunderstorms are the generators of the fair weather circuit requires us to measure currents flowing from storms. The circuit from Earth to the electrosphere is often broken into three regions: below, inside, and above the storm. The *Maxwell current density* (often shortened to *Maxwell current*) below storms can be expressed by

$$J_M = J_{PD} + J_{CV} + J_P + J_L + J_E + \varepsilon \frac{\partial E}{\partial t} , \qquad (1.26)$$

where the subscripts refer to: PD, point discharge; CV, convection; P, precipitation; L, lightning; and E, conduction. It has been suggested that the storm generator can be monitored through the Maxwell current density. All the terms will be discussed in subsequent chapters. The reason we do not do so here is that, historically, measurements to determine storm generator current have been conducted over the top of the storms, where the Maxwell current density is

$$J_M = J_E + \varepsilon \frac{\partial E}{\partial t}. \qquad (1.27)$$

This measurement is not only simpler to interpret but, except for having to get up over storm top, is simpler to do. In their often cited early work, Gish and Wait (1950) measured the potential gradient and conductivity. From each pass over the storm, they calculated a conduction current, from which they got a storm average. They obtained an average conduction current of 0.5 A at 12 km above 21 storms. Other early studies include Stergis et al. (1957) and Vonnegut et al. (1966). The latter suggested that Wilson's proposed conduction current can only flow over active convective areas that remove the screening layer.

Blakeslee et al. (1989) used a U2 airplane instrumented with two Gerdien conductivity sensors and two field mills to make measurements above storms. They reported on flights made at 16–20 km for which they found both conduction current and displacement current densities, allowing them to calculate the Maxwell current density. Typically, at least half the Maxwell current density was from conduction current. An example of their data is shown in Fig. 1.24. In this example, they calculated a conduction current of 3.2 A and an area-integrated Maxwell current of 4.3 A. The average conduction current of 1.7 A for 15 flights, each with multiple passes over the storm, was nearly the same as reported by Gish and Wait. Blakeslee et al. showed that the average conduction current appeared to level off at 3.5 A and they suggested that this indicated the generator efficiency was inversely related to lightning flash rate. Storms with low flash rates were more efficient, although their total current was less. This interpretation

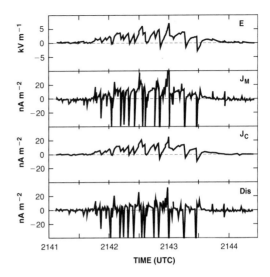

Fig. 1.24. Maxwell current and its components from a horizontal pass over an active storm on 14 July 1986. (U2 ground speed was about 200 m s^{-1}; from Blakeslee et al. 1989: American Geophysical Union, with permission.)

was based on the lightning flash rate being the indicator of generator "strength," a relationship supported by another finding by Blakeslee et al. that the total current (from integration of the Maxwell current density) was linearly proportional to the lightning flash rate (shown in Fig. 1.25).

Determining the global total of currents from the top of thunderstorms requires knowledge of the total storms in progress at any time. The numbers in the literature

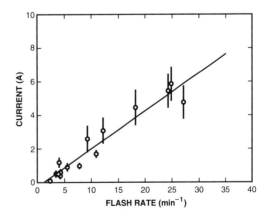

Fig. 1.25. Total Maxwell current (from integration of J_M) as a function of total lightning rate for 15 storm traverses. The integration of J_M neglects the signal directly associated with the lightning discharge. The error bars are standard deviations. (From Blakeslee et al. 1989: American Geophysical Union, with permission.)

range from about 1000 to 2000 storms. Mühleisen (1977) proposed that these numbers are too large because other nonthunderstorm generators such as rain clouds contribute also. Downward looking photographs and electrical measurements with a high-altitude U2 by Vonnegut et al. (1966) revealed that the largest current flow was only above the penetrating (i.e., highly convective) cloud tops. We know of no one who has tried to assess the relative contribution to the global current from quiescent versus deeply convective clouds.

2

Electrified Nonthunderstorm Clouds

2.1. BACKGROUND

Although we can consider any cloud as electrified because the presence of a cloud in the free atmosphere reduces the conductivity and causes charge flowing as fair weather conduction current to accumulate on cloud boundaries, clouds that are not thunderstorms have often been thought of as nonelectrified. It turns out that several cloud genera (i.e., types) have electric fields greater than can be explained by a cloud of reduced conductivity being immersed in the fair weather current. In this section, we examine such clouds, which can be highly electrified. We exclude here cumulus clouds that are in their early stages of electrical development on their way to becoming a thunderstorm. Although those clouds are nonthunderstorm in their early stages, and some cumulus clouds electrify without becoming thunderstorms, we address them later in the chapters on thunderstorm electrification. Here, we will consider the electrification of several other genera of clouds: stratus, stratocumulus, cirrostratus, altostratus, and nimbostratus.

A fundamental consideration is whether the presence of nonthunderstorm clouds can affect the air-earth current density. Certainly the development of a dielectric medium (i.e., the cloud) in the current flow alters the cloud by capture of ions flowing in the conduction current. The effect of these clouds is, however, not just local, as they are also generators, with more charging capability as the cloud thickness increases (Whitlock and Chalmers 1956). Additional evidence of their possible global effect may be seen in the work summarized

in this chapter. Burke and Few (1978) made continuous measurements of conduction near the ground and found that clouds (of unspecified genera) creating overcast caused a reversal in the conduction current. They concluded that charge generation occurs in such clouds; the clouds are not merely passive dielectrics in a conducting atmosphere. Thus, these cloud genera may play a significant role in the global circuit.

In the early 1980s, owing to the increased vulnerability of new generation airplanes to electrical disturbances and lightning, there was renewed interest in electrified nonthunderstorm clouds that do not produce lightning naturally. In the United States, the problem intensified with the loss of a rocket and its payload of a communication satellite due to lightning triggered just after launch from the Kennedy Space Center (Christian et al. 1989). Worldwide, however, there has been comparatively little research on nonthunderstorm clouds for decades (a fact obvious in the sparsity and age of most references cited in this section).

Reiter (1958) conducted extensive research over several years on stratified clouds in the Alps of Germany, where he had several sites at different altitudes. The measurements are useful, even though they were made at the ground and were affected by the presence of the ground, even when the ground was inside the cloud. Much of what is known about the electric structure above ground inside nonthunderstorm clouds comes from measurements made by Soviet scientists working in the USSR using instrumented airplanes. Summaries of this research are in Imyanitov and Chubarina (1967) and Imyanitov et al. (1972). Similar, more recent research was reported by Brylev et al. (1989).

First we examine stratified clouds from the perspective of vertical profiles of the electric field, E, obtained with instrumented airplanes. Over a few decades, Imyanitov and his colleagues studied stratified clouds in the USSR, mainly at three locations with different latitudes: St. Petersburg (Leningrad) at 60° N, Kiev at 50° N, and Tashkent at 41° N. We will concentrate mostly on the researchers's electric field measurements. The data we summarize here came from about 900 spiraling ascent soundings up to 6 km MSL through clouds. Imyanitov et al. (1972) reported each data point to be an average of 100 m vertically from a climb rate of about 4 m s^{-1}, and 3 km horizontally from a horizontal speed of about 50 m s^{-1}. From horizontal flight paths in stratus (St) clouds and stratocumulus (Sc) clouds that were not precipitating, they noted a horizontal uniformity in E. Even so, they felt the horizontal averaging from the spiral path was useful in reducing (smoothing) inhomogeneities. Thus, the E was not totally homogeneous. Their work shows that, in general, the maximum field, E_{max}, increases as the cloud genera move from St to Sc to altostratus (As) clouds to nimbostratus (Ns) clouds. Within a genera, thicker clouds tend to have larger E_{max}. For all genera, the southernmost clouds had larger E_{max} than the northernmost clouds. Those in between (i.e., at Kiev) had the smallest E_{max}, so the dependence of E_{max}, and thus a measure of electrical strength, on latitude is unclear from these data. The average E_{max} for these genera were larger in the summer than in the winter by factors ranging from about one to seven (at 60°N).

Almost no profiles of E versus height (or temperature) are available in Imyanitov and Chubarina (1967) or Imyanitov et al. (1972), as they displayed the electric field as a function of *reduced height,* obtained by dividing the cloud layer into 10 equal thicknesses, with a data point for each. Likewise the subcloud and above-cloud (to 6 km) layers had five intervals. The researchers justified their technique with their observation that only a "certain number" of E-profile patterns existed. (More recent literature citing in situ and laboratory measurements suggested temperature may be key in cloud charging.) Unfortunately with only a few exceptions, temperature measurements are unavailable in the use of reduced height. Because their data set is the largest (and maybe only) substantial collection of vertical profiles of E in nonthunderstorm clouds, we will summarize their results and use a few of these reduced profiles. In the following sections, it will be helpful to remember that the polarity of a dipole is denoted by the charge that is on top and that charge layers were calculated using a one-dimensional form of Gauss's law (see Sec. 6.2.4). Also, charge layers depicted in these clouds did not seem to include screening layers (see Sec. 2.2.). If such layers were in the data, they were apparently ignored or lost by using reduced-height plots. In a later section we will review various cloud genera, give a brief list of the characteristics of each, and summarize pertinent electrical characteristics.

First, however, we examine airborne measurements of E as a function of the depth of the cloud and the complexity of the charge distribution. Note that since the measurements were made aloft, our term for the magnitude of the maximum E, $|E_{max}|$, is equivalent to the maximum $|E_{aloft}|$. It is clear from Table 2.1 that $|E_{max}|$ increases with increasing cloud thickness. It is equally significant that the inferred electric structure becomes more complex with increasing cloud thickness, as shown in Table 2.2 below. The complexity of the charge distribution inferred by Imyanitov et al. (1972) ranged from simple monopoles to multiple layers of charge.

2.2. EXTERNAL CHARGING OF CLOUD BOUNDARIES AND SCREENING LAYERS

As noted in the last section, simply inserting a cloud into the fair weather currents will cause the cloud to become electrified at its boundaries, but observations indicate that internal charging mechanisms exist inside nonthunderstorm clouds. Before examining observations of electrical characteristics of nonthunderstorm clouds further, it will be instructive to consider how electrified a cloud can become without invoking any internal charging mechanisms.

Table 2.1 Mean of the absolute values of the maximum E (i.e., $|E_{max}|_{avg}$ in V m^{-1}) for varying thicknesses of different cloud genera from measurements near St. Petersburg.

Genera	Thickness Range (m)					
	0–200	200–500	500–1000	1000–2000	2000–4000	>4000
St	200	250	400	100	—	—
Sc	200	200	200	200	—	—
As	200*	200*	300	800	1200	5200
Ns	400*	400*	500	800	2000	3100

*are for a single thickness of 0–0.5 km. The dashes (—) indicate no data (essentially by definition of St and Sc clouds).

From Imyanitov and Chubarina (1967).

Table 2.2 Average cloud thickness, Δz_{avg} (m), for different charge structures showing increasing electrical complexity with increasing thickness in each genera, near St. Petersburg, 1958–59.

Genera	Positive Monopole	Negative Monopole	Positive Dipole	Negative Dipole	Multi-Layers
St	200	200	450	450	700
Sc	260	250	400	450	700
As	650	700	800	900	1,500
Ns	650	700	950	1,600	2,000

From Imyanitov et al. (1972).

If a cloud suddenly is placed in the atmosphere, which has a roughly uniform, vertical fair weather current, there will be a discontinuity in the ohmic currents across the upper and lower surfaces of the cloud, because the conductivity of clear air is greater than the conductivity of cloudy air. (Little data exist on the conductivity of cloudy air aloft, but Rust and Moore (1974) found that the conductivity of cloudy air was 10% that of clear air at the same altitude.) As charge continuity requires charge to collect in any region in which electric currents vary spatially, a surface charge will form on the top and bottom boundaries of the cloud, as cloud hydrometeors preferentially capture charge of one polarity. The surface charge that forms on the boundary will retard currents outside the cloud and increase them inside the cloud. Equilibrium will be attained when the magnitude of the electric field inside the cloud is so much larger than the magnitude of the field outside that it compensates for the reduced conductivity of cloudy air, and the ohmic currents are once more uniform across the cloud boundary. In clouds that are electrified internally, the internal charge affects the value of surface charge needed to attain equilibrium across the boundary, but the applicable physics will be essentially the same.

It is simple to calculate the equilibrium value of surface charge density, σ_q, at a cloud boundary in the absence of other electrification mechanisms. Remember that the ohmic current density is given by σE (plain σ denotes conductivity). We start with the simple case of the cloud forming an infinite horizontal layer between ground and some altitude, z, in fair weather; then the equilibrium condition can be written as

$$\sigma_{clr}(z)E_{clr}(z) = \sigma_{cld}(z)E_{cld}(z), \qquad (2.1)$$

where $\sigma_{clr}(z)$ and $E_{clr}(z)$ are, respectively, the conductivity and electric field in clear air just above the cloud boundary, and $\sigma_{cld}(z)$ and $E_{cld}(z)$ are, respectively, the conductivity and electric field in cloud just below the boundary. It was shown (Sec. T.5.3) from Gauss's law that an infinite horizontal sheet of surface charge density, σ_q, will create a vertical electric field of $\sigma_q/2\varepsilon$. If the charge is positive, the electric field is directed upward above the sheet and downward below the sheet. This electric field adds to the ambient electric field in which the cloud is embedded to give total electric fields in the clear and in the cloud of

$$E_{clr}(z) = E(z) + \frac{\sigma_q}{2\varepsilon} \text{ and}$$
$$E_{cld}(z) = E(z) - \frac{\sigma_q}{2\varepsilon} \quad , \qquad (2.2)$$

where $E(z)$ is the ambient E at height z. Substituting these expressions in Eq. 2.1 and solving for σ_q gives

$$\sigma_q = -2\varepsilon \frac{\sigma_{clr} - \sigma_{cld}}{\sigma_{clr} + \sigma_{cld}} E(z). \qquad (2.3)$$

Thus, we see that the surface charge density is a function of the ambient E. The surface charge will cause electric current to be continuous across the cloud boundary.

This result will be modified if the cloud also has a lower boundary above the ground. The discontinuity in conductivity at each cloud boundary will create a surface charge that establishes current continuity, so the surface charge at each boundary must satisfy Eq. 2.2. However, the situation is more complicated, because the surface charge at the lower boundary contributes to the electric field at the upper boundary, and vice versa. With an upper boundary having surface charge density $\sigma_{q\,top}$ at height z_{top} and a lower boundary having surface charge density $\sigma_{q\,bot}$ at height z_{bot}, the electric fields just inside and outside the cloud boundaries are given by

$$E_{clr}(z_{top}) = E(z_{top}) + \frac{\sigma_{q\,top}}{2\varepsilon} + \frac{\sigma_{q\,bot}}{2\varepsilon}$$
$$E_{cld}(z_{top}) = E(z_{top}) - \frac{\sigma_{q\,top}}{2\varepsilon} + \frac{\sigma_{q\,bot}}{2\varepsilon}$$
$$E_{cld}(z_{bot}) = E(z_{bot}) - \frac{\sigma_{q\,top}}{2\varepsilon} + \frac{\sigma_{q\,bot}}{2\varepsilon} \qquad (2.4)$$
$$E_{clr}(z_{bot}) = E(z_{bot}) - \frac{\sigma_{q\,top}}{2\varepsilon} - \frac{\sigma_{q\,bot}}{2\varepsilon} .$$

The first two expressions must satisfy Eq. 2.1 for the upper boundary; the second two, for the lower boundary. Solving these two equations simultaneously for $\sigma_{q\,top}$ and $\sigma_{q\,bot}$ gives fairly complicated expressions for each. However, the expressions are much simpler if we consider a cloud layer thin enough so that the ambient E and conductivity are approximately equal at the upper and lower boundaries. Then,

$$\sigma_{q\,top} = -\frac{\sigma_{clr} - \sigma_{cld}}{\sigma_{cld}}\ \varepsilon\,E(z)$$

$$\sigma_{q\,bot} = \frac{\sigma_{clr} - \sigma_{cld}}{\sigma_{cld}}\ \varepsilon\,E(z). \qquad (2.5)$$

The surface charge density at each boundary is greater than that of the cloud with a single boundary above ground, because of the interaction of the two boundaries. However, $\sigma_{q\,top} = -\sigma_{q\,bot}$, so the net electric field outside the cloud is the same as if the cloud were absent. Although the ratio of external field and internal field is the same as is the case with only one cloud boundary above ground, the internal field is larger in the case with two boundaries. It can be shown that inside the cloud

$$E_{cld}(z) = \frac{\sigma_{clr}}{\sigma_{cld}}\,E(z) \qquad (2.6)$$

for the case of an infinite horizontal cloud layer, which models an extensive stratus cloud, versus

$$E_{cld}(z) = 2\,\frac{\sigma_{clr}}{\sigma_{clr} + \sigma_{cld}}\,E(z) \qquad (2.7)$$

for the case of a single boundary. Since $\sigma_{cld} < \sigma_{clr}$, the E_{cld} with two boundaries is larger by a factor of $(\sigma_{clr} + \sigma_{cld})/2\sigma_{cld}$ than E_{cld} with a single boundary.

When the stratus layer thickens so that the ambient electric field and conductivity are different at the two boundaries, then $|\sigma_{q\,bot}| \neq |\sigma_{q\,top}|$, and there will be space charge within the cloud, satisfying Gauss's law. However, it turns out that the expressions for surface charge at the boundaries are still fairly simple if we assume that the ratio of conductivity in clear air to that in cloudy air is constant with altitude. To visualize this, remember that the charge above and below the cloud and at the cloud boundaries will not contribute to the vertical gradient of E inside the cloud, because we have assumed that these charges are in infinite layers, which produce E that are constant with height outside the layers. Therefore, within the cloud layer all variation in E will be from the variation in conductivity with height, and current continuity requires that the vertical gradient of E will have the same magnitude as the vertical gradient of conductivity when equilibrium is established. The assumption that the ratio of clear to cloudy conductivity is independent of height implies that the E inside the cloud is a constant factor of the E which would exist at the same height in the absence of cloud. This factor is the same as the ratio of conductivities in clear and cloudy air. Since $E_{cld}/E = \sigma_{clr}/\sigma_{cld}$ at the cloud boundary is the same as the expression for the change in E across the surface charge at the cloud boundary when there is no vertical gradient in E and σ, the effect of the vertical gradients on the surface charge

at one of the boundaries must be equivalent to the opposite screening layer being located just inside the boundary being analyzed:

$$\sigma_{q\,top} = -\frac{\sigma_{clr}(z_{top}) - \sigma_{cld}(z_{top})}{\sigma_{cld}(z_{top})}\varepsilon\,E(z_{top})\ \text{and}$$

$$\sigma_{q\,bot} = \frac{\sigma_{clr}(z_{bot}) - \sigma_{cld}(z_{bot})}{\sigma_{cld}(z_{bot})}\varepsilon\,E(z_{bot}). \qquad (2.8)$$

It can be shown that the total of these two surface charges approaches zero as z_{bot} approaches z_{top}. (Use the approximate empirical relationship that conductivity and electric field vary exponentially.) Furthermore, the total charge from a stratus cloud layer, including these surface charges at the cloud boundaries and internal charge due to the gradient of E in the absence of other electrification mechanisms, is the same as the charge that would have existed in the absence of the cloud, since $\sigma_{q\,int}$, the vertical integral of charge density through the interior of the cloud, is given by

$$\sigma_{q\,int} = \varepsilon[E_{cld}(z_{top}) - E_{cld}(z_{bot})] \qquad (2.9)$$

and

$$\sigma_{q\,top} + \sigma_{q\,bot} + \sigma_{q\,int} = \varepsilon[E(z_{top}) - E(z_{bot})]. \qquad (2.10)$$

2.3. ELECTRICAL ASPECTS OF DIFFERENT CLOUD GENERA

Excerpts of descriptions of several cloud genera from the World Meteorological Organization in WMO (1969) will be given. Note that any single characteristic of these clouds is usually inadequate to define them.

2.3.1. Stratus clouds

Stratus (St) clouds are typified by a grey cloud layer with a nonuniform base, which may give drizzle, ice prisms, or snow grains. They are composed of small droplets or small ice crystals, depending on temperature. Stratus clouds usually mask the sun and moon, but where they are not too thick, they can produce corona (same word but different definition for an electrical discharge—here it means small diameter colored ring(s) around the sun or moon from diffraction by water droplets). Stratus clouds are associated with calm or light winds and are generally found between 0 and 2 km in mid-latitudes. Their electrical characteristics are summarized at the end of this chapter in Table 2.4.

2.3.2. Stratocumulus clouds

Stratocumulus (Sc) clouds are typified as grey and white in patches or layers with nonfibrous rounded masses composed of water droplets, sometimes with rain or snow pellets. Rarely do Sc clouds have snow crystals and flakes. Sc cloud structure is highly variable, including parallel rolls with clear spaces interspersed. Sometimes they have low-intensity rain or snow. Their main difference from St clouds is that Sc show elements, separate or merged. These clouds are also mostly at altitudes from 0–2 km in mid-latitudes.

Mountaintop (2 km MSL) measurements by Allee and Phillips (1959) showed droplets in Sc clouds with an average diameter of ≈ 7 µm had an average charge, q_{avg}, of about 1×10^{-18} C and occurred in the presence of a positive E ≈ 500 V m^{-1}. (This is an example of the mirror image effect, which is discussed in Sec. 7.10.) They also found conductivity in Sc to be 0.05 to 0.3 the conductivity in clear air at the same altitude. Phillips and Kinzer (1958) found from their measurements at the ground inside clouds that q_{avg} in Sc clouds increased with increasing droplet diameter. They proposed that the charge was due to ionic diffusion and that the charges were still below the maximum possible.

Averaged data from Imyanitov et al. (1972) showed simple charge configurations in warm (presumably entire cloud with T > 0°C) Sc clouds at 60° N latitude. Within the few kilometers of the horizontal extent of the airplane's spiral ascent, E was homogeneous. Note that this did not preclude inhomogeneities, which could be observed sometimes in horizontal traverses at constant altitudes. About two thirds of Sc clouds had dipole (either polarity) structures. Five percent of Sc clouds were multiply charged, i.e., having more than two charge layers. In winter Sc clouds with mixed phase precipitation, Brylev et al. (1989) found both polarities of the vertical component of E with values ≈ 30 kV m^{-1}. The lapse rate in the middle of the cloud was 8°C km^{-1}. Additional electrical characteristics are summarized in Table 2.4.

2.3.3. Cirrostratus clouds

Cirrostratus (Cs) clouds are transparent white veils composed mostly of ice crystals, which usually produce halos. They are thinner than As clouds and easily confused with St clouds. In temperate regions, Cs clouds are at 5–13 km MSL (3–8 km in polar regions). Few E profiles in this type are available. They are noteworthy in that they can be multiply charged with four layers of charge. Although it is uncertain, the cases cited as electrified may have been Cs clouds formed from cumulonimbus anvils or thinning As clouds, thus having their charge generated while in another genera. Because Cs that are formed directly are generally thin, it seems unlikely that they will often have a multicharged structure.

Also, the average cloud base is 5500 m, just high enough for the level of Cs clouds defined by the WMO for temperate regions, but well within the layer for polar regions (>60°N latitude). Additional electrical characteristics are summarized in Table 2.4.

2.3.4. Altostratus clouds

Altostratus (As) clouds are a grey sheet or layer composed of water droplets and ice crystals with raindrops and snowflakes. Precipitation is usually virga. If precipitation reaches ground, it is often continuous rain, snow, or ice pellets. Altostratus clouds have large horizontal extent (up to hundreds of kilometers), with a vertical extent of few kilometers. They tend to occur at altitudes between 2 and 7 km in midlatitudes, but can extend higher than 7 km. The sun can be seen only vaguely, if at all, and no halo from ice crystals (either as a colored arc from refraction or white arc from reflection) can be observed. Formation most often is from widespread slow lifting. Altostratus clouds are more uniform than Sc, have no breaks, and are never white.

In all the above cited reports, the researchers found significant deviations of individual clouds from their "average one." This indicated an increase in the complexity of electrical structure with increasing cloud size. Zones of precipitation were found. Altostratus clouds with a dipole structure tended to have charge layers near cloud boundaries, with lower charge densities deeper inside the cloud. We cannot discern whether the charge layers near boundaries were screening layers because of the reduced-height data format. All bases of the As clouds were at about 3 km. Additional electrical characteristics are in Table 2.4. Of significance, however, is that E can be high enough to cause hazards, and the E and the space charge densities approach those of thunderclouds.

2.3.5. Nimbostratus clouds

Nimbostratus (Ns) clouds are grey, often dark, with nearly continuous rain or snow that generally reaches ground. They contain: (1) water droplets (sometimes supercooled ones) and raindrops, (2) snow crystals and snowflakes, or (3) a mixture of liquid and solid particles. Their formation is usually by slow lifting of a large area of air. They always blot out the moon or sun. According to the WMO, if there is confusion with As, a cloud is classified as Ns if precipitation reaches the ground. Ns clouds occupy altitudes of 2–7 km and often extend both below and above these altitudes.

In England, Chalmers (1956) analyzed Ns from which precipitation had fallen for ≥ 1 h and noted that the electric field and currents arriving at the ground were different for continuous rain and continuous snow. With rain, he found that in all but one 4.5-min interval, in about 1400 such intervals, the net charge on the rain at Earth was positive. The |E| was ≤ 800 V m^{-1} and usually between 100 and 200 V m^{-1}. During continuous snow, the limits of |E| were the same, but regardless of the polarity of E, the charge on the snow was negative in all but one of 990 intervals. Note that precipitation charging or modification of E at the ground by point discharge from the ground (see Ch. 4) was not a factor, as the E value was too small.

In only slight contrast, Reiter (1958, 1965) made measurements in Germany at several ground observatories located in the Alps between 700 and 3000 m MSL. All the following are observations when |E| ≤ 1000 V m^{-1}, so point discharge ions were absent. He

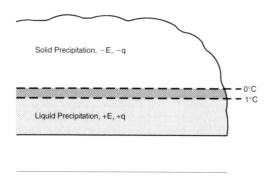

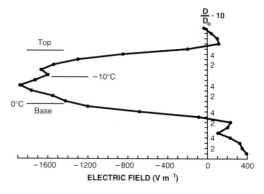

Fig. 2.1. Depiction of the relationship between polarity of charge, q, carried by precipitation and E in Ns clouds found by Reiter (1965). The transition seemed to occur consistently at temperatures between 0° and +1°C.

Fig. 2.2. Composite E profile for Ns clouds containing mixed-phase precipitation. The profile is plotted as a function of reduced height reported by Imyanitov and Chubarina (1967).

studied As clouds just prior to and during their transition to Ns. While still As (no precipitation), their base had a net negative charge density that increased in magnitude with the transition to Ns. When steady snow or rain occurred from Ns clouds, Reiter reported that a consistent phenomenon ensued: Whether inside or outside below the Ns cloud, E and the dominant polarity on snow were negative at temperatures below freezing; the situation reversed in a melting zone at 0°–1°C with E and the dominant polarity on rain being positive at temperatures warmer than +1°C (Fig. 2.1). Reiter stated that these relationships held in >80% of his observations, independent of altitude. As an aside, we note that he found that the above was not true for shower clouds, which he defined as clouds in unstable stratification. Under this latter condition, there was no conclusive or consistent melting layer effect or E polarity tied to temperature. He attributed this to turbulent zones and effects occurring close to the recording sites.

Reiter (1958, 1956) noted that as As clouds transformed into Ns clouds, the negative E was reduced, indicating that the net charge in the cloud became more negative (although not always enough to reverse the polarity of E). He found that As and Ns always had a net negative in the base when the cloud was not precipitating (both during lulls in precipitation or after it was over). He also found net negative charge in altocumulus, Ac, mammatus clouds and in short virga below As clouds.

The Soviet research on Ns clouds can also be divided into rain and snow categories. Positive and negative dipoles were equally likely in Ns clouds, whether it was raining or snowing. The researchers defined a mixed Ns cloud as one having the 0°C isotherm within the cloud, although the precipitation could be either rain or snow. By dividing the reduced height into two sectors, cloud base to 0°C and 0°C to cloud top, the Soviet scientists inferred that Ns clouds tend to be more complex, with four layers of charge. They observed, as

has Reiter (1965), that the region higher than freezing (T < 0°C) was typified by negative E. Negative space charge density was also present, but not exclusively.

The charge distribution in Ns clouds was usually a dipole or multiply charged (again, remember this was based on a reduced height analysis, and all charge layers may not have been delineated). The Soviet researchers reported that a small positive charge was beneath and in the lower part of the cloud, a large negative charge above that, and a large positive above the negative. These features can be seen in Fig. 2.2. They also reported that both negative and positive charge could reside higher than the freezing level. Examples of their E profiles in Ns clouds with different microphysics is shown in Fig. 2.3. Imyanitov et al. (1972) also inferred that charging occurs inside Ns clouds. They did not address whether it is different when an inverted (i.e., negative) dipole occurred. The maximum fields in Ns clouds ranged from 16 kV m⁻¹ at 60° N latitude to 135 kV m⁻¹ at 41° N. We cannot discern from their data whether part of the difference in electric structure was seasonal.

Andreeva and Evteev (1974) presented measurements made by Imyanitov and colleagues in which Ns clouds were found to have |E| up to 45 kV m⁻¹. Andreeva and Evteev noted that the horizontal continuity of the vertical component of E could exist up to about 20 km. They interpreted changes in the polarity of E at some particular altitude (where measurements were made) as opposite polarities side-by-side in the cloud. As these, as well as much of Imyanitov's other data, were not from vertical profiles of the cloud, sloped, charged layers cannot be ruled out. If they existed, a misinterpretation of side-by-side opposite polarity could result. Many other horizontal soundings in electrified clouds of all types suggest that polarity of charge can change in the horizontal. (We show in Ch. 8 vertical E profiles in stratiform regions of large storm systems that indicate horizontal continuity of charge.)

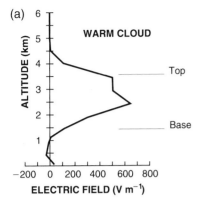

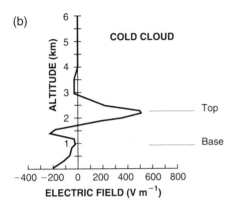

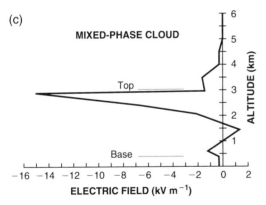

Fig. 2.3. Examples of E profiles in Ns cloud with different microphysics due to differing temperature regimes. The warm cloud has T > 0°C throughout; the cold cloud has T < 0°C throughout, and the mixed-phase cloud has temperatures above and below freezing accompanied by liquid and frozen hydrometeors. (From Imyanitov and Chubarina 1967.)

In their studies of nonthunderstorm clouds, Brylev et al. (1989) used an airplane instrumented with four field mills to obtain mostly horizontal profiles of E through As and Ns clouds in the Soviet Union. While the main emphasis of Brylev et al. was to study the haz-

ards to aircraft, we can glean basic understanding from their work. They used the concept of electrically active zones (EAZs), which were defined as heterogeneities in stratiform clouds characterized by increases in liquid water content (LWC), radar reflectivity, and |E|. They did not make a direct connection to the term inhomogeneities used by Imyanitov et al. (1972) to describe the variation of E through clouds, but it seems they are related. EAZs carry with them the connotation of hazardous electric fields. Brylev et al. stated that EAZs originate from and are caused by cellular convection. EAZs were usually in layers under the most convective cells, not in the maximum convection; this was determined from lightning strikes to airplanes, not E measurements. Thus where the aircraft flew, instead of the electric structure relative to convection, may have been the dominant, but undetermined, variable. They reported a study by Chubarina (1977) of 47 clouds (33 were Ns) in which the dimensions of EAZs with a threshold E ≈ 10 kV m^{-1} were 500–800 m in the vertical and ≥600 m in the horizontal. Magnitudes of E_z of about 40 kV m^{-1} were measured in Ns clouds. Horizontal dimensions where these high E occurred ranged from a few to 10 km. In another flight in an Ns cloud near St. Petersburg (61° N), a vertical E of 70 kV m^{-1} was recorded in mixed precipitation of rain, sleet, and snow at T ≈ 0° C just before a lightning strike to the airplane. Table 2.3 summarizes EAZs.

Although not quantifiable, the studies of Brylev et al. (1989) indicate that lower E are expected in clouds containing only liquid water with typical values of the vertical $|E_{aloft\,max}| \approx 6$ kV m^{-1}, while in mixed-phase clouds $|E_{max}|$ up to an order of magnitude greater are expected. Brylev et al. stated that EAZs needed cellular convection and that LWC had to be ≈ 1 g m^{-3} for significant electrification. Also, EAZs were usually in weak, and rarely in moderate, turbulence. EAZs were enhanced by large-scale ascent of the air. In stratified clouds in the Ukraine, for example, the higher LWC was associated with greater lapse rates. Lapse rates of 4°–6°C km^{-1} had LWC ≈ 0.2 g m^{-3}, and lapse rates

Table 2.3 Characteristics of electrically active (hazardous) zones, EAZs in stratiform clouds.

Horizontal dimensions	several kilometers
Vertical dimensions	>1 km
Lifetime	<100 min
Precipitation intensity at ground	Usually moderate, sometimes weak
Phase state	Mixed
Water content	≈1 g m^{-3}
Temperature	0° to 4°C (melting region) −6° to −11°C (freezing region)
Turbulence	Weak, rarely moderate
Electric intensity	10 kV m^{-1}

Adapted from Brylev et al. (1989).

Table 2.4 Summary of measurements on stratified clouds near St. Petersburg.

Genera	No.	Δz Range (m)	Δz Avg. (m)	z_{avg} Base (m)	$E_{80\%}$ ($V\,m^{-1}$)	E_{min} ($V\,m^{-1}$)	E_{max} ($V\,m^{-1}$)
St	116	100–1000	500	350	−300, 100	−500	1500
Sc	357	100–1800	500	1000	−400, 200	−1400	1600
Cs	48	—	1100	5500	−400, 200	−2000	900
As	218	—	950	3400	−600, 100	−6450	1450
Ns	155	—	2100	900	—	−18,000	12,000

Δz is the thickness of the clouds; z_{avg} is the average altitude of the cloud base; $E_{80\%}$ means 80% of all measurements were between the two values shown. The extreme values are E_{min} and E_{max}. Unavailable data are denoted by —.

From Imyanitov and Chubarina (1967).

of 6°–8°C km^{-1} had LWC $\approx$ 1 g m^{-3}. EAZs also appeared in stratified clouds when LWC $\approx$ 1. Their description of widespread Ns clouds with some embedded cumulonimbus (Cb) clouds makes it unclear to us if the most intense E were in the Cb clouds, related to their presence, or neither. There are similarities between these structures and those of mesoscale convective systems (see Ch. 8).

What is clear from the above is that stratiform clouds are electrified, sometimes highly, and that they affect the global circuit, pose hazards to vehicles penetrating them, and offer an opportunity to test some cloud electrification theories in conditions less chaotic than those for clouds with stronger convection. Table 2.4 summarizes values of electric parameters in different stratiform genera.

3

Introduction to the Electrical Nature of Thunderstorms

3.1. INTRODUCTION

We define a *thunderstorm* as a cloud that produces thunder. Some official organizations and some scientists restrict the term to storms during the warm season and specifically exclude winter storms, because winter storms tend to produce less thunder and because many winter storms differ in other important respects from warm season thunderstorms. However, this subverts the original distinctive characteristic of thunderstorms, the production of thunder, and forces an additional layer of meaning on the term that might be better served by adding modifiers, as in the term warm season thunderstorm. In this text, we use the broader definition, which also emphasizes the electrical phenomena associated with thunderstorms. Because thunder is produced by lightning, defining a thunderstorm as a cloud that produces thunder means that it is also a cloud that is electrified enough to produce lightning.

Since the remainder of this text could be considered an introduction to the electrical nature of thunderstorms, you might reasonably ask why basic information on this topic needs to be encapsulated now in a single chapter. For a reference book or monograph, this organization would be inefficient and inconvenient. For a textbook, however, there is an overriding pedagogical motive for choosing this organization: A student of cloud electrification needs a framework of basic concepts and observations into which more detailed information can be placed. In this chapter, therefore, we will examine simple conceptual models of thunderstorm charge structure, thunderstorm currents, and

mechanisms for charging cloud particles and precipitation. This will provide the framework we need to examine the range of electrical behavior of which thunderstorms are capable.

3.2. THE GROSS CHARGE STRUCTURE OF THUNDERSTORMS

In the eighteenth century, Benjamin Franklin established that negative charge usually was present in thunderstorms, though sometimes positive charge was observed. Because these measurements were made beneath thunderstorms, they tended to be dominated by charge in the lowest part of storms. Although in the late 1800s some investigators suspected that positive and negative charge coexisted in thunderstorms, it was not possible to establish even the gross distribution of positive charge and negative charge in thunderstorms until new instruments were developed early in the twentieth century. C.T.R. Wilson (1916, 1920, 1929), a famous and influential scientist, made measurements at the ground of both the electric field of thunderstorms and the electric field changes of lightning. Based on systematic variations in the polarity of the field and field changes with range, he suggested that thunderstorms typically have positive charge above negative charge, a configuration now called a *positive dipole*.

To see how surface electric field values at several ranges can provide this information, consider a positive dipole, as shown in Fig. 3.1. Remember that the vertical component is the only nonzero component of E at

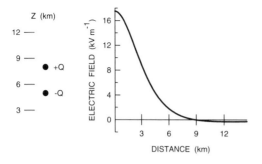

Fig. 3.1. The electric field at the surface as a function of distance from the axis of a positive dipole distribution. The height of the charges is shown in the left plot. The charge magnitude for this example is 40 C, but the magnitude of the charge does not affect the reversal distance.

the ground. If the dipole axis is aligned with the z-axis, then the electric field on the ground at an arbitrary distance D from the dipole axis can be given as

$$E_z = \frac{1}{4\pi\varepsilon}\left[\frac{2Qz_N}{(D^2 + z_N^2)^{1.5}} - \frac{2Qz_P}{(D^2 + z_P^2)^{1.5}}\right],$$
(3.1)

where z_N and z_P are the heights of the negative and positive charge, respectively, and Q is the charge magnitude in each pole of the dipole. As shown in Fig. 3.1, the electric field is zero at D such that

$$D = \left(\frac{z_N^{\frac{2}{3}}z_P^2 - z_N^2 z_P^{\frac{2}{3}}}{z_P^{\frac{2}{3}} - z_N^{\frac{2}{3}}}\right)^{\frac{1}{2}}$$
(3.2)

and is positive at closer distances and negative at farther distances. The distance at which the electric field passes through zero as it reverses polarity is called the *reversal distance*. The change in polarity occurs because the relative magnitude of the vertical component of the field decreases less rapidly with distance for the higher positive charge than for the negative charge, and this eventually outweighs the effect of the closer proximity of the negative charge. A similar expression can be written for the field change of cloud flashes that neutralize charge in each pole of the dipole, and the polarity of the field change from cloud flashes also changes with increasing distance from the axis of the charges. The net effect of ground flashes, however, is to neutralize only one polarity of charge, so the polarity of the resulting field change at the ground remains the same at all distances from the storm.

Wilson's hypothesis generated considerable controversy. The distribution of charge in thunderstorms is complicated enough that measurements made at the surface, remote from the thunderstorm charges, provided confusing evidence. As an example of the possible confusion, consider a case in which there is a small positive charge below the negative charge in the positive dipole of a storm. If there is enough lower positive charge, the electric field at the surface underneath the storm would be negative and would reverse polarity

twice with distance from the dipole axis. Some investigators suggested that the lowest charge in many thunderstorms was positive and that the typical charge distribution of thunderstorms consisted of negative charge over positive charge, i.e., a negative dipole. However, by the early 1930s, electric field and field change measurements had been made for many thunderstorms and provided considerable evidence that the polarity of thunderstorm charge distributions usually was that of a positive dipole. This conclusion was bolstered by measurements, made around 1930, showing that ground flashes typically lowered negative charge to ground, although some flashes lowered positive charge (e.g., Sporn and Lloyd 1931; Jensen 1933).

Despite the widespread support for a positive dipole, some electric field measurements and the ground flashes that lowered positive charge appeared to be inconsistent with a positive dipole. For example, Wormell (1930, 1939) concluded from many observations of E at the surface that a positive dipole existed in the vast majority of clouds, but that a negative dipole appeared to exist in a few clouds and polarity could not be determined in some clouds. To provide more definitive measurements of the polarity of the thunderstorm dipole and to investigate the observed discrepancies, Simpson, a scientist who had argued that the dipole was negative, made in situ measurements of the electric field. Simpson and Scrase (1937) and Simpson and Robinson (1941) flew balloons carrying alti-electrographs (described in Sec. 6.2.3) to record corona caused by the vertical component of the electric field. Based on these measurements, they concluded that the main charges in a thunderstorm typically formed a positive dipole. However, they also found that two vertically separated charges were insufficient to account for the electric field profiles observed in many storms. A third charge, a smaller positive charge, was needed below the negative charge in many storms. Modern in situ measurements of electric field profiles and of the charge on individual precipitation particles have confirmed that lower positive charge centers occur in many thunderstorms (e.g., Marshall and Winn 1982, Marshall and Rust 1991).

The simple conceptual model of the gross charge structure of thunderclouds thus became that shown in Fig. 3.2. This conceptual model, which we will refer to as *dipole/tripole structure,* has prevailed to the present, as may be seen in the comprehensive review of the charge distribution of storms by Williams (1989). Note that in this context, dipole/tripole does not refer to point dipoles, point tripoles, or specific terms of the multipole expansion of the electrostatic field of a charge distribution. Instead, it refers to vertically separated, oppositely charged regions. To conform to strict usage in physics, the terms should mean that the net charge in the dipole or tripole is zero. However, in thunderstorm research, the terms frequently are used to refer to the number of charges stacked vertically in the gross thunderstorm charge distribution, even if overall charge neutrality is not satisfied.

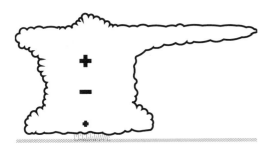

Fig. 3.2. Depiction of the charge structure of a thunderstorm as derived from measurements by Simpson and colleagues. A thunderstorm is described as a positive dipole (positive above negative charge) or a tripole, which is shown here. The lower positive charge center in this simple model may not always be present.

Although dipole/tripole structure is frequently all that is discussed explicitly when considering the gross charge distribution of storms, most investigators concede that there typically is yet another charge on the upper cloud boundary. As will be discussed in Section 3.5.4, the dominant charge in the upper part of a thunderstorm causes a layer of charge of the opposite polarity to form on the upper cloud boundary, and this layer often is called a *screening layer*. The screening layer was expected theoretically and has been inferred many times from electric field sensors flying above or through the upper cloud boundary (e.g., Vonnegut et al. 1962, Marshall et al. 1989, Marshall and Rust 1991). As a paradigm of the gross charge distribution of thunderstorms, the dipole/tripole structure plus an upper screening layer is a reasonable working hypothesis for many applications and will be the basis for our present discussion of thunderstorms.

Keep in mind, however, that we use the dipole/tripole plus screening layer only as a simple description of the minimum number of vertically stacked charges

needed to match the most prominent features of typical electric field measurements. Actual thunderstorm charge distributions usually are more complex than a simple dipole/tripole. For example, charges of opposite polarity can occur at the same altitude, and electric field soundings can have greater complexity than would be expected from a vertical path through a simple dipole/tripole with screening layer. From measurements by several investigators, Moore and Vonnegut (1977) and Krehbiel (1986) hypothesized that the detailed distribution of charge around and in small thunderstorms might appear similar to the distribution shown in Fig. 3.3. Numerical cloud models also produce more complicated charge distributions, such as the distribution in a large severe storm modeled by Ziegler and MacGorman (1994) and shown in Fig. 3.4.

Some electric field profiles measured by vertical soundings through storms have been so complex that it is doubtful that even the gross structure of the charge distribution should be approximated as a dipole or tripole; additional vertically separated charges appear to be required to explain the profiles (Marshall and Rust 1991, Stolzenburg and Marshall 1994). In particular regions of some types of storms, this greater complexity has appeared to be the rule more than the exception. When we return to the topic of charge structure as we examine different types of thunderstorms in later chapters, we will further examine situations in which the charge structure of storms often appears to be more complicated than the dipole/tripole paradigm. Since thunderstorm charge distributions typically include a screening layer, and more than four vertically stacked charges have often been inferred in some situations, Rust and Marshall (1996) suggested that the paradigm of the gross thunderstorm charge distribution should be modified to acknowledge departures of the gross structure from a dipole/tripole plus screening layer. In both isolated storms and storm systems, for example, Stolzenburg (1996) found that the dipole/tripole plus screening layer applied to regions of updraft, but that

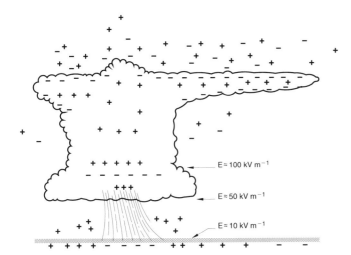

Fig. 3.3. Charge structure of thunderstorms inferred from measurements and theory as depicted by Krehbiel (1986). The electric fields are approximate values for the locations shown.

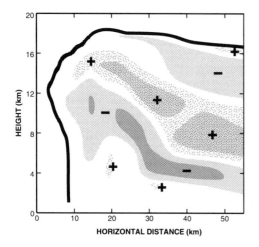

Fig. 3.4. Charge distribution of a large, severe thunderstorm from a numerical cloud model. (Adapted from Ziegler and MacGorman 1994, with permission.)

as one moved away from updrafts, there often appeared to be more vertically stacked charges, up to a total of six.

Through the years, many values have been given for the charges in the thunderstorm dipole/tripole. They are summarized in Table 3.1, and come from measurements of E at the ground and aloft and from analyses of lightning field changes.

Although the number of in situ measurements is too small to document what is responsible for the considerable variations in charge structure observed from storm to storm, some features, including some departures from dipole/tripole structure, tend to be present in many storms. The following are typical characteristics of the overall charge structure of thunderstorms:

1. Negative charge usually dominates in the lower region of thunderstorms, usually in some temperature range warmer than roughly $-25°C$ and sometimes warmer than $-10°C$.
2. A positive charge region is typically just above the negative charge (probably <1 km higher). A large body of data supports the existence of a net positive charge in the upper region of thunderstorms and in their anvils (e.g., Gish and Wait 1950, who measured conductivity and E above storms and consistently found positive charge flowing upward).
3. E measured during horizontal traverses through thunderstorms often indicates that different charge polarities are side-by-side. Models also indicate that charge can be side-by-side, as when an upper positive charge extends downward outside of the updraft, to one side of negative charge.
4. Vertical profiles of E_z often indicate more than three vertically stacked charge regions in the storm.
5. There often is a layer of charge, called a screening layer, up to a few hundred meters thick at cloud boundaries, especially at the quiescent top boundary, where the screening layer typically contains negative charge.
6. Most small-ion charge below thunderstorms is produced by point discharge, the current induced at the ground from pointed objects, such as grass and trees, under the influence of the thunderstorm's electric field.
7. Precipitation often carries predominately positive charge below the cloud; the charge on precipitation near ground is affected significantly by ions produced by point discharge.

In a typical thunderstorm, the electric field magnitude and space charge density observed throughout most of the storm are much smaller than the maximum observed values. Maximum charge densities inferred from electric field measurements typically are $\leqslant 10$ nC m^{-3} (10 C km^{-3}) and are detected over a vertical distance of less than 1 km. The sparse measurements of

Table 3.1 Altitude, z, and charge, Q, inferred or assumed for the lower positive charge center, LPCC, and the main negative and positive charges in thunderstorms. A "—" indicates no available information.

Reference	z_{LPCC} (km)	Q_{LPCC} (C)	z_- (km)	Q_- (C)	z_+ (km)	Q_+ (C)
Wilson (1920)	—	—	1	-33	10	33
Simpson and Scrase (1937)	1.7	—	2.7	—	$\geqslant 4.6$	—
Simpson and Robinson (1941)	1.5	4	3	-20	6	24
Gish and Wait (1950)	—	—	3	-39	6	39
Huzita and Ogawa (1950)	3	24	6	-120	8.5	120
Kuettner (1950)	1.5	—	3	—	6	—
Malan (1952)	2	10	5	-40	10	40
Wait (1953)	—	—	3	-39	6.1	39
Tamura (1955)	—	—	5	-120	7	80
Stergis et al. (1957)	—	—	5	-20	8	37.5
Kasemir (1965)	—	50	—	-340	—	60
Tzur and Roble (1985)	—	—	5	-50	10	50

particle charge now available suggest that the largest charge on individual precipitation particles typically is ≈ 200 pC ($\approx 2 \times 10^{-10}$ C). However, even when large net charge densities are inferred and large charges are measured on some individual particles, the charge usually is below detectable limits on many individual precipitation particles, and both polarities of charge often coexist in the same region, regardless of the polarity of net charge in the region. The largest electric field magnitude typically observed at cloud base is 10–50 kV m^{-1} and inside the cloud is 70–200 kV m^{-1}. The electric field at the ground is affected by the charge that pointed objects on the ground emit in response to the electric field of the storm. The resulting charge in the subcloud region reduces the electric field near ground (Standler and Winn 1979, Chauzy and Raizonville 1982, Soula and Chauzy 1991, Soula 1994), so the largest electric field magnitude at the ground typically is 1–10 kV m^{-1}.

3.3. CURRENT FLOW IN THUNDERSTORMS

We move from the basic charge structure of clouds to the currents that flow in the storm environment. The polarity convention for current must be assigned. The convention in this book is chosen to be consistent with our convention for a positive electric field. Remember that a positive vertical electric field is chosen to be a field that causes a positive charge to move upward. The current caused by a positive electric field (i.e., positive charge moving upward) comprises a positive current. A *positive current* is taken to be a flow of positive charge in the positive coordinate direction (east, north, and up for Earth-relative Cartesian coordinates) or a flow of negative charge in the opposite direction.

Currents in the storm environment include those in the internal generator, those from the storm top, cloud flashes, ground flashes, precipitation, corona, and those from storm-relative air flow (both vertical and horizontal) near and in the cloud. Until recently, strategies aimed at determining the total current budget of a thunderstorm were mostly attempts to measure every conceivable current source, to estimate those not easily measurable (e.g., wind-driven currents), and to total them all to derive the net current from a single storm or to get an annual total.

3.3.1. Currents from cloud tops

Wilson (1920) proposed a current flowing from the tops of thunderstorms to the upper atmosphere to supply the fair weather current. Above the tops of storms, a net positive current flows toward the electrosphere. This conduction current arises from the response of small ions in clear air to the electric field due to the net positive charge in the upper part of a storm. Gish and Wait (1950) made the first comprehensive measurements and reported that the horizontal distance over

which appreciable current flowed was ≈ 20 km. They estimated an average current of 0.5 or 0.8 A, depending on whether they included their largest value of 6.5 A. Stergis et al. (1957) determined an average current of 1.3 A. Blakeslee et al. (1989) found conduction currents averaging 1.7 A, with a maximum of 3.7 A.

3.3.2. Corona or point-discharge current

The terms corona current and point-discharge current are used interchangeably. Corona current is created by the breakdown of air in the high electric field around pointed conductors below storms, such as grass and trees. Historically, it has been measured by instrumenting a single point, tree, or sod bed. Extrapolation of such measurements to derive a net current over an area affected by a storm of tens of square kilometers involves great uncertainty. For example, the corona measured from an isolated tree depends not only upon the electric field and wind, but on the size of the tree, its proximity to other vegetation, the presence and type of leaves, etc. Thus, it is not surprising that there is a wide range of current magnitudes reported in the literature. A meaningful storm-relative measurement of corona current is difficult, if not impossible. In a stand of trees covering 3×10^4 m^2, Stromberg (1971) found currents of a few microamperes from two spruce trees, one at the edge and one inside. The outside tree transferred 25% more charge to the atmosphere than did the inner. The trees transferred only 6% as much charge as a metal point outside the trees. This calls into question the use of metal points to determine natural corona currents. Corona current density is typically on the order of nanoamperes per square meter. This suggests that for an isolated storm, total current flow is a few tenths of an ampere. Studies over periods of months (e.g., Whipple and Scrase 1936, Chalmers 1967) showed that corona current transferred more negative than positive charge to Earth by a factor of about 1.5–3.0.

3.3.3. Precipitation current

A famous early measurement of precipitation current density is that by Simpson (1949). He noted the "mirror-image" between the current on precipitation and the potential gradient at the ground. By this, he meant that the curves had similar trends with opposite polarities. He also proposed that point discharge dominated the charge on rain at the ground, because rain collected lots of point-discharge produced ions as the rain fell. The precipitation current at the cloud base is usually not the same as precipitation current at the ground, because of this modification of the precipitation charge. Typical values of precipitation current at the ground were a few tenths to several nanoamperes (nA) per square meter. Both polarities occur, with typical magnitudes at storm cloud base and at the ground of 2–3 nA m^{-2} (Rust and Moore 1974). In the same type of storms at Langmuir Laboratory in central New Mexico, Gaskell et al. (1978) made measurements and obtained -3 to

-12 nA m^{-2} when flying horizontally just inside the cloud base. Higher in the storm, Moore (1976) calculated negative current densities of >20 nA m^{-2} from particle charge measurements for most of a vertical sounding. He also measured an equally large positive current during one period. Whether this positive current is common or was due to a nearby lightning flash is unknown. Marshall and Winn (1982) found precipitation current densities inside storms over Langmuir Lab of up to 30 nA m^{-2}.

3.3.4. Lightning current

Lightning produces currents that can be divided into those that directly transfer charge to Earth via ground flashes and those that are internal to the storm generator via cloud flashes. Livingston and Krider (1978) estimated that ground strikes produced an average current density of 3 nA m^{-2} during a Florida storm, resulting in a total current of about 3.5 A. Krehbiel (1981) estimated each cloud flash involves about 50–100 nA m^{-2} or a total current of 0.1–0.7 A.

3.3.5. Maxwell current

Krider and Musser (1982) proposed determining the total current flow in a thunderstorm by measuring the Maxwell current density, J_M, which in the storm environment is given by Eq. 1.26. They also suggested that J_M is directly coupled to the meteorological structure and/or the kinematics of the cloud. Krider and Blakeslee (1985) made simultaneous measurements at the ground of electric field, Maxwell current density, and conductivity. For one isolated storm they showed ≥ 0.25 nA m^{-2} over about 50 km^2, which yielded total current estimates for the storm ranging from 0.05–0.45 A.

3.4. BASIC CONCEPTS OF CLOUD ELECTRIFICATION

Determining how thunderstorms become electrified has been a goal of laboratory experiments and field observations for several decades. Investigators have made substantial progress evaluating various electrification processes and, as we will discuss, have found one type of process that appears capable of producing maximum electric field magnitudes comparable to those observed in thunderstorms. However, many of the factors influencing this process are poorly understood. Furthermore, observations suggest that at least some charged regions in thunderstorms may be produced by other processes, some of which have received little consideration thus far. Observations of myriad electrical properties of thunderstorms remain unexplained, as we will describe in later chapters.

One reason for the lack of definitive explanations is that it is nearly impossible to sample adequately all processes important to thunderstorm electrification.

Relevant distance scales range at least from the microscale (e.g., properties of ions and water particles) to storm scale (e.g., wind patterns and particle trajectories). Storms are complex and can change drastically in a few minutes, so differences in the time and location at which various properties are measured often interfere with analyses of causality. Furthermore, access for measurements is limited, because clouds are remote from the ground and many regions inside storms are hostile to instrumentation, aircraft, and balloons.

Another reason for the lack of understanding is the multitude of mechanisms by which clouds can become electrified. Naturally occurring frictional or thermal effects can generate enough electricity to cause lightning without *hydrometeors* (defined as liquid or frozen water particles) in sandstorms and volcanic plumes. Various properties of water, such as its surface properties and its polarizability, provide many more mechanisms for thunderstorms. The preexisting electrical properties of the troposphere also may have an effect on cloud electrification. From the multitude of mechanisms that have been identified as candidates, one would have reason to conclude that the main difficulty is not to find mechanisms that can electrify hydrometeors, but to determine which mechanisms dominate in thunderstorms. Of course, important mechanisms still may be unknown or poorly understood.

The remainder of this chapter will introduce basic concepts of thunderstorm electrification processes, including discussion of several possible electrification mechanisms. Electrification of sandstorms, volcanic plumes, and nuclear explosions is peripheral to this book, and so will not be discussed. Additional viewpoints and information are available in many reviews of thunderstorm electrification mechanisms (Mason 1971; Stow 1969; Moore and Vonnegut 1977; Pruppacher and Klett 1978; Magono 1980; Latham 1981; Illingworth 1985; Williams 1985; Krehbiel 1986; Beard and Ochs 1986; Saunders 1993, 1994).

3.4.1. Hydrometeor terminology

The references to hydrometeors in the rest of this book often use special terminology for particles having different combinations of size, state, and density. *Cloud water droplets,* also called simply *droplets,* are liquid particles that are too small to have an appreciable fall speed (droplet radius is $\lesssim 100$ μm). Similarly, *cloud ice particles* and *ice crystals* are small ice particles. Particles that are large enough to have an appreciable fall speed ($\gtrsim 0.3$ m s^{-1}) are called *precipitation.* Liquid precipitation is divided sometimes by size and terminal fall speed into *drizzle* (radius roughly 0.1–0.25 mm) and *rain* (radius >0.25 mm). Solid precipitation often is subdivided by density and fall speed into *snow* (lowest density, with fall speeds of ≈ 0.3–1.5 m s^{-1}), *graupel* (intermediate density, with fall speeds of ≈ 1–3 m s^{-1}), and *hail* (greatest density and possibly larger, with fall speeds up to 50 m s^{-1}). Graupel sometimes is called *soft hail.* The density of graupel particles ($\lesssim 0.6$

g cm^{-3}) is less than that of hail, because much of graupel growth is by *riming,* a process in which accreted particles freeze rapidly to the graupel as distinct particles. When particles are collected fast enough so that the latent heat of freezing warms the graupel to 0°C, liquid water coexists longer with ice and can fill or spread over the ice structure (a mode of growth called *wet growth*). This makes the density of the graupel particle closer to the bulk density of ice (0.9 g cm^{-3}) when the liquid finally freezes, and so can cause graupel to become hail.

Graupel growth can be divided into two more regimes besides wet growth. When the concentration of water vapor at a given temperature is sufficient to make air saturated with respect to ice, graupel particles grow by deposition. If liquid hydrometeors are present also, graupel can grow both by accreting supercooled liquid water and by deposition. However, the amount of vapor needed for saturation depends on the temperature of the ice. As a graupel particle accretes liquid, latent heat released by the freezing liquid warms the surface of the graupel. If liquid water is accreted fast enough, the surface of graupel is warmed to the point that the graupel begins to sublimate, although it continues to grow by accretion. Growing graupel whose surface has been warmed enough by accretion to sublimate, but not enough to undergo wet growth, is said to be in the *sublimation growth regime*. Graupel cold enough to grow by deposition is in the *deposition growth regime*.

3.4.2. Some definitions and cautions concerning storm electrification

In popular literature, the term *strong electrical storm* usually means a storm that produces frequent lightning flashes. In meteorological literature, the term *electrical storm* usually is supplanted by *thunderstorm,* because all thunderstorms produce lightning and because the term *electrical storm* is used also for an electrical phenomenon due to disturbances in the solar wind interacting with Earth's magnetosphere. *Strong thunderstorm* does not refer specifically to electrical strength, but can refer to any of several aspects of the storm, including heavy precipitation, large hail, violent wind, large storm size, or long lifetime, as well as frequent lightning. *Strongly electrified* refers specifically to electrical aspects of thunderstorms, but can mean either that a storm produces frequent lightning flashes or that it has a large electric field somewhere inside. Both of these criteria are qualitative, and it is not always clear which is meant. Usually, when referring to the electric field, strongly electrified connotes an electric field that approaches what is needed to produce lightning. Frequent lightning flashes imply that the electric field magnitude is large enough to produce lightning and that the storm quickly regenerates the electric field after a lightning flash. (Some flashes may increase the electric field magnitude in part of the storm and trigger a subsequent flash. However, the overall tendency is for flashes to limit or reduce the maximum electric field magnitude in a storm.)

Going beyond these qualitative descriptions to quantify many specifics about the electric field and charge distributions required to produce lightning is difficult, and quantitative criteria still are controversial. In part, the difficulty is that we do not completely understand how the charge and electric field distributions of a storm influence the lightning that is produced. However, even what we do understand is difficult to use in any pseudo-realistic conceptual model of a thunderstorm because of the complex interactions in a thunderstorm.

There are several potential pitfalls when considering electrification. Some of these will be discussed as we examine various topics, but two are basic enough to warrant discussion now. First, there is a tendency sometimes to consider ingredients for producing lightning as though there is one recipe that applies uniformly to the entire thunderstorm. However, many storm parameters, such as temperature, electrical conductivity, and hydrometeor type, vary widely within a thunderstorm, and these variations may be important to electrification processes and lightning production. Some parameters, such as air's increasing electrical conductivity and decreasing mass density with increasing height, are expected to affect lightning initiation, as we will discuss in later chapters.

The second pitfall concerns a concept common to many electrification mechanisms: Although separating two oppositely charged particles after they collide is discussed often as if this process definitely increases the degree of electrification of a storm, real thunderstorm charge distributions are complex enough that this is not always the case. Separating the two particles is necessary if they are to contribute to electrification, but separation itself is insufficient to insure that they contribute. Whether the motion of any particular charged particle increases either the local electric field magnitude or the electrical potential energy of the storm depends on how the change in its location relates to the location and motion of all other charged particles in the storm. For the electric field, this is a consequence of the superposition principle. While it is simple to calculate the change in the contribution of a moving charged particle to the electric field at a particular point, the polarity of the field contributed by the particle may or may not be the same as the polarity of the net field from all other particles. Thus, determining whether the change increases or decreases the magnitude of the total electric field at that point requires knowledge of the evolving contribution from all other charges having a significant influence at that point.

The interrelationship of individual charges may not seem quite as obvious when considering changes in the electrical potential energy of a storm, but Weinheimer (1987) showed the interrelationship explicitly. If the self energy to create individual charges is ignored, the electrical potential energy, *U,* due to the configuration of a system of point charges, q_i can be written as

$$U = \frac{1}{2} \sum_{i=1}^{N} \sum_{\substack{j=1 \\ j \neq i}}^{N} \frac{q_i q_j}{4\pi\varepsilon r_{ij}} \qquad (3.3)$$

where r_{ij} is the distance between the *ith* and *jth* charges and ε is the permittivity of air. Weinheimer pointed out that the change in energy due to the motion of particles is then given by

$$\frac{dU}{dt} = -\frac{1}{2} \sum_{i=1}^{N} \sum_{\substack{j=1 \\ (j \neq i)}}^{N} \frac{q_i q_j}{4\pi\varepsilon r_{ij}^2} v_{ij}$$

$$= -\frac{1}{2} \sum_{i=1}^{N} q_i \sum_{\substack{j=1 \\ (j \neq i)}}^{N} |\vec{E}_{ij}| v_{ij} \qquad (3.4)$$

$$= -\frac{1}{2} \sum_{i=1}^{N} q_i \vec{E}_i \cdot \vec{v}_i \left(\frac{\sum_{\substack{(j=1) \\ (j \neq 1)}}^{N} |\vec{E}_{ij}| v_{ij}}{\vec{E}_i \cdot \vec{v}_i} \right),$$

where $|\vec{E}_{ij}|$ is the electric field at the location of the *ith* particle due to the charge of the *jth* particle, v_{ij} is the radial component of the relative velocity between the two particles, $\vec{E}_i$ is the electric field at the *ith* particle from all other charges, and $\vec{v}_i$ is the velocity of the *ith* particle relative to a fixed reference frame. For thunderstorms, image charges due to the conducting Earth must be included in these sums. In the general case, neither v_{ij} nor r_{ij} can be assumed to be the same for all i and j, so neither parameter can be factored out. Only if the bracketed term in the last line of Eq. 3.4 is ≈ 1 can the energy be calculated as the motion of individual particles against a background electric field (i.e., by $q_i \vec{E}_i \cdot \vec{v}_i$), as sometimes is done in simple calculations. If the geometry of a thunderstorm charge distribution can be approximated as consisting of only a few spheres, each filled uniformly with co-moving charges, the calculation of energy is much easier, because the number of terms in the summation is much smaller. This may be an acceptable approximation for relatively simple situations, such as possibly during initial electrification. However, as the charge distribution becomes increasingly complex, more caution is needed when making inferences about the result of moving any particular group of particles.

Regardless of the difficulty in determining the effect of charge motion on a storm's energy and electric field, it remains true that thunderstorms must produce regions of net positive charge and regions of net negative charge to become strongly electrified. Processes, such as lightning, that effectively neutralize charge must be offset eventually by charged particle motions that create regions of net charge if a thunderstorm is to remain strongly electrified.

3.4.3. Basic characteristics of electrification theories

Any theory of thunderstorm electrification must include two components. First, charge must be placed on hydrometeors, so that its ability to move under the influence of electric forces is drastically reduced. Second, charge of one sign must either be kept in isolation from charge of the opposite sign (if positive and negative charges are ingested and captured in different regions) or be moved by wind and gravity to create regions of net charge (if positive and negative charges are placed on pairs of particles that interact).

The limited ability of the surface of hydrometeors to withstand electric forces restricts the amount of charge that can be placed on a single hydrometeor. When the outward electric force at the drop's surface equals the restraining surface tension, the hydrometeor will break apart (Rayleigh 1882a). The charge on the drop necessary to cause disruption is called the *Rayleigh limit*. By equalizing the force per unit area for electric and surface tension forces for drops behaving as a charged spherical conductor (i.e., $0.5\varepsilon E^2 = 2\zeta r^{-1}$, where r is the drop radius and ζ is the surface tension), the Rayleigh limit can be shown to be given by

$$q_{Ra} = 8\pi(\varepsilon\zeta r^3)^{\frac{1}{2}}. \qquad (3.5)$$

The maximum charge that a drop or hailstone can retain under different conditions has been examined in more detail by Bourdeau and Chauzy (1989). Drops measured in clouds generally have much less charge than the Rayleigh limit. However, the Rayleigh limit has been attained in the laboratory by evaporating charged drops (e.g., Doyle et al. 1964, Abbas and Latham 1967, Roulleau and Desbois 1972, Dawson 1973), and the drops were observed to eject mass explosively upon reaching the Rayleigh limit. This suggests that particle charge could approach the Rayleigh limit in regions of storms where hydrometeors are evaporating.

(i) Categorization of particle charging mechanisms. Mechanisms by which charge is placed on hydrometeors can be categorized in several ways. One popular categorization is based on whether or not a mechanism requires a preexisting electric field to polarize hydrometeors. Mechanisms requiring an electric field to induce charge on the surface of hydrometeors are called *inductive mechanisms*. If a hydrometeor previously was uncharged, the electric field induces charges of opposite polarity on opposite surfaces of the hydrometeor (Fig. 3.5), and the hydrometeor is said to be *polarized*. An obvious problem with inductive mechanisms is that they cannot be invoked to explain the initial intensification of the electric field in storms unless they can be effective in the much weaker electric field characteristic of fair weather. Mechanisms that do not require hydrometeors to be polarized by the ambient electric field are called *noninductive mechanisms*.

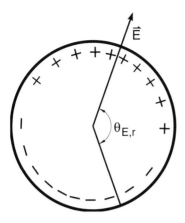

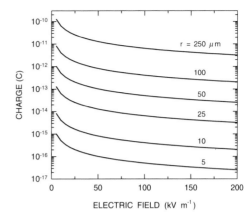

Fig. 3.5. Charge induced by an electric field on a spherical drop.

Fig. 3.6. Charge required to levitate drops of different radii as a function of the ambient electric field.

Another categorization scheme can be defined also:

1. Mechanisms in which gaseous ions (either small or large) are captured or emitted by a hydrometeor we will call *ion-particle charging mechanisms.* Ion-particle mechanisms can distribute charge of one sign systematically in different regions of thunderstorms, so the affected region can have a net charge without invoking further motions of hydrometeors. However, such systematic placement of charge is not provided by all ion-particle mechanisms.

2. Mechanisms in which equal and opposite charges are exchanged by hydrometeors that interact (e.g., by the collision of a pair of hydrometeors or by small particles breaking from the surface of a hydrometeor) we will call *particle-particle charging mechanisms.* Because equal and opposite charges are placed on particles as they interact and separate, regions in which this class of mechanism operates remain electrically neutral until negatively charged hydrometeors move systematically away from positively charged hydrometeors. Thus, particle-particle mechanisms create a neutral reservoir consisting of a mixture of both signs of charge, from which two oppositely charged regions can emerge. Some force, such as gravity or convection, must separate the oppositely charged particles before particle-particle mechanisms can contribute to storm electrification. All particle-particle mechanisms suggested thus far separate the oppositely charged particles by requiring one of them to have a larger terminal fall speed than the other. Because by definition falling hydrometeors are precipitation, particle-particle mechanisms often are called *precipitation mechanisms.* We use both terms in this book.

These two categorization schemes for particle charging mechanisms can be combined, as they are based on completely different particle properties. For example, a mechanism can be an inductive, ion-particle mechanism if it requires hydrometeors to become polarized and to either capture or emit gaseous ions.

(ii) Forces affecting the motion of charged hydrometeors. Once a hydrometeor gains charge, it moves under the influence of several forces and thereby transports the charge. The forces that normally have been considered in electrification theories are electric attraction or repulsion, gravity, and convection. Convection in this context refers to both vertical and horizontal motions of air driven by buoyancy (free convection) and other forces (forced convection).

Usually, electric forces on hydrometeors are much smaller than gravitational and convective forces, but in regions of large electric field magnitude (>100 kV m^{-1}), the electric acceleration of small charged hydrometeors can be comparable to gravity. Figure 3.6 shows the particle charge at which electric acceleration equals the acceleration of gravity as a function of particle radius and electric field magnitude. The charge values required in a weak electric field are unrealistically large for all particle sizes; for particles with radius greater than 100 μm, the charge even exceeds the Rayleigh limit. The charge values required in a strong electric field are large, but realistic, for particles with small radii, and become increasingly unlikely with increasing radius above 50 μm. Thus, the primary effect of electric force on particle motion applies to small particles in regions of large electric field. However, even larger particles will be affected to some extent when the magnitude of the electric field is large and the particle is charged. For example, the electric force on a raindrop that is 1 mm in diameter and has 5 pC (1 pC = 10^{-12} C) of charge in an electric field of 100 kV m^{-1} is approximately 10% of the gravitational force. Electric forces have a negligible effect on particle motions during initial stages of the electrification of storms, when the electric field and hydrometeor charges have the much weaker values characteristic of fair weather.

To the extent that electric force moves ions and charged hydrometeors inside thunderstorms, its immediate effect typically opposes most electrification mechanisms, because the motion that it induces dissipates electric energy (i.e., its effect is to move charge so that the electric energy of the total charge distribution is reduced). In fact, electric force would quickly neutralize much of the charge in thunderstorms if hydrometeors were removed and the charge were left on small ions, which move rapidly in response to a large electric field. The primary roles that have been proposed for macroscopic electric forces in electrification theories are to enable inductive charging mechanisms to occur and to drive gaseous ions toward the cloud, where they are captured by hydrometeors. As we already have mentioned, systematic attachment of ions of one sign at the cloud boundary tends to increase the electric field inside the storm near the boundary. Besides these effects on storm electrification, electric forces also can have significant effects on properties such as the coalescence of liquid water drops and droplets, as will be discussed in Chapter 10.

The relative roles of gravity and convection in the electrification of storms are difficult to separate. Gravity, of course, acts on all hydrometeors in every region of a thunderstorm. The process of hydrometeors falling through an air parcel under the action of gravity is called *sedimentation*. Gravity traditionally has been considered essential for particle-particle charging mechanisms, because particle-particle mechanisms place equal but opposite charges on a pair of particles with different fall speeds. (A charging interaction also can place charge of one sign on a larger particle and equal but opposite net charge on a group of small satellite particles, but the same process of particle separation applies in this case.) After the two particles interact, charge is separated vertically by positive charge falling either faster or more slowly than negative charge. In this context, falling refers to a reference frame moving with the air. Because hydrometeors are embedded in air that moves relative to the ground, falling particles may move either upward or downward relative to the ground.

Although sedimentation requires gravity, it is misleading to refer to particle-particle mechanisms as gravitational mechanisms, as they often are labeled, because the motion of all particles would be accelerated at the same rate, regardless of size and density, if gravity were the only force acting on them. However, particles moving through air also experience bouyancy and a resistive force called *drag*. As particles begin falling, the drag that the air exerts on them increases with increasing fall speed until it balances gravity at a fall speed called the *terminal velocity*. Pruppacher and Klett (1978) estimated that particles whose diameter exceeds 0.25 mm reach terminal velocity after falling a distance of the order of millimeters. The rate at which two oppositely charged particles move apart soon after

they collide is approximated reasonably well by the difference of their terminal velocities.

Sedimentation plays a role in particle-particle charging mechanisms primarily because the fall speed at which drag balances gravity is affected by properties such as the size, shape, and density of particles. Terminal velocities of the smallest cloud particles are typically <1 cm s^{-1}, so the particles move essentially with the wind. In air near ground at mean sea level, terminal fall speeds are as large as a few meters per second for low-density graupel particles, up to almost 10 m s^{-1} for raindrops, and up to roughly 50 m s^{-1} for large hailstones. Measurements of the terminal velocity of raindrops at surface air pressure by Gunn and Kinzer (1949) have been fitted statistically by Atlas et al. (1973) to give the approximate relationship

$$v_T = 9.65 - 10.3\, e^{-600D} \qquad (3.6)$$

for raindrops whose diameter, D, is between 0.6×10^{-3} and 6.0×10^{-3} m. (Raindrops larger than this become unstable and break apart at typical fall speeds.) Terminal fall speeds tend to increase with increasing altitude, because the drag acting on a particle moving through air decreases as the density of air decreases. (Chapter 10 of Pruppacher and Klett 1978 thoroughly discusses terminal fall speed.)

Determining whether gravity or convection does the initial work of separating two colliding particles that have exchanged charge can be confusing. Note that here we are not dealing with work that changes the kinetic energy of the particles. (At terminal velocity, forces on each particle are balanced so that its kinetic energy does not change.) We are dealing with the positive work done by forces in the direction of particle motion that act against electric force to create regions of net charge and increase the stored electrostatic energy of the storm.

Even in a very weakly electrified storm, what dominates the electric force against which each of the two particles moves is not the force from the other particle, but rather the force from the electric field due to all other charges. (At ranges ≥ 1 m from a precipitation particle, the electric field from even a 200 pC particle charge is at least two orders of magnitude smaller than the field throughout most of a weakly electrified storm.) As noted before, the motions of some charged particles increase neither the local net charge nor the electrostatic energy of the storm, but we assume for this discussion that each particle moves toward a charged region of the appropriate polarity to increase electrostatic energy.

We consider a pair of separating particles in three situations: (1) in updrafts strong enough that both particles move upward; (2) in downdrafts and calm air, in which both particles move downward; and (3) in updrafts weak enough that one particle moves downward, and the other moves upward. (Because gravitational potential energy is mgh, where m is particle mass, g is the acceleration of gravity, and h is height above ground, a reference frame fixed to Earth is the natural

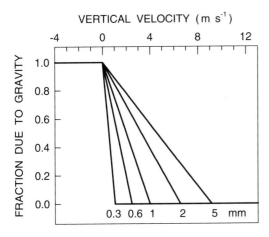

Fig. 3.7. Fraction of the work of separating a charged raindrop and a charged cloud particle due to gravity, as a function of vertical wind and raindrop diameter at a pressure of 1000 mb. The cloud particle has negligible fall speed.

reference frame for evaluating work done by gravity.) In the first situation, gravity cannot do any positive work involved in moving the two particles; in fact, the gravitational potential energy of each particle increases. Instead, the work is done by convection: Some of the convective force is transmitted to the particles by friction in a way that moves each particle toward a charged region of the appropriate polarity. In the second situation, in which both particles move down, air drag—which is directed upward, opposite to particle motion—can do no positive work on either particle. Instead, gravity moves each particle toward the appropriate charged region. In the third situation, in which updrafts cannot lift the faster falling particle, convection does some of the work, and gravity does the rest.

The fraction of the positive work attributable to convection or gravity soon after collisions, when both particles experience virtually the same wind and electric field, is simply the fraction of the relative motion of the two particles in the direction of convection or gravity, respectively. The fraction due to gravity is shown in Fig. 3.7 as a function of updraft speed and drop size for raindrops colliding with cloud particles. In regions where gravity does all of the work of charge separation, the height of the hydrometeors limits the electrostatic energy that can be generated by precipitation mechanisms. In updrafts strong enough that convection does the work of separating charged particles, gravitational potential energy is irrelevant to the generation of electrostatic energy.

As the two particles move farther apart, differences in the wind at the particles' respective locations are more likely to contribute to their relative motion. For typical thunderstorms, the relative motion due to wind

shear can become much larger than the relative motion due to different terminal velocities. When the relative motion against electric forces is dominated by wind shear, convection does most of the positive work involved in moving the particles, regardless of whether the particles are in updrafts or downdrafts.

Specific charging mechanisms are described in the rest of this chapter. We begin by examining the properties of water that can cause charge separation and then consider what appear to be some of the most promising mechanisms for charging thunderstorms.

3.4.4. Some electrical properties of water

(i) Polar molecular structure, dielectric properties, and induction.   A water molecule, shown schematically in Fig. 3.8, has a permanent dipole moment, and so is a *polar molecule*. Two characteristics of the water molecule are responsible for the dipole moment: (1) The distribution of electrons results in a net positive charge at the locations of the hydrogen nuclei and a net negative charge at the location of the oxygen nucleus; and (2) The molecular geometry is bent, not linear, with a net negative charge to one side of a line connecting the two net positive charges. (If the geometry were linear, the vector addition of the dipole moments of the two halves of the molecule would cancel each other to give zero dipole moment.)

Because water molecules have a permanent dipole moment, an external electric field tends to align the molecules, although thermal agitation prevents perfect alignment. In a hydrometeor, the partial alignment causes a surplus of negative charge on the surface of one side and a surplus of positive charge on the surface of the opposite side along the direction of the electric field (also see Sec. T.10). When an electric field induces opposite charges to form on opposite surfaces of a substance, as shown in Fig. 3.5, the substance is said to be *polarized*. A substance in which an electric field causes polarization by aligning molecular dipoles throughout the substance is called a *dielectric*, and a dielectric consisting of polar molecules, such as water, is called a *polar dielectric*. (A dielectric also can consist of molecules that form a dipole moment in response to an electric field, instead of molecules that have a perma-

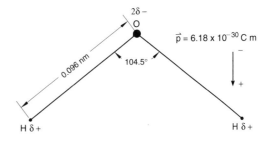

Fig. 3.8. Schematic of a water molecule. The numerical values shown are typical for liquid water.

nent dipole moment.) The effect of polarization is to reduce the electric field inside hydrometeors and increase it outside hydrometeors.

The polar molecular structure of water is responsible for some interesting behavior of water in the presence of electric fields. For example, a charged rod held near a thin stream of liquid in air attracts the liquid and deflects the motion of the stream if the liquid consists of polar molecules, but does not affect the motion if the liquid does not consist of polar molecules. Similarly, the partial alignment of dipoles enables a stream of water to resist breakup into a fountain of droplets when a charged rod is held near the orifice from which an almost vertical stream of clean water flows, a phenomenon often referred to as a Rayleigh fountain (Rayleigh 1879, 1882b). Rayleigh showed that a stream of water normally consists of drops that collide with each other, rebound, and scatter, but that the mutual attraction induced by an electric field causes the drops to coalesce instead of rebounding.

Although an electric field also causes a conducting particle to become polarized, there are some significant differences between a dielectric and a conductor. For a dielectric, the polarization charge is bound to molecules, while for a good conductor, the induced surface charge consists of free charge (i.e., charge that moves readily through the substance from molecule to molecule or from one group of molecules to the next). In a dielectric, surface charge is produced by the electric field aligning molecules. In a conductor, surface charge is produced by an electric field driving electric currents along the surface. Also, polarization simply reduces the electric field inside a dielectric, but currents flow along the surface of a conductor until the resulting charge distribution reduces to zero the internal electric field of the conductor.

However, when dielectrics have a strong tendency to become polarized in an electric field, some effects are similar to those for conductors. For example, the surface charge density, σ_q, induced on the surface of a spherical conductor in a uniform electrostatic field, $\vec{E}$, is given by

$$\sigma_q = 3\varepsilon|\vec{E}|\cos\theta_{E,r}, \qquad (3.7)$$

where $\theta_{E,r}$ is the angle from the electric field vector to the point being considered on the surface of the sphere (Fig. 3.5), and $|\vec{E}|$ is the magnitude of $\vec{E}$. For a dielectric sphere in a uniform field, the induced surface charge density is given by

$$\sigma_q = 3\varepsilon\left(\frac{\varepsilon_{diel}-1}{\varepsilon_{diel}+2}\right)|\vec{E}|\cos\theta_{E,r}, \qquad (3.8)$$

where ε_{diel} is the dielectric constant of the substance in the sphere. (Remember that the dielectric constant is a measure of the tendency of a substance to become polarized. For a derivation of Eqs. 3.7 and 3.8, see an electricity and magnetism textbook, such as pages 33–35 and 113–116 of Jackson 1962.) For $\varepsilon_{diel} \gg 1$, the factor in parentheses approaches 1, and Eq. 3.8 approaches Eq. 3.7. Because the dielectric constant of

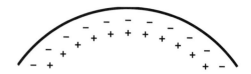

Fig. 3.9. Concept of the electrical double layer. Here a portion of a water drop is shown with its electrical double layer at the interface with air.

water is large (≈ 80), water drops frequently have been treated as conductors when considering the charge induced by an electrostatic field. Integrating over a hemisphere, then, gives the charge magnitude on each half of the sphere:

$$q = 3\pi\varepsilon|\vec{E}|r^2, \qquad (3.9)$$

where r is the radius of the sphere. Magono and Koenuma (1958) investigated the charge on each half of spherical drops that were polarized by an electric field and found it to be consistent with Eq. 3.9.

(ii) Electrical double layer. Hypotheses for many electrification mechanisms assume there is an electrical double layer at interfaces between water and air, ice and air, or water and ice, as shown schematically in Fig. 3.9. An *electrical double layer* is defined as a dipolar layer just inside an interface between two substances. For all cases in this book, at least one of the substances is water or ice. In Fig. 3.9, the electrical double layer is inside a water drop, at its interface with air. From a theoretical treatment, Fletcher (1962, 1968) concluded that it was thermodynamically advantageous for molecules at the surface of pure water in air to be oriented with their negative vertices outward. Experimental evidence for the existence, sign, and magnitude of the double layer is somewhat ambiguous. In a review of the literature on electrical double layers, Iribarne (1972) reported that the electric potential across the water-air interface varied from $+0.4$ V to -0.5 V, and that most studies found negative charge outside of positive charge. Takahashi (1969) reported a potential of ± 20 mV to $+100$ mV across water-ice interfaces at 0°C, with positive charge usually left behind on ice when the liquid water was removed.

There are various ways for the double layer to cause charge separation between drops. In all of them, more of the outer charge is removed from the double layer than the inner charge, so an excess of the inner charge is left behind. For example, if there are bubbles of CO_2 in liquid or ice and a bubble rises to the surface and bursts through, an escaping droplet will carry away the charge on the outermost part of the layer, leaving a surplus of the charge sign residing in the lower layer (see Fig. 3.9). This is thought to mean that negative charge will be carried away and an excess of positive charge will be left behind. A second possible mechanism involving the double layer occurs when two ice surfaces with different properties are rubbed together. If one particle shears more charge from the upper layer than from the lower layer of the other particle, a net charge could be transferred from the outer layer.

Several scientists, beginning with Faraday (1860), have suggested that the ice-air interface behaves like a very thin film of quasi-liquid water on an ice particle. Baker and Dash (1994) further suggested that particles will exchange material from this layer if the thickness of the films for two particles are different, with mass flowing from the thicker to the thinner layer, and that charge can be transferred during this mass exchange because of the double layer. Mass will tend to flow from warmer to colder surfaces, from regions of high surface curvature to lower surface curvature, and from regions of high vapor growth to regions of lower growth or evaporation. The results of Furukawa et al. (1987) and indirectly the results of Dong and Hallett (1992) suggest that the quasi-liquid layer occurs only for ice at temperatures $\geq -4°C$. This would eliminate processes involving the quasi-liquid layer as an explanation for hydrometeor charging over much of the temperature range in which charging is observed to occur. However, Dash (1989) cited laboratory experiments such as those by Golecki and Jaccard (1978) and Elbaum et al. (1992) that demonstrated the existence of a very thin quasi-liquid layer at temperatures extending to colder than $-30°C$. Dash noted that the layer was too thin at temperatures $\leq -4°C$ to be detected by the technique used by Furukawa et al.

(iii) Thermoelectric effect. Another microphysical property that may be responsible for charge transfer is a thermal effect, first studied extensively in ice by Workman and Reynolds (1948, 1950). In water, some molecules separate into H^+ (called cations) and OH^- (called anions). If these have different mobilities, they will diffuse at different rates across thermal gradients, and so will separate charge within a particle on which a temperature gradient is imposed. Although the mobility of H^+ is greater than that of OH^-, because H^+ has much less mass, this is not generally applicable to liquid water. In liquid water, anions and cations attach themselves to clusters of water molecules, so the difference in mobility between them is overwhelmed by the mobility of the cluster to which each is attached, and there appears to be little systematic difference in the mobility of clusters containing anions and cations.

In ice, however, the mobility of H^+ is roughly an order of magnitude greater than the mobility of anions. If a temperature gradient is maintained across a piece of ice, H^+ ions will diffuse more rapidly into colder ice and leave a surplus of negative charge in the warmer end. Latham and Mason (1961) theoretically predicted that the resulting potential difference across the temperature gradient would be $2\delta T$ mV, where δT is the temperature difference across a distance, δx. The maximum charge separation occurs at times of roughly 5–10 ms. Latham and Mason also made measurements that verified their predicted relationship between temperature gradient and potential difference. Besides anions transporting charge in ice, it also is possible for lattice defects to move under thermal gradients and to carry charge with them, thereby causing a thermoelectric effect. Takahashi (1966) concluded from laboratory experiments that lattice defects were the primary charge carrier at temperatures warmer than $-10°C$. The thermoelectric effect in ice is relatively slow on time scales relevant to particle collisions, and so probably is not a factor if particles exchange charge during collisions. However, it may still be a factor in other processes, such as small ice spicules melting or evaporating and breaking from riming graupel.

(iv) Dislocations in ice structure. Holes or dislocations in the ice lattice structure create net charge in the associated molecules, and these dislocations can move through the ice and thereby transport charge. McCappin and Macklin (1984) found that the dislocation concentration in rime ice increased as temperature decreased, and McKnight and Hallett (1978) found that slowly growing ice crystals had small dislocation densities. Because the density of dislocations depends on temperature and deposition rate, Keith and Saunders (1990) suggested that positively charged dislocations might be responsible for charge transfer during collisions of rimed graupel and ice crystals. Although this process is not well understood, the particle having the greatest concentration of dislocations has tended to gain negative charge when two ice particles collide.

(v) Contact potential. The contact potential of ice is analogous to the contact potential of metals. (The contact potential of metal is what governs the potential difference and current flow between two different metals when they touch.) Buser and Aufdermaur (1977) suggested that ice formed in different ways might also have different contact potentials, and Caranti and Illingworth (1983) and Caranti et al. (1985) suggested that a potential difference can exist between a rimed surface and an unrimed surface. Caranti et al. observed that a rimed ice surface had a negative contact potential relative to unrimed ice. The magnitude of the contact potential increased with decreasing temperature to approximately $-20°C$ and then became steady.

3.5. CLOUD ELECTRIFICATION MECHANISMS

The magnitude of charge that electrification mechanisms place on a precipitation particle can range from zero to more than 100 pC. The following is a brief explanation of a few key hypothesized electrification mechanisms. Many other mechanisms have been examined by various investigators, as described in the reviews referenced at the beginning of Section 3.4.

3.5.1. Ion capture mechanisms

One of the earliest suggestions of a mechanism by which charge can be placed on hydrometeors was an

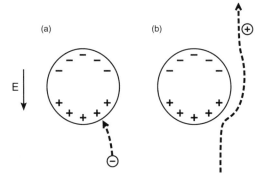

Fig. 3.10. Selective ion capture. (a) Collision between an uncharged water drop polarized in an electric field and a negative ion. (b) Deflection of a positive ion from an uncharged polarized drop.

inductive, ion-particle mechanism suggested by Wilson (1929). Although he discussed the mechanism in terms of raindrops, the same process can work for both liquid and frozen precipitating hydrometeors, so we will include both in our discussion. Wilson suggested that precipitating hydrometeors would become polarized in an electric field. If the polarized hydrometeor falls relative to ions moving under the influence of both the wind and electric field, ions of the same sign as the bottom of the hydrometeor are repelled, and ions of the opposite sign are attracted and captured. In the example in Fig. 3.10, negative ions are captured on the bottom of the drop.

For this mechanism to work, ion motions driven by the electric field must be slower than the fall speed of the hydrometeor. If the ion drift velocity is faster than the fall speed, ions of the same polarity as the bottom of the polarized hydrometeor can be captured on the top of the hydrometeor. Thus, both signs of ions would be captured, and net charging of the hydrometeor would be reduced or nullified. In Fig. 3.10, if positive ions have a drift velocity faster than the drop fall speed, those ions will be captured on the top of the drop. If the positive ions are slower, they will not be captured. This mechanism is usually called the Wilson selective ion capture mechanism, because a suitable combination of electric attraction and particle fall speed enables a particle to capture a single polarity of ions.

As a hydrometeor gains charge by selective ion capture, it reduces the attraction for additional charge. When ions of the same polarity as the top of the polarized hydrometeor are captured, the demarcation between opposite charges on the surface of the polarized hydrometeor moves downward. Eventually, the electric field is no longer sufficient to polarize the hydrometeor, so no further charging can occur. Whipple and Chalmers (1944) showed that if the velocity of the drop is much greater than the ion drift velocity, the maxi-

mum magnitude of charge acquired by a drop of radius r in a vertical electric field is given by

$$q_{max} \equiv 12\pi\varepsilon \, |\vec{E}| \, r^2. \tag{3.10}$$

In principle, if Wilson selective ion capture occurred between the positive and negative regions of a thunderstorm, precipitation particles would tend to capture negative charge, which could then increase the amount of charge in the negative region of the thunderstorm under suitable conditions. In most thunderstorm regions having precipitation, however, ion densities are insufficient to allow this mechanism to charge precipitation particles quickly enough to make a substantial contribution to electrification. One place where there might be enough ions to make a significant contribution is in the subcloud region, where ions are provided by point discharge beneath the storm.

Note that the Wilson selective ion capture mechanism is a subset of mechanisms involving ion capture by hydrometeors. However under most other circumstances (e.g., when $E = 0$ V m^{-1}), hydrometeors gain roughly equal amounts of positive and negative charge. There are two obvious exceptions: (1) Although the number of gaseous ions of opposite polarities are roughly equal in most regions of a storm, in some situations, there is a surplus of one sign of ion. This can occur beneath the storm, when point discharge currents from the ground release ions of one polarity in response to the thunderstorm electric field. It also can occur in regions where lightning deposits ions in a storm. The space charge deposited by lightning is quickly collected by hydrometeors, and small cloud particles tend to collect most of the charge, because they are much more numerous and collectively have a much larger cross section than precipitation particles. Because both of these situations occur only in active thunderstorms, further discussion of them will be deferred to Section 9.4.5; and (2) Discontinuities in conductivity create a divergence of current that causes a region of charge to form. The boundary of clouds provides a discontinuity in conductivity that causes charge to form on the cloud surface. Ion capture by hydrometeors causes both the discontinuity of conductivity and the formation of a region of charge. As discussed in Section 2.2, this mechanism has an effect on fair weather clouds, as well as thunderstorms. However, because it can be treated as a macroscopic effect of current and conductivity instead of ion and particle microphysics, we discuss it in a separate section (Sec. 3.5.4).

Most investigators have suggested that ion capture by hydrometeors is unlikely to produce thunderstorms in the absence of other mechanisms, because fair weather ion densities are too small to electrify a thunderstorm within the lifetime of storm cells. (However, see Sec. 3.5.4 for further discussion.) Numerical cloud models (e.g., Chiu and Klett 1976, Takahashi 1979) suggest that the mechanism may be important in weakly electrified storms, but can produce a maximum electric field magnitude of only ≈ 10 kV m^{-1}, roughly an order of magnitude smaller than is typical of thunderstorms. We discuss these models and present a more generalized treatment of ion capture by hydrometeors in Chapter 9.

3.5.2. Inductive charging of rebounding particles

Another of the earliest mechanisms considered was an inductive particle-particle mechanism first suggested by Elster and Geitel (1888) and revised extensively in subsequent papers. In the form in which it is usually considered today (Fig. 3.11), a precipitation particle is polarized by the ambient electric field. Sartor (1954) suggested that the cloud particle will be polarized also, thereby enhancing the charge transfer. When the falling precipitation collides with a cloud particle, some of the charge on the bottom of the precipitation particle transfers to the cloud particle. If the cloud particle rebounds, it will carry away the charge it has gained and leave behind an excess of the sign of charge that is on the top of the particles. When scientific literature simply refers to the inductive mechanism, it usually means the rebounding collisions of two polarized hydrometeors.

Equation 3.7 gave the surface charge density on a spherical conducting hydrometeor in an electric field. The treatment becomes more complicated when two hydrometeors collide, because the induced charge on each is modified by the approach of the other. Furthermore, if either hydrometeor already has a net charge, the preexisting charge also modifies the induced charge. From the work of Latham and Mason (1962), Davis (1964), and Paluch and Sartor (1973a), Chiu (1978) gave the following expression for the charge, δq, gained by the smaller of two conducting spheres that collide in an electric field:

$$\delta q = 4\pi\varepsilon\gamma_1 |\vec{E}| r_{small}^2 \cos\theta_{E,r} + A\,Q_{large} - B\,Q_{small}, \quad (3.11)$$

where r_{small} is the radius of the smaller of the two particles; $\theta_{E,r}$ is the angle between the electric field vector and the radius to the impact point on the surface of the larger particle; Q_{small} and Q_{large} are the charges already on the small and large particles, respectively; A and B are given by

$$A = \frac{\left(\dfrac{r_{small}}{r_{large}}\right)^2 \gamma_2}{1 + \left(\dfrac{r_{small}}{r_{large}}\right)^2 \gamma_2} \quad (3.12)$$

$$B = \frac{1}{1 + \left(\dfrac{r_{small}}{r_{large}}\right)^2 \gamma_2}\;;$$

r_{large} is the radius of the larger particle; and γ_1 and γ_2 are dimensionless functions of the ratio of the radii of the two particles (r_{small}/r_{large}), as shown in Table 3.2. The larger hydrometeor gains charge of the same mag-

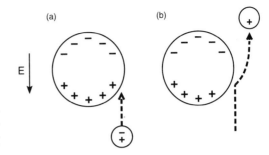

Fig. 3.11. Inductive charging of rebounding particles.

nitude but opposite polarity. Note that the effect of preexisting charge on a particle is to increase the charge transfer if the preexisting charge is of the same polarity as the charge transferred from the particle and to decrease the charge transfer if the preexisting charge is of opposite polarity. Censor and Levin (1973) and Al-Saed and Saunders (1976) showed that colliding raindrops that formed a long filament between them as they coalesced and broke apart transferred more charge than predicted by Eq. 3.11, because of the elongated geometry of the filament.

Three conditions must be met for the inductive mechanism to be effective. First, the colliding particles must separate. There has been considerable controversy about efforts to determine how often rebounding occurs under various circumstances. Most combinations of particle types separate at least occasionally under thunderstorm conditions. However, it appears that all small liquid cloud droplets coalesce with raindrops in electric fields strong enough for the inductive mechanism to be effective (Jennings 1975), so the mechanism usually is dismissed for collisions between these two particle types.

The second condition is that the contact time must be long enough so that the charge has time to flow from one particle to the other. Equivalently, the conductivity of the particles must be large enough that the charge can be transferred during the contact time. Charge transfer has been observed for collisions between two liquid water particles and for collisions between a liquid water particle and an ice particle. However, Latham and Mason (1962), Aufdermauer and Johnson (1972), Buser and Aufdermauer (1977), Takahashi (1978), and

Table 3.2 γ_1 and γ_2 as a function of the ratio of the radii of the two particles.

r_{small}/r_{large}	0.0	0.2	0.4	0.6	0.8	1.0
γ_1	$\pi^2/2$	3.90	3.10	2.55	2.06	$\pi^2/6$
γ_2	$\pi^2/6$	1.36	1.21	1.11	1.04	1.0

From Latham and Mason (1962); analytical expressions for γ_1 and γ_2 are given by Latham and Mason.

Gaskell (1981) showed that little charge was transferred when ice crystals collided with and rebounded from larger, unrimed ice particles. The conductivity of ice is smaller than the conductivity of liquid water, so it was suggested that the conductivity of ice was too poor for charge to be transferred in the short time available during collisions. This was confirmed by Illingworth and Caranti (1985), who compared the behavior of unrimed ice particles typical of a thunderstorm environment with that of unrimed ice particles that were doped to make the surface a better conductor. In collisions between doped ice particles, charge was transferred in agreement with theory. For natural ice particles, no charge transfer could be detected at temperatures $\leq -10°C$. However, Saunders (1993) cited recent, unpublished laboratory experiments in which charge was transferred during collisions between ice particles and a charged graupel particle that was actively riming. He suggested that the surface conductivity of actively riming graupel is greater than that of dry ice, and so suggested that the inductive mechanism is plausible and should be investigated further for riming graupel.

Third, the inductive mechanism cannot increase the amount of charge in two neighboring, oppositely charged regions unless collisions transfer positive charge to a particle that subsequently moves toward positive charge and negative charge to a particle that subsequently moves toward negative charge. Usually this means that particles moving upward must collide with and separate from the lower surface of particles falling downward, so that upward moving particles will add to the upper charge in the storm and falling particles will add to the lower charge in the storm. McTaggart-Cowan and List (1975) and Low and List (1982) found that colliding raindrops separate, but do so by first merging and then breaking apart, with smaller drops and droplets emerging from the top of the larger particle. Canosa and List (1993) examined charge transfer for colliding raindrops in an electric field and found that it agreed with inductive theory for collisions at the top of the larger particle, where the particles separated. Similarly, Latham and Warwicker (1980) showed that a raindrop colliding with a hailstone will rotate around the ice surface and separate from the top of the hailstone. Again, for collisions with hailstones the charge transfer agreed with inductive theory for collisions at the top of the hailstone. Thus, in both types of collisions, the inductive mechanism would not be expected to enhance electrification. Instead, if it had any effect, it would be to reduce the amount of charge separated by thunderstorms in the affected regions.

Note that this discussion assumes that one of the colliding particles moves downward and at least one moves upward. Paluch and Sartor (1973b) pointed out that the inductive mechanism is most effective when collisions occur in an updraft that makes the upward velocity of the smaller particles approximately equal to the downward velocity of the larger particles. If up-drafts or downdrafts carry both particles together into one of the charged regions before the particles separate very far, then the inductive mechanism can cause regions of net charge to move farther apart or can reduce the amount of net charge in a given region.

Today, the only type of collision for which the inductive mechanism usually is considered a plausible electrification mechanism is collisions between frozen precipitation and supercooled liquid droplets (i.e., droplets colder than 0°C). A laboratory study by Aufdermaur and Johnson (1972) suggested that such collisions rebounded infrequently (only 1–10 rebounds per 1000 collisions, giving separation probabilities of 0.001–0.01). However, in the presence of electric fields of at least 10 kV m^{-1}, they concluded that enough charge was transferred by this fraction of collisions to keep thunderstorms electrified. Note that 10 kV m^{-1} is much greater than the fair weather electric field, so Aufdermaur and Johnson's results would be applicable only after a cloud has become initially electrified. Other investigators have found larger separation probabilities: Gaskell (1981) found rebounding for up to 0.1 of collisions, and Brooks and Saunders (1994) found rebounding for 0.004–0.02 of collisions.

Theoretical modeling of the inductive mechanisms must consider other complications. For example, the dependence of surface charge density on $\cos\theta_{E,r}$ in Eq. 3.11 means that in a vertical electric field, the charge transferred will be largest for collisions on the top and bottom of a particle and will decrease to zero on the sides. Aufdermaur and Johnson (1972) concluded that most rebounding collisions were glancing collisions. Since the impact was on the side of the large particle, at a large angle from the electric field vector, Moore (1975) suggested that little induced charge would be transferred. Furthermore, he noted that, on the large particle, the surface area having induced charge of the proper polarity for increasing thunderstorm electrification by the inductive mechanism will shrink as the particle gains charge, and that this will limit the inductive mechanism further.

The inductive mechanism would be more effective if particle asymmetry or roughness modified either the probability of separation or the distribution of induced charge on one or both of the colliding particles. Sartor (1981) showed that droplets can rebound from any impact point on a rough graupel particle, and Mason (1988) suggested that the relative velocity between droplets and low-density graupel is small enough to permit rebounding from the bottom, as well as the side, of graupel particles. The results of laboratory experiments by Brooks and Saunders (1994) broadly supported the position of Sartor and of Mason that rebounding from the roughened surface of rimed graupel occurred for collisions on the bottom of the graupel, as well as the sides. However, the fraction of droplets rebounding from graupel was approximately a factor of two smaller than suggested by Mason.

Because the inductive mechanism has appeared most likely to have a significant effect when the preexisting electric field is substantially larger than the fair weather electric field, the role usually hypothesized for the mechanism has been to intensify the electrification initially achieved by other mechanisms. Some investigators have suggested that thunderstorms cannot produce sufficient hydrometeor charge to cause lightning unless the inductive mechanism plays this role. However, estimates of the relative importance of the inductive mechanism for thunderstorm electrification still are controversial, and several investigations since 1985 have found that another mechanism (the noninductive, graupel-ice mechanism discussed in the next section) is capable of producing enough charge to cause thunderstorms by itself. Furthermore, in situ measurements have detected more charge on some individual particles than could be achieved by the inductive mechanism with the maximum measured electric field (e.g., Gaskell et al. 1978, Marshall and Winn 1982), so it appears that there is at least one other mechanism capable of producing enough charge. We will discuss the role of the inductive mechanism further when we examine modeling studies of electrification in Section 9.5.1.

3.5.3. Noninductive mechanisms

As noted previously, any mechanism that does not require polarization of hydrometeors by an electric field can be considered a noninductive mechanism. This category encompasses many mechanisms. For example, Workman and Reynolds (1950) examined charge separation across the ice-water interface of freezing water and suggested that this effect could be responsible for thunderstorm electrification if water is shed when glaze ice forms on solid precipitation particles. However, Reynolds (1953) showed that this process is unlikely to occur at temperatures colder than $-10°C$. Charge also is separated by collisions between graupel and liquid water droplets and between graupel and ice particles in the absence of liquid water (Reynolds et al. 1957, Takahashi 1978), as well as by evaporation from or deposition on ice particles in the absence of collisions (Findeisen 1940, Dong and Hallett 1992). However in laboratory experiments, Reynolds et al. (1957) demonstrated that graupel pellets gain much more charge per collision when graupel is growing by riming and collides with ice crystals. In this text, we usually refer to the mechanism examined by Reynolds et al. as the *noninductive, graupel-ice mechanism.*

Of the various types of noninductive mechanisms that are possible, the graupel-ice mechanism is the only one thus far that detailed laboratory and modeling studies have suggested is capable of causing clouds to become electrified enough to be thunderstorms, although other mechanisms also may make significant contributions. Furthermore, the observed dependence of graupel-ice charging on environmental parameters appears to explain qualitatively in almost all cases examined thus far why some storms are thunderstorms and others are not. As we shall discuss, however, much remains to be learned about the graupel-ice mechanism, and other mechanisms probably are needed to produce strong electrification in at least a small fraction of thunderstorms.

(i) Empirical description of the graupel-ice mechanism. The microphysical basis for the noninductive, graupel-ice mechanism is poorly understood, so most of our knowledge of this mechanism is empirical. Laboratory experiments have found that noninductive, graupel-ice charging depends on several parameters. Takahashi (1978), for example, found that the magnitude and sign of the charge deposited on a graupel particle depended on temperature and liquid water content, as shown in Fig. 3.12. The size of the liquid water droplets had no discernable effect when the liquid water content was kept the same. Subsequent work in Great Britain (e.g., Jayaratne et al. 1983, Keith and Saunders 1990) confirmed that the charging depended on these parameters, and also found that it depended on the size of the ice crystal colliding with graupel (Fig. 3.13), on the impact velocity, and on contaminants in the water particles. Note that the increase in charge transferred per collision slows with increasing ice crystal size at the larger diameters in Fig. 3.13. Keith and Saunders (1988) provided evidence that corona occurs between ice particles as they separate, if they carry enough charge, and suggested that corona would limit the charge that could be gained by colliding particles.

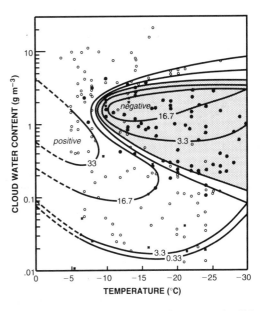

Fig. 3.12. Charge (in fC) gained by rimed graupel colliding with ice particles as a function of temperature and liquid water content. Open circles indicate that rimed graupel gained positive charge, solid circles indicate that it gained negative charge, and x's indicate that no charge was transferred. (Adapted from Takahashi 1978, with permission.)

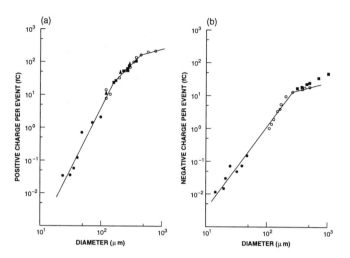

Fig. 3.13. Charge per event versus crystal size. Solid circles are data points from Jayaratne et al. (1983); triangles, squares, and open circles are data points from Keith and Saunders (1990). Squares denote particles composed of bullet clusters (i.e., clusters of bullet-shaped ice crystals). (a) Positive charge per event; (b) negative charge per event. (From Keith and Saunders 1990, with permission.)

Our understanding of the dependence of charge transfer on some parameters has been refined in subsequent work. For example, Jayaratne and Saunders (1985) and Baker et al. (1987) demonstrated that charging depended on the presence of liquid droplets on the graupel surface and that liquid particles not colliding with graupel had no detectable effect on charging. Therefore, instead of tabulating charging in their laboratory experiments simply as a function of liquid water content, Saunders et al. (1991) and Saunders and Brooks (1992) adjusted the liquid water content in their analysis to compensate for droplets being swept around the graupel by aerodynamic forces instead of colliding with it. They called the adjusted liquid water content the *effective liquid water content,* given by the liquid water content times the fraction of droplets in the graupel path that collided and adhered to the graupel.

However, the amount of liquid water collected by graupel is affected by more than just the liquid water content and aerodynamic forces. Brooks et al. (1997) pointed out that increasing the velocity of the riming target increases two parameters of importance to the charge transfer per ice crystal collision: (1) the impact velocity of ice crystals, and (2) the rate of collisions with cloud droplets. Keith and Saunders (1990) had examined the effect of impact velocity on charge transfer, but had reduced the liquid water content in their laboratory environment to keep the rime accretion rate constant. Under these conditions, the charge transfer was proportional to impact velocity raised to the 2.5–2.8 power, but the polarity of charge transfer was unaffected. When Brooks et al. kept the liquid water content constant so that the number of collisions also increased as the velocity of the riming particle increased, both the magnitude and polarity of charge could be changed by changing the velocity.

To explain the change in polarity, Brooks et al. (1997) noted that several previous laboratory studies by their British group had found the rime accretion rate

to be a parameter that controlled charge transfer. The use of effective liquid water content was one way of incorporating this underlying relationship for a constant riming rate. However, the increase in the rime accretion rate caused by an increase in graupel velocity for a constant liquid water content had the same effect on charge transfer as an increase in the rime accretion rate caused by increasing the effective liquid water content.

The effect on polarity from changing the liquid water content can be seen in Fig. 3.14, which shows the polarity of charge transfer found by Saunders et al. (1991) and Saunders and Brooks (1992), superimposed on the results of Takahashi (1978). (To superimpose the two sets of results, we assumed, as suggested by Saunders et al., that roughly half the droplets in the path of graupel actually collided with it.) At any temperature in the figure, the polarity of charge transfer can be changed by changing the liquid water content appropriately. The results of Brooks et al. (1997) meant that the same reversal of charge polarity could be obtained by changing rimer velocity to make an equivalent change in the rime accretion rate. Brooks et al. found that the effect of changing the velocity of the riming particle could be obtained from the results of Saunders et al. by multiplying the effective liquid water content in Saunders et al. by one third the graupel velocity. Details of the parameterization of graupel-ice charge transfer proposed by Saunders et al. are presented in Section 9.4.3.

Avila et al. (1995) suggested that liquid water content affects charge transfer primarily through its effect on the surface temperature of graupel. Remember that droplets freezing on a graupel particle release latent heat which warms the surface. By directly controlling the heating of graupel, Avila et al. were able to produce similar charging of rimed graupel regardless of whether liquid water droplets were present. Thus they suggested that the temperature contrast between a graupel particle and the surrounding air is more directly relevant to charge transfer than is the liquid water content, although liquid water content affects the temperature contrast for naturally occurring graupel. Although several previous independent laboratory experiments (e.g., Reynolds et al. 1957, Takahashi 1978, Jayaratne et al. 1983) have observed much less charge transferred to graupel when liquid water droplets were absent, as mentioned above, it is possible that graupel's resulting lack of elevated temperature was responsible, instead of some other process involving the droplets. The relationship underlying the dependence on elevated temperature suggested by Avila et al. is very similar to the dependence on rime accretion rate suggested by Brooks et al (1997), but further work is needed to establish Avila et al.'s result under a broader set of conditions mimicking previous experiments.

The results of Takahashi (1978) and Saunders et al. (1991) shown in Fig. 3.14 are broadly consistent over much of the region shown. Note that several aspects of the laboratory experiments, including the velocity of the riming targets, were different in the two studies (9 m s^{-1} for Takahashi versus 3 m s^{-1} for Saunders et al.). The rime accretion rate, the impact velocity, and the fraction of ice crystals colliding with graupel are all functions of the velocity of the riming target, and they influence charge transfer, as we have discussed. Other aspects of the laboratory experiments, such as differences in their droplet size spectra and in their methods of measuring liquid water content, may also have influenced results for some combinations of temperature and liquid water content. It would be very difficult now to deconvolve the interacting effects of all these factors from the results shown in Fig. 3.14 to examine more than broad agreement of results and, as stated above, the broad agreement is reasonably good.

However, two discrepancies still appear prominent when allowances are made for these interacting influences. The first, which is not apparent in Fig. 3.14, is for charge transfer at liquid water contents large enough to cause wet growth of graupel. Results for very large liquid water contents are not shown by Saunders and Brooks (1992), because they observed a reduction of an order of magnitude or more in charge transfer when conditions for wet growth existed. They hypothesized that the lack of charge transfer occurred because ice crystals rarely rebounded from wet graupel. It is difficult to reconcile this with the charge transfer shown by Takahashi (1978) in the upper left portion of Fig. 3.14, where liquid water content apparently was large enough to cause wet growth (see Jayaratne 1993b and Fig. 1 of Williams et al. 1994).

To better understand this discrepancy, Brooks and Saunders (1995) attempted to duplicate Takahashi's (1978) experimental procedure. They found that two previously suggested explanations were incorrect: Neither the centrifugal force of target rotation nor the type of agent introduced to seed the cloud for ice crystals had any effect on the polarity of charge. What they found, instead, was that the liquid water content was overestimated when they followed Takahashi's proce-

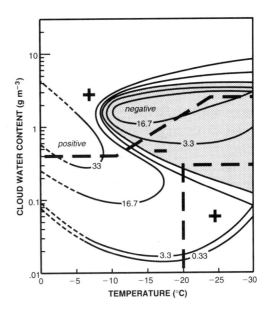

Fig. 3.14. The polarity of charge gained by graupel as a function of temperature and liquid water content for the laboratory experiments of Takahashi (1978) (as in Fig. 3.12) and Saunders et al. (1991). Bold dashed lines delineate the values of temperature and liquid water content (LWC) at which the charge Saunders et al. measured on graupel changed polarity. Saunders et al. plotted their results against effective liquid water content, which is the LWC times the fraction of cloud liquid in a graupel particle's path that the graupel is able to collect. Since Saunders et al. noted the effective LWC was typically half the actual LWC, we have used this fraction to plot their results on those of Takahashi. However, this strategy does not account for possible variations in the appropriate fraction that are a systematic function of LWC or temperature.

dure. Continuous seeding kept ice crystal concentrations large, and many ice crystals were captured by the target. Since Takahashi measured liquid water content by weighing the rime on the target and assuming it was all from collisions with liquid droplets, the large number of captured ice crystals increased the estimate of liquid water content. When comparing with the liquid water content that existed before seeding, Brooks and Saunders found that liquid water content was overestimated by up to a factor of six. Furthermore, the overestimate was actually larger, because the large number of ice crystals grew at the expense of droplets and so reduced the liquid water content during the experiment below what had been measured before seeding began, which was the value they used in computing the overestimate. Thus, Brooks and Saunders concluded that the liquid water content tabulated by Takahashi was overestimated and that wet growth was probably not achieved while graupel was charging.

The second discrepancy is apparent in regions of Fig. 3.14 with small liquid water content and warmer temperatures, in which Takahashi (1978) found that rimed graupel gained positive charge, but Saunders et al. (1991) found that it gained negative charge. Brooks and Saunders (1995) noted that, at very low liquid water content, the charge transfer appeared to be a strong function of the amount of supersaturation with respect to ice and that even the polarity of the transfer could be affected. This dependence on supersaturation has not yet been quantified in published studies. However, Saunders (personal communication, 1996) noted that the sensitivity to supersaturation made it difficult to get good laboratory results at low liquid water content for temperature $> -10°C$, although he believed the negative polarity of graupel charging measured by the British experiments was correct.

The refinements that have been made in our understanding illustrate the difficulty in separating the subtly interrelated processes responsible for the graupel-ice mechanism. Investigators obviously must exercise extraordinary care to control as many parameters of their laboratory experiments as possible, but even when this is done, subtle technical problems can interfere. We have already mentioned the difficulty in determining liquid water content and the dependence of some results on the contaminants in hydrometeors and on the degree of supersaturation within the chamber where collisions occur. Furthermore, Avila et al. (1995) noted that if the elevated temperature of graupel governs the charge it gains, then laboratory measurements of graupel charging may be contaminated by the heat conducted from a simulated graupel particle by the apparatus used to support it.

Jayaratne (1993b) pointed out yet another way in which a laboratory apparatus might affect results when he addressed a concern raised by Williams et al. (1991). Williams et al. had argued that the water vapor provided by droplets during riming at lower liquid water contents would affect too little of the surface of graupel to affect the polarity of the mean charge transfer to graupel, as proposed by Baker et al. (1987) and others. Jayaratne admitted the validity of this point. However, he pointed out that laboratory results at low liquid water contents may still have been affected by water vapor from droplets, because aerodynamic effects from the fixed orientation of a cylindrical rod used to simulate graupel created preferred zones for collisions, as noted by Baker et al. Although natural graupel can have a preferred orientation, it is unlikely to fall at a fixed orientation. Thus, Jayaratne suggested, laboratory results at low liquid water content may not be applicable to natural graupel, and natural graupel may gain negative charge instead of positive charge at low liquid water content.

The following list summarizes significant areas of agreement concerning graupel-ice charging from several studies: (1) Significant charging occurs only when the larger particle is rimed and at least small amounts of cloud liquid water are present; (2) If liq-uid water contents are large, graupel becomes positively charged; (3) At lesser liquid water contents, graupel tends to become negatively charged. (This part of the diagram in Fig. 3.14 is what is often hypothesized to cause the main charged regions of thunderstorms.); (4) At freezing temperatures near 0°C, graupel charges positively for most of the range of liquid water content; (5) The amount of charge transferred during a collision increases with increasing ice crystal size, but the increase approaches a limit at large radii; (6) The results of an individual collision between an ice crystal and graupel particle can be considerably different from the mean behaviors described by Jayaratne et al. (1983), Takahashi (1978), and others. Even the sign of the charge transfer to graupel can be different from one collision to the next with the same graupel particle; the charge transferred depends on local conditions at the site of the collision. Therefore, the noninductive charging process should be regarded as stochastic, although most experiments describe only the mean behavior; and (7) Results can be strongly affected by contaminants (primarily various salts) in the ice.

(ii) Hypothesized mechanisms for noninductive charging.

There is not yet a consensus on the microphysical mechanism responsible for the observed noninductive graupel-ice charging, although several studies have examined possible mechanisms. These mechanisms typically explain variations in the polarity of charge transfer by invoking changes in the surface properties of ice. Without invoking a particular microphysical process, Williams et al. (1991, 1994) hypothesized that variations in charge transfer between colliding graupel and ice particles can be explained by variations in ice surface properties characteristic of different graupel growth regimes. Part of the motivation for the hypothesis was the observation by several investigators (e.g., Findeisen 1940) that, in the absence of liquid water, growing ice particles tend to gain positive charge, and sublimating ice particles tend to gain negative charge. Similarly, Rydock and Williams (1991) had observed that ice surfaces growing by vapor deposition ejected negatively charged fragments and became positively charged. (Note that the reported charge gain in the absence of liquid water has usually been at least an order of magnitude less than that from noninductive graupel-ice interactions.)

The primary evidence offered by Williams et al. (1991, 1994) for their hypothesis was an apparent correspondence between the growth regimes and charge polarity in Takahashi's (1978) results: Graupel gained negative charge in what Williams et al. estimated was the sublimation growth regime and positive charge in what they estimated were the deposition and wet growth regimes. Williams et al. noted one discrepancy: Takahashi observed that graupel gained positive charge at warm temperatures in a region identified by Williams et al. as a sublimation growth regime. Furthermore, they noted that their hypothesis was inconsistent with those laboratory results of Jayaratne et al. (1983) and Saunders et al. (1991) that differed from Takahashi's.

Opposing this interpretation, Saunders and Brooks (1992) found that graupel could gain either positive or negative charge while in the deposition growth regime and could gain positive charge while in either the deposition or sublimation growth regime. Brooks and Saunders (1995) and Brooks et al. (1997) pointed out that if Takahashi (1978) overestimated liquid water content, as they suggested, then his results also contradicted the hypothesis of Williams et al. (1991, 1994). Furthermore, Jayaratne (1993b) performed laboratory experiments to estimate the parameters used by Williams et al. (1991) in calculating the growth regimes for Takahashi's results. Using these values in the equation given by Williams et al. to calculate graupel growth regimes, he found no correspondence between the growth regimes and the polarity of charge gained by graupel in Takahashi's original results.

Alternatively, several investigators (e.g., Gaskell and Illingworth 1980, Jayaratne 1993b, Baker and Dash 1994) have suggested that local conditions at the impact site on the graupel surface may govern the polarity of charge transfer. One set of experiments that demonstrated this point clearly were conducted in Argentina by Avila et al. (1988), Caranti et al. (1991), Avila and Caranti (1994), and Avila et al. (1995), who examined the charge transferred by individual collisions of ice particles with rimed graupel. (Most studies by other investigators examined the average charge transferred by many collisions.) For their experiments, they used ice spheres 100 μm in diameter colliding with a fixed cylindrical rimed target. They found that under most environmental conditions either polarity of charge could be transferred to the simulated rimed graupel by a given collision, but that one polarity tended to prevail, on average, for many collisions under a given set of conditions. Furthermore, they observed that the collisions broke off frost or rime structures from the surface of the simulated graupel, and the broken structures carried net charge away from the graupel and left the opposite polarity of charge behind. When surface structures broke off, graupel growing by vapor deposition gained positive charge, while graupel whose surface was sublimating gained negative charge.

Caranti et al. (1991) and Avila and Caranti (1994) hypothesized that thermal gradients in the rime and frost structures were responsible for the observed charging. In the absence of active riming, they hypothesized that the tip of a structure on graupel growing by deposition is warmed as water vapor releases latent heat during deposition. Thus, the tips are more negative than the base, due to the thermoelectric effect, and carry away negative charge when they break off. Similarly, the tips of structures on sublimating graupel are cooled when ice molecules take away latent heat as they sublimate. A sublimating tip, therefore, is more positive than the base of the structure and carries away positive charge when it breaks off. The situation is complicated by riming, because there are competing effects. Liquid droplets warm the surface of graupel and the base of structures relative to the tip, but also provide a source of water vapor that increases the tendency toward deposition. If enough water vapor is provided to cause or maintain deposition at the site of a collision, the resulting warming of the tip could be greater than the warming of the base.

The thermoelectric effect (e.g., Latham and Mason 1961), however, is too small by itself to explain the magnitude of graupel charging observed in the Argentine studies. Thus, Avila and Caranti (1994) suggested that the charge transfer is increased by other effects, such as a quantum mechanical effect associated with the fracture. They hypothesized this effect would increase the number of protons on the colder side of the fracture, but quantitative and theoretical descriptions of this mechanism remain to be worked out. In support of their hypothesized mechanism, Avila et al. (1995) showed that the magnitude and polarity of the mean charge that was gained by graupel depended on the temperature difference between the graupel surface and the air around the graupel.

At first glance, this dependence on the temperature difference might appear to be contradicted by Marshall et al. (1978), Gaskell and Illingworth (1980), and Jayaratne et al. (1983), who observed that the charge gained by graupel did not depend on the temperature difference between interacting graupel and ice crystals. However, it is not clear that there was a contradiction. The relevant temperature gradient for Avila et al.'s (1995) hypothesis is across rime or frost structures, not between colliding particles. Furthermore, the lack of dependence observed by other investigators was based at least in part on observations when the temperature of graupel was depressed below the environment, whereas Avila et al. used elevated temperatures to simulate heating by riming.

The observations and hypotheses of the Argentine investigators have been tested independently in three other studies. By examining the charge gained by graupel when pieces of frost or rime were removed by a jet of air, Saunders et al. (1993) confirmed that graupel gained positive charge when it was growing by deposition and negative charge when it was sublimating. Jayaratne and Griggs (1991) examined the charge gained by graupel during the fragmentation of rime structures by air or changes in graupel momentum and presented evidence that the charge was not transferred during the fracture itself, but occurred when the rime fragment subsequently collided with the rimed graupel surface and rebounded. Jayaratne (1993a) examined the charge gained by rimed graupel as a function of the temperature difference between the outer edge of the rime and the surface of the simulated graupel particle below the rime. Using vapor-grown ice crystals that collided with rimed graupel in the absence of liquid droplets, he found that the polarity of charge gained by graupel appeared to be unaffected by the temperature gradient across the rime. However, the charge gained per collision was more than an order of magnitude smaller than that observed by Avila, Caranti, and their colleagues, consistent with the greatly reduced charge per collision found by Jayaratne et al. (1983) in the absence of liquid water. Jayaratne (1993a) suggested that the small, vapor-grown crystals that he used were unable to fracture appreciable rime structures. Thus, he could not rule out charging due to thermal gradients in the experiments of Avila et al. (1995), in which rime or frost structures were fractured by ice particles that had much more momentum than the small ice crystals he used.

Another microphysical mechanism proposed to explain noninductive graupel-ice charging involves the quasi-liquid layer and electrical double layer on the surface of ice and also depends on local conditions at the point of impact on the surface of the two colliding particles. Baker et al. (1987) suggested that the particle growing more rapidly from vapor deposition tended to gain positive charge. Baker and Dash (1994) suggested that the comparison of the propensity for growth should be evaluated at the point of contact, where the details of particle geometry and surface conditions for both particles were hypothesized to affect the direction of mass transfer between the quasi-liquid layers of the colliding particles. Baker and Dash further hypothesized that the mass transferred between the quasi-liquid layers carries charge from the upper part of the electrical double layer and that the direction of mass transfer between the colliding particles governs the polarity of charge transfer. From a theoretical evaluation of the direction of mass transfer under various conditions, they concluded that the charge that would accompany mass transfers was consistent with the graupel charging observed by Baker et al. as a function of temperature and liquid water content.

Other microphysical mechanisms also have been proposed to explain noninductive charging of ice particles, although these mechanisms are capable of explaining the graupel-ice mechanisms under only a subset of the conditions that have been examined. These mechanisms still are considered plausible because different mechanisms may govern charge transfer between graupel and ice under different conditions. For example, Buser and Aufdermaur (1977) found that the charge transferred between ice crystals and various metal targets depended on the contact potential of the metal. Furthermore, they suggested that ice formed in different ways might also have different contact potentials. Caranti and Illingworth (1983), and Caranti et al. (1985) suggested that a potential difference can exist between a rimed surface and an unrimed surface and that the charge transferred depends on both the area of contact and the potential difference between the surfaces. Caranti et al. observed that a rimed ice surface had a negative contact potential relative to unrimed ice, and the magnitude of the contact potential increased with decreasing temperature to $-20°C$. Their evaluation of charge transfer by this mechanism suggested that it was able to explain negative charging of graupel during graupel-ice collisions, but was unable to explain positive charging of graupel.

As discussed in Section 3.4.4, Saunders (1994) and Keith and Saunders (1990) have suggested that positively charged dislocations in ice structure may be responsible for charge transfer during collisions of rimed graupel and ice crystals, because the dislocation densities in ice depend on temperature and deposition rate, with colder temperatures and faster deposition rates causing larger concentrations of dislocations. However, the detailed process by which dislocations are transferred between particles was not specified. Charge separation has been observed to occur when ice fractures, and this presumably is caused by some process involving dislocations. Illingworth (1985) suggested that, similarly, charge separation might be caused by plastic deformation of ice when graupel and ice particles collide. He noted that Petrenko and Whitworth (1980) observed charge separation due to the movement of charged dislocations when monocrystals of some compounds were subjected to plastic deformation.

The last mechanism that we will discuss was proposed to explain the well-established observation that ice particles growing by vapor deposition in the absence of liquid water tend to gain positive charge, and sublimating particles tend to gain negative charge. This charging occurs in the absence of collisions with ice, but tends to be much smaller than the charging that can occur from the graupel-ice mechanism. Although the hypothesized mechanism may not apply directly to the graupel-ice mechanism, we present it because it illustrates some factors that may influence hydrometeor charge.

From their studies of charging in the absence of collisions, Dong and Hallett (1992) hypothesized that a combination of the thermoelectric effect and the surface properties of ice were responsible for the charging. They pointed out that in sublimating ice particles the surface is colder than the interior, so positive ions migrate toward the surface. Because the energy binding H_3O^+ ions to the surface is weaker than the energy binding neutral water molecules, positively charged molecules escape from the surface through evaporation more readily than neutral molecules. The escaping positive ions tend to leave a negative charge on the ice, unless the negative ion concentration in the surrounding air, which exchanges molecules with the ice surface in equilibrium, is large enough to dominate the effect of escaping positive ions. In particles growing by deposition, Dong and Hallett hypothesized that positive ions are incorporated more readily into ice than negative ions, because positive ions are better able to become attached to water molecules, and water molecules can be incorporated more readily than other molecules into ice. Thus, ice particles growing by deposition would be expected to become positively charged. However, this charge transfer process remains to be proven.

3.5.4. Electrification by external currents

(i) Screening-layer charge. As discussed in Section 2.2, conduction current flowing across the discontinuity in conductivity at the boundary of clouds causes a layer of charge to form on the boundary. Although an electric field drives the current, the formation of the charge layer does not depend on hydrometeors being polarized by the electric field, so we classify the mechanism as a noninductive, ion-particle mechanism. Besides being produced by the fair weather electric field, as discussed in Section 2.2, the charge layer also can be produced by a thunderstorm's electric field. The physics and mathematical treatment are basically the same in both situations. For both, the condition for equilibrium (Eq. 2.1) for a thunderstorm is that the component of current normal to the boundary must be constant across the boundary.

To maintain a constant current across the boundary, the layer of charge increases the thunderstorm electric field just inside the boundary and decreases it outside the boundary just enough to compensate for the discontinuity in conductivity. Because the layer of charge reduces the electric field outside the storm, it often is called a *screening layer*. The ratio of the electric field just above and below the screening layer at the cloud boundary is given from Eq. 2.1 as

$$\frac{E_{clrn}}{E_{cldn}} = \frac{\sigma_{cld}}{\sigma_{clr}}, \qquad (3.13)$$

where E_{cldn} and E_{clrn} are the normal components of the electric field just inside and outside the screening-layer charge, respectively, and σ_{cld} and σ_{clr} are the conductivities of cloudy and clear air, respectively. Although there have been relatively few measurements of the ratio of conductivity inside and outside storms, the measurements of Rust and Moore (1974) suggest that the electric field outside a storm should be a factor of 5–10 times smaller than the electric field inside the storm, after a screening layer forms.

As will be derived in Section 9.4.6, the time required for the screening-layer charge to reach $1 - e^{-1}$ of its equilibrium value is $2\varepsilon/(\sigma_{clr} + \sigma_{cld})$. Normally a period of 4–5 times the e-folding time is assumed necessary to reach conditions close to equilibrium. This means that it takes anywhere from tens of seconds at the top of thunderstorms to tens of minutes near ground for screening layers to form. If the electric field at some point fixed with respect to the changing cloud boundary evolves much more rapidly than the screening layer can form, the electric field at that point will not be screened, and a recognizable screening layer may not be able to form. Because other cloud boundaries may or may not have screening layers, and the geometry of the boundaries is not simple, it is often easiest to treat screening-layer calculations by considering only the local boundary. The effect of other screening-layer charges is included in the net electric field at the boundary. Then the boundary can be treated as a single infinite plane, and the current driven by the component of the net electric field normal to the boundary is the current that must be constant. At equilibrium, the screening surface charge density at a given point on the boundary then is given by Eq. 2.3.

Electric field measurements from aircraft flying above or through storms and from balloons flying vertically through storms have provided considerable evidence of the existence of screening layers. Vonnegut et al. (1962) found screening layers at the boundaries of small cumuli whose interiors had been charged artificially. Furthermore, during flights over the tops of developing thunderclouds that were not modified, they measured no significant E unless the top was actively growing at the time of the pass. Significant electric field magnitudes were detected only from rapidly growing turrets, not from the flat cloud top corresponding to the buoyant equilibrium level. Vonnegut et al. suggested that the screening-layer charge was rapidly advected away from the growing turret. It is also possible that the boundary of the rapidly growing cloud turret did not have enough time to accumulate much charge in a screening layer. Either way, the charge in the screening layer of the cloud turret would have been small, so E outside the cloud from charge inside the turret and from charge in the upper part of the storm near the turret was not diminished much by the screening layer. Balloon-borne soundings of the electric field also often have shown evidence of a screening layer at the top of the storm (e.g., Marshall and Rust 1991). The evidence for screening layers on the sides of thunderstorms is inconclusive. Although Gunn (1957) often observed an increase of a factor of about five in the magnitude of the electric field as an aircraft penetrated a storm, some investigators have suggested that at least part of the increase was a result of the aircraft becoming charged inside the cloud.

Measurements of E in thunderstorm anvils (Secs. 7.8, 8.1.3) clearly show the presence of screening layers, often on both the upper and lower boundaries. To explain electric field profiles with E ≠ 0 near ground (i.e., E is not fully screened outside the anvil), Marshall et al. (1989) showed that, even though screening layers form quickly, E changes much more slowly at the ground. They suggested that the slower response at the ground was caused by space charge from point discharge beneath the anvil, which shields the ground from the effect of screening layer formation on the base of the anvil cloud. In a case that they examined with an analytical model, they found that $E_{gnd} = 2.5$ kV m^{-1} even after 20 min.

Sensors penetrating cloud base typically observe a rapid decrease in the conductivity over a distance of a few meters, and this decreased conductivity is capable of producing a screening layer at cloud base. However, a relatively long time is required for the screening layer to form at lower altitudes. Evidence for screening layers at the base of convective clouds is mixed. In the sounding by Winn et al. (1981), which went into a precipitation shaft, they inferred that the downdraft and rain removed any definite cloud base. Thus, they concluded that the negative charge that was located just above the apparent base (i.e., where visual contact was lost) was much higher than the updraft base. Byrne et al. (1983) called their lowest positive regions screen-

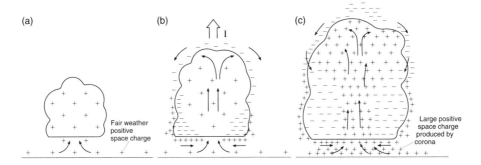

Fig. 3.15. Convective charging mechanism. (a) As a small growing cumulus cloud forms, it ingests positive space charge and forms a negative screening layer at the cloud boundary. (b) Organized transport of negative charge occurs as cloud particles on the boundaries move down the sides toward the cloud base. Inflow of positive charge continues in the updraft, and current continues to flow to the top of the cloud to adjust to the newly ingested positive charge and to the loss of screening-layer charge from the top due to transport down the sides of the cloud. (c) The electric field from negative charge in the lower part of the cloud becomes strong enough to produce corona at Earth's surface, thereby increasing the flux of positive charge into the cloud base. Transport of negative charge continues down the sides of the cloud from cloud top. (B. Vonnegut, personal communication, 1996, with permission)

ing layers (three of four began near cloud base). Dye et al. (1989) did not detect a screening layer in spirals at cloud base, and they attributed this to the updraft not allowing the layer to form. (The charged particles are swept upward continually as long as the updraft occurs.) From the location of the lowest charge in six of eight soundings for which a cloud base was noted by Marshall and Rust (1991), it appears as if the bottom charge could have been a screening layer, but the measurements were inadequate to rule out other possible mechanisms for producing the observed charge.

Large storm systems, in which the horizontal extent of charged regions can exceed 100 km, have been observed to provide an exception to the tendency for a screening layer to occur on an upper quiescent cloud boundary. The exception occurs because the electric field is constant with vertical distance from a horizontally infinite charge layer. Thus, when charge is in an infinite layer above an infinite horizontal plane consisting of a perfect conductor, the image charge nullifies the contribution of the charge layer to the electric field above the layer. When the electric field of a region of the storm can be approximated as being due to vertically stacked, infinite horizontal layers, therefore, the thunderstorm charge distribution makes a negligible contribution to the electric field above the highest charge layer within the storm. The screening layer on the upper cloud boundary due to thunderstorm charge then also becomes negligible, although it is possible that the relatively weak fair weather screening layer could form there (positive charge instead of negative charge locally on the upper cloud boundary). This scenario applies only when the horizontal extent over which charge layers are approximately uniform is large

compared with twice the height of the highest layer, because that is the effective vertical distance from the layer to its image charge.

Numerical models to describe screening layers in thunderstorms were published by Brown et al. (1971) and Hoppel and Phillips (1971). These two studies were followed by an analytical treatment by Klett (1972). Numerical cloud models used to study thunderstorm electrification include screening-layer charge either by parameterizing ion production and capture (e.g., Chiu 1978, Helsdon and Farley 1987) or by parameterizing the bulk properties of screening-layer formation (Ziegler et al. 1991). These last two treatments are described in Sections 9.4.5 and 9.4.6.

(ii) Thunderstorm electrification requiring screening-layer charge. Grenet (1947) and Vonnegut (1953) independently proposed the only conceptual model of thunderstorm electrification thus far that does not involve precipitation charging mechanisms. In their conceptual model, illustrated in Fig. 3.15, the screening layer and fair weather space charge play prominent roles. As described in Section 1.4.5, fair weather space charge is positive. Both Grenet and Vonnegut noted that the positive fair weather space charge is carried by updrafts into developing clouds, where it is quickly captured by hydrometeors. The resulting hydrometeor space charge then causes a negative screening layer to form on the cloud boundary. Cooling and convective circulation cause the cloud boundary to move downward, and the downward motion of the boundary carries the screening-layer charge. Vonnegut (1953, 1955, 1994), Vonnegut et al. (1962), and Moore and Vonnegut (1977) suggested that negative screening-layer charge can be transported to the lower part of the storm and then into the cloud's interior, thereby creating a positive dipole in the thunderstorm and attracting more positive charge from below the cloud. This conceptual model frequently is called the *convective mechanism,* because of its obvious reliance on convective motions. However, we avoid this term because convective motions also play an important role in conceptual models based on precipitation mechanisms. Instead, we refer to this conceptual model as the *Grenet-Vonnegut mechanism.*

Key processes involved in the Grenet-Vonnegut mechanism have been observed to occur in storms and probably make significant contributions to the charge distribution of thunderstorms. The primary point of controversy has been whether the mechanism can account for the major features of the overall thunderstorm charge distribution typically observed shortly before the first lightning flash and can transport enough charge to cause lightning to occur. Moore et al. (1989) attempted to demonstrate the feasibility of the mechanism by releasing large amounts of negative space charge beneath clouds. Their hypothesis was that, if a cloud ingested negative space charge and eventually became a thunderstorm, the gross structure of its charge distribution would be a negative dipole instead of a positive dipole. They presented data from one thunderstorm in which negative space charge was ingested into a cloud that then formed positive charge at lower altitudes and produced ground flashes that lowered positive charge to ground. Winn et al. (1988) examined data from electric fields measured by balloons and aircraft in clouds treated by Moore et al. Although they also found positive charge in the lower part of clouds, they found that the polarity of the overall charge distribution was normal. Similarly, Dye et al. (1988) concluded from aircraft data in one cloud treated by Moore et al. that the upper region of the storm had positive charge, as in untreated thunderstorms. Thus, while Moore et al. appeared to be able to influence the charge distribution, it is not clear that the effect was enough to reverse the overall charge distribution.

Whether or not the Grenet-Vonnegut mechanism is capable of explaining the entire thunderstorm charge distribution for all thunderstorms, we believe that most of the basic concepts of the mechanism are applicable to thunderstorms and are valuable in emphasizing electrification processes that often are neglected, as we will discuss shortly. However, several investigators have raised valid issues concerning specific details of the mechanism, and we will discuss these first. For example, Latham (1981) pointed out that: (1) The space charge density in air drawn into the updraft often is too small to account for observed thunderstorm charge densities; (2) The time scale for the convective motions initially needed to electrify the storm is comparable to the lifetime of thunderstorm cells; and (3) The mechanism has no process for forcing negative charge to be in a similar temperature range from storm to storm, as has been observed for diverse types of storms dur-

ing all seasons. In regard to the first point, the space charge density appears large enough in some cases, but not others. Measurements on mountaintops in New Mexico suggest that the fair weather space charge density is large enough to provide charge for mountain thunderstorms. However, fair weather space charge decreases rapidly with height above ground (see Sec. 1.4.5), so for some storms, such as high-based thunderstorms that form above a stable layer of the lower atmosphere (e.g., Baughman and Fuquay 1970), the fair weather space charge density is too small. Thus, it appears that there are at least some thunderstorms whose electrification cannot be produced completely by the Grenet-Vonnegut mechanism.

The second issue, the time scale for convective motions, could conceivably be addressed by invoking the succession of convective cells that often are observed to precede cells that become thunderstorms, if the only requirement were for the convective scale motions eventually to produce a thunderstorm. However, the time and the electric field distribution required by convective transport to establish a typical thunderstorm charge distribution appear incompatible with the rapid increase of the maximum electric field magnitude observed at middle levels of many storms a few minutes before they produce the first lightning flash (e.g., Dye et al. 1989, Winn 1992).

A related problem is that highly systematic convective motions are required by the mechanism. The negative charge must be transported all the way from the upper part of the storm to the height at which negative charge commonly is observed (heights where storm temperatures typically are 0°C to −20°C). Entrainment, evaporation, and diffusion will greatly reduce the negative charge on the cloud boundary during transport and may prevent effective transport to the bottom of a storm. Furthermore, tracer measurements of entrainment found that the upper cloud boundary was mixed into the central region of the upper part of the storm, instead of being transported down the side of the cloud (Blythe et al. 1988, Stith 1992). With entrainment, therefore, it appears unlikely that much charge on the upper cloud boundary would be transported to the lower part of the cloud.

An additional problem with the convective motions required by the Grenet-Vonnegut mechanism was noted by Winn (1992). For the cloud boundary to transport enough charge to achieve observed charge densities and charging rates in small thunderstorms, he calculated that the electric field between the inner positive charge and the negative screening-layer charge would become large enough to cause lightning between them, before the negative charge could reach the lower regions of the storm. Support for this interpretation was provided by the observations of Taylor et al. (1984) and Mazur et al. (1984): They observed lightning high in clouds, and suggested that the flashes occurred between positive charge in the upper part of the storm and negative screening-layer charge near the top of the storm.

Even if sufficient negative charge survived transport down the side of the cloud, it still would be necessary for negative screening-layer charge to be folded into the storm at precisely the height range that would lead to the observed charge distributions inside storms. It seems improbable that the transport into the lower part of the storm would occur systematically enough to produce negative charge in a similar temperature range for all the types of thunderstorms in all the environments that have been observed thus far. Wind patterns and the distribution and types of precipitation and cloud particles vary too much relative to the height of temperature isotherms for systematic transport of much of the cloud surface into a particular temperature range to be plausible.

For all of these reasons, it seems unlikely that the Grenet-Vonnegut mechanism is universally able to produce the major features of the charge distribution for most types of thunderstorms. However, the mechanism still may produce most of the charge in some thunderstorms, though additional work is needed to demonstrate that this is possible. No known electrification mechanism can provide a universal explanation of the electrification of all thunderstorms that have been observed. Mechanisms involving ice are commonly thought to be important in most thunderstorms, but are incapable of explaining observations of some tropical thunderstorms that appear to have little, if any, ice (see Sec. 8.4.8). Although numerical cloud models have been able to produce only weak electrification with the Grenet-Vonnegut mechanism (see Sec. 9.5.1), it may be that the mechanism is effective under specific circumstances that are not yet understood.

Furthermore, the Grenet-Vonnegut mechanism is important in highlighting the significance of convective motions and external charge and currents for electrification. During the period that the mechanism was formulated and for many years afterward, the contribution of convection to charge transport tended to be ignored in favor of the contribution of gravity. As discussed in Section 3.4.3, the relative motion of charged particles due to convection can be much larger than the relative motion due to sedimentation, and even sedimentation is not wholly due to gravity.

At least part of the reason for the importance of external charges and currents should be apparent from our discussion of screening layers. If screening-layer charge contributes to the production of some lightning flashes, as suggested by Taylor et al. (1984), Mazur et al. (1984), and Winn (1992), then it makes a significant contribution to the electrification of storms. Furthermore, MacGorman et al. (1996) suggested that flashes between the screening layer and upper positive charge of a storm help foster the subsequent production of ground flashes by allowing more charge to accumulate in the lower negative charge than in the upper positive charge. If smaller electric field magnitudes are required to initiate lightning near the top of storms, then flashes involving the screening layer may be among the first flashes produced by at least some storms. It is unclear whether the space charge produced underneath storms plays a significant role in the production or characteristics of lightning, but it appears important for the charge distribution in the subcloud region (Standler and Winn 1979, Standler 1980, Chauzy and Raizonville 1982, Soula and Chauzy 1991, Soula 1994), and possibly in the lower part of storms (e.g., Moore et al. 1989).

No process that is capable of influencing the net charge of a sizeable region of the storm or its immediate vicinity should be ignored. Because the electric field is inversely proportional to the square of the range, a region of net charge has a strong influence on the electric field in its vicinity, and so may also influence the rate, type, or location of lightning produced by a storm. Furthermore, the effect of charge can be exerted at a distance through the electric field, so the production of charge in one region may cause feedback effects elsewhere. Screening-layer charge is an obvious example of feedback, and additional possibilities also were suggested by the Grenet-Vonnegut mechanism.

3.5.5. Charging during melting

Noninductive charging as ice melts was first studied by Dinger and Gunn (1946) and later by Magono and Kikuchi (1965), Kikuchi (1965), and others. There are some conflicting reports, but usually ice gains positive charge as ice melts. The charging appears to be caused by CO_2 bubbles in the ice being freed and bursting through the wet surface of the ice. The outer layer of the bursting water film is thought to carry away negative charge from the electrical double layer of the melt water. The ice particle is then left with an excess of positive charge. The amount of charge gained during melting is larger when there are more CO_2 bubbles. However, the amount and even the polarity of charge transferred is sensitive to impurities in the water.

Another mechanism involving melting is an inductive mechanism proposed by Simpson (1909). As frozen precipitation falls and melts, it can shed droplets. In the presence of an electric field, two separating particles become polarized and exchange charge as described in Section 3.5.2. Although droplets separate from the top of the precipitation, and so would not usually be expected to enhance electric fields or increase the amount of charge in thunderstorms, this process may be important in the lower part of storms, particularly below preexisting charge regions. If negative charge is above the melting layer, for example, such a process could produce negative rain topped by a region of positive charge on droplets near the melting layer. Such a process may help explain some observations of the charge structure of large stratiform clouds that trail behind squall lines (see Sec. 8.2.4.iii).

3.5.6. Other miscellaneous mechanisms

Hydrometeors can gain charge in many other ways. For example, liquid hydrometeors gain negative charge when growing by condensation and positive charge when evaporating. Snow gets positive charge when growing by riming and negative charge when it is sublimating or stable. Water drops also become charged when they splash, an effect called the *Lenard* (1892) *effect* or waterfall effect. Larger drops usually become charged positively, and the smaller spray droplets negatively. However, none of these mechanisms is expected to generate enough charge under conditions that exist in thunderstorms to play a significant role in thunderstorm electrification.

4

Corona and Point Discharge

4.1. INTRODUCTION

Electric corona—often referred to as point discharge—is a somewhat ambiguous term, but usually refers to any electric discharge less violent and energetic than an electric spark. Thus, a number of successively more energetic phenomena fit this description. If all parameters except the electric field are held constant, transition to a more energetic form of corona is caused by increasing the electric field. Naturally occurring corona in the atmosphere can be divided into two groups: (1) corona from vegetation at the ground, and (2) corona discharges from hydrometeors in clouds. In this chapter, we will concentrate initially on a basic description of corona processes. We will then examine corona from hydrometeors and point discharge at the ground in more detail. Note that the terms corona and point discharge refer to the same basic class of phenomena. Which term is used is a matter of personal choice or convention in a particular field of study. We present only basic concepts and information about corona that we have drawn from an overview of corona in the atmosphere by Latham and Stromberg (1977) and an extensive treatment of corona by Loeb (1965). More details are available in these and other sources.

Corona, sparks, and lightning can occur when the electric field magnitude is large enough to cause the air to fail catastrophically as an insulator and to become ionized enough to act as a good conductor. In the process of becoming ionized, the air through which these electric currents pass becomes luminous. The luminosity of lightning is often obvious, especially at night, but corona is much less luminous. It requires nearly total darkness to be visible in the form of a brushlike or branched violet discharge. Corona, as lightning, also emits electromagnetic radiation.

Corona occurs near conductive points in an ambient electric field smaller than required for breakdown in the absence of points, because conductive points (i.e., regions where the radius of curvature is very small) considerably enhance the electric field locally. One property of a conductor is that external electric field lines must terminate perpendicular to its surface. Any component of electric field parallel to a conducting surface would cause charge to rearrange itself rapidly on the surface until the field from the charge nullified the parallel electric field. To satisfy this boundary condition on the surface of an elongated conductor, the ambient field lines must be distorted as shown in Fig. 4.1, so that they are densest (meaning there is a greater electric field magnitude) around the smallest radius of curvature.

Corona can occur from conductors in a variety of locations and situations whenever there is an electric field above a corona-onset threshold. For example, it occurs on pointed objects connected to the ground and from cables and wires running above ground. It occurs from rain and possibly wet ice particles in regions of large electric field magnitudes in thunderstorms. It occurs from airplanes and ships. The electrons emitted by corona form regions of space charge in and around thunderstorms, and some sources of corona make a significant contribution to the electrical nature of the atmosphere.

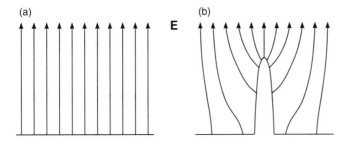

Fig. 4.1. Enhancement of electric field by pointed conductor in ambient electric fields near a positively charged ground. Ambient electric field (a) without a pointed conductor and (b) with a pointed conductor.

Most of the information we have on the physics of corona comes from laboratory studies of corona from metal points. Corona is often studied by charging two metal electrodes with a high-voltage source to provide an ambient electric field. Parallel plates, cylinders, and spheres are most commonly used. With parallel plates, a needle is often inserted between the plates to provide a region of intense electric field for corona. For other geometries, one electrode has much greater curvature than the other. To study corona from water drops, the drops are injected between electrodes and observed as they fall or are suspended in a wind tunnel at terminal velocity.

To measure the magnitude and polarity of point-discharge currents underneath thunderstorms, investigators have used both natural and artificial points. Perhaps the easiest technique is to measure the current that flows from a metal point through a simple amplifier circuit such as discussed in Section 6.7. Isolated trees or sections of turf have also been used successfully, but entail more problems. For example, trees suffer more than metal points from microscale charges in their leaves or needles, thus affecting the corona threshold value. On a more macroscale, trees in leaf produce less point discharge than the same, but leafless, trees under the same electric field and wind.

4.2. CORONA INITIATION

The nature of corona discharge is basically the same whether it occurs on metal points, hydrometeors, or natural points such as trees, grass, and bushes. In all cases, corona is initiated near the points when the local electric field is strong enough to accelerate free electrons to kinetic energies that will ionize molecules before the electrons collide with an atmospheric particle or molecule. The onset field (sometimes called the critical field), E_{on}, of corona is highly variable. Some low values of about 1 kV m^{-1} have been quoted, but the heights of the points were not given. Stromberg (1971) reported that E_{on}, which was measured in a clearing 40 m to the side of a grove of trees, for a tree 18 m tall and on the edge of the grove was 6 kV m^{-1}. From the measurements reported by Winn and Moore (1972) at Langmuir Laboratory (at an atmospheric pressure of about 700 mb), we infer an E_{on} for a treeless plot of grass

on the mountain ridge to be 4 kV m^{-1}. Later, we will examine the corona onset threshold from hydrometeors. Onset thresholds from hydrometeors are substantially greater than those from exposed, grounded conductors.

Corona is different from ohmic currents (i.e., currents that follow the relationship $\vec{J} = \sigma \vec{E}$, where σ depends on the preexisting ion population). In corona current, the number of charge carriers (primarily free electrons) increases in response to the electric field. This assumes, of course, that initially there is at least one free electron to start the discharge. As discussed in Section 1.4, electrons in the troposphere are usually produced by cosmic rays and by decay of radioactive substances from the ground. In later stages of corona development, photons emitted by the corona itself can photoionize air molecules to provide initial electrons for corona in new regions.

To develop an expression for the growth in the number of electrons, we can begin with the condition that an initial electron must gain enough energy from the electric field to ionize a molecule with which it collides. Then each electron creates a second electron when it collides; those two create four, and so on. In reality, some electrons will be captured by molecules and ions, some electrons will have too little energy to ionize molecules because of statistical fluctuations, and some will heat gas molecules or stimulate photon emission. If we neglect losses, however, and assume that there are n electrons moving in the same direction for a distance, dx, then the number of new electrons created by collisions can be given by

$$dn = \alpha\, n\, dx, \qquad (4.1)$$

where α is the number of electrons created per unit length by an initial electron. In the troposphere, α generally increases with increasing electric field and decreasing atmospheric pressure. Integrating this equation over a distance, x, gives

$$n = n_0\, e^{\alpha x}, \qquad (4.2)$$

where n_0 is the number of electrons at $x = 0$. Similarly, the current due to the increasing number of electrons is given by

$$i = i_0\, e^{\alpha x}. \qquad (4.3)$$

This process, in which the number of electrons increases exponentially under the influence of large elec-

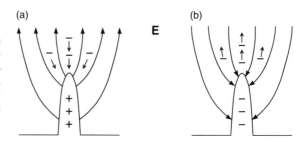

Fig. 4.2. Schematic diagram of positive and negative corona. (a) For positive corona, an electron avalanche propagates into increasing electric field magnitudes, denoted by an increasing density of electric field lines. (b) For negative corona, an electron avalanche propagates into regions of weaker electric field magnitudes.

tric fields, is called an *electron avalanche*. As corona advances outward from its origin, it can develop into a streamer.

4.3. POSITIVE AND NEGATIVE CORONA

The *polarity of corona* and point discharge is defined as the sign of charge on the object from which it originates. This is also the polarity of charge leaving the point. Thus, under the influence of a positive E from an overhead storm, a positive point-discharge current flows upward from the ground toward the cloud base. Some have defined the polarity as that of the charge flowing into the point, so be aware of the definition of polarity in individual papers you read. We will note when this latter definition is used in this text. Electrons are the principal charge carriers in both cases. The properties of positive and negative corona are similar, but have significant differences. The most elementary difference is the direction in which the electron avalanche moves. If the corona is positive, electrons move from the front of the streamer towards the point. If it is negative, electrons move outward away from the point.

Because electric fields are initially enhanced at a point or at a streamer head as it propagates, the electron avalanche is moving into stronger electric fields for positive corona and into weaker fields for negative corona. These two situations are shown schematically in Fig. 4.2. Because of this difference in direction relative to the convergent E around a point, positive corona can begin and propagate in weaker E than negative corona. The reason is that the negative corona moves into diverging, weaker fields that are less likely to be able to support a continuing avalanche.

When the E initially increases to the onset threshold for positive corona, the discharge begins with short pulses referred to by various terms (e.g., *burst pulses* or *onset streamer*). These faintly luminous streamers advance away from the point, photoionizing oxygen molecules in front and thereby creating new avalanches. If the avalanches reach a critical size, they can propagate into lower ambient E, where avalanches would not start, because the streamer tends to increase the field in front of it. The streamers pulse because negative space

charge collects behind the streamer and eventually weakens the electric field enough to quench the streamer until the charge can be cleared away by the electric field. As E becomes stronger, the pulses become more frequent. Eventually, if the electric field is large enough, equilibrium is established between the space charge and the internal electric field, and the discharge becomes a steady positive glow with no pulses. With further increases of the electric field, pulsed streamers occur once more. These are prebreakdown streamers in the sense that sparking will occur if the negative electrode is reached.

Negative corona begins as pulses (called *Trichel pulses*) whose frequency of pulsing range from 1–100 kHz. Luminosity occurs only near the negative electrode and is similar in appearance to the positive glow discharge. Negative prebreakdown streamers occur only at very high E.

4.4. POINT DISCHARGE IN THE ATMOSPHERE

Corona in the atmosphere, which is more frequently called point discharge, is an important phenomenon because of its relevance to both the global circuit and thunderstorm processes. From 1900 almost until World War II, it was one of the most extensively measured parameters of thunderstorms, as scientists tried to use it to find the electric field, occurrence of lightning, currents flowing in the thunderstorm generator, net contribution of corona to the global circuit, etc. (e.g., Wormell 1927). From long-term measurements, investigators determined there is a net transfer of negative charge to Earth via point discharge. They found a ratio of positive-to-negative point discharge of about two (see summary in Latham and Stromberg 1977). Point discharge inside storms may be important because of its possible effects on charge distribution, hydrometeor growth, and lightning initiation. In this section, we will learn that estimating the contribution of point discharge to the thunderstorm current budget is inherently imprecise and difficult. Because the ions formed by point discharge move under the force of E, ions of the polarity being repelled by E away from the point of origin move out into lower E. These ions form space charge around the point and reduce the external E that provided the driving force for the discharge. This eventually stops the point discharge unless the space charge is removed. So whatever the meteorological conditions, a basic aspect is that the magnitude of the point-discharge current is a function of the removal of the ions from near the points creating them. Obviously, therefore, the wind affects point discharge.

4.4.1. Point discharge from single and multiple points

Whipple and Scrase (1936) derived the regression relationship between the point-discharge current from a single point and the potential gradient, ∇V,

$$I_{PD} = a(\nabla V^2 - \nabla V_0{}^2), \qquad (4.4)$$

where I_{PD} is in μA, ∇V is in kV m^{-1}. When ∇V is positive, $a = 0.08$ and $\nabla V_0 = 0.78$; when ∇V is negative, $a = 0.1$ and $\nabla V_0 = 0.86$. Note that their equation gives the polarity of charge going into the point. Typically, the current that flows from a single elevated point ranges from a tenth of a microampere to a few microamperes. (Although we use the terms corona and point discharge interchangeably, the variables in equations will be subscripted with *PD*.) Jhawar and Chalmers (1967) compared point discharge from single points and small trees. They found point discharge from a single point was proportional to V^2, but point discharge from multiple points (the trees) was proportional to the cube of the potential at the tip, V:

$$I_{PD} \propto V(V - V_{on})^2, \qquad (4.5)$$

where V_{on} is the onset potential for the point discharge.

Ette and Utah (1973) compared point discharge from metal points, trees, and grass, and found that the characteristics of the points affect the threshold E at which point discharge starts. The nature of the points also affects the ratio of the negative to positive charge released during a storm. They confirmed earlier studies that had indicated there is a tendency to approach a constant value for the ratio of charges. They found that when $|E_{gnd}| > 2.5$ kV m^{-1}, the point-discharge current deviated from the widely used relationship in Eq. 4.4. Other investigators have noted that there are large uncertainties in extrapolating from single measurements, whether from trees, sod, or metal points, to calculate the net current over an area of tens of square kilometers affected by the storm. Standler (1980) summarized the problems by noting that only in regions where a single plant species is regularly spaced and of the same height is such a relationship plausible.

4.4.2. Effect of wind on point discharge

Since removal of the ions created by the point-discharge process is key to the magnitude of the current, investigators have tried to find the relationship between the point-discharge current and wind speed. At lower wind speeds, corona current is proportional to the wind speed, W, and the square of E. At high speeds, I_{PD} is more linear with E. Large and Pierce (1957) found that in wind of less than a few meters per second, the point-discharge current is not a function of the wind speed, while at winds greater than this, I_{PD} is linearly proportional to the wind speed. Various relationships have been derived; Large and Pierce derived the following one, which has been widely quoted as showing the dependence of point discharge on wind speed,

$$I_{PD} = b(V - V_0)(W^2 + c^2V^2)^{\frac{1}{2}}, \qquad (4.6)$$

where b, V_0, and c are constants, and W is the wind speed. In a small laboratory device, Jhawar (1968) used air flow up to 245 m s^{-1} and found several different, but similar, equations relating I_{PD}, W, and V. He also reported that b and c in a similar equation were not constants, but a function of V.

Point-discharge current is a complicated function of several variables and 'constants' that are not really constant. Thus in nature, defining those for a large area to make quantitative use of such data appears impossible. Nevertheless, much has been learned about point discharge from the Earth under storms, and the use of point-discharge sensors as simple indicators of E has merit in certain applications.

4.4.3. Net point-discharge currents below storms

Ette and Utah (1973) reported that when the lightning flash rate was high, the total point discharge current was less than when the flash rate diminished, usually toward the end of a storm. At the Kennedy Space Center (KSC) in Florida, Livingston and Krider (1978) calculated point-discharge currents from their measurements of E. Point discharge typically averaged about 0.6 μA per point under storms with frequent lightning and 2.4 μA during the final phase of storms, when lightning was infrequent. The reason is that times of frequent lightning often tend to have a smaller E at the ground because E does not have as long to build up between flashes. Lightning also affects the total point-discharge current by the displacement current it causes.

The equation(s) for the point-discharge current can be converted to a current density, J_{PD}, as noted by Chauzy and Raizonville (1982):

$$J_{PD} = aE(E - E_{on})^2, \qquad (4.7)$$

where a is a constant and E_{on} is the onset electric field for point discharge. This relationship does not, however, contain the dependence of J_{PD} on the wind. Latham and Stromberg (1977) estimated that storms produced about 1 nA m^{-2} by point discharge. They also inferred, in part from the work of Ette and Utah (1973), that a typical thunderstorm produces a total current of about 0.6 A from point discharge, similar in magnitude to current measured above storms.

In an attempt to surmount the difficulty in extrapolating a single measurement to the larger area of a thunderstorm, Standler (1980) proposed using a network of field measuring devices. The following equation describes the concept:

$$J_{PD}(0, t_s) \approx \varepsilon \left. \frac{\partial E(0, t)}{\partial t} \right|_{t_0} - J_i(0, t_s)$$

$$- J_{CV}(0, t_s) + J_{CV}(0, t_0), \qquad (4.8)$$

where J_i is the conduction current density, and J_{CV} is the convection current density. There are assumptions, estimates, and constraints that must be applied when using this equation. Standler argued, based on observational evidence, that J_i and J_{CV} can be neglected in calculating J_{PD}. Standler used E measurements at the surface and found that $J_{PD} \approx 0.4$–3.5 nA m^{-2}. This range encompasses that reported by Latham and Stromberg (1977) and agrees with those reported by Standler and Winn (1979). Standler proposed that J_{PD} is difficult to measure directly, but Eq. 4.8 provides the means to calculate it using routinely measured parameters. Use of a network of field measuring devices probably increases the representativeness of the calculated J_{PD} for an entire storm.

4.4.4. Space charge beneath storms

As mentioned earlier in this section, the charge emitted by point discharge creates regions of space charge beneath thunderstorms. A major consequence of this space charge is its effect on E at the ground. Beneath storms over land, $|E_{gnd}|$ is almost always less than 10 kV m^{-1}, considerably smaller than would occur in the absence of point discharge (see Section 4.4.5). The sign of the charge that an electric field draws from the ground via point-discharge current is always such that the magnitude of E_{gnd} is reduced. For example, if at some point on the ground the net effect of all thunderstorm charges is equivalent to a negative charge overhead (i.e., there is a positive E_{gnd}), the charge emitted by point discharge will be positive. The electric field produced at the ground by this space charge will be negative, so the superposition of this electric field on the thunderstorm electric field will result in a smaller positive E_{gnd}.

In an attempt to quantify the effects of point discharge on the electric field, Standler and Winn (1979) made simultaneous measurements of E_{gnd}, point discharge from a sod bed and small trees, and precipitation currents. They measured E_{gnd} with an upward facing, flush-mounted field mill and also E at various heights up to 300 m by raising and lowering a tethered balloon carrying a balloon-borne electric field meter (EFM). Measurements were made on the mountaintop at Langmuir Laboratory in New Mexico and over the flat terrain of the eastern coast of Florida at KSC.

Figure 4.3 shows E_{gnd} and the electric field aloft, E_{aloft}, from a storm over Langmuir Laboratory and gives an example of the systematic relationships found. As the balloon-borne EFM was raised, the magnitude of E_{aloft} increased; as it was lowered, the magnitude of E_{aloft} decreased. Over both the mountainous terrain of New Mexico and the flat terrain of the Florida coast, E_{aloft} followed the same trend as E_{gnd}. Figure 4.4 shows data from another storm that illustrates how the difference between E_{aloft} and E_{gnd} depends on point-discharge current as E_{gnd} slowly decreases. Initially, when point-discharge current was largest, $|E_{aloft}| >> |E_{gnd}|$, and E_{aloft} increased with height in the lowest hundred meters or so. During the period labeled a–b, the EFM was kept at 150 m, and E_{gnd} slowly decreased. As E_{gnd}

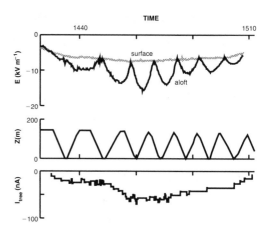

Fig. 4.3. Simultaneous measurements of E_{gnd}, E_{aloft}, and point discharge below a mountain storm. The height of the balloon-borne EFM varied from the surface to 200 m and is indicated by z. The surface wind speed measured about 350 m from the balloon winch was approximately 5 m s^{-1}. (From Standler and Winn 1979, with permission.)

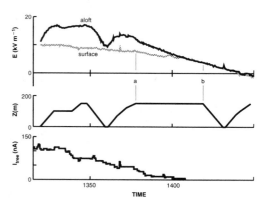

Fig. 4.4. Simultaneous measurements of E_{gnd}, E_{aloft}, and point discharge below a mountain storm. The height of the balloon-borne EFM from the surface to 200 m is indicated by z. Note the decay in E between a and b, with E_{aloft} becoming approximately equal to E_{gnd} as point discharge subsided. (From Standler and Winn 1979, with permission.)

became small, the point-discharge current decreased to zero, and E_{aloft} (within 100 m) approached and then equaled E_{gnd}. When point discharge stops, the screening, space charge layer dissipates under electric and wind forces.

The space charge from point discharge beneath thunderstorms can have other significant effects. For example, it can cause E_{gnd} to change sign during a lightning flash, while E_{aloft} does not. If lightning reduces $|E_{aloft}|$ to a relatively small value, the electric field from nearby space charge can dominate the field from the thunderstorm enough to change the polarity of E_{gnd}. Often, however, E_{gnd} soon recovers to its preflash value and polarity.

Space charge from point discharge also can affect the charge on precipitation beneath storms. The relevant charging mechanism is the Wilson selective ion capture mechanism. As discussed previously, a vertical electric field beneath storms polarizes raindrops so that they selectively attract a particular sign of charge. If enough charge is captured, it can reverse the polarity of net charge on falling precipitation. Since the sign of charge attracted electrostatically to raindrops for a particular polarity of vertical electric field is the same as that emitted by point discharge, point-discharge current can contribute to the efficacy of the selective ion capture mechanism beneath storms.

4.4.5. Point discharge over water

Point discharge is less (and probably often nonexistent) at the calm surfaces of lakes and oceans because of their lack of sharp, elevated objects. Thus a larger E should be present over water, because there will not be as much space charge as over land. Toland and Vonnegut (1977) used a floating point-discharge instrument on a lake 500 m from shore and measured E_{max} ranging from 38–130 kV m^{-1} in seven storms. Vonnegut (1974) found that in the presence of 1-m-high waves, $|E|$ is enhanced by a least a factor of two at 0.5 m above the crests, as compared to that over the wave troughs. This suggests point discharge is more likely over rough seas (and also rough fresh water). Point discharge from water surfaces also occurs when precipitation splashes and ejects a Worthington jet, with the E_{on} inversely proportional to drop radius to the 3.5 power (Griffiths et al. 1973). The E_{on} is independent of whether the water is salt or fresh. No direct measurements or estimates of total point-discharge current over large water surfaces have been published, so any effects on the global circuit or storm electrification are unknown.

4.4.6. Point discharge from precipitation

Point discharge from precipitation particles is important to our attempts to understand how lightning starts. The basic question is whether such hydrometeor-initiated corona can become a lightning flash. Dawson (1969) showed that point discharge occurs from water

drops and depends on the size and impurity of drops and the atmospheric pressure, P. He also found that the mechanics of electric discharge from the drop varied with pressure. At higher pressure, point discharge occurred after distortion and disruption of the drop. At lower pressure, "pure corona" occurred without drop disruption. He called the demarcation altitude the "transition altitude." The upper limit of the transition altitude for positive corona began at $\approx$4.5 km MSL and at $\approx$7 km MSL for negative corona. In the region between these two, the corona-onset fields were different, and the result of any discharge was the production of negative drops and positive ions. Drop-surface E that produced these phenomena ranged from about 2000–16,000 kV m^{-1}. Thus, point discharge from individual drops in storms seemed unlikely.

Crabb and Latham (1974) investigated point discharge caused by colliding drops. They used a pair of drops of radii 2.7 mm and 0.65 mm; their relative fall velocity was 5.8 m s^{-1}. During glancing collisions, they photographed a water filament between the drops (Fig. 4.5). The total length of filaments ranged from 6–25 mm. Point discharge was produced when E reached 250–500 kV m^{-1}, depending on the length of the drop-pair in the direction of E at separation. These E magnitudes are in the upper range of those measured in thunderstorms. Furthermore, with significant charge on the drops, the onset E will be reduced. Thus colliding raindrops in storms likely produce point discharge and may initiate lightning.

Griffiths and Latham (1974) examined point discharge from ice crystals. They found both polarities of point discharge occurred simultaneously from opposite ends of the crystal. In contrast to small wire (highly conductive) samples, there was no clear relationship of the onset E and the crystal length. Smooth, spherical surfaces had a higher E_{on} than crystals or spheres with spikes of >1 mm. With large charges (30–600 pC) on the ice, the E_{on} was reduced only 5–20%. Point discharge from ice is dominated by its surface, not bulk, conductivity. Point discharge seemed likely in thunderstorms with E > 400 kV m^{-1} and T > $-$18°C. At colder temperatures, no significant point discharge occurred. They concluded that point discharge from ice may play a role in the initiation of lightning, but measurements of $|E_{aloft}| \approx$ 400 kV m^{-1} are exceedingly rare.

Recently, Coquillat and Chauzy (1994) combined their earlier modeling study of raindrops experiencing both electric and aerodynamic forces (Coquillat and Chauzy 1993) with calculations of corona from drops to find the field for corona and disruption at various altitudes. They found that corona could occur at altitudes lower than found by Dawson. Coquillat and Chauzy concluded that at E = 100 kV m^{-1}, corona from large negatively charged drops would be impossible, as they would have to be at $\approx$8 km MSL, where such drops are unlikely to exist. They further concluded that their results and reported in situ measurements of drop charge

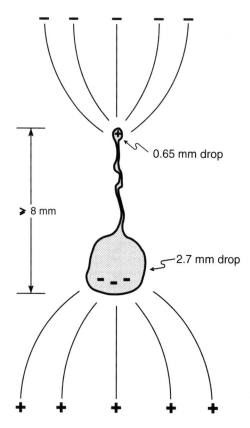

Fig. 4.5. Sketch of electric field concentration on a stretched filament produced by a glancing collision of 2.7- and 0.65-mm drops. The total length here was about 8.2 mm. (Sketch from a photograph in Crabb and Latham 1974, with permission.)

indicated that corona from drops in such E are low-probability events and unlikely to influence lightning. Their calculations showed that corona propagation depended on drop size and charge and that the transition altitude was variable.

4.4.7 Unusual observations of corona-like phenomena

Numerous reports of unusual optical phenomena in the atmosphere that may well be associated with point discharge are in the popular and scientific literature. One is the report of multicolored glows lasting several seconds reported in a nighttime storm over the southern Great Plains of the United States by McGinley et al. (1982). All the authors have been involved professionally with research on storms, and their observation seems credible. In the absence of lightning, they first saw a glow like a neon light that extended over a large sector and lasted 45 s. That was followed by multicolored cloud lightning, after which they saw aquamarine and reddish-orange glows that lasted up to 4 s. Pulsations of the latter color were observed at a rate of about 20 min^{-1} as the storm receded. They do not speculate on a mechanism, but point discharge from precipitation is a candidate for the nonlightning portion of the unusual luminosity.

4.5. UNDESIRABLE EFFECTS OF POINT DISCHARGE ON MEASUREMENTS

When making measurements of electrical parameters in storm environments, we must be concerned with point discharge, not to find its value, but to understand that it can alter the parameter we are measuring. For example, there can be effects on the measurement of E and particle charge. If point discharge occurs from an electric field mill, the instrument indicates a value of E different from the true ambient value. Therefore, it is essential to use care in the design and construction of field mills and to determine the limits of reliable measurement as part of the calibration process. Similarly, if an instrument for measuring particle charge emits point discharge, the ions emitted from the instrument can be captured by falling precipitation, thus altering its charge and creating an error. Again, care is needed in instrument design and calibration to avoid or reduce these problems as much as possible and to determine conditions under which they will occur.

5

Lightning

5.1. INTRODUCTION

The observation of lightning has been an important part of the history of humankind, undoubtedly because of the impressive and dangerous nature of the phenomenon. Through the years, lightning has been a major concern for a diverse segment of the engineering and scientific communities. The role of lightning within meteorology has recently gained even greater interest. A major reason is the availability of lightning ground strike locating data that are accessible in real time, cover large areas, and are sufficiently accurate to be a useful adjunct to radar, satellite, and other data. In this chapter on lightning fundamentals, we strive to provide a broad background of quantitative and qualitative details of lightning. This information will enhance understanding of research issues facing those using lightning data in research and applications to try to discern storm evolution, severity, etc., as presented in later chapters. We want to emphasize that we treat the electromagnetic aspects of lightning sparsely because of the context in which we expect this book to be used; it is in no way intended to send the incorrect message that knowledge of the physics of lightning is not critical. On the contrary, such knowledge is critical to the advancement of science and to solving the severe problems that can result from the interaction of lightning radiation signals with our increasingly technological world. The study and application of such information are too large a segment of science and engineering for us to cover adequately. The interested reader is referred to the book by Uman (1987), reviews such as Beasley

(1995), and the myriad papers in the scientific and engineering literature.

Lightning is referred to by various names. In the nonscientific literature the use of diverse, nonspecific, and even contradictory terms is common. We will define the terms here that are used in this book and in much of the scientific literature. First, we consider a *lightning flash* to be an entire lightning event. The term *discharge* can and often is used as the equivalent to flash. There are two fundamental types of lightning: *cloud flashes* and *ground flashes*. The obvious difference between the two is that a cloud flash does not strike Earth. Note that in much of the literature, *intracloud (IC) flash* is used to mean cloud flash. The term *cloud-to-ground (CG) flash* is often used for those flashes that do strike Earth; we will mostly use the equivalent term, *ground flash*.

5.2. TYPES OF CLOUD FLASHES

There is confusion when intracloud flash means those confined to inside the cloud instead of all cloud flashes. Types of cloud flashes include cloud-to-air and cloud-to-cloud. Their meanings should be as the words indicate, but often cloud-to-cloud is a confusing term, as it is when used for all cloud flashes. For example, if several storm cells are joined at low and midlevels and lightning crosses through clear air from one cell to another at higher altitudes, is that cloud-to-cloud or intracloud? An *air discharge* is a cloud flash with visible channels that propagate out into the air around the storm, but do not strike the ground. Mostly we use the terms cloud flash and intracloud flash in this book.

Terms often found in nonscientific publications include sheet, heat, and rocket lightning. Schonland (1964) described *sheet lightning* as due to the appearance of lightning embedded within a cloud, which lights up as a sheet of luminosity during the flash. A related term is *heat lightning*. The name comes from flashes in storms so distant that observers cannot see channels to ground and cannot hear thunder. The reason it is called heat lightning seems to be that such observations are common during the hot weather storm season (Uman 1971). *Rocket lightning* is a long air discharge whose horizontally propagating channels remind one of a rocket in flight. The progression appears slow, but since channel propagation is $\geq 10^4$ m s^{-1}, it may be that the long distances over which they move make them appear slow. Such discharge channels are slow, of course, when compared with the return stroke in a channel to ground (described later). The term rocket lightning is not commonly used in the scientific literature. A newer term that appears to label at least a subset of, if not all, rocket lightning is *spider lightning* (Mazur et al. 1994), which refers to the long, horizontally propagating flashes that often are seen on the underside of stratiform clouds such as can accompany mesoscale convective systems.

Upward discharges extending very high into the clear air from the tops of storms have been reported sporadically over many years. The phenomenon was often described as a blue shaft of light, but channels, including those that spread out horizontally at some height, have also been reported. Such phenomena have also been reported as rocket lightning from above the tops of storms (e.g., Everett 1903). It was not until early 1993 that the number of recorded images of such phenomena became so large that the community realized that they might be common, rather than rare, as previously supposed. Intense interest led to a vigorous response by the community to obtain more quantitative observations. Now the documentation is convincing, but the explanations are incomplete. A brief description of "red sprites" and "blue jets" is in Section 5.15, even though we are not sure that they are lightning flashes in the conventional sense. A possible relationship to E_{aloft} is given in Section 8.2.2.vii.

5.3. TYPES OF GROUND FLASHES

Terms that are used to describe a flash to ground include *forked lightning,* which denotes the presence of branches from a nearly vertical channel to ground (Schonland 1964) or multiple channels striking the ground. Ground flashes also can take a form termed *ribbon lightning* because the horizontal displacement of the channel by the wind appears as a series of ribbons. A photograph of ribbon lightning has been used to calculate the shear in the low-level wind (Orville 1977a). Many photographs purported to be ribbon

lightning are the result of camera movement during exposure. Only photographs of ribbon lighting made with a rigidly mounted, stationary camera are to be trusted.

Sometimes the decaying channel of a ground flash will break into a series of bright and dark spots, which has brought about the term *bead lightning*. Photographing bead lightning is difficult, because a time exposure will have the entire channel illuminated early in the flash, and thus the beads occurring at the end of the stroke are not seen in the photo. There is, however, a sequence of frames from a high-speed movie camera of a lightning strike to the water plume resulting from the explosion of a depth charge (Fig. 5.1). We and other observers of large storms on the plains have seen that some storms produce many bead lightnings. Some have been documented with video cameras. So far those video images have been made with cameras whose sensor has persistence from field to field (a field is scanned by the camera in 16.7 ms). In such cases, the images of bead lightning have not been reproduced as clearly and dramatically as the original observations seen by the human eye.

Uman (1987) summarized *ball lightning* as a luminous sphere that is associated with the location of a ground strike. Nonspherical shapes have also been reported. The usual size is given as a few tens of centimeters across. Reports of ball lightning are many, but reliable documentation is not. It is believed to exist, but its physics is not well understood. Barry (1980) published a book containing a large collection of reports, theories, etc., regarding ball lightning.

5.3.1. Categories and polarity of ground flashes

There are two categories of ground flash: *natural* and *artificially initiated* or *triggered*. By the first we mean any ground flash that occurred naturally because of electrification in the environment and without the aid of any man-made structures or intervention. Artificially initiated lightning includes strikes to very tall structures, airplanes, and rockets. The work of Berger (1977) is usually acknowledged as having produced the subdivision of ground flashes into four types, depending on their direction of propagation and polarity of charge effectively lowered from the cloud to Earth (Fig. 5.2). These have long been accepted, but recently Mazur and Ruhnke (1993) challenged the correctness of types 2R and 4R (in the figure). We present their arguments later. Also, we make little attempt to try and relate the charge structure of the storm itself with the charge carried on the channels. That relationship is not clearly understood, but is addressed later.

By far the most common ground flash is depicted in panel 1 of Fig. 5.2; it is a downward propagating negative leader, followed by an upward propagating return stroke. The net effect of this flash is to lower negative charge from the cloud to the ground. Keep in mind that the actual flow of electricity is always electrons, so here

Fig. 5.1. Lightning strike to a water plume from a depth charge. This flash is unusual for two reasons: It apparently was triggered by the plume and developed into bead lightning.

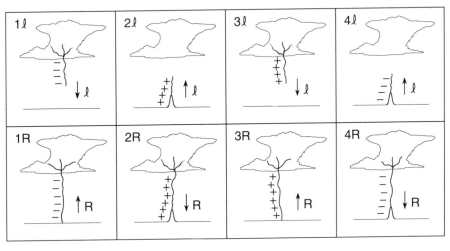

Fig. 5.2. Types of cloud-to-ground flashes. 1ℓ is a downward propagating, negatively charged leader followed by an upward return stroke (1R) that effectively lowers negative charge from the cloud to the ground. Similarly, 2ℓ is an upward positive leader followed by a downward stroke (2R) that lowers negative charge; 3ℓ is a downward positive leader followed by an upward stroke lowering positive charge; and 4ℓ is an upward negative leader followed by a downward stroke (4R) lowering positive charge. (After Berger 1977, with permission.)

the electrons flow from the channel to Earth. Type 2 is an upward propagating positive leader that is followed by a downward propagating negative return stroke. Again this type of ground flash, which generally is artificially initiated, lowers negative charge to Earth. Types 3 is a downward propagating positive leader followed by an upward return stroke that lowers positive charge to Earth. Until about the last decade, this type of flash was thought to be exceedingly rare. Several investigators have found that they are numerous and that such positive ground flashes seem linked to certain severe storms (e.g., Rust et al. 1981, 1985). Type 4 is an artificially-initiated positive ground flash composed of an upward negative leader followed by a downward positive return stroke. This is common in strikes to tall structures. We will return to these flash types in various sections on lightning and on lightning in the context of storm behavior.

5.4. Lightning initiation

Details of the origin of lightning within storms and the processes that govern lightning initiation remain unclear. In this section, we define breakdown phenomena and begin putting lightning in the context of the storm producing it. We start with a discussion of what little is known about how and where lightning starts. Much of the physics of breakdown that is accepted as relating to lightning can be found in Loeb (1965). We will use his definitions for the initial breakdown. Loeb described *corona* as luminous glows, spots, halos, branches, and streamers; all have low current. Loeb stated corona is "the general class of luminous phenomena appearing associated with the current jump to some microamperes at the highly stressed electrode preceding the ultimate spark breakdown of the gap." Similarly, the term *corona threshold* describes the potential of an electrode at which the current increases suddenly just before the observation of corona. Corona thresholds have been subdivided to describe specific types of corona. A significant one in lightning initiation is the streamer threshold, because streamer initiation and propagation are fundamental to the lightning process. In the atmosphere, the corona (or point discharge) threshold is often given as the ambient electric field at which corona onset occurs. These differences in usage will cause us little problem, as we will not delve into the detailed small-scale physics of corona.

Initially, corona is often intermittent and pulsed in nature. With increasing potential, it can become steady. Loeb defined the change from intermittent to steady corona as the *onset of steady corona*. At higher potentials, the corona closely follows Ohm's law, $V = IR$. At still higher potentials, the current increases faster than the potential (and is thus no longer following Ohm's law). Continued increase in corona current eventually leads to *spark breakdown*.

There are at least two different approaches to understanding the lightning initiation process. One starts from observations and theoretical treatments of electrical breakdown by using basic physics. The other use measurements, such as the electric field, to define the different aspects of a lightning flash. Though these approaches have been used successfully to treat some lightning processes, no set of treatments describes lightning comprehensively from the first corona, to flash initiation, and through to the end of the flash. We summarize descriptions of breakdown phenomena that apply specifically to lightning in clouds.

Although small-scale studies in the laboratory using points, planes, etc., have been applied to lightning initiation, Loeb (1966) said the comparison of a lightning stroke to a spark between metal electrodes is illusory. He believed that an acceptable laboratory model of a natural lightning was a spark moving between metal electrodes through a quasi-uniform E. Pulsed or static potentials can initiate an acceptable laboratory spark model. In their widely referenced works, Phelps and Griffiths (1976) applied voltage pulses to a corona point recessed at or below one of two parallel-plate electrodes to create a uniform E across a gap. The laboratory studies have been useful, but extrapolating them to lightning in clouds needs to be attempted with caution. Inside the clouds there are no electrodes tied to a power supply (or large source of charge) that can feed current directly into a channel once it is established.

In Section 4.4.6, we discussed the initiation of point discharge from precipitation and its possible role in lightning initiation. The type(s) of precipitation particles or particle interactions that result in lightning have not been fully delineated, and it is still unknown if lightning starts from point discharge off particles in high E. The value of the ambient E necessary for and present at initiation of lightning is also not known. Based on measurements in storms, we expect the initiation usually occurs in large-scale, background E < 300 kV m^{-1}. There may be very small regions of higher fields, (i.e., breakdown fields). Such small regions never seem to be sampled, thus raising the question of whether they exist. Mazur (1986) reported on high frequency discharges that he interpreted were unsuccessful attempts of streamers to become full-fledged flashes. The length of these streamers was <900 m, which he interpreted as an indication of the size of the region in which lightning is initiated. We do not really know what the threshold E for a lightning flash is or even if there is single threshold. We certainly expect it to have an atmospheric pressure (air density) dependence.

Other hypotheses about lightning initiation of electric field magnitudes required to sustain a discharge are more than an order of magnitude smaller than those required to initiate breakdown (e.g., Phelps 1974, Phelps and Griffiths 1976) in clear air. One mechanism that appears plausible is initiation by the energetic secondary electrons produced by cosmic rays, which tend to penetrate the atmosphere everywhere sporadically. McCarthy and Parks (1992), Gurevich et al.

(1992, 1994), and Roussel-Dupre et al. (1992) suggested that *energetic electrons* (defined as having an energy of ≈ 1 MeV, million electron volts) produce a sustained electron avalanche that can lead to lightning at electric field magnitudes much smaller than needed for initial breakdown, in fact at magnitudes that often are even smaller than needed to sustain a streamer using the more plentiful low-energy electrons. The threshold electric field magnitude necessary for the average kinetic energy of an energetic electron to remain constant as it loses energy to collisions is called the *breakeven electric field*, E_{be}. It decreases with altitude, as shown in the following equations that contain a model of the vertical distribution of the mass density of air, ρ_A, in kilograms per cubic meter:

$$E_{be}(z) = \pm 167 \, \rho_A(z)$$
$$\rho_A(z) = 1.208 \exp\left(-\frac{z}{8.4}\right), \qquad (5.1)$$

where E_{be} is in kilovolts per meter and the altitude is in kilometers. Marshall et al. (1995) compared values of the vertical profile of E_{be} with the measured E profiles from soundings in 23 storms. In most cases, the observed electric field magnitudes were bound by E_{be} at all altitudes, and observed relative maxima were similar in magnitude to the E_{be} at corresponding altitudes. Their soundings fit with, but did not prove, the theory that lightning initiation occurs when energetic electrons are accelerated by the large scale E in thunderstorms.

Regardless of how lightning starts, the discussion of lightning initiation next moves to streamer propagation. Historically, conceptual models have employed either unipolar or bipolar streamers. The former is by far the most widely used and evolved, partly because it can explain the change in E from lightning. However, there continues to be growing concern that the physics of lightning initiation makes sense only in the context of bipolar streamers.

Loeb (1966) stated that a bipolar streamer, as proposed by Wagner (1966), breaks down the air by its action. Loeb (1968) proposed that when the electron density in point discharge reaches 100 cm^{-3}, an avalanche of electrons occurs. When the ensuing streamer tip gets 10^8 electrons, it becomes self-propagating, with both positive and negative streamers occurring simultaneously, as proposed by Wagner. The negative streamer has high-electron density, which causes self-repulsion of the electrons. The high mobility of electrons tends to dissipate the streamer, but the E keeps it going. The positive streamers proceed via photoelectron influx from ahead of the tip. These do not attenuate as they propagate, except when a branch occurs occasionally. Loeb also noted that the existence of an ionizing wave of potential gradient makes brighter, stepped advances at about 10^6 m s^{-1}.

A basic model of positive streamer propagation was developed in Loeb's laboratory by Dawson and Winn (1965). Their theory explained the propagation of the streamer in a zero-field. The *primary streamer* (first

one to break down the air) was modeled as a sphere with 10^8 ions that moved ahead of a trail of nonluminous, low-conductivity streamer channel of radius 30 μm. The streamer did not carry the potential of the origin with it. The high E necessary for continued breakdown was a result of the positive charge in the spherical tip.

Phelps (1971) noted differences in the propagation behavior of negative and positive streamers. Negative ones advanced in a diffuse and self-dissipating manner, as Loeb (1968) had noted. Positive streamer propagation was dominated by the ambient E, and the distance of propagation increased monotonically with E. Until there was a highly conductive leader established, only positive streamers could propagate long distances from their source. Phelps found that positive streamers would propagate in a critical field, E_{crit}, of about 500 kV m^{-1}. Phelps (1974) verified that positive streamers intensified as they propagated if they were in an E > E_{crit}. Phelps stated that his laboratory study supported the Dawson-Winn (1965) model of lightning initiation (i.e., the channel itself has little role in the propagation of the streamer tip). Phelps extended the Dawson-Winn model to include propagation in a uniform E in the direction of propagation. Certainly a nonzero E is more realistic for all but a comparatively few places in a storm environment. He concluded that streamers can propagate and intensify in an E ≈ 300 kV m^{-1}. Even this reduced value must be reconciled with the typical, smaller measured E and with lightning propagating over long distances, especially in the horizontal. An E ≈ 300 kV m^{-1} should be commonly observed if it is required for lightning propagation over distances of kilometers. Also, as observations of E at the ground are generally 10–15 kV m^{-1}, lightning flashes of both polarities obviously come to ground in E << 300 kV m^{-1}. (See Chs. 7 and 8 for observations of E in storm environments.)

Phelps (1974) suggested that lightning can start by streamer initiation in strong E below the negative charge via point discharge from a single drop or elongated drop-pair after collision. The channels created by this process would be negatively charged. This is a version of a unipolar streamer lightning model. Subsequent breakdown could occur owing to the increased negative charge deposited by this process, until finally negative streamers also could occur.

The onset electric field, E_{on}, to create lightning is a function of pressure (Phelps 1971). Phelps and Griffiths (1976) found that the $E_{on} \propto P_a^{1.65}$, where P_a is the pressure of dry air and E_{on} is measured at the origin of the flash (i.e., the rear of the streamer). Their model was a conical-shaped advancing positive streamer, which deposited negative charge behind. Values for E_{on} ranged from 150–250 kV m^{-1} between 6.5 and 3.5 km, respectively. These values tend toward the higher E measured inside storms. Phelps and Griffiths (1976) also hypothesized that the only viable mechanism for explaining lightning that taps the highly dispersed electrostatic energy of a cloud invokes the occurrence of long positive streamers. This hypothesis has received experimental support from the analyses of the mapping of cloud flashes by Richard et al. (1985) and Mazur (1989b).

Proctor (1981) found that first streamers of in-cloud lightning and first leaders of ground flashes did not behave differently. That suggested the breakdown process was the same for both types of flashes. From the channel paths he mapped and the accompanying change in E, Proctor calculated the total charge that was removed from the originating region by the streamers. (Note that this is equivalent to the opposite charge being deposited in the region.) He found that the total charge did not depend heavily on whether the streamers were modeled as a point charge, dipole, or line charge. However, when he compared the change in E calculated from the models with the measured change in E, he got agreement only when the charge calculations were from a model with charge deposited along the channel behind the advancing streamer tip. Also he found that a streamer model having a tip with charge isolated in it did not yield good agreement with measurements. This conclusion contradicted the interpretations of the laboratory work by Phelps and Griffiths (1976). Based on E changes associated with the mapped flashes, Proctor concluded there was excess negative charge within the 20-dBZ (decibels of reflectivity factor, Z; see Sec. 7.1.2) reflectivity. He also concluded that of the 165 flashes for which he had field change data, only one did not carry negative charge from its origin.

Kasemir (1983) reproposed his hypothesis of a bidirectional leader with no net charge as the mechanism by which lightning forms. He had put forth the idea three decades earlier (Kasemir 1950). There are three basic aspects to his hypothesis: (1) The developing streamer carries no net charge; (2) The streamer does not collect charge from the cloud or precipitation particles; and (3) The energy source for the lightning is the electrostatic field. We will look at lightning theory in Section 5.10. Interestingly, in their widely referenced work, Ogawa and Brook (1964) interpreted their electric field measurements as a net result from a unidirectional streamer, but stated that the actual intracloud lightning may develop as a bidirectional streamer. Many authors have continued to assume unipolar streamer development, which was originally proposed by Schonland (1938). One exception is Loeb (1966), who stated that a ground flash starts by an upward moving positive streamer coincident with a downward moving negative streamer, in apparent agreement with Kasemir's theory.

Mazur (1989a, 1989b) analyzed observations in the context of Kasemir's (1983) concept of lightning initiation. Mazur's experimental data were direct in-channel measurements provided by strikes to an instrumented airplane and coincident mappings of the cloud lightning. For the airplane, he interpreted the data as showing that the negative point discharge occurs first (because of its lower onset field). This charges the air-

plane, so positive point discharge can occur. The resulting positive streamer then creates an E for a negative streamer from the airplane to occur simultaneously. Mazur (1989b) has argued that this process is the same for naturally occurring lightning (i.e., when the airplane is not there). To draw this conclusion, Mazur analyzed mapping images of cloud flashes obtained with an interferometric mapping system (Richard et al. 1985).

5.5 LOCATION OF LIGHTNING ORIGINS

Systems for mapping lightning in three dimensions (see Sec. 6.10) have been used to determine the origin of lightning. In their analysis of mapped lightning from four small, severe thunderstorms in Oklahoma, Rust et al. (1985) used the first mapped point from each flash as the point of origin. They reported that the origins of lightning can be distributed bimodally in altitude. For the lower group of flashes, they found the origin averages just below 7 km (T ≈ $-14°$C). Ground flashes originated lower than cloud flashes, but the difference was not statistically significant at the 0.05 level. The upper group of flashes, which were all intracloud, originated at about 10 km (T ≈ $-38°$C). These temperatures were from a sounding, which was made in the clear air in the same region as the storms and during the lightning observations.

Proctor (1991) also investigated where in thunderstorms lightning originates. He analyzed nearly 800 flashes that he mapped in portions of 13 storms. He calculated a centroid from the first six to ten mapped points and concluded that the origins of flashes tend to cluster in regions a few kilometers in the horizontal and at one or two regions in the vertical. He found a lower region at an average altitude of 5.3 km, corresponding to 540 mb, and an inferred temperature of about $-3°$C. When there was an upper region, it was at an average altitude of 9.2 km (317 mb, $-28°$C). Possible reasons for differences in the altitude of origins determined in these two studies were not addressed.

Proctor (1991) found >90% of the flashes began within 300 m (inside or outside) of 20-dBZ radar reflectivity. Only 3 of his 13 storms had a high percentage of high-altitude origins. The other 10 had <40% of the flash origins in an upper region. There was a tendency for both the height of the lower origin of lightning and for the percentage of high flashes to decrease with storm age. He attributed exceptions to most flash origins being clustered in space to the possible influence of midlevel winds.

Proctor (1991) also reported that flashes often begin in "holes" in the reflectivity. Although Vonnegut and Moore (1965) related such holes to nucleation of ice from thunder, Proctor noted that many holes were there prior to the flash. Thus, they may represent discontinuities in the charge density which could be preferential for initiating lightning. The size of these holes ranged from 250 to >1000 m.

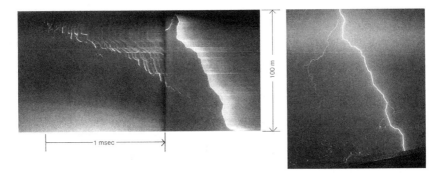

Fig. 5.3. Moving- and still-film images of the same ground flash. The stepped leader is the series of discrete downward moving channels to the left of the very bright return stroke. (From Berger and Vogelsanger 1966, with permission.)

We emphasize as we leave this section that we still do not have a clear understanding of the patterns of breakdown and the distributions of the space charge that causes the breakdown E and the ensuing discharges.

5.6. VISIBLE CHARACTERISTICS OF GROUND FLASH DEVELOPMENT

Much of what is known about the structure and time evolution of ground flashes was determined by high-speed photography. Summarized here is the evolution of a typical lightning flash to ground. The features described have been documented also with electric field measurements. We will present only the basics and refer you to Uman's (1987) book for details. Given here will be a typical flash to ground, which lowers negative charge.

Berger (1977) documented lightning processes in his years of experiments at Mount San Salvatore near Lugano, Switzerland. The initial discharge processes within the cloud cannot be photographed (hence the value of the lightning mapping systems). After the initial breakdown inside the cloud, the typical ground flash emerges from the cloud and heads to the ground. Figure 5.3 contains moving- and still-film images of the same flash. Discrete streaks of light indicate that the flash initially comes to ground in a series of steps as the *stepped leader*. The luminosity is brightest at the tip of each step. The speed of stepped leaders determined from photographs ranges from $1-25 \times 10^5$ m s^{-1}, with the average about 2×10^5 m s^{-1}. With time, the steps become fainter, but continue to just above the ground where an upward propagating steamer attaches to the downward moving stepped leader. This process is not seen in the figure, but an example of rare photographs of upward streamers is shown in Fig. 5.4. These processes are too faint and fast to be resolved by the human eye.

The return stroke occurs next and is the brightest part of the lightning; it is clearly visible unless deeply embedded in precipitation. In the typical negative flash to ground, the *return stroke* is caused by a wave front traveling rapidly up the channel as it drains negative charge to ground. The motion is continuous, without steps, and very fast. The speed is the greatest near the bottom of the channel and can exceed 2×10^8 m s^{-1}. The optical wave front slows with height to about 1×10^8 m s^{-1}. This can be the end of the flash if it has only a single stroke. There are two other processes that often occur. If the flash is multistroke, another leader comes to ground. Sufficient ionization can remain in the channel so the *dart leader* moves without steps and faster than the stepped leader. The dart leader has a typical speed of about 1×10^7 m s^{-1}. The speed varies as a function of the time from the last stroke because of channel cooling and decay of its ionization. Because of

Fig. 5.4. Still photograph of an upward streamer from a tree. To its right, another streamer that connected to the downward moving leader became part of the return stroke channel, which extends off the top of the photo. (With permission © Johnny Autery).

channel cooling, there is often a mixture of leader processes. As the leader nears the ground, it can begin to step, but stays in the original channel. Another variation is that the leader creates an entirely new lower channel and contact point to the ground. In both cases, the process is called a *dart-stepped leader,* as long as the upper part of the leader did not step.

Another visible process that can occur after any stroke is *continuing current.* On a moving-film photograph it shows up as a continuous smear of luminosity lasting tens to hundreds of milliseconds (Fig. 5.5). As its name implies, it is caused by the continuous flow of current in the channel of a few hundred amperes. This process is the primary one for starting forest fires, as the current can flow to the tree for several tenths of a second and raise the wood to its combustion temperature.

A human observer can often detect multiple strokes in a flash from an apparent flicker in the channel. We cannot unambiguously determine continuing current by eye, but its presence can be inferred if a steady, bright channel persists. People report seeing ground flashes travel upward between the ground and the cloud; they also report the direction as downward.

Fig. 5.5. Moving-film photograph of a positive ground flash. The continuing current is indicated by the continuous smearing of luminosity to the right from the return stroke channel (on the left side of luminosity). The small horizontal bars of light at the bottom are a time code to denote 10 ms resolution. (From Rust 1986, with permission.)

Because the return stroke is too fast for the direction of travel to be resolved by eye, our observations of direction are illusions. However, the direction of the original breakdown can be determined by the direction of branching. Typically the branches are pointed toward the ground, showing that the flash began in the cloud and moved downward.

5.7. RETURN STROKE VELOCITY MEASUREMENTS

Most measurements have come from high-speed photography with a streaking camera as originated by Schonland and Collens (1934) and continued by Orville et al. (1978), Idone and Orville (1982), and others. More recently, optical detectors behind multiple slit arrangements have also been used (see Sec. 6.9.3). Regardless of technique, the measured velocity often seems to be a function of the channel length sampled. This is true for return strokes where the velocity is not necessarily constant with height, as well as in other discharge processes. We begin our summary with results of measurements on return stroke velocity from naturally occurring (as opposed to triggered) lightning. There has been considerable study of the velocity of triggered lightning channels, which we also present.

We consider first the most common ground flash: a downward propagating negative stepped leader followed by upward moving negative return stroke(s). Idone and Orville (1982) analyzed the velocity near the ground of 63 strokes whose channel lengths were ≤ 1300 m, except for one channel of 2000 m. For all strokes, they found $v_{avg} = 1.1 \times 10^8$ m s^{-1}. The peak in the distribution was at $v \approx 0.9 \times 10^8$ m s^{-1}. The return stroke velocity almost always decreased as the channel propagated upward. The average reduction for the 27 strokes with both short and long channel segments was about 25%, with the v_{avg} decreasing from 1.4×10^8 m s^{-1} near the ground to 1.1×10^8 m s^{-1} higher up. They recommended that return stroke models need to incorporate the variation in velocity to obtain realistic currents. Most later measurements support this.

A nonphotographic technique has been to place optical detectors behind slits in the focal plane of a lens (e.g., Hubert and Mouget 1981, Nakono et al. 1983). The photoelectric device has the advantage of reduced size and cost. Mach and Rust (1989a) used such a device on a mobile laboratory. Included in their assessment of the technique was a comparison of velocities from their instrument and those from a streak camera for the same triggered flashes; the measured velocities were the same for comparable channel lengths (Idone et al. 1987).

Mach and Rust (1989b) divided the velocities from 86 natural flashes into two groups: short channels (<500 m long) and near the ground, and long channels (>500 m long) and starting near the ground. Short channels had an $v_{avg} = 1.9 \pm 0.7 \times 10^8$ m s^{-1}, which

was statistically significantly different at the 90% level from the long channels with an $v_{avg} = 1.3 \pm 0.3 \times 10^8$ m s^{-1}. It is accepted from the results of several investigators that negative stroke velocity slows with height above the ground. A division of the data into first and subsequent stroke velocities showed no significant difference. Comparison of the two largest data sets (Idone and Orville 1982 and Mach and Rust 1989b) showed that the long segment velocities from each were not significantly different. Mach and Rust found that short segments had velocity differences between natural and triggered flashes. The data set available on natural positive ground flashes is very small. Mach and Rust (1993) found that the velocity of seven natural positive flashes was about half that of negatives at a significance level of 99%. Furthermore, they found no slowing of the stroke velocity with height. Example results from several investigators are given in Table 5.1. Several researchers found a decrease in stroke velocity with height. As pointed out by Idone and Orville (1982), the decrease in velocity with height could be a manifestation of using two-dimensional velocities. For example, if tortuosity changes with height, the upper lengths would be underestimated. However, tortuosity does not seem to vary systematically with height (Idone and Orville 1988).

The velocity of channel propagation has been of particular interest for return strokes owing to its presumed tie to current flow in the channel and its relationship to the frequency of radiated signals. Idone et al. (1993) noted that several papers had shown a contradiction regarding the dependence of current on return stroke velocity, and that a resolution of the issue was not possible owing to lack of measurements in the first tens of meters to hundred meters of channel above the ground.

5.8. VELOCITY OF OTHER GROUND FLASH PROCESSES

We turn now to the measurement of the velocity of propagation for processes other than return strokes. The experimental values have been determined optically, from mapping of radiation sources, and from radar. For years such information was provided by the outstanding early work by Schonland and his colleagues (reviewed by Uman 1987). Here we restrict the summary to more recent measurements to take advantage of improved instrumentation. We do not include the more indirect determination of velocity from field change measurements and an analytical model. The tables are divided by process. The only data we know of for a streamer moving upward to meet a downward leader are also given in the following Table 5.2 under the term connecting leader.

All of the following measurements on dart leaders given in Table 5.3 are from streaking cameras. The number of events, n, can be more than the number of flashes.

Table 5.1 Examples of measured return stroke velocities. Natural and triggered strokes are depicted by N and T, respectively, along with stroke polarity (i.e., the charge effectively lowered to ground). The dimensionality of the velocity is d. Length, L, is the total measured. The number of first strokes is n_1, and the number of subsequent strokes is n_s. First and subsequent strokes did not always come from the same flashes. A single velocity was used for each stroke here, but the original reported may have been an average for a stroke. If it was available or attainable, we give the standard deviations, even though the distribution of velocities was not necessarily normal. Speeds are average values or/and ranges are given in { }; values separated by commas give the specific values available. If the standard deviation was reported, it is given after a ±.

Investigators	Type/ Pol	d	Length (m)	n_1	First Stroke (m s^{-1})	n_s	Subsequent Stroke (m s^{-1})
Hubert and Mouget (1981)	T/−	3	≤800	2	$0.6 \pm 0.1 \times 10^8$ $\{0.6 \times 10^8\}$	—	—
Hubert and Mouget (1981)	T/−	3	≤800	—	—	11	$1.1 \pm 0.4 \times 10^8$ $\{0.4-1.7 \times 10^8\}$
Idone and Orville (1982)	N/−	2	≤1300	17	$1.0 \pm 0.5 \times 10^8$ $\{0.3-2.4 \times 10^8\}$	—	—
Idone and Orville (1982)	N/−	2	≤1300	—	—	46	$1.2 \pm 0.4 \times 10^8$ $\{0.6-2.2 \times 10^8\}$
Beasley et al. (1983a)	N/−	1	470–690	17	$\{0.3-3 \times 10^8\}$	—	—
Beasley et al. (1983a)	N/+	1	470–690	3	$\approx 1 \times 10^8$	—	—
Idone et al. (1984)	T/−	3	≤575	—	—	56	$1.2 \pm 0.3 \times 10^8$ $\{0.67-1.7 \times 10^8\}$
Idone et al. (1987)	T/+	2	50	1	$\approx 1 \times 10^8$ $0.93, 1.0 \times 10^8$	—	—
Mach and Rust (1989b)	N/−	2	<500	25	$1.7 \pm 0.7 \times 10^8$	43	$1.9 \pm 0.7 \times 10^8$
Mach and Rust (1989b)	N/−	2	>500	25	$1.2 \pm 0.6 \times 10^8$	54	$1.2 \pm 0.2 \times 10^8$
Mach and Rust (1989b)	T/−	2	<500	—	—	39	$1.4 \pm 0.4 \times 10^8$
Mach and Rust (1989b)	T/−	2	>500	—	—	40	$1.2 \pm 0.2 \times 10^8$
Mach and Rust (1989b)	N/+	2	1300, 2300	2	$1.0, 1.7 \times 10^8$	—	—
Mach and Rust (1993)	N/+	2	<500	4	$0.8 \pm 0.3 \times 10^8$ $\{0.6-1.2 \times 10^8\}$	—	—
Mach and Rust (1993)	N/+	2	>500	7	$0.9 \pm 0.4 \times 10^8$ $\{0.6-1.7 \times 10^8\}$	—	—

Finally we summarize velocity measurements from other processes. In Table 5.4, they come from standard video, Brantly (1975); VHF mapping, Proctor (1981); radar, Mazur and Rust (1983); and VHF mapping plus high-speed video, Mazur et al. (1994).

5.9. VISUAL ASPECTS OF CLOUD FLASHES

We can obviously see lightning that does not strike ground, but much of the luminosity is within the cloud, rendering channel structure indistinguishable. However, there have been a few successful studies of physical processes in cloud discharges using visual techniques. For example, an observer can see air discharges propagate over long distances. An observer usually can correctly determine the general direction of propagation in two dimensions (Fig. 5.6). These air discharges propagate at speeds of 10^3-10^5 m s^{-1} (Brantley et al. 1975).

There also are faster current pulses that propagate in a direction opposite to the initial channel development and cause the channel brightness to flicker. They are called *K changes* or *recoil streamers*. In images of cloud discharges, Ogawa and Brook (1964) identified recoil streamers coming back along previously created channels. In video images of lightning channel attachment to an instrumented aircraft, Mazur (1989b)

Table 5.2 Measured velocities of stepped leaders. Type refers to natural and triggered lightning flashes, respectively, N and T. Polarity refers to polarity of charge brought to ground by the flash. The dimensionality of the velocity is d. Distance, D, is above the ground, and length, L, is the total leader length measured. If it was available or attainable, we give the standard deviations, even though the velocities are not necessarily distributed normally. Speeds are average values and/or ranges are given in { }; values separated by commas give the specific values available. If the standard deviation was reported, it is given after a ±.

Investigators	Phenomenon: Type/Polarity	d	Distance or Length	n	Speed (m s^{-1})
Orville and Idone (1982)	Stepped leader: N/−	2	D < 200 m	3	1.1×10^6
Beasley et al. (1983b)	Stepped leader: N/−	1	D < 100 m	3	{$0.8–3.9 \times 10^6$}
Proctor et al. (1988)	Stepped leader: N/−	3	L = 6.4 ± 2.3 km	66	$1.6 ± 0.7 \times 10^5$ {$0.3–4.2 \times 10^5$}
Idone (1990)	Upward connecting leader: T/−	2	L < 20 m	9	$1.3 ± 0.6 \times 10^7$ {$0.15–2.1 \times 10^7$}
Idone (1992)	Downward stepped leader: T/−	2	—	1	3.1×10^5
Idone (1992)	Upward stepped leader: T/+	2	L = 530 m	1	3.6×10^5 {$1.2–6.5 \times 10^5$}
Idone (1992)	Step pulses: T/+	2	L = 600, 400 m	1	$5, 7 \times 10^7$

showed that the cloud flash channels often have recoil-streamer current pulses superimposed on continuous current flow. Mazur et al. (1995) analyzed correlated observations of ground flashes with a high-speed video system, which had a 1 ms image resolution, and a VHF radio interferometer, and concluded that K changes and *M streamers* are the same—a fast negative streamer—except that the M streamers contact a conducting channel to ground.

5.10. OVERVIEW OF LIGHTNING THEORY

The occurrence of a lightning flash entails highly complicated processes. Although there is a large body of research addressing numerous aspects of lightning, a complete theory of a flash does not exist. A comprehensive theory of lightning requires understanding the physical processes in electrical breakdown, which are

Table 5.3 Measured velocities of dart leaders. Type refers to natural and triggered lightning (N and T). Polarity is the reported polarity of charge that moved. The dimensionality of the velocity is d. Distance, D, is from the ground, and length, L, is the total leader length (or segment, if noted) measured. If it was available or attainable, we give the standard deviations, even though the distribution of velocities was not necessarily normal. Speeds are average values and/or ranges are given in { }; values separated by commas give the specific values available. If the standard deviation was reported, it is given after a ±.

Investigators	Phenomenon: Type/Polarity	d	Length	n	Speed (m s^{-1})
Orville and Idone (1982)	Dart leader: N/−	2	D < 800 m	21	1.1×10^7 {$0.29–2.3 \times 10^7$}
Idone et al. (1984)	Dart leader: T	3	{segment L = 15–90 m}	32	2.0×10^7 {$0.95–4.3 \times 10^7$}
Jordan et al. (1992)	Dart leader: N	2	{L = 700–1400 m}	9	$1.5 ± 0.5 \times 10^7$ {$0.8–2.4 \times 10^7$}
Jordan et al. (1992)	Dart leader: T	2	{L = 400–700 m}	36	$1.6 ± 0.9 \times 10^7$ {$0.6–3.4 \times 10^7$}
Jordan et al. (1992)	Dart-stepped leader: N	2	{L = 700–1400 m}	2	$0.54, 1.2 \times 10^7$
Orville and Idone (1982)	Dart-stepped leader: N	2	D < 800 m	4	{$0.2–0.46 \times 10^7$}

Table 5.4 Measured velocities of various processes. Type refers to natural and triggered lightning (N and T). Polarity is the reported charge that moved. The dimensionality of the velocity is d. Length is the total measured. If it was available or attainable, we give the standard deviations, even though the distribution of velocities was not necessarily normal. Speeds are average values and/or ranges are given in { }; values separated by commas give the specific values available. If the standard deviation was reported, it is given after a ±. The phenomenon is given as reported and can include more than one identifiable process (e.g., speed can include steps, pauses, etc.). The number, n, of events may be larger than the number of flashes.

Investigators	Phenomenon: Type/Polarity	d	Length (km)	n	Speed (m s^{-1})
Brantley et al. (1975)	Horizontal channels: N	2	—	13	{0.56–1.1 × 10^4}
Proctor (1981)	Initial streamer: N/−	3	—	25	1.3 ± 0.3 × 10^5 {0.9–2.1 × 10^5}
Proctor (1981)	Initial streamer: N/+	3	—	1	5 × 10^5
Proctor (1981)	Q noise: N	3	{0.1–4.4}	19	2.5 × 10^7 {0.25–4.4 × 10^7}
Proctor (1983)	In-cloud channels: N/both	3	{1–90}	32	1.4 ± 1.2 × 10^5 {0.4–7.7 × 10^5}
Mazur and Rust (1983)	Gross-in-cloud propagation: N	1	{a few to ≥ 20}	1055	1 × 10^5 max. 2.5 × 10^5
Proctor et al. (1988)	Q noise: N	3	{0.3–3}	78	8.7 ± 6.8 × 10^7 {0.2–30 × 10^7}
Proctor et al. (1988)	Horizontal retrogression: N	3	—	48	2.2 ± 1.3 × 10^4 {0.1–7 × 10^4}
Mazur et al.	Horizontal: N/−	3	—	1	{2–4 × 10^5}

microscale processes, and in flash propagation, which is a macroscale process that can cover hundreds of kilometers of path length in a single flash. We will examine only simple physical approaches to the main aspect of the lightning leader process, specifically two electrostatic models for a vertically oriented leader: the source-charge model and the bidirectional leader model. We present a summary of and the salient analytical expressions in each theory.

We remind you that in much of the literature, authors have used the potential gradient polarity convention. A careful determination of the polarity convention throughout each paper is imperative if quantitative understanding is the goal. Electromagnetic theory related to that used in this section is in the Tutorial.

5.10.1. Fundamentals of the source-charge model

In the *source-charge model* (Fig. 5.7a), which was proposed by Schonland (1938), a unidirectional leader emanates from a charge region, and a unipolar charge flows from the source into the leader channel, where it distributes itself uniformly along the length of the channel. Uman (1987) presented a review of this theory. The incremental E measured at the ground, which is assumed to be a flat conductor, from an element of charge along an incremental channel length, dz, aloft is

$$dE = -2\lambda_q(z) \frac{z\,dz}{4\pi\varepsilon(z^2 + D^2)^{\frac{3}{2}}}, \qquad (5.2)$$

where the 2 results from the image charge. The line charge density, λ_q, is constant. In these equations, the polarity of the leader is assigned to λ_q, so for a negative leader, λ_q is negative and dE is positive.

With the leader moving downward, there is a decrease of the charge in the source region; this decrease also contributes to the change in electric field at the ground. The total field change produced by a negative leader moving downward and the decrease of charge in

Fig. 5.6. Air discharge. (Photo courtesy of D. W. Burgess.)

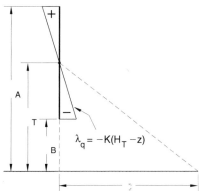

(a) Source Charge Model (b) Bidirectional Leader Model

Fig. 5.7. Conceptual models of a lightning leader. The ambient electric field is uniform in both. Image charges are not shown but are in the models. (a) The source-charge leader moves only downward from its origin at a height (in the z direction) of H_T and has a nonzero net charge. H_B is the height of the leader's lower tip. (b) A bidirectional leader moves upward and downward from its origin at H_T and has a net charge of zero. H_A and H_B are the heights of the tips of the positive and negative parts of the bidirectional leader, respectively. (After *Mazur and Ruhnke* © 1993 American Geophysical Union.)

the source is

$$\Delta E = -\frac{\lambda_q}{2\pi\varepsilon} \qquad (5.3)$$

$$\cdot \left[\frac{1}{(H_B^2 + D^2)^{\frac{1}{2}}} \frac{1}{(H_T^2 + D^2)^{\frac{1}{2}}} - \frac{(H_T - H_B)H_T}{(H_T^2 + D^2)^{\frac{3}{2}}} \right],$$

where the Hs and D are defined in Fig. 5.7. The two left-hand terms are from charge on the leader, and the term on the far right is from the decrease in the source charge.

5.10.2. Fundamentals of the bidirectional leader model

The electrostatic model of a bidirectional leader with a net charge of zero was proposed by Kasemir (1950, 1960). However, his model went mostly unnoticed, and the source-charge model was not extensively challenged. Research on lightning-aircraft interactions during the 1980s provided compelling evidence that a flash begins as simultaneous, oppositely propagating leaders. Both airborne recordings and ground-based radar data of the lightning showed such bidirectional propagation of positive and negative streamers that originated at the opposite extremities of the airplane (Mazur et al. 1984, Mazur 1989a). Mazur (1989b) compared flashes triggered by the airplane to naturally occurring cloud flashes and concluded that they were similar. Mazur and Ruhnke (1993) reviewed the electrostatic theory of lightning development and compared the source-charge and bidirectional leader models. Their findings are summarized here.

In the *bidirectional leader model*, the leader is bipo-lar with positive and negative parts that move in opposite directions from its origin at H_T (Fig. 5.7b). Although the total net charge on the bipolar leader is zero, the distribution of charge on the leader is determined by the ambient potential distribution. In a simple treatment, the ambient electric field is assumed constant. The line charge density for a bidirectional leader with downward moving negative and upward moving positive parts is

$$\lambda_q(z) = -K(H_T - z). \qquad (5.4)$$

The dE from an elemental line of charge is still as given in Eq. 5.2, and the line charge density changes linearly with a slope, K, that is governed by the ambient E. Mazur and Ruhnke (1993) assumed that the positive and negative parts of the leader move at the same velocity, so the total electric field change for a negative ground flash can be obtained from integration after combining Eqs. 5.2 and 5.4:

$$\Delta E = \frac{K}{2\pi\varepsilon} \left\{ \frac{H_A - H_T}{(H_A^2 + D^2)^{\frac{1}{2}}} + \frac{H_T - H_B}{(H_B^2 + D^2)^{\frac{1}{2}}} \right.$$
$$- \ln\left[H_A + (H_A^2 + D^2)^{\frac{1}{2}} \right] \qquad (5.5)$$
$$\left. + \ln\left[H_B + (H_B^2 + D^2)^{\frac{1}{2}} \right] \right\}.$$

When the leader connects with the ground, $H_B = 0$ and $H_A = 2H_T$, and ΔE becomes

$$\Delta E_0 = \frac{K}{2\pi\varepsilon} \left\{ \frac{H_T}{(4H_T^2 + D^2)^{\frac{1}{2}}} + \frac{H_T}{D} \right.$$
$$\left. - \ln\left[2H_T + (4H_T^2 + D^2)^{\frac{1}{2}} \right] + \ln D \right\}. \qquad (5.6)$$

Shown in Fig. 5.8 are comparisons between the

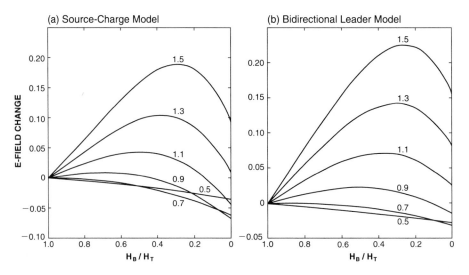

Fig. 5.8. Electric field change per unit charge for (a) source charge model with a downward moving negative leader and (b) bidirectional leader model with the downward half of the leader being negative. Each curve is labeled with its value of H_T/D. In both panels, H_B is the height of the lower tip of leader, and H_T is the top of bidirectional leader or initiation height. (After Mazur and Ruhnke, 1993: American Geophysical Union, with permission.)

electric field changes for the two models. Mazur and Ruhnke (1993) calculated the height-to-distance ratio, at which the $\Delta E_0 = 0$, to be $H_T/D = 1.27$ for the source-charge model and an $H_T/D = 0.98$ for the bidirectional leader. Thus for the same ΔE and D, the calculated initiation height, H_T, for a flash is 30% higher with the source-charge model than with the bidirectional leader model. This suggests a solution to the problem of trying to reconcile calculations with the source-charge model and measurements that had been noted by several investigators.

5.10.3. Combined effect of leader and return stroke

We consider now what happens when the return stroke is included in the models. The return stroke was envisioned by Malan (1963) as discharging the charge along the leader channel. Thus, he assumed that the field change from the stroke was the same magnitude as, but of opposite polarity to, that from charge on the leader. To get the resultant field change at the ground for the entire process, the expression for the return stroke is added to that of the leader (Eq. 5.3) with the proper signs. The resulting total field change at the ground is

$$\Delta E_{L+RS} = -\frac{Q}{2\pi\varepsilon} \frac{H_T}{(H_T^2 + D^2)^{\frac{3}{2}}}, \tag{5.7}$$

where the total charge $Q = \lambda_q H_T$. This expression is equivalent to that obtained using Eq. 5.2 if we consider a ground stroke as depositing a positive point charge

within the cloud and thus neutralizing a part of the source charge of the leader. Application of such a point charge analysis is described in Section 6.10.1.

In the bidirectional leader model, the field change of the return stroke is produced by a ground potential wave that moves from the ground to the top of the leader (H_A in Fig. 5.7). For the case of a negative ground flash, Mazur and Ruhnke (1993) would have envisioned the effect of the return stroke as neutralization of the negative charge in the bottom half of the leader and the doubling of positive charge in the top half of the leader. The negative stroke then puts a positive charge, $Q = KH_T^2$, in the bidirectional leader channel. Assuming equal propagation velocities for the leader halves so $H_A = 2H_T$, Mazur and Ruhnke found the field change at the ground for the bidirectional leader and return stroke to be

$$\Delta E_{L+RS} = \frac{Q}{4\pi\varepsilon H_T^2}\left\{\frac{-2H_T}{(4H_T^2 + D^2)^{\frac{1}{2}}}\right.$$
$$\left. + \ln\left[2H_T + (4H_T^2 + D^2)^{\frac{1}{2}}\right] - \ln D\right\}. \tag{5.8}$$

Although both the bidirectional leader and the source-charge models produce realistic waveforms of the electric field for leader-return stroke combinations, Mazur and Ruhnke (1993) concluded that the bidirectional leader concept makes sense physically, whereas the unipolar leader from a source charge does not. For example, the source-charge model violates the requirement that a conductor in an electric field will have an induced charge distribution dictated by the ambient potential distribution. Furthermore, Kasemir (1983)

(a)

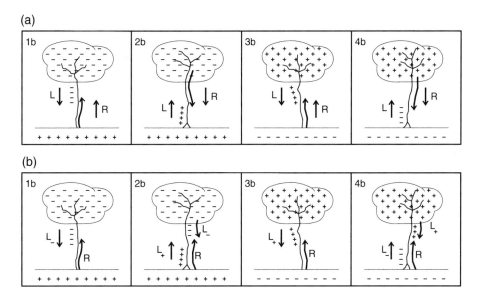

(b)

Fig. 5.9. (a) The four types of ground flashes, which are composed of a leader, L, and a return stroke, R, as proposed by Berger (1977). (b) Proposed modifications to two of the four types of ground flashes by Mazur and Ruhnke (1993: American Geophysical Union, with permission.) The change is that a return stroke in types 2 and 4 can flow only from the ground upward as a ground-potential wave. The hypothesized sequence in these two flash types is upward leader, downward leader, and then return stroke.

noted that no known physical law can make a distributed charge concentrate itself by flowing into a lightning channel. Rather, Kasemir showed that the energy stored in a storm's electric field induces charge on the bidirectional leader.

Mazur and Ruhnke (1993) also proposed modification to two of the four types of ground flashes introduced by Berger (1977) (see Sec. 5.3.1). They based this proposal on analyses of physical processes common to the development of both natural and triggered flashes. For convenience, we illustrate both sets of proposed descriptions of each type in Fig. 5.9, Berger's descriptions summarized in Fig. 5.9a and Mazur and Ruhnke's descriptions summarized in Fig. 5.9b. Berger (1975) had noted difficulty in determining the direction of return strokes from his photographic streak images of 2b and 4b, and the streak image of type 4b shown by Berger had a time resolution of only 1 ms, which is inadequate to determine the direction of the stroke from the image. Basically, Mazur and Ruhnke argued that a return stroke can move only from the ground to the cloud, not the reverse.

To make types 2b and 4b conform to the principle, Mazur and Ruhnke (1992) proposed the following descriptions: In type 2b, a positive leader initially moves upward into the cloud; a dart leader then moves back down from the cloud to Earth; and when the dart leader reaches Earth, a return stroke occurs. This scenario was meant to describe rocket-triggered ground flashes that involve the negative charge in summer thunderstorms

and is consistent with Hubert et al.'s (1984) interpretation of data from rocket-triggered flashes. In the revision of type 4b, a positive leader descends from the cloud toward Earth and triggers a negative leader from a tall structure or rocket; after the two leaders meet, a return stroke occurs. This scenario was intended to describe the lightning sometimes observed under winter storms when positive charge is overhead. Both of these two scenarios revise Berger's versions by adding a second leader.

5.11. TEMPERATURE, PRESSURE, ELECTRON NUMBER DENSITY, AND THUNDER

The electrical resistance of air at ambient environmental temperatures is large. Consequently, air is heated when the large current densities of lightning flow through it. When large current surges such as return strokes pass through a lightning channel that is relatively cool (e.g., before the first return stroke), they rapidly heat the air in the channel to temperatures greater than 20,000 K, causing air molecules to become ionized. The heating continues until the ionization has created enough free electrons to make the channel a good conductor. This process typically requires <10 μs. (A neutral gas that is ionized to the point that it is a good conductor is called a *plasma*.) Although currents continue to flow in the channel, further heat input to the lightning channel becomes small as the electrical resistance decreases to values typical of good conductors. Uman and Voshall (1968) showed that once a channel is heated, it cools relatively slowly by radiation and heat conduction until it again becomes nonconducting, typically at temperatures of 2000–4000 K, roughly 50 ms after energy input to the channel ends.

Besides ionizing the air, rapid heating increases the pressure and luminosity of air inside the lightning channel. The large pressure of the channel causes rapid expansion into the surrounding air, initially in the form of a shock wave, which starts moving outward from the hot core at roughly ten times the speed of sound (i.e., at $\approx 3 \times 10^3$ m s^{-1}). In a shock wave, both the pressure and the radial expansion speed are larger than is characteristic of sound waves. Because the time that lightning current deposits significant heat in a channel is short, the region of high pressure typically expands no more than 1–2 cm during heating, so theoretical treatments of thunder often consider the heating to be instantaneous compared with the time for hydrodynamic processes. As the shock front propagates through the surrounding air, it does work by raising the pressure of air it overtakes, so the outward expansion slows, and the pressure of the disturbance decreases until it becomes a minor perturbation of ambient pressure. At some point, the pressure disturbance becomes small enough to become a sound wave. The sound generated by a lightning flash is called *thunder.*

5.11.1. Results from spectroscopy

The intensity and frequencies of light emitted by a lightning channel in air depend on the temperature, pressure, and electron number density of the channel in such a way that estimates of these three properties can be made from spectroscopic measurements. A discussion of the theory and measurement techniques for making these estimates involves at least a basic level of quantum mechanics, which is beyond the intended scope of this text, so we will give only the pertinent results. For information on the theory and measurement techniques of lightning spectroscopy, see a text on atomic spectra, such as Kuhn (1969), and treatments of lightning spectroscopy by Orville (1968a–d, 1977b), Uman (1969, 1987), and Salanave (1980).

From an analysis of spectral lines of singly ionized nitrogen (i.e., nitrogen that has lost one electron) in 10 return stroke channels below cloud, Orville (1968a–c) estimated that the peak temperature occurred within the first 10 μs after the return stroke and was typically 28,000–31,000 K, more than twice as hot as the surface of the Sun. Because of limitations imposed by the time resolution of the spectra and the characteristics of the nitrogen line that was used, it is possible that higher peak temperatures occurred during the initial few microseconds, but were masked by the measurement. The temperature was below 30,000 K after 10 μs and dropped to roughly 20,000 K or less after 20 μs. Brook et al. (1985) found similar temperatures for lightning

channels in the upper part of clouds, although they were not able to tell whether the channels belonged to cloud or ground flashes.

Orville (1968a–c) also estimated that electron number densities averaged over the first 5 μs after return strokes were approximately 10^{24} m^{-3} and decreased to 10^{23} m^{-3} after 20–30 μs. These estimates agreed reasonably well with theoretical treatments of a nitrogen plasma (Drellishak 1964) and with measurements of a spark in air (Orville et al. 1967). According to Drellishak's theoretical treatment, the density remains approximately 10^{23} m^{-3} until the channel temperature drops below 13,000 K. With these electron densities and temperatures, the pressure of the channel was estimated to be 8 atm in the first 5 μs and to decrease to 1 atm in 10–20 μs. Because the initial measurements were averages over the first 5 μs, the peak pressure probably was even higher. However, since even 8 atm is greater than ambient pressure (≤ 1 atm), it is clear that the heated channel expands and thereby produces thunder.

5.11.2. Thunder

Lightning produces thunder by electrostatic forces and by the heating described above. Audible thunder is due primarily to heating, but electrostatic forces are thought to be a major source of *infrasonic thunder* (i.e., thunder at frequencies too low to be heard). We begin with thunder caused by heating and then move to electrostatic production of infrasonic thunder.

Figure 5.10 shows examples of thunder signatures of individual lightning flashes. Thunder spectra almost always peak at frequencies below 150 Hz and typically below 100 Hz (e.g., Holmes et al. 1971). The duration of sound from a single lightning flash is typically 10–60 s, and the amplitude varies considerably during

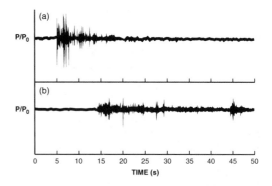

Fig. 5.10. Thunder signatures of two lightning flashes. Microphone output amplitude is plotted as pressure perturbation, P/P_0, and shows the claps (large spikes) and rumbles of thunder from (a) a nearby cloud-to-ground flash and (b) a long horizontal flash. Time = 0 is the time of lightning. (Adapted from Few 1974: American Geophysical Union, with permission.)

that period, from sudden, loud claps to quieter, more gradually varying rumbles.

The duration of thunder is a function primarily of the geometry of a lightning flash and its location and orientation relative to the observer. The order in which channel segments form and are heated has relatively little effect, because thunder propagates at the speed of sound (roughly 340 m s^{-1} at sea level) while lightning channels propagate three to five orders of magnitude faster (typically 10^5–10^7 m s^{-1}). Thunder from the initial breakdown of a flash typically travels only 50–500 m by the time the flash ends, so the flash can be considered to be instantaneous compared with the time it takes thunder from different channel segments several kilometers apart to reach an observer. The first thunder heard, therefore, is from the section of the lightning channel closest to the observer, and the last is from the section farthest away. The duration of thunder is just the difference in the times required for thunder to propagate these two distances.

As shown in Fig. 5.11, this is equivalent to saying that the duration of thunder defines the thickness of a hemispherical shell, centered on the observer, that completely contains all thunder-producing channels in a flash. Furthermore, the thickness of the shell gives the minimum length of the lightning flash. (It is a minimum estimate, because there can be considerable channel structure parallel to the shell without affecting duration.) Because the speed of sound is approximately 1/3 km s^{-1}, you can estimate the minimum length of a flash in kilometers by dividing the duration of thunder in seconds by three. The range in kilometers to the closest part of the channel is given by counting the number of seconds from the time you see a flash until the time you first hear thunder and then dividing by three. (You can use the time you see lightning as the time to start counting, because light travels six orders of magnitude faster than sound in the troposphere.)

Reviews of thunder by several investigators (Hill 1977a, Few 1982, Uman 1987) have agreed that lightning channel tortuosity is necessary to produce the characteristic sound of thunder, as suggested by Few (1969, 1974). To see this, first consider the sound produced by a single linear acoustic source. Using both

laboratory measurements and theory, Wright (1964) and Wright and Medendorp (1967) studied the sound wave overpressure and waveform as a function of angle from the axis of a 1-cm spark. They found that the peak overpressure, P(θ), produced by the spark is a steep function of angle, θ, from the axis

$$P(\theta) = P(90°) \left[1 - \left(\frac{L_s \sin|90° - \theta|}{c_s \tau_s} \right) \right], \quad (5.9)$$

where L_s is the length of the spark; c_s is the speed of sound; and τ_s is the period from the beginning of the compression pulse to the end of the rarefaction. Few (1974) computed from this that 80% of the acoustic energy from a short spark is radiated within ±30° of the plane perpendicular to the spark.

Furthermore, Wright (1964) and Wright and Medendorp (1967) found that at angles away from the perpendicular, the compression and rarefaction become separated. Uman et al. (1968) showed that this separation is a characteristic of acoustic radiation from a straight line source. On the perpendicular at distances from the source much greater than its length, the radiation from all parts of the source arrive at essentially the same time; the compression and rarefaction that is radiated by each infinitesimal element reinforce each other to produce a single wave whose pressure trace has an N shape, and so sometimes is called an *N-wave* (Fig. 5.12). Off the perpendicular, however, the compression and rarefaction of internal, infinitesimal elements interfere with each other in such a way that pressure perturbations are received only from the ends of the linear source. The initial compression is received from the end of the linear source closest to the observer, and then there is a period of quiet until the rarefaction is received from the farther end. The interval between compres-

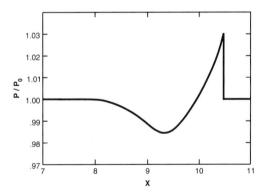

Fig. 5.12. Overpressure versus distance in the weak shock regime for an N-wave generated by a single acoustic source. The values shown in the plot have been based on spherical expansion, input energy of 10^5 J m^{-1}, $P_0 = 1$ atm, and spherical divergence. The N shape is reflected about the vertical axis in this plot. (Adapted from Few, 1968, with permission.)

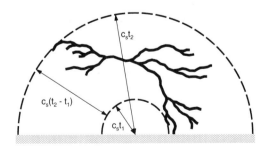

Fig. 5.11. Schematic diagram of the relationship between minimum length, $c_s (t_2 - t_1)$, of a lightning flash and thunder duration.

sion and rarefaction is largest along extensions of the long axis of the spark, because that orientation maximizes the difference in the distances from the observer to the two ends.

Even sharp claps of thunder are of much longer duration than the initial compression pulse from a single linear source, and typically, there is not a period completely without sound immediately after thunder begins or after every clap. Instead, tortuosity breaks up the lightning channel into a series of acoustic sources. This was shown experimentally by several investigators, as mentioned by Brook (1969) and described by Hill (1977a) and Uman (1987). Short lengths of explosive line, called Primacord, were used for the acoustic source. When the lengths were strung together in a straight line, the result was what is described above for a linear source. When the lengths were arranged to simulate a tortuous lightning channel, features of the sound were qualitatively similar to thunder. Similarly, measurements of the pressure pulses from a tortuous 4-m long spark by Uman et al. (1970) showed multiple compression pressure pulses, not just the single compression pulse that is characteristic of a simple cylindrical source. Also, Ribner and Roy (1982) and Few (1974) were able to produce thunder-like sounds numerically by constructing tortuous channels from short channel segments, each having the acoustic characteristics of short, linear sparks determined by Wright and Medendorp (1967).

It is less clear what size of channel segment should be considered an independent acoustic source. If lightning tortuosity was caused by straight channel segments in only a narrow range of lengths long enough to act as independent acoustic sources, the length of the linear acoustic sources would be obvious. However, tortuosity exists over a broad range of distance scales; to delineate segments of a tortuous channel on a given distance scale, segments are defined by changes in direction on the chosen scale, and each segment is fitted through tortuous structure on smaller scales. Evans and Walker (1963), Hill (1968), and Salanave (1980) showed that lightning channel tortuosity has similar statistical properties on distance scales from 10 cm to hundreds of meters. Hill showed that the change in direction from one channel segment to another on several scales can be described as a random walk with a mean deviation from the preceding segment's forward direction of $16°$. Therefore, the scales of tortuosity provide no clue to the size of independent acoustic sources.

Instead, the size of individual acoustic sources must be governed by the channel heating process and the hydrodynamics of rapidly heated air embedded in ambient air. As pointed out by Few (1969, 1982), the expanding pressure pulse from adjoining channel segments on the smallest distance scales will tend to merge together under the high temperatures and pressures of the channel when it is initially heated, so these small channel segments will produce a single n-wave. As the distance scale increases, it takes longer for pressure pulses from the extremes of the segment to reach the

expanding shock front from the middle of the segment. Therefore, when these pulses arrive at the shock front, the temperature and pressure of the shock will be smaller, and the propagation velocity of the pulses themselves will have slowed. As the length of a channel segment continues to increase, eventually pressure pulses from the extremes of the segment will not be able to travel fast enough to overtake the shock front and merge into a single n-wave with pressure pulses of disparate phases from other parts of the segment. The distance scale of channel segments whose pressure pulses no longer must merge will be the scale of tortuosity that separates segments into independent acoustic sources. For input energies typical of lightning, Few (1969, 1974, 1982) suggested that the length of quasi-linear segments that form independent acoustic sources is roughly 1 m. Although Hill (1977a) agreed that the characteristic sound of thunder is a result of tortuosity, he disputed aspects of Few's theory affecting the estimate of source length.

Using concepts described in the previous paragraph, Few (1969, 1974, 1982) defined three scales of tortuosity: (1) Microtortuosity involves small channel segments on distance scales where the shock waves of adjoining segments merge together; (2) Mesotortuosity involves channel segments that behave as individual linear acoustic sources when the shock wave makes the transition to an acoustic wave. The energy from an individual linear source will tend to be beamed perpendicular to the axis of the channel segment, as described above; and (3) Macrotortuosity involves channel segments consisting of many acoustic sources (i.e., of many mesotortuous segments). The orientation of mesotortuous segments within a macrotortuous segment tend to be similar, because there is a strong bias in the forward direction for changes in orientation from segment to segment.

Mesotortuosity and macrotortuosity each play a role in producing the characteristic sound of thunder. As pointed out by Few (1974, 1982), much of the perceived variation in the amplitude of thunder is caused by the orientation of macrotortuous channel segments relative to the observer (see Fig. 5.13). Loud, sudden claps of thunder are heard by observers along or near a perpendicular to macrotortuous segments. Few described three effects of channel geometry that contribute to the loudness of claps: (1) When an observer is near a plane perpendicular to a long (macrotortuous) channel, the rate at which the individual acoustic waves from the mesoscale segments reach the observer is greater than for any other orientation, because the difference in ranges to the closest and farthest segments is minimized; (2) Since the orientation of mesotortuous segments within a macrotortuous segment tends to be similar, as discussed above, and each mesotortuous segment radiates most of its acoustic energy near the plane perpendicular to it, macrotortuous segments will tend to radiate most of their acoustic energy near the perpendicular to the segment, too; and (3) When acoustic waves are added, waves that are in phase have

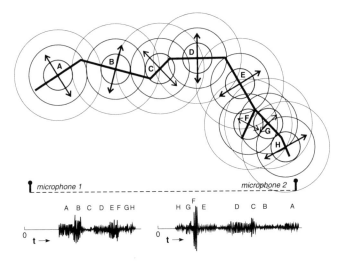

Fig. 5.13. Schematic diagram of the effect of macrotortuous channel orientation on thunder amplitude perceived by an observer. The thunder signal recorded at two locations is shown on the bottom. The time of the flash is t = 0. Letters indicate corresponding thunder and channel segments. Note the different order of letters and different relative amplitudes at the two locations. (From Few 1975, with permission.)

a larger amplitude than waves that have random phases or are out of phase. Since the mesotortuous segments in a long channel are heated at essentially the same time, the acoustic waves of neighboring segments will be nearly in phase perpendicular to the macrotortuous segment. Evidence for this was found by Uman et al. (1970), who showed that the number of pressure pulses decreased with distance from a 4-m laboratory spark, as the distance to different segments of the spark (and, hence, the phases of acoustic waves from the segments) became more nearly equal. Few (1982) quoted a discussion with Plooster to note that in-phase acoustic waves from the individual sources within a macrotortuous segment can merge together as a result of nonlinear processes, so that they propagate as a single, larger wave instead of propagating independently in their original directions.

Few (1974, 1982) suggested that rumbling (i.e., the part of a thunder signature in which the amplitude is smaller and more gradually varying) is produced by the erratic amplitudes and phases of the many acoustic waves arriving from macrotortuous channel segments more parallel to the line of sight of an observer. The reasons for the quieter sound are analogous to the reasons for the loudness of claps: for signals arriving from directions more nearly parallel to the macrotortuous segment (1) the number of acoustic sources whose signals arrive per unit time is smaller, (2) less energy is radiated at angles away from the perpendicular, and (3) the relative phases of individual waves arriving at a given time are more likely to vary erratically. The duration of rumbling tends to be longer than the duration of claps because an orientation more along an observer's line of sight increases the difference in the distances to the closest and farthest segments.

Mesotortuosity plays a role by allowing sound to be heard from the internal part of a long channel segment that is not perpendicular to the line of sight of an observer. Remember that if the channel segment were straight, the only sound received at angles away from the perpendicular to the segment would be from the end points. Tortuosity introduces erratic variations of phase and amplitude that prevent the superposed waves received from internal segments of a long channel from consistently nullifying each other.

There has been little research examining whether thunder generation varies significantly for different flashes or for different channel segments within a flash, apart from the effects of channel orientation relative to the observer. Holmes et al. (1980) presented data from Langmuir Laboratory suggesting that thunder generation can vary considerably. Their data consisted of simultaneous recordings of thunder, electric field changes, and radar returns from lighting channels. Little or no thunder was detected from what appeared to be a network of weaker lightning channels above an altitude of 6 km MSL, but large amplitude thunder was received from lightning channels below 6 km. Note that MacGorman et al. (1981) reported many thunder sources as high as 12 km MSL from storms in other geographic regions, so the lack of thunder reported by Holmes et al. was not caused simply by the lower atmospheric pressure at altitudes above 6 km. These reportedly weak channels at high altitudes may have been similar to the almost continuous lightning flashes observed in the upper part of storms by Taylor et al. (1984), Mazur et al. (1984), and Rust et al. (1985).

Another case in which little thunder appeared to be generated by lightning channels was observed by one of us (MacGorman). On a couple of days when long, horizontal flashes propagated along the bottom of a stratiform cloud slowly enough that the eye could follow their motion, no thunder could be heard at the ground, although several flashes were almost directly overhead, and the environment was not particularly windy or noisy. Since it would be difficult to explain this observation simply by propagation effects, it may be that the channels did not generate thunder loud enough to be heard at the ground.

Detailed theoretical treatments of thunder generation have been somewhat controversial. Few (1969, 1982), for example, provided a theoretical treatment of thunder generation in which propagation was divided into three regimes: strong shock, weak shock, and acoustic. To delineate the strong and weak shock regimes, he estimated the radial distance over which energy input to the lightning channel would be expended in thermodynamic work done by the shock wave on the atmosphere, i.e.,

$$E_L = P_{amb} \, \pi R_C^2$$
$$E_T = P_{amb} \, \frac{4}{3} \pi R_S^3,$$
(5.10)

which gives

$$R_C = \left(\frac{E_L}{\pi P_{amb}} \right)^{\frac{1}{2}}$$
$$R_S = \left(\frac{3E_T}{4\pi P_{amb}} \right)^{\frac{1}{3}},$$
(5.11)

where R_C is the radius for cylindrical expansion, R_S is the radius for spherical expansion, P_{amb} is the ambient environmental pressure, E_L is the energy input per unit length of channel for cylindrical expansion, and E_T is the total energy input into the center of a spherical expansion. For input energy in the range that has been considered likely for lightning ($E_L = 1 \times 10^4$ to 5×10^5 J m^{-1}), R_C varies from 0.2 m to 1.3 m at sea level and from 0.3 m to 2.3 m at a height of approximately 9 km.

Few (1969, 1982) suggested that the shock wave decays to the weak shock regime when it reaches a radius approximately equal to R_C. This has been demon-

strated by numerical simulations of the shock wave as it propagates outward. Figure 5.14 shows two plots of pressure as a function of radius during expansion of the disturbance. One is a detailed analysis close to the channel in the strong shock regime, although a relatively small input energy was assumed. The other is a simulation of the transition into the weak shock regime. In the strong shock regime shown, the pressure is greater than atmospheric pressure throughout the expanding volume, and there are pressure fluctuations from internal waves that move back and forth between the origin and the shock front. The internal waves appear to organize internal properties of the expanding region to allow steady progression of the front.

When, as shown in Fig. 5.14, the shock wave expands beyond a radius approximately equal to R_C or R_S, depending on the appropriate geometry, pressure exceeds 1 atm only in the leading shock front; the momentum of the expanding gas carries it outward and forces the central pressure to drop below 1 atm. As the weak shock continues to propagate outward, the central pressure returns to 1 atm, and the shock wave decouples from the hot channel. The propagating pulse consists of a compression followed by a rarefaction, and so is an n-wave. The pressure pulse eventually weakens to the point that it can be treated by linear perturbation theory, and it begins to propagate as a sound wave.

There is still disagreement about the details of energy input for thunder generation and about whether spherical or cylindrical geometry is more appropriate for treating divergence of the shock wave from tortuous lightning channels. Few (1969, 1982), for example, suggested that independent acoustic sources (i.e., mesotortuous segments) should be modeled as a series of short cylindrical sources close to the channel, but that at a dis-

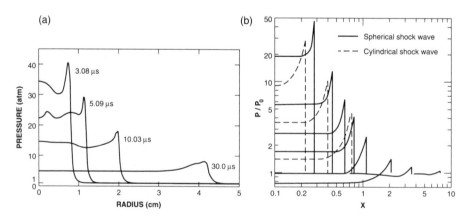

Fig. 5.14. Numerical simulations of overpressure versus distance for different times in the expansion of a shock wave. (a) Strong shock development with cylindrical divergence and input energy of 1.5×10^4 J m^{-1}. Note internal waves to the left of the shock front. (From Hill, 1971: American Geophysical Union, with permission.) (b) Expansion of spherical and cylindrical shock waves from strong shock regime to weak shock regime. X is a normalized distance equal to r/R$_c$ or r/R$_s$. (From Few, 1968, with permission.)

tance greater than roughly the length of the mesotortuous segment, further divergence would become approximately spherical. He also suggested that the distance at which this transition occurs is roughly equal to R_C.

Several measurements appear to support Few's (1969, 1982) theory. Dawson et al. (1968a) found that overpressures measured by Hill and Robb (1968) at a range of 0.55 m from artificially initiated lightning appeared to be consistent with transition from cylindrical to spherical expansion if the input energy was 10^5 J m^{-1}—a typical estimate of input energy from other measurements (e.g., Uman 1987). Uman et al. (1970) specifically tested Few's theory by comparing it with measurements of the input energy and pressure of a shock wave from a 4-m long spark. They reported that, at distances greater than R_C, the measured pressure was consistent with spherical divergence and was smaller than would be predicted for cylindrical divergence. Furthermore, Few (1969) used his theory to relate the peak frequency of thunder spectra to the input energy. The resulting input energies inferred by Holmes et al. (1971) from their measurements of thunder spectra (with the exception of cases having infrasonic peaks) appear consistent with estimates of input energy made by other means. Similarly, Few's theory correctly reproduced the relationship between peak acoustic frequencies and input energy measured by Dawson et al. (1968b) for a 4-m laboratory spark.

However, other interpretations are possible. Plooster (1971a,b) and Hill (1971, 1977b, 1979) presented numerical simulations of thunder generation using detailed lightning channel physics and cylindrical divergence and pointed out that Uman et al.'s (1970) measurements were consistent with cylindrical divergence, if the effective input energy was an order of magnitude smaller than measured. Plooster (1971a) argued that there was an order of magnitude error in Uman et al.'s measurement of input energy. Hill (1979) stated that only 10% of the energy goes into the shock wave; the remainder is deposited diffusely around the spark channel and produces no sound. However, Hill's (1977a) main point was not that cylindrical symmetry is correct; his main point was that there are enough uncertainties in our understanding and measurements of processes involved in thunder generation by a tortuous lightning channel that the quantitative aspects of Few's (1969, 1982) theory are suspect. Few (1982) argued that Uman et al. (1970) did not make an order of magnitude error in their measurements, that Hill (1971) used a current rise time an order of magnitude slower than typically has been measured with modern instruments, and that it is unreasonable to expect silent dissipation of 90% of the input energy. However, there have been no published experiments specifically designed to test Plooster's, Hill's, or Few's hypotheses further.

Besides being affected by the characteristics of lightning channels, thunder also is affected by propagation through the atmosphere. Few (1982) discussed the primary effects in considerable detail. We will list several of these, but will not present a detailed description.

The spectrum of thunder from a given flash tends to be shifted toward lower frequencies as the distance of propagation increases. Several processes cause this shift, and their effects can vary with range and with the characteristics of the medium through which thunder propagates:

1. The wavelength of large amplitude thunder increases as it propagates. Few (1982) estimated that amplitudes of thunder large enough to cause this lengthening occur primarily within 1 km of a lightning flash, and he estimated that the resulting decrease in frequency is no more than a factor of 2.
2. Viscosity, heat conduction, and molecular absorption attenuate acoustic waves, but attenuation by molecular absorption dominates viscosity and conduction and is a function of humidity, decreasing with increasing humidity. Over ranges at which thunder typically is audible, attenuation is insignificant at frequencies below 100 Hz, but increases at higher frequencies. Few (1982) estimated that thunder at 1 kHz is attenuated by a factor of 2 over a range of 10 km under typical atmospheric conditions. Bass and Losely (1975) estimated that thunder is attenuated by a factor of 3 at 400 Hz, 50% relative humidity, and 20°C at a range of 5 km, and that attenuation at 400 Hz increases by approximately a factor of 3 over a range of 2 km when relative humidity increases from 20% to 100%.
3. Scattering from turbulent eddies also increases with increasing frequency, but the increase with frequency is much less rapid than for attenuation from molecular absorption (Few 1982, Brown and Clifford 1976).
4. Cloud hydrometeors interact with thunder. First, they increase viscosity and heat conduction and so also increase thunder attenuation. Second, acoustic pressure perturbations produce small changes in the equilibrium rates of condensation and evaporation of hydrometeors. The resulting release or absorption of heat causes pressure changes within an acoustic wave opposite to the acoustic pressure perturbations, and so diminishes the wave amplitude. Third, hydrometeors change the speed of sound and the acoustical impedance of cloudy air and so may cause partial reflection at cloud boundaries. Few (1982) estimated that the first two effects of hydrometeors can increase attenuation by a factor of 10 or more over attenuation in cloud-free air, depending on the type, size, and number density of cloud particles. Few also pointed out that attenuation of acoustic waves at low frequencies is caused primarily by turbulent scattering and interactions with cloud hydrometeors.

Another important propagation effect is refraction, which typically governs the maximum range at which thunder can be heard from a lightning flash. Acoustic rays in the troposphere are curved upward by the ver-

Fig. 5.15. Acoustic rays tangent to the surface for an atmosphere with a lapse rate of 7.5°C km^{-1} and no wind. S$_1$ and S$_2$ are acoustic sources. As a ray propagates to the right from the acoustic source, it curves upward until it becomes tangent to the surface at the corresponding $\mathcal{T}$. Beyond $\mathcal{T}$, sound cannot be heard at the surface from that source.

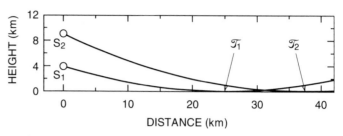

The ray tangent to the surface from a higher source is tangent at a longer range, so higher sources can be heard farther away. Rays emitted in other directions from the sources either hit the surface before reaching $\mathcal{T}$ or never reach the surface.

tical gradient of temperature. An acoustic ray originating at some height will become tangent to the surface of Earth at some point and curve back into the atmosphere at greater ranges; thunder from a source at a particular height cannot be heard beyond the range at which the ray becomes tangent to the surface (Fig. 5.15). Fleagle (1949) showed that sound rays originating at or below a height of 4 km cannot be heard beyond a range of 25 km when the temperature decreases with height by 7.5°C km^{-1}. However, Fleagle also showed that a layer of temperature inversion can cause the range to be extended substantially. Wind shear can produce even larger ray curvature than the temperature lapse rate (Fleagle, Teer 1973, Few and Teer 1981), and so can further limit the range at which thunder can be heard. A theoretical treatment of ray curvature produced by temperature lapse rate and wind shear is given in Section 6.10.3. Measurements of the maximum range at which thunder can be heard suggest that the maximum usually is no more than 25 km, but can occasionally be up to 100 km (e.g., De L'Isle 1783, Veenema 1920). In windy conditions, the maximum range thunder can be heard is something much less than 25 km; for example, Ault (1916) reported one instance in which thunder could not be heard beyond 8 km.

5.11.3. Electrostatic generation of infrasonic thunder

Although Holmes et al. (1971) found that the peak of the Fourier spectrum of thunder can be at infrasonic frequencies, they were unable to detect an infrasonic peak in thunder from many flashes. Even when a flash did have an infrasonic peak, the frequency of the peak changed with time and was infrasonic for only part of the flash's thunder signature. Similarly, Bohannon et al. (1977) and Balachandron (1983) found that infrasonic thunder was observed preferentially underneath thunderstorms, and it arrived as a discrete pulse.

According to Few (1969) and Holmes et al. (1971), channel heating requires more input energy than is available in lightning channels to produce thunder with a peak at infrasonic frequencies. Instead, Holmes et al. and Dessler (1973) suggested that infrasonic thunder can be generated by the electrostatic energy stored in a region of charge. Benjamin Franklin was the first to

suggest that electrostatic processes could produce thunder. Wilson (1920) estimated the pressure perturbation caused by the change in electrostatic forces when lightning neutralizes a volume of charge, and he suggested that it could make a substantial contribution to thunder.

Initial forcing for the electrostatic mechanism is provided by the mutual repulsion felt by charged particles and ions of the same polarity in a region of net charge. The particles transmit this force to air through their frictional drag on the air. The resulting outward expansion leaves behind a pressure deficit at the center of the charged region and continues until the inward pressure exerted by the atmospheric pressure gradient balances the outward pressure exerted by the electric force. Dessler (1973) calculated the pressure due to electrostatic force for particular charge geometries and found that, at the midplane of a flat disk of charge Q of radius R, the pressure P_E is given by

$$P_E = \frac{Q^2}{8\,\pi^2\varepsilon\,R^4} \; . \tag{5.12}$$

Only the charge neutralized by lightning need be considered, because the effect of other charges is the same before and after the flash, and so is not involved in creating thunder.

When charge is neutralized by lightning, the outward electric force produced by the charge vanishes, and the opposing pressure gradient force causes the volume to collapse, creating an acoustic rarefaction wave. The collapsing volume tends to overshoot equilibrium and produce a subsequent compressional wave before reaching pressure equilibrium, although the compression may be weak. According to Dessler (1973), the time needed to restore equilibrium in the collapsing volume is of the order of 1 s, so the resulting acoustic wave also will have a period of the order of 1 s. Furthermore, Dessler showed that, if the horizontal dimension of the neutralized charge is much greater than the vertical dimension, the electrostatic mechanism will tend to beam acoustic energy vertically. The infrasonic wave then will be detected at the ground only by sensors almost directly beneath the flash, and so will be observed sporadically. This prediction was confirmed experimentally by Bohannon et al. (1977) and Balachandron (1983).

The main problem with this theory is that it predicts an initial rarefaction wave, while all reported infrasonic pulses have begun with a compression. Bohannon (1980) suggested that the initial compression could be generated by releasing an attractive electrostatic force between two regions of opposite charge neutralized by a lightning flash. Few (1985) suggested that lightning neutralizes a region of charged hydrometeors by producing an extensive network of small electric discharges that permeate and heat the region, resulting in the observed initial infrasonic compression. Both Bohannon (1980) and Few (1985) suggested that the subsequent rarefaction is caused by the mechanism modeled by Dessler (1973).

5.12. ELECTROMAGNETIC RADIATION FROM LIGHTNING

5.12.1. Brief overview

There is a large body of literature that describes several decades of measurements of the signals radiated from lightning. A concise review of electromagnetic signals from natural lightning is found in Uman and Krider (1982); in addition there is a large body of detailed literature (summarized in Uman 1987). We summarize the basics here; there are additional aspects in the description of lightning mapping techniques (Sec. 6.10). A lightning flash is a broadband electromagnetic radiator. Ground flashes have a strong, low-frequency component, which is one reason why a broadcast band AM radio tuned to its lower end works as a rudimentary lightning detector. The peak in frequency of a return stroke is around 10 kHz. The amplitude of the different frequencies varies during a flash and is dependent upon the process(es) under way at any instant. A long standing rule of thumb is that above 10 kHz, the frequency amplitude decreases linearly on a log-log plot of signal amplitude versus frequency normalized to a given distance (e.g., 10 km) and a given bandwidth (e.g., 1 kHz).

This seems true at least out to about 10 MHz. Until recently, high-frequency measurements were seldom obtained close to lightning, so the historic data likely are biased against frequencies above 1 MHz, whose amplitudes attenuate with distance much more than lower frequencies.

Krider (1992) corrected the theory describing radiation of lightning to handle the return stroke propagation at a large fraction of the speed of light. At distances where the radiation field dominates the measured return stroke signal, a correction factor is needed in the radiation term. Krider reaffirmed the result by Krider and Guo (1983) that the total power in the limit where the velocity is much less than the speed of light is

$$P_{tot} = \frac{4\pi R^2}{3} \frac{E_{pk\,S}^2}{\mu c}, \tag{5.13}$$

where R is the range and $E_{pk\,S}$ is the peak radiation field on a hemispherical surface, S, above the strike. From their measurements and this equation, Krider and Guo and Krider found peak powers of $1–2 \times 10^{10}$ W for first return strokes. Using the geometry shown in Fig. 5.16 and analysis of the transmission line model (TLM) in Le Vine and Willett (1992), Krider's more general version of the above equation was expressed in terms of the ratio of the velocity of the channel to c, $\beta = v/c$. Then the total power is

$$P_{tot} = \frac{Z_0}{8\pi} I_{pk}^2 \left[\frac{1+\beta^2}{\beta} \ln \left(\frac{1+\beta}{1-\beta} \right) - 2 \right], \tag{5.14}$$

where $Z_0 = (\mu/\varepsilon)^{0.5}$ and I_{pk} is the peak current in the return stroke. Krider gave a typical example with $I_{pk} = 30$ kA and $\beta = 0.5$ that yielded a $P_{tot} = 1 \times 10^{10}$ W.

Krider also derived the following for the peak electric radiation field, E_{pk}:

$$\vec{E}_{pk}\left(t_{pk} + \frac{R}{c}\right) = Z_0 \frac{I_{pk}t_{pk}}{4\pi R} \left[\frac{\beta \sin\theta}{1 - \beta\cos\theta} + \frac{\beta \sin\theta}{1 + \beta\cos\theta} \right] \hat{\theta}$$

$$= Z_0 \frac{I_{pk}t_{pk}}{2\pi R} \left[\frac{\beta \sin\theta}{1 - \beta^2\cos^2\theta} \right] \hat{\theta}, \tag{5.15}$$

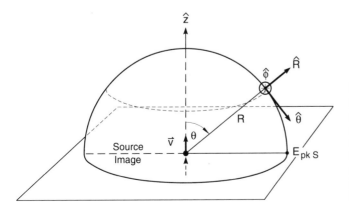

Fig. 5.16. Depiction of hemisphere throughout which a return stroke in its early stage radiates. The strike point is the dot at the source-image boundary, i.e., a conducting flat plane. The encircled + on the hemispherical surface is where the radiation or power is calculated. (From Krider, 1992: American Geophysical Union, with permission.)

where t_{pk} is the time of the peak and R is the distance from the strike point. Similarly the peak magnetic radiation field is

$$\vec{H}_{pk}\left(t_{pk} + \frac{R}{c}\right) = \frac{I_{pk}t_{pk}}{2\pi R}\left[\frac{\beta \sin \theta}{1 - \beta^2 \cos^2 \theta}\right]\hat{\theta}. \text{ (5.16)}$$

This equation is valid for the TLM of a stroke early in its development. The stroke radiates into the hemisphere above the strike point, which is on an infinite, flat, conducting Earth. That the model is generally applicable, at least to some degree, is supported by observations showing that the largest radiation often occurs at the beginning of the first return stroke and that radiation seems reasonably described by the TLM. Krider noted that different correction factors than his may be needed to handle other situations; the relevance here is that calculation of the radiation term from lightning depends on several factors that include acceleration of the stroke, simultaneous wave fronts, and discontinuities in the channel geometry. Aspects of the validity of the TLM were addressed by Le Vine and Willett (1992).

A comparison of a close, severe ground flash with the electromagnetic energy from an exoatmospheric nuclear explosion by Uman et al. (1982) showed that the electric field of a return stroke channel just above the ground exceeded that of a nuclear electromagnetic pulse for frequencies below 1 MHz. Examples of why high-frequency radiation is important include coupling into flight-critical control systems in modern airplanes and space vehicles and damage and substantial economic losses for sophisticated equipment in ground-based facilities (e.g., modern weather sensor systems).

5.12.2. Measurements of lightning waveforms

The electric field change for a lightning flash is a fundamental measurement made in many lightning and thunderstorm studies because the field change for the whole event can often reveal important physical parameters as well as determine if it is a cloud or ground flash. We will show examples and introduce the various component parts of a lightning flash (for a detailed review, see Uman 1987). Such measurements can be made with simple electronic circuits. More information is given in Section 6.9.1, but we introduce you to the terms *slow antenna* and *fast antenna*. The slow antenna reproduces the total field change from an entire flash, whereas a fast antenna is designed to reproduce the faster field changes that are characteristic of some of the component parts of a flash. As the examples will show, we can determine some types of flashes from the slow antenna recording. Please note that in this section we reintroduce some terms for parts of a flash that you have seen before.

(i) Electric field change from a negative ground flash.

We begin with an example of a ground flash whose effect is to lower negative charge from the cloud to the ground. We reproduce two examples of field changes and photographic recordings by Kitagawa and Brook (1960) that show the manifestation of leader and return stroke(s), which are seen in Fig. 5.17. In most of the literature, the field change for a negative return stroke has been shown as positive, which is really the change in potential gradient. We show the lightning field changes in our usual electric field polarity. The following refer to aspects of ground flashes such as are seen in Figs. 5.3 and 5.17. Perhaps the most obvious are the large, rapid changes in E for the return strokes (marked R). Such a negative increase in E is the result of the flow of a negative monopole of charge from the cloud to the ground. The slope of the leader field change, which is also negative in the figure, depends on the distance from the leader to the sensor of the field change. As the leader gets closer to the sensor, the slope of the field change becomes less, and when the leader occurs within a few kilometers (horizontally) from the sensor, the slope reverses to positive. The direction of the return stroke field change is unaffected by distance, as it is equivalent to the lowering of a point charge. If there are subsequent return strokes before the first channel decays, they are initiated by a dart leader. The dart is faster than the stepped leader. The entire interval in between return strokes are in-cloud processes called the J (junction) change, which manifests itself as a slowly varying field change between strokes. The J change is a result of more breakdown inside the cloud—both new channels and recoil streamers. The slope of a J change depends on channel orientation and distance from the sensor. If we examine the field change recorded by a fast antenna, we see there are rapid pulses within the J changes. Such changes are called K changes and have been attributed to recoil streamers coming back along previously ionized channels. On a slow antenna record, the summation of many fast field changes can make the appearance of a slowing varying change. The ΔE after the final return stroke has been called the F change. In recent years, various of these processes have been studied in detail with VHF mappers, optical detectors, and video as reviewed by Mazur and Ruhnke (1993). From his analysis of natural and airplane-triggered flashes, Mazur (1989b) hypothesized that recoil streamers retrace the positive leader channels. This is supported by the VHF mapping data that show them moving toward the initiation place of the flash. Dart leaders and K changes were postulated to be identical processes, except that the dart reaches the ground. This has received confirmation in observations by Mazur et al. (1995).

From a practical viewpoint, a very important parameter is shown in the ground flash in the lower part of Fig. 5.17. It is labeled C for continuing current. Kitagawa and Brook (1960) called a flash a *hybrid* if it contained continuing current, and *discrete* if it did not. As the smearing of luminosity from a moving-film camera clearly shows, the continuing current makes the channel stay illuminated a much longer time than the return stroke. Continuing current carries tens to a few hundred

DISCRETE FLASH (Flash No.109, 19 km distant)

PHOTOGRAPHIC RECORD

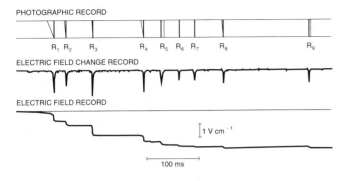

R_1 R_2 R_3 R_4 R_5 R_6 R_7 R_8 R_9

ELECTRIC FIELD CHANGE RECORD

ELECTRIC FIELD RECORD

$\updownarrow$ 1 V cm^{-1}

|——— 100 ms ———|

HYBRID FLASH (Flash No.106, 20 km distant)

PHOTOGRAPHIC RECORD

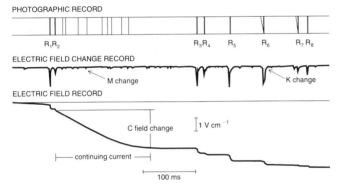

$R_1 R_2$ $R_3 R_4$ R_5 R_6 $R_7 R_8$

ELECTRIC FIELD CHANGE RECORD

M change   K change

ELECTRIC FIELD RECORD

C field change $\updownarrow$ 1 V cm^{-1}

|——— continuing current ———|

|——— 100 ms ———|

Fig. 5.17. Sketch of simultaneous photographic record from streak camera and field changes from fast and slow antennas of two ground flashes. Other typically named processes are shown. Flash at top ≈ 19 km distant. The lower record shows a flash with continuing current, C. (After Kitagawa et al., 1962: American Geophysical Union, with permission.)

amperes (comparable to an electric arc welder). The duration of a continuing current ranges from a few tens to a few hundreds of milliseconds. Thus the channel can stay attached long enough to bring combustibles such as trees up to ignition temperature and often to melt through metals. Continuing currents are a major cause of forest and grassland fires. So the risk from a lightning strike is twofold: (1) explosive from the rapid heating of the high-current, very short duration return stroke, and (2) burning and melting from the lower amplitude, but longer duration continuing current. During continuing current, there can be transient increases in E that show in the fast antenna record. These are M changes that flow within the continuing current. Their sensors showed them to be the same in appearance as the transient, higher current pulses that flow during the return stroke each time the upward wave reaches a branch that was created by the stepped leader, which were called M components. Mazur et al. (1995) used high-speed video and a VHF interferometer to study M changes. They found M events consist of a fast negative streamer that propagates in and brightens the return stroke channel. They also found that M streamers are the same as in-cloud K events and dart leaders ex-

cept that the M streamers contact the conducting, continuing-current channel and propagate to ground.

(ii) Electric field change from a positive ground flash. Flashes that lower positive charge to Earth are now known to occur routinely, although usually as only a small percentage of all ground flashes. It is now widely accepted that they occur a higher percentage of the time in winter thunderstorms than in summer thunderstorms. In the warm season, they can be a high percentage of the ground flashes in the stratiform region of mesoscale convective systems and in certain scenarios of severe convective storms. The determination that a flash to ground was positive is often not possible with just a slow antenna record. Before we knew that positive flashes were common, the distinction between ground and cloud flashes from a slow antenna was often easy. Now, however, we know that a positive ground flash has a field change that is usually devoid of conclusive evidence of a return stroke (Fig. 5.18). It was not until combined electric field change, optical, and video data were analyzed by Beasley et al. (1983a) that we even had conclusive proof that naturally occurring positive ground flashes really contained a re-

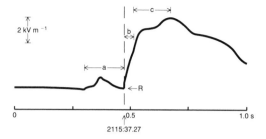

Fig. 5.18. Slow antenna field change from a positive ground flash about 4 km distant. The calibrated field change scale is shown. Interval 'a' is the preliminary breakdown and leader; R denotes the return stroke, which was confirmed from a fast antenna record and a video image. The interval 'b' is the minimal duration of continuing current (see Fig. 5.5), and 'c' is a longer interval of additional continuing current or in-cloud breakdown. (After Rust et al. 1985, with permission.)

turn stroke. An important practical advancement has been that the lightning ground strike mapping systems were modified to identify positive ground strikes.

A typical field change of a positive flash consists of a slow ΔE of tens to hundreds of milliseconds prior to a return stroke that is itself followed by additional slow change. The latter slow change apparently often results from continuing current and more extensive in-cloud breakdown, which may be inferred in Fig. 5.18. The return stroke field change is a small fraction of the total electric field change. The overall field change is easily confused with a cloud flash—thus the difficulty in identifying a positive ground flash from a slow antenna alone. Even though a fast antenna record often clearly indicates a return stroke, its occurrence can be confused with other fast processes (Beasley et al. 1983a). Most positive ground flashes are single stroke.

(iii) Electric field change from a cloud flash. We show in Fig. 5.19 an example of field change recordings from a cloud flash. The ΔE can be of either polarity, and often has slopes of both polarities within a single flash. This can due to the horizontal or vertical movement of charge as the flash evolves. Visualize an example as having your slow antenna being ap-

proached by a negative in-cloud streamer that moves toward, over, and then away from it. The slopes before and after it moves overhead will be opposite. In their original work, Kitagawa and Brook (1960) divided the cloud discharge into initial, very active, and J-type portions. Krehbiel (1981) concluded that K changes could occur throughout the duration of a cloud flash. Both polarities of K change are observed. Although Fig. 5.19 has been used extensively in the literature as an example of a cloud flash, we now know that the appearance of such a record is a function of the bandwidth of the instrument and recorder used. For example, Bils et al. (1988) noted that such displays do not show the microsecond-scale features. Villaneuva et al. (1994) also divided the cloud flash into two segments only. From their analysis of several of the flashes recorded by Kitagawa and Brook, Villaneuva et al. concluded that a cloud flash should be divided only between an early and a late stage. Further clarification likely awaits comparison of high-resolution (time and space) discharge mapping coincident with wideband waveforms. Such results are becoming available.

5.13. ELECTRICAL CURRENT FLOW IN LIGHTNING

5.13.1. Estimating current from theoretical calculations

When using lightning measurements to solve a variety of problems, it is necessary to be able to calculate key physical parameters in the lightning based upon the remote detection of the flash (i.e., measurement of the electric or magnetic field changes produced by the flash). Historically, this has been particularly important for lightning strikes to ground, but both cloud and ground flashes can have deleterious effects. We shall give only the basics from this large area of work. Here we move from being able to use theory applicable to static and only slowly varying changes in lightning to theory that describes the fast changing signals generated by lightning. The starting place is, not surprisingly, Maxwell's equations. From them are derived the time-varying electric and magnetic fields as a function of coordinates and time-varying current; the case considered is for a vertically oriented channel above an infinite plane, which Earth approximates. We show here the equations for the field changes from lightning developed by Master et al. (1981):

$$d\vec{E}(r, \phi, z, t) = \frac{dz'}{4\pi\varepsilon}\left\{\left[\frac{3r(z - z')}{R^5}\int_0^t i\left(z', \tau - \frac{R}{c}\right)d\tau\right.\right.$$
$$\left.\left. + \frac{3r(z - z')}{cR^4}\, i\left(z', t - \frac{R}{c}\right)\right.\right.$$

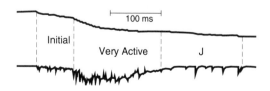

Fig. 5.19. Field changes of a cloud flash from slow and fast antennas. (After Kitagawa and Brook 1960: American Geophysical Union, with permission.)

$$+ \ \frac{r(z - z')}{c^2 R^3} \ \frac{\partial i\left(z', t - \frac{R}{c}\right)}{\partial t} \Bigg] \hat{r} \quad (5.17)$$

$$+ \left[\frac{2(z - z')^2 - r^2}{R^5} \int_0^t i\left(z', t - \frac{R}{c}\right) d\tau \right.$$

$$+ \ \frac{2(z - z')^2 - r^2}{cR^4} \ i\left(z', t - \frac{R}{c}\right)$$

$$\left. - \ \frac{r^2}{c^2 R^3} \ \frac{\partial i\left(z', t - \frac{R}{c}\right)}{\partial t} \right] \hat{z} \Bigg\}$$

for the electric field change, and for the magnetic field change:

$$d\vec{B}(r, \phi, z, t) = \frac{\mu dz'}{4\pi}$$

$$(5.18)$$

$$\cdot \left[\frac{r}{R^3} i\left(z', t - \frac{R}{c}\right) + \frac{r}{cR^2} \frac{\partial i\left(z', t - \frac{R}{c}\right)}{\partial t} \right] \hat{\phi},$$

where the symbols are defined previously and the notation is shown in Fig. 5.20. The expression $\tau = t - R/c$ is called the *retarded time,* which is the time required for the radiation to propagate from the source point to the point where it is measured. In the electric field expression, there are three components. The component containing the integral of current is the electrostatic field, which is from the movement of charge. The electrostatic component has the strongest dependence on range and tends to dominate at very close range. The component containing current is the next most depen-

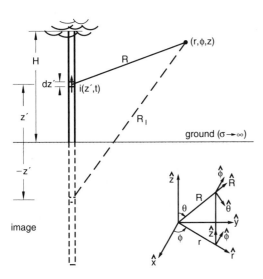

Fig. 5.20. Geometry and coordinate system for field calculations. (From Master et al. 1981: American Geophysical Union, with permission.)

dent on range and is the induction component. The component least dependent on range is dominant at far distances; it is called the radiation component and contains the terms with the partial derivative of the current. For the magnetic field, the component containing the current is called either the induction or the magnetostatic. The term with the derivative of current is the magnetic radiation and dominates far from the source. These equations are applicable above Earth's surface, as well as in the special case at ground level.

Most measurements have been and continue to be made at Earth's surface (i.e., at $z = 0$). For such measurements, the equations describing the electric and magnetic fields are

$$\vec{E}(r, \phi, 0, t) = \frac{1}{2\pi\varepsilon} \left[\int_{H_B}^{H_T} \frac{(2z'^2 - r^2)}{R^5} \right.$$

$$\cdot \int_0^t i\left(z', \tau - \frac{R}{c}\right) dt \ dz' \qquad (5.19)$$

$$+ \int_{H_B}^{H_T} \frac{(2z'^2 - r^2)}{cR^4} \ i\left(z', t - \frac{R}{c}\right) dz'$$

$$\left. - \int_{H_B}^{H_T} \frac{r^2}{c^2 R^3} \ \frac{\partial i\left(z', t - \frac{R}{c}\right)}{\partial t} \right] \hat{z}$$

and

$$\vec{B}(r, \phi, 0, t) = \frac{\mu}{2\pi} \left[\int_{H_B}^{H_T} \frac{r}{R^3} i\left(z', t - \frac{R}{c}\right) dz' \right.$$

$$\qquad (5.20)$$

$$\left. + \int_{H_B}^{H_T} \frac{r}{cR^2} \ \frac{\partial i\left(z', t - \frac{R}{c}\right)}{\partial t} \ dz' \right] \hat{\phi}$$

(Uman 1987, p. 137). As above, the three terms in $\vec{E}(r, \phi, 0, t)$ are the electrostatic, induction, and radiation components. In $\vec{B}(r, \phi, 0, t)$, the two terms are magnetostatic and radiation. An early development to place such calculations in the time domain was made by Uman et al. (1975).

From such analytical expressions as are shown above come models that are used to determine the current or velocity of channel propagation from measured E or B changes. There are several models, both analytical and numerical, which we summarize later in this chapter.

5.13.2. Measurements of current in lightning

The direct measurement of the current that flows in lightning has long been important for application to the design of lightning protection for all sorts of equipment, structures, facilities, and aircraft. For years, measurements on return strokes dominated, which is where we begin this summary. Widely used current measurements, which are still in use today in many lightning protection schemes, were made by K. Berger and associates at Mount San Salvatore near Lake Lugano, Switzerland. He used photography and towers instrumented with resistive shunts to compile statistics on the distribution of peak current and waveforms. His measurements led to the concept of the four types of ground flashes presented earlier in this chapter; Berger published several reviews of the extensive work at Mount San Salvatore (e.g., Berger 1967, 1977, Berger et al. 1975). Most measurements have been on negative ground flashes, but there are measurements on positive flashes. A more recent and extensive set of measurements with resistive shunts on instrumented towers has come from Italy by Garbagnati and Lo Piparo (1982). A different technique was used by Eriksson (1978) in South Africa. He connected an isolated 60-m high tower through an open-core transformer to ground. As he intended by design, many strokes measured by Eriksson were from downward negative leaders (i.e., what we call natural lightning). In his data, the time for the current to reach its peak in subsequent strokes often was ≤0.2 μs, which was the limit of his system. He questioned the validity of current risetimes recorded with slower techniques that were prevalent in the literature. His measured maximum current rise rate was 180 kA μs^{-1}, which may still be one of the fastest ever measured on a tower. An interesting and likely significant contrast is found in the calculated mean current from electric field by Weidman and Krider (1984). They found an average the same as Eriksson's extreme value and a maximum about twice as large. Thus it seemed as if the measurement on a tower can dramatically reduce the apparent maximum severity of a flash, at least as far as rate of rise is concerned. Later, Leteinturier et al. (1990) found more evidence that natural lightning might be more severe than indicated by tower measurements and also found an excellent correspondence between their calculated I_{pk} from integration of di/dt and the directly measured values of I_{pk}. The correlation was 0.91 for the slope of the fitted curve of 1.05 (1.0 would be a perfect correspondence). Results from a model by Diendorfer and Uman (1990), which is discussed later, indicated that a possible reason was that with an assumed stroke velocity of 1 × 10^8 m s^{-1}, the TLM overestimated the current rise rate. This issue was not resolved.

The distributions of peak currents from Mount San Salvatore are shown in Fig. 5.21. Those distributions were the foundation for believing that positive strokes were more damaging as a category, but it was long thought that this applied only to triggered positives. There is now evidence to the contrary. Garbagnati and Lo Piparo (1982) found that downward leaders followed by negative return strokes yielded an average of 33 kA for first strokes and 18 kA for subsequent strokes. Their distribution for negative flashes was about the same as Berger et al.'s (1975). They also reported two categories of positive flashes: seven flashes

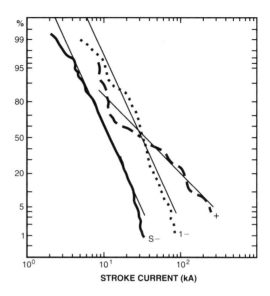

Fig. 5.21. Distribution of stroke currents measured on instrumented towers at Mount San Salvatore, Switzerland. Curve 1− is from negative first strokes, S− is subsequent negative strokes, and + is positive strokes. The straight lines are least squares fit to the points. (From Berger et al. 1975, with permission.)

with peak currents <10 kA and five flashes with ≥30 kA, which are the ones in their table and are presented here. Berger put all 26 of his positive flash current measurements in the same data set, and he had values <4.5 kA. Whether this was due to the existence of two categories or to the small sample size is not clear.

Shown in Table 5.5 are example parameters from lightning return stroke currents.

The charge in a return stroke can be determined by integrating the current over time. A decision has to be made about whether the charge is due just to the stroke impulse current or whether any following current is included. Examples are given in Table 5.6.

The values from Berger's measurements in both tables are not necessarily the same as those derived from curves he published in the widely used Berger (1967); the values here came from his analysis of data covering 1963–71.

With the coming of lightning ground strike locating systems (see Sec. 6.10.7) has come the opportunity to use models to calculate the peak stroke currents for a huge data base such as that from the U.S. National Lightning Detection Network (NLDN). An example of trying to calibrate the NLDN for such use is in Orville's (1991) comparison of direct measurement of 18 triggered flashes in Florida with six nearby magnetic direction finders. He used the TLM, which we describe later. For such networks working at high gain, Orville found the peak current was related to the so-called normalized LLP units (see Section 6.10.7) that the NLDN

Table 5.5 Lightning peak currents measured with instrumented towers and rocket-triggered lightning. The polarity is that on the leader and of the charge effectively lowered to ground in the stroke. The values for Berger are medians, whereas those for Garbagnati are means and logarithmic standard deviations (and do not include positive strikes with <10 kA). Eriksson's values are medians with a log standard deviation. Values by Fisher are means and logarithmic standard deviations from all strokes in triggered lightning flashes.

Investigators	Polarity	n_1	First Stroke (kA)	Extreme 5% level (kA)	n_s	Subsequent Stroke (kA)	Extreme 5% level (kA)
Berger et al. (1975)	−	101	30	80	135	12	30
Berger et al. (1975)	+	26	35	250	—	—	—
Garbagnati and Lo Piparo (1982)	−	42	33 ± 0.25	87	33	18 ± 0.22	33
Garbagnati and Lo Piparo (1982)	−	61	7 ± 0.23	16	142	8 ± 0.22	19
Garbagnati and Lo Piparo (1982)	+	5	{30–160}	—	—	—	—
Eriksson (1978)	−	11	41 ± 0.7	165	—	—	—
Fisher et al. (1993)	−	—	—	—	45	12 ± 0.28	29

uses to record the peak magnetic field. The *LLP units* were normalized to 100 km by the system. The relationship he found was

$$I_{pk} = 2.3 + 0.19 \times (LLP), \quad (5.21)$$

where I_{pk} is in kiloamperes. He noted that further simplification to $I_{pk} = 0.2 \times (LLP)$ resulted in an error of 6% or less. Because of the limitations of his small data set, he cautioned against using the relationship for $I_{pk} > 60$ kA, as only lower currents went into the regression analysis.

In their subsequent examination of the calibration of the NLDN, Idone et al. (1993) used a data set of 56

triggered flashes from Florida and found the relationship

$$I_{pk} = 4.20 + 0.171 \times (\overline{LLP}), \quad (5.22)$$

where $\overline{LLP}$ is the mean of the normalized signals from all direction finders that recorded the flash. This regression was based on data between 15 and 60 kA, within which the uncertainty in predicted I_{pk} was <15%. Below 15 kA, the errors increased rapidly; for values above 60 kA, they stated that the errors likely remained at <15%. Their objective was to provide an independent verification of the initial work by Orville (1991) and to extend the data set to include new cases of direct

Table 5.6 Impulse charge (except where noted) in return strokes calculated from current measurements with towers and rocket-triggered lightning. The polarity is that on the leader and of the charge effectively lowered to ground in the stroke. The values for Berger are medians. Those for Garbagnati are means and logarithmic standard deviations and for charge flow either to peak current or for 500 μs. Values in { } are the measured range and are used when no median or mean was published. The total charge in the stroke, which is impulse plus that in continuing current, is only slightly larger in the Berger data except for positive strokes. [†]The total charge (as opposed to impulse charge) in positive first strokes had a median value of 80 C and a 5% extreme value of 350 C. [‡]Total charge from stroke impulse plus continuing current. Values by Fisher are from all strokes in triggered lightning flashes.

Investigators	Polarity	n_1	First Stroke (C)	Extreme 5% (C)	n_s	Subsequent Stroke (C)	Extreme 5% (C)
Berger et al. (1975)	−	90	4.5	20	117	0.9	4
Berger et al. (1975)	+	25	16	150[†]	—	—	—
Garbagnati and Lo Piparo (1982)	−	42	2.8 ± 0.35	10	33	1.4 ± 0.37	10
Garbagnati and Lo Piparo (1982)	−	61	0.5 ± 0.30	1.3	142	0.6 ± 0.32	1.4
Garbagnati and Lo Piparo (1982)	+	5	(15–50)	—	—	—	—
Fisher et al. (1993)	−	—	—	—	65	2.5 ± 55[‡]	15

current measurements on ground flashes. The resulting equation gave only slightly different results from the first.

Idone et al. (1993) normalized each value of the signal from the direction finders to 100 km using the one-over-range dependence of the radiation component of the signal from each stroke. They found only a negligible difference when they varied the dependence on distance from $R^{-0.85}$ to $R^{-1.4}$, which showed that the R^{-1} normalization predicted by theory was adequate. Idone et al. also renormalized their data using an exponential distance dependence from another propagation model used in Canada by Herodotou et al. (1992). Again they found no statistically significant improvement over an R^{-1} dependence on distance to the stroke. Including a nonlinear variation of peak current with return stroke velocity was not supported by Willett et al.'s (1988, 1989) results. Whether this will still be the case for signal propagation over terrain other than that in Florida or when an even larger data set is processed is unknown. They noted that sources of uncertainty included 10% in the measurement of I_{pk}, a limited range of I_{pk}, the distribution of errors in I_{pk} and $\overline{LLP}$ were unknown, and only 56 data points went into the regression.

While we wrote this, changes were being made to the U.S. NLDN. National networks of lightning ground strike locators are an evolving but already important tool in a variety of operational applications. We will need to be mindful of network changes, calibration issues, etc. as dependence on and diversity of use of these data in applications and operational meteorology increases throughout the world. It is important to recognize that if we can calibrate many of these systems, there will be an excellent data base upon which to base the design and verification of lightning protection for susceptible structures and facilities that incorporates year-to-year and seasonal variations for any desired location.

5.13.3. Overview of models and comparisons with measurements

Several models have been developed over the years to deal with the various aspects of the physics and engineering applications of lightning. Most have tried to depict the return stroke, in large part due to its obvious devastating capability. The concept of using a model can be described in several ways. One is that given the stroke current as a function of time as a model input, the model produces the electric and magnetic field or the channel velocity. An obvious corollary is to model peak current flow in the channel from remote measurements of the electric or magnetic field. We briefly describe only a few of the models. For more detail, refer to Uman (1987), Nucci et al. (1990), and others. The most widely used has been the transmission line model, TLM. In its simplest form, the TLM depicts current flow up a vertical channel at a velocity of v_{TLM}. More complicated models can be made using nonvertical channel segments, variable velocity, etc.

A model that has produced good results is that by Lin et al. (1980). It contained three adjustable parameters: an initial breakdown pulse described by the TLM, corona current from charge on the leader, and continuous current. While physically realistic in several aspects, it was complicated and had more adjustments than desirable in a universally usable model.

There have been a few tests of the TLM that included comparison with directly measured current or channel velocity, either of which can be obtained from the model if other parameters are known. The model can be expressed as

$$I_{pk}(t) = -\frac{2\pi\epsilon c^2 R}{v} E_{pk}\left(t + \frac{R}{c}\right) \quad (5.23)$$

for all times before the wavefront reaches the end of the line (i.e., the top of the channel). The distance from the strike point is assumed to be measured along a flat conducting plane (Earth). The equation can also be written in terms of the magnetic field in place of E. Another form of the model, which is used to calculate the velocity of strokes, is

$$v_{TLM} = 2\pi\epsilon c^2 R \frac{E_{pk}}{I_{pk}} . \quad (5.24)$$

The derivative form, which replaces the three variables with their peak time derivatives, can also be used. If there are measurements of the field and current and their time derivatives, then the two calculations of velocity can be compared. The form used to calculate the stroke velocity for comparison with direct measurements by Willett et al. (1988, 1989) was

$$E_z(R, t) = -\frac{v}{2\pi\epsilon c^2 R} i\left(0, t - \frac{R}{c}\right), \quad (5.25)$$

where the height of the channel is much less that the distance to the measurement. The current as a function of time is

$$i(z, t) = i\left(0, t - \frac{z}{v}\right) \qquad t > \frac{z}{v}$$
$$i(z, t) = 0 \qquad t \le \frac{z}{v} , \qquad (5.26)$$

when just the radiation field is present. In their first test, which used triggered subsequent strokes, Willett et al. (1988) found a reasonable agreement for the first two microseconds between the measured velocities and those calculated from the TLM. They also suggested changing the model to include two wavefronts, initially in opposite directions starting just above ground. The downward one would soon reflect from the ground and move upward. This improved the agreement between measured and calculated wavefront propagation speed, apparently because it took into account the connecting leader.

In their follow-up study, Willett et al. (1989) had the added capability of unambiguous identification of the measured stroke velocities with their measured current and electric field and their time derivatives. They iden-

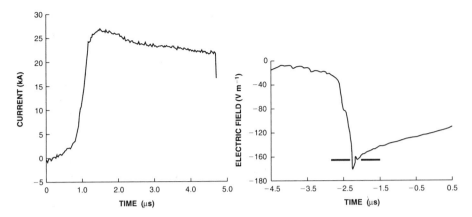

Fig. 5.22. Waveforms of current and electric field from triggered, subsequent stroke. The horizontal line below the peak E denotes the modified electric field peak from reflection at the ground. (From Willett et al. 1989: American Geophysical Union, with permission.)

tified a modified electric field peak value (Fig. 5.22), which they postulated came from the partial reflection of the downward stroke wavefront. Their original hypothesis was for a connection of the flash to the ground a few tens of meters above ground, which then resulted in wavefronts moving down and up. They concluded that for triggered strokes, the TLM gave reasonable fit to their data during the first two microseconds if the modified E_{pk} was used to account for the reflected wave at the ground. They also found linear relationships between the electric field and the current and between their derivatives, dE/dt and di/dt. The correlations between the model and the measured stroke velocities were poor. They concluded that use of the TLM for calculations on natural, subsequent return strokes likely would be accurate to about 20%.

Diendorfer and Uman (1990) tried to keep their model physically meaningful while reducing mathematical complexity. They incorporated a spatially and temporally varying current that flowed up a vertical channel. After the time when the stroke reached a charge height, the channel discharged exponentially. Use of their model led them to offer explanations for the discrepancies between directly measured currents on towers and those calculated from measurements of the electric field and discrepancies between modeled and measured return stroke velocities. Furthermore, they noted that triggered lightning may result in faster discharging in the channel than natural lightning, in agreement with others who had previously suggested various differences between natural and triggered strokes. Their model indicated that the peak currents from the TLM could be too small by as much as a factor of two. That underestimate could be offset if the peak electric field used came from elevated strikes such as to trees, towers, etc. The model showed that for strike point heights of only 5–20 m above Earth, the TLM current could be at least doubled.

That model was advanced by Thottappillil et al. (1991) with a more straightforward derivation and modifications to incorporate a return stroke velocity varying as a function of height. Their model also allowed for variable-speed, downward waves, instead of the previously used constant value, the speed of light. For the same current into the channel base, the changing speed of a stroke did not alter significantly the peak electric field, but the wave shapes of the field were changed at both far and very close distances. This affected the radiation and electrostatic components of the field.

Various methods to test the models have been employed. Nucci et al. (1990) tested five models by using the same current wave shape injected at the channel base. They noted that the fundamental differences in return stroke models can be stated in terms of (1) whether it treats the wave front in the stroke channel as a discontinuity traveling up the channel or a fast risetime assumed for the current specified at the base, and (2) the removal of charge from the leader in time and space. The TLM does not address the second. They concluded that all models gave reasonable approximations to direct measurements of fields in the literature for the first 5–10 µs, and all but the TLM had reasonably correct approximations for durations up to 100 µs.

5.13.4. Measurements of current aloft

It is important to know the values for currents flowing in channels that are not attached to the ground, but it is a difficult measurement to make. The most successful have come from instrumented airplanes. An obvious requirement was to 'harden' the airplanes to survive the lightning without putting crew members or the airframe at unacceptably high risk. Examples include the instrumented F-106 jet-powered fighter converted to this use by NASA and a turboprop-powered CV-580 flown for the FAA. In those programs, the current flow in the channels attached to the airplanes aloft was greater for dart leaders and recoil streamers than for the return stroke channel, at least when a return stroke channel could be definitely determined. In some cases, (e.g., see Mazur and Moreau 1992), the dart leader and return stroke for the same flash were obtained. The dart had a peak current of about 1.1 kA, and the stroke had about 0.6 kA. Another example had a dart of 3 kA and a stroke of about 1.8 kA. They showed several examples with multiple recoil streamers whose amplitudes ranged from hundreds of amperes to >1.5 kA, which was the upper limit of the system on the CV-580 for that measurement. No tabulation of peak current values was given, but they concluded that recoil streamers were usually much larger than those of negative stepped leaders during lightning initiation. Two larger currents of >20 kA were the greatest measured during ground flashes, but it was not known if those peaks were from return strokes (Mazur and Fisher 1990). Analysis of hundreds of current pulses to the CV-580 by Burket et al. (1988) revealed that most common peak currents at altitude were in the range of 1–3 kA, only about a dozen were >5 kA, and an extreme of 25 kA was recorded.

5.14. SCHUMANN RESONANCES

Schumann (1952) suggested that the Earth-ionosphere system forms a cavity that has resonant electromagnetic radiation frequencies. A simplistic way to visualize this is the Earth and ionosphere forming a waveguide within which electromagnetic waves of certain characteristics resonate. The largest peaks in the spectrum are in the extremely low frequency, ELF, band at about 8, 15, and 20 Hz, with two smaller peaks at 27 and 32 Hz. These resonant frequencies are called the *Schumann resonances.* Lightning around the globe is the source of most of the radiation in these spectral peaks. Although originally the source was thought to be the return stroke channel, Pierce (1963) concluded that cloud flashes are comparable sources at the two lowest resonant frequencies, but that ground flashes dominate at the higher resonances. Polk (1969) stated, however, that the vertical return strokes are the main producer at ELF. We can find no report of an observational comparison of the relative contribution of ground and cloud flashes to Schumann resonances. Such observations would be difficult, as the Schumann resonances are the summation of numerous, randomly occurring sources at different locations around Earth. Data are often averaged for ≈10 min to obtain the amplitudes of Schumann resonances. (Measurements by Boccippio et al. (1995), discussed in Section 5.15, matched a few individual ELF signals with what appears to be a Schumann resonance spectrum.)

We discuss the propagation of radiation in the Earth-ionosphere cavity in general terms. Detailed, quantitative explanations are found in reviews such as *Polk* (1982). There are two possible modes of propagation of the spherical ELF waves produced in the waveguide by lightning. One mode is transverse electric; this means the component of E directed along great circles radially away from a vertical lightning channel is zero. The amplitude of waves in this mode are reduced (attenuated) by a factor of about 10^{10} per 1000 km of travel. The other mode is transverse magnetic, which means the radial component of the magnetic field is zero. In this mode, signals are attenuated by a factor of 1.06 per 1000 km—they propagate almost without loss. This mode is the dominant one for Earth's waveguide. The wavelength of the fundamental ELF resonance is ≈ 4×10^4 km, the circumference of Earth. The spacing between the cavity walls (i.e., between Earth's surface and the ionosphere) affects propagation characteristics of the waveguide. The height of the ionosphere, and hence this spacing, varies diurnally and with latitude.

Pierce (1963) stated that because the global circuit is largely dominated by the transfer of charge through the circuit from ionic currents and because the total lightning activity is indicative of total thunderstorm activity, Schumann resonances should in turn be indicative of total thunderstorm activity. Others have also assumed that lightning activity itself is related to the diurnal variation in the global-circuit (e.g., Sentman and Fraser, 1991). It is not uncommon to find statements of the links between the global-circuit E and "thunderstorm activity." Sentman and Fraser noted the theoretical link between the fair weather E (the zeroth order transverse magnetic (TM) mode) and Schumann resonances (the higher order modes of the cavity). They stated that all such modes track the intensities of the same global lightning source function.

Sensors to measure Schumann resonances in the TM mode respond to the vertical component of E and the horizontal component of B. These measurements require much care and are useless unless obtained at electromagnetically quiet sites. The instrumentation and data processing issues were discussed by Polk (1982) and by Sentman and Fraser (1991).

The location of distant storms can be estimated from measurements of Schumann resonances, because the ratio of the vertical component of E and the horizontal component of B is affected primarily by the range of propagation. This technique has useful applications, but it sometimes also has been used with controversial assumptions to derive a measure of global thunderstorm activity or "intensity." Intensity in this context can have various meanings, including the number of storms or the lightning flash rate.

Whether such derived measures are meaningful remains to be demonstrated. Even the simplest approach of using the current in a typical return stroke to estimate the number of ground flashes may be misleading, because little is known about variations in the contribution from individual ground flashes or from vertical channels of cloud flashes. Extending estimates of ground flash rates to total flash rates is even more problematic, as cloud flashes comprise a majority, though a variable percentage, of all flashes. Futhermore, if one then tries to estimate the global number of storms, it is not clear how a storm unit should be defined. For example, is a 500-km long squall line having many cells counted as one storm or many? It might be reasonable to count individual storm cells instead of whole storms, but then one must cope with the considerable variability, and some troublesome systematic seasonal and regional tendencies, in the flash rates produced by individual cells. The list of problems is long.

However, there are problems interpreting any global storm index simply, and one can argue that a parameter tied to lightning is a better measure for thunderstorm climatology than many other parameters, such as those based on satellite cloud imagery. Though we believe that the assumptions used to convert Schumann resonance measurements to other measures have been too simplistic to yield reliable results, such conversions are not required to obtain a global index. Instead, an index based more directly on the Schumann resonance measurements themselves would be useful for monitoring effects on global lightning activity and aspects of global climate change.

Clayton and Polk (1977) used 6 days of Schumann resonances from each month of a 9-month period to derive lightning activity. The days were grouped in three pairs across each month; the days had clear skies and no thunderstorms within 1000 km of their site, which was in the northeast United States. These requirements limited their analysis to 9 months and excluded the summer thunderstorm season. From their recordings of the vertical E and horizontal B, they expressed "lightning intensity" in terms of an integrated charge moment squared per second (C^2 m^2 s^{-1}) for the entire planet, as they assumed that the average charge moment was better established than that of the current moment. From the calculated intensity, they had to assume parameters for a typical flash to get the world-wide activity. We wonder whether use of a typical flash yields meaningful results—but that is the hope. They noted the location of the lightning was not their primary objective. As they also noted, however, their procedure to determine world-wide lightning activity required the source locations, as the calculation of E requires values of the angular separation between the source and the receiver.

To determine the location of the ELF sources, they assumed the Earth is a perfect conductor and that the bottom of the ionosphere forms the other side of the spherical cavity in which ELF propagates. Since the ionosphere does not have a sharp boundary owing to the changing conductivity with height between Earth's surface up through the ionosphere, they used a conductivity profile that stratified the ionosphere. Their model had the lowest layer from 0–40 km. Other layers, whose profiles were selected based on other data, covered the 40–90 km interval. Another difficulty is handling nearly simultaneous sources. Simultaneous lightning at different locations are a distributed, rather than a point, source. To reduce such errors, they calculated the charge moments across the entire 12–21 Hz band instead of at each spectral peak. Individual peaks were used to find the source-receiver separations; the resulting errors in charge moments ranged from tens of percent to as low as one percent. They did not use data from the large ELF peak at about 8 Hz, as it is often contaminated with extraterrestrial sources.

They concluded that use of a typical flash of 20 C transferred over a 3-km path yielded lightning activity of 42–608 s^{-1} (i.e., flashes per second), while increasing the moment to 90 C km yielded a lightning rate of 19–270 s^{-1}. Note that these estimated rates vary by more than a factor of ten and represent only those flashes contributing to the Schumann resonances.

Sentman and Fraser (1991) showed that there is a significant modulation of the global signal from the local height of the ionosphere over the receiving site. They measured the vertical E and the horizontal components of B at two sites separated about 14,800 km in great-circle distance and 8.6 h in local time. Diurnal intensity patterns in the signal power received had a day-to-day variability that was similar to that in the global number of thunderstorms (as estimated by Whipple and Scrase, 1936). They stated that the positive correlations, albeit moderate, were evidence that the variability in the intensities of the Schumann resonances at the two sites came from a common source. They concluded that raw Schumann resonance data from a single station would not seem a feasible way to monitor global activity. Sentman and Fraser showed a significant global coherence in the variations in intensity of the Schumann resonances when the local ionospheric height variations were considered. Furthermore, they believed that the day-to-day variations in the diurnal activity might correspond to daily and seasonal variations in global lightning activity. They believed that such measurements supported suggestions that Schumann resonances can be used to define a geoelectric index such as proposed by Holzworth and Volland (1986). Sentman and Fraser reported a strong global coherence in their data for times of <1 h that led them to conclude it was from global lightning activity (Fig. 5.23).

Williams (1992) recently proposed Schumann resonances as a means for monitoring global warming. He stated that Sentman and Fraser (1991) provided convincing evidence of global representation from a single site. Presumably he was referring to the data after processing to include the effects of the local height of the ionosphere. The premise of his paper was that differences of ≈1°C in the surface air temperature cause sufficient increases in the bouyant energy available for convection to affect the number of thunderstorms or the average lightning flash rates of storms. His interpretation was based in part on the assumption that the noninductive graupel-ice mechanism dominates thunderstorm electrification. Williams stated that about two thirds of all lightning occurs in the tropics (i.e., within ±23° latitude of the equator). He calculated monthly mean values of the surface wet-bulb temperature (to

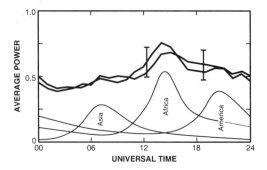

Fig. 5.23. Average power from Schumann resonances and world-wide thunderstorm activity on three continents from Whipple and Scrase (1936). Schumann resonances were corrected for local height of the ionosphere and their 24-h averages are from 14–21 April 1990. The curves are for measurements at Table Mountain, California, and Norwest Cape, Australia. (From Sentman and Fraser, 1991: American Geophysical Union, with permission.)

account for daily variations in temperature and moisture, both of which contribute to convection and lightning). He concluded that the farther from the equator, the less the lightning is related to seasonal variations in wet-bulb temperature. His hypothesis awaits further tests.

5.15. RED SPRITES AND BLUE JETS

As mentioned at the outset of this chapter, substantial documentation of above-cloud discharges has been obtained in the last few years. These phenomena have stimulated considerable interest recently, and theoretical and observational studies are proliferating. We provide a very brief summary of the observational data. Recent recordings of luminous phenomena above

storms include images from the space shuttle (Vaughan and Vonnegut 1989, Vaughan et al. 1992, Boeck et al. 1992), a single image above a storm 250 km away by Franz et al. (1990), about 250 video images during the summer of 1993 over mesoscale convective systems (MCSs) by Lyons (1994), and the observations of two phenomena called "red sprites" (Sentman and Wescott 1993, Sentman et al. 1995) and "blue jets" (Wescott et al. 1995). These last two studies used synchronized video cameras and other optical sensors on two aircraft to determine the location, dimensions, and visual characteristics of sprites and jets.

Sprites appear as transient vertically elongated spots of luminosity suspended above cloud top, usually in clusters as shown in Fig. 5.24, but sometimes alone. Sentman et al. (1995) observed the following characteristics of sprites: Individual sprites are 5–30 km wide, and the brightest part is at an altitude of ≈70 km MSL; from the brightest region, a wispy red glow splays outward and upward to an altitude of 88 ±5 km MSL (the tallest sprites extend to >95 km MSL), and tendrils extend downward to ≈40 km MSL; sprites are mostly red in color, changing to blue at the lowest extremities of the tendrils; all sprites in a cluster become luminous simultaneously (to within 16 ms); the time for luminosity to decay is uncertain but may be roughly 100 ms; the estimated total optical energy radiated by individual sprites is 1–5 kJ, with instantaneous radiated power of 0.5–2.5 MW; and their preferred location is over stratiform regions of MCSs, in the vicinity of positive ground flashes.

Boccippio et al. (1995) inferred a link between sprites and positive ground strokes with large currents detected by the National Lightning Detection Network. They studied two mesoscale convective systems and found ≥80% of the sprites were coincident with positive ground strokes in the stratiform region of the MCSs. The peak currents of these positive strokes ranked in the upper 3–15% of the current distributions estimated from ground strike mapping networks. ELF transients of the appropriate polarity for positive

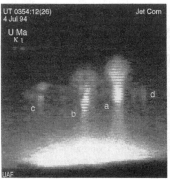
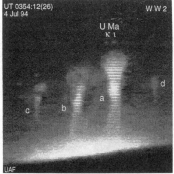
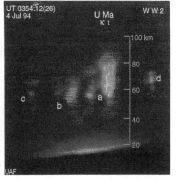

Fig. 5.24. Video image of cluster (a–d) of "red sprites" on 4 July 1994, 0354:12 UTC. Sprite "a" reached an altitude of ≈ 85 km. (From Sentman et al., 1995: American Geophysical Union, with permission.)

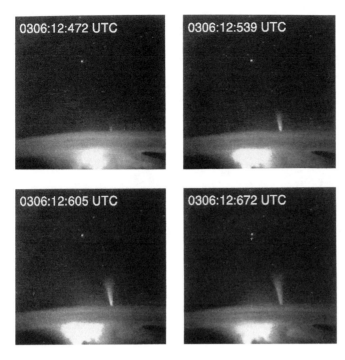

Fig. 5.25. Sequence of video images showing development of "blue jet" on 4 July 1994, 0306:12 UTC. Each image is separated in time by 67 ms. (From Wescott et al. 1995: American Geophysical Union, with permission.)

ground flashes appeared to be coi ident with sprites and their associated positive ground strokes. Boccippio et al. invoked the common observations of extensive horizontal lightning and of extensive horizontal reservoirs of positive charge in MCS stratiform regions to support their notion that flashes in the stratiform region create large-scale electrostatic stresses in the atmosphere that lead to sprite formation. They stated that their data supported Wilson's (1925, 1956) proposal that a quick electric field change can cause the mesosphere to break down electrically. Finally, Boccippio et al. proposed that large bandwidth recordings of ground stroke transients and measurements of E_{aloft} over MCS stratiform regions were needed for conclu-sive confirmation of an electrostatic triggering mechanism for sprites.

Blue jets appear to emerge from the top of storms, as shown by the sequence of video images in Fig. 5.25. Sometimes a blue jet is preceded by a lightning flash with well-defined channels that propagate upward from cloud top, but do not extend as high as jets, and sometimes two sequential jets emerge from the same location. Westcott et al. (1995) observed the following characteristics of blue jets: They look like a narrow cone that fans out at the top and is mostly blue in color; they extend upward at $\approx 10^5$ m s^{-1} to a maximum altitude of 40–50 km MSL; their duration is $\geq$200–300 ms; and luminosity decays simultaneously along the entire jet.

6

Instruments

This chapter is designed to give you key elements of the measurement techniques used in storm electricity research, which is largely an observational science. Understanding several key instruments is important to conducting or evaluating experimental research. It is obviously helpful to remember that most of the time the measurements of electrical and other parameters are made in a harsh environment. Further exacerbating the situation is that the parameter to be measured often has small values (i.e., we are challenged to sense a small signal in a very noisy background). Please note that we include circuit diagrams for some key instruments. They are examples provided for those with some knowledge of electronic circuits. It is not essential that they be included in the teaching or the learning of this course. The fundamental issues of making the measurements can be grasped without knowledge of how the electronic components and circuits in the instruments work.

Although we discuss techniques individually, most measurements are made in a field project or at a facility that includes multiple instruments. As the knowledge of storm electricity has grown, it has become more critical to try to surround the storms with a substantial array of instruments in order to place individual measurements in the larger context of the storm creating them and thus to acquire comprehensive data that provide observational constraints to new theories and models of how storms work. Mention of some facilities and the measurements made at them appear throughout this book.

6.1. ELECTRIC FIELD MILLS

Among the devices that measure the electric field are cylindrical and rotating-vane field mills and other field meters. Although field mills have been used for decades, evolving technology has contributed significantly to new designs and applications. We begin with a single field mill, such as is frequently used to measure E_{gnd} and then consider systems to measure E_{aloft} with rockets, airplanes, and balloons. First, we look at the basics of a fundamental circuit to make the measurement. In modern instruments to measure E (and charge), the circuit often used is a solid-state operational amplifier connected as a charge amplifier (Fig. 6.1). The circuit is simple, yet useful in many sensors. We will discuss features common to measurement of E and of charge, q. The behavior of an operational amplifier in a charge-amplifier configuration is described by a simple expression—at least to a first approximation. The charge measured is given by

$$q = -\frac{V_o}{C_F}, \qquad (6.1)$$

where V_o is the voltage out of the amplifier that causes a current to flow around the feedback loop, and thus through the feedback capacitor, C_F, and resistor, R_F. In doing so, it keeps the input of the circuit at virtual ground, which turns out to be a nice feature of circuit performance. When used in a field mill, the charge measured by the circuit can be related directly to the

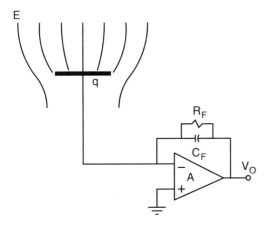

Fig. 6.1. Charge amplifier circuit used in many sensors of atmospheric electricity. In this example, the electric field, E, induces charge on the sensor plate. The decay time of the instrument is defined by $R_F C_F$. This time constant needs to be several (e.g., ≥ 5) times the duration of the event to be measured to reproduce faithfully the signal from that event.

electric field. Because the q induced on the stators (Fig. 6.2) is a function of the electric field (from Gauss's law), the relevant relationship for a field mill becomes

$$E \propto -\frac{V_o}{C_F}. \qquad (6.2)$$

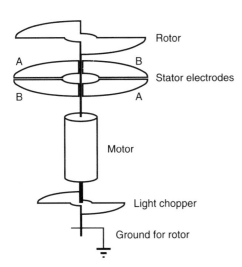

Fig. 6.2. Sketch of major mechanical parts of typical rotating vane field mill. The stators are connected in pairs, A and B. Each pair is connected to the input of a charge amplifier. The light chopper provides a reference signal to determine polarity of E. (After Winn 1993: American Geophysical Union, with permission.)

The amplifier can produce both polarities of voltage, so both polarities of E (or q) can be measured. The maximum V_o is slightly less than the amplifier's power supply voltage. The desired maximum charge to be measured can be used to determine the value of C_F. The minimum detectable charge is a function of the electronic noise but, in a storm environment, transients from lightning and corona can be major noise sources. Especially if the circuit is being used to measure charge on particles, the minimum detectable charge of the sensor system (not the circuit itself) can change with environmental conditions, but it can be estimated from the data.

6.1.1. Rotating-vane field mills

The charge amplifier in a rotating-vane field mill senses charge induced on different electrodes, called *stators,* that are alternately shielded and exposed to the atmospheric E by a grounded rotor (Fig. 6.2). A simplified circuit that uses charge amplifiers for such a field mill is shown in Fig. 6.3. Geometrically opposite sensor plates (stators) are connected to each other, and each pair is connected to the inverting input of an amplifier. The amount of charge induced on the stators depends on the magnitude of the local E. In a charge amplifier, the signal is not a function of the rotation frequency of the rotor if it exceeds about five times the electronic frequency of the feedback amplifier, $f = 1/(R_F C_F)$. If the frequencies are about the same, as in this example, there is only a 2% effect on the signal amplitude. The key in this example mill is that the rotation does not vary significantly, which would cause unknown variations in E (unless the data were corrected for the frequency change). In this circuit, the rotation can vary by two, and the change in uncorrected E would be only about 4%. Once the rotation rate exceeds the electronic frequency by five, the effect of changing rotation rate is negligible. These design parameters make the measurement of E at Earth's surface straightforward. A triangle wave is the theoretical signal out of each amplifier, but electric field fringing effects at the rotor edge as it crosses a stator cause the actual wave shape to be more like a sinusoid (V_A and V_B in Fig. 6.3b). The two amplifier signals are 90° out of phase and are sent through solid state, analog switches for combination into a single wave. Switches are driven by a reference square wave that is generated by a black plate that is the same size as the rotor and that chops a light beam shining on a detector. Mechanical design allows the relative position of the chopper to the rotor to be changed, which is the process of *phasing* the mill. The reference square wave, V_C, is used to drive the analog switches synchronously with the E signals to determine the polarity of the signal and thus of E. Phasing yields the desired, full-wave rectified wave form, V_3, and is used to set the desired polarity out of the circuit for a particular polarity of E. The fully rectified sine wave is filtered at a few Hertz, and the final field mill output is a vary-

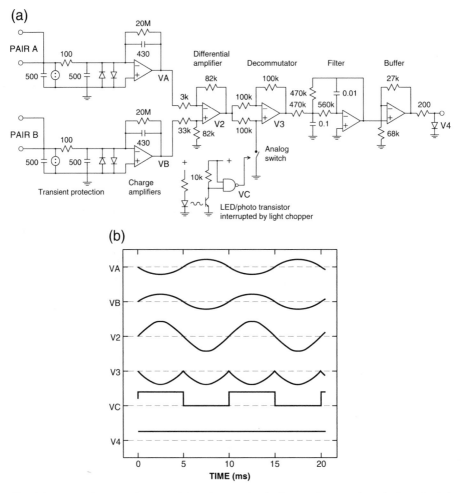

Fig. 6.3. Circuit for a rotating-vane field mill. (a) Circuit diagram for one channel of sensitivity. Resistances are in ohms; capacitances >1 are in picofarads; C <1 are in microfarads. Amplifiers have field-effect-transistor inputs. (b) Waveforms of voltages at corresponding points in the field mill circuit. (From Winn 1993: American Geophysical Union, with permission.)

ing direct current (d.c.) voltage, V_4, whose polarity and magnitude are directly tied to the atmospheric electric field. More complicated designs and circuits exist. For example, field mills in a network at the Kennedy Space Center have several refinements, including circuitry to allow placing a calibration voltage on the rotor. This can be done remotely from a monitoring site and yields a check on proper operation and calibration of the mill.

6.1.2. Field mill calibration

A rotating-vane field mill is calibrated directly by placing it in a uniform E made by applying known voltages, V, across a parallel plate capacitor, separated by a known distance, d. An example setup has plates 1 m square separated by a distance d = 10 cm; the plate spacers are a dielectric material. Batteries or a d.c. power supply is used to apply several values of both polarities on the upper plate; the calibration uses the relationship E = V/d. (Although not strictly necessary as part of the calibration when V = 0, the plates should be electrically shorted to ascertain that the output at E = 0 is correct, i.e., there is no significant offset in the mill output). Since the field mill circuit is very linear, only a few points are required to get a good calibration.

In use, one configuration is to mount the field mill flush with Earth's surface and facing upward. This simplifies the calibration, as there is minimal perturbation of electric field lines, and the geometrical form factor of the instrument is unity. Then the output voltage, denoted by V_4 here to coincide with Fig. 6.3b, can be directly calibrated to the atmospheric electric field E by

$$E = GV_4, \qquad (6.3)$$

where G is the circuit calibration factor for a particular gain setting. (Field mills often have switchable gains or more than one channel of output sensitivity.)

Field mills for ground-level recording are often designed to measure from <100 V m^{-1} (fair weather) to about 20 kV m^{-1} (intense thunderstorm overhead). Electric fields at Earth's surface generally do not exceed 15 kV m^{-1} because they are limited by space charge from point discharge. Sometimes other installation configurations besides flush mounted are needed; in particular, the mill can be mounted a meter or so above ground, facing downward (Fig. 6.4). This arrangement is beneficial if measurements are to be made in snow and very intense rain during which electronic shorting of the sensors can occur and charged rain adds noise. Such a mounting configuration distorts E because of the geometrical form factor, $\mathcal{F}$. Although the field mill itself is still calibrated as before, the system is calibrated (i.e., $\mathcal{F}$ is determined) by comparison with an upward facing, flush-mounted mill operated nearby. Equation 6.3 is modified to

$$E = \mathcal{F}GV_4. \qquad (6.4)$$

If, for example, the enhancement of E for a particular mill installation configuration is 2.5, then $\mathcal{F} = 0.4$. Major practical problems when using a field mill and means of overcoming them include: (1) Drifting electronics can cause d.c. shifts and change the output voltage for the true E = 0. An 'electronic zero' can be obtained by disconnecting the stators. Temperature stable circuits and components should be used; (2) Leakage currents can develop across the stator insulators (generally Teflon or KEL-F) from dirt, spider webs, or blades of vegetation, especially in high humidity, and cause the circuit output to drift or saturate. Good design, smooth surfaces from proper machining, and routine cleaning with a low residue solution such as denatured alcohol, mitigate the problem; (3) Errors in the electric field measurement can be caused by charges on nearby dielectric surfaces such as paints and plastics. (Therefore, do not cover a field mill with a plastic cover to keep the rain off.); (4) Point-discharge ions created in large E from the mill parts or nearby vegetation or objects will change the local E and can cause erroneous readings. To decrease this effect, the rotor, stators, and other nearby metal edges must be smooth and rounded for larger radii of curvature. If possible, the mill should be placed in an area that is clear of point-discharge sites such as grass and is several times the diameter of the mill, or the height of the mill above ground, whichever is larger; (5) The same material should be used to fabricate stators, rotor, and nearby housing to avoid an E arising from differences in contact potential. Such local fields limit the minimum measurable field and add noise as a resultant d.c. signal at the instrument output. Stainless steel is a preferred material;

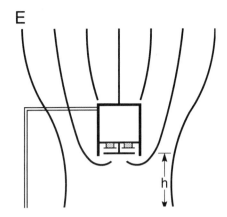

Fig. 6.4. (top) Sketch of downward facing field mill installation. The height, h, is not critical, but is often about 1 m. The ground beneath the mill should be level and without tall grass, etc. A flat conductive base plate is advisable. The distortion of E lines is depicted and indicates why there is a geometrical form factor, i.e., perturbation in the local E by this mounting configuration. (bottom) An example of a new mill used at the Kennedy Space Center. (Photo courtesy M. Stewart and L. Barnum)

and (6) Nearby structures and tall vegetation should be minimal or, better yet, nonexistent, to avoid perturbing the local E.

6.2. INSTRUMENTS TO MEASURE E_{aloft}

The airborne platforms that carry instruments to provide in situ measurements of E in thunderstorms are rockets, airplanes, and balloons. We give attributes of each, as well as describe sample systems. A rocket carrying electric field measuring devices through clouds can produce nearly instantaneous profiles of the measured components of E along its trajectory. However, there are only a few locations within the continental United States where airspace restrictions allow the firing of rockets of adequate size to lift even small payloads part way up through thunderstorms. In spite of these limitations, rockets have provided useful data. They have been used to obtain nearly vertical profiles. Airplanes and balloons often can be operated without a restricted airspace, but have other operational constraints.

Many envision airplanes as ideal platforms, because they seem as if they can cover large parts of the clouds quickly. There are tremendous challenges to overcome, however, to make reliable E measurements, which we describe soon. Constraints that impact the usefulness of airplanes include operating cost; inability to operate in hail; no storm penetrations after nightfall owing to hazards from flash blindness; and difficulty in obtaining vertical profiles owing to climb limitations and, often, maximum flight altitude that is lower than storm tops. Airplanes are better suited for horizontal profiles. Large airplanes have the significant advantage of carrying observers and more and heavier instrumentation than rockets or balloons.

Balloons cannot be guided, but do provide a convenient platform for obtaining vertical profiles of E. Tethered balloons are also used; they are suitable for low heights, but the tether is problematic in large E and windy conditions. Although we often think of planes as fast and free-flying balloons as slow, obtaining vertical profiles of E occurs at about the same rate—if the airplane can even do a complete vertical profile in storms. An exception would be a high-performance jet aircraft that could fly nearly vertically. Balloons and rockets can be flown without exposing personnel to aircraft hazards associated with hail, wind shear, etc.

There are irreconcilable differences and trade-offs with these platforms. Optimally, coordinated combinations of the platforms would be used to obtain more complete spatial and temporal coverage within storms. Specific examples of how these platforms are used follow.

6.2.1. Rocket-borne E sensors

A rocket-borne electric field meter was developed by Winn and Moore (1971) for measuring the component

of E perpendicular to the longitudinal axis of the rocket. The folding-fin aircraft rocket motor used was a military Mark 40, Model 1, which had a solid-fuel motor; the body of the rocket was 70 mm in diameter and about 1.5 m long when equipped with the E-measuring nose. The rockets were launched within the restricted airspace at Langmuir Laboratory in New Mexico.

As Winn and Moore (1971) stated, care must always be exercised when making measurements of E within clouds and caution used in interpreting the data. They discussed two possible sources of error in their rocket instrument: charge on the device and point discharge from the rocket. For the first, any charge that was distributed on the rocket body resulted in a d.c. offset of the sinusoidal signal that was produced by the external field. Thus, they could separate the signals from the field produced by charge on the rocket from the external E. For the second, Frier (1972) warned of possible vitiation of the measurements if point discharge occurred at the front of the rocket. Winn and Moore (1972) said that even in the unlikely event of point discharge from the dielectric nose ahead of the sensors, the resulting E could be distinguished. An exception would have required that the point discharge produce a sinusoidal varying space charge. The more likely place

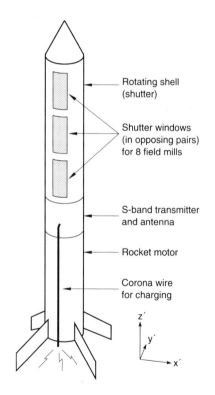

Fig. 6.5. Rocket-borne electric field mill (REFS). Eight mills are arranged as four pairs. A single pair is located 90° from the middle pair. Three windows on one side are shown. (After Willet et al. 1992, with permission.)

for point discharge to occur, they contended, was at the fins, with the only result normally being spike-type noise on the telemetry record. This was shown on several of their recordings and was distinguishable from signals due to the external E.

A new version of a rocket E-measuring system called REFS has been developed by Willett et al. (1992); (also described in Appendix B of Marshall et al. 1995b). As did the earlier instrument, this one uses surplus military rocket motors (Mark 40, Model 4). A major improvement in the instrument is its ability to measure all components of E through use of eight field mills (Fig. 6.5). Another innovation is an outer rotating shell with shutter windows that serve as the field mill rotor. The sensor stators are thin electrodes glued to the rocket body. Each revolution of the shell exposes and covers all eight stators twice in each revolution of the shell by the three pairs of shutter windows. Both transverse components of E are measured. Signals from opposing pairs are differenced to subtract any E due to charge on the rocket, E_Q, and the longitudinal E. The result is three independent measurements of one transverse component of E and one measurement of the other transverse component. When summed, opposing mill signals cancel the transverse components to derive the longitudinal one and E_Q. These are combined in a calibration matrix to yield the components of E and E_Q. The longitudinal and transverse components can be coupled with the rocket inclination angle, which changes with time, to calculate the atmospheric E components, E_x, E_y, and E_z.

Just as with other asymmetrical vehicles such as an airplane, the rocket platform must be chargeable in a controlled manner in flight to handle the E due to self-charge. A 10 kV supply and outboard corona point allow this. The power supply is turned on for 100 ms every 1.5 s during flight. The self-charging allows the necessary determination of calibration matrix coefficients plus a relative calibration of the mills during flight. Each field mill signal is telemetered to ground to reduce electronics needed in the rocket and to provide diagnostic information for each mill. Other key design aspects are the inclusion of a pressure transducer for determining the altitude, an accelerometer to determine when the motor stops burning and to calculate trajectory, and a simple magnetometer that senses Earth's magnetic field to measure rocket rotation. Calibration of the rocket was done both theoretically and in the parallel capacitor system at Langmuir Laboratory (see Sec. 6.2.3). The dynamic range in storms of REFS is about 1–900 kV m^{-1}, with an absolute accuracy of no more than $\pm 10\%$ of the measured E. As of 1993, about four flights had been made through thunderstorms. Vertical profiles of E made with the REFS agreed with those in the same storm made with a balloon-borne electric field meter (Marshall et al. 1995a, b). That result added to the credibility of both. Key constraints for using the rockets are the necessity of a restricted airspace and the maximum ceiling of the rockets of <8 km MSL, even when launched from Langmuir Laboratory at 3.2 MSL.

6.2.2. Instrumented airplanes for measuring E_{aloft}

Both field mills and radioactive probe devices have been used to measure E_{aloft}. An example of the latter is found in Vonnegut et al. (1961). For measurements in thunderstorms and high E, field mills are preferred. Here we address characteristics of field mills on recent example airplanes.

(i) Airborne rotating-vane field mill systems. An excellent example of the use of a set of rotating-vane field mills to make measurements in storms is the Special Purpose Test Vehicle for Atmospheric Research (SPTVAR), a powered sailplane operated by the New Mexico Institute of Mining and Technology (Fig. 6.6). While there have been a few other well instrumented airplanes, this one has been in research service since 1976. Its instrumentation has evolved with available technology and advancing thunderstorm research problems. Until recently, a total of five mills were used, one pair mounted vertically and another pair horizontally on the fuselage. The fifth mill faced aft.

Individuals mills are calibrated as described previously. However, in an airplane system, changes in field mill output due to geometrical form factors, aircraft charge, and contributions from all three components of E must be determined. Charge acquired by the airplane is a big problem, as large local E can be generated by this charge. Jones (1990) studied the SPTVAR, the NCAR/NOAA sailplane, and an NCAR King Air. He found that the two piston-powered airplanes charged when impacting liquid hydrometeors. Also, induction charging of the SPTVR occurred in liquid water when an external E was present. Furthermore, self charge came from corona when it occurred from the airplane. He did not study the even more complicated charging process by ice crystals.

As corona ions flow off surfaces such as propellers, wing tips, and other conductors with small radii, and as charged engine gases exhaust, they all form plumes of charge. These plumes of space charge cause a local E that very adversely affects the measurement of the fore-aft component of E (Jones et al. 1993). They found that through linear combinations of the field mill signals, they could judge the effect on the measurement of E. They also assessed inductive charging as precipitation impacts the airplane in the E from the storm and determined that this process can generate a charging current even greater than corona-ion plumes. The aft-facing mill on the SPTVAR was the most affected. Other installation configurations will respond differently, but these problems will still need addressing on any airplane.

Even without the plume problem, the E caused by aircraft charge, E_Q, must be removed from the data to determine the undisturbed components of the atmospheric electric field near the airplane. We begin with a description of the often used technique of charging the airplane aloft in fair weather. Preferably, charging is done several kilometers above ground, where the am-

Fig. 6.6. Photograph of field mills on fuselage of New Mexico Tech SPTVAR.

bient E is only a few volts per meter. There, essentially all the E will be due to airplane charge. Airplane charging is accomplished by changing from cruise to nearly full engine power, which increases charge emission, or by 'pumping' charge overboard with a corona point attached to a high voltage d.c. power supply. The second technique is preferred, as it is controllable, easily done, and some airplanes do not charge much from engine exhaust. Placing the corona point in the air stream allows point-discharge ions of one sign to flow away (as opposed to coming back to the aircraft skin), leaving the opposite sign of charge on the airplane.

Each field mill can have all three components of E plus E_Q, the electric field due to airplane charge, present in its output signal. The following is an example of the signals in field mill number one:

$$E_1 = a_1E_x + b_1E_y + c_1E_z + d_1E_Q, \qquad (6.5)$$

where a–d are coefficients that must be determined by the calibration procedure. Such an equation exists for each field mill, and finding the Es requires solving a matrix equation. The implicit assumption is that the aircraft surface near the field mills is conducting and free from paint, antenna radomes, and other dielectrics. The presence of such dielectrics can result in charge being trapped in nonuniform and nonreproducible distributions, which, in turn, cause nonrepeatable E_Q that contaminates the measurement of the ambient E (e.g. see Jones 1990).

The form factor of the airplane can be determined by three methods: direct calibration, a small-scale model to which charge is applied, and a numerical model. If one coefficient is known along with the relationships (ratios) to others, then all coefficients can be found. In direct calibration, the form factor affecting E_z is found by flying just a few meters above a calibrated field mill at the ground. Then other flight maneuvers such as pitch-up, pitch-down, and banking turns at known angles to the vertical allow calibration of the horizontal components, E_x and E_y, by comparison with the vertical component. To be most useful, this requires that the field mills be sensitive enough to measure fair weather E. Another procedure is to fly the airplane by a balloon-borne field meter for an independent measurement of E. Use of an electrically conductive, small-scale model or a numerical model of a particular airplane are appealing because they do not require flights. Large uncertainties occur, however, if the actual mill locations are near small but significant curves in the airplane skin that are not in the model. Irrespective of calibration procedure, an often used technique has been to choose geometrically symmetrical locations for pairs of field mills to reduce the unwanted components of E. They cannot be entirely eliminated (e.g., Mazur et al. 1987).

The signals from all mills can be combined electronically or processed analytically to determine the three E components and E_Q by simultaneous solution of the set of equations like that shown above for mill number one. The analytical technique is much preferred because the data from each mill are recorded without processing to facilitate real-time monitoring of mill performance. Furthermore if a coefficient were to need correction, it could be applied in postprocessing of the data.

In a new approach to the problem of airplane calibration, Winn (1993) used the concept of self-calibration in radio astronomy. He used data obtained when the SPTVAR made two turns while flying in an otherwise constant direction near a thunderstorm. This procedure carries two constraints: the field mills must not be near dielectrics or moving parts such as a propeller, and the vector electric field must be uniform over a larger volume than that occupied by the airplane if the plane were not there. Slowly varying E during the turns can be accounted for if necessary. A major advantage of this new procedure is that the field mills do not need to be at symmetrical locations, but can be placed optimally to minimize exposure to charged plumes from the airplane. In the mid-1990s, the SPTVAR was equipped with additional field mills, including new versions of cylindrical mills, in response to the self-calibration results. Winn's innovative approach, mentioned above, is not constrained as was the much earlier use of cylindrical field mills by Kasemir (1951, 1972), which follows. Also, Winn uses a different design and modern circuit components, but as his mills are not yet documented in the literature, we will describe this type of mill using Kasemir's original design and usage.

(ii) Cylindrical field mill. Another type of field mill designed for use on airplanes is the cylindrical mill (Fig. 6.7), which was developed by Kasemir (1951, 1972). It was unique in that it measured two orthogonal components of E with a single mill. Although its mechanical design is quite different from the rotating vane type, major circuit elements can be the same. When mounted on the nose of an airplane, the vertical, E_z, and wingtip-to-wingtip components of E are measured. A second cylindrical mill mounted perpendicularly and on the top or underside of the fuselage measures both horizontal components, E_x and E_y, of the horizontal field, E_h. The two mills give a redundant measure of the wingtip-to-wingtip component.

The sensing elements of Kasemir's (1951, 1972) mill consists of two hemicylinders. One is grounded and insulated by Teflon from the other. The external E induces an alternating charge on the sensors as the mill rotates. This results in an alternating current that flows from the insulated sensor through a rotating capacitor to circuits where it is amplified, rectified, and filtered. (Mechanical slip rings can be used to send signals from the rotating to the fixed parts of the mill.) Two sinusoidal reference signals, 90° out of phase, are produced by a signal generator driven by the shaft of the field mill motor. These are used in the phase-sensitive rectification circuits so that the two orthogonal components of the external E may be determined. After phase-sensi-

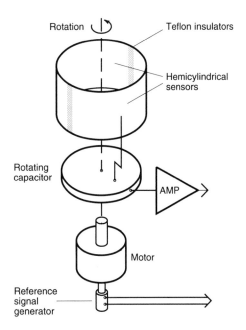

Fig. 6.7. Major mechanical parts of cylindrical field mill. One hemicylinder was grounded through the motor brushes; the other provided the input signal. The rotating capacitor coupled the signal to the first-stage amplifier, which was located in the field mill head. All other electronic circuits were located away in a control console. (After Rust and MacGorman © 1988, Oklahoma Univ. Press, with permission.)

tive rectification, the signals are filtered to give d.c. outputs proportional to E. Use of several gains allowed Kasemir to measure E from about 10 V m^{-1} to 500 kV m^{-1}, the latter being an extrapolated E_{max} from calibration data. Possible loss of reliability from corona in very high E was not reported.

Kasemir and his colleagues calibrated the cylindrical field mills in the laboratory between parallel plates, and then they calibrated the field mill system on the airplane in flight as described above for a rotating-vane field mill installation. He compensated for the effects of charge on the airplane by using smoothly rounded metal bosses called "hump rings," which were mounted on the cylindrical housing just below the sensing elements (see Fig. 6.8). Procedures to charge the airplane in flight are as described above. The hump ring was rotated and translated until the charging procedure caused no noticeable outputs from the field mill, even on sensitive ranges. Done manually, this was a slow and tedious task, as each adjustment of the hump ring required a take off, flight to a few kilometers altitude for charging, and return to ground for the next adjustment. A remotely controlled motorized set of spherical bosses, mounted symmetrically about the mill body

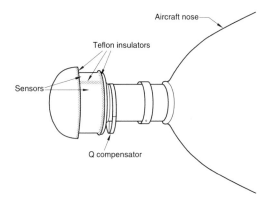

Fig. 6.8. Cylindrical field mill on nose of airplane. The conductive hump ring was used to null out the E from charge on the airplane. The aircraft nose was made conductive to make charge near the mill collect in a reproducible way (i.e., there was no nearby dielectric paint surface). (After Rust and MacGorman © 1988, with permission.)

was later used successfully, with the adjustment taking only a single calibration flight to produce the charge nulling.

Cylindrical mills have excellent performance when mounted on a conductive airplane nose, but this precludes use of a nose-mounted radar or a single engine airplane. The mill on the top or bottom of the fuselage to measure E_h does not work well in cloud, probably because of some of the problems discussed by Jones et al. (1993). (Winn's (1993) new installation scheme likely gets rid of these problems.) Most information in the literature has been only on E_z and the one component of E_h that were obtained with a nose-mounted cylindrical mill. To distinguish it from the total E that we normally use, we call the field measured with Kasemir's (1972) device, E_K, where

$$|E_K| = (E_z^2 + E_H^2)^{\frac{1}{2}}. \tag{6.6}$$

The field was often dominated by the vertical component, E_z.

6.2.3. Balloon-borne sensors of E_{aloft}

Through the years, several instruments have been developed for use on balloons. Some others have yielded interesting information, but because of cost, size, etc., were not widely used. Most have been designed for flight on free balloons. A term we use is *radiosonde*, by which we mean any balloon-borne device that transmits its data back to ground. More specifically, we usually are referring to a meteorological sonde that measures temperature, pressure, and humidity. The term *rawinsonde* is a radiosonde that also provides a measure of the winds aloft. Originally the complete term was rawinsonde system, denoting the sonde plus the tracking system from which winds were calculated. Now in some systems, the wind is provided by

the sonde without a tracking antenna through use of navigational signals from LORAN, OMEGA, or GPS. OMEGA (and maybe LORAN) is to be decommissioned; the global positioning system, GPS, is fast becoming the navigation system of choice and is now in some rawinsondes.

(i) Alti-electrograph. The most significant early balloon-borne instrument was the alti-electrograph developed by Simpson and Scrase (1937). It responded to point-discharge current and was designed to reveal the polarity of E_z. We describe it because of its historical prominence and the frequent references made to its results. The recording mechanisms were clock-driven disks of aluminum. One disk had a smoked surface on which a stylus scratched traces of the relative humidity from a human-hair hygrometer and pressure from an aneroid. The second disk served as a turning platter for a paper disk that was chemically treated to show current polarity. Two iron electrodes moved over the paper. One electrode was connected to the metal case, which had several wires projecting above the sensor. These formed the upper electrode of the point-discharge sensor. The other electrode was isolated from ground and connected to a 20-m long copper wire of about 0.3-mm diameter. The wire hung below the recording box during flight. The iron pens were set 3 mm apart with no signal from point-discharge current flowing from the electrodes. Positive current flow into an electrode caused that electrode to make a blue trace; negative current made a trace that was not blue. The polarity of the point-discharge current, and thus the polarity of the E, could be determined. Fortuitously, the researchers found that wider traces resulted when more point discharge flowed. From the trace width, they obtained a crude estimate of the magnitude of the potential gradient, but only up to 10 kV m^{-1}, where the instrument began sparking and became useless for estimating E_z. Simpson and Scrase flew their instrument below a parachute, which allowed them a chance to retrieve the alti-electrograph data. The instrumentation did not measure temperature, so they estimated altitude from the pressure trace plus assumed lapse rates below and inside the cloud. They noted that the hygrometer did not clearly show passage through cloud boundaries. (Modern soundings also often do not show abrupt changes in humidity at cloud boundaries.)

It was flights of the alti-electrograph that formed the basis of our concept of the gross charge structure of thunderstorms. The influence of measurements made with the alti-electrograph has been substantial and long-lived. However, aspects of the original data analysis technique and resulting conclusions were challenged by Marshall and Rust (1991). They argued that the method of using zero crossing of point-discharge current versus altitude to identify a charge region was flawed and tended to make Simpson and colleagues exclude regions of charge. Although Simpson and Scrase (1937) assumed they could find charge regions from zero-crossings in their corona current data, Simpson and Robinson (1941) noted that this method was "defective." Corona current does not have to be zero in the middle of a charge, a result of E from other charge regions in the cloud. Also, the thickness of the regions could not be determined. Another important contribution, which came from Simpson and Robinson's flights of alti-electrographs, was their convincing the scientific community that the storm structure was not a dipole but, rather, in many cases a tripole, which became the paradigm for thunderstorm charge structure.

(ii) Radioactive probes.   Moore et al. (1958) used tethered balloons to fly modified radiosondes that had radioactive (polonium) probes for measuring the potential gradient. They found the sonde reliable in clouds—at least in low to moderate magnitudes of E. Takahashi (1983) combined sensors in one instrument to measure particle charge, identify graupel, and measure the vertical component of E with polonium probes. He flew the instrument on free balloons, which he had to be recover to retrieve the particle size data.

(iii) Coronasondes.   Another widely used technique to measure E was with corona points, which were usually added to conventional rawinsondes. Because they are inexpensive compared to field mills, their simplicity, and their use in many field projects, we describe one such coronasonde. As its name implies, it uses corona points to cause point discharge to flow through an electronic circuit. The resulting signal is then multiplexed with the other sonde data telemetered to the ground. Usually one or two points aligned vertically are used to measure the vertical component of E. If the instrument swings as a pendulum, the variations in E sometime allow an estimate of the horizontal component of E.

Byrne et al. (1986) built such a coronasonde into a commercially available rawinsonde. Winds aloft and balloon location were from radio tracking of the instrument. They found that point-discharge current, I_{PD}, flowing in the sonde circuit depended on atmospheric pressure, temperature, wind velocity past the point, and the water vapor in the air. A high impedance resistor in series with the points tended to linearize the relationship between I_{PD} and E. They calibrated their coronasonde using an environmental chamber that contained a high-voltage supply to create E up to 100 kV m^{-1}. From their calibrations, they obtained this linear least squares relationship among I_{PD} and the other major variables:

$$I_{PD} = A \left(\frac{P}{P_0} \right)^B \left[E - E_0 \left(\frac{P}{P_0} \right)^C \right]. \quad (6.7)$$

This equation uses I_{PD} in microamperes to get E in kilovolts per meter ($P_0 = 1000$ mb). They fit the data above and below E = 30 kV m^{-1} separately; above the coefficients were $A = 0.086$, $B = 0.015$, and $E_0 = 14.0$.

If a water drop hangs from the bottom point, I_{PD} is affected until the drop falls off, which should happen quickly owing to the balloon rise of ≈ 5 m s^{-1} relative to the air. Freezing of supercooled drops on the points would be more adverse since the dielectric ice would dramatically affect I_{PD} from a point, but it requires a collision with the point by the supercooled drop. They could identify no such effect in their data. Finally, we note that care must be taken in designing the points due to the chance, albeit slight, of striking someone upon descent. For example, a flexible wire with a point is much less hazardous than a rigid one.

(iv) Balloon-borne field mill.   An electric field mill is a better instrument for use on balloons flown into storms. A variety of designs for balloon-borne instruments using field mills have been developed. For example, a spherical field meter, which used two small rotating-vane field mills to measure the total vector electric field was developed by Rust and Moore (1974) and flown beneath tethered balloons into thunderstorms over Langmuir Laboratory. Although reliable, its large mass kept it lower than a few hundred meters above cloud base and made it usable only in a restricted airspace. Chauzy and Raizonville (1982) combined a cylindrical field mill with two rotating-vane mills in a single balloon-borne instrument to measure the total E. Chauzy et al. (1991) also flew a large tethered balloon with several field meters, each with vertically opposed, dual field mills to obtain E$_z$ at several heights.

(v) Balloon-borne electric field meter.   The largest number of published vertical profiles of E have come from a low-mass balloon-borne electric field meter (EFM) which was invented by Winn and Byerley (1975). We examine some of its details and its calibration. The original EFM measured only the horizontal component of E, but the next version could measure the total vector electric field as the EFM ascended through storms and back down on a parachute (Winn et al. 1978). In the late 1980s, the EFM was redesigned with digital signal processing circuits. This version of EFM (Fig. 6.9) has been flown extensively in thunderstorms, mesoscale convective systems, and severe (and tornadic) supercells.

The sensing elements consist of two 15-cm diameter aluminum spheres. The spheres are also the transmitting antenna. A 400-MHz crystal-controlled radiosonde transmitter, analog circuits, analog-to-digital convertor, microprocessor, and batteries are inside the spheres. The smooth, spherical sensors inhibit point discharge and thus reduce the likelihood of adverse effects on the measurement of E. A motor powered by batteries at one end of the fiberglass tube rotates the spheres in the vertical plane at a rate of about 2.5 Hz. (There is a bearing at each end of the fiberglass tube.) The electric field causes opposite polarity of charge to be induced on the spheres, so the spin causes a sinusoidal signal at the spin frequency. As the EFM spins, a mercury switch in the tube between the spheres switches on and off and provides a pulse to determine which sphere is up. When this is compared with the rectified sinusoidal signal, the polarity of E$_z$ can be determined, just as from the reference signal in a rotating-vane field mill. Airflow past the styrofoam, rhomboid-shaped vanes causes the

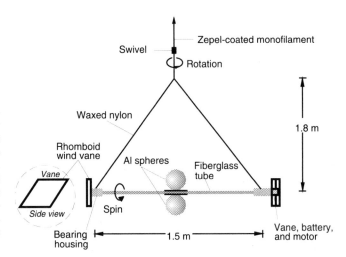

Fig. 6.9. Updated version of balloon-borne electric field meter, EFM, originally designed by Winn Byerley (1975). The aluminum spheres are the sensors of the electric field and the transmitting antenna. Electronics are housed in one sphere and their battery power source in the other. The dimensions shown are typical, but not critical. The vanes are low-density styrofoam. (From Marshall et al. 1995b: American Geophysical Union, with permission.)

instrument to rotate in the horizontal plane. This rotation about the vertical axis allows a determination of the horizontal component of E, E_h. There are two ways to consider charge placed on the dielectric parts of the instrument by precipitation (or any other means). One is to assume that charge on the EFM will be distributed symmetrically since the EFM spins, and that, "on average," charged particles impacting and leaving the EFM result in a uniform surface charge density on the dielectric. The second is that even if the previous assumption were not true, static charge that is on the dielectric parts and spinning with them will not be seen in the data. Only if precipitation charging could occur in a way that mimics the sinusoidal signal of the spinning and slowly rotating EFM would it give an apparent E. Spikes riding atop the E data from impacting and departing precipitation are not obvious, and they would be ignored even if there. Thus, no significant error in E will result from precipitation charging.

The EFM measures $\vec{E}$ and E_z; E_h must be calculated. An example of "raw" data voltage from the EFM is in Fig. 6.10. The components of the vector electric field, $\vec{E}$, are determined from

$$|\vec{E}| = (E_z^2 + E_h^2)^{\frac{1}{2}}. \qquad (6.8)$$

Often the total $|E|$ is plotted with its polarity defined by that of E_z. There is no polarity associated with E_h, but its direction can be determined from the output of the magnetic field sensor inside the spheres. The direction information in models up through 1996 was probably no better than 30°, and sometimes worse owing to noise on the magnetic field data. However, the sensor was useful for approximate direction of E_h and to confirm EFM spin and reversals in vertical motion. The latter is seen as the more slowly varying sinusoidal magnetic field signal goes to zero anytime the balloon stops moving up or down. Although the mercury switch provides the automatic detection of spin, the mercury freezes at about $-40°C$, and the switch quits operating. Then spin is confirmed with the magnetic field signal.

Although not always evident in the usual plots of E versus z in journals, the occurrence of lightning shows in the raw EFM data when it is close or makes a large enough ΔE to affect the calculation. Lightning field changes are easy to discern in the raw data (e.g., Fig. 6.11).

A typical instrument train for a balloon-borne EFM is shown in Fig. 6.12. A stainless steel fishing swivel with a small dielectric rain shield inhibits twisting of

Fig. 6.10. Example of 'raw' voltage in digital units from EFM showing change in magnitude due to nonzero horizontal E, E_h during rotation. Simply number and digital units are tied to absolute time and E via calibration and data processing. All of the E_h is at the rotational position as marked, which is when the azimuth of the EFM is in the plane of $\pm E_H$. When the EFM rotates 90°, about 2 s later, only E_z is detected. One measured E, E_m, can be calculated from each adjacent peak and trough during each spin. (From Stolzenburg 1993, with permission.)

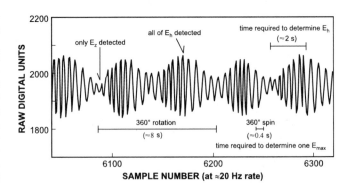

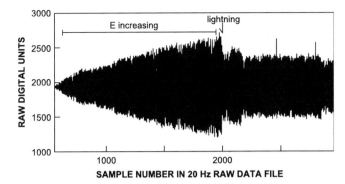

Fig. 6.11. Raw E data showing increasing E followed by a field change from lightning.

the line and facilitates rotation of the EFM. The line to the balloon is nylon monofilament treated with a water repellent solution of Nalan W and Zepel to inhibit water absorption, a leakage current, and perturbation of the E measurement. Waxed nylon is tied to the instrument because it is easier than the monofilament to tie and handle during launch preparations. Jonsson (1990) tested the treated lines. His test showed that the waxed nylon line has water-shedding properties at least equal to treated monofilament, with both having ≥1000 less electrical leakage than untreated nylon monofilament. No problems in the data have been apparent from line wetting. Marshall et al. (1995b) compared their balloon soundings with rocket profiles in the same storms and concluded that the data strongly indicated that the rigging lines in the balloon-borne instrument train do not adversely impact the measurement of E.

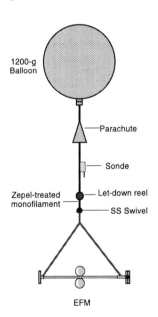

Fig. 6.12. Balloon-borne instrument train showing complement of sensors to measure E, thermodynamic variables, winds, and location. (From Marshall et al. 1995a: American Geophysical Union, with permission.)

This particular EFM has been used extensively in mobile ballooning operations in which an instrumented mobile laboratory facility has been used to make E soundings in a variety of thunderstorms and weather systems, including severe storms. The mobile laboratory portion grew out of the necessity to put instruments in desired locations of severe and possibly tornadic storms (Rust 1989). Then the development of a launch tube enabled the inflation and release of instrumented balloons in very high winds (Rust and Marshall 1989). The final piece of the facility is a box van for logistical support, including traveling with a fully inflated balloon. This led to extensive mobile ballooning to obtain soundings of E, thermodynamics, wind (Rust et al. 1990), and other parameters such as particle size and charge and x-rays; aspects of these will be presented in appropriate sections of later chapters.

(vi) EFM Calibration at Langmuir Laboratory. An EFM was tested and calibrated in a parallel plate capacitor at Langmuir Laboratory (altitude ≈ 3.2 km, P_{atm} ≈ 700 mb) as shown in Fig. 6.13 (see Marshall et al. 1995b). This large calibration setup also has been used to calibrate rocket and other balloon-borne instruments and to conduct other tests in simulated thunderstorm E, such as with water drops. The capacitor consists of a ≈3-m square aluminum electrode suspended 1.0–1.5 m above an aluminum plate of larger area lying on the ground. Known d.c. voltages up to 200 kV from a power supply were applied to the upper plate, yielding E as high as 200 kV m^{-1} according to

$$E_z = -\frac{V}{d}, \qquad (6.9)$$

where V is the voltage of the top plate relative to the bottom one, and d is the plate separation. However, if the instrument being calibrated is large compared with the capacitor dimensions, it can perturb E. Marshall et al. found that E from Eq. 6.9 was within a few percent of that directly measured for the EFM calibrations.

During calibration, the EFM hung from its waxed nylon line. Usually, E_z in the capacitor was recorded with a flush-mounted, rotating-vane field mill located near the center of the ground plane. Estimated errors in the mill values were ±10% owing to: uncertainty in the

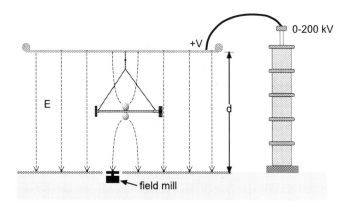

Fig. 6.13. Parallel plate capacitor and d.c. (direct current) high-voltage power supply to produce large E for instrument calibration such as for the EFM shown. Use of the flush-mounted field mill allows comparison with E = V d^{-1} for a particular instrument. The distortion of E around the instrument is indicated by the set of curved field lines.

V values, the unknown effect of the EFM on the field, and the fringing at the edges of the capacitor. In addition, there are errors in the EFM electronics: The output of the EFM charge amplifier, V_o, is proportional to the feedback capacitance, C_F (Eq. 6.2). The measured error in C_F is $\approx$ 3%, and the digitization error in V_o is $\lesssim$1%. Thus the error due to the electronics is <4%.

Calibration in positive E_z with d = 1.43 m showed that the EFM values were within 5% up to E_z = 123 kV m^{-1}. The consistency of the difference between calculated and EFM fields throughout the test indicated that the EFM values were biased downward by about 4%. There was also a calibration for negative E_z with the same electrode spacing. The flush field mill was in place. The EFM values were consistently lower, as for positive E_z, but within 3% of the field calculated from Eq. 6.9. The average error increased slightly with field magnitude and was 1.8% for negative E up to $-$133 kV m^{-1}.

Based on their full set of calibrations, Marshall et al. (1995b) concluded that errors in the EFM measurements are <10% for E values between $-$200 kV m^{-1} and +125 kV m^{-1}. Furthermore, all the evidence suggested that the EFM is linear across its dynamic range of about $\pm$220 kV m^{-1}. An extensive error analysis by Stolzenburg (1993) showed possible underestimates of $\vec{E}$ and E_z of up to $\pm$8%. This is from the combined effects of the rotation rate of 0.125 Hz and the instrument digitization frequency of about 20 points s^{-1}. Major overestimates of E_h up to >60% can occur. This problem was discussed in detail in Stolzenburg (1993) and Stolzenburg and Marshall (1994), but can be remedied in part by a modified EFM design to increase the data rate.

6.2.4. Determining space charge density from E

Knowledge of the electric field obviously is an important parameter in many aspects of atmospheric electricity. In addition, whatever instrumented platform is used to measure the electric field, another important parameter, the charge density, ρ, can often be calculated using Gauss's law,

$$\rho = \varepsilon \, \vec{\nabla} \cdot \vec{E}, \qquad (6.10)$$

where

$$\nabla \cdot \vec{E} = \frac{\partial E_x}{\partial x} + \frac{\partial E_y}{\partial y} + \frac{\partial E_z}{\partial z} . \qquad (6.11)$$

Recall that the coefficient ε is the permittivity of air, 8.86 $\times 10^{-12}$ F m^{-1}. There is currently no instrumented platform that allows simultaneous measurement of all the derivatives. Since we often talk about charge density contributions from different sources (e.g., ions, precipitation, cloud particles, etc.), we use the term *total space charge density*, ρ_{tot}, to denote specifically that all charges through which an instrument moves contribute to the calculated profile of charge density.

(i) Application to vertical soundings of E$_{aloft}$. Some investigators have calculated the horizontal ρ from transverses made with an airplane, but more extensive use of Gauss's law has been made to deduce vertical charge structure of storms from vertical soundings of E (we drop the vector notation for E unless specifically needed). The data from vertically moving instruments allow an adequate estimate only of the partial derivative in the vertical. Thus, we use a one-dimensional analysis, and only an approximation of ρ_{tot} can be obtained. The one-dimensional approximation in the vertical is

$$\rho_{tot} \approx \varepsilon \, \frac{\Delta E_z}{\Delta z} . \qquad (6.12)$$

The one-dimensional approximation does not mean that E_x and E_y are necessarily small, but it does assume that their horizontal derivatives, $\partial E_x / \partial x$ and $\partial E_y / \partial y$, are negligible compared with $\partial E_z / \partial z$, as they are when the charge structure is horizontally homogeneous and extensive. Fortunately in this context, "horizontally extensive" can mean spanning only a few kilometers around the measurement location. For example, Marshall and Lin (1992) studied dying thunderstorm anvil clouds, which tend to be horizontally stratified. They compared the approximation (Eq. 6.12) with a model having finite disks of charge plus image charges to assess the error in ρ_{tot} if the charge is not horizontally infinite. With disks of 12-km diameter, the calculated E profile was nearly identical to that measured; even with disks only 2 km across, the agreement was fair. Subsequently, Stolzenburg and Marshall (1994) tested the charge structures in a three-dimensional model that used Coulomb's law to compute $\vec{E}$ directly from the charge layers, which were calculated with Eq. 6.12. The modeled E_z profiles and the observed E_z profiles agreed well, indicating that they were not eliminating significant charge layers in the analysis, nor were they including any layers that were not significant. They showed that the one-dimensional approximation gives reasonable results if a charge region is only 6×6 km in horizontal extent. Their more general results were that the profile can be 80% of the distance from the charge center to its closest edge, and still Eq. 6.12 will yield reasonable values of ρ_{tot}.

Because vertical soundings measure E as a function of time, it is assumed that temporal changes in the measurement are dominated by the changing vertical position of the EFM and the cloud's overall E structure is quasi-steady during the sounding, except for transient lightning field changes. It is necessary to address what happens if the charging rate of the storm, which produces the E, varies while the vertical profile is made. The simple answer for balloons and airplanes (and maybe even rockets) is that we cannot determine all possible time-dependent changes from the measurements. However, ρ_{tot} can be calculated reliably in several situations. One often valid condition is that the charge in the layer or region does not change appreciably during the short time needed for the instrument to traverse that layer or region. A second valid condition is that the electrical evolution of the storm after it is electrified and producing lightning varies much more slowly than we might believe based on large ΔE and visually spectacular lightning transients. An exception can occur if lightning permeates the region near the EFM. In the balloon-borne EFM data, most lightning flashes show distinctly. Often when lightning transients happen, the slope of the vertical gradient of E stays the same, which shows that the same charge is still in the region. Thus, ρ_{tot} can still be found from the total change in E over the depth it occurred unless the lightning changes the charge density in the layer containing the EFM when the flash occurs. However, significant changes can, and probably often do, occur in charge regions during the time necessary to acquire the full vertical profile. There is no solution to this problem except making vertical soundings with vehicles that go fast vertically through the storm.

You can easily interpret vertical profiles of E and E_z qualitatively, as significant changes in E denote significant space charge density whose polarity is the same as the slope of E on its plot versus altitude. The greatest quantitative use of E to calculate ρ_{tot} has come from soundings of E made with balloons. Various aspects of applying Eq. 6.12 to balloon data were discussed in several papers (e.g., Schuur et al. 1991, Marshall and Rust 1991, Marshall and Lin 1992). A critical step in analysis is to define the minimum significant value of $\Delta E/\Delta z$ acceptable for a space charge calculation. Stolzenburg et al. (1994) defined a significant slope of E_z as one that yielded a $\rho_{tot} \geq 0.1$ nC m^{-3}. In their stratiform cloud soundings, they also required charge layer depths ≥ 200 m. Once these two thresholds were applied, they then eliminated any layers for which $\rho \Delta z <$ 50 nC m^{-2}. Hence, if a layer had low-charge density, it had to be relatively deep to be included; if a layer was shallow, it must have had relatively high-charge density to be included. These guidelines minimize, if not prevent, spurious charge layers from being included in the inferred storm structure. Note in addition that $\Delta E/\Delta z$ large enough to infer charge using the one-dimensional version of Gauss's law can occur when the measuring sensor moves toward or away from (as opposed to through) charges as occurred in a modeling study by Ziegler and MacGorman (1994). Thus, we must always be concerned about overestimating the number of charges using this technique. This is not likely to be a problem if the measurements are in the midst of horizontally extensive charges.

6.3. INSTRUMENT FOR CURRENT DENSITY CARRIED BY PRECIPITATION AT GROUND LEVEL

Mostly because of its historical importance, we discuss briefly the measurements of precipitation charge arriving at Earth. Such measurements started the process of scientists trying to understand why precipitation is charged and what effect falling, charged precipitation has on the cloud, but the measurements are inadequate to explain what is happening in the cloud (see Sec. 7.10). The earliest measurements of charge on precipitation were apparently by Elster and Geitel (1888); they measured charge by collecting many particles and calculating a current density (i.e., the "flow of charge" carried on the precipitation). Lenard (1892) documented that splashing drops cause drops to become charged. His findings are referred to as the *waterfall effect,* because he observed the phenomenon around water falls. This finding implies that for reliable measurements, precipitation must not splash out of or into the collection container, because it will change the current measured.

The basic elements of a ground-based instrument are seen in Scrase's (1938) design (Fig. 6.14). An electrometer has typically been used to measure the small currents. Electrometers, whether old or newer solid-state versions, are electronic devices that can be configured for measuring charge, current, voltage, and resistance over a wide range of values. Their merit in many applications of atmospheric electricity is their ability to measure very small charges and currents. Of course, the smallness of the signal to be measured means that shielding the collecting surface from the ambient electric field is critical for meaningful measurements. The tall, vertical shield and the grounded conical section inside and just above the funnel are used to shield the sensor from

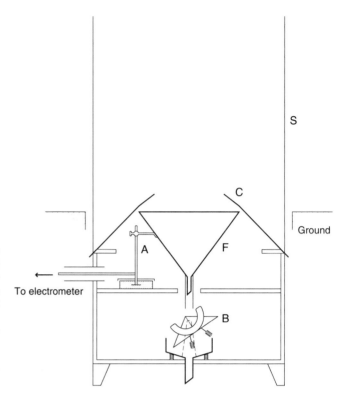

Fig. 6.14. Sketch of device used to measure precipitation current density (charge per time per area) at Earth. The tipping bucket, B, below the funnel measures the total rainfall volume per time and allows a calculation of rain rate. F is the collecting funnel connected by A to the electrometer, S is the electrostatic shield, and C is a conical shield. (From Scrase 1938 © Crown copyright. Reproduced with the permission of the Controller of Her Majesty's Stationery Office.)

the external electric field, including lightning field changes, and to prevent particles from charging by induction if they collide with the apparatus and then are measured. Aspects of the design of the shielding and entry orifices are given later. Depending on how the electronics were configured, investigators have measured the total charge on a given volume (e.g., one tip of the bucket) or the precipitation current density, which was determined from summation (integration) of the total charge per unit time per area. The area used in the calculation is the smallest cross-sectional opening above the collecting funnel. Perhaps the first self-recording precipitation current instrument was that developed by Simpson (1909). It was housed in a metal hut >2 m square. Scrase found that his gauge for measuring current collected only about half as much rain as a standard gauge; the tall electrostatic shield above the collector caused this by keeping out some nonvertically falling precipitation. The problem obviously comes from wind at the collector and is worse for smaller drops (or snow). The larger the opening, the smaller the amount of discrimination of particles and the smaller the percentage error from particles colliding with the edge and getting into the sensor with their altered charge. Finally, it is important that the water leave the instrument in a field-free region to avoid inductive charging of the exiting water, which would produce an erroneous current.

To calculate the precipitation current density from the output of such a device, Scrase (1938) and others

have divided the current density by the measured rain rate. The rain rate has units of velocity, but it is tied to the total amount of precipitation. What was calculated is sometimes called the "*specific charge density*," which means that the current density takes into account the volume of the precipitation and not the volume of the atmosphere within which that precipitation fell, so the volume is just that of, and therefore specific to, the water. Thus, this calculation of precipitation current density yields values that are about four to six order of magnitude greater than that calculated from both precipitation current measuring devices and from instruments that sense individual particles that are not calibrated in specific charge density (see Sec. 6.4.3).

6.4. CHARGE ON INDIVIDUAL PARTICLES

The charge on individual particles provides important information about the electrical state of a cloud, especially when measured inside the cloud. Gschwend (1922) made measurements of individual precipitation charge by using a collector of 12 cm^2 that was connected to an electrometer and on which he placed a chemically-treated paper to allow an imprint of the particle, and thus its size, to be obtained. Gunn (1947) devised a novel approach to making the measurement without touching or capturing the particle. He con-

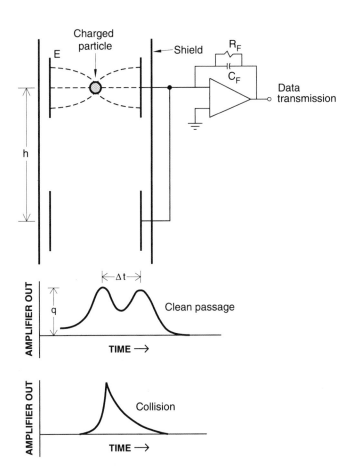

Fig. 6.15. Concept of instrument to measure charge, q, and fall speed, $h(\Delta t)^{-1}$, of charged particles with induction cylinders. The fall speed, coupled with assumptions, allows an estimate of particle size. Waveforms from the amplifier show the difference between clean passage and collision with an induction cylinder.

nected an electrometer to a hollow cylinder through which the particles fell. Interestingly, this design also allowed Gunn and Kinzer (1949) to measure terminal fall velocity of drops in air as a function of size. By artificially charging drops and allowing them to fall through two vertically separated cylinders, they could measure the particle velocity. A schematic of this technique is shown in Fig. 6.15. It works by induction of charge on the cylinder—hence the name *induction cylinder* (sometimes called induction ring). An outer conducting cylinder is needed to shield the sensitive induction cylinder(s) from external electric fields, field changes, and point discharge. Typically this shield extends beyond both cylinders. The circuit shown is a sensitive charge amplifier. Shown in the figure are two different waveforms. In a clean passage, each induction ring produces a Gaussian-shaped pulse; the time difference between the two is used to calculate the fall speed, if desired. If a particle collides with a cylinder, charge is transferred to the cylinder. This causes a rapidly rising pulse that is followed by an exponential decay of the waveform back toward zero. The decay is easily identifiable as the $R_F C_F$ time constant of the charge amplifier (Fig. 6.1) connected to the cylinder.

The cylinder radius and length affect the number of field lines that terminate on its inner surface from the charged particle and thus contribute to the determination of q. For example, a cylinder with a length to radius ratio of 1.0 collects 85% of the field lines from an on-axis particle; this increases to 95% for a ratio of 2.0 (Weinheimer 1988). The $R_F C_F$ time constant is chosen to be long (e.g., ten times) compared to the slowest particle transient time to minimize loss of signal pulse amplitude. Particle charge measurements have been made at the ground, but those aloft are of more importance, so we concentrate on measurements aloft. The data are transmitted to ground if the sensor is on a balloon and usually recorded on board for a similar system, but with horizontally aligned induction cylinder(s), on an airplane. Modern instruments usually convert the analog signal to a digital form, either before or after transmission to the ground.

Calibration of such charge measuring devices can be done with falling charged water drops, ball bearings, and charged pellets shot from a gun. Unless high-speed particles are required, ball bearings are the most convenient (Fig. 6.16) by easily allowing variation of charge and size. Winn (1968) described the technique and derived an equation relating the charge on the spherical bearing:

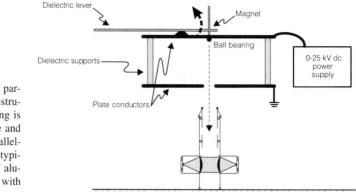

Fig. 6.16. Calibration technique for particle charge (and size) measuring instruments. The charge on the ball bearing is controlled by varying the magnitude and polarity of voltage applied to the parallel-plate electrodes. The ball bearing is typically of steel, and the plates are of aluminum. (After Bateman et al. 1994, with permission.)

$$q = 1.83 \times 10^{-10}\, E r^2, \qquad (6.13)$$

where E is the electric field between the plates, and r is the radius of the ball bearing.

We now look in more detail at two crucial aspects of design: detecting a collision, which is often called a splash, and avoiding instrument-induced artifacts. Usually a collision cannot be detected directly if it occurs on any part of the instrument or the vehicle carrying it ahead of the measurement sensor. An exception is airplanes, on which investigators have reported success in identifying such collisions from the different speed of the particle as it went through two induction cylinders after colliding with the shield. Collisions with an induction cylinder are easy to detect, as seen in Fig. 6.15. The best procedure is to configure the instrumentation so that collisions from the transporting vehicle are highly unlikely. On airplanes, the instruments have been mounted on the nose and under the wings and far enough away so propeller spray is not a problem. An advantage to proper placement under the wing is that it helps to electrostatically shield the instrument and reduces the likelihood of point discharge at its entrance. Point discharge has two deleterious effects: It makes the signal noisy from the sensors from which it emanates, and it releases ions that could alter the charge of particles headed through the instrument. Both lead to errors, with the second creating charges that are not indicative of the undisturbed cloud. On balloons, the optimal sensor placement is on top of the balloon—impossible for most ballooning operations. More practical is the use of a dielectric boom to hold the instrument out from under the balloon. Particles falling from the undersurface of the balloon will not go through the instrument, and it is highly unlikely that a particle can bounce off the balloon and still be able to get through the instrument cleanly. Having the sensor out on the boom reduces any possible problems associated with currents that might flow in response to a high E, most likely in the line to the balloon. Quantifying errors resulting from a particular system configuration seem impossible in a practical sense owing to the large number of variables. Good design dictates using as many optimal procedures as possible.

A quasi-constant E does not cause a signal from the induction cylinder, as the circuit time constant is much less than the slowly changing E. This is not the case during lightning; then a rapidly changing charge can be induced on the cylinder. Even during such a transient, charge on precipitation can still be measured if the combined signal does not saturate the electronics.

A key to evaluating a technique for measurement of particle charge is that if a particle can collide and then be measured by the device without the collision being noticed, serious errors can result. It is then impossible to trust the data. A design innovation apparently first used by Marshall and Winn (1982) is to place an induction cylinder as the leading edge of the whole instrument. If no particle collisions are seen, then it is obvious that the particle did not come off the sensor entry and have its charge modified. As the first cylinder is exposed to the ambient electric field, its amplifier time constant often is made shorter to reduce the duration of any saturation during large field changes from lightning. However, instruments without leading detectors can give useful results if they have multiple entry orifices and other design features that increase the likelihood that a particle that hits the instrument will not get through all its sensing elements and that allow identification of such a particle as contaminated. The careful reader of the literature will pay particular attention to whether or not the instruments used to gather the data are carefully designed, well documented, and if their calibration is clearly described.

6.4.1. Measurement of particle charge with an airplane

Gunn (1947) mounted two, horizontally separated induction cylinders of different diameters in a grounded conical shield under the fuselage of an airplane to measure charge on precipitation as a function of altitude. He reported that the few drops that splashed on the sensor could be detected by their different, characteristic waveforms. He did not say how, but he stated that he could detect those striking the shielding cone and breaking up ahead of the induction cylinder. He calibrated the instrument using charged shot from an air rifle. The shot was captured in a Faraday cup connected to an electrometer set up to measure charge, and thus it provided an absolute measurement of the particle charge. In what has became a widely used configuration, Latham and Stow (1969) used two cylinders of the same size within a shielding cylinder. They found that the dual cylinders helped identify artifacts in the data. Also, particles that collided with the shield were unlikely to go through both induction cylinders. Laboratory tests showed identical instrument operation when covered with ice. External electric fields were shielded, with the problem of noise occurring only when the airplane went into point discharge. Their calibration via calculation, not direct measurement, may have been in error (Weinheimer 1988). Nevertheless, they recognized key points of good instrument design.

We turn now to modern instruments on aircraft. One instrument, designed by Saunders et al. (1988), measures only charge. It uses an outer shield, but they stated that it is critical to keep particles that collide with the leading edge from making it through the two induction cylinders for measurement. They also designed the instrument so particles could not go between the wall and the induction cylinder by imbedding the sensor in a solid dielectric (Teflon) spacing between the shield and the cylinders. The shield and tubes are de-iced with heaters. The amplifier on the second induction cylinder has a time constant 40 times longer than the first. The time constants are less than the transient time of the particle through an induction cylinder, so bipolar pulses are produced. Any charged particles that touch a cylinder cause a unipolar pulse and are identified as spurious. The second tube has more particle collisions than the first, but provides validity of the pulse from the first. The detection range is ± 2 to ± 200 pC.

Weinheimer et al. (1991) described a similar design for a charge- and size-measuring instrument (Fig. 6.17) for use on the NCAR/NOAA instrumented sailplane. A sailplane has a major advantage over engine-powered aircraft in that, if hail is not present, it can often spiral upward in the updraft and obtain vertical profiles of the particles and other parameters. Weinheimer et al. tested the induction cylinders ahead of and behind the laser beam of a two-dimensional optical array spectrometer with grey scale to determine the size of many of the particles. They concluded that behind the beam is better to avoid many false triggers of the optical probe. To prevent excessive ice buildup, they also found it necessary to put a heated entry orifice just ahead of the first induction cylinder. Because the induction cylinder has a larger cross section than the size-measuring beam, charge can be measured on particles that miss the beam. Particles that go through both are identified by the time between sensors, and the charge and size of individual particles are determined. The key to this technique is a requirement that the time intervals between the laser beam and each cylinder are correct for the airspeed of the sailplane. In this way, particles that collide

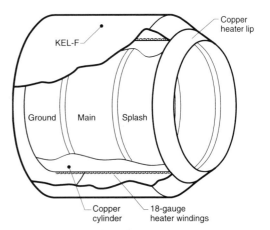

Fig. 6.17. (a) Sketch of induction cylinders for measuring particle charge built into the front of a PMS 2D probe, shown in (b). The detected charge can often be attributed to a particular particle image, whose size, shape, and state often can be determined when the particle goes through the beam of the probe. (See Weinheimer et al. 1991; figure provided by D. Breed, with permission.)

with the laser beam support or cylinder heater are removed from the data. An exception to this might be if a frozen particle collides at such a shallow angle that it does not slow but still goes through the cylinder. Whether such a glancing collision will last long enough for the particle charge to change is not known, but the percentage of such events in the data base is likely very small. The researchers then calibrate the instrument using charged conducting spheres (BBs) shot from an air gun and collected in a Faraday cup. They estimate their error in measuring charge at about 20%. The particle-size detection is 50 µm to 8 mm. The response of the probe to particles <200 µm is poor and is a function of shadowing, which they describe. The minimum detectable charge and size vary with the noise in the cloud. Sometimes they could not detect a charge of 3 pC, but generally the minimum detectable is 1–3 pC. The maximum detectable charge with this system is about 500 pC. They state that determining charge and size for particles <1 mm is problematic. They provide a detailed description of analysis procedures and their impact on the results. Most spurious associations of charge and size are going to be on particles of <1 mm.

6.4.2. Measurement of particle charge with a balloon

In Fig. 6.18 is the sensing part of the balloon-borne instrument developed by Marshall and Winn (1982) to measure precipitation charge and fall velocity. They used two 15-cm diameter aluminum spheres to house the sensors, electronics, and batteries, making the device similar in exterior design to that of the balloon-borne field meter made from two spheres. They purposely designed their instrument to have an induction cylinder as the entry port. The top cylinder was used to detect the presence of charged particles that collide; the two better shielded cylinders were used to measure charge and fall speed. All cylinders were 8.3 cm in diameter. The overriding advantage was that a particle could not easily get into the sensing area after colliding with the instrument. Nearly all colliding particles were going to hit the outer induction cylinder and be identified as a collision. The unlikely exception could be a particle that collided with the shield and then bounced into the instrument without colliding with an induction cylinder. The device measurement sensitivity was ±10 pC to maxima of −350 pC and +450 pC; the asymmetry was due to amplifier design. The instrument had the advantages of light weight, small size, and design features to inhibit point discharge. Shortcomings included the high minimum detectable charge of 10 pC and using fall speed to determine particle size. The upper- and lower-most surfaces were covered with a dielectric material to inhibit point discharge from the instrument. In general, the ability to see point discharge

on the data depends on signal processing. The signal from point discharge manifests itself as a high-frequency noise and is discernible unless signal processing filters it out.

Bateman et al. (1994) added to the design of Marshall and Winn with an inexpensive optical technique for measuring size (Fig. 6.19). It solves the problems of using fall speed to obtain particle size. The induction cylinders are 6.6 cm in diameter; the size sensor is 5.5 cm in diameter and oriented horizontally to intersect the particle after it passes through the induction cylinders. Size is determined from the amount of light blocked as a particle passes through the beam. Because the particle is not imaged, its size is given as equivalent diameter (or cross sectional area) from a calibration using spheres. Thus, nonspherical particles are sized with their equivalent diameter. Less that 1% of the particles with charges did not also pass through the size-measuring beam. Detection ranges of the instrument are ±4 to ±400 pC and 0.8–8 mm diameter. This instrument is one of the few whose onset threshold for point discharge has been reported in the literature; point discharge is not discernible until $|E| \geq 100$ kV m^{-1}. An updated instrument with greater low-end sensitivity and dynamic range is being developed.

In a follow-up comment expressing concerns similar to those voiced about the balloon rigging line by Jonsson (1990), Jonsson and Vonnegut (1995) questioned whether the measurements of particle charge could be in error from current flow in the line. They were concerned, in part, because measurements by Marsh and Marshall (1993) and Marshall and Marsh (1993) (called the MM papers here) were made with untreated rigging lines. Jonsson and Vonnegut stated that those results subsequently appeared to be contradicted by results in Bateman et al. (1994), who used a similar instrument, but one with rigging lines treated to be water repellent (Bateman et al. 1994). We note possible confusion in that one of the results of Bateman et al. was that the total charge density in the cloud was not accounted for by precipitation. On the other hand,

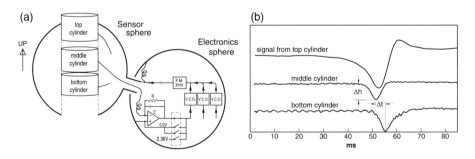

Fig. 6.18. (a) Sketch of balloon-borne instrument designed with its foremost part (assuming particles come in from top) to be a charge sensing element. This optimizes the identification of particles and reduces chances of measuring any contaminated particles; (b) data waveform examples for one charged particle going cleanly through the instrument. From calibration data, Δh yields q and Δt gives the fall velocity. (From Marshall and Winn 1982: American Geophysical Union, with permission.)

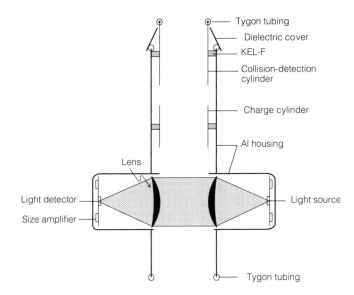

Fig. 6.19. Sketch of balloon-borne sensor that measures charge with an induction cylinder and equivalent diameter with optics. The latter allows uncharged precipitation to be measured. (From Bateman et al. 1994, with permission.)

Marshall and Marsh (1995) countered by pointing out that the values of individual charges and histograms of charge were similar among the studies. Noted in defense of the MM measurements were (1) their similarity to measurements made with a different type of instrument on the NCAR sailplane, but also at Langmuir Lab, and (2) that the majority of the data questioned by Jonsson and Vonnegut were collected at temperatures too cold for the problem to exist according to Jonsson's test results. To date, no definitive information has shown that the MM or similar balloon-borne measurements are contaminated to any significant degree, but rather that they appear to be reasonably self-consistent, at least regarding the distribution of charges carried on individual particles. Such a debate serves as an important reminder and highlights the fact that making such measurements in the harsh environment of a thunderstorm is fraught with problems.

6.4.3. Calculation of current density from individual particles

Current density can be calculated from individual particle data. First, the precipitation charge density for increments of charge magnitudes is calculated by summing individual charges. For horizontal flight paths, as when using airplanes, the precipitation space charge density, ρ_P, is calculated over the desired length of flight path using the expression

$$\rho_P = \frac{\sum_i N_i q_i}{Au\Delta t},$$ (6.14)

where A is the "acceptance area" (about 75% the geometric area) swept out by the induction cylinder device, u is the airplane speed, Δt is the time of the summation,

N_i is the number of particles with charge q_i. The acceptance area is smaller than the geometric area because of the particle fall speed (i.e., the particle does not go through horizontally, but drops vertically as it traverses from its entry to its exit). The particle must enter high enough in the cylinder opening to pass through cleanly. Then the precipitation current density can be calculated from ρ_p with

$$J_P = \rho_P \bar{v},$$ (6.15)

if we assume an average fall velocity, $\bar{v}$, which can be based on observed particle size. The calculation assumes all particles are falling. For precipitation-size particles this is often, but not always, true. Instruments that operate in the vertical rather than in the horizontal do not have this ambiguity, although they may be able to measure only the particles going in one direction. Thus, a different J_P can result from a sample of the same cloud's precipitation population, depending on how the measurement is made.

In the case of balloon-borne measurements, the individual particle and fall velocity data can be combined to obtain both ground-relative and balloon-relative current densities. It is easier to do balloon-relative calculations. We summarize the details, which Marshall and Winn (1982) provide, of such a calculation of ρ_p. The precipitation charge density from the contribution of each ith charge increment can be calculated from

$$\rho_P = \sum_i \frac{q_i}{Av_i \Delta t},$$ (6.16)

where A is the cross-sectional area of the induction cylinder, and Δt is the time of the summation. The precipitation current density can be calculated from

$$J_{P_{bal}} = \sum_i J_i,$$ (6.17)

where each increment of current density is $J_i = \rho_{P_i} v_i$. The direction of the velocity is used to determine the precipitation current density. Recall that falling net negative charge gives a positive J_P in our polarity convention (often in the literature the polarity is that of the polarity falling).

6.5. MEASUREMENT OF SMALL, CLOUD PARTICLES ALOFT

The charge on small, cloud-size particles is as important as that on precipitation particles. Although there is no absolute demarcation size, we follow McDonald (1958) and consider that droplets of diameter <200 μm are cloud droplets. The measurement problems are very difficult owing to the small magnitudes of both the charge and size of these particles. Very few such measurements have been made, and most of these have been at the ground. The measurements at the ground can be useful in limited research (e.g., Takahashi and Craig 1973, Saunders 1978). Two serious problems exist: (1) Point discharge vitiates any meaningful interpretation of measurement at the ground in thunderstorm environments; and (2) Thus far, the inability to make the measurement through the cloud has not allowed a determination of the complete charge distribution inside thunderstorms. We limit our discussion to instruments used to measure the cloud particle charge above the ground and inside clouds.

Moore et al. (1961) measured net charge in given ranges of small particle sizes and types using filtration techniques. They presented convincing evidence, however, that, inside all clouds, airborne measurements using filters are invalid. An important part of their proof involved changing the dominant polarity of measured space charge by changing from a dielectric to a conductive intake orifice. Their definitive results remove all such measurements from our consideration.

If the charge and size of individual particles are desired, they have been obtained typically by letting particles fall between vertical parallel plates to which a voltage (direct or alternating) is applied. Both the mass and charge can be calculated. Colgate and Romero (1970) described an instrument that was flown on a tethered balloon into the bases of thunderstorms over Langmuir Laboratory. They used an oscillating acoustic wave to measure mass and parallel plates with a d.c. voltage to measure charge. Their calculated space charge density of 16 nC m^{-3} is high. We cannot determine if this calculation is unrealistically high, as there are no precipitation or E data with which to compare or to estimate the total space charge density.

The instruments that, perhaps, came closest to achieving the goal of measuring cloud particle charge aloft were those developed by *Takahashi* (1975) to measure charge and size. He described a sonde that measured the charge of raindrops with a cup and cloud droplets with another cup having a wire mesh bottom. During calibration, he simulated air flow in the as-

cending sonde by using a pump to pull droplet-laden air into and through the cup with the mesh bottom. The minimum detectable cloud particle charge was 0.3 fC (femtocoulombs) within a size range of 5–150 μm, although the size of individual cloud droplets was not determined. He reported that the instrument produced no spurious cloud particle data when large raindrops splashed into the top cup. This test also was done with airflow equivalent to flight conditions. In another sonde, he used an isolated dish in a shielding cylinder to capture particles and measure their charge. He used an induction cylinder above a microphone to detect charge and size. The limits of detection were about 7 fC and 300 μm diameter, which made it sensitive enough to detect very small raindrops with small charges.

6.6. MAXWELL CURRENT SENSOR

A fundamental parameter of storm electrification is its production of current. In practice, this current is measured as a current density at a particular site. There are two ways to measure the current density in the storm environment: (1) The direct method uses an isolated bed of sod connected to appropriate electronics. Blakeslee and Krider (1985) used such a sod bed (Fig. 6.20) along with field mills and conductivity sensors at Kennedy Space Center. The sod bed insulators were unheated, but a conducting cylinder surrounded their upper portion to keep water from running down the insulators. The conducting pan containing the sod was connected to a solid-state electrometer circuit; (2) The second method of measuring the Maxwell current den-

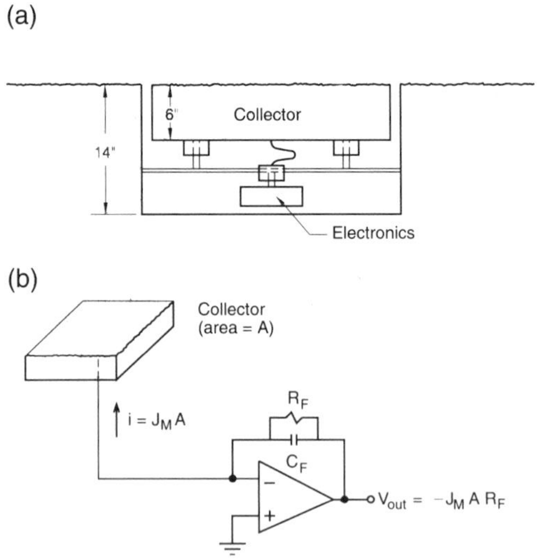

(a)

(b)

Fig. 6.20. (a) Sod bed for measuring Maxwell current from the ground beneath thunderstorm. (b) Solid-state electrometer circuit to measure current. (After Blakeslee and Krider, 1985, with permission.)

sity is to estimate its value from the slope of the output of a field mill (i.e., the electric field change $\Delta E/\Delta t$) when $E \approx 0$. This technique has been verified by comparison with direct measurement of the Maxwell current density by Krider and Blakeslee. There was usually good agreement between the measured and inferred Maxwell current density values as described in Section 7.11. Apparently, a single field mill near a thunderstorm is adequate to obtain a good estimate of Maxwell current density for the storm.

6.7. POINT-DISCHARGE SENSOR

The measurement of point discharge is perhaps the oldest and simplest measure of the electrical state of the thunderstorm environment. In his brief historical review, Chalmers (1967) noted that point discharge was measured as early as 1826 by Colladon (1826). Many of the measurements were analyzed to provide estimates of the net current flow in the atmosphere. They have also been calibrated to provide a measure of the electric field. Since there is a turn-on, or onset, electric field, E_{on}, for a point to emit corona, points have been elevated or even raised to a high potential by a high-voltage supply to keep them emitting point discharge even in low ambient E to facilitate their use in measuring E quantitatively. They offer an inexpensive technique for warning of electrical development in clouds overhead, but field mills are more widely preferred for quantitative measurements.

A point-discharge instrument consists of a point, a way to sense electronically the current that flows, and a data display device. A simple instrument could be a point connected directly to a nonrecording indicator such as a microammeter. A more typical, simple instrument has an amplifier between the point and the display device (Fig. 6.21).

The sensor can be calibrated by applying known currents, which are usually in the range of microamperes, to the point. This calibrates the instrument's current response, but does not calibrate its output relative to its placement in the atmosphere. The geometry of the point and its height above ground, affect its calibration.

Nor is its output yet calibrated in terms of E. That is more complicated, as the current is a function of E, the form factor, and the wind speed past the point. Several investigators have tried to define the relationship. Note that we have left the expressions (Eqs. 6.18–6.20) in their original potential gradient form. One early and widely used equation is the empirical formula developed by Whipple and Scrase (1936), which they expressed as

$$I_{PD} = a(F^2 - M^2), \qquad (6.18)$$

where F is the potential gradient at ground nearby, a = 0.0008 and M = 7.8 V cm^{-1} for positive F and a = 0.0010 and M = 8.6 V cm^{-1} for negative F; here I_{PD} is in microamperes. They did not take into account effects of the wind. Much of the laboratory assessment of point discharge has used the potential of the point. For the measurement of E, the current in the sensor must be related to E during some type of calibration. For example, Milner and Chalmers (1961) used a single point at a height of 15 m and further explored Chalmers's (1957) finding that

$$I_{PD} = G(V - V_0)(W^2 + a^2 V^2)^{\frac{1}{2}}. \qquad (6.19)$$

Milner and Chalmers reported the constant G = 6.5 × 10^{-11}, V_0 = 900 volts, and a$^2 \approx$ 2.5 × 10^{-9}. This indicates that the point-discharge current from the point is a function of the square of the wind speed, W. Rewriting in terms of the ∇V at the ground and $V = h\nabla V$ for a height, h, we get

$$I_{PD} = Gh(\nabla V - \nabla V_0)(W^2 + a^2h^2\nabla V^2)^{\frac{1}{2}}. \qquad (6.20)$$

A novel version of a point-discharge sensor was developed by Toland and Vonnegut (1977) to determine the maximum E at the surface of large bodies of water, where we expect E to be higher owing to much reduced or even an absence of point-discharge created space charge. The point was attached to a small motor that was activated by point discharge when E > 35 kV m^{-1}. The motor retracted the point by an amount determined by the maximum E (as the point became shorter, a larger E was required to activate it). The instruments

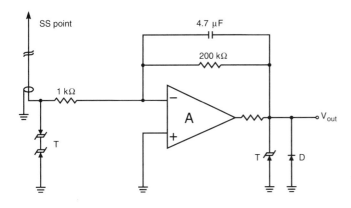

Fig. 6.21. Schematic of point-discharge sensor. SS denotes stainless steel point to resist weathering. (A hypodermic needle can be used.) The T and D denote transorbs and diode, respectively, for protection from lightning-induced transients.

were calibrated before and after use. They measured a surface-level $E_{max} = 130$ kV m^{-1} on a lake.

6.8. IONIC CONDUCTIVITY

Recall (Sec. 1.4.3) that conductivity of the atmosphere is due to small-ion movement under the influence of E. Its definition is

$$\sigma = nq_e k, \quad (6.21)$$

where n is the ion number density, q_e is ionic charge, and k is the mobility. Small ions are taken as carrying a single elementary charge per ion. For multiple mobilities, the equation is expressed as a summation over all the appropriate mobilities. There is an equation for each polarity. The total conductivity of the atmosphere can be written as

$$\sigma_{tot} = \sigma_+ + |\sigma_-| . \quad (6.22)$$

The methods to measure conductivity fall into two categories: dissipation and aspiration. Dissipation techniques use an isolated charged conductor exposed to the air. Conductivity is determined from the charge decay with time,

$$Q_\pm = Q_{0\pm} \exp\left(\frac{-\sigma_\mp t}{\varepsilon}\right), \quad (6.23)$$

where Q_0 is the initial charge on the object and is opposite in polarity to the conductivity to be determined. It is the second method, aspiration, that is almost always used. The apparatus is called a Gerdien (1905) capacitor since it was first used by him in 1905 to measure conductivity. The device also can be configured as an ion counter, and we begin our treatment of it in this mode.

The apparatus can be understood as a hollow cylinder with a smaller coaxial solid cylinder. Air is drawn through the hollow cylinder and across the inner one by a fan. A constant potential difference, V, is maintained between the cylinders. Ions move under the influence of the electric field produced by the V. The current flow that results from maintaining the constant V is measured. Configurations of Gerdien cylinders are shown in Fig. 6.22. Typical design values are a = 5 mm, b = 5 cm, and L = 20 cm.

The theory of operation was summarized by Israël (1970) and Chalmers (1967). Using Gauss's law on a long cylindrical capacitor (i.e., L ≫ b), calculating the potential difference, and then solving for the electric field inside the cylinder, E_{cyl}, yields

$$E_{cyl} = \frac{V}{r \ln\left(\dfrac{b}{a}\right)}, \quad (6.24)$$

where r is a radius measured from the central axis. Under the force provided by E_{cyl}, an ion of mobility, k, moves an incremental radial distance, dr, given by

$$dr = kEdt. \quad (6.25)$$

The air speed through the cylinder is u. In a cylinder, u = u(r), but later we will see that the conductivity is not a function of velocity. For ions starting at the maximum radius, b, and moving inward, the time to go from b to a is

$$
\begin{aligned}
t &= \int_b^a \frac{dr}{kE} \\
&= \frac{\ln\left(\dfrac{b}{a}\right)}{kV}\int_b^a r dr \quad (6.26) \\
&= \frac{(b^2 - a^2)\ln\left(\dfrac{b}{a}\right)}{2kV} .
\end{aligned}
$$

During t, the air moves a distance, ut. If $ut < L$, then all ions moving toward the central electrode will be collected. The condition that is met when the velocity is

$$
\begin{aligned}
u &< \frac{L}{t} \\
&< \frac{2kVL}{(b^2 - a^2)\ln\left(\dfrac{b}{a}\right)} \quad (6.27)
\end{aligned}
$$

is the condition in which the instrument operates as an ion counter.

To measure ionic conductivity, the device needs to be set up such that $u \gg L/t$. Then only a fraction of the

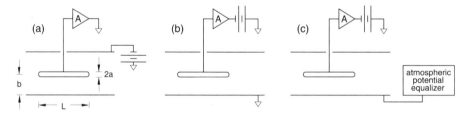

Fig. 6.22. Configurations for use of Gerdien cylinder. The battery creates the accelerating potential, V, between the central electrode and the outer cylinder. The electrometer amplifier circuit is denoted by the symbol A. (After Anderson and Bailey 1991, with permission.)

ions will reach the inner cylinder. They will be the ions within a critical radius r_{crit}, which is $<b$, given by

$$u(r_{crit}^2 - a^2) = \frac{2kVL}{\ln\left(\dfrac{b}{a}\right)}. \qquad (6.28)$$

If n is the number of ions per unit volume, the number entering the cylinder within the critical radius per unit time is

$$n_{crit} = \pi(r_{crit}^2 - a^2)nu, \qquad (6.29)$$

which, from the previous equation, is

$$n_{crit} = \frac{2\pi nkVL}{\ln\left(\dfrac{b}{a}\right)}. \qquad (6.30)$$

The charge arriving at the central cylinder per unit time (current) can be expressed as

$$\frac{dq}{dt} = \frac{2\pi nkVLq_e}{\ln\left(\dfrac{b}{a}\right)}. \qquad (6.31)$$

As nkq_e is the conductivity, σ, the current flow in the device, i, is proportional to the ionic conductivity.

$$i = \frac{2\pi VL\sigma}{\ln\left(\dfrac{b}{a}\right)}. \qquad (6.32)$$

While this can be used to measure σ, an alternative expression is often used. The capacitance, C, of a cylindrical capacitor is, from Gauss's law,

$$C = \frac{2\pi\varepsilon L}{\ln\left(\dfrac{b}{a}\right)}. \qquad (6.33)$$

From the last two equations

$$i = \frac{CV\sigma}{\varepsilon}. \qquad (6.34)$$

Finally the general equation for conductivity for each polarity can be expressed as

$$\sigma_\pm = \frac{\varepsilon i_\pm}{CV_\pm}, \qquad (6.35)$$

where the subscripts apply to the polarity of the conductivity, the potential of the outer cylinder relative to the inner, and the polarity of the current in the device, which is related to the instrument output voltage. (Note that the amplifier or full circuit output polarities could be reversed depending on the design of the electronic circuits.) The units for conductivity used now are siemens per meter ($S\ m^{-1}$), which replaces mho per meter or $\Omega^{-1}m^{-1}$, which, until recently, was often used in the atmospheric electricity literature.

6.8.1. Practical considerations

The Gerdien cylinder does not lend itself to direct calibration; there is no easy way to produce air of known electric conductivity. Therefore, we have to use theory and supporting information. The capacitance of the system can be measured directly with two measurements: one with and one without the center cylindrical electrode. The part of the support for the electrode that extends into the volume surrounded by the outer cylinder is part of the capacitance and also captures some of the ions. So the measurement of C without the central electrode is made with a shorter stem that does not extend into the volume. The difference in the measured two capacitances is the capacitance of the collecting electrode. A significant problem is the determination of zero conductivity, because $V = 0$ out of the circuit does not necessarily occur at $\sigma = 0$. Three techniques can be used (Kraakevik 1958). Each can have a different value of the circuit output, which supposedly corresponds to $\sigma = 0$: (1) Short the input of the electrometer to ground; (2) Reduce the flow, u, to zero; and (3) Switch the accelerating potential to zero and leave the rest of the system operational, which is called a *dynamic zero*. A qualitative example of the outputs for these three is depicted in Fig. 6.23. The dynamic zero is most accurate because it accounts for the current flow due to impaction of ions on the collector, even with no accelerating voltage applied to it. The equation for finding conductivity converts to the differences between the value during the dynamic zero and when V is applied. The equation to calculate conductivity in practical use thus becomes

$$\sigma_\pm = \frac{\varepsilon\Delta i_\pm}{C\Delta V_\pm}. \qquad (6.36)$$

Other aspects of the practical use of a Gerdien cylinder have been studied by Anderson and Bailey (1991). They obtained their experimental data by simultaneously comparing the measurements from their control Gerdien cylinder configuration with several others. The tests were conducted in a large underground clean room. Figure 6.22 shows their depiction of configurations for applying the accelerating voltage relative to the electrometer circuit and the outer cylinder. The optimum (Fig. 6.22c) has V applied to the inner electrode to avoid repulsing the polarity of ions that are the same as the po-

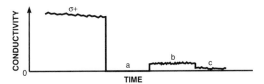

Fig. 6.23. Schematic of signals for determining the value of zero conductivity and a simulated data portion. The various techniques for determining $\sigma = 0$ are explained in the text. The dynamic zero is the preferred technique.

larity of V on the outer cylinder. This procedure holds the outer cylinder entrance at the same potential as the atmosphere to avoid changing the local ion density near the intake. If the accelerating potential is applied to the outer cylinder, large errors will result from ion repulsion, unless the V is small or the flow rate very high.

If tubes are used to duct the air into the cylinder, there are losses by diffusion to the walls. Anderson and Bailey (1991) found expressions for the loss of each polarity. In each of their semi-empirical expressions, n_0 is the original concentration of ions, Φ is the volume flow rate, and the other variables are as given before. For positive ions, the fractional loss is

$$\frac{n_+}{n_{0+}} = \exp\left(-3.8 \times 10^{-3} \, d^{-\frac{7}{8}} \, \Phi^{-\frac{1}{8}} X\right), \quad (6.37)$$

and for negative ions, the expression is

$$\frac{n_-}{n_{0-}} = \exp\left(-4.5 \times 10^{-3} \, d^{-\frac{7}{8}} \, \Phi^{-\frac{1}{8}} X\right), \quad (6.38)$$

where d is the diameter and X the length of tube connected to the Gerdien cylinder. Since it is common to use bends in the tubing, they checked a 90° bend and found no ion loss if the right-angle tube is not less than the diameter of the outer cylinder.

It is important to obtain an i versus V curve for a particular instrument setup to ascertain that the apparatus is operating in the conductivity mode. An example of such a curve with the conductivity regime marked is in Fig. 6.24. In addition, Anderson and Bailey (1991) recommended that V be a factor of three lower than the maximum, which can be calculated from

$$V = \frac{\varepsilon \Phi}{C k_{crit}}, \quad (6.39)$$

where k_{crit} is the *critical mobility*, which is the lowest mobility fully collected by the apparatus.

Use of a Gerdien cylinder in precipitation has the added problem of getting water into the cylinder, which

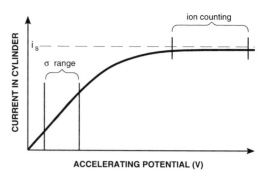

Fig. 6.24. Current versus accelerating potential curve for a Gerdien cylinder at a given flow rate. In the quasi-linear, ohmic region—σ range—is where conductivity is measured. (Ion counting occurs where the current is saturated, i_s, and does not change with changing accelerating potential.)

creates noisy signals from impaction of low-mass particles and can easily wet the insulator and short out the very low-current signal to the electrometer. To obtain measurements aloft in thunderstorm conditions, Rust and Moore (1974) used a specially designed Gerdien cylinder (Fig. 6.25) flown on a captive balloon at Langmuir Laboratory. It was calibrated as above and corroborated by placing it on the ground next to a flush-mount system. In the balloon-borne instrument, the accelerating potential was only 9 V, which minimized the problem of altering the ion concentrations near the intake. Furthermore, the spherical intake (and the rest of the conducting cylinder) tended toward the atmospheric potential. The design of the intake and the use of a guard ring (Keithley Instruments 1972) made the Gerdien cylinder functional in precipitation and high electric fields. The insulating material around the guard ring (really a short cylinder) and the support for the center electrode was made with KEL-F. This removed the problem of piezoelectric signals found in tests in a

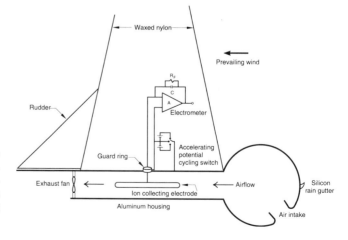

Fig. 6.25. Balloon-borne conductivity instrument. A small motor switched the accelerating V polarities and had a dynamic-zero position. The RC time constant of the instrument was 3.7 s. (From Rust and Moore 1974, with permission.)

cold chamber due to changing temperature that occurred with Teflon for the insulator. The rudder kept the intake aimed into the ambient wind and decreased the likelihood of inadvertently modified air being sampled.

Gerdien cylinders have been put on airplanes and used successfully to measure clear air conductivity (e.g., Kraakevik 1958), but they have not been able to make reliable measurements in precipitation. One recent example of a sophisticated conductivity device is that designed for use in a research program conducted by French scientists (Gondot 1987). Another is a parallel plate capacitor system flown on a high-altitude aircraft to measure conductivity in the clear air above storms by Blakeslee et al. (1989). For balloon-borne measurements in the clear, low air densities of high altitude, Rosen and Hofmann (1981a,b) devised a modification to maintain the volume flow up to very high altitudes.

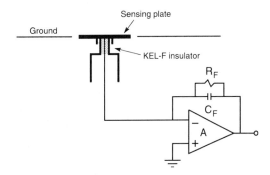

Fig. 6.26. Schematic of slow antenna for measuring field change from lightning. Typically used is an operational amplifier, A, with FET (field-effect transistors). The slow antenna is shown flush-mounted in the ground.

6.9. LIGHTNING PARAMETERS

6.9.1. Field changes

For years, measurements have been made of the electromagnetic fields generated when a lightning flash occurs. Of the various instruments and frequency bands covered, we present only the fundamentals of instrumentation to measure electric field change, ΔE, and optical emission. The terminology changes when discussing lightning. It is the electric field that is still measured; we often use the term electric field change to denote the electrostatic component (i.e., the slower portion lasting milliseconds to seconds) of the change in E. The term electric field waveform is used to denote the radiation component (i.e., higher frequencies lasting microseconds to milliseconds) of lightning. We alert you, however, that these terms are used interchangeably in the literature and even in this book outside this section. It is very common for electric field change to mean a portion or all of the change in E during a flash. When examining lightning ΔE, we are usually not concerned with the quasi-static value of the atmospheric E prior to or after the flash. The electrostatic field change is the portion of radiation that is measured close to lightning (within a few tens of kilometers).

Modern sensors are conceptually quite simple, with one widely used version shown in Fig. 6.26. The sensing plate is a conducting disk connected to a charge amplifier. The insulator around the input connection should be KEL-F or Teflon. In addition, the insulator should have wide grooves or depressions machined in it to increase the path length along it, thus reducing the possibility of leakage currents and their attendant fluctuating circuit outputs and long-lasting amplifier output offsets. Keeping the insulator clean is critical to avoid leakage across the insulator. For example, a common problem is a strand of a spider web that, although it may not be seen, causes a leakage current to saturate the circuit in high humidity. While a ΔE circuit responds to an impulse transient on the order of a microsecond, it is the slower components of the field change that are electrostatic. The circuit must allow decay of the signal; otherwise the instrument would saturate and remain so. The decay time is selected to be five to ten times longer than the phenomenon of interest. Since lightning generally lasts <1 s, a decay time of 10 s allows faithful recording of the electrostatic field change throughout the event. (Too short a time constant would cause the electronics to bring the output signal quickly back toward zero, and the relatively slowly varying components of the field would not be recorded.) These instruments are often called *slow antennas,* owing to their characteristic slow decay time.

The field change instrument can be calibrated by applying pulses of known amplitude through a precision capacitor connected to the input (i.e., the sensing disk) or by flush-mounting the instrument in a parallel-plate electrode system to which pulses are applied. The plates should be several times the disk diameter to assure a uniform E between them. This calibrates the slow antenna if it is flush-mounted in the ground (as shown in Fig. 6.26). If the instrument is not to be mounted flush with the ground, the calibration must include the geometrical form factor, $\mathscr{F}$, of the particular installation. The $\mathscr{F}$ is determined by comparison with an identical instrument that is mounted in a flat ground plane at the surface of Earth. The electric field change for components of a flash (e.g., a return stroke) or the entire flash is determined from

$$\Delta E = \frac{\mathscr{F}CG}{\varepsilon A}\,\Delta V, \qquad (6.40)$$

where $\mathscr{F}$ is determined from the ratio of the voltage outputs recorded for the same lightning transient by the flush-mounted instrument to the other instrument and A is the area of the sensor. The other symbols are as defined before for field mills. For a specific instrument and installation, we can simplify the equation to $\Delta E = KG(\Delta V)$, where K is a constant containing all but the gain factor for a given installation. With a network of such instruments that have been carefully calibrated, the location of individual charges neutralized during ground flashes and the dipole moments of intracloud lightning processes can be determined and compared with storm structure (Krehbiel 1981).

The same instrument can be used to observe the higher frequency components of flashes, as well. This is often done merely by reducing the time constant, with common values being 100 μs–1 ms. Since saturation is less likely owing to the fast decay, these instruments can be set to higher gains than a slow antenna for the same flash, allowing smaller pulses to be recorded. These instruments have been called *fast antennas,* although they are also referred to simply as field change instruments. A time constant of at least 1 ms allows the major aspects of a return stroke waveform to be recorded faithfully, which is necessary if the measurements are to be used quantitatively.

With the advent of high resolution (≥ 12 bit), fast (>1 MHz) analog-to-digital converters and digital memory, both the electrostatic and radiation components of a flash can be obtained with the same instrument, especially if you are willing to record only flashes that activate the storage of the digital data (Brook et al. 1989, Kawasaki et al. 1991). What we have described are only basic lightning field change sensors. Faster and more sophisticated instrumentation is available for looking at the fine structure of lightning.

6.9.2. Optical waveforms

The electronic sensing of the light output from lightning is a simple, yet useful, technique to detect lightning, even when it is totally inside a cloud in bright daylight. A simple detector can have an audio or flashing light indicator and can be made for a few dollars with parts from a consumer-grade electronics store. Photoelectric devices are now incorporated into commercial hand-held lightning detectors and are used to acquire the waveforms from optical emissions of lightning. Simple detectors have been flown on the space shuttle (Vonnegut et al. 1985) and on high-altitude airplanes (Brook et al. 1980, Goodman et al. 1988). Optical detectors can have very fast response times (we are actually just addressing a specific part of the electromagnetic spectrum). Observation of unique optical characteristics of lightning is also possible. For example, the all-azimuth optical detector that was designed by Guo and Krider (1982) views a small segment of leader and return stroke channels.

6.9.3. Channel velocity

An early attempt to study lightning with high-speed photography was conducted by Boys (1926), who developed a two-lens streak camera. His camera, and its subsequent improvements, became widely used and allowed measurements of the velocity of propagation of return strokes and leaders. A modern version of a streaking camera was described by Idone and Orville (1982),

who modified the technique by adding a sensor to trigger the camera before or during the first return stroke to allow photographs of subsequent return strokes in daylight.

Nonphotographic techniques have been developed to measure return stroke velocity. Most had photodiode detectors behind slits in the focal plane of a lens (e.g., Hubert and Mouget 1981, Nakono et al. 1983). Mach and Rust (1989) designed and built such a device (Fig. 6.27) for use on a mobile laboratory. It incorporated a video camera with the same field of view to provide images of the lightning to be used with the velocity information. They intercepted storms and obtained return stroke velocities while the mobile laboratory was moving down roads, as well as when stationary. The reason that was possible is that the duration of the optical transients is so short that there is no significant relative motion between the lightning source and the detector on the mobile lab, even if it is turning a corner when the flash occurs. Measurements of the velocity of the same strokes with their instrument and with a streak camera yielded the same results (Idone et al. 1987). There are compromises with either velocity measuring technique. The nonphotographic technique has the advantage of more easily capturing all strokes in a flash, significantly reduced size and cost, not requiring a stable site for operation of a high-speed camera, and, with analog tape or adequate digital storage, having only occasional data gaps rather than the large down times that occur during film changes. Both the photographic and the optoelectrical techniques remain viable.

6.9.4. Triggered lightning facilities

Lightning physics research has benefited tremendously from the ability to trigger lightning strikes to specific measurement devices or to items to be tested for lightning susceptibility. Triggered (sometimes called artificially initiated) lightning has been measured for

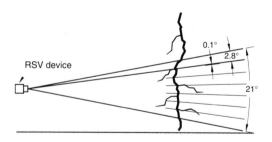

Fig. 6.27. Concept of measuring channel velocity with a optoelectric detector. An optical pulse comes from each 0.1° slit across which the lightning image moves. Using the time difference between corresponding points on the waveforms allows the angular velocity to be calculated. If the distance is known (or estimated), the two-dimensional velocity can be determined. (From Mach and Rust 1989, with permission.)

decades with instrumented towers. The usefulness of using a system to trigger flashes from a thunderstorm, as opposed to having to wait for their occurrence to a desired place, led Neuman (1958) to suggest firing a small rocket trailing a wire from his research schooner on open water. He was successful at doing so (Newman 1965). For many years, however, no one had such success over land. The breakthrough in the development of a technique for land-based triggering was in France. Fieux et al. (1975, 1978) described their successful capability using small rockets carrying fine wire aloft.

Triggering lightning has evolved and is now successfully and routinely done at a few places around the world. The various sites have similarities, but there are important differences in purpose. For example since the 1970s, such sites often have been used to trigger lightning in winter storms produced by the Sea of Japan on Japan's west coast to study positive, as well as negative, ground flashes (e.g., Nakamura et al. 1992, Horii et al. 1996). Extensive rocket triggering has been a component of several research programs in Japan. In the United States, a few sites have had the capability to trigger lightning routinely. For example, at Langmuir Laboratory in the mountains of New Mexico, triggered lightning comes to Earth on the top of an underground room with metal walls, floor, and roof. This room forms a Faraday cage where people and instruments can safely reside to measure the current and other properties of lightning. The triggering operations are part of a larger research effort at the Laboratory to study thunderstorms and lightning. For several years, rocket triggering was done at the Kennedy Space Center (e.g., Fisher et al. 1993). That site was recently decommissioned, and a new one was established at Camp Blanding, Florida. At this site (Uman et al. 1996), there is considerable emphasis on the effects of lightning on overhead and buried utility cables along with basic, lightning channel physics.

At all sites, cameras and other instruments are typically placed at various distances from the triggering location to record the lightning. At some places, two types of triggering have been done. One consists of a rocket spooling out a grounded wire as it rises. The other method is for the wire to have an insulating line below the wire, so it is not grounded. An overview of triggering activities though the mid-1980s is found in Uman (1987) and more recent developments are found in papers such as the examples above. The evolution of rocket triggered lightning has contributed significantly to new knowledge of lightning, and results of triggered lightning projects are found in several chapters in this book.

6.10. LIGHTNING-LOCATING TECHNIQUES

Knowing where lightning occurs in storms is essential to gaining an understanding of storm electrification, particularly of relationships between lightning and its storm environment. Until the early 1970s, however, the technology for locating lightning was limited to radar, to magnetic direction-finding that used very low frequency (VLF) electromagnetic radiation (often called *sferics*), and to measurements of electric field changes for estimating the charge 'neutralized' by a lightning flash. (The term 'neutralized' implies that positive and negative charge recombine to produce neutral particles. Although charge often is said to be neutralized by lightning, we do not know that lightning physically recombines positive and negative charge in a storm. Other hypotheses have been suggested. The same effect could be obtained simply by moving the charge to the ground or by adding the opposite sign of charge to the same region to mask the previous charge, without the charge physically recombining to produce neutral particles.) Beginning about 1970, advances in electronics and computers were coupled with significant new insights to improve existing methods and to develop new methods. Lightning can now also be located from thunder received by microphone arrays and from sferics received by antenna arrays. In this section, we discuss several mapping techniques.

6.10.1. Charge center analysis

Lightning can be located by recording electrostatic field changes at several stations simultaneously. The measured field changes are inserted into equations for the electrostatic field changes of lightning, which then are solved to estimate the coordinates and net magnitude of charge involved in the flash. To measure the field changes, Koshak and Krider (1989) used a field mill network and Krehbiel et al. (1979) used a network of slow antennas. The time resolution of the field mill network was adequate only to determine a single charge center or dipole for an entire ground or cloud flash, respectively. The better time resolution of the slow antennas allowed analysis of flash development.

To infer the location and magnitude of charge from measurements of electric field changes at the ground, it is necessary to model the charge geometry. Two geometries commonly have been used for this analysis. The first, a point charge model, applies only to flashes lowering charge to ground. If it is assumed that the charge to be lowered is spherically symmetric, or that its dimensions are small compared with its height above ground, and if the charge is lowered to a perfectly conducting plane, the resulting field change for a station at coordinates x_i, y_i, z_i, is

$$\Delta E_i = \frac{Qz}{2\pi\varepsilon[(x - x_i)^2 + (y - y_i)^2 + (z - z_i)^2]^{\frac{3}{2}}},$$

(6.41)

where x, y, z are the coordinates of the charge center and Q is the charge being lowered. Because there are four unknown variables, measurements of ΔE_i at four locations are sufficient to determine the coordinates and charge, but redundant measurements at additional sites provide more confidence in the solutions.

The second model, a dipole model, is assumed to describe field changes from nonreturn stroke processes. We will call these intracloud processes. In this context, however, intracloud includes ground flash processes other than return strokes and leaders to ground, as well as cloud flash processes. If an intracloud process neutralizes equal, but opposite, point charges above a perfectly conducting plane, the electric field change at the ground is the superposition of the field changes from each point charge, and is given by

$$\Delta E_i = \frac{Q}{2\pi\varepsilon} \left\{ \frac{z_+}{[(x_+ - x_i)^2 + (y_+ - y_i)^2 + (z_+ - z_i)^2]^{\frac{3}{2}}} \right.$$
$$\left. - \frac{z_-}{[(x_- - x_i)^2 + (y_- - y_i)^2 + (z_- - z_i)^2]^{\frac{3}{2}}} \right\}, \tag{6.42}$$

where x_+, y_+, z_+ are the coordinates of the positive charge and x_-, y_-, z_- are the coordinates of the negative charge. In this situation, seven measurements of E_i are needed to determine the six coordinates and the charge. If only six measurements are available, the expression for ΔE_i can be reformulated in terms of the dipole moment $\vec{p}$,

$$\vec{p} = Q[(x_+ - x_-)\hat{i} + (y_+ - y_-)\hat{j} + (z_+ - z_-)\hat{k}]. \tag{6.43}$$

If all the measured field changes are small because the charge magnitude or the separation between the two charges is small, it also may be necessary to use the dipole moment to avoid large errors in the solution. However, solving only for the dipole moment obviously provides less information about the charge involved in a flash.

When there are more measurements than the minimum required for a solution, it is necessary to reconcile the resulting redundant solutions. Two optimization techniques have been used: a scatter plot of redundant solutions (Krehbiel et al. 1979) and nonlinear, least-squares fitting of the measurements (first applied by Jacobson and Krider 1976). Krehbiel et al. (1979) found that results from the scatter plot and least-squares analysis were in good agreement with each other. As the least-squares technique gives a centroid more directly and objectively, it became preferred over the scatter plot.

In least-squares fitting, the analysis seeks to minimize the chi-squared statistic, which measures the deviations between the measured and derived values of the field and is given by

$$\chi^2 = \sum_i \frac{(\Delta E_i - \Delta E_{mi})^2}{\sigma_{Ei}^2}, \tag{6.44}$$

where ΔE_{mi} is the measured field change at the ith location, ΔE_i is the field change calculated from the model charge distribution, and σ_{Ei}^2 is the variance in the measurement ΔE_{mi}. Since ΔE_i is a nonlinear function of the coordinates of the centroid, an iterative, non-

linear algorithm, such as the Marquardt algorithm described by Bevington (1969), is needed for the least-squares analysis.

The success of the charge centroid analysis in finding solutions has been cited in arguing that a point charge can model the actual geometry of charge in lightning flashes reasonably well. However, the charge center analysis is relatively insensitive to the assumption that charges are spherically symmetric, as long as there is a vertical axis of symmetry in the charge geometry. For example, in a computer simulation of a thin disk of charge it was found that the point-charge analysis determined Q and the horizontal coordinates of the charge centroid accurately, when given values of ΔE_{mi} computed from the disk of charge (E. Williams 1984, personal communication). The vertical coordinate was the most strongly affected by the departure from spherical geometry, and even it was within 0.5 km of the actual centroid.

Koshak and Krider (1994) developed a third technique that avoids minimizing a nonlinear expression and provides an estimate of the distribution of charge instead of simply using point charges. Instead of minimizing χ^2 for a dipole or monopole model of the lightning charge, Koshak and Krider's technique uses a general volume charge distribution on a large grid. A linear set of equations relates the measured ΔEs to the charge at each grid point, and a Landweber iterative search method (Landweber 1951) is used to solve the system of equations to determine the charge distribution. By solving calculated ΔEs for known charge distributions of model lightning flashes, Koshak and Krider found that this technique tended to find a reasonable solution, as long as the flash was over the network of electric field sensors. For such flashes, typical errors in the location of the centroid of charge were 1–2 km and in the charge magnitude were 20–35%.

The constraints imposed on electric field change measurements by charge analyses are fairly strenuous. The instruments should be spaced only a few kilometers apart and in a relatively level region. Krehbiel et al. (1979) indicated that to measure the location of a charge centroid to within 100 m it is necessary to measure electric field changes to within 1% at the different stations. Because of the dependence of the form factor on the position of the sensor relative to its surroundings, this corresponds to variations in the position of the sensor of about 2.5 cm. Other sources of error in their system included the electronic circuitry, the telemetry, and the analog recording system. Environmental effects such as varying space charge and precipitation impacting the sensing disk also introduce errors, though the errors can be subtracted if their period is much slower than the field change being analyzed. Krehbiel et al. estimated that a typical error in their measurement of field changes was approximately 2%.

Similar information on the distribution of charge lowered to ground by lightning can be provided by measurements of the electric field change by a network of field mills. The frequency response of a typical field mill is inadequate for resolution of the field change of return strokes, but a good field mill does measure the total field change from the flash. From the total field change, one can find the equivalent total charge neutralized during a flash. This degrades the accuracy of solutions, since the return strokes are mixed with in-cloud field changes, and the geometry of the neutralized charge is likely to be larger and to have a less symmetric shape than for individual strokes or intracloud processes. Constraints on the network are the same as those on the network of field change sensors. A network of field mills has been operating for a number of years at the Kennedy Space Center. Analysis of network data (Jacobson and Krider 1976, Lhermitte and Williams 1985, Koshak and Krider 1989, Krider 1989, Nisbet et al. 1990) and comparison with independent data (e.g., Uman et al. 1978, Krehbiel 1981) have shown the technique to be useful in ascertaining the heights of charge centers relative to ambient temperature and storm structure.

6.10.2. Lightning location with radar

Although the observation of lightning with weather radars was first published in the 1950s, only recently have radars been used specifically to locate lightning, determine physical characteristics of channels, and relate lightning to storm evolution. Here we briefly examine examples of how radar is used to locate lightning. See Williams et al. (1990) for a full review of radar studies of lightning. Radar allows a type of lightning mapping from very close ranges out to a few hundred kilometers. One simple way to observe lightning is to aim a stationary (nonscanning) radar antenna at an active region of a storm (Fig. 6.28). Spatial resolution of the lightning location in the plane orthogonal to the radar beam decreases with range. Lightning detection results vary widely depending on the antenna beam pattern. For example, the narrow conical beam in the figure likely will detect fewer lightning flashes than a cosecant beam, which is narrow in azimuth but exten-

sive (e.g., 0°–60°) in the vertical. The longer the wavelength, the easier the detection of lightning in precipitation. In a practical sense, the wavelength needs to be at least 10 cm to avoid excessive precipitation masking of the echoes from lightning. At 10 cm, lightning will be detectable in precipitation outside of the most intense cores. Longer wavelengths of up to a meter or two reduce the precipitation return even more. Use of polarization-diversity radar techniques can suppress the precipitation echo and enhance lightning detection at a given wavelength.

The reason radars can be used to map lightning is that the channels are highly reflective. The plasma condition of a lightning echo is overdense for hundreds of milliseconds and has a temperature >5000 K (Williams et al. 1989). When the lightning channels are overdense, we can consider the lightning to be like wires or conductive cables in the cloud; they are easy targets to detect if the precipitation echo does not overpower them. As channels cool, they become underdense and reflect less powerfully. The reflectivity power often decays at 0.2 dB ms^{-1} (Holmes et al. 1980).

A few different uses for radar mapping of lightning follow. If an antenna is pointed at low elevation angles at a storm, the velocity of propagation of lightning along the radar beam can be determined. Mazur et al. (1985) used a vertically pointing Doppler radar and other instrumentation (Fig. 6.29) to capture radar echoes of lightning and precipitation in the same sample volume(s). The radar had a wavelength of 10.5 cm, an antenna beamwidth of 3°, and a Doppler velocity resolution of 1 m s^{-1}. The antenna was in a microwave absorbing shroud to minimize side lobes from what they would have been had the dish antenna just been pointed vertically. This provided a well-defined sample volume. There were 16 observation volumes; each was 45 m deep and separated by 300 m. The entire column could be moved up and down to increase chances of recording lightning. This technique provided more information on channel characteristics than on lightning location in the storm context. It also provided a

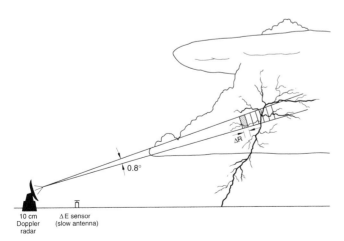

Fig. 6.28. Remote observations of lightning inside storms using a pointable, stationary antenna. There are numerous observational volumes that are defined by the antenna beamwidth (here 0.8°) and the length of the transmitted pulse (shown as ΔR). The ΔE sensor allows comparison of the time evolution of the lightning field change and its echo, identification of flash type, etc. (From Rust et al. 1981, with permission.)

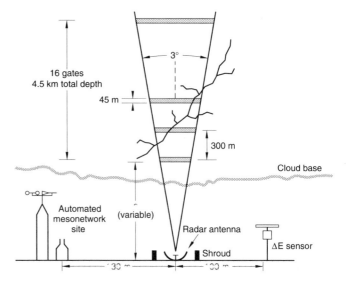

Fig. 6.29. Vertically pointing Doppler radar and other instrumentation to study lightning echoes and precipitation overhead. (After Mazur et al., 1985: American Geophysical Union, with permission.)

spot measurement of the vertical air velocity, since the channel initially moves with the air. A depiction of the precipitation and lightning velocities is given in Fig. 6.30. Radars have also been used to look for changes in lightning channels and precipitation immediately following the lightning flash. This has included pursuit of evidence for the rain gush phenomenon, such as found by Szymanski et al. (1980). Their observations are discussed in Section 10.11. Radar location of lightning has been conducted in severe storms using simultaneous observations with a 10-cm Doppler radar to locate mesocyclones (Sec. 8.1.5) and a 23-cm wavelength radar with a cosecant antenna to record the total lightning (ground plus cloud flashes) activity in the mesocyclone region (Fig. 6.31).

As noted by Williams et al. (1990), a radar suitably configured to map lightning in three dimensions would have an advantage over passive techniques because the channels reflect whether or not their electromagnetic radiation is detectable by a passive system, and passive systems tend to miss certain lightning processes such as positive streamer propagation. What is required for optimal radar mapping in three dimensions throughout a storm is a radar system capable of high spatial and time resolution. As often happens, a less than optimal radar system can provide useful lightning information, especially when used with other mappers and instrumentation.

Although not lightning mapping in the sense being discussed here, we alert you to the interesting use of radar techniques to detect remotely the alignment of ice crystals caused by the electric field of storms. Detecting particle alignment is a promising new technique to provide short-term prediction of when lightning will occur. See Section 10.10 for examples and references,

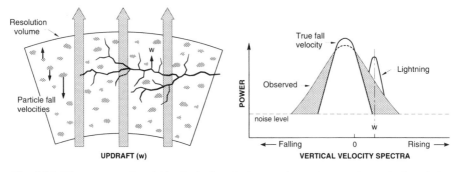

Fig. 6.30. The spectrum of vertical velocity in a single range interval (gate) of a vertically pointing Doppler radar. The lightning initially moves with the vertical air velocity, w. The observed fall velocity can be corrected to the true one through use of the independent measure of w from the lightning echo.

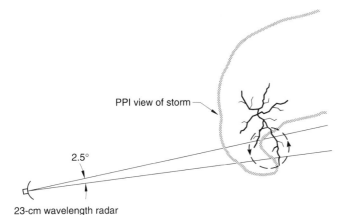

Fig. 6.31. Lightning observations with radars in a severe thunderstorm. The dashed circle denotes the mesocyclone, which was detected in real time with a 10-cm radar. The 23-cm wavelength radar antenna beam was cosecant-squared in the vertical, so it recorded all lightning within the storm that had at least part of its channel in the 2.5° azimuthal beam. The antenna was stationary except for occasional reaiming to track the mesocyclone. (From Rust and MacGorman © 1988, Oklahoma Univ. Press, with permission.)

and especially note Krehbiel et al.'s (1996) overview of theory and observations with this technique.

6.10.3. Acoustic mapping techniques

Acoustic mapping techniques for tracing the propagation of sound back to its source, originally developed for underwater applications (sonar) and oil exploration, have been adapted to locate lightning channels from the thunder that they generate. Since the development of a lightning flash is very fast on the time scale applicable to acoustic propagation, the collection of mapped acoustic sources derived from the thunder recording of a flash basically gives a snapshot of the hot, explosive channels producing thunder, with no information about their sequence of development. Although acoustic techniques are limited to a range of roughly 20 km for audible frequencies, and the analysis of data is complicated by the sensitivity of sound propagation to normally encountered atmospheric variations, the required instrumentation is relatively simple and can be deployed easily.

Acoustic mapping locates the source of a thunder pulse by determining the pulse's range and direction. Range can be determined by measuring the propagation time of the acoustic pulse, since the speed of sound, c_s, can be determined accurately from atmospheric soundings. Because electromagnetic signals travel roughly a million times faster than acoustic signals in the troposphere, the propagation time of the electromagnetic field change of a flash is negligible compared with that of acoustic signals from the same flash. Thus, the acoustic propagation time to a site from the lightning channel segment that generated a particular pulse is given by the time delay between the electric field change of the flash and the arrival of the acoustic pulse at the site. Sensors for recording electric field changes are described above. Any rugged outdoor microphone with a reasonably flat frequency response between a few hertz and a few hundred hertz can be used to record the arrival of a thunder pulse.

There are two techniques—thunder ranging and ray tracing—for determining location of an acoustic source. Thunder ranging, the simpler of the two, was evaluated first by MacGorman (1977) and later, in greater generality, by Bohannon (1978). Ray tracing was developed by Few (1970) and evaluated by Few and Teer (1974). Each technique has advantages and disadvantages.

(i) Thunder ranging. Thunder ranging requires at least three microphones separated by 1–3 km in a non-collinear array. The ranges of the source of a thunder pulse to each of the microphones are computed as the speed of sound times the propagation time described above. (The ranges can be modified in an iterative procedure to account for propagation effects, as described by Bohannon 1978.) The intersection of the three ranges then gives the location of the acoustic source in three-dimensional space. The equations for the coordinates of the intersection are simplest if the microphones are arranged in a right triangle with one of the legs aligned north as shown in Fig. 6.32. In this arrangement the coordinates are given by

$$x = \frac{R_3^2 - R_2^2 - D_{23}^2}{2D_{23}}$$

$$y = \frac{R_2^2 - R_1^2 + D_{12}^2}{2D_{12}} \qquad (6.45)$$

$$z = (R_2^2 - x^2 + y^2)^{\frac{1}{2}} \,,$$

Where D_{ij} is the distance between microphones i and j, R_i is the range from the ith microphone to the acoustic source (estimated from the pulse's time of arrival), x and y are the coordinates of the source east and north from microphone 2, and z is the height of the source above the plane that contains the three microphones.

To perform a thunder ranging analysis, the thunder waveforms from all microphones were plotted along with the electric field change record on a common time scale on the order of 1 cm s^{-1} and were examined by eye to find features common to all. Because the thunder waveform is influenced by propagation effects and by the different position and orientation of a lightning flash relative to each microphone, the analysis should be based only on features that are fairly prominent in waveforms from all the microphones.

(ii) Ray tracing. The second technique, ray tracing, uses an array of three or more microphones separated by only a few tens of meters. A cross-correlation analysis such as that described by Bendat and Piersol (1971) is used to identify corresponding thunder pulses at all microphones and to compute the propagation time of the pulse from one microphone to another. Each cross correlation uses a 0.1–0.5 s segment from the data record. To avoid computational problems, the segment should be about twice the maximum expected propagation time, but needs to be short to minimize confusion caused by thunder from multiple sources arriving at about the same time. The maximum expected propagation time is set either by the time required to traverse the baseline being analyzed or by the time lags between a clearly defined feature in the thunder waveform.

The values found by the cross-correlation analysis for the propagation time between each pair of microphones are used to compute the direction from which the thunder pulse arrived. For the same geometry of baselines as in Fig. 6.32, but with the legs of the triangle equal in length, the azimuth, ϕ, and elevation, θ, from which the pulse arrived are given by

$$\phi = \arctan \frac{t_{23}}{t_{12}}$$

$$\theta = \arccos \frac{c_s}{\dfrac{L}{t_{23}} \sin\phi - u\cos(\phi' - \phi)} \quad (6.46)$$

$$= \arccos \frac{c_s}{\dfrac{L}{t_{12}} \cos\phi - u\cos(\phi' - \phi)} ,$$

where c_s is the speed of sound, L is the length of the north-south and east-west baselines, t_{ij} is the propagation time between microphones i and j (positive when the pulse arrives at microphone i first), u is the surface wind speed, and ϕ' is the azimuth from which the wind is blowing.

To locate the source of the thunder pulse, an acoustic ray with this initial azimuth and elevation must be propagated backward through a model atmosphere. The total path length is determined from the speed of sound and the measured time between occurrence of the lightning flash and arrival of the acoustic pulse at the microphones. If the atmosphere had uniform temperature and humidity and no wind, the path would be a straight line, and the computation of the coordinates of the thunder source would be simple. However in the real atmosphere, paths are curved, and this must be approximated in the computation. For a continuously varying, stratified atmosphere with wind,

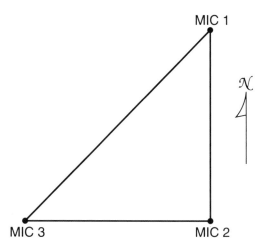

Fig. 6.32. Geometry of microphone (MIC) array to record thunder for flash location. (From Rust and MacGorman © 1988, Oklahoma Univ. Press, with permission.)

a form of Snell's law describes how the direction angles change along the ray path:

$$\cos\alpha = \cos\alpha_0 \left[\frac{1 + \Gamma_c dz}{1 - (\Gamma_{ux} + \Gamma_{uy})dz} \right]$$

$$\cos\beta = \cos\beta_0 \left[\frac{1 + \Gamma_c dz}{1 - (\Gamma_{ux} + \Gamma_{uy})dz} \right] \quad (6.47)$$

$$\cos^2\gamma = 1 - \cos^2\alpha - \cos^2\beta ,$$

where α, β, and γ are the new direction angles to the x, y, and z axes, respectively; α_0 and β_0 are the direction angles for the previous step along the ray path; and

$$\Gamma_c = \frac{1}{c_{s0}} \frac{\partial c_s}{\partial z}$$

$$\Gamma_{ux} = \frac{\cos\alpha_0}{c_{s0}} \frac{\partial u_x}{\partial z} \quad (6.48)$$

$$\Gamma_{uy} = \frac{\cos\beta_0}{c_{s0}} \frac{\partial u_y}{\partial z} ,$$

where u_x and u_y are the components of the local wind, c_s is the local speed of sound, and c_{s0} is the value of c_s for the previous step. The ray path can then be approximated as a series of steps of length, ds, the components of each step being given by

$$dz = \cos\gamma\, ds$$

$$dx = \frac{c_s \cos\alpha + u_x}{c_s} ds \quad (6.49)$$

$$dy = \frac{c_s \cos\beta + u_y}{c_s} ds .$$

The time increment for each step is given by

$$dt = \frac{ds}{c_s} \, . \qquad (6.50)$$

Steps are added until the sum of the time increments equals the measured propagation time.

(iii) Location errors in acoustic mapping techniques.

Inherent errors in the two acoustic techniques are of comparable magnitude. The source of the largest errors in ray tracing is the difficulty in modeling wind shear. Few and Teer (1974) estimated that differences in wind shear between the model and real atmospheres would cause errors typically within 10% of the range to the source. Thunder ranging is much less sensitive to wind shear than ray tracing, but is more sensitive to other sources of error. MacGorman (1977) found that when the range of lightning from the microphone array is ≲5 km, errors in reconstructed locations from thunder ranging are dominated by the different orientation and range of lightning structure relative to each site. Typical errors in this region are 15% of the range. When lightning is >5 km outside the microphone array, errors in thunder ranging are dominated by the uncertainty in identifying a particular short pulse within bursts of pulses in the thunder signature; errors in this region are typically 10% of the range. Much larger errors can result from incorrect subjective choices of corresponding pulses in the thunder signatures from different microphones.

MacGorman (1977) examined whether results from the two techniques were consistent. Three flashes were reconstructed by ray tracing with two independent arrays and by thunder ranging with three microphones separated by 2–3 km. Agreement between the two ray-tracing analyses and between the ray-tracing and thunder-ranging analyses was generally within the predicted errors, as shown by the example in Fig. 6.33.

Thunder ranging is most useful for focusing attention on interesting data before ray tracing or for studying large numbers of lightning flashes to determine the general location of lightning channels. When these techniques were in use, thunder ranging was much faster than ray tracing, requiring a total analysis effort of a few hours per flash versus 20–40 h for ray tracing. However, thunder ranging provides much less detail in the reconstructed lightning structure, and the analysis is somewhat more subjective, since the choice of corresponding features in the thunder signatures is made by eye. Therefore, ray tracing is more appropriate when greater detail or more reliable locations are needed. If the microphone array is small enough (probably a few meters) so that thunder signals are highly coherent between microphones, it may be possible to automate the calculation of direction angles for ray tracing; the time required for analysis might then be less than that required by thunder ranging. While these techniques are still viable, they have largely been replaced by VHF mapping, which we describe next, because VHF mapping provides greater range of coverage and information on the initiation of flashes and the subsequent development of lightning channel structure.

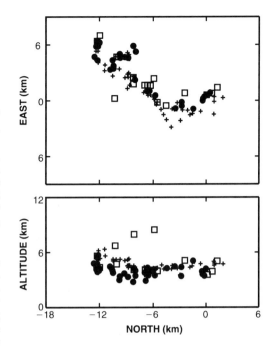

Fig. 6.33. Comparison of three acoustic reconstructions of the same flash. Sources mapped by ray tracing from two arrays are shown as +s and dots to indicate array used. Squares denote sources determined by thunder ranging. (From MacGorman 1978, with permission.)

6.10.4. VHF radio mapping

A technique for mapping the three-dimensional structure of lightning by using the radio frequency signals that it radiates was first formulated by Oetzel and Pierce (1969). Several techniques have since been developed to map the location and development of lightning from the electromagnetic radiation it produces. These techniques often use frequencies in the very high frequency (VHF) band (30–300 MHz), but sometimes use other frequencies to emphasize particular lightning processes or to provide specific design advantages for the system. The effective range of techniques for determining the three-dimensional location of lightning is generally ≲60 km, because of signal attenuation and decreased spatial resolution in the mapping analysis at longer ranges. Beyond these common factors, the techniques vary considerably.

Each technique can be described in terms of the size of its antenna array and its method of determining locations. Small arrays have baselines of the order of 1–100 m and determine only direction angles to the radio sources. Range can be determined if there are two such arrays for triangulation. Large arrays have baselines on the order of 10 km and determine both direction angles and range. Short-baseline arrays determine direction angles either by measuring the difference in times of arrival of the radio signal between pairs of antennas or by using an interferometric technique. All long-baseline arrays developed thus far determine locations by measuring the difference between times of arrival. The following sections describe specific systems of each type.

(i) Long-baseline, time-of-arrival technique. The first VHF system to locate noise sources on lightning channels in three dimensions was a long-baseline, time-of-arrival system developed by Proctor (1971). Initially, the chief disadvantages of the system were that data could be recorded continuously for only 250 ms, and analysis of the data was extremely laborious, requiring 6 months to determine source locations for a single 250-ms record. With the addition of new devices as described by Proctor (1983), these problems were alleviated, though analysis was still fairly tedious.

In Proctor's (1983) system, radio signals were sensed by five broadband, vertically polarized antennas arranged to form two crossed baselines (Fig. 6.34). The signal from each antenna was fed into a crystal-controlled receiver operating at a center frequency of 355 MHz. The output of each outlying receiver was transmitted to the central station over a frequency-modulated, 3-cm wavelength microwave link, and care was taken to avoid signal contamination by sferics or by cross coupling signals from other stations. Coordinates of the outlying stations were determined relative to the central station to an accuracy of ≤ 10 cm, so the propagation time between stations was known much more accurately than required for the time-of-arrival measurements. The time at which a signal arrived at an outlying station was given by the time the telemetered receiver output arrived at the central station minus the propagation time of the telemetry between the two stations. Data were recorded at the central station on long strips of photographic film by an analog, laser-optical recorder. Up to 20 min of continuous data from five channels could be recorded with an interchannel, root mean square (rms) timing error of 82 ns and a bandwidth of 6 MHz per channel.

Determining the location of the source of a sferic involved, first, determining the differences in the times at which the sferic arrived at different stations and, second, using these time differences to compute the source location. Differences in the arrival times were measured visually by a person who used a custom-built analyzer to inspect signals in 1-ms segments of data. By displaying the data with increasingly better time resolution, signals could be lined up in 62.5-ns increments. Once signals were aligned, the time delays between stations were provided automatically by the machine.

The resulting delays were sent to a computer, which calculated the coordinates of the source of the signal and checked that the calculated location was consistent with all the redundant time delays. Each time delay between stations defined a hyperbolic surface of constant time delay, and the VHF source for the signal was somewhere on that surface. To calculate the source location, equations for three of the hyperbolic surfaces

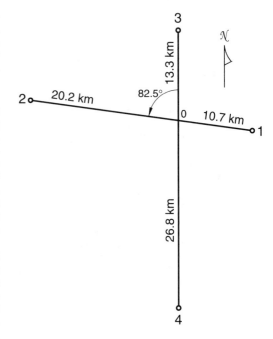

Fig. 6.34. Geometry of a long baseline VHF array. The receiving locations are labeled 0–4. (After Procter, 1981: American Geophysical Union, with permission.)

were solved to find the point at which the surfaces intersected. For a coplanar array of stations arranged in the same order as shown in Fig. 6.34, but with orthogonal baselines, one of the redundant solutions can be written as

$$x = \frac{A_1 L_1 + A_2 L_2}{2(A_2 - A_1)}$$

$$\cdot \left\{ -1 \pm \left[1 - \frac{c^2 (A_2 - A_1)(\tau_2^2 A_2^2 - \tau_1^2 A_1^2)}{(A_1 L_1 + A_2 L_2)^2} \right]^{\frac{1}{2}} \right\}$$

$$y = \frac{A_3 L_3 + A_4 L_4}{2(A_4 - A_3)} \tag{6.51}$$

$$\cdot \left\{ -1 \pm \left[1 - \frac{c^2 (A_4 - A_3)(\tau_4^2 A_4^2 - \tau_3^2 A_3^2)}{(A_3 L_3 + A_4 L_4)^2} \right]^{\frac{1}{2}} \right\}$$

$$z = \left[\frac{1}{4} (L_1^2 - c^2 \tau_1^2) \left(\frac{4x^2}{c^2 \tau_1^2} - 1 \right) - y^2 \right]^{\frac{1}{2}},$$

where $\tau_i = t_i - t_0$, t_i is the time of arrival at the ith station, L_i is the length of the baseline between stations i and station 0, c is the speed of light, and the As are defined by

$$A_i = \frac{L_i^2}{c^2 \tau_i^2} - 1 . \tag{6.52}$$

The choices of sign in the expressions for x and y are positive when $\tau_1\tau_2 > 0$ and $\tau_3\tau_4 > 0$, respectively. See Proctor (1971) for additional information about the computations.

Because there was no vertical baseline, the standard error in z was much larger than standard errors in x and y. Within a plan quadrilateral having the four outlying stations as vertices, the error in x and y was approximately 25 m. The error in z, however, was a function primarily of elevation angle, varying from approximately 100 m directly over station 0 to 300 m at $(-3, -3, 4)$ km and to 1000 m at $(-6, -6, 2)$ km.

A second long-baseline time-of-arrival system, the Lightning Detection and Ranging system (LDAR), was developed at the Kennedy Space Center for use in rocket launch operations (Poehler and Lennon 1979). This system operated at a center frequency of 63 MHz and consisted of two synchronized independent networks of antennas. Lightning signals were processed automatically in real time from each network. To be accepted by the system as a valid VHF source location, signals had to be detected in coincidence by both networks, and the resulting independent source locations had to agree within prespecified limits. The redundant data provided by the two networks were important in eliminating contamination from noise, as discussed by Krehbiel (1981). The processing speed of the system severely limited the number of locations that could be produced in real time, but digitized raw sensor data also were recorded. Investigators could use either the real-time locations or could reprocess the raw data to produce more detailed reconstructions of lightning flashes (Uman et al. 1978, Lhermitte and Krehbiel 1979, Krehbiel, Lhermitte and Williams 1985, and Nisbet et al. 1990).

A new version of the LDAR system replaced the original system and has been used in some investigations (e.g., Maier et al. 1995b, Mazur et al. 1995). The new LDAR system, described by Lennon and Maier (1991) and Maier et al. (1995a), consists of seven, circularly polarized, broadband, receiving antennas that communicate by microwave links to a central processor. Six of the antennas are located roughly at the vertices of a hexagon centered on the seventh antenna. Each site sends analog data continuously to the central processor. The system has a 6 MHz bandwidth and can operate at either 60–66 MHz or 222–228 MHz. When a signal from the center antenna exceeds an adjustable threshold, the system opens an 80-μs window and records the peak amplitude and the time of the peak amplitude within the 80-μs window for each antenna. These data are then transferred to a computer workstation that calculates the source location. The system can process a maximum of 10^4 pulses per second.

The times at which a signal arrives at four antennas are required to compute a source location. The central station must be one of the four sites used in solutions, so the seven sites provide up to 20 redundant solutions. Initially, the system uses solutions from the two combinations of stations that provide the best distribution of location errors, and it checks to see if corresponding coordinates from each of the two solutions agree within 5% or 350 m, whichever is greater. If so, the average of the two solutions is accepted as the solution. If not, solutions are computed from all twenty combinations, and each of the twenty combinations is tested to find how many other solutions have coordinates within 5% or 350 m. The solution that has the largest number of other solutions within this range then is chosen as the system's solution.

Maier et al. (1995a) examined location errors from LDAR by comparing the location of a transmitter on an aircraft tracked by the Global Positioning System (GPS) with the source location found by LDAR. They found that the random timing error of signals from each LDAR antenna appeared to be ±50 ns. The resulting location error varied with the range and height of the source, but was typically 50–100 m for sources at altitudes greater than 3 km within the perimeter of the network. Errors were slightly larger at lower altitudes, and increased with range to a mean error of 1 km at a range of 40 km. Most of the location error at 40 km was along the horizontal radial from the center antenna to the source.

A third long-baseline, time-of-arrival system was developed by Thomson et al. (1994), again at the Kennedy Space Center. The system consisted of five stations, with one station at the center and four stations located in different directions approximately 10 km away. Signals from each station were sent to a central processor over microwave or analog fiber optic links. Broadband measurements of dE/dt were used, with signals down no more than 6 dB from 800 Hz to 2 MHz at the central station and from 800 Hz to 4 MHz at the remaining four stations. Absolute timing was provided with 1-μs resolution, and the relative time between stations was adjusted to within 400 ns from television synch pulses. Relative times were then determined as part of the analysis to within better than 50 ns. Signals from each station were digitized by a processor at the central station at 2×10^7 samples per second in 204.8-μs segments each time a signal triggered the system. The system could record up to 25 data segments per flash, with 40 μs of dead time between sequential segments. When 25 segments filled the digital data buffer, the system required 1.5 s to transfer the buffered data to memory in a personal computer before the data could be digitized again. Data were recorded on a 1.5-gigabyte disk and then archived on optical disk.

The time of arrival was determined for three features of the dE/dt waveform at each station: (1) the rising half peak, (2) the peak, and (3) the descending half peak. The time of arrival of the signal was estimated as the mean of these three times, which also were used to measure the similarity of pulse shape between stations and to estimate timing errors. Measurements of signals from any four of the stations were enough to determine the source location and time of occurrence, so the five-stations provided an overdetermined set of equations. Two different optimization techniques were used to determine solutions, and both produced comparable errors. The location error was expected to be less than 100 m for sources over the network.

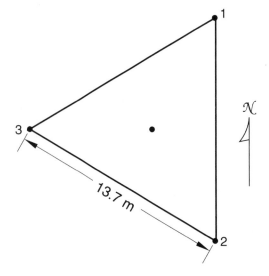

Fig. 6.35. Layout of full-hemisphere VHF array. The interior dot indicates location of vertical baseline antennas, where one antenna is 13.7 m above the one on the ground. (From Rust and MacGorman © 1988, Oklahoma Univ. Press, with permission.)

(ii) Short-baseline, time-of-arrival technique. Taylor (1978) developed the first VHF short-baseline, time-of-arrival system. His system located radio sources radiating in the 20–80 MHz band. The system was upgraded in 1984, but is no longer in use. As shown in Fig. 6.35, five antennas were used in the upgraded system: three at the vertices of an equilateral triangle, 13.7 m on a side, and two forming a vertical 13.7-m baseline at the center of the triangle. Coverage was divided into seven contiguous sectors: Each of six sectors spanned 60° in azimuth by 60° in elevation, and a seventh sector covered the upper 30° of elevation at all azimuths. Each VHF impulse that was accepted (i.e., that had a sufficiently fast risetime and an amplitude above the system threshold) was assigned to the appropriate sector from the sequence of arrival at the antennas. This sector was encoded along with the time, azimuth, and elevation of the VHF source.

The azimuth and elevation to a VHF source were determined by the difference in the time of arrival of the VHF impulse between pairs of antennas. The time difference was measured to an accuracy of 0.5 ns, sufficient to give elevation and azimuth to within approximately 0.5°. For the six sectors with elevation angles ≤60°, the elevation to a VHF source was calculated from the measured propagation times for the vertical baseline by using the equation

$$\theta = \arcsin\left(\frac{c\,\tau_v}{D}\right), \qquad (6.53)$$

where τ_v is the positive time lag between the top and bottom antennas, c is the speed of light, and D is the

distance between antennas. The choice of antennas for the azimuth calculation depended on the sector from which the VHF source arrived. For azimuths between 60° and 120°, the azimuth was calculated by using the antennas numbered 1 and 2 in Fig. 6.35 and was given by

$$\phi = \arccos\left(\frac{c\,\tau_{12}}{D\cos\theta}\right), \qquad (6.54)$$

where τ_{ij} is the lag between antennas i and j, positive when the signal reaches i first. The azimuth for a VHF source from a different sector with an elevation angle ≤60° was calculated from a similar expression in terms of the time lag across another horizontal baseline.

For VHF sources from elevation angles ≥60° (i.e., the overhead sector), the VHF system measured time lags of a signal along two horizontal baselines of the array, and these time lags were used to determine azimuth and elevation. For the array geometry shown in Fig. 6.35, azimuth is given by

$$\phi = \arctan\left(\frac{2\,\tau_{13} - \tau_{12}}{\sqrt{3}\,\tau_{12}}\right), \qquad (6.55)$$

and elevation is given by

$$\theta = \arccos\left(\frac{c\,\tau_{12}}{D\cos\phi}\right) \qquad (6.56)$$

$$= \arccos\left[\frac{2c}{\sqrt{3}\,D}(\tau_{12}^2 - \tau_{12}\,\tau_{13} + \tau_{13}^2)^{\frac{1}{2}}\right],$$

where D is the distance between antennas.

With two of the VHF systems separated by 15–50 km, it was possible in later analysis to triangulate the real-time directions to locate sources of VHF signals. The first step of this analysis was to identify which VHF signal in data from one array corresponded to a particular signal received by the other array. Identifying common VHF signals was difficult and required the time between stations to be synchronized to within approximately 10 μs. (Taylor 1978 analyzed the data from both arrays to synchronize them more accurately than provided by the available timing signal, but adequate synchronization is now readily available by using clocks synchronized to global positioning system (GPS) signals.) Once the same VHF signal was identified in data from both arrays, the coordinates of the source of the signal were calculated from the direction angles measured by each array by using the following equations:

$$x = L\tan\phi_A\,\frac{\tan\phi_B\cos\psi - \sin\psi}{\tan\phi_B - \tan\phi_A},$$

$$y = L\,\frac{\tan\phi_B\cos\psi - \sin\psi}{\tan\phi_B - \tan\phi_A} \qquad (6.57)$$

$$= \frac{x}{\tan\phi_A},$$

$$z = (x^2 + y^2)^{\frac{1}{2}}\tan\theta_A,$$

where Ψ is the azimuth of the baseline from array A to array B, L is its length, and array A is the origin.

With a baseline of 40 km, it usually was possible to determine three-dimensional source locations for only 20–30% of the signals detected by a single array. The percentage was low for three reasons: (1) The logic in the circuitry for the mapping system had thresholds that a signal had to exceed before being accepted. Because the amplitude of electromagnetic radiation decreases linearly with range, pulses that satisfied thresholds at one array could fall below the thresholds at the other array; (2) The lightning channel segments that generate the signals are anisotropic radiators. Thus, again there could be different amplitudes at the two arrays even if ranges were similar, because the orientation of the individual radiating segments relative to the two stations was different; and (3) One of the arrays was occasionally processing another signal at the time a signal detected by the other array arrived. Failure to locate the source of a radio signal was usually the result of one of the first two problems.

It was impossible to place absolute confidence in any single, three-dimensional source location, because it was always possible that signals coincident within a reasonable time window were not actually from the same source. However, clusters of located sources had greater reliability, because it was unlikely that many locations in a cluster were identified with falsely associated impulses. In addition, greater confidence could be gained in a particular source by comparing the measured time delay between arrays with the time delay that would have resulted from the calculated location.

(iii) Short-baseline, interferometric technique.

Interferometry has been used for a number of years in radio astronomy. It was adapted by Warwick et al. (1979) and Richard et al. (1986) to map centroids of VHF sources generated by lightning. In its simplest form, an interferometer consists of two identical antennas some distance apart, each connected to a receiver. The outputs of the two receivers are sent to a phase detector that produces a voltage level that is a function of the phase difference between the signals at the two antennas. The phase difference, in turn, is a function of the direction angle of the VHF source relative to the baseline between the antennas. To determine the azimuth and elevation to a VHF source, an interferometer with two orthogonal baselines is needed; to locate the source in three dimensions, two such interferometers are needed several kilometers apart.

The dependence of the measured phase difference on the direction angle is determined by the geometry shown in Fig. 6.36. For a signal of wavelength λ arriving from a direction angle θ the phase difference α between two antennas separated by a distance D is given by

$$\alpha = \frac{2\pi L}{\lambda}, \qquad (6.58)$$

where

$$L = D \cos \theta. \qquad (6.59)$$

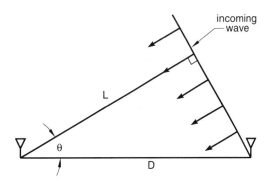

Fig. 6.36. Geometry for determining dependence of phase difference on the direction angle in an interferometer system. D is baseline; L is the distance the wavefront propagates as it traversed D; θ is the direction angle along which the wavefront propagates. (From Rust and MacGorman © 1988, Oklahoma Univ. Press, with permission.)

Then the direction angle, θ, is given by

$$\theta = \arccos \left(\frac{\alpha \lambda}{2\pi D} \right). \qquad (6.60)$$

Measurements of α typically are averaged for a period ranging from one to tens of microseconds to reduce random noise fluctuations.

Because the output of the interferometer is a trigonometric function of the phase difference and so has a cycle of 2π radians, α cannot be uniquely determined if it varies by more than 2π as L varies from $+D$ to $-D$. Thus, the measurement of α, and hence of θ, is ambiguous for any $D > 0.5\ \lambda$. This can be shown explicitly if we rewrite Eq. 6.60 as

$$\theta = \arccos \left[1 - (\alpha_0 - \alpha) \frac{\lambda}{2\pi D} \right], \qquad (6.61)$$

where $\alpha_0 = 2\pi D/\lambda$ is the value of α at $\theta = 0°$. As θ increases from $0°$, the measured phase difference α changes until $\alpha_0 - \alpha$ reaches an integral multiple of 2π, at which point the system measures the same phase difference as at $\theta = 0°$. Thus, all values of α such that

$$\theta = \arccos \left(1 - \frac{n\lambda}{D} \right) \qquad (6.62)$$

correspond to the same interferometer output as $\theta = 0°$. Figure 6.37 shows the multiple values of θ corresponding to any given value of measured phase difference for an interferometer with a baseline 4λ long. These cycles in the phase difference are often referred to as *fringes*. For a given baseline, there are $2D/\lambda$ fringes.

Although the discussion above suggests that a short baseline is desirable, the lessening in number of ambiguities for shorter baselines is countered by a corresponding increase in angular errors. Warwick et al. (1979) assumed that a lightning source generates band-limited white Gaussian noise, and they estimated the

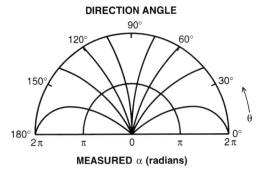

DIRECTION ANGLE

MEASURED α (radians)

Fig. 6.37. Direction angle, θ, as function of measured phase difference, α, for an interferometer with a 4λ baseline. The radial axis of this polar plot is α. There are eight direction angles corresponding to a given α. (From Rust and MacGorman © 1988, Oklahoma Univ. Press, with permission.)

resulting error in the direction angle to be

$$\delta\theta = \frac{\sqrt{2}\,\lambda}{\pi D(B\,\tau)^{\frac{1}{2}}}, \qquad (6.63)$$

where B is the bandwidth of the system, and τ is the time over which the phase measurement is averaged. (Examinations of interferometric data and other estimates of the error indicate that this equation overestimates the error; however, the functional dependence on D, B, λ, and τ is similar in all estimates). Thus, the error is inversely proportional to the baseline length. To circumvent balancing angular ambiguity versus error, two collinear baselines can be used, one with D = 0.5λ, which has no ambiguities but poor resolution, and one with a value of D such that the interval of direction angles over which the phase difference makes a complete cycle equals δθ for the 0.5λ baseline.

There are a number of techniques for determining the phase difference of a signal between antennas, and interferometers can be classified by the techniques they use. The interferometer designed by Warwick et al. (1979) is a simple multiplying interferometer. In this system, the radio frequency signal received by each of the antennas is amplified and mixed with the output of a local oscillator to produce a signal at an intermediate frequency; the amplitude of the intermediate frequency signal is proportional to the amplitude of the radio frequency signal. The signals from the two antennas are then amplified and multiplied together to produce a voltage level that is directly proportional to the power of the combined signals. This voltage is a function of the relative phase difference and the direction angle, such that for a point scource

$$V = K \cos \alpha$$
$$= K \cos \left(\frac{2\pi D \cos \theta}{\lambda} \right), \qquad (6.64)$$

where K is a constant of proportionality.

If analog recording is used with the interferometer, it is expedient to output a time varying sinusoidal signal instead of a d.c. voltage level. Warwick et al. (1979) accomplished this by offsetting the frequencies of the local oscillators mixed with each of the antenna outputs by f_o. The relative phase and the output to the recorder then beat at a frequency equal to f_o, i.e.,

$$V = K \cos \left(2\pi f_0 t + \frac{2\pi D \cos \theta}{\lambda} \right), \qquad (6.65)$$

where t is time. As can be seen from this relationship, a result of offsetting the two mixed frequencies is that a stationary point source gives a time varying output equivalent to that of a moving source for an interferometer in which the mixed frequencies are identical. This apparent motion of a stationary source can also be viewed as equivalent to moving the fringe pattern of an interferometer across the sky along the direction of its baseline. Since the value of t is unknown, the measured phase differences, and hence the direction angles, are only relative. To determine the direction to a source, it is then necessary to provide a source of periodic signals at a known location to provide a reference direction for the relative direction angles.

The interferometer described by Hayenga and Warwick (1981) operated at 34.3 MHz and had two orthogonal baselines, each 2λ long and having four fringes. There was not a second site, so all VHF sources were located in only two spatial dimensions. Data were recorded on analog magnetic tape. The local oscillators for each pair of antennas were offset 200 kHz, so the output of the system had a center frequency that was also 200 kHz. The bandwidth was 3.4 MHz, and the signal averaging time was 2.5 μs. From Eq. 6.63 the resulting angular uncertainty for the direction angle to each baseline was 4°. An estimate that appeared to be more consistent with the measured data was 2.5°, based on an error of $1/B\tau$ radians in the measurement of phase.

A new version of the system, described by Rhodes et al. (1994), Shao et al. (1995), and Shao and Krehbiel (1996), consists of five antennas arranged to form a long and a short baseline along each of two orthogonal directions. One antenna is located at the intersection of the two longer baselines and is shared by each of the two shorter baselines. Rhodes et al. and Shao et al. used 4λ and λ/2 baselines, so there would be no angular ambiguity in the measured direction angles. Shao and Krehbiel used 4.5λ and 1λ, because avoiding baselines that were integral multiples of each other improved immunity to noise in the results and because the increased length of the short baseline reduced systematic errors caused by interactions between antennas. The two fringes in direction angles from their 1λ baseline were far enough apart that the ambiguity did not affect results.

The new interferometer has a bandwidth of 6 MHz centered on 274 MHz. Because the characteristics of an interferometer baseline depend on its length measured in multiples of the wavelength, the higher frequency of the new interferometer results in shorter baseline dimensions. The signal output of each baseline of the interferometer is averaged over a running window 1 μs long and is sampled every 1 μs. With these values, the angular error calculated from Eq. 6.63 is 2.5°, but this appears to overestimate typical errors. Rhodes et al. (1994) estimated that the random error is 1° at high elevation angles and increases as $1/\sin\Theta$ (where Θ is the elevation angle) at lower elevation angles. Signal dispersion between antennas has no effect for signals arriving from bearings orthogonal to the baseline. The maximum effect, which occurs when signals propagate along the length of the longer baselines, is to increase the error by a factor of $2^{0.5}$. Systematic errors are caused primarily by interactions between the antennas on the short baseline and by the finite conductivity of the ground.

Data from each baseline are recorded on a 28-track, digital, high-density recorder capable of recording 6 Mb s^{-1} for up to 15 min. Rhodes et al. (1994) and Shao et al. (1995) used an interactive display driven by a computer workstation to remove fringe ambiguities manually from the recorded data. With the modified baselines used by Shao and Krehbiel (1996), it became possible to remove ambiguities automatically. Electric field changes from lightning also are digitized and recorded on one of the tracks of the high-density recorder. Shao et al. pointed out that limited information about the three-dimensional structure of a flash can be provided by combining information from the electric field change data with the two-dimensional VHF source locations from the interferometer.

Another adaptation of the interferometric technique was developed in France beginning in 1979 and was subsequently commercialized as the SAFIR system by the French company Dimensions (Richard et al. 1986, 1988; Richard 1990, 1992). SAFIR is used in meteorological operations for the European space program and is the only commercial system now available that is designed to map both cloud and ground flashes. The system receives signals in a selectable band 1 MHz wide between 110 and 118 MHz. The band used by a particular installation is chosen to avoid local noise sources. A narrow band is used also to reduce signal dispersion between antennas; the design philosophy was that this improved the accuracy of the system more than using a broadband signal.

Each station in the system has an array of antennas and uses interferometry to determine the direction to a lightning channel source of radiation and sends that information to a central processor. In systems designed to map the three-dimensional location of sources, a station has two independent arrays of antennas, one array of 8 horizontally separated antennas to determine azimuth and an array of 16 horizontally and vertically separated antennas to determine elevation. In systems designed to map only the plan location of lightning, a station may have only a single array of antennas to determine azimuth. The central processor determines the location of a VHF source by triangulating the direction angles measured by two or more interferometric stations, much like a short-baseline, time-of-arrival system. The central processor can record data at up to 4×10^4 pulses per second.

A typical SAFIR system has three interferometric stations separated by 10–100 km. To map lightning in three dimensions, baseline lengths approaching the lower limit are needed to provide sufficient vertical resolution. When longer baselines are used, the system provides only two-dimensional, plan maps of lightning. VHF signals propagate essentially along a straight line, so a VHF interferometer cannot detect signals from sources in the lower region of storms at long ranges from the stations because of the curvature of the Earth's surface. Furthermore, few signals received in the VHF band from lightning more than 160 km away remain above the noise level of ground-based radio receivers. At a range of 160 km, a SAFIR system with a 100-km baseline between stations has a typical error of 5–8 km in the plan location of lightning channels that it maps. Within the perimeter of the network, typical location errors with a 100-km baseline are approximately 1.6 km. Smaller location errors can be obtained within the perimeter of the network by using smaller baselines.

To make a preliminary test of SAFIR operation, Kawasaki et al. (1994) examined the cross-correlation of lightning flash density and radar reflectivity at a height of 2 km for three storms. The cross-correlation coefficient was typically about 0.6. Furthermore, Kawasaki et al. found that the visual appearance of the spatial distribution of lightning typically appeared similar to the reflectivity pattern for thunderstorms.

6.10.5. Comparison of three-dimensional lightning-locating techniques

Each technique for locating lightning can contribute something unique to our understanding of lightning, but each has shortcomings. The equivalent charge center analysis, for example, is the only technique that estimates directly the net charge involved in lightning processes. If the analysis is performed on sequential segments of the field charge, as by Krehbiel (1981), it is even possible to study the development of lightning by studying how the net charge and the coordinates of the charge centroid vary. However, the required electric field measurements are difficult to make, and the analysis gives no indication of lightning channel structure or of the geometry of the charge involved in a particular short segment of the field change. Mapping horizontally extensive flashes, such as commonly occur in the decaying stage of large thunderstorms and in extensive stratiform regions of mesoscale convective systems, produces large errors, if such mapping is possible at all, when using the low-resolution measurements that give the field change only for the entire flash. Mapping horizontally extensive flashes requires that the field-change measurements have enough time resolution to analyze the development of the lightning flash.

The acoustic technique can define in detail the structure of the volume permeated by explosive lightning channels, and measurements for the analysis are relatively easy to make. Furthermore, in the limited number of instances in which an infrasonic thunder signal is received, the location of the infrasonic source corresponds to the location of charge neutralized by the flash (see Sec. 5.11.3). However, the acoustic technique gives no direct indication of charge magnitude or of the development of a lightning flash; the analysis is fairly time-consuming, and errors are sensitive to atmospheric conditions. There is also some indication that the acoustic technique may sometimes be biased against some lightning channels at higher altitudes. Although high-altitude thunder sources have been reported by MacGorman et al. (1981) and Rust et al. (1981), Holmes et al. (1980) reported that they observed few thunder sources in the upper region of storms, where there were many strong radar echoes from lightning.

The radar technique is the only technique that actively probes lightning channels instead of passively receiving signals. It provides the location and development of lightning channels directly out to long ranges, with accuracy defined by the resolution of the radar being used. Furthermore, because radar responds to ionization, it is possible to study how the ionization of a channel increases and decays. The primary disadvantage of radar is that it observes lightning channels only within its beam. Radars with mechanical antenna drives cannot scan a storm within the time interval that ionization is typically maintained by lightning channels (0.1–1 s) unless the angular field of view of the radar encompasses the entire storm. Phased-array antennas used with UHF radars would be capable of extremely fast scanning, but practical design constraints make them marginal for mapping lightning flashes at ranges less than about 100 km. Even if entire lightning flashes cannot be mapped reliably by radar, important details on portions of individual flashes, as well as statistical data indicative of characteristics in the whole storm, can be obtained.

Techniques for locating sources of radio noise have a unique capability as a class to observe how lightning develops in time and space throughout a storm. However, some uncertainty persists about precisely what is being observed. Anything that creates a suitable change in the current moment of a lightning channel generates a signal. Proctor (1981) and Hayenga and Warwick (1981) observed that radio sources usually occur in new regions, instead of continuing to be generated in approximately the same place, and so suggested that the sources are created by breakdown processes.

Although the three types of radio systems for mapping lightning respond to the same parameter of lightning and so have many similarities in their data, there may be significant differences in the lightning processes that they map because of the different frequency bands and processing techniques that are used. Most time-of-arrival systems, for example, have processed primarily well-defined pulses and the leading edge of more noise-like pulse trains; they cannot readily identify features within pulse trains to tag the time of arrival. Interferometers, on the other hand, would be expected to excel in processing pulse trains, in which the phase of signals is relatively slowly varying, but to do less well processing pulses that are short compared with the window (typically of the order of 1 μs) during which signals are averaged. Interferometers also are unable to process signals of comparable strength arriving at a station at the same time (i.e., within the averaging window) from different sources; coincident signals produce interference that causes phase variations unrelated to the direction of propagation.

In a preliminary study, Mazur et al. (1995, 1997) compared lightning locations from the LDAR time-of-arrival system and the SAFIR interferometer, both of which were designed to be operational systems and so were not necessarily optimized to extract as much data as possible from a lightning flash. They found systematic differences in the data from the two systems that appeared related to the differences in signal processing. VHF pulses typically occur throughout the duration of a lightning flash, but pulse trains occur sporadically. Thus, the LDAR system detected signals continually during a flash, while SAFIR detected signals sporadically. The vertical distribution of VHF sources mapped by the two systems also differed. Lightning structure mapped by LDAR typically extended higher in the storm than structure mapped by SAFIR, but LDAR detected much less structure in lower parts of the storm, particularly during processes such as dart leaders and return strokes. The two systems also measured different lightning propagation velocities: Lightning development measured by LDAR progressed at velocities of 10^4–10^5 m s^{-1}, while development measured by SAFIR progressed at velocities of 10^7–10^8 m s^{-1}. Mazur et al. interpreted these differences as indicating that LDAR responded better to initial breakdown at the tips of leaders, while SAFIR responded better to fast negative streamers and dart leaders. Because Mazur et al. used data that were already available to analyze these differences, the data sets were not ideal for comparing the systems. It is important to pursue more systematic investigation of differences between the systems to determine the information that each provides about lightning.

Only a few total-lightning mapping systems exist around the world, but the mapping technologies and scientific research from their use have progressed steadily. Data rates and timing accuracies of both interferometric and time-of-arrival systems have improved, and the improvements have provided a more complete picture of lightning development. However, further work is needed to better understand what lightning processes are responsible for the signals detected by existing systems, and different approaches may provide a more complete picture of lightning than present systems. For example, cross-correlating signal waveforms to determine differences in the time-of-arrival (instead of using the arrival time of particular well-defined features) should provide detection of both isolated pulses and pulse trains, unlike present time-of-arrival systems.

6.10.6. Satellite lightning mapper

For more than a decade, there have been efforts to get a lightning mapper in geostationary orbit. The sensor, which was developed by NASA, is a CCD (charge coupled device) optical array with electronics capable of detecting transient light from lightning, even during the day. It is designed to detect lightning from geostationary altitudes during daylight and darkness, with spatial resolution of 10×10 km and with timing resolution of 1 ms (Davis et al. 1983). This sensor has been packaged with lenses to provide coverage of much of North America, including all of the contiguous United States ocean areas, Central America, South America, and the intertropical convergence zone (Christian et al. 1989). From work with prototypes, NASA estimates that the system will detect 90% of the flashes that occur in daylight. Such a system would enable mapping of both cloud and ground lightning continuously on continental scales. No reliable means have yet been found to distinguish between ground and cloud flashes with the satellite's optical sensor, but identification of flash type can be provided by combining satellite data with data from a ground-based network, such as described in the next section.

These data are expected to be extremely valuable for thunderstorm and climate research and for lightning and thunderstorm detection and warning. As has been reported by Robertson et al. (1984), lightning measured from space could be examined relative to storm-top development, growth rates, precipitation, and storm and larger scale wind fields. If such data were made available routinely to the scientific and operations communities, it would increase our understanding of such topics as the global electric circuit, relationships between storm development and electrical activity, and thunderstorms over oceans. There are no definite plans at this time to launch a geostationary lightning mapper, although a prototype has been completed and existing satellite platforms would be suitable. It has proven difficult to get an allocation of space on a geostationary satellite and funding for the lightning sensor at the same time. However, modern optical detectors that were derived from the design of the geostationary mapper are being flown on satellites in lower orbits than required by geostationary satellites (Christian et al. 1992, Christian et al. 1996, Mach et al. 1996).

6.10.7. Cloud-to-ground lightning location

Systems for locating ground flashes have been available for decades (e.g., Horner 1954). However, early systems were cumbersome and not accurate enough for studies of many of the relationships between lightning strikes and storms. Two developments greatly enhanced the usefulness of this mapping technology. First, advances in electronics provided fast and reliable communication from remote sites and speeded computations to the point where data could be processed automatically in real time in a compact unit at reasonable cost. Second, Krider et al. (1976) realized that, if the plan location of lightning were measured at the peak in the return stroke waveform, the location would correspond to a point on the lightning channel within a few hundred meters of ground and nearly directly above the strike point. Currently, the accuracy and range of systems that determine locations from this well-defined point in the lightning waveform are comparable with weather radar capabilities.

Besides being easier to locate accurately, the ground strike point is the part of a lightning channel that is most important to those concerned with the lightning hazard to life and property. Thus, in addition to being used for studies of storms, lightning strike locating systems have been used to manage or avoid lightning hazards in forest fire detection (Krider et al. 1980), management of electric power distribution lines, employee safety, and space vehicle operations. Data also can be incorporated in climatological studies to yield maps of lightning flash density for engineering design (e.g., MacGorman et al. 1984, Orville et al. 1987, Orville 1994).

Commercially available lightning strike locating systems utilize either magnetic direction-finder technology, time-of-arrival technology, or a combination of both. Because the two technologies were developed separately and still can be used separately, we will describe each of them in the following sections, before discussing a hybrid system, as well as a different system used by the United Kingdom. Keep in mind that the technology for locating lightning strikes continues to evolve. The descriptions in the following sections portray the technologies as they were in 1995; subsequent versions may differ in some details.

(i) Direction-finder system. The direction-finder system is an outgrowth of a technology that is decades old. The basic sensor is a crossed-loop antenna that consists of two vertical loops mounted perpendicular to each other, one oriented north-south, and the other, east-west. The technique assumes that the lightning channel being located is a vertical channel, because the magnetic field produced by a vertical channel has only an azimuthal component; the radial and vertical components are zero. Horizontal channels, which produce nonzero radial or vertical components, cause errors in the measured bearing to a channel. The signal induced in each vertical loop depends on the electric current in the vertical lightning channel, the range from the channel, and the cosine of the angle between the plane of the loop and a bearing to the lightning channel. When the loop points toward the lightning channel, the signal produced in the loop is the maximum possible at a given range and for a given lightning current. When the loop is orthogonal to the bearing toward the channel, no signal is induced in the loop. To obtain a signal that is independent of range and lightning current, a direction-finder system uses the ratio of the signals induced in two orthogonal loops; the ratio of the signal in the north-south loop to that of the east-west loop depends only on the tangent of the bearing to the lightning channel.

By triangulating the bearings measured by two or more direction-finder stations, the location of the channel can be calculated, as shown in Fig. 6.38. Thus, direction-finder systems require a minimum of two stations. However, location errors are distributed more uniformly around the network of stations if there are three or more stations in a noncollinear arrangement.

Krider et al. (1976) made a major improvement in direction-finder technology that was soon incorporated into a commercial system by Lightning Location and Protection, Inc. (LLP, Tucson, AZ). (LLP eventually became GeoMet Data Services and later merged with Atmospheric Research Systems, Inc., under Global Atmospherics, Inc., to provide a hybrid system.) Direction-finder systems work properly only for vertical lightning channels, so LLP designed direction-finder systems that used only the part of a lightning signal generated by the cloud-to-ground lightning channel within roughly 100 m of ground. Because lightning channels are predominantly vertical near ground, the errors caused by departures from vertical orientation tend to be small. Of course, to use the signal from a lightning channel near ground, it is first necessary to make sure that the signal is generated by a ground flash. The LLP system uses signals in the radio frequency band below about 100 kHz, where ground flashes radiate much more energy than cloud flashes. LLP also developed several tests for the shape of signal waveforms in this band to identify signals from ground flashes and reject signals from cloud flashes.

The largest sources of error in the direction-finder system are terrain features and man-made structures that reradiate lightning signals. Errors from these sources are called *site errors*. The site errors of a given station vary with azimuth, but are essentially the same for all lightning flashes occurring at the same azimuth. Also, site errors usually are reasonably constant in time. Therefore, once site errors are determined, measured bearings to flashes can be corrected in real time.

An LLP direction-finder network typically consisted of at least three direction-finder stations that sent lightning data to a central analyzer, called a position analyzer. If the position analyzer received data from two or more direction-finder stations within a preset time window, it assumed that the stations detected the same flash and calculated a strike location. Eventually, some networks became very large (e.g., Orville et al. 1987), and it became common for most flashes to be detected by several stations.

When more than two stations detect the same flash, they provide redundant data, although errors in the data cause triangulation of different pairs of bearings to give multiple solutions. In a normal calculation of strike location, the position analyzer used spherical trigonometry and a least-squares technique to determine an optimal solution based on bearings from all stations detecting the flash. A nonlinear least-squares technique (Bevington 1969) determines the strike location by minimizing χ^2, given by

$$\chi^2 = \sum_i \frac{(\phi_i - \phi_{mi})^2}{\sigma_{az\,i}^2} , \qquad (6.66)$$

where ϕ_i is the bearing of the calculated strike location from the ith station; ϕ_{mi}, the bearing measured by the ith station; and $\sigma_{az\,i}$, the expected azimuthal error in the measurement by the ith station. If only two stations detect a ground strike, a location is calculated by triangulation using spherical geometry. If the measured bearings are along or near the baseline between the two sites, triangulated locations have large errors, so range to the strike is calculated from the measured signal amplitudes by utilizing the 1/R dependence of amplitude on range.

There have been several evaluations of the LLP direction-finder system (e.g., Krider et al. 1976, Hojo et al. 1989, MacGorman and Rust 1989). Typically, direction-finder systems detected 60–90% of the ground flashes that occurred. At least part of this spread of values was caused by variations in lightning characteristics from one season or geographical region to another. Evaluations also found that the standard deviation of the random error in the measured azimuths was 0.5°–1.0°, with the smaller end of this range being typical of more recent systems that have had technological improvements. The error in strike locations varied with the orientation and length of the baselines between stations relative to the strike location. However, within the interior of a network having three or more stations separated by 200–400 km, random errors typically were ≤5 km. Better accuracy could be obtained with smaller baselines.

(ii)   Time-of-Arrival System. The Lightning Position and Tracking System (LPATS) for mapping lightning was first developed and manufactured in the

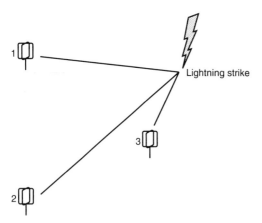

Fig. 6.38. Concept of direction angles (in this case, 3) from detection sites to a ground strike.

early 1980s by Atlantic Scientific Corp. (ASC later became Atmospheric Research Systems, Inc. (ARSI) and eventually merged with GDS as Global Atmospherics, Inc.) This system uses a time-of-arrival technique: Each station identifies the time of arrival of a vertically polarized electromagnetic signal from lightning, and a central processor determines the differences in the times the lightning signal arrives at four to six stations. The difference in the time of arrival for a pair of stations defines a locus of constant time difference that passes through the lightning strike location. For stations on a flat plane, the locus is a hyperbola, but for ranges at which the curvature of Earth must be considered, the locus is distorted from a hyperbola. A third station provides a second independent time difference, and the locus of this second time difference intersects the first locus at the point at which the lightning strike occurred (Fig. 6.39). However, for flashes in some regions, there are two plausible intersections. A fourth station, which provides a third independent time difference, is required to eliminate the ambiguity. For cases in which the ambiguity does not need to be considered, LPATS determines the location of the lightning by solving for the intersection of two hyperbolas. This location is verified, if possible, by checking it against redundant data.

Although a ground flash generates the largest signals in the radio frequency band used by LPATS, an individual LPATS station accepts signals from both ground and cloud flashes, unlike stations in the LLP system. Most versions of LPATS have used only a simple waveform criterion to evaluate whether a signal is from a ground flash. An unpublished study by one of the authors found that flashes identified by LPATS as negative ground flashes were usually correctly identified. However, the majority of flashes identified as positive ground flashes appeared to be cloud flashes: A comparison of LPATS data with ground-truth television video recording of lightning channels to ground failed to find corresponding images of channels to ground for most flashes identified by LPATS as positive ground flashes within the region of the study. The most recent version of LPATS can include extensive waveform storage capabilities, which can be used to implement criteria for identifying flash type.

Each LPATS station consists of a vertical antenna to receive reference timing signals and sense electric field changes from lightning, a time signal receiver, a time signal generator, and a lightning stroke detector. The critical task for a time-of-arrival system is to accurately time the arrival of the same part of a lightning signal at four or more stations. When stations are more than 200 km apart, the specified performance of an LPATS network can be maintained by synchronizing time between stations within $<10^{-6}$ s. ARSI developed three techniques for synchronizing stations. Originally, LPATS used the signal from a single LORAN-C navigational transmitter. The most recent version of the system uses timing from clocks synchronized to signals from the Global Positioning System (GPS).

The feature identified by LPATS in a lightning signal to determine the time of arrival has varied. The most recent version processes the signal to identify the beginning of the pulse. For ground flashes, this should correspond to the return stroke. Identifying the same return stroke signal at four stations is not usually difficult, because the propagation time of the signal between stations is much smaller than the time interval between return strokes, even in a multistroke flash. When there is too much local radio noise at a site, however, the LPATS station's processing capacity is overloaded. Noise problems can be caused either by manmade noise or by frequent radio impulses from nearby cloud flashes. Sites can be chosen to eliminate most problems with man-made noise, but thunderstorms can occur near any site. LPATS turns a station off if the number of signals detected by the station is large enough that thunderstorms are likely to be nearby.

Even when this has been done, however, data rates sometimes have been large. Evaluations (e.g., MacGorman and Rust 1989) have found that a large LPATS network in the United States sometimes detected $<40\%$ of ground flashes that occurred. It now appears that this happened when data rates exceeded the available communications bandwidth. Reducing the data rate and increasing the bandwidth of communication links apparently eliminated, or at least minimized, the problem.

Errors in lightning locations from LPATS are caused by anything that affects the determinations of the time of a lightning signal. The actual location error is a complicated function of range and azimuth. However, the largest errors generally are caused when time is synchronized incorrectly or the wrong part of the signal is

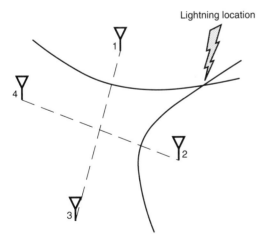

Fig. 6.39. Location of a ground strike by a time-of-arrival system. The difference in times that a signal arrives at two stations (numbered antenna symbols) defines a hyperbola on which the flash is located. The hyperbolas from different pairs of stations intersect at the location of the lightning channel.

chosen at one of the stations. These errors usually can be detected and discarded by newer models of LPATS (models after Series 2). The next largest errors are caused by changes in a lightning signal as it propagates from a flash to a station. The effect of propagation on location errors usually decreases with distance outside the network and is largest within the boundary of the network. In the newest version of LPATS, the location error from all sources is expected to be typically <1 km.

(iii) Hybrid system. By 1991, two separate lightning strike locating networks covered the contiguous United States, one a direction-finder system and the other a time-of-arrival system. In late 1993, the two national networks were consolidated under the same parent company (Global Atmospherics, Incorporated), and work on merging the two technologies into one network was begun. By the end of 1994, the two technologies had been merged, and a new U.S. national network had been installed. As of 1995, the resulting U.S. National Lightning Detection Network had more than 100 stations. Roughly 60% of the stations employed only time-of-arrival, and the remainder employed both time-of-arrival and direction-finder sensors. The location of ground strikes in the combined system is computed by an optimization technique similar to the technique that had been used in the direction-finder network. The algorithm searches for a location and time of the lightning strike that minimizes χ^2, now given by

$$\chi^2 = \sum_i \frac{(\phi_i - \phi_{mi})^2}{\sigma_{az\,i}^2} + \sum_j \frac{(t_j - t_{mj})^2}{\sigma_{tj}^2}, \quad (6.67)$$

where ϕ_i, ϕ_{mi}, and $\sigma_{az\,i}$ have the same meaning as in Eq. 6.66, and where t_{mj} is the time at which the signal was observed to arrive at the jth station, t_j is the time of arrival at the jth station from the trial solution (given from the trial flash time by adding the propagation time from the trial location of the flash to the jth station), and σ_{tj} is the expected error in the time measurement. The network was designed to detect 80–90% of ground flashes and to have a typical location error of 500 m (Cummins et al. 1995). The new network provides data both for individual strokes and for flashes. As defined by the system, a flash includes strokes that occur within 10 km and 1 s of the first stroke in the flash.

(iv) VLF arrival-time-difference system. The British Meteorological Office developed the arrival-time-difference (ATD) system (Lee 1986a, 1986b, 1989a) to replace the manual direction-finder system that it had used for decades. The primary goal was to develop an automatic system for accurately locating thunderstorms over ranges of thousands of kilometers, a task made more difficult by most of the range being over oceans in some directions. The operating frequency of the ATD system was chosen to be 2–23 kHz, in the very low frequency (VLF) radio band, because ground

flashes radiate strong signals in this band, and because VLF signals can propagate thousands of kilometers. Detecting the majority of lightning flashes was not a requirement; the operational system was designed to analyze only ≈400 flashes per hour. Lee (1986b) noted that a previous, unpublished experiment had found that detecting 50–100 flashes per hour was sufficient to detect 94% of storm areas reported by observers in the United Kingdom.

The British ATD system determines the difference in the times at which a lightning signal arrives at different pairs of stations and uses the arrival-time differences to calculate the location of lightning. The technique is the same in this respect as the time-of-arrival technique, but a different nomenclature was chosen because the ATD system differs considerably from existing time-of-arrival systems in the way that time delays are measured. Instead of timing the arrival of only a single feature of the signal waveform, each station in the British ATD system stores the entire waveform and the time it was received and then transmits these data to a central analyzer. The central analyzer performs a cross-correlation analysis of this waveform with the waveform received by another station, to determine the arrival-time difference between the stations.

Two stations provide a single arrival-time difference that defines a locus of constant arrival-time difference passing through the lightning flash that produced the sferic. As for other long-baseline, time-of-arrival systems, the locus is a hyperbola within ranges where the Earth can be approximated as flat, but becomes distorted from a hyperbola when extended over a spherical surface. A third station provides a second arrival-time difference and a corresponding locus that intersects the first locus at two points, although one intersection sometimes is so far from any of the stations that it would be implausible as a solution. To eliminate ambiguities, a fourth station, which provides a third independent arrival-time difference, is required. The British system has seven stations, five in the United Kingdom, one at Gibraltar, and one on Cyprus. Employing three stations more than the required minimum provides redundant data, which are used to improve the accuracy of computed lightning locations. Baselines between neighboring stations are 250–3300 km to provide the desired accuracy over the region of coverage.

Most published evaluations of the British ATD system have been statistical analyses of ATD data alone, although there have been some comparisons with independent storm data. Published estimates (Lee 1986b, 1989b) have indicated that typical location errors are <2 km over the United Kingdom, <5 km throughout most of western Europe, and increase to roughly 15 km at the western limit of the system's service area, approximately 2400 km west of the United Kingdom. Lee (1990) suggested that errors due to systematic propagation effects could be largely eliminated to reduce typical errors to 1.2 km over most of western Europe.

7

Observations of the Electrical Characteristics of Thunderstorms

I. General Characteristics

7.1. INTRODUCTION

In this chapter and the next, we present the observational information that comprises the heart of our understanding of storm electricity. It will serve as the foundation upon which conceptual and numerical models of storms can be built and evaluated. In this chapter, we try to cover general characteristics of thunderstorms by concentrating on nonsevere, isolated, warm season thunderstorms over both mountains and flatter terrain, inland and coasts. In Chapter 8, we present electrical characteristics of several specific types of storms: severe and supercell storms, mesoscale convective systems, winter thunderstorms, tropical thunderstorms, and tropical cyclones. Many of the electrical characteristics of these types overlap, but there are enough differences that we find the divisions are helpful, even if artificial. To avoid awkward gaps in our treatments of topics in these two chapters, there is some redundancy with the brief introduction to storm electricity in Chapter 3.

Within our discussion of each type of storm, we will present overviews of what is known about their electric field structure, charge structure, and lightning production. Discussion of charge will include the charge on cloud particles and precipitation, because these hydrometeor charges are tightly tied to assessments of charging mechanisms and the resulting spatial electrical structure of storms. The goal is to understand storm electrification as a whole, not just isolated pieces of it. You will learn, however, that details of a broad understanding elude us, owing in large part to inadequate

observations with which to compare to theory. Nevertheless, it is in this context and toward such a goal that we must strive if we are to develop the understanding necessary to describe quantitatively how clouds electrify and why they have the electrical structures they do.

As you read this chapter, be aware that it is difficult to ascertain to what extent any particular results are "typical" of storms overall or even of storms in a given region. For example, Raymond et al. (1991) documented that the relatively small mountain thunderstorms that occur over Langmuir Laboratory can have different dynamical behaviors, even when their radar reflectivity structures have only minor differences. Thus, it is unclear whether observations of a single storm or a few storms are representative of thunderstorm behavior or whether such observations could be representative of any class of storms.

Numerical cloud models have been developed that should eventually allow rapid tests of hypotheses to supplement what can be learned from measurements, which often are difficult to obtain and have severely limited continuity in time and space. Models can provide complete sets of observable properties and can allow tests with controlled changes in many variables, resulting in much more hypothesis testing than could ever be done solely with observations. However, models cannot replace observations if reality is to be checked. Since even the results of models must be based on observations, the quality of observations is key to assessing the believability and value of our hypotheses. Keep in mind the major features of the instrumentation and be especially mindful that the measurements often require very sensitive sensors, which are embedded in a large electric field and in the electrically noisy environment associated with nearby lightning.

An important concept in this chapter is *initial electrification*, for which no universally accepted definition exists, as far as we know. Some researchers use the change in polarity of E from the fair weather polarity to the foul weather (i.e., negative to positive E). Others use a particular threshold of E (e.g., 1 kV m^{-1}). In some papers, initial electrification is defined as the occurrence of the first lightning flash. We will revisit this issue in relevant sections, but it is wise to be alert to the lack of a simple and widely held definition. We do not try to provide a single quantitative definition, as it is not clear to us what would be most meaningful. For example, should a criterion for identifying initial electrification be that the cloud goes on to produce a lightning flash, or just that the internal E (or even E$_{gnd}$) of the cloud attains some threshold value?

7.1.1. Basic concepts of thunderstorm morphology and evolution

A major thrust of this chapter is to present electrical observations in the context of storm evolution or structure. To understand material concerning the storm context, however, the reader must have at least a basic understanding of some core concepts and related terminology concerning the production and morphology of storms. In this section, we summarize the necessary material for students unfamiliar with this topic. Our discussion will tend to be qualitative and will omit many details, some of which are important, because a more in-depth understanding is not necessary to the rest of this chapter, and more thorough quantitative descriptions are readily available (Ludlam 1980, Kessler 1986, Houze 1993, Bluestein 1992).

Thunderstorms involve strong convection, i.e., vertical motions that transport heat to bring the troposphere locally into stable equilibrium. Convection exists in the troposphere on many distance scales, with convection on larger distance scales tending to have weaker vertical motions. Updrafts in isolated thunderstorms are toward the small end of the scale of atmospheric convection (a few kilometers in diameter) and tend to have larger magnitudes (≥ 10 m s^{-1}). However, the relationship between size and speed is not simple. Thunderstorm updrafts with small cross sections typically are weakened because the surrounding dry environmental air becomes mixed with the updraft air (i.e., the dry air becomes *entrained*), while larger thunderstorm updrafts have a core that is protected from entrainment.

(i) Instability. Thunderstorm convection has two key ingredients, buoyancy and water, although other factors also can have a strong influence. Buoyancy occurs because a parcel of warm air expands until its pressure is nearly equal to environmental pressure. The expansion causes the parcel to become less dense than the surrounding air, in accordance with the ideal gas law. The downward gravitational force on the parcel then no longer balances the upward pressure gradient force (which is approximately balanced over most of the troposphere), so the parcel is accelerated upward. This acceleration can be given by

$$\frac{dw}{dt} = \frac{g(\rho_a - \rho_p)}{\rho_p}$$
$$= \frac{g(T_{Vp} - T_{Va})}{T_{Va}} \tag{7.1}$$

where w is the vertical velocity of the parcel, g, the acceleration of gravity, ρ_a, the density of the environmental air at the same level, ρ_p, the density of the parcel, T_{Va}, the virtual temperature of the air, and T_{Vp}, the virtual temperature of the parcel. (Virtual temperature is the temperature to which a moist parcel would have to be heated for its pressure to be equal to that of a dry parcel.) As the parcel rises under buoyant forces, it expands and cools. As long as the temperature of the parcel remains warmer than the environmental temperature as it rises, the parcel will remain buoyant and continue to be accelerated upward. This occurs as long as the temperature lapse rate (the rate at which temperature decreases with height) of the parcel is less than the temperature lapse rate of the environment. When the temperature of the parcel cools to the environmental temperature, it is no longer considered buoyant. The height at which this occurs is called the *equilibrium level* or the *level of neutral buoyancy*.

If the parcel contains water vapor and cools enough so that it becomes saturated with respect to liquid water before it stops rising, water vapor condenses to form water droplets. As the water vapor condenses, it releases its latent heat of vaporization, thereby increasing the buoyancy of the parcel. Largely because of the release of latent heat, the temperature lapse rate of a saturated parcel is smaller than the lapse rate of dry air. If the parcel rises past the 0°C isotherm, some of the water particles begin to freeze, but others (typically smaller particles) remain liquid as they become colder than 0°C. These cold liquid particles are called *supercooled liquid water particles*. At some still colder temperature, all remaining liquid particles freeze. This temperature is no colder than −40°C, but typically most liquid particles have frozen by −30°C and frequently at warmer temperatures. The region in which liquid water and ice coexist is called the *mixed phase region*. When particles freeze, they release their latent heat of fusion, and this additional heat again increases the buoyancy of the parcel.

To estimate the likelihood and strength of convection, the environmental temperature profile of the troposphere is compared layer by layer with the estimated profile of temperature in a parcel that rises adiabatically through the atmosphere. If the temperature lapse rate of a parcel having no moisture is smaller than the envi-

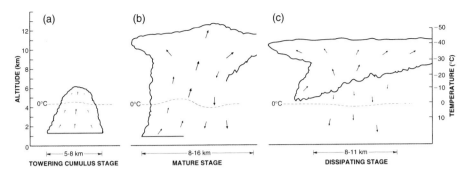

Fig. 7.1. The stages of a thunderstorm cell: (a) the towering cumulus stage, (b) the mature stage, (c) the dissipating stage. (After Byers and Braham 1949, with permission.)

ronmental lapse rate above some altitude in the troposphere, so that a rising parcel starting at that altitude is always warmer than the environment in some substantial altitude range, the environment is said to be absolutely unstable. If this condition is not satisfied, but the temperature lapse rate of a parcel having environmental moisture is smaller than the lapse rate of the environment once condensation begins (i.e., at the *lifting condensation level*), the environment is said to be conditionally unstable. If a parcel is conditionally unstable, it must be lifted by some forcing mechanism to the lifting condensation level before it becomes unstable. The altitude at which a parcel becomes unstable to vertical motion, whether it is absolutely unstable or conditionally unstable, is called the *level of free convection*.

A useful measure of the potential for convection is provided by the *convectively available potential energy* (CAPE), defined as the total buoyant energy available per unit mass to an air parcel that rises from the level of free convection to the level of neutral buoyancy for a given atmospheric sounding. *CAPE* can be expressed by

$$CAPE = \int_{LFC}^{LNB} (T_{Vp} - T_{Va})R_d\, d\ln P, \quad (7.2)$$

where *LFC* is the level of free convection, *LNB* is the level of neutral buoyancy (i.e., the equilibrium level), T_{Vp} is virtual temperature of a parcel, T_{Va} is the virtual temperature of the environmental air, and R_d is the gas constant for dry air. If we assume that all potential energy is converted to kinetic energy, the updraft velocity at the equilibrium level is given by

$$w = (CAPE)^{\frac{1}{2}}. \quad (7.3)$$

(ii) Thunderstorm morphology. Convection in thunderstorms has cellular structure, with each thunderstorm cell having a strong updraft ($\geq 10\ \mathrm{m\ s^{-1}}$), typically a few kilometers in diameter, extending through much of the vertical depth of the troposphere. At some

point in its life cycle, which typically takes 30–60 min to complete, a thunderstorm cell produces precipitation and a downdraft. Thunderstorms can include a single isolated cell, several cells (called a *multicell storm*), or a long-lived cell with a rotating updraft (called a *supercell storm*), which we will describe further in Chapter 8. In multicell storms, the individual updrafts can coexist or be sequential but, typically, multicell storms consist of a group of coexisting cells at various stages of evolution (Marwitz 1972, Browning et al. 1976).

Whether isolated or in a multicell storm, thunderstorm cells tend to evolve in characteristic patterns that were first detailed by The Thunderstorm Project (Byers and Braham 1949) for single-cell storms (see Fig. 7.1). When a cell first forms, a single updraft completely fills the cell and produces small water particles and ice particles. This period is called the *cumulus stage*. As the updraft rises through the troposphere, particles grow in the updraft by a combination of collection (large particles fall relative to small particles and collect them) and condensation or deposition. Eventually, some particles grow large enough that their terminal velocity exceeds the updraft velocity, so they begin to fall toward the ground. The drag of falling particles and cooling from evaporation and sublimation of particles cause downdrafts to form.

The *mature stage* begins when precipitation reaches Earth's surface. During this stage updraft magnitudes increase as the cell continues to grow vertically, with the largest updraft speeds typically occurring early in the mature stage. Also during this stage, the cell reaches its maximum height, and the horizontal area of precipitation in the lower part of the cell increases. Downdrafts, which began at middle and low altitudes, grow horizontally and vertically. When the downdraft reaches the ground, it diverges to produce outflowing wind at the surface.

The *dissipating stage* begins when the area of the downdraft extends completely across the lower level of the cell. During the dissipating stage, the region of downdrafts continues to grow. Updrafts may still exist in the upper part of the storm, but they are cut off from air below the storm, and they weaken. Because updrafts weaken, condensation and deposition decrease, so precipitation growth slows. This reduces the contribution of precipitation to downdraft acceleration, so downdrafts also eventually weaken.

7.1.2. Radar reflectivity of hydrometeors

Because some sections of this chapter will examine electrification relative to storm characteristics measured by radar, we will describe very briefly how radar detects hydrometeors, before examining observations of the electric field, charge, and lightning. For more information about radar meteorology, see a textbook such as Doviak and Zrnić (1993).

A weather radar radiates an electromagnetic pulse and detects the signal scattered back to the radar receiver by a volume of cloud. The signal received by the radar is a function of several parameters, including the wavelength of the radar signal, the radiated power, the gain of the antenna, the range to the cloud, and the number, size, and type of cloud and precipitation particles. All else being equal, the scattered signal increases rapidly with increasing particle size, so a few large precipitation particles can reflect more power than orders of magnitude greater numbers of small cloud particles. The *radar reflectivity factor, Z,* often called simply *reflectivity* in the literature, is the parameter from which many other dependencies have been removed to indicate primarily the effect of the number density and size of particles. For sufficiently small particles ($D < \lambda/16$, where D is particle diameter and λ is signal wavelength), the scattering is called Rayleigh scattering, and Z is given by

$$Z = \sum_i n_i D_i^6, \qquad (7.4)$$

where n_i is the number of particles of diameter D_i per unit volume. Scientists often use a logarithmic scale $dBZ = 10 \log_{10} Z$, where Z is in units of $mm^6 \ m^{-3}$. Other parameters, such as mean particle diameter and rainfall rates, can be estimated from Z.

Features may be added to a radar to enable it to measure other properties of the particles that scatter the radar signal. Doppler radars measure Doppler shifts in the frequency of the scattered signal to calculate the velocity of particles toward or away from the radar. Polarimetric radars (also called multiparameter radars) process additional parameters of signals, such as the phase shift of scattered signals at two orthogonal signal polarizations, to estimate other properties of particles, such as asymmetries in their shape. Considerable work has been done in recent years on inferring the dominant type of particle (e.g., rain, hail, graupel) in a given region of a storm from polarimetric radars.

Characteristics of a radar, such as its wavelength, pulse rate, beamwidth of the radiated signal, antenna rotation speed, and receiver noise level, affect how well Z can be measured. Short wavelength (≤ 3 cm) radars

are well suited to detecting smaller cloud particles, but often have problems measuring thunderstorm structure, because the radiated power is attenuated strongly by precipitation over signal pathlengths of only a few kilometers. Radars used to study thunderstorm structure typically have a wavelength of 5–11 cm.

Differences among radars probably account for some of the variations in results of electrification studies, particularly in studies from earlier periods, when calibration procedures were not as standardized as in more recent years. Even with modern equipment, it is difficult to calibrate a radar to an accuracy of 1–2 dB and to maintain its calibration (see Doviak and Zrnić 1993). Uncertainties of 3–4 dB are not unusual in well-executed, modern field experiments. Furthermore, early radar experiments frequently relied on photographic records of storm displays in which reflectivity was depicted by a continuous gray scale, and the photographs themselves posed additional sources of error. Photographs of gray shades were sensitive to the brightness of the display, camera exposure settings, focus, film characteristics, and the film developing process. Even though great care often was used in making and calibrating the photographs, it appears unlikely that errors from the photographs could have been made consistently negligible, because of the multitude and sensitivity of sources of error.

7.2. TYPICAL HISTORY OF E$_{gnd}$ BENEATH A THUNDERSTORM

We begin with measurements of electrical aspects of thunderstorms made on and above mountaintops, where orographically induced convection often results in isolated thunderclouds during the summer. Much impetus for the comprehensive observational study of initial cloud electrification has come from the research on such thunderstorms in the mountains of central New Mexico. From the early work in the 1940s and 1950s, it became obvious that a permanent mountaintop facility was needed. This led Drs. E.J. Workman, Marvin Wilkening, and Marx Brook of the New Mexico Institute of Mining and Technology to plan and oversee the building of the Irving Langmuir Laboratory for Atmospheric Research. The Laboratory is named in honor of Noble laureate, Dr. Irving Langmuir, who along with Drs. Vincent Schaefer and Bernard Vonnegut had conducted cloud seeding experiments in New Mexico. Langmuir Laboratory is at 3.2 km MSL on the crest of a small mountain range called the Magdalena Mountains, and it continues to be operated as a division of the New Mexico Institute of Mining and Technology.

Perhaps the measurement considered most fundamental, historically, has been that of the electric field at the ground, E$_{gnd}$, largely because the ground is the easiest location to make measurements. To assess charging mechanisms, scientists looked for correlations between the evolution of precipitation and electrification. Reynolds and Brook (1956) placed a field mill atop Mount Withington (west of the Magdalena Mountains and the site of a temporary predecessor to Langmuir Laboratory) and monitored clouds over it with a radar. They found from 75 h of data on clouds without precipitation that such clouds did not electrify. From thunderstorm data, they suggested that the presence of precipitation was a necessary, but not sufficient, condition for cloud electrification and that rapid vertical growth of the precipitation echo was required for the cloud to electrify. Their definition of initial electrification seems to have been the deviation of E_{gnd} from its usual fair weather state. Although the growth of E_{gnd} under a developing thunderstorm looked exponential, they noted that it deviated from a true exponential relationship. In their few cases, the first lightning occurred as quickly as 11 min after the onset of electrification in E_{gnd}. Moore and Vonnegut (1977) showed a record of E_{gnd} that is typical of those measured underneath thunderclouds (Fig. 7.2). The significant features include: the reversal in polarity and rapid increase in magnitude of E_{gnd} at initial electrification; a maximum E_{gnd} that limits often at $<10 \, kV \, m^{-1}$; polarity reversals due to lightning; an excursion in E_{gnd} to the opposite polarity lasting a few minutes and associated with a burst of precipitation called a *field excursion associated with precipitation*, FEAWP; and an *end-of-*

storm-oscillation, EOSO. This record applies only to observations beneath the storm. These features have been corroborated in other studies. For example, the FEAWP was found in Alabama. There, Williams et al. (1989) linked it not only to falling precipitation but to microbursts.

Comparisons of E_{gnd} and E_{aloft} show that the initial reversal of E from fair weather polarity to foul weather polarity occurs within about 1 min at the two locations (Breed and Dye 1989). If initial electrification is defined as such a polarity reversal, this observation suggests that E_{gnd} is a useful detector of initial electrification if it is measured very close to or beneath the developing cloud. The tendency for $|E_{gnd}|$ to have a limiting value between lightning and the polarity reversals from lightning are both evidence of a space charge layer just above the ground. As verified by Standler and Winn (1979), the space charge is the result of point discharge from grass, trees, buildings, etc. at the ground (Sec. 4.4.4).

As mentioned in the instrumentation chapter, E_{gnd} can be very large if point discharge cannot occur. Recall that Toland and Vonnegut (1977) measured a maximum $E_{gnd} = 130 \, kV \, m^{-1}$ at the surface of a lake. Whether the general trends of the time variation of E_{gnd} remain the same for storms over water as for storms over land is not known because no life histories of E_{gnd} have been published for storms over water.

7.2.1. End-of-storm oscillation in E_{gnd}

The term *end-of-storm-oscillation* defines a comparatively slow oscillation in the polarity of E_{gnd} that is often from positive to negative to positive and finally back to fair weather negative values (Fig. 7.2). In their depiction of the storm during an EOSO, Moore and Vonnegut (1977) suggested that the swing to negative E occurred when the lower negative charge was gone or being moved aside, which exposed the positive charge above it. Their whole-sky, time-lapse movies made below storms supported this view. They proposed that the final change to enhanced positive E is from negative charge that was a lower screening layer. During an EOSO, they often found a radar *bright band*, which results when frozen hydrometeors fall through the 0°C level and begin to melt (e.g., see Battan 1973). Whether the bright band is indicative of processes that affect E_{gnd} is unknown. Just before the first excursion of the EOSO, lightning strikes that lower positive charge can occur. Positive ground flashes can occur also in the first excursion of the EOSO (to fair weather polarity). Usually only a few, if any, positive ground flashes occur, and they are often the last flashes of the storm.

Marshall and Lin (1992) added new information about EOSOs with their in situ profile of E during the thundercloud's dying stage (Fig. 7.3). The accompanying E_{gnd} (Fig. 7.4) showed that the profile was from the final part of the EOSO and that the storm was clearly dissipating, since the last lightning of the storm was 20

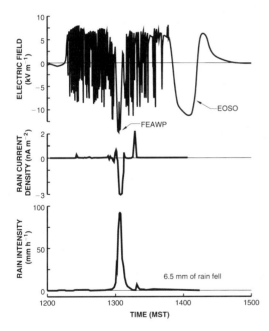

Fig. 7.2. Typical record of the electric field at the ground and the coincident rain rate and current (negative current denotes fall of net-positive precipitation) beneath a thundercloud. An excursion in E_{gnd} associated with increased rainfall, FEAWP, and the end-of-storm oscillation, EOSO, are often prominent features beneath an isolated storm. (After Moore and Vonnegut 1977, with permission.)

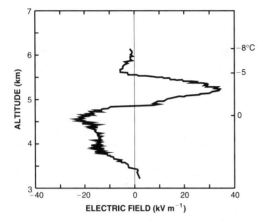

Fig. 7.3. Profile of E through a dying thunderstorm over Langmuir Laboratory. Selected temperatures measured aloft are on the right axis. (From Marshall and Lin 1992: American Geophysical Union, with permission.)

min before balloon launch. The profile indicated an internal negative charge in the cloud bounded on top and bottom by screening layer charges. Their hypothesis for the EOSO was similar to that of Moore and Vonnegut (1977). Marshall and Lin, however, proposed that negative precipitation falls out to drive E_{gnd} positively. After the precipitation has fallen to the ground, a negative E_{gnd} occurs from a positive screening layer on the lower cloud boundary that now dominates. Finally, E_{gnd} decays to fair weather negative values as the screening layer is rescreened and the cloud continues to dissipate.

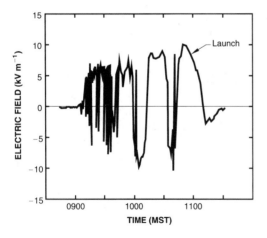

Fig. 7.4. E_{gnd} beneath a small thunderstorm over Langmuir Laboratory during a sounding to measure the E profile in the dying stage of the storm. The balloon launch was during an EOSO. (From Marshall and Lin 1992: American Geophysical Union, with permission.)

Radar and E_{gnd} data were obtained by Williams et al. (1994) for storms near Orlando, Florida. In their data, they found an apparent link between the mesoscale vertical velocity and E_{gnd}. As E_{gnd} started its oscillation to fair weather values, the updraft that had been in the mixed phase region diminished and became a downdraft above the 0°C level. A few large field changes, which they suggested could have been from positive flashes to ground, occurred. The oscillation back to foul weather was followed by the return of updraft above 5 km. They attributed the change to fair weather as evidence of an inverted dipole, but they had no in situ measurements of E aloft to confirm it. They noted the similarity between stratiform convection in a mesoscale convective system, MCS (Sec. 8.2) and the decaying stage of a thunderstorm during an EOSO. Their apparent claim that the charge structure of such stratiform clouds is a dipole is in disagreement with in situ observations in MCSs (e.g., see Sec. 8.2.2). Thus, their model of the dipole inverting to give the EOSO may not be correct if the similarity between decaying deep convection and MCS stratiform regions is true. However, their proposal also involved effects of evolving kinematics and microphysics that appear to be viable: They suggested that the observed formation of a downdraft during the storm's decaying stage and the inferred associated changes in microphysics would produce positive charge in the lower region of the storm. This process would reverse the polarity of E_{gnd} temporarily to produce the EOSO.

7.3. SMALL-ION SPACE CHARGE DURING FAIR WEATHER AND CONVECTION

We begin our consideration of space charge with fair weather. Here we consider *fair weather* as clear air or clouds that are not electrified enough to reverse the polarity of E_{gnd}. The role of fair weather space charge in cloud electrification has not been clearly delineated. However, repeatable and reasonable measurements have been made in clear air at the ground and aloft beneath developing cumuli. To be specific, we are talking about space charge carried on ions. Vonnegut et al. (1962) used a high-voltage supply connected to an elevated wire to create space charge near the ground. They then measured the artificially produced space charge in the clear air around growing cumuli (Fig. 7.5). Vonnegut et al. concluded: (1) Space charge at the ground was convected into the cloud by updrafts; (2) The space charge that entered the cloud did not come out its top; this was interpreted to mean that charge laden air diverged at the top of the cloud and moved downward along its sides; and (3) The polarity and concentration of space charge determined the polarity and electrical intensity of small, nonprecipitating cumuli.

Moore et al. (1958) and Vonnegut et al. (1959) measured space charge beneath developing cumuli with a Faraday cage atop Mount Withington and found a vary-

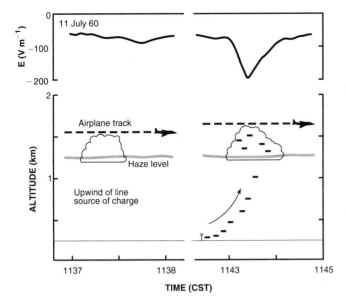

Fig. 7.5. Depiction of an experiment to trace fair weather space charge from ground into growing cumulus. The left side is upwind of the point of release of point-discharge ions into the atmosphere. The right side shows that negative charge released at the ground was ingested into the small cumulus cloud. The E shown was measured by the airplane over cloud top. (After Vonnegut et al. 1962: American Geophysical Union, with permission.)

ing positive charge often ≥ 0.1 nC m^{-3}. They used the one-dimensional Gauss's law and computed the space charge aloft using data from four field-measuring sondes flown on a tethered balloon before and during early convection. They inferred negative charge at the base of cumulus clouds before significant convection. With increased updrafts, they inferred increased negative charge just inside the cloud and increased positive charge just below the base. Magnitudes of both polarities were ≈ 0.1 nC m^{-3} and were about the same as at mountaintop. If convection subsided, the negative charge in the cloud base also dissipated. The surface air was laden with positive space charge prior to cloud electrification. After cloud electrification (i.e., when E reverses polarity) positive ions are attracted upward and leave negative space charge at the surface. In similar observations, Moore et al. (1989) repeated the direct measurements of space charge and again found positive ions were carried upward by convection before cloud development. (An example of the variation of space charge at the ground beneath a developing thundercloud may be seen in Sec. 7.7.1 and Fig. 7.9.)

Measurements of radon gas (Rn^{222}) over Langmuir Laboratory show that surface-level air goes into the developing cumulus clouds over Langmuir Laboratory with little dilution of the Rn^{222} (Wilkening 1970). Since both radon and space charge are carried in the same air parcel from near ground, this is further evidence that the positive fair weather space charge can get into the developing cloud. Additional measurements and analysis reported in Raymond et al. (1991) showed that upslope flow was a significant contributor to the air moving upward into the cloud. Little doubt remains that the fair weather space charge in boundary layer air can get into cumuli that develop over a moun-

tain. However, it is not yet clear what effect this space charge has on the overall electrification of a cloud.

7.4. SMALL-ION SPACE CHARGE IN AND NEAR ELECTRIFIED CLOUDS

The space charge near and within developing cumuli that become thunderstorms is a parameter of cloud electrification. Moore et al. (1958) raised the point that the initial weak fields, which can result from space charge flowing into the cloud, may play a significant role in the formation of drops by coalescence (Sec. 10.8). In attempts to verify the role of space charge in cloud electrification, Moore et al. (1989, 1991) tried to create electrically upside down clouds over Langmuir Laboratory by releasing negative charge into the subcloud air (in contrast to the natural fair weather, positive charge). They found that they would need a larger charge-source array to dominate the natural space charge in the subcloud air, and no definitive test has yet been possible.

We now consider the space charge after a natural cloud is electrified. Standler and Winn (1979) measured the E profile in the subcloud air 200 m above the mountaintop. From their data (e.g., Figs. 4.3 and 4.4), we compute that space charge of both polarities and up to 0.5 nC m^{-3} existed. Soula and Chauzy (1991) used a tethered balloon with field mills to measure E continuously at five different levels to determine space charge under a storm in Florida. The measurements reported were in a rain-free region, where they found that space charge ranged from about 0.1–0.8 nC m^{-3} up to 600 m above ground.

The net small-ion space charge can also be calculated from conductivity using

$$\rho_{net} = \frac{\sigma_+}{k_+} + \frac{\sigma_-}{k_-} , \qquad (7.5)$$

where k is ionic mobility and the subscripts denote ion polarity. A value of $k \approx 10^{-4}$ m s^{-1} per V m^{-1} can often be used for both ion polarities without significant error. In the following form, algebraic errors from sign confusion associated with σ_- are removed, and the approximation is

$$\rho_{net} = \frac{\sigma_+ - |\sigma_-|}{k} . \qquad (7.6)$$

Layers in which the conductivity is approximately constant must be picked for the equation to work correctly. Uncertainty in assuming constant conductivity exists, especially if a single Gerdien cylinder is used to cycle and measure both positive and negative conductivity. Rust (1973) measured the conductivity profile over the mountain ridge (Fig. 7.6) at Langmuir Laboratory and calculated a small-ion $\rho_{net} \approx +0.5$ nC m^{-3} from 150–500 m in the layer beneath a storm after positive point discharge had been occurring for an hour.

These studies showed that space charge layers of similar magnitudes occur above the vastly different terrains of the mountain ridge and coastal flatland. The studies also showed that space charge rises from the ground toward an electrified cloud. Other aspects of space charge around fully developed storms follow in our consideration of conductivity.

7.5. CONDUCTIVITY NEAR AND IN STORMS—THEORETICAL DISCUSSION

Vonnegut (1963) expressed concern that conductivity may be an invalid concept within a thunderstorm. Krehbiel (1969) showed theoretically that the conductivity would be nonohmic in clouds and would affect current flow for $|E|$ ranging from 1–100 kV m^{-1}, which we will see is typically measured in storms. Though the small-ion conductivity might be low, there are also larger charged particles that move under the influence of air motions. Conductivity measurements have yet to be made in regions of the largest electric fields in storms. In these high-field regions, charge transport may vary in a nonlinear way with the electric field, and then $J \neq \sigma E$. The reason is that E could affect not only the motion of charged particles but also the population of the particles because of particle charging processes.

Whether the conductivity inside storms is greater or less than clear air values at the same altitude has been controversial. For example, Phillips (1967) calculated that the conductivity inside a thundercloud would be from <0.05 to 0.1 its value in clear air. In contrast, Hoppel and Phillips (1971) argued for the possibility of conductivity inside thunderstorms being at least as high as that of the surrounding clear air. Their reasoning was that the cloud particle population would be dramatically reduced in convective updrafts because of

precipitation hydrometeor growth. They termed this region *electrically thin,* meaning that ions can move through with less likelihood of being captured. Then in the presence of ionization of at least five times that due to cosmic rays, the conductivity in the cloud could be about the same as in the clear air. Sources other than cosmic rays inside storms include point discharge from precipitation, drop disruption, lightning, and ions moved in by the airflow and E. Griffiths et al. (1974) concluded from their theoretical treatment of conductivity in clouds that it is much less than in clear air. They also concluded that the total electric conductivity of the storm medium is not a simple concept.

Kamra (1979, 1981) proposed that charged precipitation and cloud particles could cause very high conductivities and questioned whether the concept of small-ion conductivity can be applied. He noted that even if the total conductivity, in which he included cloud and precipitation particles, is large, it does not preclude the small-ion conductivity of cloudy air from being low. The impact of low conductivity is that the relaxation time of cloudy air still would be quite slow. Weinheimer and Few (1981) commented that Kamra's estimates of space charge densities were too high and could affect the conclusion of large total conductivity. As Kamra (1981) pointed out, the necessary, reliable data on space charge inside storms are lacking for the spatial and temporal scales needed to address this issue definitively with observations.

7.6. CONDUCTIVITY NEAR AND IN STORMS—MEASUREMENTS

Almost no measurements of conductivity aloft within clouds exist in the refereed literature. A major reason appears to be that measurements of conductivity using the most likely platform, an airplane, are fraught with problems, even in fair weather (Kraakevik 1958). We know of no reliable in-cloud measurements made with an airplane. We turn to other techniques to get data inside clouds. Measurements made at the tops of mountains embedded in nonstormy clouds show that the conductivity drops and is 0.05–0.3 of its fair weather value (Allee and Phillips 1959).

Scott and Evans (1969) reported on a balloon-borne Gerdien cylinder that they designed to measure conductivity upon descent after release at altitude (i.e., it functioned as a dropsonde). They designed it for order of magnitude measurements of conductivity at least as large as in clear air at the ground. They tested for effects of many typical errors associated with conductivity measurements, but they did not state whether the piezoelectric property of Teflon affected the measurement as the temperature changed during a flight. A flight in fair weather produced results comparable with those by Woessner et al. (1958), which suggests the instrument may have worked correctly in clear air over a wide temperature range. They then made a stormy weather flight. The conductivity was never less than the

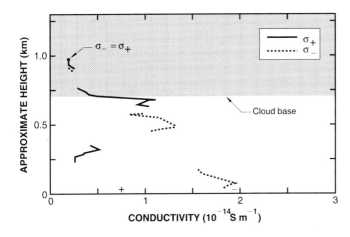

Fig. 7.6. Electric conductivity during ascent below and inside a weakly electrified cloud over Langmuir Laboratory, 9 Aug 1972, 1200–1215 MST. The + and − denote values of positive and negative conductivity, respectively, at the mountaintop. Height is the approximate distance above the mountain, which is at 3.2 km MSL. (From Rust and Moore 1974, with permission.)

clear air value, and they measured peak values of both polarities of 2×10^{-12} S m^{-1}. In a cloud they identified as a storm cell, they measured only the positive conductivity and were confident of values of 3.5×10^{-14} S m^{-1} above 6 km. Their conclusions included identification of instrumentation design problems and the finding that conductivity was not lower inside clouds than in the clear air.

Evans (1969a) reported on a different dropsonde that measured E and allowed calculation of the conductivity to an accuracy of about 50%. He reported conductivities in thunderstorms up to ten times that of clear air. He found these in E ≤ 30 kV m^{-1}. Vonnegut (1969) argued that the instrument might not have functioned reliably owing to asymmetry and enhancement of E, which would affect the calculation of conductivity. He also restated his earlier concerns about whether the concept of conductivity inside a storm is valid. Evans (1969b) countered with laboratory tests showing no point discharge from the instrument until E ≈ 50 kV m^{-1}, which was greater than the measured E, and noted other aspects of the instrument designed to keep self-charging from affecting the measurements. Scott and Evans (1969) did not continue their instrument development, and these conductivity measurements may be the only possibly reliable ones from deep within the interior of thunderstorms.

Conductivity measurements do exist below and in the bases of thunderstorms. Rust and Moore (1974) used a tethered balloon to carry an electric field meter and a Gerdien capacitor designed for measuring conductivity in the presence of large E (Fig. 6.25). The sounding in Fig. 7.6 was made into the base of a weakly electrified cloud. During this time, $E_{gnd} \approx 600$ V m^{-1}, while inside the cloud base the maximum $E_{aloft} \approx 710$ V m^{-1}. There was no point discharge at the ground. The conductivity data showed a net negative space charge at ground level, which is typical under an electrified cloud. The conductivity dropped rapidly with penetration into the first several meters of visible cloud. Inside the cloud base, the net small-ion space charge

was about zero. The reduced conductivity and space charge were likely the result of ion capture by cloud particles (see 'screening layers' in Sec. 3.5.4). During an ascent beneath a thunderstorm with an $E_{gnd} \approx 8$ kV m^{-1} (except during lightning), a different space charge distribution occurred. Positive point discharge began 1 min before balloon ascent. The conductivity data showed that the subcloud air contained a net positive small-ion space charge. After positive point discharge had been in progress for about an hour, the descent sounding (Fig. 7.7) showed a three to five times greater enhancement of the positive space charge beneath the storm.

The results from nine soundings into the bases of storms are summarized in Fig. 7.8. For the data in the figure, the mean conductivities and standard deviations

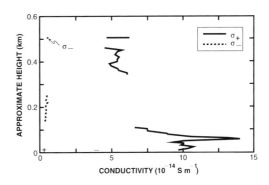

Fig. 7.7. Profile of small-ion conductivity below a highly electrified thundercloud whose cloud base was ≈1 km above a ridge at Langmuir Laboratory, 22 July 1972, 1735–1740 MST. The + and − denote values of positive and negative conductivity, respectively, at the surface of the mountaintop. Height is the approximate distance above the mountain, which is at 3.2 km MSL. The peak in positive conductivity at ≈70 m corresponds to ρ ≈ 0.5 nC m^{-3}. (From Rust 1973, with permission.)

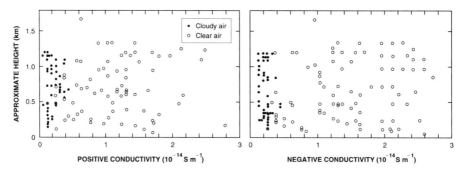

Fig. 7.8. Scatter plots of representative measurements of electric conductivity in clear air and inside nonelectrified and weakly electrified clouds from nine captive balloon flights at Langmuir Laboratory. (From Rust and Moore 1974, with permission.)

in clear air were $\sigma_- = 1.45 \pm 0.67 \times 10^{-14}$ S m^{-1} and $\sigma_+ = 1.28 \pm 1.11 \times 10^{-14}$ S m^{-1}. For cloudy air, the means were $\sigma_- = 0.23 \pm 0.08 \times 10^{-14}$ S m^{-1} and $\sigma_+ = 0.21 \pm 0.10 \times 10^{-14}$ S m^{-1}. The data, limited in clouds to locations where $|E| < 2$ kV m^{-1}, indicated that conductivity of cloudy air is about 0.2 that of clear air at the same altitude. Rust and Moore (1974) noted that the conductivity of cloudy air was about 1/6 that of the adjacent clear air and about 1/10 that of clear air at the same level in the absence of any cloud. Another way to look at the conductivity of the cloudy air is in terms of its relaxation time constant. These measurements gave a relaxation time constant of about 4000 s, which agrees with theory (e.g., see Pruppacher and Klett 1978). However, the data are not adequate for a full assessment of conductivity because there are no in-cloud conductivity data in the presence of large E.

The small-ion conductivity inside storms must be low in many regions, for it is difficult to understand how clouds could separate charge and produce a large electric field if conductivity were high and thus tending to short circuit the charge separation process. Still, no one has adequately addressed the problems delineated by Vonnegut (1969) or by Scott and Evans's (1969) tentative acceptance of their own measurements. Conductivity within the heart of storms remains an unknown, which is unfortunate because of its fundamental importance in any analytical formulation of cloud charging.

7.7. ELECTRIC FIELD ALOFT

Measuring E in storms is critical to our understanding their electrical activity. The electric field is linked with storm charging mechanism(s), the charge carried by particles and ions, and the resulting overall charge structure of clouds. Several significant tasks of studies of E include: (1) defining initial electrification, (2) determining $|E|$ and the rate of change of E during initial electrification, (3) measuring $|E_{max}|$ in storms, and (4)

ascertaining links between E and hydrometeor growth and electrification.

Most in situ measurements aloft of E, E_{aloft}, have been made with airplanes and balloons. Although perhaps obvious, it is worth emphasizing that the information obtained from these two platforms can be very different. Usually airplanes have extensive horizontal records at a few selected altitudes, and balloons have a vertical profile without systematic measurements in the horizontal. However, some investigators have spiraled an airplane down or up trying to obtain a vertical sounding. Recently, this has been particularly successful in the early stages of thunderstorms through use of an instrumented sailplane to spiral in the updraft (e.g., Dye et al. 1988, 1989). A few partial vertical profiles of E have been obtained with rockets.

None of the E measurements meet the ideal, which would be something like vertical and horizontal profiles of E obtained every few minutes throughout the life of the storm. We begin with E data from the initial stages of cumuli. A long-running measurement program was undertaken by the Air Force Cambridge Research Laboratory. Intermittently from the 1950s through the mid-1970s, various aircraft were instrumented to make electric field and other electrical measurements. Unfortunately, few results from the huge data set were published in scientific journals, but we can glean some of the results for cumuli from Fitzgerald and Byers (1958). They found from their investigation of about 700 trade wind cumulus clouds with no frozen particles that $|E_{max}|$ ranged from 100–1000 V m^{-1}. Only a single cloud had a bigger electric field, and it contained frozen precipitation.

7.7.1. E_{aloft} and particle charge during initial electrification

We are aware of no extensive, published measurements of hydrometeor charge during the initial electrification of thunderstorms. There are several reasons: Initial electrification is a short-lived phase and obtaining the in situ measurements in the right places is very difficult; to date, the entire set of particle measurements during initial electrification totals only a few minutes. The minimum detectable charge often has been about 5 pC, which seems too large for initial electrification studies; there are no measurements of charge on small cloud particles during this stage. Usable measurements of the electric field do exist.

When comparing initial electrification studies, be aware that different authors often used different criteria to define when a cloud became electrified. For example, Moore et al. (1958), who examined the growth of foul weather E_{aloft} up to 1 kV m^{-1}, defined initial electrification as E becoming positive inside the base of a cloud. However, Dye et al. (1989), who measured $|E_{aloft}|$ from 100 V m^{-1} to $\approx$100 kV m^{-1}, defined initial electrification as $|E_{aloft}|$ becoming $\geq$1 kV m^{-1}. Clouds can meet both criteria and yet never produce lightning. Other definitions also are possible.

The observations at Mount Withington described in Moore et al. (1958) and Vonnegut et al. (1959) indicated that E in the cloud changes before E_{gnd} (Fig. 7.9). Their data, like that of others, showed that electrification is linked to vertical development of the cloud. For measuring the vertical field, they sometimes flew captive balloons with two devices separated vertically by about 800 m inside the cloud. The devices measured

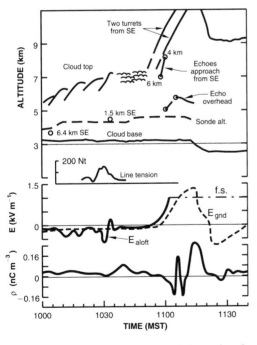

Fig. 7.9. Electrical parameters and cloud observations for a developing thundercloud over a mountaintop observatory at Mount Withington in New Mexico, 20 Aug. 1957. E_{aloft} was measured with a tethered balloon system. (After Moore et al. 1958, Vonnegut et al. 1959, with permission.)

the potential gradient up to 1 kV m^{-1}, which was their full-scale saturation level. Thus, their instruments had high sensitivity for detecting changes from the fair weather state. The upper instrument was often at a level just warmer than 0°C ($\approx$5 km MSL); the lower was about 400 m above the cloud base. Initial electrification (defined here as E reversing from fair to foul weather polarity) was detected first at the lower device in the cloud. E_{gnd} reversed polarity next. The upper device continued to show an enhanced fair weather E prior to the first lightning. E_{aloft} usually reversed polarity several minutes before the first radar echo was detected (e.g., 5 min before in Fig. 7.9) and increased rapidly to 1 kV m^{-1} before or as the first echo appeared. From their radar calibration, they estimated that their minimum detectable radar signal was equivalent to a median drop size of $\approx$150 µm at distances of 2–3 km.

Attempts to repeat Vonnegut and Moore's accomplishments observing initial electrification with tethered balloons have not been successful. Most of what has been measured has recently come from instrumented aircraft. One result of measurements with aircraft has been the acquisition of in situ particle data. This removes total reliance on radar detection, with its inherent uncertainties about particle size distribution for a particular reflectivity. Radar, however, appropriately remains an important measurement tool.

Apparently, the most comprehensive case study with in situ measurements inside an electrifying cumulus cloud since the original studies by Moore and Vonnegut is that described for a small thunderstorm in Montana by Dye et al. (1986). The full life cycle of the storm was observed with the NCAR/NOAA sailplane and other instrumented airplanes as part of the Cooperative Convective Precipitation Experiment (CCOPE). The cloud was in a weakly sheared environment with moderate thermal instability. The initial radar observations indicated a maximum reflectivity of −5 dBZ when E_{aloft} < 50 V m^{-1} (i.e., the cloud still was not electrified). Within the updraft, there initially was no precipitation except in the top 200 m of the cloud. The updraft was about 5 m s^{-1} at cloud base and as much as 15 m s^{-1} at 7 km, which was about the maximum altitude of the observations. The organized vertical motions of the cloud were tied with the organized increase in electrification. The converse was also true: When the motions decreased, so did the electrification.

A synopsis of E_{aloft} is that it initially remained at <100 V m^{-1}; once it started increasing, it took 4 min to reach 800 V m^{-1}; and in the next 5 min it reached 8 kV m^{-1}. Adding credence to these particular E values is that a second airplane making repeated traverses at 4.5 km (−6°C) observed E to increase from 500 V m^{-1} to 15 kV m^{-1} in 4 min. The increase in E lagged the development of first precipitation by $\approx$8 min. Though it is not readily apparent in their figures, Dye et al. reported that E_{aloft} reversed to foul weather and remained at 100–200 V m^{-1} for 11 min before it intensified. This reversal occurred 4 min before the radar echo reached 5 dBZ. One interpretation of these observations is that precipitation development is required for substantial growth in E, but whether it is required for the initial polarity reversal is still unclear. Although the radar reflectivity was quite low, in situ measurements showed that graupel as large as 0.6 mm diameter existed at the $-15°C$ level when E had foul weather polarity, but a small magnitude. The inferred region of initial charge was near $-20°C$. As is typical of such clouds, this one had an updraft-downdraft interface, which favors precipitation growth, at the $-15°C$ level.

Dye et al. (1989) analyzed the E_{aloft} data from twenty developing thunderstorms in which the sailplane spiraled up to 7 km and the SPTVAR usually made horizontal passes at 4–5 km over Langmuir Laboratory. This study had the added advantage of using three radars, all of whose reflectivity measurements agreed well during comparisons. Dye et al. could detect $|E| < 100$ V m^{-1}, which was well below the threshold they chose to delineate initial electrification (1000 V m^{-1}). While the threshold value of E for initial electrification might be debated, the evolution of E from fair weather magnitudes to substantial, foul weather values certainly denotes electrification of the cloud; such can be seen in their data. Fourteen storms became electrified during convective growth, and three became electrified at the time of the highest cloud top. Many periods of convection occurred without a measurable increase in electrification. The cloud tops at initial electrification ranged from 8–12 km, and at the time of the first lightning flash they were ≥ 9.5 km.

Dye et al. (1989) found that precipitation had to be evident for tens of minutes before $|E_{aloft}|$ exceeded 200 V m^{-1}. They suggested that electrification of >1 kV m^{-1} is tied to radar reflectivities at 6 km MSL ($\approx -10°C$) being >40 dBZ and that electrification occurs concurrently with or just after convective growth. The observed time from the first 10-dBZ echo at 6 km until initial electrification ranged from about 15 min to 1 h. If the reflectivity was caused by rain having a Marshall-Palmer (1948) size distribution, 40 dBZ translates to a median volume diameter of 1.5 mm. (*Median volume diameter* is a drop diameter such that half the water volume in the cloud is in larger drops; e.g., see p. 226 in Doviak and Zrnić 1993). The precipitation size at initial electrification was in marked contrast to the above results of Vonnegut et al. (1959), but the contrast seems likely to have been dominated by the different definitions of initial electrification used by the two studies.

After Dye et al. (1989) detected $E_{aloft} = 1$ kV m^{-1} in a storm, the time to the first lightning flash was as short as <1 min. Of course, the E they measured was not necessarily the maximum in the cloud, but all studies have shown that, once started, the growth of E is often rapid. For example, Breed and Dye (1989) examined the growth rate of E_{aloft} in electrifying clouds. The growth during initial electrification was exponential. The e-folding time (i.e., the time to increase by a factor of e = 2.718) was 50–100 s. When E was larger, the growth rate slowed to ≤ 300 V m^{-1} s^{-1}.

A final note about the difficulty of defining initial storm electrification is that even with the threshold as high as 1 kV m^{-1}, Dye et al. (1989) observed that four of 18 clouds whose E reached this threshold did not go on to produce lightning. It seems to us that a minimum E of 1 kV m^{-1} might be useful as a threshold beyond which we would expect the cloud has a good chance of becoming a thunderstorm. Furthermore, when E_{aloft} is as large as 1 kV m^{-1}, it is highly probable that precipitation exists in the cloud.

7.7.2. E_{aloft} after initial electrification

We examine examples of in situ measurements of E_{aloft} and the size of the maximum E_{aloft} with data spanning four decades. To simplify notation in this section, we note that unless otherwise specified, we are talking about E_{aloft}, so $|E_{max}|$ means the maximum $|E_{aloft}|$. Airplanes have long been appealing as a measurement platform for in situ cloud measurements. As discussed in Section 6.2.2, reliable measurements come only from great care and effort. It is also worth keeping in mind that airplanes that flew at only a few altitudes within a storm were likely to miss the E_{max}. This may be seen in the often shallow depths over which the most intense E have been measured in vertical soundings, as will be discussed later.

In several hundred continental cumulus congestus clouds with mixed phase precipitation, Fitzgerald and Byers (1958) found that $|E_{max}|$ was 10 to 100 times greater than in all-liquid clouds of the same type. We find the measurements reported (mostly in conferences) by Fitzgerald and colleagues showed $|E_{max}| < 200$ kV m^{-1}. Although in individual storms, his storm penetrations were often in limited altitude ranges, over the years he made numerous penetrations at different altitudes. His $|E_{max}|$ agree well with others reported here.

Imyanitov et al. (1969) instrumented an airplane to measure all three components of E and the charge on the plane. In their summary of flights through 300 shower clouds and thunderstorms, they reported an $|E_{max}| = 400$ kV m^{-1} in storms near St. Petersburg (then Leningrad), Russia. In another summary report, Imyanitov et al. (1972) listed an $|E_{max}| \approx 280$ kV m^{-1} from numerous measurements in storms.

Chapman (1953) used balloons to carry a coronasonde into storms. The largest field he found in a few flights was in what he termed a "mild thunderstorm," in which he measured an $|E_{max}| \approx 80$ kV m^{-1}. Winn et al. (1974) used a rocket-borne E meter (Sec. 6.2.1), which measured the component of E perpendicular to the rocket body. Thus, the data consisted almost entirely of the horizontal component of E. Near rocket apogee, the trajectory caused a significant vertical component of E to be measured, but the vertical and horizontal components could not be separated. The small values of E_{max} found in many of their profiles (see Fig. 7.10) likely were the result of being able to measure only the horizontal component of E for most

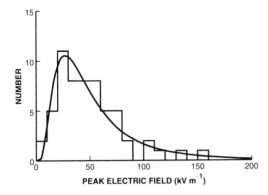

Fig. 7.10. Histogram of maximum $|E_{aloft}|$, mostly the horizontal component of E, from rocket measurements over Langmuir Laboratory, 1970–72. All relative peaks on each flight were included in the histogram. The smooth curve is a log-normal density distribution from the histogram and used a median E = 43 kV m^{-1}. (From Winn et al. 1974: American Geophysical Union, with permission.)

tive cloud, since the layer will be swept away as it forms. This would be similar to the removal of an upper screening layer from the convective cloud turret, which has been documented by flying over active convection. However, this study and other evidence indicate that lower screening layers sometimes exist.

One sounding by Byrne et al. was through a storm whose base was at the mountaintop level. Positive charge was indicated in the bottom part of the cloud, and a negative region started at a temperature of about +1°C (Fig. 7.12). This sounding is important because it has the same apparent stack of charge regions as others, but seems to have them at warmer temperatures. No radar data were presented to evaluate or compare with the other storms. Assuming these data are correct, any proposed charging mechanism(s) should account for this profile as well as the others displaced upward to colder temperatures.

Balloon soundings of E were also made by Marshall and Rust (1991) who used balloon-borne electric field meters. They showed six profiles of E inside isolated storms: five at Langmuir Laboratory and one from the

of each trajectory. In two storms, Winn et al. measured $|E_{max}|$ = 300–400 kV m^{-1}, but they questioned the validity of these two measurements and so omitted these values from Fig. 7.10. After switching from rockets to a balloon-borne electric field meter (EFM) to measure the total vector E, Winn et al. (1981) found an $|E_{max}| \approx 140$ kV m^{-1} in a small thunderstorm over Langmuir Laboratory (Fig. 7.11).

Other early free-balloon soundings of E over Langmuir Laboratory were by Weber et al. (1982), who launched their balloon electric field sensor about 6 min after E_{gnd} reversed to foul weather. The balloon initially tracked next to the main region of precipitation and at 8 km ($\approx -20°C$) recorded an E = -110 kV m^{-1}, which was the maximum in the profile. They interpreted their sounding as fitting the tripole model of a thunderstorm with calculated values of $\rho_{tot} \approx -0.6$ to -4 nC m^{-3} in the main negative charge center. In followup studies by the same Rice University group, they switched to coronasondes to measure the electric field. They obtained an interesting set of four profiles at Langmuir Laboratory in 1981 (Byrne et al. 1983). Three of the four storms appeared to have simple electrical structure.

The maximum $|E_z|$ they measured was 130 kV m^{-1}. The charge regions that they calculated from their E profiles had magnitudes of total charge density ranging from a few tenths to <2 nC m^{-3}. Byrne et al. (1983) interpreted positive charge just above cloud base as lower screening layers in two of the storms. Two of the four soundings obtained data through the cloud top, and the presence of an upper screening layer of negative charge was indicated. There are other discussions in the literature in which it is proposed that screening layers cannot form on the lower boundary of a convec-

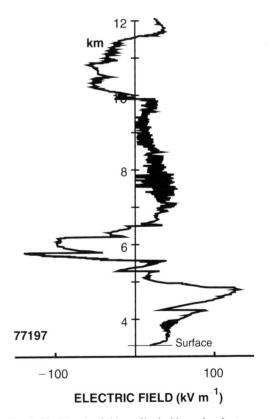

Fig. 7.11. Electric field profile inside a thunderstorm over Langmuir Laboratory, 16 July 1977, obtained from a sounding with a balloon-borne electric field meter and a meteorological sonde. (After Winn et al. 1981: American Geophysical Union, with permission.)

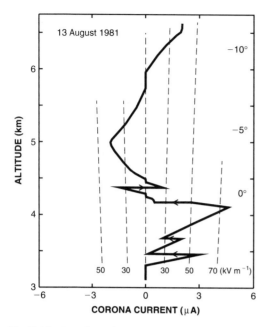

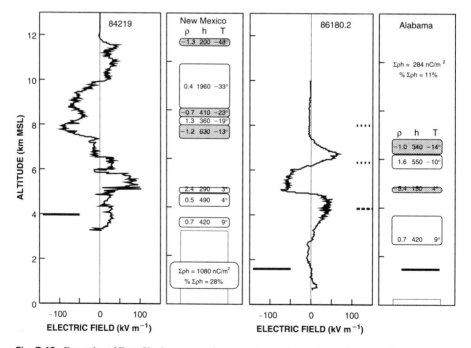

Fig. 7.12. Sounding of E derived from a coronasonde through a storm whose cloud base was at mountain top level at Langmuir Laboratory, 13 Aug. 1981, 1538 MST. Arrowheads denote major changes in E from lightning. Measured temperatures are on the right. (After Byrne et al. 1983: American Geophysical Union, with permission.)

southeast United States. They found an $|E_{max}|_{avg} = 87 \pm 16$ kV m^{-1}, with the largest $E_{max} = 125$ kV m^{-1}. Figure 7.13 contains examples of their soundings. In Chapter 8, we will show that the values of $|E_{max}|$ measured in big storms are comparable to the values in these smaller storms, as would be expected if $|E_{max}|$ is limited by the threshold E for lightning initiaton. Still, big storms can sustain much larger lightning flash rates, and this implies a larger charge generation rate, possibly caused by their stronger updrafts and more plentiful hydrometeors. We will return to such differences later.

Heinz Kasemir and colleagues used cylindrical field mills (see Sec. 6.2.2) during the early 1970s in storms in Colorado and Wyoming and also over Langmuir Laboratory. The cylindrical mill on the airplane nose measured E_z and the along-wing component, E_x, reliably. The measurement of the fore-aft component in clouds was noisy and unusable, so the researchers often reported the magnitude of the vertical and one horizontal component of E (E_K in Eq. 6.6). They specifically searched for the highest fields in the lower half of the storm. Figure 7.14 shows that values of E at cloud base are greater than the maximum at the ground. For 67 storms, $|E_{K max}|_{avg} = 78 \pm 50$ kV m^{-1}. The distribution of $|E_{K max}|$ inside storms (Fig. 7.14) had an average of 148 ± 67 kV m^{-1} for measurements usually just inside the visual cloud base. During flights in storms at Kennedy Space Center, the investigators

Fig. 7.13. Examples of E profiles in storms at Langmuir Laboratory in New Mexico (left) and in northern Alabama (right). The h denotes the thickness of the layer, Δz. The temperature in 0°C is at the center of the charge layer. (From Marshall and Rust 1991: American Geophysical Union, with permission.)

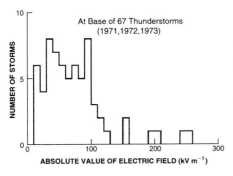 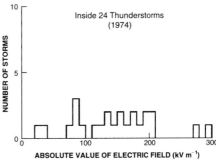

Fig. 7.14. Magnitude of the vertical and one horizontal component of E, E_K, observed at or just inside the base and inside the lowest part of storms in the Rocky Mountain region. Only a single value of $|E_{K\,max}|$ obtained in each storm is used. (Unpublished results from H. Kasemir and colleagues.)

found that substantial E did not occur within the cloud until its top was >5 km MSL (Fig. 7.15) (Cobb 1975). From this, they inferred that the development of E was related to cloud top height. However, the two small E_{max} above 7 km suggest that growth of E is likely not solely a function of cloud top height. Furthermore, the apparent dependence of E on cloud top height may be caused by an underlying dependence on other processes

(e.g., the formation of coexisting liquid and solid hydrometeors) influenced by cloud top height.

A conclusion from all the measurements is that electric fields >100 kV m^{-1} exist but occur in a relatively small fraction of the total cloud volume. The vertical profiles from balloons suggest that the vertical thickness of regions of large E usually is only on the order of hundreds of meters and is no more than a few kilometers. We remind the reader that the distributions of charge that cause the electric field are impossible to determine unambiguously from an E profile.

7.7.3. E_z above storm top

Gish and Wait (1950) made flights over the tops of several thunderstorms. They measured the vertical electric field and found both simple and complicated patterns in the traverses. They explained many simple patterns by a positive dipole in the storm beneath their instrumented B-29 airplane, but some required other charge distributions. They reported that 2 of their 21 overflights had such low tops that no E_z was observed. No specific cloud top heights were given. They showed an example above a storm having a smooth increase from zero to a peak $E_z = 3$ kV m^{-1} at an altitude of 13.1 km. A major reason for the flights was to determine the currents that flow above the storm (see Secs. 1.4.9 and 3.3.1).

Using balloons, Stergis et al. (1975) reported on measurements of the electric field at altitudes of $\approx$21–27 km MSL and above storms across Florida with tops from 9.1–18.2 km. They found E_{max} from 0.15 to >0.5 kV m^{-1}, which was instrument saturation. The largest value occurred during an over-storm flight at 23 km when the balloon went directly over a line of storms rather than a few kilometers to the sides of maximum storm tops as in other flights. They concluded that E patterns showed the storms were not always a simple dipole, but rather a more complicated configuration.

Vonnegut et al. (1966) measured the vertical field and took downward looking photographs from an in-

Fig. 7.15. $|E_{K\,max}|$ inside developing storms in Florida as a function of cloud top height. The temperatures came from environmental soundings and are approximate. (From Cobb 1975, with permission.)

strumented U-2 airplane. They discovered that significant deviations from E ≈ 0 came only in the region over penetrating convective tops. The remainder of the storm top that was quiescent (i.e., no noticeable vertical motions) had little E. They found peak E_z ≈ 0.4 kV m^{-1}. They hypothesized that the convection in the penetrating tops swept aside the screening layer and thus exposed the cloud's internal charge. Their records also showed transient increases in $|E_z|$ that they attributed to lightning. The lightning on one traverse changed E_z from ≈0.2 to ≈ −0.2 kV m^{-1}. Vonnegut et al. hypothesized that lightning inside a screened storm could neutralize some of the storm's internal upper positive charge, which would in turn cause the flash to reverse the polarity of E above storm top.

Blakeslee et al. (1989) showed records from storm overflights in which the lightning both decreased and increased E_z a few kilovolts per meter. In their figures, the maximum E_z above the storm top was approximately 7 kV m^{-1} at about 20 km MSL. They reported frequently finding $|E_z|$ > 5 kV m^{-1} over cloud tops and at 18–20 km MSL. In one case, they found such a significant electric field for a horizontal distance ≈20 km.

In two flights above thunderstorms at Langmuir Laboratory, Marshall et al. (1995a) observed E_{aloft} ≲ 1 kV m^{-1} within 5 km of the electric cloud top. This top was based on E decreasing in magnitude rapidly and staying low, as would occur if the sensor had traversed the uppermost charge—a screening layer—in the storm. They recorded increases in E_{aloft} caused by lightning. These changes in E were up to 10 kV m^{-1} followed by a decay back to the low ambient field in 50–125 s. The field and field changes above storm top will be revisited in our consideration of discharges above the top of the stratiform region of mesoscale convective systems.

7.8. ELECTRIC FIELD AND PARTICLE CHARGES IN ANVILS

Historically, little has appeared in the refereed literature on the electrical structure of the anvil cloud of thunderstorms. Although the demarcation between thunderstorm core and anvil is a gradual one, the often substantial extent of the anvil horizontally out from the core makes the anvil an entity of interest and practical importance. For example, questions have been raised concerning the likelihood that an airplane, rocket, or space shuttle will trigger a lightning flash by flying through or above an anvil. The recent confirmation of blue jets emanating from cloud top may be relevant to this threat, but the relevance is unknown. In the United States, the anvil specifically has been considered in the rules for launch and return of the space shuttle and launch of unmanned space vehicles.

Unpublished measurements of E_z within small thundercloud anvils in Florida by Kasemir and associates indicated that E of a few to several tens of kilovolts per meter were associated with anvils. At 10–20 km

away from the core of Florida thunderstorms, Bly and Nanevicz (1977) occasionally found $|E|$ > 15 kV m^{-1}. Other measurements showed that values of $|E|$ > 1 kV m^{-1} are scarce far out in the anvils of small thunderstorms. Against this background, Marshall et al. (1989) reported on E profiles in two anvil clouds. In the anvil of a small thunderstorm over Langmuir Laboratory, they found an $|E_{max}|$ = 70 kV m^{-1}. A significant part of the charge of anvils, and thus a reason for the high E there, was the presence of screening layers that formed in response to charge inside the anvils. The maximum total charge density ρ_{tot} estimated by Marshall et al. for the anvil of the small storm was less than half that which they estimated for a severe storm anvil. Because of the sparsity of measurements, the typical values of ρ_{tot} in anvils are not known.

Particle charge measurements in anvils are nearly nonexistent. Takahashi (1970) reported on two soundings of particle charge. Each was from the anvil of a summer thunderstorm near Tokyo that was producing lightning in the distance (toward the storm core) during the flights. Beneath the anvils, he found $|E_{gnd}|$ < 200 V m^{-1} during both flights. He observed that both storms had tops to about 15 km, with one top confirmed via radar. In the first storm, he found few charged precipitation particles below 8 km; above that, particles were almost exclusively positively charged with a q_{max} = 6 pC. In the second storm, charged precipitation was shown from 5–15 km, with most of the particles negatively charged and a q_{max} = −3 pC. Takahashi was unable to address this surprise finding of negatively charged particles and pointed out the need for additional soundings. In the only other similar measurements, Marshall et al. (1989) reported that the charges on precipitation particles below and inside the anvil of a storm at Langmuir Laboratory were <10 pC, the minimum detectable charge of their instrument. (More particle measurements have been made at Langmuir Laboratory by Bateman and colleagues; these data await analysis.)

7.9. VERTICAL PROFILES OF TOTAL SPACE CHARGE DENSITY

Much of the sounding data described above and in other sections contained depictions of the total space charge calculated from the one-dimensional Gauss's law (Eq. 6.12). Some were compared with precipitation charge density. It is useful to recall the information about use of Gauss's law and the E sounding data to derive ρ_{tot}. The number of charge regions and the magnitude of total space charge in a layer one calculates from a sounding is dependent on the criteria one chooses for defining a charge region. We averaged the magnitudes of approximately 50 calculations for total space charge in small thunderstorms from several papers referenced in this section and obtained a $|\rho_{tot}|_{avg}$ = 1.2 ± 0.7 nC m^{-3}.

From the stack of space charge layers, we get an indication of the overall charge structure of a storm. For example, E profiles obtained by Winn et al. (1981) and Byrne et al. (1983) seem to require at least one additional charge more than a tripole—an upper negative screening layer. In the storm investigated by Winn et al., the balloon may have gone through a screening layer on the underside of the anvil and entered one on the top side before the ascent was terminated at 12 km. It also seems typical of many profiles that the lowest (in height) negative charge region within the cloud is shallow, often <1 km in depth. It is this negative charge that seems to come closest to being the "main negative charge," although in more complicated soundings it is sometimes not possible to identify unambiguously a single lower main charge. The upper positive charge apparent in the simple sounding profiles is often deeper and less dense than the negative, and this also fits the classic dipole/tripole model.

In the soundings shown by Marshall and Rust (1991), the dipole/tripole structure of the storm was often not evident (within the assumptions of the required one-dimensional calculation). Even small storms such as those over Langmuir Laboratory showed four charges, which could be interpreted as a tripole structure plus a screening layer at the cloud top. Some storms had additional regions of charge.

Along with a discussion of the difficulties of calculating charge regions, Marshall and Rust (1991) reviewed the types of data on which the historical acceptance of the dipole/tripole model of a thunderstorm evolved: field changes from lightning, point discharge and E_{gnd} measurements, and balloon soundings by Simpson and Scrase (1937) and Simpson and Robinson (1941). (Hereafter we write "Simpson and colleagues" when we refer to both papers.) Recall that the tripole paradigm consists of a vertically-oriented positive dipole above a lower, positive charge (i.e., positive, negative, positive). In Section 6.2.3 we discussed how Simpson and colleagues developed and flew on balloons an instrument called the alti-electrograph to record corona current (and thus the polarity of the potential gradient) and atmospheric pressure. To interpret their corona current data, Simpson and Robinson chose a tripole charge distribution and then determined simulated alti-electrograph profiles for various hypothetical balloon paths around and within their tripole model.

Marshall and Rust (1991) questioned the correctness of the long-existing tripole paradigm of the electrical structure of thunderstorms. They showed that the evidence supporting the paradigm can be separated into remote measurements and in situ measurements. Then they pointed out that although tripole charges represent the simplest interpretation of many remote measurements, the remote measurements do not preclude the presence of more than two or three charges. Furthermore, they noted that charge regions were often inferred from the remote measurements after first assuming the paradigm. They concluded that in situ measurements, rather than remote ones, were needed to determine more accurately the number of main charge regions inside thunderstorms.

To examine the tripole paradigm further, Rust and Marshall (1996) reanalyzed the E profiles of Simpson and colleagues. Rust and Marshall noted that many of the profiles were consistent with a more complex thunderstorm charge distribution. Similarly, Stolzenburg (1996) analyzed many modern E profiles and suggested that some were caused by more than three vertically separated charges. Since knowledge of the vertical distribution of charge regions within thunderstorms is key to understanding cloud electrification and modeling the electrical aspects of storms, this subject is one that should receive continuing study.

7.10. PRECIPITATION CHARGE AND CURRENT AT THE GROUND

We now examine examples of measurements of the charge and current density associated with precipitation. The early measurements of precipitation charge were made at the ground rather than aloft, because suitable electronics to make instruments for measurements aloft were unavailable. Like the inconsistent definition of the polarity of the electric field, which we have discussed, the polarity of precipitation current also has been defined inconsistently among studies. It is often defined as the polarity of the dominant charge received. To be consistent with the standard conventions of physics, our convention for the polarity of *precipitation current* and *precipitation current density* is such that falling precipitation that carries a net positive charge would produce a negative precipitation current (since it falls in the negative z direction as it moves toward the ground.)

The early work sought to understand the role of falling precipitation in cloud electrification. For example, Simpson (1909) combined measurements and laboratory studies to develop a theory of cloud electrification. Wilson (1929) discussed problems in cloud electrification, including his perception that the charged rain measured at the ground was the result of, not the cause of, the cloud electrification.

7.10.1. The mirror image effect

Simpson (1949) documented the "mirror image" effect between the dominant precipitation polarity and the potential gradient at the ground. Simpson defined the *mirror image effect* in a broad sense to imply that changes in potential gradient are accompanied by similar changes, but of opposite sign, in precipitation polarity. He showed the mirroring was not necessarily across zero, but it could be symmetrical about a nonzero value of the parameters. Because our polarity conventions for E and the precipitation current density, J_p, reverse the polarity conventions used by Simpson for both parameters, the mirror image effect does not depend on which convention is used.

In an attempt to document the mirror image effect more fully, Ramsay and Chalmers (1960) found three patterns in the effect. They reaffirmed that the mirror

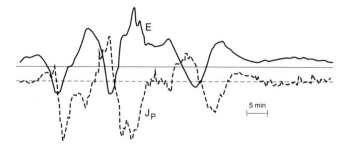

Fig. 7.16. Records of electric field, E, and precipitation current density, J_P, showing the mirror image effect. (From Ramsey and Chalmers, 1960, with permission.)

image effect does not require the parameters to cross zero, only that the superposition of the two traces shows a mirroring. About half their data showed the mirror image effect with no time delay between the current and E (Fig. 7.16). Their remaining data showed a clear mirror image pattern, but with a delay between the peak (and zero crossings) of precipitation current and E, split almost equally between current leading and lagging E. The magnitude of the lag times was not always the same; no complete explanation was reported for this. When E changes before current, it may be because approaching charged precipitation is affecting E. This seems to occur in the field excursion associated with precipitation described by Moore and Vonnegut (1977). Though no systematic study has examined the timing of mirror image patterns in this situation, field excursions associated with precipitation have been observed in many storms (e.g., Fig. 7.2).

Measurements show that precipitation polarity at the ground is not always tightly coupled with the polarity of E_{gnd}. However, when precipitation falls during sustained point discharge, the observations do show a link. Then, precipitation arriving at Earth's surface likely has been electrically modified since leaving the cloud. The degree of that modification can range from small to large; the polarity of the charge carried can even be reversed in the subcloud layer. Wilson (1929) proposed that the prevailing positive charge on rain at Earth's surface is the result of the capture of positive ions within the subcloud region and is driven by the commonly occurring positive E there. He stated that this is true even if E is not large enough to create point-discharge ions, and he suggested that selective ion capture is responsible (see Sec. 3.5.1).

Using simultaneous measurements of precipitation at the ground and several hundred meters aloft near cloud base, Rust and Moore (1974) found reversals in the polarity of charge on precipitation between cloud base and ground when point discharge from Earth was sustained for several minutes (Fig. 7.17). They described calculations that showed that the Wilson ion-capture process is a viable candidate to explain the charging of precipitation in the subcloud region and the mirror image effect at the ground. Subsequent measurements also were made at the ground at Langmuir Laboratory by Stow (1980). He found greater magnitudes of particle charge, which he attributed to differences in instrument sensitivity or meteorological conditions. His particle instrument had a range of 6–200 pC. He found that his particle spectra did not fit that expected from the Wilson ion-capture process and concluded that the origin of large charges was different from that of small ones for precipitation at the ground. He proposed that a measurement range of 0.05–500 pC is needed for charge and a range of 0.2–5 mm is needed for size to determine the charge and size spectrum of precipitation particles.

Also in contrast to the Rust and Moore (1974) measurements, Magono (1980, pp. 72–73) cited their research (from original papers in Japanese) in which they found both the Wilson process and other ion-capture processes during point discharge to be inadequate to explain the mirror image effect. So this apparently simple issue does not seem to be resolved to the satisfaction of all—testimony to the complexity of electrical processes even in the subcloud air. Snow seems more likely to be charged below cloud base and to show the

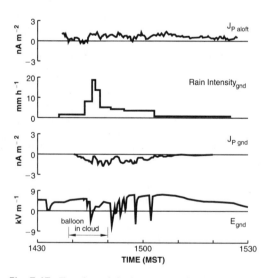

Fig. 7.17. E and precipitation current density measured simultaneously at cloud base and the ground at Langmuir Laboratory. Precipitation current density (labeled J) was defined as positive when negative particles fell downward. (From Rust and Moore 1974, with permission.)

mirror image effect than liquid precipitation, according to the observations by Asuma and Kikuchi (1987). Finally, a mirror image effect is often not seen aloft.

7.11. MAXWELL CURRENT

To determine a fundamental parameter of the storm generator, Krider and Musser (1982) used the electrical current density, J, as defined by one of Maxwell's equations,

$$\vec{\nabla} \times \vec{H} = \vec{J} + \frac{\partial \vec{D}}{\partial t} , \qquad (7.7)$$

where $\vec{H} = \mu_0 \vec{B}$ ($\vec{B}$ is the magnetic field). Remember that $\vec{D} = \epsilon \vec{E}$ and that the second term on the right side of Eq. 7.7 is the displacement current density. If the storm is considered a current source, the Maxwell current density for the storm can be depicted as in Figure 7.18 and explained as follows: *The Maxwell current density, $\vec{J}_M$, is given by*

$$\vec{J}_M = \vec{J} + \epsilon \frac{\partial \vec{E}}{\partial t} . \qquad (7.8)$$

We can expand this to show individual contributors to the total current density

$$\vec{J}_M = \vec{J}_{PD} + \vec{J}_{CV} + \vec{J}_P + \vec{J}_L + \vec{J}_E + \epsilon \frac{\partial \vec{E}}{\partial t}, \qquad (7.9)$$

where the following are the individual current density terms: J_{PD} is point discharge; J_{CV} is current density due to transport of charge by air motion such as convection;

J_P is precipitation current density; J_L is transient lightning current density; J_E is the conduction current density due to the electric field, and the last term is the displacement current density.

Krider and Musser (1982) attempted to determine J_M from measurements made at the ground, so the horizontal components of E were always zero. At times when a lightning flash is not occurring and when $E_z = 0$, as often occurs as E relaxes back to ambient values after a flash, then J_E, J_L, and J_{PD} are zero. In the absence of precipitation at the field mill, only the air transport mechanism remains. Krider and Musser assumed that the current density due to both point discharge and convection are widely distributed in the storm environment and vary slowly with time. Under the assumption that $J_{CV} << J_M$, the Maxwell current density becomes

$$J_M \approx \epsilon \left(\frac{\Delta E}{\Delta t} \right) \bigg|_{E \approx 0} , \qquad (7.10)$$

where the vector notation is dropped, because at the surface only the vertical components of E and J_M are involved.

To evaluate this technique, Krider and Blakeslee (1985) estimated J_M from surface measurements of E_Z and compared these estimates with direct measurements of J_M. For the direct measurements, they used an isolated section of natural turf connected to ground through an electrometer to measure current (see Sec. 6.6). Shown in Fig. 7.19 are the various J_M observations throughout a storm at Kennedy Space Center. Not only was the theoretical prediction of Eq. 7.10 shown, but they found that the 2.5-min averages of J_M changed slowly throughout the storm, on time scales similar to those of storm development. They also found that J_M was not altered significantly by lightning, thus supporting the suggestion of Krider and Musser (1982) that J_M is an electrical parameter that may be correlated directly with storm evolution. Notice the good agreement between the measured J_M and those inferred from Eq. 7.10. The two peaks are due to two cells. Larger, stationary storms tended to have one peak, while smaller ones had one or more small peaks, each lasting 15–20 min. However, the lightning flash rate had only one peak. Areal integration of J_M provides an estimate of the lower limit of total charging current aloft. Krider and Blakeslee reported that a lower bound on total charging current of one storm was ≈ 0.5 A.

Deaver and Krider (1991) compared the lightning character of small and large storms at Kennedy Space Center. They defined a small storm as having flashes that caused E_{gnd} to recover "back, in a quasi-exponential fashion, for a few tens of seconds to a value close to the predischarge field, and then another discharge occurs." They defined large storms as having flash rates that can exceed 25 min^{-1} and with E recoveries that were usually linear with a slope proportional to flash rate. Deaver and Krider confirmed that the displacement current dominated J_M except when precipitation was falling at the measuring site, when E was large and

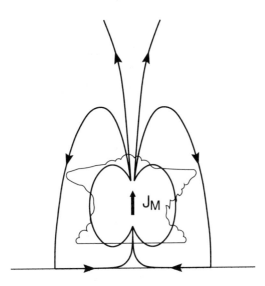

Fig. 7.18. Conceptual sketch of the Maxwell current density, J_M, flowing in the thunderstorm environment. (After Blakeslee and Krider, 1985, with permission.)

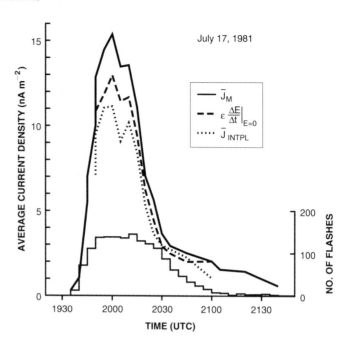

Fig. 7.19. Maxwell current density, J_M, during the lifetime of a thunderstorm at Kennedy Space Center, Florida, 17 July 1981. The J_M data points are 2.5-min averages of the directly measured J_M between lightning flashes, and $\Delta E/\Delta t$ is the mean displacement current obtained by averaging the slopes when $E \approx 0$ during the 2.5 min periods. J_{INTPL} is the interpolated estimate from the KSC field mill network data. Flashes were determined via automatic data processing that used a field change of ≥ 350 V m^{-1} at two or more field mill sites. (From Krider and Blakeslee 1985, with permission.)

steady, or when E was too noisy to use. They also found that J_M was not quasi-steady under small thunderstorms, and this means that earlier assumptions of a constant J_M between flashes may not be valid under small storms. In these cases, departures from quasi-steady J_M probably were not caused by the generator in the small storms, but were caused by current produced in the atmosphere in response to lightning field changes. Because of the different pattern in J_M records under larger storms with frequent flashes, they still believed J_M to be quasi-steady and therefore a good measure of the generator current in larger storms.

7.12. SIMULTANEOUS MEASUREMENTS OF E$_{aloft}$ AND PRECIPITATION CHARGE

Many measurements of the precipitation charge and current at the ground exist, but to understand the distribution of charge in storms, we would benefit from what is, in a practical sense, impossible to obtain—high time resolution, in situ vertical and horizontal distributions of particle charge and E. Because of the difficulty of the measurement and the time it takes to traverse storms with either airplanes or balloons, we have only small amounts of data. We will group them by whether they were acquired or presented along mostly horizontal or mostly vertical paths through the storms. We begin with quasi-stationary measurements made with a tethered balloon system.

Rust and Moore (1974) used tethered balloons to measure precipitation and E in and below thunderstorms at Langmuir Laboratory. The precipitation instruments measured precipitation charge; particle fall

speed, from which one can estimate particle size; and precipitation current density. Also flown was an electric field mill system. They usually did not get vertical distributions, since they tried to raise the captive balloon as high as possible in the developing stage of a cloud over the mountain and then keep the balloon parked there. They thus obtained a time series measurement at a point in space, which usually changed slowly as the balloon moved due to downdrafts and precipitation loading. Their precipitation measurements were typically below cloud base and up to the lowest few hundred meters of the cloud because of the lift limitation of the tethered balloon. Their instrument could measure both polarities of charge between 0.15 and 140 pC. They found that nearly all precipitation particles at and below cloud base had <5 pC. Aloft they found that the polarity of precipitation charges was almost invariably opposite that of E_z.

7.12.1. Horizontal distributions of E$_{aloft}$ and precipitation charge

Gunn (1950) made horizontal passes through a storm core with an induction cylinder device under the wing of a B-17 airplane. Little information was given about E, which he obtained from only two vertically looking field mills whose calibration was not discussed. He described the thunderstorm as prefrontal and typical of midsummer in Minnesota. Lightning occurred close to the airplane several times inside the storm. The horizontal traverses through the storm were about 10 km across and at four altitudes with temperatures from +15°C to −10°C. Precipitation often had both polarities of charge at the same altitude. The net space charge on precipitation, ρ_P, was as large as 30 nC m^{-3} over horizontal distances of ≈ 2 km. His maximum values of precipitation charge density are a factor of two larger than those reported by others. Also, each traverse had at least one, and up to three, changes in the polarity of net charge. We have found no challenge in the literature to the validity of his large values of precipitation charge density, and the differences remain unexplained; they may be correct.

Latham and Stow (1969) used a nose-mounted cylindrical field mill, induction cylinders, thermometers, continuous hydrometeor samplers, and a liquid-water content sensor to get data inside developing cumuli before the first lightning. They studied mountain thunderstorms in the southwestern United States (Arizona). The particle size measuring instruments were separate from the charge measuring device, so charge versus size data were not obtained. As seen in Fig. 7.20, which is a pass at $T \approx -16°C$, a small convective cloud in its early mature stage had a simple horizontal profile of E (presumably E_z). Although a single polarity of particle charge was occasionally dominant over a few hundred meters, positive and negative particles were mixed together in most regions, especially at altitudes where T $< 0°C$.

They compared a cumulus cloud that had been seeded with silver iodide and was glaciated to an adjacent, unmodified and unglaciated cloud of about the same size. The traverses were at $-16°C$. The unglaciated cloud had a higher liquid water content of 0.7–1.5 g m^{-3}, only a few ice crystals, and an E ≤ 1 kV m^{-1}. The seeded, fully glaciated cloud had low liquid water content and an E ≈ -6 kV m^{-1}. They did not compare the particles in each. They penetrated another set of clouds at T $= -11°C$, one unglaciated and another that contained a mixture of supercooled water, ice crystals, and rimed aggregates. The mixed phase cloud had numerous charged particles, while the unglaciated cloud had none. Unfortunately, neither E nor calibrated precipitation charge density were given for those clouds. They did show examples of data with particle charges

up to sensor saturation of 330 pC. Their findings included: (1) Numerous big charges on precipitation required the coexistence of supercooled droplets, ice crystals, and rimed aggregates: (2) Less electrification occurred when aggregates and crystals coexisted, but droplets were absent; (3) Negligible cloud electrification occurred when aggregates and droplets coexisted, but in the absence of ice crystals. Their work leaves little doubt about the close connection between the microphysical and the electrical state of clouds as manifested in E, but as is typical of all case studies, not all connections and causalities are delineated.

We now summarize measurements of E and precipitation made via horizontal penetrations of storms with the powered sailplane, SPTVAR (Sec. 6.2.2). The data were collected in thunderstorms in New Mexico by Gaskell et al. (1978) and by Christian et al. (1980). In both studies, the SPTVAR made horizontal passes through the clouds at a few different altitudes. (It is worth recalling here that an airplane flying at only a few altitudes can always be vertically displaced from the region of, and thus not measure, the maximum E in the storm.) For these cases, precipitation reports by the pilot indicated mixed phase precipitation in the clouds.

Gaskell et al. (1978) found a maximum $E_z \approx 30$ kV m^{-1} in their traverses at temperatures of 4 to 1°C. Mostly, there was a single, dominant polarity during each cloud penetration. One traverse had a particle charge reversal coincident with a change in the polarity of E_z. Gaskell et al. found that both polarities of charged precipitation particles coexisted, but most often the precipitation had a net $\rho_P \approx -5$ nC m^{-3}. Only occasionally was net positive ρ_P recorded. Precipitation of d < 1 mm contributed most to the charge density. The calculated rain rates during the traverses were 1–10 mm h^{-1}. Gaskell et al. did not find any evidence of a relationship between the size and the charge of individual particles.

Christian et al. (1980) observed two thunderstorms. The first had two sequential cells: One cell had no lightning, and the second cell produced only 7 flashes, which occurred within a span of 9 min. The horizontal traverses were at temperatures of $+5$ to $-10°C$. The first cell had $|E_z| \leq 50$ kV m^{-1} and calculated precipitation charge densities, $\rho_P \approx -0.5$ nC m^{-3}. The cell that produced the 7 flashes had measured $|E_z| \leq 50$ kV m^{-1} and $\rho_P \approx -1$ nC m^{-3}.

In the second storm, Christian et al. (1980) measured both polarities of electric field, with maximum $|E_z| \approx 100$ kV m^{-1}. They found that many smaller particles had a charge greater than the inductive-charging limit, and the precipitation charge density sometimes changed polarity during a horizontal pass. As in their first storm, the calculated precipitation rate was about 10 mm h^{-1}, but in the second, the magnitude of precipitation charge density was 5 nC m^{-3}. They found predominately negative precipitation charge density existed on particles with d ≤ 2 mm, and there was no charge versus size relationship in the data. Charges on individual particles were within ± 250 pC, with $q =$

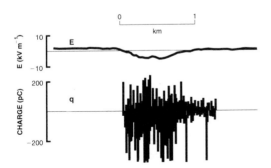

Fig. 7.20. Sketch of the vertical component of E and of particle charges, q, from the early stage of a developing storm. The horizontal traverse was made at $-16°C$. (From Latham and Stow 1969, with permission.)

±50 pC common. Both polarities of particle sometimes coexisted, but usually one dominated. Christian et al. concluded, as had Gaskell et al. (1978), that inductive charging was not the dominant charging process.

To learn the relative contributions of all hydrometeors, we need measurements over broader size and charge ranges. Keep in mind that all the published calculations of precipitation charge density are incomplete—they are biased because a large fraction of particles can carry charge below the detection threshold. For example, on one pass Christian et al. (1980) noted that 95% of the particles had <25 pC, the detection threshold. This problem exists in all studies thus far, and we cannot determine the magnitude of the error.

7.12.2. Vertical soundings of E_{aloft} and precipitation charge

Marshall and Winn (1982) studied precipitation in the lower region of thunderstorms with free balloons carrying an electric field meter, a rawinsonde, and a precipitation charge measuring device whose dynamic range was about ±10 to ±400 pC (see Sec. 6.4.2). They reported on two lower positive charge centers, which they described as "transient and localized regions consisting solely of positively charged particles." Charged precipitation particles were 1–3 mm in diameter and generally had charge magnitudes of 10–200 pC. Marshall and Winn found a few charges of 200–400 pC. Of course, all particles of <10 pC were undetected. The vertical extent of the charge in the lower positive charge centers was about 1 km, with a total calculated charge of 0.4 C and 2 C. They concluded that lightning caused the second of the lower positive charge centers, and offered no explanation for the other.

Whatever their cause, lower positive charge centers offer a reasonable explanation for the field excursions associated with precipitation (Sec. 7.2). As it falls, the precipitation from a lower positive charge center makes E more negative on the ground beneath it. However, its overall effect at the ground appears small; for example, Marshall and Winn found that the total charge conveyed to ground from lower positive charge centers was at least an order of magnitude less than that delivered by lightning. They found precipitation particles had the following polarity of charge versus temperature for this one flight: Particle charges were mostly positive for T warmer than −7°C, a mixture of both polarities from −9°C to −11°C, and mostly negative from −12°C to −15°C, where a lightning strike ended the ascent. (Their balloon-borne instrument was one of several that have been struck by lightning just as E was increasing rapidly to negative values greater than 100 kV m^{-1}. It seems likely that some part of the instrument train triggered these lightning flashes.)

Weinheimer et al. (1991) mounted an induction cylinder behind the laser beam for size imaging in a PMS 2D probe and flew the instrument on a sailplane (see Sec. 6.4.1). They reported on 6 min of data obtained as the sailplane spiraled from 7.5 up to 10.8 km

MSL (−13°C to −32°C) inside a large multicell storm at Langmuir Laboratory. Lightning occurred during the data collection, so the storm was highly electrified. All particles with d > 300 μm were graupel, with no dendritic or hexagonal structure seen. This seems typical of other measurements at Langmuir Laboratory in storms with liquid water content more than a few tenths of a gram per cubic meter. Though all the charged particles were graupel, not all graupel had detectable charge. Positive charge densities were up to 6 nC m^{-3}. Peak negative charge densities were −12 nC m^{-3}, with a typical range of −2 to −6 nC m^{-3}. The net charge became more negative with increasing altitude. As with some earlier investigators, Weinheimer et al. found that larger particles tended to have larger charges. They found about 0.1% of the detected particles had charge magnitudes >1 pC. Other investigators have found a few percent of particles had charge. This percentage varies with altitude and perhaps other parameters.

In what is probably the most complete vertical sounding of precipitation, Marsh and Marshall (1993) acquired precipitation and electric field data from the ground up through a storm top at Langmuir Laboratory. Their minimum detectable particle charge was 10 pC. The balloon was launched shortly after the cloud produced a foul weather $E_{gnd} = 1$ kV m^{-1}. Three specific aspects of Marsh and Marshall's sounding may be significant and contribute to its being somewhat atypical: (1) It appears that most of the flight occurred when the cell (the first of two) was dissipating. A related interpretation is that the first cell was going through an EOSO or equivalent, which caused enhanced negative E_{gnd}. The E profile, however, does not match any other from dying storms. (2) Negative charge was being released at the mountaintop as part of an experiment to test the convective hypothesis of electrification (Moore et al. 1986, 1989); and (3) Marsh and Marshall suggested that the lower positive charge center may have been unusually extensive horizontally.

An example of the storm structure derived from multiple Doppler radars is shown in Fig. 7.21. The E sounding (Fig. 7.22) showed a lower positive charge center, above which were three charge regions. The lower positive charge center could not have been from lightning, since there had been none. At cloud top in the anvil there was a negative screening layer. The vertical distribution of total space charge density, as calculated from the E profile, appeared to be carried on the precipitation except in the upper regions. There, Marsh and Marshall (1993) inferred that the main positive charge and the negative charge in the screening layer were on cloud particles. Total and precipitation charge densities in the storm had magnitudes of ≤3 nC m^{-3}.

Marshall and Marsh (1993) also presented a complete sounding of E and precipitation charge with the same instrumentation through another small thunderstorm at Langmuir Laboratory. In the gross vertical scale, the storm also exhibited four main charge regions, including a negatively charged screening layer at the cloud top. The average particle charges through

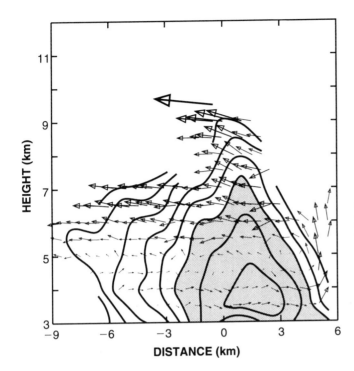

Fig. 7.21. Reflectivity and wind structure of a thunderstorm through which E and precipitation data were obtained at Langmuir Laboratory, 15 August 1984. Distances are east-west from Langmuir Laboratory. During the radar scan, the balloon was at 6.3 km altitude and at distance -1.3 km. The hatched reflectivity begins at 30 dBZ; reflectivity contour intervals are 5 dBZ. The wind vectors are ≈ 6 m s^{-1} per apparent length of 1 km. (From Marsh and Marshall 1993: American Geophysical Union, with permission.)

the storm are shown in Fig. 7.23. A comparison of particle data with the ρ_{tot} from E shows the following: (1) In the lower positive charge center, $\rho_P \approx \rho_{tot}$; (2) In the lower negative charge region, $\rho_P > \rho_{tot}$, which implies that some positive charge was on cloud particles or on precipitation, but was below the minimum charge magnitude detectable by the instrument; (3) In the upper positive charge region and in the screening layer, $\rho_P < \rho_{tot}$, implying that the dominant charges were carried on cloud particles; (4) Magnitudes of ρ_P ranged from 0.1–2 nC m^{-3} in the storm; and (5) The average particle charges were a few tens of picocoulombs.

Table 7.1 summarizes selected parameters of precipitation measurements from several studies.

An important use of precipitation charge measurements is in assessing whether or not the motion of precipitation increases or decreases the local E and generates or dissipates the storm's electric energy. For example, Christian et al. (1980) found that the precipitation acted to dissipate E at their flight level of 5.2 km ($-3.5°$C). A complete assessment requires data on the relevant parameters throughout the storm. The relationship between J_p and E can, in principle, be used to calculate the energy generation rate, i.e., power, P, from

$$P = -\vec{J} \cdot \vec{E}. \qquad (7.11)$$

However, Weinheimer (1987) showed that calculating energy generation is difficult because charged particles often move with a spectrum of velocities in a given region and because the change in the contribution of a specific particle depends on the motion of other

charged particles, as well as on the particle's own motion. We cannot determine the role of precipitation in cloud electrification solely from its effects in a few measurements; the issue is more complicated.

7.13. INTRODUCTION TO LIGHTNING OBSERVATIONS IN THE CONTEXT OF THUNDERSTORMS

In the rest of this chapter, we will examine lightning in isolated thunderstorms. In Chapter 5 we already have discussed the physics of lightning channels, a topic that will receive little attention in this chapter. Instead, we will concentrate on observations concerning when and where lightning occurs in storms. Discussion of trends that are characteristic of lightning in several specific types of thunderstorms or storm systems will be deferred until Chapter 8. However, sometimes we will refer in this chapter to studies of specific storm types when they reinforce more broadly observed relationships. We specifically exclude in this chapter lightning-like phenomena that occur above thunderstorms; these are discussed in Chapters 5 and 8.

Before we begin to examine lightning in the context of thunderstorms, it might be helpful to review briefly what must occur to initiate lightning. As noted in Chapter 5, lightning occurs in a thunderstorm when the electromagnetic forces from charge in and around the storm accelerate electrons in some small region of air to the point that suddenly the air is forced to become

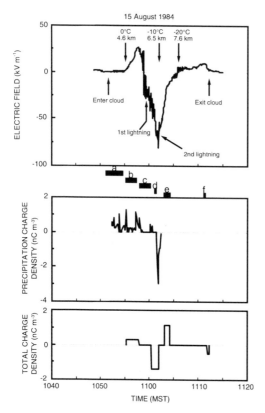

Fig. 7.22. Vertical sounding of electric field and precipitation parameters through a mountain thunderstorm at Langmuir Laboratory. (From Marsh and Marshall 1993: American Geophysical Union, with permission.)

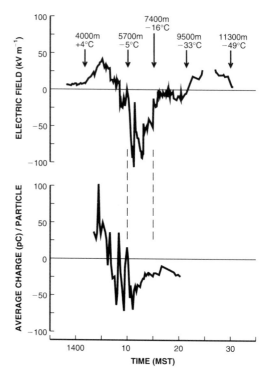

Fig. 7.23. Vertical sounding of the electric field and the average precipitation particle charge for a thunderstorm over Langmuir Laboratory. The particle charge averages were calculated every 12 s. From 1406–10, the averages are not meaningful. During other times, the particles nearly all had the same polarity, and the averages are meaningful. (From Marshall and Marsh 1993: American Geophysical Union, with permission.)

Table 7.1. Examples of the magnitude of precipitation current density, J_P, and charge density, ρ_P, inside thunderstorms. The sounding types from aircraft are horizontal, H, and spiral, S, and from balloons are vertical, V. Most values come from published graphs and are approximate. The maxima of different quantities in a particular study did not necessarily coincide in time or space. T is the temperature(s) at which the data were acquired; — indicates not available.

Reference	Type	T (°C)	max $|J_P|$ (nA m^{-2})	max $|\rho_P|$ (nC m^{-3})	max $|E|$ (kV m^{-1})	max $|q|$ (pC)
Gunn (1950)	H	15 to −10	—	30	—	90
Latham and Stow (1969)	H; S	10 to −30	—	—	15	>330
Gaskell et al. (1978)	H	4 to 1	12	5	30	200
Christian et al. (1980)	H	5 to −10	20	5	90	250
Marshall and Winn (1982)	V	0 to −15	30	5	130	400
Gardiner et al. (1985)	H	−5	—	0.5	15	50
Dye et al. (1988)	H	−12	—	—	0.6	12
Weinheimer et al. (1991)	S	−13 to −35	—	12	—	460
Marshall and Marsh (1993)	V	4 to −49	15[a]	4	110	430
Marsh and Marshall (1993)	V	5 to −24	30[a]	1	80	270

[a]These particular values are from the same storms, but in Marshall (1993).

a good conductor. This catastrophic failure (called *electric breakdown*) of the insulating properties of air becomes visible as a transient spark which conducts large currents to reduce the local electromagnetic forces that created it. Because the electric field dictates the force that a test charge will experience, the condition that electromagnetic forces be large enough to create a spark is equivalent to a condition on the magnitude of the electric field.

However, what constitutes "large enough" is still controversial. Maximum values of the electric field measured inside thunderstorms are much smaller than is thought to be needed to cause electrical breakdown of cloudy air. Thus, either electric field magnitudes sufficient to cause breakdown occur only in regions so small they are almost never sampled, or some other process aids or triggers flash initiation. Electric field magnitudes required to sustain a discharge are more than an order of magnitude smaller than those required to initiate breakdown in clear air (e.g., Phelps 1974, Phelps and Griffiths 1976, Griffiths and Phelps 1976a, Williams et al. 1985).

Several mechanisms have been suggested which might allow lightning initiation at smaller electric field magnitudes than needed for clear air electric breakdown. An example of such a mechanism is electric field enhancement around the edges or points of ice particles (e.g., Griffiths and Latham 1974, Griffiths and Phelps 1976b). Another mechanism, suggested only recently, is initiation by the energetic secondary electrons produced by energetic cosmic rays, which are sporadic, but pervasive throughout the atmosphere. In this context,

"energetic electrons" means, at a minimum, that the electrons are moving fast enough to exhibit effects of special relativity. McCarthy and Parks (1992), Gurevich et al. (1992, 1994), and Roussel-Dupre et al. (1992) have suggested that energetic secondary electrons produce a sustained electron avalanche that can lead to lightning at electric field magnitudes much smaller than needed for initial breakdown—in fact, at magnitudes that sometimes are even smaller than needed to sustain a streamer using the more plentiful low-energy electrons. (Obviously, lightning cannot propagate beyond the region in which energetic secondary electrons are available unless the electric field at the streamer tip in that region is large enough to sustain a streamer without the energetic electrons.) The threshold electric field magnitude in the presence of energetic secondary electrons is called the *breakeven electric field,* and it decreases with altitude (Eq. 5.1). Marshall et al. (1995b) compared values of breakeven electric field as a function of height with the magnitude of relative maxima in the electric field observed during vertical soundings of storms (e.g., see Fig. 7.24). In most cases, the observed electric field magnitudes approached and were bound by the breakeven electric field at all altitudes.

7.14. GLOBAL FLASH RATE AND PERCENTAGE OF FLASHES OVER LAND AND WATER

The majority of lightning occurs in storms near or over land, not over open oceans. Note that this trend may be

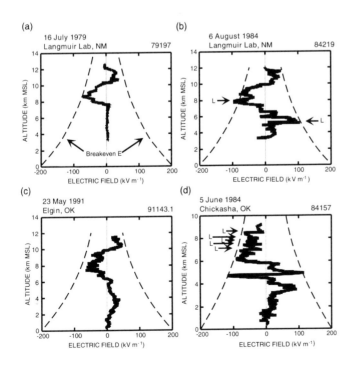

Fig. 7.24. Vertical soundings of thunderstorm electric fields superimposed on a vertical profile of the breakeven electric field. The breakdown field is greater than the maximum E on the graph at all altitudes. (From Marshall et al. 1995: American Geophysical Union, with permission.)

less pronounced for rainfall, particularly in the tropics. Zipser (1994) noted that some observing stations in the tropical Pacific received large amounts of rain from oceanic storms, though no lightning was detected.

The affinity of thunderstorms for land has been accepted for many years on the basis of records of *thunderstorm days* (e.g, Brooks 1925, World Meteorological Organization 1956), defined as the number of days during which observers at official sites hear thunder. Although the resulting isokeraunic maps (with contours in units of thunderstorm days) give a rough indication of the distribution of thunderstorms, they can be misleading, as noted by MacGorman et al. (1984) and Orville and Henderson (1986). For example, a thunderstorm day does not distinguish days with few flashes from days with many flashes. Furthermore, to be recorded, thunder must be heard by an official observer, and detection is dependent on the range of audibility of thunder, which commonly varies from 8 km to 25 km (more than a factor of three), even without interference from local sounds and barriers. Radio receivers have been used to measure the global or hemispherical distribution of sources of sferics (electromagnetic radiation from lightning) (e.g., International Radio Consultative Committee 1964, Freeman 1977). However, these estimates of the relative number of flashes over land and oceans were suspected of bias favoring land areas, because the sensors were not distributed uniformly over the globe.

Satellite sensors eliminated the potential problem of predominately land-based sensors biasing the measured oceanic and continental fractions. Both sferics and optical sensors have been used. Kotaki et al. (1981) used four sferics receivers to measure the global distribution of lightning from the equator to latitudes of $\pm 70°$ in bins of 10° longitude by 10° latitude. The circular satellite orbit was at an inclination of 70° from the equator and precessed relative to local time so that all local times in a particular bin were sampled approximately every 2 months for low latitudes and every 4 months for high latitudes. Data from a 2-year period (Fig. 7.25) indicated that, during all seasons, the largest discharge rates were in bins near or over land, while the smallest were in bins over oceans. The maximum discharge rate was always between $\pm 40°$ latitude, and the most widespread minimum was always at longitudes through the middle of the Pacific Ocean. The minimum in the Pacific was $10^{-8}\,\mathrm{s}^{-1}\,\mathrm{km}^{-2}$ or less during all seasons. The location and value of the largest discharge rate varied seasonally; the magnitude ranged from at least $5 \times 10^{-7}\,\mathrm{s}^{-1}\,\mathrm{km}^{-2}$ in most seasons to at least $10^{-6}\,\mathrm{s}^{-1}\,\mathrm{km}^{-2}$ in the northern fall. Typical values over land within $\pm 40°$ of the equator were greater than $10^{-7}\,\mathrm{s}^{-1}\,\mathrm{km}^{-2}$ during all seasons. Note that these values are smaller than those found by some local measurements (e.g., Piepgrass et al. 1982), possibly because the global study averaged rates over much larger regions. However, as one might expect, these values are larger

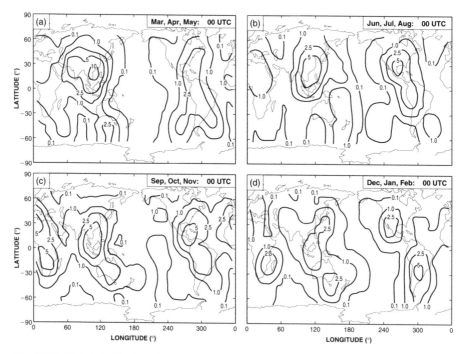

Fig. 7.25. Global distribution of lightning discharge rates at 0000 UTC by season from satellite-borne sferics sensors. Contour intervals are in units of 10^{-5} flashes per 100 km². (From Kotaki et al. 1981, with permission.)

than those found by Orville (1994), because they indicate total lightning activity, while Orville tabulated only ground flashes.

Several studies have used optical sensors on satellites, but all optical studies thus far have been limited to specific local times. The first and most restrictive studies (e.g., Vorpahl et al. 1970, Sparrow and Ney 1971) were limited to within $\pm 35°$ of the equator near local midnight during the new moon (lightning could be detected only when Earth was dark). These studies found that storms producing lightning at midnight were much more likely to occur over land than over oceans. Data from Vorpahl et al., for example, indicated ten times as many thunderstorms over land as over oceans and fifty times as many flashes.

Several more recent studies have used optical sensors on polar orbiting satellites from the Defense Meteorological Satellite Program (DMSP). The orbit of these satellites maintained a fixed local time, typically dawn-dusk or midnight-noon for lightning studies. Using a sensor that detected the brightest 2% of lightning flashes, Turman and Edgar (1982) estimated that, at dusk, 15% of flashes were from oceanic storms and 85% were from continental storms, while at dawn, 37% were oceanic and 63% were continental. The larger percentage over oceans at dawn is consistent with observations by Maier et al. (1984), who showed that, in the vicinity of Florida, the diurnal variation of ground flash activity over the ocean was much smaller than over the land.

The best spatial resolution (approximately 4 km by 100 km) was obtained by Orville and Henderson (1986), who analyzed transient bright streaks on cloud imagery taken at local midnight by a DMSP satellite. (Better resolution is now available from new optical sensors on satellites in high-inclination orbits; for details see Christian et al. 1996 and Driscoll et al. 1996.) The resulting monthly ratios of continental flashes to oceanic flashes are shown in Table 7.2 and range from 2.2 to 4.2. Orville and Henderson pointed out that because the surface area of oceans is 2.4 times the surface area of land, the ratio of the average densities is a factor of 2.4 larger than the ratio of the number of flashes. Note that different studies have sometimes used different rules for classifying oceanic regions near coastlines as land or ocean, but these differences have been insufficient to change the basic conclusion that the majority of lightning occurs near or over land.

Most satellite studies that have examined the number of flashes in zonal (i.e., constant latitude) bands as a function of latitude have found that the latitude of the maximum varies seasonally, but usually is somewhere between 20°N and 20°S, as shown in Fig. 7.26. The number of flashes at high latitudes typically is more than an order of magnitude smaller than the peak, so the decrease cannot be explained simply by the decrease in Earth's surface area with increasing latitude. (The surface area at the highest analyzed latitudes is a factor of 2–3 times smaller than at the equator.) The peak in equatorial regions has been observed for lightning from the entire day (Kotaki et al. 1981) and for lightning at dusk and midnight (Orville and Spencer 1979, Turman and Edgar 1982). However, Turman and Edgar found that the peak at dawn was at 40°–60°S latitude during late southern spring and north of 40°N during northern spring. They attributed these peaks at higher latitudes to enhanced thunderstorm activity from polar fronts at dawn during these seasons.

Table 7.2. Summary of midnight lightning by month. Mean values of the flash ratio and density ratio were 3.2 and 7.7, respectively.

Month	Number of flashes detected	Land: Ocean flashes[a]	Land: Ocean flash density[b]
September 1977	1826	2.2	5.3
October 1977	2181	3.3	7.9
November 1977	2174	3.3	7.9
December 1977	2891	2.6	6.2
January 1978	2440	2.5	6.0
February 1978	1890	3.4	8.2
March 1978	2113	3.5	8.4
April 1978	3947	3.0	7.2
May 1978	3433	3.1	7.4
June 1978	3369	3.4	8.2
July 1978	3481	4.2	10.1
August 1978	2517	3.7	8.9

[a]Ratio of the number of flashes over land to the number over oceans;
[b]Ratio of flash density over land to flash density over oceans.

From Orville and Henderson (1986) with permission.

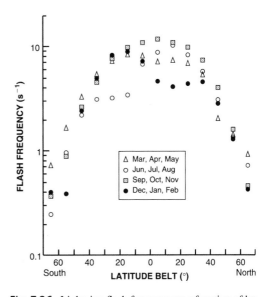

Fig. 7.26. Lightning flash frequency as a function of latitude belt and season. (From Kotaki et al. 1981.)

Satellite studies also can be used to revise estimates of global flash rates. As discussed in Chapter 1, Brooks (1925) had made a crude estimate of 100 s^{-1}. From sferics measurements, Kotaki et al. (1981) estimated that the global flash rate (integrated over the globe and averaged across the diurnal variation discussed in Ch. 1) was 63 s^{-1} for the entire year and ranged from 54 flashes s^{-1} in northern winter to 80 s^{-1} in northern fall. Estimates of global flash rates from DMSP data were sometimes larger than those found by Kotaki et al., but uncertainties in those estimates were considered to be at least a factor of two. Orville and Spencer (1979) estimated 123 s^{-1} at dusk and 96 s^{-1} at midnight, with a seasonal variation of 10%–20% of the mean. Turman and Edgar (1982) estimated that global flash rates were somewhere in the range of $40–140 \text{ s}^{-1}$, with a seasonal variation of 10%.

All three studies agreed that the minimum global flash rate occurred during winter in the northern hemisphere. (An error in one statement by Turman and Edgar (1982) attributed winter's flash rate to late spring.) There was disagreement on the season of the maximum. Kotaki et al. (1981) found a large maximum in northern fall, with a much smaller secondary maximum in northern spring. From data presented by Orville and Spencer (1979) and Turman and Edgar, it appears that the maximum occurred around May; in both studies, there was less variation in global rates from April through September than found by Kotaki et al., and global rates from October through November were somewhat lower than average values during northern summer.

Part of the variation in results among studies may have been caused by the different sensitivities of lightning detection, the different times of observation, and the very sparse sampling available for each study. Geostationary satellites would be needed to address the limited sample rate. However, the satellite-based estimates of average global flash rates agree reasonably well with each other. They also agree remarkably well with the crude estimate by Brooks (1925).

7.15. RELATIVE PROPORTIONS OF DIFFERENT FLASH TYPES

7.15.1 Cloud flash percentage

Cloud flashes usually outnumber ground flashes, both climatologically and for individual storms. The relative proportion of cloud and ground flashes is given often in the literature as the ratio of cloud flashes to ground flashes ($r_{cld/gnd}$). This statistic has the unfortunate property of being nonlinear, because ground flashes are counted in the denominator, while cloud flashes are counted in the numerator. Thus, the ratio can be infinity and can have large changes in value for small changes in the number of ground flashes. Instead of this ratio, we will use the percentage of all flashes that are cloud flashes (f_{cld}), in which the denominator is the to-

tal number of flashes, so that changes in the number of ground flashes will cause changes in the statistic comparable to those caused by changes in the number of cloud flashes. In many cases, the values (in percent) that we give in this section are converted from values of $r_{cld/gnd}$ in the literature by using

$$f_{cld} = \frac{100 \, r_{cld/gnd}}{1 + r_{cld/gnd}} \, . \tag{7.12}$$

Not surprisingly, climatological values of the cloud flash percentage are more characteristic of a given region than values for individual storms. Climatological measurements have been made, for example, by Mackerras and Darveniza (1994), Mackerras (1985), Prentice and Mackerras (1977), and Pierce (1970) using flash counters (Mackerras 1985). Pierce and Prentice and Mackerras found that when the lightning flashes of all storms in each region were combined, the percentage of flashes identified as cloud flashes decreased with increasing latitude. Prentice and Mackerras found that the mean value of the percentage varied from 85% between 2°N and 19°N to 64% between 52°N and 69°N. Pierce suggested that the percentage was given by $90 - 25 \times \sin\phi_{lat}$, and he hypothesized that the variation was caused by the systematic decrease in the mean height of the 0°C isotherm with increasing latitude, as suggested previously by Gunn (1935). This decrease was assumed to cause a corresponding decrease in the height of negative charge in thunderstorms, which then was expected to increase the proportion of ground flashes (with a corresponding decrease in the cloud flash proportion).

However, Mackerras and Darveniza found a much smaller variation with latitude than found by the other two studies (see Table 7.3). Although all sites having cloud flash percentages greater than 75% were at latitudes less than 30°, some sites at all latitudes had cloud flash percentages less than 65%. Furthermore, the variation in cloud flash percentage from site to site at similar latitudes within 30° of the equator was at least as large as any linear decrease with latitude in the data. Mackerras and Darveniza suggested that earlier measurements of large cloud flash percentages at low latitudes were in error, because those measurements used simpler flash counters and visual identification of ground flashes and so tended to underestimate ground flash counts when large lightning flash rates were observed.

Mackerras and Darveniza (1994) also noted that cloud flash percentages for individual storm days spanned the entire range from 0–100% at most sites, but extreme values tended to be associated with relatively few lightning flashes. To show the cumulative distribution of daily values of cloud flash percentage for several sites listed in Table 7.3, they plotted the fraction of all flashes occurring on days with a cloud flash percentage greater than or equal to the value on the ordinate (see Fig. 7.27). Only Bogota had a cloud flash percentage less than 50% for days having more than half of the lightning activity at the site.

Table 7.3. Cloud flash percentage (f_{cld}), positive ground flash percentage (f_+), and mean annual total flash density (v_f) as a function of latitude.

Site	Latitude	Period (y)	f_{cld} (%)	f_+ (%)	v_f (flashes $km^{-2} y^{-1}$)	ID in Fig. 7.27
Singapore	1.2°N	4.5	77	2	61	—
Bogota, Columbia	5.0°N	1.0	33	2	19	1
Lae, PNG	6.4°S	0.6	57	3	—	—
Darwin, Australia	12.2°S	3.0	58	2	10	2
Stanwell, Australia	23.5°S	4.0	62	2	5.6	—
Gaborone, South Africa	24.4°S	1.1	52	4	16	—
Ackland, Australia	27.3°S	0.3	79	9	—	—
Brisbane, Australia	27.3°S	4.0	71	3	4.5	3
Kathmandu, Nepal	27.4°N	2.0	79	28	11	4
Gainesville, Florida, USA	29.4°N	0.2	60	3	—	—
Tel Aviv, Israel	32.1°N	3.0	67	15	3.2	—
Toronto, Canada	43.4°N	1.0	58	2	3.3	—
Berlin, Germany	52.3°N	5.0	50	12	0.73	5
Uppsala, Sweden	59.9°N	4.4	60	9	0.66	—

Adapted from Mackerras and Darveniza 1994, with permission.

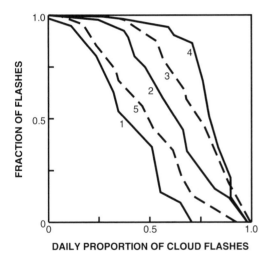

Fig. 7.27. Fraction of flashes occurring on days with a cloud flash percentage greater than or equal to the value on the horizontal axis for selected sites. The number labeling each curve corresponds to the number assigned to the site in Table 7.3. Curve 1 is Bogatá; curve 2, Darwin; curve 3, Brisbane; curve 4, Kathmandu; and curve 5, Berlin. (After Mackerras and Darveniza 1994: American Geophysical Union, with permission.)

Cloud flash percentages for individual storms in other studies, many of which analyzed data from lightning mapping systems, have shown similar variability. Figure 7.28 shows the cloud flash percentage for many of the storms documented by various investigators (only storms producing at least 15 flashes are included). Most storms from which data in Fig. 7.28 were taken occurred in the United States, but some are included from other countries. The distribution in the figure should not be taken as representative of storms globally or in any particular region. Even on climatological time scales, there are considerable regional variations in typical values of the cloud flash percentage, as discussed above, so the choice of locations represented in Fig. 7.28 probably affected the distribution. Furthermore, some studies specifically focused on storms with cloud flash percentages ≥90%, and so may have biased the distribution in that bin.

This last possible bias may be offset, however, by a probable undercounting of cloud flashes in many storms. Because most flashes that produce weak electromagnetic signals at the ground near thunderstorms are cloud flashes, the measured cloud flash percentage can be a function of the sensitivity of the measurements. For example, Rust et al. (1985) obtained a much higher cloud flash percentage by counting small electric field changes, all of which were from cloud flashes. Maier and Krider (1986) and Nisbet et al. (1990) also noted an increase in cloud flash percentage when small electric field changes were included. This may be one reason why most storms shown in Fig. 7.28b that produced more than 1000 flashes had a relatively large cloud flash percentage, although few of the storms had that many flashes. Therefore, even with the biases it appears likely that in a majority of the locations represented in Fig. 7.28, most storms with >15 flashes have cloud flash percentages >50%, and few have cloud flash percentages <40%.

Mackerras (1985) noted that the variability in cloud flash percentage from storm to storm and among sites having similar latitudes cannot be explained, as had

Fig. 7.28. (a) Distribution of the number of reported cases observed within a given range of cloud flash percentages; (b) Cloud flash percentage versus number of flashes for the same storms plotted in panel a. Plotted points are taken or estimated from Baughman and Fuquay (1970), Brook and Kitagawa (1960), Driscoll et al. (1994), Funaki et al. (1981b), Fuquay (1982), Goodman et al. (1988a, 1989), Krehbiel (1981), Livingston and Krider (1978), MacGorman et al. (1989), Mackerras (1985), Maier and Krider (1986), Mazur and Rust (1983), Nakano (1979a), Nakano et al. (1984), Norinder and Knudsen (1961), Pierce and Wormell (1953), Pierce (1955a), Prentice (1960), Reynolds and Neill (1955), Richard (1991), Rust et al. (1981a, 1985), Rust (1989), Schonland (1928), Taylor et al. (1984), Teer and Few (1974), Williams et al. (1989), and Workman et al. (1942). (All with permission.)

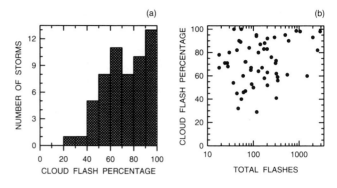

been hypothesized, by variations in the height of the 0°C isotherm; such variations tend to be relatively small for storms at a given latitude or site during seasons when thunderstorms are prevalent. Price and Rind (1993) suggested that, instead, observations support the hypothesis that the cloud flash percentage increases with the depth of the cloud above the 0°C isotherm. This depth tends to increase with decreasing latitude, but it also can vary markedly from storm to storm, and so possibly can account for daily variations in f_{cld}. Rutledge et al. (1992), Price and Rind (1993), and Carey and Rutledge (1996) argued that the number of cloud flashes increases as the volume of the upper part of the storm increases, and that a systematic increase in the number of cloud flashes at low latitudes can increase f_{cld} as much as the systematic decrease in the number of ground flashes hypothesized by Gunn (1935) and Pierce (1970).

7.15.2. Positive ground flash percentage

Most ground flashes in isolated storms—indeed in most types of storms—lower negative charge to ground. For example, annual flash statistics compiled for the United States from a 6-year period by Orville (1994) indicated that less than 10% of all ground flashes in the United States lowered positive charge to ground. Table 7.3 shows climatological values of the positive ground flash percentage tabulated by Mackerras and Darveniza (1994) for various locations around the world. Note that all four tropical sites in Table 7.3 had very low positive ground flash percentages, and that some locations at higher latitudes had much larger values. The site with the largest climatological value, Kathmandu, has been studied in greater detail by Baral and Mackerras (1993).

There also tends to be a seasonal variation in the positive ground flash percentage. In most regions, the percentage increases during cooler seasons, often with a peak in the winter (e.g., Takeuti et al. 1977, 1978, Orville et al. 1987, Reap 1991). In Kathmandu, how-

ever, Baral and Mackerras (1993) found that the percentage increased during the spring and summer and was smallest during the winter, when it was 20–21%.

Usually, even when storms produce positive ground flashes, the positive ground flash percentage is small. However in some storms, the majority of ground flashes can be positive either in the whole storm or during some stage, or in some component of the storm. This tends to occur in one of five storm situations: (1) in some winter storms (see Sec. 8.3.4), (2) in the stratiform precipitation region of mesoscale convective systems (see Sec. 8.2.3), (3) in some relatively shallow thunderstorms of any type, including isolated thunderstorms, the convective region of squall lines, and rainbands (e.g., Bosart and Sanders 1986, Takagi et al. 1986, Engholm et al. 1990, Morgenstern 1991), (4) in some severe storms (see Sec. 8.1.10), and (5) during the dissipating stage of many isolated storms (e.g., Pierce 1955b, Krehbiel 1981, Fuquay 1982). This last observation has been suggested as a possible signature of imminent storm dissipation. Fuquay observed positive ground flashes on 16 of 48 storm days, usually during the last 30 min of a storm's lightning activity, after negative ground flashes had stopped occurring.

7.16. LIGHTNING HEIGHT AND SPATIAL EXTENT

7.16.1. Height distribution of lightning inside clouds

Lightning channels are not distributed uniformly with height in thunderstorms, but tend to occur most often in layers considerably smaller than the depth of the storm. This usually is true also of the charge involved in flashes, an observation contrary to the early hypothesis of Malan and Schonland (1951). On the basis of single-station surface measurements of electric field changes from lightning in South Africa, they had suggested that ground flashes tap negative charge from increasingly higher regions in a tall vertical column of negative charge. However most studies, even those in which there were not enough measurements to unambiguously determine the location and magnitude of charges involved in lightning, have suggested that the negative charge effectively neutralized by flashes typically was within a shallower region that was more horizontal than vertical (e.g., Pierce 1955b, Tamura 1958, Hatakeyama 1958, Ishikawa 1961, Ogawa and Brook 1969).

(i) Lightning charge centers and dipoles. Experiments that had enough measurements to determine the location of negative charge centers neutralized by cloud flashes and individual strokes of ground flashes in New Mexico storms (e.g., Workman et al. 1942, Reynolds and Neill 1955, Krehbiel et al. 1979) typically found that the negative charge centers tended to occur in a layer 1–3 km thick, with a lower boundary at some temperature between 0°C and −10°C, as shown by the example in Fig. 7.29 (open circles are negative). A similar result was found in Florida storms for negative charge centers neutralized by strokes (Uman et al. 1978, Krehbiel 1981) and entire flashes (Jacobson and Krider 1976, Maier and Krider 1986, Koshak and Krider 1989, Krider 1989, Nisbet et al. 1990). The combined results of the above studies indicate that the negative charge centers neutralized by lightning tend to occur at similar heights during the evolution of an individual storm, from storm to storm, and from region to region. Jacobson and Krider (1976) and Proctor (1991) have reviewed results from studies of the height of negative charge centers.

Studies evaluating the location of both positive and negative charge centers neutralized by individual ground strokes and cloud flashes (Workman et al. 1943, Workman and Reynolds 1949, Reynolds and Neill

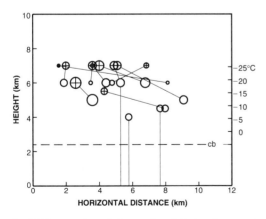

Fig. 7.29. Location of charges for all flashes from an 8-min period during a New Mexico storm; cb denotes cloud base. (After Workman et al. 1942, with permission.)

1955, Krehbiel 1981, Nisbet et al. 1990) found that the negative charge tended to occur in a relatively compact region, while the positive charge was spread through a larger region mostly above the negative charge, but sometimes also below or to the side. In New Mexico storms, both Workman et al. and Reynolds and Neill found that the mean location of positive charge centers typically was 0.5–1 km above the mean location of negative charge centers, which usually occurred somewhere between 0°C and −20°C (Fig. 7.29). However, the heights of some positive charge centers were the same or lower than the heights of some negative charge centers. In fact, in two storms analyzed by Workman et al., the majority of positive charge centers were at or below the height of negative charge centers for several minutes, but these two cases appear to be atypical.

Krehbiel (1981) and Nisbet et al. (1990) found more vertical separation between lightning charge centers in Florida, possibly because the top of storms in Florida tended to be colder than those in New Mexico. Typically, negative charge centers in Florida were between −10°C and −25°C, and positive charge centers were between −30°C and −60°C. Because Krehbiel found that the negative charge involved in cloud and ground flashes occurred in approximately the same region, he inferred that the negative charge sources of both types of flashes coincided with the main negative charge region of the thunderstorm. The charge transfer of sequential processes in a single cloud flash often appeared to form an upwardly divergent sequence of dipoles from a relatively compact region of negative charge into a larger region of positive charge (Krehbiel 1981). By comparing the lightning charge centers and dipoles with channel structure from a VHF mapping system for similar flashes (Fig. 7.30), Shao and Krehbiel (1996) inferred that the sequential, upwardly diverging dipoles were not caused by separate, predominately vertical channels. Instead, they were caused by the growth of horizontal lightning channels at the level of the positive charge centers, which were joined by a single vertical channel to the negative lightning charge centers.

Similarly, those studies treating the charge neutralized by cloud flashes as a point dipole (e.g., Jacobson and Krider 1976, Maier and Krider 1986, Koshak and Krider 1989, Krider 1989) showed that most point dipoles in Florida storms were within a layer above the negative charge neutralized by ground flashes (Fig. 7.31). However, the height of the dipoles varied in different storms. For example, Krider reported that negative charge centers occurred at similar heights in both a small storm and a large storm. However, the point dipoles were a mean distance of 1.6 km above the negative charge centers in the small storm, but 4.6 km above the negative charge centers in the large storm.

As shown in Fig. 7.31, almost all point dipoles above the negative charge pointed downward, most at the same level as the negative charge pointed horizontally, and almost all below the negative charge pointed upward. Krider noted that in three dimensions the

Fig. 7.30. Charge centers (circles) neutralized by successive processes in a cloud flash in a Florida storm superimposed on channel structure (dots) from a flash with similar structure mapped by a VHF system from another Florida storm. Letters indicate point dipoles (dots with arrows) or corresponding positive (open circles) and negative (shaded circles) charge centers for the same flash process. The size of the circles is adjusted to indicate the relative amount of charge neutralized by each process, and the length of a dipole is proportional to its dipole moment. The charge analysis shown left and right is for the same flash. The

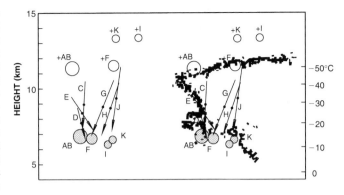

heights of charge centers tend to agree with the heights of horizontal channel structure in similar storms, and the channel structure suggests that the upwardly diverging charge dipoles are caused by horizontal channel development away from a single vertical channel. Temperatures are from a typical environmental sounding. (From Shao and Krehbiel 1996: American Geophysical Union, with permission.)

dipoles tended to be distributed around the negative charge centers and to point toward them. With the exception of one storm observed by Workman et al. (1942), the observations of lightning charge centers and dipoles have been consistent with a conceptual model suggested by Workman and Reynolds (1949), in which cloud flashes involve positive charge distributed around and above negative charge, with most positive charge being located in the upper part of the storm. Several studies (e.g., Koshak and Krider 1989) have suggested that cloud flashes involving positive charge below negative charge were between the main negative charge and the small lower positive charge in a tripolar thunderstorm charge distribution. Some of these studies have noted also that the tripolar distribution is a gross oversimplification of the actual charge distribution of the thunderstorm.

Although many winter storms are much shallower and colder than warm season thunderstorms, the charge centers involved in lightning, including many positive ground flashes, still occur within a range of temperature smaller than the range over the depth of the storm (e.g., Takeuti et al. 1978, Funaki et al. 1981a, Brook et al. 1982). Typically, the negative charge centers neutralized by ground flashes were at temperatures similar to those found for negative charge centers in summer storms, and positive charge centers were often above, but sometimes below, the negative charge centers. Takeuti et al. and Brook et al. reported that the main positive charge was above the main negative charge in winter thunderstorms, as observed in summer thunderstorms, but the main positive charge also tended to extend farther horizontally beyond the main negative charge in the winter storms that they observed.

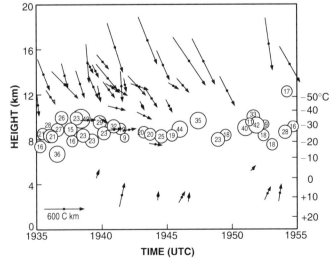

Fig. 7.31. Height and time of occurrence of charge centers of negative ground flashes and dipoles of cloud flashes from a large Florida storm on 6 July 1978. The angle from the vertical at which dipoles are plotted is the angle in the vertical plane that completely contains the dipole. Each charge center is labeled with the magnitude of charge (in C) neutralized by that negative ground flash. (From Koshak and Krider © 1989: American Geophysical Union, with permission.)

(ii) Lightning channel segments. Because lightning channel structure has been mapped by several different technologies based on different lightning characteristics (thunder, electromagnetic radiation, and radar reflectivity), we will use the term *lightning channel segment* to mean the part of a lightning channel mapped from a particular received signal, regardless of the technology that was used. We do this in part to emphasize that all systems map some subset of the channels of a lightning flash (although possibly a large fraction of the whole set) and in part for the convenience of having a term that applies to all technologies. In the scientific literature and in this section, these channel segments sometimes also are called *sources* of whatever type of radiation was detected by the mapping system (e.g., acoustic sources, VHF sources).

Systems that map lightning channel segments have shown that the channel segments tend to occur in layers 2–3-km thick. For example, MacGorman et al. (1981) showed that acoustically reconstructed lightning channels tended to occur in one or two layers in storms in Arizona, Colorado, New Mexico, and Florida (Fig. 7.32). When there were two layers, the lower layer was approximately 2-km thick and located somewhere between 0°C and −25°C; the upper layer was 2–3-km thick and began 1–2-km above the lower layer. (The thickness of a layer was measured between the uppermost and lowermost boundaries of the continuous height interval in which the number of acoustic sources in each bin was at least half the peak value.) In all cases the lower boundary of the upper layer was −18°C or colder. When there was only one layer, it appeared to correspond approximately to the lower layer of storms having two layers; its temperature range was similar, but it tended to be somewhat thicker (2–4-km thick).

Similar results were obtained from VHF and radar

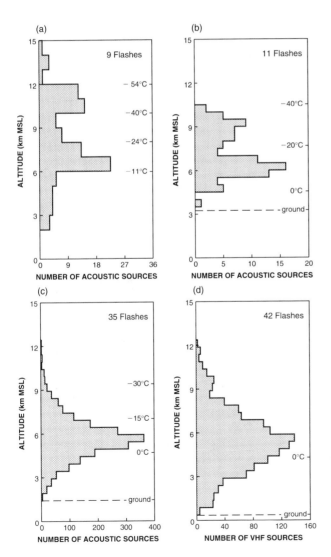

Fig. 7.32. Vertical distribution of reconstructed channel segments from acoustic (panels a, b, and c) and VHF (panel d) mapping systems: (a) a Florida storm, (b) an Arizona storm, (c) a Colorado storm, and (d) an Oklahoma storm. (From MacGorman et al. 1981: American Geophysical Union, and from MacGorman et al. 1983. Both with permission.)

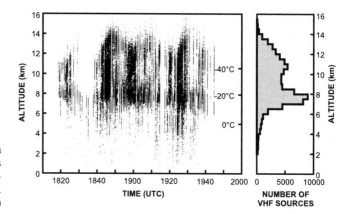

Fig. 7.33. Height of lightning in a Florida storm: (left) height of VHF sources versus time of occurrence; (right) vertical distribution of VHF sources for the entire storm. (From Maier et al. 1995, with permission.)

systems for mapping lightning. Lhermitte and Krehbiel (1979), MacGorman et al. (1983), Taylor et al. (1984), and Maier et al. (1995) found that lighting channel segments radiating in the VHF band tended to occur in one or two layers, as shown by the examples in Figs. 7.32 and 7.33. Furthermore, Mazur et al. (1984, 1986) found that lightning channel segments detected by radar tended to occur in one of two layers for storms in Virginia. Although these vertical distributions of channel segments often showed layering, an even stronger impression of layering sometimes was given by the cumulative lightning structure of several flashes plotted in a vertical plane, as done by Proctor (1983) and shown in Fig. 7.34, because systematic variations in height as a function of horizontal location were separated. When two layers were observed, the distribution of lightning activity often appeared somewhat mushroom shaped, with a broad upper region overlying a narrower lower region (Krehbiel 1981, MacGorman et al. 1981, Taylor et al. 1984, Maier et al. 1995).

A comparison of studies of the vertical distribution of lightning suggests that the layers of lightning activity tend to be 3–5°C colder in storms on the Florida coast than in storms in New Mexico, Colorado, and Oklahoma, as suggested by Jacobson and Krider (1976) and MacGorman et al. (1981). However, this hypothesis needs further testing, since the number of storms in the sample is marginal for determining a characteristic difference in heights between regions. If the difference exists, it probably reflects some systematic regional difference in other storm characteristics that affect lightning, possibly microphysical differences caused by the more maritime environment of the Florida coast.

The proportion of lightning channel segments in each layer can vary as a cell evolves. Mazur et al.'s (1986) data can be interpreted to indicate that new, growing cells tended to have more lightning in the upper layer, near storm top. As the cell matured and cores of high reflectivity descended to lower heights, the lower layer became more dominant. However, such evolution of the height distribution of channel segments may not occur in all cells. Mazur et al. indicated that lightning in some storms may occur only in either the low or high layer and can have different patterns of evolution. MacGorman (1978) and MacGorman et al. (1981) showed two storms with a pattern of evolution different from that reported by Mazur et al. Acoustically

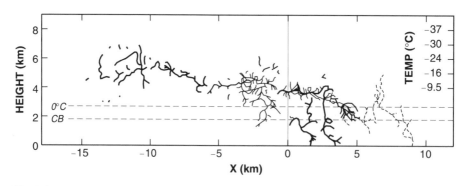

Fig. 7.34. Composite of seven sequential flashes mapped by a VHF mapping system during a 52-s period of a South African storm. (From Proctor 1983: American Geophysical Union, with permission.)

reconstructed channels were found only at lower altitudes during the earlier stages of each storm and then spread to the upper part of the storm as it matured and dissipated.

To some extent, such differences may be a result of the different mapping technologies used by MacGorman et al. (1981) and Mazur et al. (1986), as suggested by differences in the maximum height of channels mapped in one New Mexico storm by radar and acoustic techniques (Holmes et al. 1980). However, it also is possible that the differences in reported evolution of lightning height were real, caused by regional differences or storm-to-storm variability. Takahashi (1984) suggested that storm microphysics can affect the evolution of the height of lightning activity (see Sec. 9.5), and MacGorman et al. (1989) suggested that the depth and magnitude of updrafts also may affect the evolution of lightning height and type (see Sec. 8.1).

There also is evidence of storm growth and dissipation in the height of the highest lightning channel segments. Krehbiel (1981) found that the height of the topmost VHF sources increased at approximately the same rate as the height of storm top measured by radar, but the minimum height (ignoring channels to ground)

remained approximately constant. Lhermitte and Krehbiel (1979), Lhermitte and Williams (1985), and Maier et al. (1995) all showed cycles in the maximum height of lightning mapped by VHF systems, similar to the cycles of storm top growth (see examples in Figs. 7.33 and 7.35). The maximum height of VHF sources increased as cells grew to greater heights, decreased as cells dissipated, and then increased again as new cells grew. Again, the minimum height was relatively constant compared with the maximum height. Changes in the upper extent of lightning structure also have been found by other mapping techniques (e.g., the study of radar mapping by Mazur et al. 1986 and Keener and Ulbrich 1995 and the study of acoustic mapping by MacGorman et al. 1981), although changes in the upper extent of lightning structure observed by acoustic mapping have not always appeared to correspond to changes in the height of storm top. Cycles in the maximum height of lightning have not been readily apparent in studies of charge dipoles neutralized by lightning (Krider 1989), but point dipoles would not necessarily be expected to reflect changes in the height of the upper and lower extremes of lightning structure.

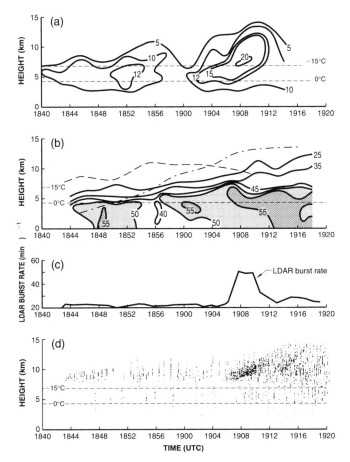

Fig. 7.35. Lightning and Doppler radar data from a Florida storm. (Adapted from Lhermitte and Krehbiel 1979 IEEE, with permission.) (a) maximum updraft speed (in m s^{-1}) as a function of height and time, (b) maximum reflectivity (in dBZ) as a function of height and time and cloud top height for two cells as a function of time, (c) burst rate (roughly equal to flash rate) received by the LDAR system as a function of time, and (d) height of mapped VHF sources versus time of occurrence.

In contrast with the findings of VHF and acoustic studies of the height distribution of lightning, one radar study has indicated that the lower extent of lightning, as well as the upper extent, appeared to be affected by storm structure. In Arecibo, Puerto Rico, Keener et al. (1993) and Keener and Ulbrich (1995) found that both the minimum and maximum height of lightning channels mapped by a vertically pointing, dual wavelength Doppler radar sometimes increased as a storm grew. As shown in Fig. 7.36, flashes tended to occur near either the upper or lower boundary of a region in which pre-

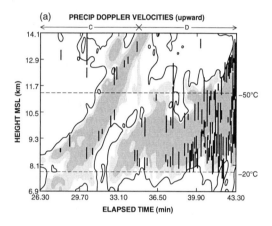

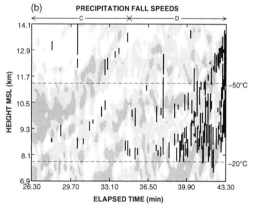

Fig. 7.36. Lightning relative to storm structure detected by a dual wavelength, vertically pointing radar in Arecibo, Puerto Rico. Vertical lines indicate the vertical location of lightning detected by the radar. (a) The height of lightning relative to simultaneously measured Doppler velocities of ascending precipitation as a function of time. Shading darkness indicates upward velocity, with thresholds at 2, 5, and 8 ms^{-1}, (b) Precipitation fall speeds after vertical air speed and the variations caused by the decrease of atmospheric density with height have been subtracted. Shading darkness indicates fall speed, with thresholds at 5 and 8 m s^{-1}. Larger fall speeds typically indicate larger, heavier particles. (From Keener and Ulbrich 1995, with permission.)

cipitation moved upward at a velocity ≥ 8 m s^{-1}. When the lower boundary of this region of faster upward velocities was relatively constant (as between elapsed times of 33 and 39 min in the figure), the height of the lower lightning remained relatively constant. When the lower boundary of this region moved higher (as between elapsed times of 29 and 33 min in the figure), the height of the lower lightning also increased.

The apparent discrepancy between this observation and the observations of a more constant lower height by other mapping studies may have been caused by differences in experimental techniques. Keener et al. (1993) and Keener and Ulbrich (1995) obtained both lightning and precipitation data simultaneously from a radar pointing directly overhead, so only a small part of lightning and storm structure were observed, although they were sampled with high temporal and spatial resolution. Other studies examined lightning structure for the entire storm volume, and so tended to concentrate on the lowest extent of flashes anywhere in the storm or to examine statistics summarizing the distribution of lightning activity as a whole. Furthermore, other studies did not observe lightning and storm structure simultaneously in the same sample volume. Since we already have found that the height of lightning channels can vary systematically with horizontal location in the storm (Fig. 7.34), it is possible that an association between the height of lightning channels and the height of a storm feature would be obscured by experiments that either emphasized the structure of whole flashes or did not simultaneously observe both lightning and storm structure.

7.16.2 The Structure and Dominant Dimension of Lightning Flashes

(i) Lightning characteristics as a function of height.
MacGorman et al. (1981) noted that individual lightning flashes could have channel structure in either or both of the layers mentioned previously (Fig. 7.37). Late in an Arizona storm, they observed that flashes often had extensive horizontal structure in each of the two layers, with little mapped structure observed between them. Similarly, late in the life of a Colorado storm, MacGorman (1978) found some lightning flashes with horizontal channels in each of two layers connected by a vertical channel (e.g., see Fig. 7.40c), although the vertical distribution of acoustically reconstructed channel segments for the storm as a whole did not show two layers. Using an interferometer in New Mexico, Scott et al. (1995) and Shao and Krehbiel (1996) also have observed individual flashes with horizontal channel structure in two layers (Fig. 7.38), in some cases with a much more pronounced vertical channel between them than observed by either MacGorman et al. (1981) or MacGorman (1978).

Some aspects of lightning in the upper part of storms can appear different from those of lightning in lower regions of the storm. Taylor et al. (1984) observed a small, steady rate of VHF sources in the upper

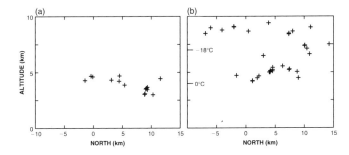

Fig. 7.37. Two flashes reconstructed by acoustic techniques in an Arizona thunderstorm: (a) a flash that was characteristic of an early period of the storm. (From Teer 1973, with permission.), (b) a flash that was characteristic of a later period in the storm. (From MacGorman et al. 1981: American Geophysical Union, with permission.)

part of a severe, multicellular storm in Oklahoma, which were unlike the large bursts of VHF sources generated by flashes at lower altitudes. He attributed these continual sources to frequent discharges of small extent, possibly initiated between the main upper positive charge and screening-layer charge on the upper cloud boundary. Similarly, Mazur et al. (1984) observed frequent discharges having relatively small radar cross sections near the top of storms in Virginia and suggested that these might be the same phenomenon observed by Taylor et al. (1984). Holmes et al. (1980) may have observed a similar phenomenon; they reported many flashes detected by radar near the top of New Mexico thunderstorms that produced little or no thunder detected at the ground. Remember that both the breakdown and breakeven values of the electric field decrease rapidly in magnitude with increasing height in a storm. Thus, all else being equal, initiation of flashes can occur at a much lower magnitude of the electric field near the top of storms than at middle levels.

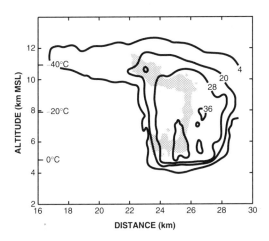

Fig. 7.38. A flash mapped by a VHF interferometer for a New Mexico thunderstorm. Contours indicate radar reflectivity, labeled in dBZ. Lightning structure is shown by the shaded region. The indicated temperature levels were not measured near the storm, but are typical values. (After Scott et al. 1995 and Shao and Krehbiel 1996: American Geophysical Union, with permission.)

Another difference was reported by Proctor et al. (1988). They noted that cloud flashes initiated at heights above 7.3 km MSL (temperature $< -15°C$) in South African storms produced smaller rates of VHF impulses ($\leq 10^4$ s^{-1} above 7.3 km MSL versus $\geq 10^5$ s^{-1} for flashes beginning at heights below 7.3 km MSL), and the individual source regions that emitted the impulses were broader (mean extent of 300 m versus a median of 60 m for lower flashes). Cloud flashes that began higher than 7.3 km also exhibited initial phases during which VHF sources clustered near the origin of the flash, while ground flashes and lower cloud flashes did not.

(ii) Vertical versus horizontal lightning geometry. The geometry of an individual lightning flash can be either predominately horizontal or vertical. The issue of vertical versus horizontal lightning geometry generated considerable controversy before the advent of systems for mapping lightning channels. Malan and Schonland (1951) had interpreted visual observations and surface electric field measurements as indicating that the structure of a typical ground flash extended vertically in a column inside clouds. Although some individual lightning flashes were observed by radar to extend long distances horizontally, and in a few cases more than 100 km (Ligda 1956, Atlas 1958), it was unclear to what extent horizontal flashes were characteristic of thunderstorms. As discussed above, studies of surface electric field measurements (e.g., Pierce 1955b, Ogawa and Brook 1969, Krehbiel et al. 1979) determined that the negative charge effectively neutralized inside storms by successive strokes of individual ground flashes tended to be displaced more horizontally than vertically from each other. Ogawa and Brook and Krehbiel et al. also demonstrated that the South African data of Malan and Schonland (1951) could be consistent with predominately horizontal structure. Furthermore, Ogawa and Brook inferred from their own data and those of Workman et al. (1942) that lightning flashes in New Mexico thunderstorms appeared to be mutually aligned, with similar horizontal orientations. However, the separation between positive and negative centroids of cloud flashes sometimes has been predominately vertical (e.g., Krehbiel 1981, Lhermitte and Williams 1985).

The channel extents mapped by acoustic and VHF systems have tended to have a larger horizontal component than that shown by the evolution of neutralized charge centers deduced from electric field change measurements as a flash develops, but the conclusion regarding the dominant dimension is the same: The dominant dimension of mapped flashes could be either horizontal or vertical. In a small New Mexico storm, Christian et al. (1980) observed predominately vertical lightning flashes with an acoustic mapping system, and Proctor (1981, 1983), Proctor et al. (1988), Maier et al. (1995), Scott et al. (1995), and Shao and Krehbiel (1996) also observed some predominately vertical lightning flashes with VHF mapping systems in South Africa, Florida, and New Mexico.

However, from acoustic reconstruction of lightning structure, Teer and Few (1974) showed that the predominant dimension of lightning flashes in an Arizona thunderstorm was clearly horizontal. Ground flashes had an average horizontal extent of 9.8 ± 3.6 km, while cloud flashes had an average horizontal extent of 11.0 ± 4.7 km. The ratio of horizontal to vertical extent averaged 2.1 ± 0.8 for ground flashes and 2.9 ± 0.9 for cloud flashes. Nakano (1973, 1976), Winn et al. (1978), and MacGorman et al. (1981) also used acoustic reconstruction from thunder recordings to demonstrate that the geometry of lightning flashes within clouds was predominately horizontal in several storms in Japan, New Mexico, Colorado, and Florida. Similar observations were made by Proctor (1981, 1983), Proctor et al. (1988), Lhermitte and Krehbiel (1979), and Maier et al. (1995), who used long-baseline, time-of-arrival VHF mapping systems in South Africa and Florida. Predominately horizontal lightning flashes also were observed by Rust et al. (1981b), MacGorman et al. (1983), Taylor (1983), and Scott et al. (1995) with short-baseline VHF mapping systems in Oklahoma and New Mexico.

(iii) Lightning geometry relative to storm dimensions. The issue of lightning geometry relative to storm dimensions concerns primarily the in-cloud structure of flashes, which we expect to be influenced by the distribution of E and charge in storms. The effect of the channel to ground on channel geometry typically is ignored, because its length is a function more of the height of the initiation point above ground than the geometry of the distribution of charge.

The extent to which the vertical or horizontal dimension dominates lightning structure during a particular period of a given storm appears to be strongly influenced by the dimensions of the storm. Since most visible lightning channels do not extend far beyond the side or top of clouds, the dimensions of a storm obviously limit the dimensions of the lightning it produces. However, the relationship between storm dimensions and lightning dimensions is not a simple proportionality. As we have discussed already, the vertical extents of many lightning flashes appear to span only a few kilometers, and even in large storm systems some flashes have horizontal extents of only a few kilometers. Furthermore, in some storms, flashes tend to occur near and inside reflectivity cores (Fig. 7.39), while in other storms, flashes are offset from reflectivity cores (Fig. 7.40) (flash location relative to reflectivity cores is discussed in Sec. 7.17). Still, the horizontal extent of a storm does appear to influence the average horizontal extent of the lightning flashes it produces.

Mapping systems have found that the horizontal extent of many flashes is comparable to the horizontal extent of regions of the host storm having moderate reflectivity at roughly the height of the lower layer of lightning activity in that storm (typically between $0°$ and $-25°C$). In fact, most lightning structure is within regions of moderate to large reflectivity. However, there are relatively little data on the specific value of reflectivity within which most lightning occurs, and the value may vary during the lifetime of a storm or from storm to storm. MacGorman (1978) and MacGorman et al. (1983) found that throughout the lifetime of a severe, multicellular Colorado storm, the horizontal extent of many acoustically reconstructed flashes was comparable to the horizontal extent of the region having reflectivity ≥ 36 dBZ at or just below the altitude of the lightning (Fig. 7.40). Toward the end of the analyzed period (Fig. 7.40e,f), however, flashes began to extend almost 10 km into weaker reflectivities in what probably was the anvil. Proctor (1983) observed that cloud flashes in a small, multicellular storm in South Africa typically remained within regions having ≥ 25 dBZ, which extended into the anvil of the storm and included cores of heavier precipitation. In a study of radar-detected light-

Fig. 7.39. Two lightning flashes mapped by the LDAR system from a storm in Florida. Contours of radar reflectivity are shown for 0, 10, 20, 30, and 40 dBZ, with the value increasing inward from one contour to the next. (a) The first flash in the storm, a cloud flash, extended from 7 to 12 km. Reflectivity is shown at 8 km MSL. (b) A ground flash that occurred approximately 45 min after the first flash. Reflectivity is shown at an altitude of 6 km MSL. (From Maier et al. 1995, with permission.)

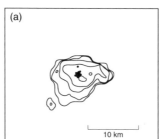
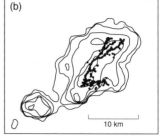

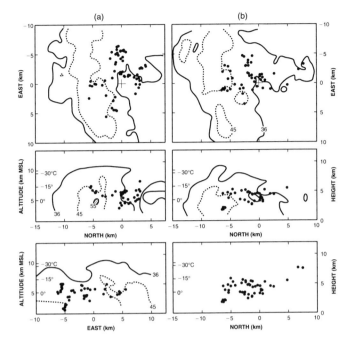

Fig. 7.40a–b. Acoustically mapped lightning channel segments relative to reflectivity structure of a Colorado storm. Reflectivity contours are labeled in dBZ. The top panels show all mapped channel segments and show radar data from scans at an elevation angle that passed through the lightning at a height of 4–5 km. The middle panels show the vertical cross section of radar data along the north-south line passing through the origin and show mapped segments occurring within ±2 km of this cross section. (a) A flash at 1709:29 Mountain Standard Time (MST). The bottom panel shows the vertical cross section of radar data along the east-west line passing 1.5 km south of the origin and shows mapped channel segments occurring within ±2 km of this cross section. (b) A flash at 1716:34 MST. The bottom panel shows all mapped channel segments. (From *MacGorman* 1978, with permission.)

ning in an Oklahoma squall line by Mazur and Rust (1983), most mapped flashes occurred within regions having ≥20 dBZ, including those flashes stretching horizontally from near the leading line of heavy precipitation back more than 60 km into the trailing region of light precipitation (see Sec. 8.2.3).

In two isolated, severe storms in Oklahoma, Taylor et al. (1984) and Ray et al. (1987) observed most lightning channel segments within or just outside regions with reflectivity ≥30 dBZ, except near the top of the storm, where channels extended into weak reflectivi-

ties above the reflectivity core or in the anvil (see Sec. 7.18). Likewise, in a Virginia storm, Mazur et al. (1986) found that the region of lightning activity typically was within regions with reflectivity ≥20–30 dBZ, but also extended into weak reflectivities over reflectivity cores. A similar result was found by Scott et al. (1995) and Shao and Krehbiel (1996) for a New Mexico storm (Fig. 7.38). Again, flashes penetrated into regions of weaker reflectivity near the top of the storm, either above the reflectivity core or in the anvil. In a Florida storm, Maier et al. (1995) found that most

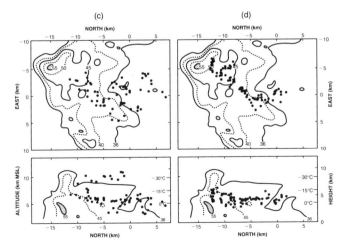

Fig. 7.40c–d. Mapped lightning channel segments relative to reflectivity structure of a Colorado storm. All mapped segments are plotted in both projections. Radar data in the top panels are from a constant height of 5 km. Radar data in the bottom panels are from a vertical cross section along a parabolic line through the main channels in the plan projection of flash d. (c) A flash at 1720:29 MST. (d) A flash at 1721:25 MST, the next flash detected after flash c. (From *MacGorman* 1978 and *MacGorman et al.* 1983, with permission.)

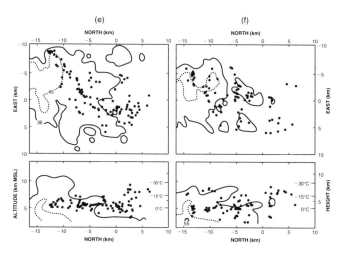

Fig. 7.40e–f. Mapped lightning channel segments relative to reflectivity structure of a Colorado storm. All mapped segments are plotted in both projections for each flash. Radar data in the top panels are from scans passing through the lightning at a height of 4–5 km. Radar data in the bottom panels are from a vertical cross section along the north-south line through the origin. (e) A flash at 1726:11 MST. (f) A flash at 1730:44 MST. (From *MacGorman* 1978, with permission.)

VHF sources occurred in regions with ≥30 dBZ; only occasional channels penetrated outward beyond the 20-dBZ contour (Fig. 7.39).

As might be expected from the above observations, the horizontal extent of lightning flashes tends to be larger when the horizontal extent of the region of moderate reflectivities is larger, whether comparing lighting from storm to storm or within a single storm at different times. This means that flashes in storms with larger horizontal dimensions tend to have longer horizontal extents than flashes in smaller storms. Cherna and Stansbury (1986) reported evidence that lightning structure in two large, severe storms in Quebec, Canada, tended to be more horizontally extensive than was typical of that region. A similar comparison can be made of lightning from different studies. Flashes in the small Florida and New Mexico storms studied by Winn et al. (1978), Lhermitte and Krehbiel (1979), Christian et al. (1980), Krehbiel (1981), Lhermitte and Williams (1985), Nisbet et al. (1990), Scott et al. (1995), and Shao and Krehbiel (1996) tended to have much smaller horizontal extents than flashes in the larger, multicellular storms studied by MacGorman (1978), MacGorman et al. (1981, 1983), Mazur and Rust (1983), Rust et al. (1981a), and Proctor (1983). Compare, for example, the flash from a small, mountain storm shown in Fig. 7.38 with the flashes from a large, multicellular storm shown in Fig. 7.40. Even longer horizontal flashes were presented by Mazur and Rust, who found flashes more than 60 km long in the extensive region of light precipitation behind a squall line, and by Ligda (1956), who found flashes more than 100 km long in a large storm system.

The dependence on storm dimensions also means that the dominant dimension of flashes can change as a storm evolves. Lhermitte and Krehbiel (1979), Krehbiel (1981, 1986), Scott et al. (1995), and Maier et al. (1995) found that lightning flashes occurring early in small storms often were predominantly vertical. As the storm became multicellular and larger, the structure

of lightning flashes became increasingly horizontal, typically spanning more than one cell. An example of this evolution is presented in Fig. 7.39, which shows two flashes from a Florida storm studied by Maier et al. The first (Fig. 7.39a) was the initial cloud flash in the storm and was predominately vertical, with little horizontal structure, as was typical of flashes early in the storm. The second (Fig. 7.39b) was a ground flash representative of many flashes during the period after the storm had formed additional cells southwest of the original cell. This second lightning flash permeated much of the region of moderate to large reflectivity and extended horizontally from the reflectivity core of newer cells to the decreased reflectivity of the old, original cell.

MacGorman (1978) and Krehbiel et al. (1979) also pointed out that flashes tended to span several cells and began to extend to a new cell as it matured (see examples from MacGorman 1978 in Fig. 7.40), although not all flashes extended to a new cell after it began producing lightning. Furthermore, if an anvil or a region of widespread, lighter precipitation formed downstream from new and mature cells, lightning flashes often began to extend into these regions (MacGorman 1978, MacGorman et al. 1983, Proctor 1983, Mazur and Rust 1983, Scott et al. 1995, and Shao and Krehbiel 1996). Examples of flashes extending into part of an anvil are shown in Figs. 7.38 and 7.40. As a result of both anvil growth and formation of new cells, the horizontal extent of typical flashes in the Colorado storm shown in Fig. 7.40 grew from roughly 10 km at the beginning of the analyzed period to ≥20 km.

It sometimes has been suggested that horizontal flashes are restricted primarily to the dissipating stage of storms. Although this appears to be true for relatively small storms (e.g., Krehbiel 1981), it is not always true. Instead, predominately horizontal flashes can occur during any stage of a storm in which the region of lightning activity is horizontally extensive. A large storm, such as those studied by Cherna and Stansbury (1986), MacGorman et al. (1983), Proctor

(1983), and Mazur and Rust (1983), often has predominately horizontal flashes throughout most of its lifetime. In the storm studied by MacGorman (1978), for example, almost all flashes were predominately horizontal, even early in the analyzed period, when convection appeared strong in several cells and the anvil and the region of dissipating cells were much smaller than they later became (Fig. 7.40).

7.16.3. Hypothesized conceptual model for layering

MacGorman et al. (1981) hypothesized that each of the two layers of lightning channel segments corresponds to a major region of thunderstorm charge, and that the two layers of lightning channel structure reflect the vertical dipole that is believed to describe the gross charge structure of many thunderstorms. As evidence to support their hypothesis, they noted that the heights of the two layers of channel structure were similar to the heights of charge inferred from in situ measurements of the electric field in the same storms and to the heights of charge neutralized by lightning in other storms analyzed by Workman et al. (1942), Reynolds and Neill (1955), and Krehbiel et al. (1979). Furthermore, they noted that the range of heights in the two layers encompassed most sources of infrasonic thunder for the same storm. (Recall from Sec. 5.11.3 that infrasonic sources are thought to be coincident with charge regions neutralized by lightning.) MacGorman et al. suggested that single layers of lightning occur when the vertical separation between positive and negative charge in a thunderstorm is small, as might occur when strong vertical shear in the horizontal wind shears storm structure horizontally with height and restricts the height of a storm outside the immediate vicinity of updraft cores.

To further support their hypothesis, MacGorman et al. (1981) offered a conceptual model that is illustrated in Fig. 7.41. They pointed out that lightning is initiated in regions of large electric field magnitudes, which will tend to drive the propagating ends (i.e., the leaders) of lightning channels toward regions having smaller electric field magnitudes. To the extent that the preexisting electrostatic field affects lightning propagation, it will drive the lightning into regions of weak electric field (thereby maximizing the potential difference between oppositely charged ends of the lightning) and will prevent subsequent propagation from moving back out into a stronger electric field (which would reduce the potential difference between the ends of the flash). Propagation in regions of weak electric field (i.e., regions of relatively uniform electric potential) will be dominated by the electric field of the leader tip. Thus, MacGorman et al. expected that most lightning structure would be within regions of charge, where the thunderstorm electric field tends to be relatively small and the potential difference with an oppositely charged region tends to be largest.

Support for this interpretation was provided by studies showing that mapped channel segments of individual flashes tended to be located in the vicinity of the charge centers neutralized by the flash (Uman et al.

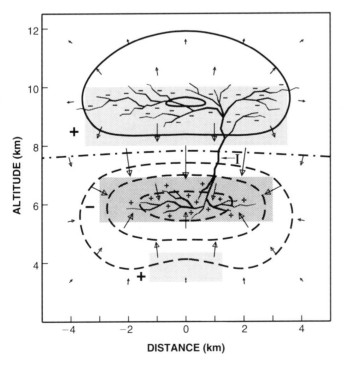

Fig. 7.41. Conceptual model of lightning initiation and development relative to electric field and charge regions of a highly idealized thunderstorm. Charge is in three vertically separated disks (shaded). Electric field lines are indicated by arrows, with the length roughly proportional to the field magnitude at the midpoint of the arrow shaft. Contours indicate equipotential lines. Lightning is initiated at I in a region of large E, and each end moves into a charge region, where it develops horizontally through the charge. Negative charge is distributed on the upper channels in the positive disk, and positive charge, on the lower channels in the negative disk. The direction of channel development into the negative disk is somewhat controversial (see Sec. 7.17).

1978, Krehbiel 1981, Lhermitte and Williams, 1985, Nisbet et al. 1990), although the acoustic system and the model of the LDAR VHF mapping system used in these studies sampled too slowly to provide much detail in the flash structure. Subsequent studies using faster VHF mapping systems did not have data on lightning charge centers. However, Shao and Krehbiel (1996) noted that the heights of the upper positive and lower negative charge centers of cloud flashes in previous studies corresponded well to the heights of horizontal channels in the two layers mapped in similar storms by newer VHF systems. A superposition of mapped channel structure and lightning charge centers for two different flashes from similar storms was shown in Fig. 7.30 to illustrate the layering of both and the apparently close relationship between charge centers and channel structure.

The conceptual model of MacGorman et al. (1981) was developed further by Williams et al. (1985), who studied spark propagation through plastic blocks, each of which was doped inside with a layer of negative charge that varied horizontally in one of several patterns. The doping process also created a layer of positive charge on the surface of the plastic. If the electric field between these regions was large enough, a spark that was created by a sudden enhancement of the electric field at a point near the surface would bridge the short gap between the surface positive charge and the internal negative charge and would propagate through the two charge layers. The spark left a visible trace of its path through the inside of the plastic (i.e., primarily through the negative charge).

For experiments in which isolated spots of high charge density were embedded in a region of low charge density, Williams et al. (1985) showed that the spark permeated spots of high charge density with extensive branching, but propagated through the regions of low charge density between the spots as relatively isolated channels having much less branching (Fig. 7.42a). Conversely, when charge density was low in isolated spots and higher in the surrounding region, the spark permeated the surrounding region with heavy branching and tended to avoid the isolated spots of low space charge density (Fig. 7.42b). By invoking scaling of parameters between plastic and the atmosphere that had been determined by previous experiments, Williams et al. argued that the sparks's behavior in seeking out regions of charge implies that lightning also will seek out regions of charge, and this supports and extends the hypothesis of MacGorman et al. (1981).

Williams et al. (1985) also showed that sparks propagated farther and faster as the charge density through which they propagated was increased. Propagation ceased when the magnitude of the electric field at the tip of the spark became too small. As the space charge density increased, the potential difference between the positive charge and negative charge increased, so the length of spark and amount of branching for which a large enough electric field could be maintained at the propagating tip also increased.

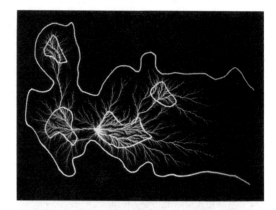

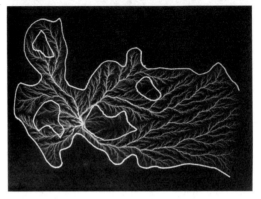

Fig. 7.42. Spark propagation through two blocks doped with negative charge in different horizontal patterns. The blocks were 12.7-mm thick, and negative charge was in a thin layer roughly midway through the block. In both blocks, the space charge density outside the outer contour was 0 C m^{-3}. (a) Space charge density was 1.2 C m^{-3} in the isolated spots and 0.6 C m^{-3} in the surrounding region within the outer contour. (b) Space charge density was 0.6 C m^{-3} in the isolated spots and 1.2 C m^{-3} in the surrounding region within the outer contour. (From Williams et al. © 1985: American Geophysical Union, with permission.)

However, properties of the medium also affect spark length and branching. Williams et al. (1985) argued that, if the critical electric field strength for sustained propagation is very close to the value needed to cause electrical breakdown of the medium, discharges will extinguish quickly. But if the critical field for sustained propagation is very small, the discharge will continue until every region of charge is neutralized. This will affect both how far a flash propagates and on how small a distance scale it will continue to branch. Williams et al. hypothesized that the ratio of neutralized charge to initial charge ($\Delta Q/Q$) depends on the ratio of the critical field for sustained propagation to the breakdown field. They suggested that a range of $\Delta Q/Q$ of 0.3–0.5, such as measured by Winn and Byerley (1975), is consistent with breakdown fields of 500–1000 kV m^{-1} and critical fields for sustained propagation of 100–150 kV m^{-1} at middle levels of storms. This range of critical fields for sustained propagation also is the range of values they inferred from laboratory experiments (Griffiths and Phelps 1976a, Uman et al. 1970, Gallimberti 1979) and from electric fields measured by instruments struck by lightning in clouds, but Williams et al. noted that the correct range of values for sustained propagation of lightning still was poorly understood.

From the above arguments, one might expect that, if the critical electric field for flash initiation is smaller than assumed by Williams et al. (1985) (as will be true if lightning is initiated at the breakeven field by relativistic electrons), then either $\Delta Q/Q$ or the threshold field for sustained propagation must be smaller than previously estimated. However, the way in which $\Delta Q/Q$ depends on the ratio of the critical field for sustained propagation to that for flash initiation can be affected by several factors, including the geometry of the thunderstorm charges causing the critical fields, the geometry of the lightning, and the location at which conditions for sustained propagation are evaluated. By making different simplifying assumptions for thunderstorms, one can infer that the above values for sustained propagation and $\Delta Q/Q$ are consistent with either the breakdown field or the breakeven field. Nonetheless, one still would expect that decreasing the threshold for sustained propagation would lead to more charge being neutralized and that lightning tends to branch more extensively and propagate farther through charge as the charge density increases.

7.16.4. Structure of positive and negative ground flashes

As might be expected from the conceptual model discussed in the previous section, many negative ground flashes reconstructed by mapping systems have predominately horizontal structure in clouds, usually in the lower layer of lightning activity, where negative charge typically is found. The channel to ground can be anywhere relative to the horizontal structure, from the middle out to one of the extremities. Thus, the appearance of flashes in a vertical plane varies from an L-shape to a T-shape (Fig. 7.43).

Only two studies have reconstructed flashes identified as positive ground flashes in warm season storms. Krehbiel (1981) analyzed the charge removed by various processes in three positive ground flashes from a Florida storm, and Rust et al. (1981c) acoustically reconstructed two positive ground flashes from an Oklahoma storm. In both studies, positive ground flashes occurred in regions of weak reflectivity and weak vertical winds, 30 km or more from deeper convection. Krehbiel found that long, horizontal channels propagated from the direction of deeper convection, which was beyond the range of the lightning measurements, and removed positive charge from a layer in the radar bright band near the 0°C isotherm, where frozen precipitation was melting. These horizontal flashes then formed a vertical channel to ground near heavier embedded precipitation and transported positive charge to ground from a region that was higher than the horizontally propagating channels.

One of the positive ground flashes reconstructed by Rust et al. (1981c) is shown in Fig. 7.44. Vertical structure extended through most of the depth of the storm at the location of the lightning. However, the horizontal extent was even larger, with channels extending at right angles to the more vertical structure to produce a horizontal extent of 20 km. The channel to ground, not readily apparent in the reconstruction, was roughly 10 km from the vertical lightning structure. (The maximum range of the reconstruction was limited by the audibility of thunder, so it would not have been possible to observe channels extending farther toward regions of deeper convection.)

Note that, in a charge distribution that can be approximated as a series of layers of charge, such as found in anvils (Secs. 7.8 and 8.1.3) and in the stratiform precipitation of mesoscale convective systems (see Sec. 8.2.2), any particular equipotential surface is at a roughly constant height. Thus, long horizontal channel propagation is consistent with the hypothesis of MacGorman et al. (1981) and the spark observations of Williams et al. (1985). Furthermore, if a more vertical region of charge or a change in the height of charge is embedded in a layered distribution, it would distort the equipotential surface of the layers and so could be responsible for causing the formation of a channel to ground. If so, then the location of a channel to ground

(a) (b)

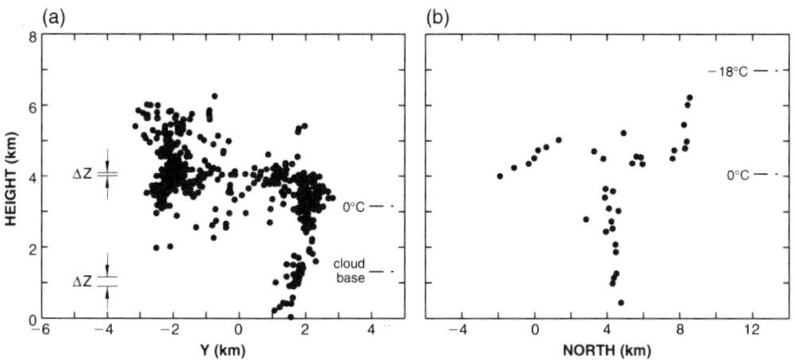

Fig. 7.43. Mapped structure of two ground flashes. Besides being a function of flash geometry, the perceived geometry tends to vary with the viewing angle from different locations on the ground. (a) An L-shaped discharge from South Africa. The Y-axis was along an antenna baseline, approximately NW-SE. (From Proctor et al. © 1988: American Geophysical Union, with permission.) (b) A T-shaped discharge from Arizona. (From Teer and Few © 1974: American Geophysical Union, with permission.)

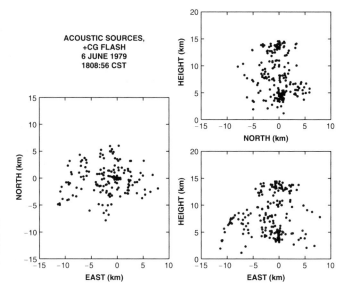

Fig. 7.44. Mapped structure of a positive ground flash that occurred in stratiform precipitation trailing an Oklahoma squall line. (From Rust et al. 1981c, with permission.)

from anvils and stratiform regions is controlled primarily by the fact that the vertical distribution of charge observed in regions of heavier precipitation in the stratiform region is different from the distribution in more stratified regions. The location of the ground channel would be affected relatively little by the vertical distribution of charge in more stratified regions through which the flash propagates horizontally.

7.17. LIGHTNING LOCATION RELATIVE TO STORM STRUCTURE

7.17.1. Lightning location inside clouds

(i) Lightning channels. As discussed in Section 7.16.2, most lightning channels at heights corresponding roughly to the lower layer of lightning activity are within regions having moderate to large radar reflectivities (i.e., $\gtrsim$20–30 dBZ), although they are not necessarily coincident with the largest reflectivities. Higher channels in the upper layer often occur in weaker reflectivities, typically over regions of larger reflectivity or in anvils. The association with reflectivity also can be inferred from Kawasaki et al. (1994), who examined the correlation of radar-inferred rainfall at a height of 2 km and lightning flash density detected by an interferometric mapping system in Japan. Typical correlation coefficients were 0.6–0.7, which indicate reasonably good correlation, but do not rule out significant differences between rainfall distribution and lightning location. From the studies mentioned in Section 7.16.2, there appears to be considerable variation in the location of lightning relative to vertical cores of heavy precipitation.

This variation in the location of lightning relative to storm structure appears to depend on the structure of the storm's internal wind field relative to updraft and reflectivity cores in a way that is consistent with the wind transporting charge from these regions. For example, in many storms, particularly larger ones, lightning is displaced from reflectivity and updraft cores. (The magnitude of reflectivity or updraft defining a core has varied, but cores often are defined as regions with reflectivity $\geq$40–50 dBZ or updrafts $\geq$10 m s^{-1}.) Nakano (1976) suggested that acoustically reconstructed horizontal lightning flashes were aligned so they extended downstream from the updraft along the direction of the upper level outflow. A similar suggestion was made by MacGorman (1978) and MacGorman et al. (1983), who noted that long horizontal lightning flashes extended from new and mature cells northward into the anvil of a Colorado storm (Fig. 7.40). Earlier in the storm, before northern cells dissipated and an anvil formed, lightning channels tended to be restricted to a region closer to, but still displaced to the north side of, reflectivity cores. Similarly, Proctor (1983) showed long, horizontal flashes extending into the growing anvil of a small, multicellular storm, while ground flashes tended to occur on the opposite side of the storm, which Proctor called the propagating side, where we would expect updrafts and new cells to form. By analyzing lightning location relative to the wind and reflectivity fields measured by Doppler radar for one storm, Rust et al. (1981a) and MacGorman et al. (1983) provided confirmation that lightning flashes in regions of weaker reflectivity were downstream from updraft and reflectivity cores (see Fig. 7.45).

However, in many storms lightning does not appear to be offset appreciably to the side of reflectivity cores, but tends to occur within and above them (Lhermitte

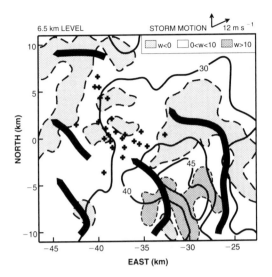

Fig. 7.45. A cloud flash mapped by a VHF system relative to reflectivity and wind fields at a height of 6.5 km. This height is through the middle of the greatest concentration of mapped VHF sources. Crosses indicate the plan location of channel segments that produced VHF noise. Contours of reflectivity are labeled in dBZ. Shading indicates vertical velocity in m s⁻¹. Arrows show horizontal winds. (From Rust et al. 1981a, with permission.)

and Krehbiel 1979, Krehbiel 1981, Taylor et al. 1984, Lhermitte and Williams 1985, Maier et al. 1995, Scott et al. 1995, and Shao and Krehbiel 1996). Examples are shown in Figs. 7.38 and 7.39. Typically in these cases,

the region of lightning activity appears mushroom shaped, with a relatively narrow region centered roughly on the reflectivity core at low and middle altitudes and a broad region in the upper part of the storm where winds begin diverging into the anvil, as described by Taylor et al.

Ray et al. (1987) illustrated how the distribution of lightning is influenced by the wind field by analyzing lightning in a multicell storm with a supercell storm that had markedly different wind fields (Fig. 7.46). Both storms were severe. In the multicell storm, storm-relative inflow sloped upward into the reflectivity core and did not penetrate beyond the core at low and middle levels of the storm. Near the top of the storm, wind from the updraft diverged fairly symmetrically into the anvil. Lightning below the level of divergence was concentrated in and near the reflectivity core. In the region of strong divergence near the top of the storm, lightning was scattered over a much larger region. The net effect was a roughly mushroom-shaped distribution centered on the reflectivity and updraft cores, which were close to each other.

In the supercell storm, storm-relative streamlines in the updraft passed through the reflectivity core at middle levels of the storm and continued into the anvil, where vertical motions were weaker. There was divergence from the updraft near the top of the storm, but upshear flow (relative to environmental winds and pointed toward the right in the figure) in this region of divergence was dwarfed by flow into the rapidly expanding downshear anvil (to the left in the figure). No lightning was observed on the inflow side of the reflectivity core until almost the top of the storm, where some VHF sources occurred in diverging winds from the updraft. Most of the lightning at all heights was far-

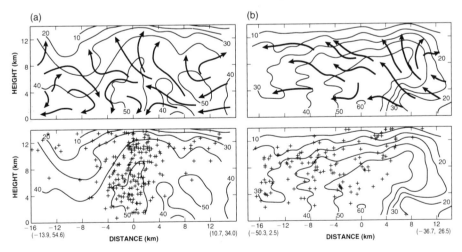

Fig. 7.46. Mapped channel segments (crosses) relative to reflectivity contours and wind streamlines for all lightning within ±15 min of the time of the radar volume scan for two Oklahoma storms. Only channel segments within ±1 km of the vertical plane were plotted. (a) A multicell storm. (b) A supercell storm. (From Ray et al. © 1987: American Geophysical Union, with permission.)

Fig. 7.47. VHF impulse density as a function of reflectivity for (a) The multicell storm and (b) The supercell storm shown in Fig. 7.45. Dashed lines indicate values of reflectivity that existed over too few grid points to give reliable estimates of impulse density. (From Ray et al. 1987: American Geophysical Union, with permission.)

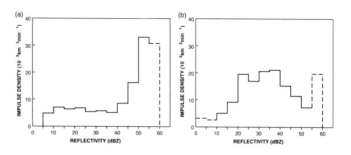

ther downshear, with the region of lightning activity beginning where streamlines were approximately midway through the reflectivity core at most heights.

The above differences in these two storms are reflected in the distribution of lightning channel segments relative to reflectivity and vertical velocity, shown in Figs. 7.47 and 7.48, respectively. The multicell storm had high concentrations of lightning at larger values of reflectivity and updraft, indicating the tendency of lightning to occur in updraft and reflectivity cores. In contrast, the supercell storm had an almost uniform distribution with respect to vertical velocity. The distribution with respect to reflectivity for the supercell had a smaller, broader peak than that of the multicell storm, and the peak for the supercell occurred at moderate reflectivities instead of large reflectivities.

The vertical velocity field in the vicinity of lightning flashes had been analyzed previously in other storms by Rust et al. (1981a) and MacGorman et al. (1983), as illustrated in Fig. 7.45. Most of the reconstructed channels of flashes in this storm tended to be downstream from reflectivity cores and were in weak updrafts, often near the boundaries of weak downdrafts. Keener et al. (1993) and Keener and Ulbrich (1995) examined the location of lightning relative to the vertical distribution of vertical winds in a tropical storm by using simultaneous measurements of reflectivity, wind, and lightning location from a dual wavelength, vertically pointing radar. They found that lightning tended to occur in regions of updraft at heights above the height of the largest reflectivities and updrafts. Lightning channels were most likely to be clustered around regions of large gradients in the vertical velocity (Fig. 7.36).

An apparently related effect of the influence of wind is suggested by another observation: Away from reflectivity cores, lightning structure often roughly parallels reflectivity contours, especially those that are elongated downstream toward an anvil (e.g., Fig. 7.40 and MacGorman 1978, Proctor 1983). This observation probably is related to earlier observations that the in-cloud structure of lightning flashes in individual storms appeared to be mutually aligned (Ogawa and Brook 1969, Teer and Few 1974, and Nakano 1976).

The reason for mutual alignment and for the tendency to parallel reflectivity contours is not well established. However, we would expect that advection by the wind would affect both the distribution of reflectivity and the distribution of charge. Although particle microphysics also continues to affect reflectivity and charge during advection, microphysical interactions relevant to charge generation should be slower away from reflectivity cores and strong updrafts. The observation that reflectivity contours and lightning

Fig. 7.48. VHF impulse density as a function of vertical velocity for (a) The multicell storm and (b) The supercell storm shown in Fig. 7.45. Dashed lines indicate values of vertical velocity that existed over too few grid points to give reliable estimates of impulse density. (From Ray et al. © 1987: American Geophysical Union, with permission.)

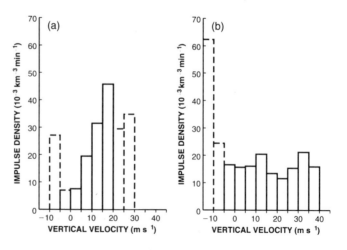

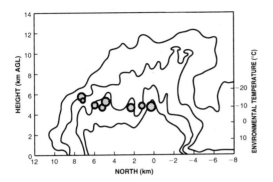

Fig. 7.49. Charge centers neutralized by a single ground flash as it developed in a New Mexico storm. Radar data are from a 3-cm radar and are not corrected for range or attenuation. Contours are roughly 20 dB apart. The size of the circles indicates the relative magnitude of charge neutralized by each stroke. (From Krehbiel et al. © 1979: American Geophysical Union, with permission.)

structure are roughly parallel approaching an anvil suggests that the effects of advection dominate local microphysical charge separation in determining the horizontal distribution of charge in that region. It is not clear whether this would continue to be true in the horizontally extensive region of light precipitation and weak vertical winds (sometimes called the stratiform region) that often accompanies mature squall lines, since dynamics and microphysics acting on time scales of hours may be as important as advection in es-

tablishing the structure and electrification of that region (see Sec. 8.2.4).

(ii) Lightning charge centers. Measurements of the negative charge neutralized by flashes find that it usually is located near or inside reflectivity cores and sometimes extends into adjoining regions. In small, isolated storms in New Mexico and Florida, Krehbiel et al. (1979) and Krehbiel (1981) found that negative charge centers for most cloud and ground flashes were near reflectivity cores, usually above the level of maximum reflectivity, at temperatures of $-10°$ to $-25°C$. An example of charge centers from successive strokes of a single ground flash is shown in Fig. 7.49. Charge for the first stroke was located above the newest cell at the northern extreme of the storm, and charges for subsequent strokes typically were taken from centroids above older cells increasingly farther south.

Similar results were observed by Lhermitte and Williams (1985) and Nisbet et al. (1990) (Fig. 7.50). Lhermitte and Williams found negative charge above and beside the upper part of a 45-dBZ core. Negative charge centers determined by Nisbet et al. typically were near or inside 30-dBZ cores and outside 40-dBZ cores, but sometimes were within regions having reflectivities between 40 and 50 dBZ. In these studies, most positive charge centers of cloud flashes were in weaker reflectivities surrounding, and usually above, the negative charges. Krehbiel (1981) reported that as each flash progressed, the region from (to) which positive (negative) charge was transferred broadened so that it was upwardly divergent (e.g., the charge centroids of Fig. 7.30).

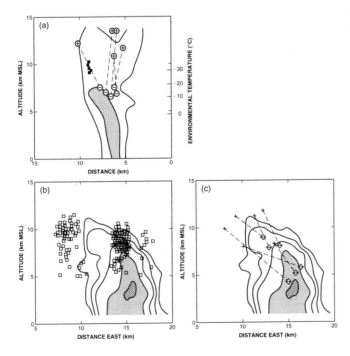

Fig. 7.50. Charge centroids and channel segments of cloud flashes superimposed on reflectivity for two Florida storms. Channel segments from a VHF mapping system are indicated by squares. Positive and negative charge centroids for a single flash are joined by a dashed line to delineate the charge dipole. (a) The first detected VHF impulse and the neutralized charge dipole for five flashes. The solid contour indicates 30 dBZ, and the shading indicates ≥45 dBZ. (After Lhermitte and Williams 1985: American Geophysical Union, with permission.) (b) The first and last VHF impulses detected for each cloud flash during a 5-min period in another storm. Only channel segments within 1 km of the vertical cross section were plotted. Radar reflectivity contours are in 10-dBZ increments from 10 to 50 dBZ. (After Nisbet et al. 1990: American Geophysical Union, with permission.) (c) Charge dipoles for six cloud flashes during the same period shown in panel b. (After Nisbet et al. 1990: American Geophysical Union, with permission.)

(iii) Flash Initiation. As discussed previously, even if flash initiation partly depends on other processes, such as cosmic rays, one would expect flashes to begin in regions of a storm that have a large electric field (approaching the breakeven or breakdown value). In charge distributions approximated by simple geometries, the electric field would tend to be largest near the boundaries of the charge regions. For example, with two vertically separated spheres or disks of opposite charge, the electric field would be largest between the charges. With three vertically stacked charges of alternating polarity, a large electric field could exist between either the lower two or upper two charges, depending on the distribution and relative magnitude of the charges. More complicated charge geometries (either in number or shape of the charge regions) could certainly produce more complicated vertical distributions of large electric field magnitudes.

The height at which a cloud flash begins often has been found to be between the heights of the positive and negative charge centers neutralized by the flash. Krehbiel (1981) and Lhermitte and Williams (1985) found that the height of the first VHF source in a cloud flash tended to be between 7.5 and 10 km MSL (approximately $-20°C$ and $-40°C$, respectively) in one storm on the Florida coast. In most of the 13 analyzed cloud flashes, the height of the initial VHF source was between the positive and negative charge centers that the flash effectively neutralized, although the initial source often was offset from a line joining the two charges (Fig. 7.50a). In another Florida storm, Nisbet et al. (1990) showed lightning charge centers with both the first and last VHF sources, so it is difficult to identify initiation points (Figs. 7.50b,c). However, plots of VHF sources and charge centers for periods having the most lightning flashes would be consistent with the observations of Lhermitte and Williams and Krehbiel if initial VHF sources of cloud flashes analyzed by Nisbet et al. tended to occur in the cluster of VHF sources near reflectivity cores, while the last source was in the cluster of VHF sources near positive charge in regions of much weaker reflectivity. Several studies (Nakano 1979b, Krehbiel 1981, Liu and Krehbiel 1985, Rhodes et al. 1994, Shao et al. 1995) have shown that individual cloud flashes usually began by developing vertically, but in many cases eventually developed horizontal structure.

Some studies have reported that the initiation heights of ground flashes have been lower than the initiation heights for cloud flashes in the same storm. For example, the only two ground flashes analyzed by Lhermitte and Williams (1985) began between roughly 5 km ($-2°C$) and 6 km MSL ($-8°C$), at least 12°C warmer than the initiation temperature of the cloud flashes and below most of the negative charge centers neutralized by lightning for that storm. Although Lhermitte and Krehbiel (1979) were unable to determine discrete charge centers for the Florida storm they analyzed, the initial VHF sources that they mapped for several cloud flashes and two ground flashes were at heights and temperatures very similar to those found by Krehbiel (1981) and Lhermitte and Williams (1985). Taylor (1983) found that the mean heights of initial VHF impulses for 28 ground flashes and 35 cloud flashes in Oklahoma were 5.9 km MSL ($-13°C$) and 7.6 km MSL ($-23°C$), respectively.

Proctor (1991) analyzed a much larger data set and was able to provide new detail about the relative heights of initiation of cloud and ground flashes, although it is unknown whether his results from South Africa differ in some respects from what would be observed at other locations. The data set consisted of 559 cloud flashes and 214 ground flashes from 13 storms (Fig. 7.51). Origins for both total flashes and cloud flashes appeared to have a bimodal distribution, with a lower mode at 1°C to $-9°C$ and an upper mode at roughly $-25°C$ to $-35°C$. The upper mode of the cloud flash distribution contained the origins of approximately 50% more cloud flashes than the lower mode. Out of 214 ground flashes, approximately 80% originated between 0°C and $-15°C$. The vertical distribution of origins for ground flashes was similar to the distribution in the lower mode for cloud flashes. Roughly half of the lower mode of the distribution of total flashes was comprised of ground flashes. Proctor noted an apparent tendency for a larger percentage of total flashes to originate in the lower mode when flash rates decreased as a storm cell aged.

Proctor also noted that 73% of high cloud flashes (initiation height >7.4 km MSL) began within 270 m of a 20-dBZ contour, 19% began farther inside regions of larger reflectivity, and only 8% began farther outside in regions of weaker reflectivity. Ground flashes were more likely to begin farther inside regions of ≥20 dBZ than cloud flashes: 54% of ground flashes began within 270 m of a 20-dBZ contour, 36% began farther inside, and 9% began farther outside. Proctor noted that many of the flashes beginning inside a 20-dBZ contour began at the edge of a reflectivity core or in regions where reflectivity gradients were large. Some of the 20-dBZ contours that were included in these statistics bounded regions of weaker reflectivity inside regions of reflectivity ≥20 dBZ. Approximately 13% of all flash origins occurred at a 20-dBZ contour bounding one of the weak echo holes, 24% occurred along the upper boundary of a region of ≥20 dBZ, and 27% occurred along the side of a region of ≥20 dBZ.

The height distributions reported by Proctor (1991) had similarities and differences compared with the distributions reported by Proctor et al. (1988) from a smaller data set of 98 cloud flashes and 47 ground flashes. The distribution of ground flash origins appeared similar in both studies, and the lower part of the distribution of cloud origins occurred at roughly the same height as ground flash origins in both studies. Proctor et al. performed a Student's t test on the cloud and ground flash distributions and found that the distribution of initiation heights for ground flashes was statistically the same as the distribution for the 78 cloud flashes that began at temperatures warmer than $-15°C$.

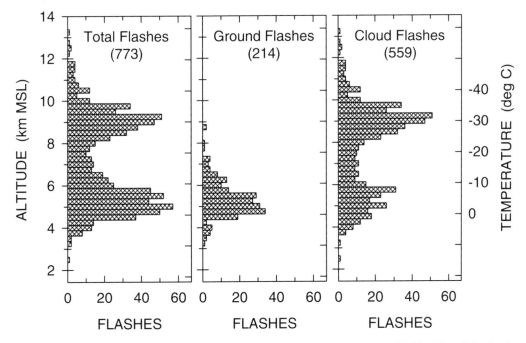

Fig. 7.51. Distributions of the height at which flashes originated for 13 storms from South Africa. The origin of each flash was taken to be the centroid of the first 6–10 VHF sources mapped for that flash. Terrain was at 1430 m MSL. The distributions for 773 flashes and 214 ground flashes were replotted from Proctor (1991). The distribution for 559 cloud flashes was obtained by subtracting the ground flash distribution from the total flash distribution. The heights at which various temperatures are shown on the right axis are typical heights for the 13 storms. In all but one of the storms, the temperature at each height was within 1–2°C of the temperature shown. For the one exception, temperatures in the upper part of the storm were up to 7°C warmer than shown.

However, Proctor's results differed from those of Proctor et al. in two respects. First, the relative proportions of cloud flash origins below and above −15°C were reversed. In Proctor et al., more cloud flash origins were warmer than −15°C than were colder, with 78 origins warmer and 20 colder. In Proctor, roughly 50% more cloud flash origins were colder than −15°C than were warmer. Second, Proctor's distribution of cloud flash origins extended to colder temperatures than Proctor et al.'s, and the uppermost of the two modal values in Proctor's distribution was roughly 5°C colder than in Proctor et al.'s distribution.

The lower mode in the distributions found by both Proctor (1991) and Proctor et al. (1988) overlaps the range of temperatures for cloud flash initiation found previously by Nakano (1979b) from a lightning charge center analysis based on multiple surface electric field measurements. Although neither Proctor et al. nor Proctor determined lightning charge centers, the lower mode of origins is within a temperature layer at which negative charge centers often are found in other storms, although positive charge centers occasionally have been reported in the colder part of this range. However, the upper mode of cloud flash origins spans temperatures at which positive and negative charge centers

have been found. Without lightning charge measurements, it is impossible to tell whether these upper cloud flashes began between lower negative and upper positive charge centers of lightning as found by Lhermitte and Williams (1985), although the temperature range is consistent with this hypothesis. Proctor (1991) hypothesized that cloud flashes beginning in the upper mode involved an upper positive charge and a negative charge in the screening layer above it.

As discussed in Section 7.16.2, Proctor et al. (1988) found that cloud flashes that began at temperatures ≤−15°C had traits (VHF pulse rates, breadth of VHF sources, and clustering of VHF sources around the origin) different from those that began at warmer temperatures lower in the storm. Although three ground flashes began at temperatures ≤−15°C, no ground flash had these different traits. Proctor et al. reported that the medium-frequency electromagnetic noise radiated by ground flashes was indistinguishable from noise radiated by cloud flashes that began at temperatures warmer than −15°C. On the other hand, Kitagawa and Brook (1960) and Shao and Krehbiel (1996) reported that ground flashes in New Mexico produced rapid initial pulse rates, while cloud flashes did not. To reconcile their observations with those from New

Mexico, Proctor et al. suggested that the relative percentages of cloud flashes beginning in the lower and upper temperature ranges might be different in some storms or in other locations. The observations from South Africa and the United States would be consistent if all cloud flashes observed by Kitagawa and Brook and Shao and Krehbiel began at the colder temperatures identified by Proctor et al. with flashes having slower VHF pulse rates. This appears to have been the case for the two Florida storms studied by Lhermitte and Krehbiel (1979) and Lhermitte and Williams (1985).

(iv) Channel development. Most studies that have used VHF mapping systems to study flash development have found that flashes often began near or above regions of greater radar reflectivity (e.g., Lhermitte and Krehbiel 1979, Lhermitte and Williams 1985), with at least one major channel moving out into areas of weaker reflectivity (e.g., Proctor 1983, Rust et al. 1981a, MacGorman et al. 1983, Nisbet et al. 1990, Maier et al. 1995, Scott et al. 1995) (Figs. 7.36 and 7.45). When more than one cell was electrically active, flashes often began near the tallest or most vigorously growing reflectivity core and propagated back into older cells (e.g., Proctor 1983, Maier et al. 1995) (Fig. 7.39). The structure of acoustically mapped lightning flashes (e.g., MacGorman 1978, MacGorman et al. 1983) is consistent with these observations (Fig. 7.40), though acoustic mapping cannot depict flash development. In studies of lightning charge centers, Krehbiel (1981) found that flashes often began near the tallest reflectivity core, above the height of maximum reflectivity. As noted previously, the negative charge neutralized by the flashes often was in a layer and was concentrated near reflectivity cores (Fig. 7.49). During cloud flash development, the outward sequence of positive charge centers typically progressed into regions of weaker reflectivity.

Krehbiel (1981), Proctor (1983), and Proctor et al. (1988) determined that channels of cloud flashes moving into weaker reflectivity usually carried negative charge to (or equivalently removed positive charge from) the region into which they propagated, although some developing channels had the opposite polarity of current. Rhodes et al. (1994) and Shao et al. (1995) made a similar observation of the charge of developing channels, but without corresponding radar data for the storm. Figure 7.52 shows typical cloud flash development mapped by a VHF system in New Mexico and Florida. All the mapped channels developed as negative leaders. Initial channel growth was upward from the initiation point to the height of the upper horizontal channel, which grew outward from the vertical channel. Channels below the initiation point were detected later, and their mapped development was retrograde, from a location away from preexisting channels back toward the initiation point. Thus, the initial and upper horizontal channels transported negative charge away from the initiation point, while the later channels at lower heights transported negative charge toward the initiation point.

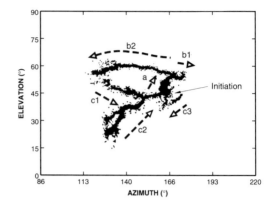

Fig. 7.52. Azimuth and elevation of selected VHF radiation sources for a cloud flash, showing the overall structure and development of the flash. The flash began at the point marked "Initiation." Letters and numbers indicate the sequence of development, and arrows indicate the direction of observed channel development. Some of the lower channels were closer to the system than the upper channels, so the vertical separation between the lower and upper channels appears smaller than it actually was. (From Shao and Krehbiel 1996: American Geophysical Union, with permission.)

The cause of the relative lack of observations of positive leaders is not well established. Some (e.g., Mazur 1989) have suggested that this is because positive leaders radiate little VHF noise. Others (e.g., Shao 1993, Shao et al. 1995) have suggested that because some positive leaders are observed, it is plausible that there are few such leaders. However, Shao and Krehbiel (1996) noted at least one example in which a channel formed by a positive leader was not detected by a VHF system until the return stroke, unlike negative leaders of channels to ground, which are readily detected.

Channel development inside clouds often was more complex than the well-defined channels seen beneath clouds. Proctor (1981, 1983) suggested that individual VHF sources in clouds were 0.1–1 km in diameter. Furthermore, Hayenga and Warwick (1981), Rust et al. (1981b), and Proctor et al. (1988) showed that lightning often did not progress simply (Fig. 7.53). Instead, progression of a lightning flash occurred as the outer envelope of a multitude of channels moved forward, and flash development proceeded at two velocities. Individual new channels moved from a location to the side and slightly ahead of the flash back toward preexisting channels at speeds of 3×10^6–1×10^8 m s^{-1} (retrograde motion); the forward edge of the envelope of channels progressed more slowly at speeds of 2×10^4–2×10^6 m s^{-1}. For more information on lightning propagation, see Chapter 5 and reviews by Uman (1987), Proctor (1988), Rhodes et al. (1994), and Shao et al. (1995).

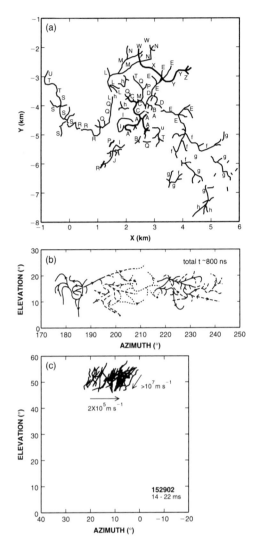

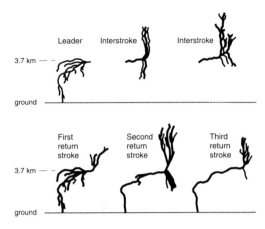

Fig. 7.54. Sketch of ground flash development due to return strokes and interstroke processes. Each leader precedes the return stroke shown below it. Return strokes tend to overshoot the end of the interstroke channels. Interstroke processes extend the flash horizontally into new regions. The VHF mapping system shows retrograde motion from the vertical channels that form during the interstroke interval back toward the previously existing channel. Although not shown in this sketch, channels that form during a given interstroke interval can be in a different direction from the flash origin than channels during previous interstroke intervals. (After Proctor et al. 1988: American Geophysical Union, with permission.)

Fig. 7.53. Detailed pictures of flash development depicted by VHF mapping systems. (a) Plan projection of a flash mapped by Proctor et al. (1988). Letters indicate sequence of development. Capital letters denote channel segments that formed during K changes. (From Proctor et al. 1988: American Geophysical Union, with permission.) (b) An azimuth elevation presentation of a flash. The different line types indicate progression of time, increasing as the flash formed generally from left to right. (From Rust et al. 1981b, with permission.) (c) An azimuth elevation presentation of a small portion of a flash. Overall flash development was from left to right at an estimated speed of 2×10^5 m s^{-1}, but vertical channel segments developed downward at more than 10^7 m s^{-1}. (From Hayenga and Warwick 1981: American Geophysical Union, with permission.)

Charge center analyses of multiple stroke ground flashes by Krehbiel et al. (1979) and Krehbiel (1981) found that subsequent channels began somewhere at the top of a preexisting flash and extended into new regions in a predominately horizontal direction (Fig. 7.49). Similarly from VHF mapping systems, Proctor (1981, 1988), Rhodes et al. (1994), and Shao and Krehbiel (1996) found that interstroke processes extended the flash into new regions, as shown in Fig. 7.54.

MacGorman (1978) and Proctor (1983) reported that sequential flashes sometimes appeared related. A flash often overlapped the volume filled by the previous flash and extended into new regions. The relative volumes of overlapping and new regions varied considerably. In some cases, MacGorman found that several kilometers of channel length from two sequential flashes overlapped, but the second flash extended a comparable distance into a new region. This was characteristic of flashes during the earlier part of the analyzed period, when the anvil and the spatial length of flashes were smaller. In other cases, much of the lengths of two sequential flashes overlapped, and the second flash extended only a few kilometers into new regions. This was characteristic of flashes later in the storm, such as the sequential flashes shown in Fig. 7.40c and d. MacGorman noted that many of the long horizontal flashes that tended to occur during later stages of the storm had very similar structure, with one end near a vertically extensive reflectivity core and the other extending by a tortuous path into the anvil.

7.17.2. Lightning strikes to ground

(i) Negative ground flashes. The horizontal extent of in-cloud channels of ground flashes can be comparable to the horizontal extent of the corresponding host storms, and there is considerable variety in the part of the flash structure connected to the channel to ground. Also, we noted in Section 7.17.1 that flashes tend to be co-located with updraft and reflectivity cores in some storms and are offset from the cores in other storms. Thus, it should come as no surprise that there is considerable variation, too, in the location at which ground flashes strike ground relative to storm structure. Most often, ground strikes of negative ground flashes are

densest near or inside contours of 20 dBZ or 30 dBZ on plan projections of the radar reflectivity structure of storms from the lowest elevation angle of a radar. However, there can be considerable variability in the location within this region, particularly with respect to whatever cores of larger reflectivity or heavy rain exist (Fig. 7.55). Sometimes strikes extend ahead (relative to the direction of storm motion) of moderate reflectivity (Shackford 1960, Rutledge and MacGorman 1988, Keighton et al. 1991), and sometimes they are absent along the leading edge and most concentrated behind cores of maximum reflectivity, along a strong reflectivity gradient (e.g., López et al. 1990). Sometimes strikes fill most of the region having moderate reflectivity (Kinzer 1974, Rutledge et al. 1990, Mohr et al. 1995); sometimes they cluster mainly around or in the cores of largest reflectivity (Rutledge et al. 1990, MacGorman and Nielsen 1991, Keighton et al. 1991, Rutledge and Petersen 1994) or only in some of the

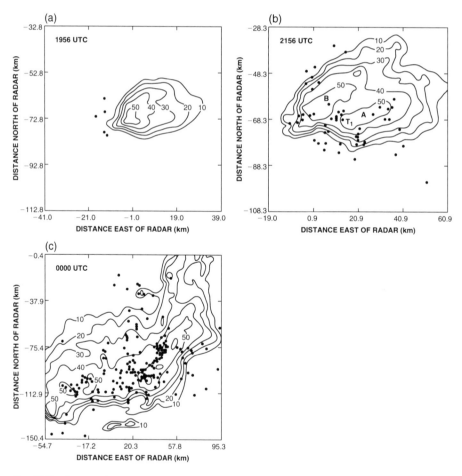

Fig. 7.55. Ground strike points (large dots) of ground flashes relative to base-scan radar reflectivity for three time periods of an Oklahoma storm system. Reflectivity contours are labeled in dBZ. (From Keighton et al. 1991, with permission.)

larger cores (López et al. 1990); and sometimes strikes occur in a region of weak reflectivity adjoining heavier reflectivity (Rutledge and MacGorman 1988, Rutledge et al. 1990, Nielsen et al. 1994) or under anvils or high-based storms, completely outside detectable reflectivity at low levels (Reap 1986, Keighton et al. 1991).

Since base-scan reflectivity is the lowest elevation angle of reflectivity measured in a storm, it is used often to infer rainfall and, in many circumstances, relationships between ground flashes and base-scan reflectivity would be expected to be similar to relationships between ground flashes and rainfall. Several studies have noted that the pattern of ground strike density produced by a storm often is similar in many ways to the pattern of rainfall (Goodman and MacGorman 1986, Rutledge and MacGorman 1988, Roohr and Vonder Haar 1994); however, there also are differences, as would be expected from the description in the previous paragraph of ground strikes relative to radar reflectivity. Ground flashes and rainfall can appear related in ways other than just spatial distribution. In Ontario, Canada, Clodman and Chisolm (1993) found that the largest rainfall and largest ground flash rates usually occurred close together in space and time; furthermore, extremes of both rainfall and ground flash rates were produced by stationary or slow moving storms that were triggered by various effects of the Great Lakes. In southern Florida, Maier et al. (1978) also found that slow moving storms tended to produce larger ground flash rates than faster moving storms but, unlike in Ontario, total rainfall was larger on days when storms moved. Other aspects of the relationship between rainfall and ground flashes will be discussed in Sections 7.18.1 and 7.19.

It is difficult to find a consistent relationship between stage or type of storm and the location of lightning strikes to ground. Kinzer (1974) suggested that ground flashes tended to cluster closer to regions having moderate to large reflectivity at low levels in squall lines than in typical isolated storms in Oklahoma. Keighton et al. (1991) observed this pattern in Oklahoma on one day that produced a supercell storm which evolved into a squall line. When the storm first began to produce ground flashes, the strike points were outside 10-dBZ contours at a height of 1 km, but were under the anvil (Fig. 7.55a). During the supercell stage of the storm, strike points continued to occur under the anvil, but also were clustered around and inside 50-dBZ contours at a height of 1 km, with many flashes in regions of large reflectivity gradients (Fig. 7.55b). During the squall line stage, the ground flash rates increased to almost 20 min^{-1}, and most strike points were concentrated within a large region having reflectivity $\geq$40 dBZ (Fig. 7.55c). Similarly, Roohr and Vonder Haar (1994) and Mohr et al. (1995) observed that in some squall lines, negative strikes were more likely to cluster under regions of moderate-to-large reflectivity and cold cloud tops during the mature stage of the squall line than during the developing and dissipating stage. However, Kane (1993a) reported on an isolated storm

near Washington D.C. in which strikes tended to cluster within large reflectivities. Furthermore, in a study of several squall lines in the central United States, Rutledge et al. (1990) observed some in which clusters of negative ground strikes occurred under regions having reflectivity <20 dBZ.

The mean distance from one strike point to its nearest neighbor is a function both of the number of flashes produced by a storm and the horizontal area over which they are distributed (i.e., related to average density). Krider (1988) found that in small, isolated Florida storms, the mean distance to the nearest strike was approximately 0.7 km. The mean distance between successive strikes was 3–4 km. Krider calculated that if all ground flashes originated from the same localized region, the random variation in the mean angle of channels from the vertical would cause a mean separation between successive strikes of 2–3 km. Thus in these small storms, the random variation in channel orientation was sufficient to explain most of the separation between successive strike points. We would not expect this to be as true in larger storms, because the charge regions producing the flashes are more horizontally extensive. However, this work provides an estimate of the extent to which variations in the angle from the vertical of the channel to ground contribute to the scatter of ground strikes.

Although the ground flash rates of storms of similar size and duration varied widely, Peckham et al. (1984) found that the average ground flash rate was larger in storms that had more convective cells. The mean ground flash rate of spatially isolated storms whose flash rate increased to a single peak and then decreased monotonically was 1.7 min^{-1}, and the maximum flash rate observed in any such storm for a 5-min period was 17 min^{-1}. Multicellular storms that had multiple peaks in 5-min lightning flash rates had a mean rate of 3.4 min^{-1} and a maximum 5-min rate of 22 min^{-1}. Storm systems consisting of a cluster of single- or multipeak storms had a mean ground flash rate of 6.8 min^{-1} and a maximum rate of 32 min^{-1}.

Still larger storm systems that occur over the central United States can produce much larger maximum rates. Goodman and MacGorman (1986) reported that some mesoscale convective complexes produced average rates greater than 50 min^{-1} for 1 h or more. Kane (1993b) observed a mesoscale convective system that produced an average of more than 60 min^{-1} for 13 hours. Note also that some isolated storms have larger rates than observed by Peckham et al. (1984). For example, MacGorman and Burgess (1994) described one large isolated storm that had multiple peaks in its ground flash rates and produced rates of more than 30 min^{-1} as it began to merge with other storms in the squall line studied by Kane. Besides being related to convective cycles of cell growth and decay, an increase in ground flash rates can occur when a storm splits into two storms or when the boundaries of two or more storms begin to merge together (Goodman and MacGorman 1986, Curran and Rust 1992, MacGorman and Burgess 1994).

Peckham et al. (1984) found that strike densities averaged over an area spanning all lightning strikes in each of their three categories of storms had a trend opposite to that for ground flash rates: The average density was smaller for categories of storms that tended to last longer and have more storm cells, while ground flash rates tended to be larger. Single-peak storms had a mean ground flash density of $18 \pm 11 \times 10^{-3}$ km^{-2} min^{-1}, multi-peak storms had a mean of $15 \pm 8 \times 10^{-3}$ km^{-2} min^{-1}, and storm systems had a mean of $10 \pm 6 \times 10^{-3}$ km^{-2} min^{-1}. The area of the lightning activity increased faster than the decrease in strike density from one category to another, so that mean flash rates increased going from single-peak storms to multiple-peak storms to storm systems, as noted above. Smaller densities are typical of some types of isolated storms in other regions. Stolzenburg (1990) reported observations of isolated storms in various locations that had flash densities of $2–8 \times 10^{-3}$ km^{-2} min^{-1} when data were corrected for detection efficiency, as done by Peckham et al. In some isolated supercell storms in the central United States, we have observed flash densities comparable to those Peckham et al. found for storm systems and multipeak storms (Seimon 1993), although many supercell storms have lower densities (inferred, for example, from data presented by MacGorman and Burgess 1994). Some mesoscale convective complexes (e.g., the squall line stage of the 28 August 1990 storm observed by MacGorman and Burgess and Kane 1993b) in the central United States have clusters of ground flashes with densities comparable to or larger than those that Peckham et al. found for single-peak storms.

Note that the results discussed in the previous three paragraphs were obtained from warm season thunderstorms and that flash rates and densities tend to be smaller in winter storms. At Sakata, Japan, for example, Ishii et al. (1981) found that although there were approximately as many thunderstorm days in winter as in summer, the number of ground flashes occurring in winter was much lower. The winter storms had much smaller flash rates, and flash densities were almost an order of magnitude smaller than in summer.

(ii) Positive ground flashes.   The location and density of strike points for positive ground flashes typically are different from those for negative ground flashes. During isolated, warm season thunderstorms, positive ground flashes often occur during the dissipating stage of storms of any size or emanate from the anvil of large storms (Pierce 1995b, Fuquay 1982, Rust et al. 1981a, b, López et al. 1990). In both instances, the maximum reflectivity above positive ground strike points tends to be smaller than is typical for negative ground strike points. Furthermore, the density of positive ground strike points in these situations tends to be at least an order of magnitude less than is typical for negative ground strike points (e.g., Stolzenburg 1990, Engholm et al. 1990). Positive ground flashes constitute a higher percentage of ground flashes at middle and

upper latitudes in winter (e.g., Takeuti et al. 1977, 1978; Orville et al. 1987) and can have densities similar to those for negative ground flashes in winter (Ishii et al. 1981). However, both the density of strike points and the reflectivity above strike points tend to be smaller in winter than in summer for both polarities of flashes. In one winter case studied by Bosart and Sanders (1986), both negative and positive ground flashes were observed, but positive ground flashes dominated in storms that were shallower.

There are two situations in which positive ground strike points can occur at densities comparable to the densities of negative ground strike points in typical isolated thunderstorms. The first is a subset of relatively shallow thunderstorms (at least by warm season standards), as observed by Bosart and Sanders (1986), Engholm et al. (1990), and Takagi et al. (1986). Although positive ground flash densities in these storms often are less than typical negative ground flash densities, sometimes they are not. We have observed several cases in the central United States, most often in late winter or early spring, in which either the northern cells in a line of storms or an entire line of shallow storms is dominated by positive ground flashes (Morgenstern 1991).

The second case in which dense positive ground flashes are observed is a subset of severe storms (e.g., Curran and Rust 1992, Branick and Doswell 1992, Morgenstern 1991, Kane 1991, Seimon 1993, MacGorman and Burgess 1994, and Stolzenburg 1994). In an extremely strong, classic supercell storm, Seimon showed that strike points of positive ground flashes sometimes clustered around the high reflectivity core on a plan projection of base-scan radar reflectivity and sometimes were concentrated throughout the high-reflectivity region. Once negative ground flashes began to dominate ground flash activity, strike points of negative ground flashes tended to cluster near reflectivity cores on the leading edge of the storm, where new cells tended to form. Winter storms, squall lines, and severe storms will be discussed further in Chapter 8.

7.18. CHARACTERISTICS OF STORMS INDICATIVE OF LIGHTNING PRODUCTION

Prior to the development of automatic lightning detection systems, the only means for identifying thunderstorms operationally was to infer their existence from measurements of other storm characteristics. For example, there have been many attempts to determine whether a storm is producing lightning by examining radar-determined storm characteristics, such as storm height, formation and location of precipitation, and the rate of storm growth. Although there are trends in relationships between lightning and other storm parameters that can be used to identify thunderstorms with varying degrees of success from nonelectric storm characteristics, geographical and storm-to-storm variability causes considerable problems in applying these relationships. Storm electrification requires interactions of too many storm parameters for the relatively simple parameter tests that have been used thus far to have a high degree of diagnostic skill. However, because many of these radar tests and in situ measurements of related parameters reveal information about the nature of storm electrification and in some cases have been widely considered for operational meteorological applications, we will examine them in this section. We also will examine studies of flash rates as a function of the same parameters. While these relationships are similar in some respects to the relationships between initial electrification and various properties of storms, which were discussed in a previous section, this section will differ from that section by concentrating specifically on lightning production, instead of the production of charged particles and a foul weather electric field.

7.18.1. Base-scan reflectivity

One of the simplest tests to determine if a cloud is a thunderstorm has been to use thresholds on the maximum reflectivity found in a storm during a radar scan at the lowest elevation angle (referred to as a *base scan*). A de facto threshold sometimes used in meteorological operations has been that storms having base-scan reflectivity of at least VIP 3 (≥ 41 dBZ) are thunderstorms. However, the extent to which the rule is valid varies from one geographical region to another. For example, Reap and MacGorman (1989) examined the fraction of storms that produced ground flashes versus their maximum base-scan reflectivity for storms in Oklahoma and Kansas. The probability of producing ground flashes increased with increasing reflectivity, as shown in Fig. 7.56a, but was never 100%. Furthermore, only 50% of cases having VIP 3 reflectivity (the operational threshold then used by the U.S. National Weather Service to identify thunderstorms based on radar data) produced ground flashes.

The relationship is different in the western United States. There, it is not unusual for mountain thunderstorms to have ground flashes when the maximum base-scan reflectivity is less than VIP 3 (Fig. 7.56b). Reap (1986) noted that in the drier environment of the southwestern United States, storms often have high cloud bases and produce little precipitation at the ground, and so have small base-scan reflectivities. Furthermore, storms that formed over mountainous terrain often were small and so only partially filled the 2° beam width of the WSR-57 radars used in Reap's study.

In the southeastern United States, on the other hand, unpublished data (personal communication, R. Holle and I. Watson) showed that the probability of lightning strikes increased with increasing reflectivity up to ≈50 dBZ, but began to decrease with increasing reflectivity above 50 dBZ. Storms in the southeastern United States tend to be dissipating when they rain heavily enough to produce very large reflectivities and either have not yet produced lightning (a shallow, heavy rain-

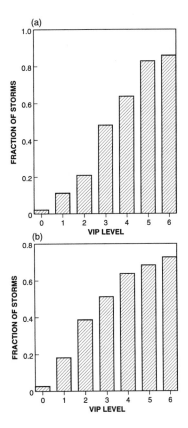

Fig. 7.56. Fraction of storms with ≥ 2 ground flashes versus maximum base-scan radar reflectivity. (a) For storms in Oklahoma and Kansas. (From Reap and MacGorman 1989, with permission). (b) For storms in the western United States. (From Reap 1986, with permission.)

shower) or are less likely to produce ground flashes during this stage.

Although the above studies concerned only ground flashes, not total lightning activity, a similar variation in regional trends would be expected for total lightning. Since precipitation at the ground is related only indirectly to any of the storm parameters that have been hypothesized to play a vital role in storm electrification, we will not examine this relationship further here. However, in Section 7.19 we discuss some quantitative observations of the relationship between precipitation and lightning production.

7.18.2. Storm height

Another storm parameter that has been examined as a predictor of lightning occurrence is storm height. Several studies (e.g., Lhermitte and Krehbiel 1979) have noted that lightning activity began when the top of the radar echo of the storm grew to an altitude where the temperature was $\leq -20°$C. However, the use of almost any threshold of storm height (or temperature) leads either to inadequate probability of detection or to false detection of some nonthunderstorm clouds as thunderstorms. For example, Jones (1950) found an increasing probability of lightning with decreasing cloud top temperatures for cloud tops colder than $-25°C$ (i.e., for taller clouds), but he also observed several storms with cloud tops between $-25°C$ and $-40°C$ that produced no lightning. Similarly, Tohsha and Ichimura (1961) reported that there was a 90% probability that storms whose radar-detected tops were $-16°C$ or colder were thunderstorms; this result presumably means that 10% of these storms were not thunderstorms. Also in a study of 20 storms, Dye et al. (1989) observed six storms with maximum storm heights of 8–9.5 km (-20 to $-30°C$) that did not become thunderstorms, although three had an electric field of more than 10 kV m^{-1}.

Holle and Maier (1982) examined the probability of ground flashes being produced as a function of the altitude of the radar-determined storm top for storms in Florida. Their results are in Table 7.4. The minimum storm height at which ground flashes were observed was 7.8 km, which corresponded approximately to the $-18°C$ isotherm. The probability of a storm producing ground flashes was less than 100% until the altitude of storm tops was 16 km MSL. Although some storms without ground flashes may have produced cloud flashes, the trend of increasing probability of ground flashes with increasing storm height and the apparent existence of a minimum storm height for ground flash production appear consistent with previously reported observations of total lightning in individual storms.

The increasing probability of lightning occurrence with increasing storm height is consistent with obser-

vations showing that the magnitude of the electric field tends to increase inside thunderstorms as the storm height increases. Mach and Knupp (1993) examined the electric field inside storms as a function of the maximum height of 10-dBZ reflectivity, and found a monotonic increase in the magnitude of the electric field as the height increased, particularly as the temperature of the cloud top became colder than $-20°C$. A similar increase of electric field with storm height was reported by Moore et al. (1958).

In addition to taller storms being more likely to produce at least one lightning flash, several studies have noted that taller storms tend to have larger flash rates (Shackford 1960, Baughman and Fuquay 1970, Jacobson and Krider 1976, Livingston and Krider 1978, Williams 1985, Cherna and Stansbury 1986, Goodman et al. 1988a, b, Rutledge et al. 1992, Price and Rind 1992, Rutledge and Petersen 1994, Watson et al. 1995). Shackford (1960) found that as the radar-detected height of summer thunderstorms increased from ≈5 km to 17 km, the maximum flash rates (whether ground, cloud, or total was not specified) observed for storms in each height bin tended to increase from roughly ≤10 h^{-1} to >1000 h^{-1}. Median flash rates also increased with height, although the increase was smaller than for maximum flash rates. However, flash rates ≤20 h^{-1} were observed in some storms in all height bins.

Williams (1985) suggested that flash rates increase as the fifth power of thunderstorm height. However, several studies since then have suggested that the relationship is more complicated. Cherna and Stansbury (1986) found that the dependence on height varied considerably from one day to another and appeared to be greatest on days with stronger vertical shear in the horizontal wind. Several investigators (Williams et al. 1992, Rutledge et al. 1992, Rutledge and Petersen 1994) have shown that some tall tropical storms (height >15 km) had little or no lightning. Zipser (1994) and Petersen et al. (1995, 1996) reported that tropical monsoon storms can produce heavy rainfall and extend to 17 km in height and yet produce no lightning. Furthermore, Goodman et al. (1988a) presented a case in which flash rates were larger than those predicted by either Williams (1985) or Cherna and Stansbury (1986). If the fifth power relationship exists, it is modulated by the effect of other parameters.

The results discussed in this section suggest that summertime storms with tops higher than 16 km might be identified reliably as thunderstorms in Florida. However, such a threshold would miss the considerable fraction of thunderstorms with warmer tops, and observations in other climatological regions have reported taller storms that produced no lightning.

7.18.3. Vigorous convective growth

One reason that a storm height threshold sometimes falsely identifies storms as thunderstorms is that tall storms can have updrafts with insufficient speed or spatial extent to drive electrification mechanisms to the

Table 7.4. Occurrence of ground flashes relative to radar-detected storm height.

Altitude (km MSL)	Number of Storms	Storms with Ground Flashes[a]	Percentage
5.0–5.9	1	0	0
6.0–6.9	11	0	0
7.0–7.9	21	1	4.8
8.0–8.9	16	2	12.5
9.0–9.9	25	4	16.0
10.0–10.9	32	12	37.5
11.0–11.9	38	18	47.4
12.0–12.9	70	48	68.6
13.0–13.9	100	76	76.0
14.0–14.9	145	125	86.2
15.0–15.9	74	68	91.9
16.0–16.9	19	19	100
17.0–17.9	1	1	100

[a]Number of cases with ground flashes within ±5 min of the height determination by radar.

Adapted from Holle and Maier 1982, with permission.

point of causing lightning. Thus, several studies (e.g., Workman and Reynolds 1949) have suggested another requirement for lightning: the presence of vigorous convective growth at temperatures colder than some value often estimated to be somewhere between −10°C and −20°C. Reynolds and Brook (1956) noted that two storms with precipitation and with a top that was steady at roughly −5°C did not produce lightning until the storm top began to grow vigorously higher. Similarly, Michimoto (1991, 1993a) indicated that storms must have rapid vertical movement of 30-dBZ radar echoes, which, he suggested, imply rapid development and motion of graupel particles. Note that the timing of initial lightning activity does not always correspond to the timing of strong convection. Dye et al. (1989) pointed out that initial electrification occasionally began as storms reached their maximum height and started dissipating. Thus, they modified the requirement for vigorous convection to indicate that electrification sufficient to cause lightning occurs during or at the end of convective growth.

Vigorous growth of storm tops and strong updrafts are linked closely, so this section will treat them together. Michimoto (1991, 1993a) noted that in winter the −10°C isotherm was at altitudes higher than 1.4 km for all thunderstorms, because storms with a lower −10°C isotherm had updrafts too weak to produce intense electrification. Based on work by Ishihara et al. (1987) and Tabata et al. (1989), Michimoto (1991) suggested that the updraft threshold is greater than 7 m s⁻¹ for Japanese thunderstorms. Zipser (1994) and Petersen et al. (1996) suggested that updraft speeds of at least 6–7 m s⁻¹ were needed at the −10°C level to produce lightning in tropical oceanic thunderstorms.

As suggested by the specification of a temperature level for the convection, the mere presence of vigorous convection somewhere in a cloud is insufficient to produce lightning. Apparently, strong updrafts are needed in the mixed phase region, where the noninductive mechanism discussed in Chapter 3 is effective. Several of the above studies indicated that strong updrafts must occur above the −10°C level to cause electrification. For example, tropical oceanic storms, which can be very tall, rarely produce lightning (e.g., Rutledge et al. 1992, Williams et al. 1992, Zipser 1994). Williams and Renno (1993) Zipser and Lutz (1994), Zipser (1994) and Lucas et al. (1995) have noted evidence that the release of latent heat during formation of cloud ice particles can increase updraft speeds in the upper region of tall oceanic storms, which tend to have weak updrafts in the mixed phase region. They suggested that the lack of strong updrafts in the mixed phase region is responsible for the absence of lightning in oceanic storms, even when larger velocities occur higher in the storms.

Increases in vigorous convective growth have been tied to increasing flash rates, as well as to initial lightning production. Lhermitte and Krehbiel (1979), for example, observed a pronounced increase in flash rates in a Florida storm when there was an increase in the rate of cloud height growth and an increase in updraft speed

to >20 m s⁻¹ at temperatures colder than −10°C (Fig. 7.35). Weber et al. (1993) observed that lightning flash rates tended to increase when updrafts were near their maximum values and top heights of 20-dBZ radar echoes were growing rapidly. Similarly, Williams et al. (1989) noted that once lightning began in Alabama storms, flash rates tended to increase, when storm tops measured by radar grew rapidly. Mazur et al. (1986) noted that the trend in the maximum flash density (defined in this case as the number of flashes per minute per kilometer along the radar beam) was similar to the trend in the maximum height of 40-dBZ reflectivity; the growth and subsidence in the height could be interpreted as reflecting the increasing depth in which large precipitation formed during strong convection, followed by sedimentation as either the updraft weakened or precipitation grew to a size too large to be supported.

The dependence of lightning rates on convective vigor is related to the observation by several investigators that days with greater conditional instability or greater CAPE at a given location, tended to produce storms with higher flash rates. In his analysis of thunderstorms with high cloud bases, Colson (1960) found that flash rates tended to be larger on days with greater instability and with more precipitable water. In analyses of tropical storms, Rutledge et al. (1992), Williams et al. (1992), and Petersen et al. (1996) noted that flash rates tended to be larger on days with larger CAPE. However, the instability discussed in these studies is conditional instability and may not be fully realized by a given storm. Rutledge et al. noted that many storms with low flash rates occurred on days with large CAPE. Williams et al. suggested that something other than CAPE must also have affected ground flash rates, because the number of flashes on days with similar values of CAPE varied by up to roughly an order of magnitude and appeared to be affected by when in a cycle of increasing and decreasing CAPE the day occurred. Zipser (1994) discussed earlier studies showing that oceanic cases having values of CAPE comparable to those of continental storms tended to have much smaller updraft velocities than the continental storms. Expanding earlier hypotheses by Rutledge et al., Williams et al., and others, Zipser suggested that the weaker vertical velocities in the mixed phase region of oceanic storms hindered the noninductive electrification mechanism by causing either a lack of large ice particles or insufficient concentrations of supercooled liquid water.

There is some disagreement about the role that strong convection plays in electrifying thunderstorms. The hypothesis suggested by most investigators has been that strong convection at temperatures of −10°C to −40°C tends to produce graupel growth leading to electrification. The strong updraft plays a dual role: It provides moisture through condensation or deposition of water vapor as air parcels rise and cool, and it supports graupel aloft long enough to have many collisions with ice crystals. Furthermore, there is a feedback effect as condensation or deposition release heat, which intensifies the updraft. Evidence for these relationships was discussed by Dye et al. (1986). Furthermore, Bringi et al. (1993a, b) provided additional evidence based on aircraft penetrations and multiparameter radar for storms on the Florida coast: Electrification increased as graupel accreted liquid particles, reflectivity in the mixed phase region increased to 40 dBZ or more, and updraft velocity increased (probably because of the release of latent heat during freezing). Lightning began within a few minutes of glaciation (i.e., freezing of liquid water particles).

Those who favor electrification by storm ingest and collection of external charge (e.g., the Grenet-Vonnegut mechanism, often called the convective mechanism) argue that strong convection also would increase charge transport for the Grenet-Vonnegut mechanism. Thus, the relationship between convection and lightning also has been cited as evidence for the Grenet-Vonnegut mechanism. However, a requirement that the convection extend higher than a particular temperature threshold to produce lightning would be based on different processes if the Grenet-Vonnegut mechanism dominates electrification instead of precipitation interaction mechanisms. Precipitation growth in itself would be of little aid to electrification by the Grenet-Vonnegut mechanism. Instead, dependence on convective growth could be based on a correlation between the height of convective storms and the speed of updrafts. Certainly the release of latent heat from freezing hydrometeors boosts updraft velocities. Furthermore, Williams (1985) and Price and Rind (1992) compiled data from several studies to show that taller storms tend to have larger updraft velocities (Fig. 7.57). Though

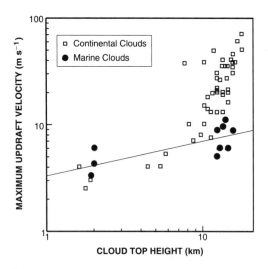

Fig. 7.57. Maximum updraft magnitude versus storm depth. The line shows the regression for marine clouds and is fit by the equation, $W_m = 2.86\ H^{0.38}$. (After Price and Rind © 1992: American Geophysical Union, based on data from other studies, with permission.)

data points are considerably scattered about any regression line for data on maximum updraft velocity versus storm height, continental storms having depths less than 2 km tend to have maximum updrafts of 2–6 m s^{-1}, while those having depths of at least 10 km tend to have updrafts of 10–60 m s^{-1}. Besides being based on the tendency toward faster updrafts in taller storms, a temperature threshold for the Grenet-Vonnegut mechanism also could be related to the larger conductivities at higher altitudes; a larger conductivity allows the upper screening layer to form faster and thereby increases the rate at which charge becomes available for transport from upper cloud boundaries.

7.18.4. Precipitation aloft

As suggested by observations discussed in previous sections, some storms with cloud tops colder than $-20°C$ do not produce lighting because they have an unfavorable vertical distribution of precipitation. Although a foul weather electric field can exist in clouds when cloud particles are too small to have appreciable terminal fall speeds, the electric field under these conditions usually is no more than a few kV m^{-1} (see, for example, data in Ch. 2 and discussion in Sec. 7.7). In all documented cases, the electric field became large enough to cause lightning only after clouds had formed precipitation and, usually, ice. Early documentation of this behavior was provided, for example, by Workman and Reynolds (1949), Byers and Braham (1949), Jones (1950), Kuettner (1950), Reynolds and Brook (1956), and Tohsha and Ichimura (1961). Many other, more recent studies have provided additional confirmation (e.g., Larsen and Stansbury 1974, Takahashi 1978, Christian et al. 1980, Dye et al. 1986, 1988, 1989, Williams et al. 1989, Michimoto 1991, Rutledge and Petersen 1994, and Zipser 1994).

Although some storms have been observed producing lightning when they were completely warmer than 0°C (discussed in Sec. 8.4.8), this has not been true in the vast majority of cases. Several studies have suggested that some threshold of precipitation usually is needed at temperatures between $-10°C$ and $-40°C$ to produce lightning, although there is uncertainty about the amount and altitude of precipitation that are required. For example, Workman et al. (1942), Reynolds and Brook (1956), and Workman (1967) observed that New Mexico storms produced lightning only after precipitation was detected at the $-10°C$ isotherm by a 3-cm wavelength radar. Larsen and Stansbury (1974) found that sferics were not received from a storm until the 43-dBZ contour of reflectivity reached a height of 7 km, where temperature was $-30°C$. Miyazaki et al. (1982) found that precipitation rates of 16 mm hr^{-1}, corresponding roughly to reflectivity of 40 dBZ, typically extended to the $-20°C$ isotherm at the time of the first lightning flash (the range of temperatures encompassing most cases at the time of first flash extended from $-10°C$ to $-40°C$). Michimoto (1991) showed that the first lightning discharge in Japanese storms occurred 5 min after radar reflectivity of 30 dBZ appeared at the $-20°C$ isotherm. Petersen et al. (1995, 1996) noted that tropical oceanic storms tended to produce lightning only when reflectivity of at least 30 dBZ extended above the $-10°C$ isotherm. Zipser (1994) and Zipser and Lutz (1994) suggested a similar requirement.

Studies that included in situ microphysical measurements provided additional information about this relationship. Dye et al. (1986), for example, made in situ electric field and microphysical measurements in a storm that produced a single lightning flash. Shortly before precipitation was first detected by radar, they measured a weak foul weather electric field of 40 V m^{-1}. For 8–10 min after the first precipitation was detected (initially 5 dBZ at -15 to $-20°C$), the electric field magnitude increased slowly (but remained <1 kV m^{-1}), as precipitation began to grow. There were few graupel particles until the end of this period, when reflectivity grew to 36 dBZ and the first 5 mm graupel particle was detected. As ice particles quickly became larger and more plentiful and reflectivity increased over the next 5–6 min, the maximum measured electric field grew rapidly to 15 kV m^{-1} at approximately the time of maximum precipitation development, which

was at the $-20°C$ level. The storm then began to dissipate, and 2–3 min later lightning occurred (implying that someplace in the storm the electric field magnitude was larger than had been measured). As the storm continued to dissipate, the measured electric field magnitude decreased. Others also have found that the first flash occurs within a few minutes of first graupel detection (e.g., Goodman et al. 1989, Bringi et al. 1993a).

Dye et al. (1989) extended Dye et al.'s (1986) observations of early electrification to a total of 20 New Mexico storms, some of which did not become thunderstorms. Results are shown in Table 7.5. Electric field magnitudes less than 200 V m^{-1} were sometimes observed before precipitation was detected. However, the electric field did not exceed 1 kV m^{-1} until reflectivity had reached ≥ 40 dBZ, at least 15 min after reflectivity of 10 dBZ had first been detected. Breed and Dye (1989) noted that the measured electric field magnitude often grew rapidly after passing the 1 kV m^{-1} threshold. Lightning often occurred within 2–7 min of the time that 1 kV m^{-1} initially was observed. This timing is similar to what was found by Hondl and Eilts (1994), who noted that lightning occurred 5–25 min (median of 15 min) after the first 10-dBZ echo was observed at the $-10°C$ isotherm in 22 Florida storms. Note that the 10-

Table 7.5. Summary of 1984 initial electrification cases from Langmuir Laboratory, New Mexico.

Lightning				Time from dBZ detection to lightning (min)				Radar Top (km)	
No. of Flashes	Time from IE[a] to first (min)	Observed[b] E_{max} (kV m^{-1})	Comment on Initial Electrification	10 dBZ	40 dBZ	dBZ$_6$[c] at IE	Maximum dBZ$_6$[c]	at IE	Maximum
0	No IE	≈0.2	None	—	—	—	15	—	8
≈50	>2	55	With top growth	>32	14	≥40	50	≈11.5	13
0	—	3	Weak electrification	27	2	≈36	≈40	≈7.5	≈9
>3	17	95	With slow growth	62	11	>40	46	>8	≈10.5
0	—	36	With slow growth	>15	13	50	52	9	9
>100	2	80	With growth	>62	2	43	50	10.5	14
0	—	60	At maximum height	27	9	41	46	9.5	9.5
≈11	5	17	With growth	>20	6	45	52	≈9	13.5
3?	?	≈80	With growth	≈25	6	51	52	10.5	13.5
6	3	40	With growth	26	12	40	43	9.5	12
0	No IE	<0.2	None	—	—	—	12	—	≈8
?	≈5	65	With growth	>23	>8	≈47	50	≈10	12
?	<7	>65	With growth	23	1	44	50	10.5	11.5
≈10	16	55	With weak growth	>16	>10	42	51	>10.5	≈12.5
18	1	70	With growth	?	?	≤55	61	≈11	12
0	—	15	Near relative maximum of height	16	4	44	45	9	9.5
6	23	28	At maximum height	26	3	38	43	10	11
≈10	27	60	Uncertain	>23	6	44	46	>7.5	?
≈50	4	40	With growth	24	6	47	55	12.5	14
30	≈0	80	With growth	25	≈6	52	53	11	12

[a]Initial electrification, defined for this study as the first time when $|E| \geq 1$ kV m^{-1} was observed; [b]Maximum value of $|E|$ measured by aircraft inside the storm; may be smaller than true maximum; [c]Reflectivity at an altitude of 6 km MSL.

Adapted from Dye et al. 1989.

dBZ radar reflectivity signature at $-10°C$ did not by itself guarantee that a storm would produce lightning: Of the 37 storms studied by Hondl and Eilts that produced 10-dBZ echoes at $-10°C$, 15 produced no lightning.

Several studies have demonstrated that lightning flash rates tend to increase as the area of high reflectivity increased at an altitude within the mixed phase region of storms. Larsen and Stansbury (1974) showed that flash rates measured from sferics were roughly proportional to the area within 43-dBZ contours of reflectivity at a height of 7 km ($-30°C$) for two Canadian storms. Flash rates were less well correlated with the height of the 23-dBZ contour and were poorly correlated with the area within 23-dBZ contours at 7 km.

Marshall and Radhakant (1978) extended the work of Larson and Stansbury (1974) to several thunderstorms on another day, but instead of the area within 43-dBZ contours, they considered the area within 30-dBZ contours at a height of 6 km, where the environmental temperature was roughly $-17°C$. Although details of the evolution of flash rates and reflectivity area had obvious differences, the two parameters were well correlated (correlation coefficient of 0.91). Once sferics rates were normalized to compensate for signal attenuation with range, the sferics rates predicted from the reflectivity area by the regression analysis were within a factor of two of observed sferics rates. This was true of individual storms as they grew and decayed and when comparing flash rates from many storms. Cherna and Stansbury (1986) expanded the previous work once again by considering the effect of both storm height and reflectivity area at a height of 7 km. We briefly discussed their observations of the dependence on height in Sec. 7.18.2. The dependence of sferics rates on the area within 30-dBZ contours at 7 km was roughly linear in most cases. However in severe storms on one day, the dependence on reflectivity area was greater: Flash rates were proportional to $A^{1.24}$, where A is the area of 30 dBZ at a height of 7 km.

Similarly, Keighton et al. (1991) found that ground flash rates were correlated with the area within 40-dBZ, 50-dBZ, and 55-dBZ contours of reflectivity at 6.4 km MSL (roughly $-15°C$) for variations sampled every 10–20 min over a period of 5 h in one storm. In the storm depicted in Fig. 7.35, Lhermitte and Krehbiel (1979) found that trends in flash rates followed trends in the integral of reflectivity over the horizontal area at a given altitude above 7–8 km MSL (i.e., at T $\leq$ $-15°C$).

MacGorman et al. (1989) found that cloud flash rates tended to increase approximately 5–7 min after the area within contours of 40-dBZ and 45-dBZ reflectivity increased at 8.4 km MSL (environmental temperature of $-30°C$), at roughly the same time as the area within contours of these reflectivities at 6.4 km MSL, and at roughly the same time as the area within contours of larger reflectivity at 8.4 km MSL. Cloud flash rates appeared uncorrelated with changes in the area of these reflectivity values at 10.4 km MSL or higher. Thus, MacGorman et al. suggested that particle

interactions leading to reflectivity growth at $-25°C$ to $-40°C$ were important for the production of cloud flashes. Ground flash rates in the supercell storm that MacGorman et al. studied increased 20–25 min after the area of large reflectivity at 7 km increased. However, the delay between peaks in cloud and ground flash rates were larger than in most storms, and MacGorman et al. argued that the unusually strong, deep updraft delayed the sedimentation of larger precipitation particles to altitudes at which the negative charge involved in ground flashes normally is found. (See Sec. 8.1.9 for details about the hypothesis.)

Shackford (1960) examined a different dependence of flash rates on reflectivity aloft in a study of New England thunderstorms. Besides the dependence of flash rates on storm height that was discussed in a previous section, he noted that flash rates tended to increase as the maximum value of reflectivity increased. However, the dependence on maximum reflectivity value was much more pronounced when the maximum reflectivity occurred at temperatures colder than $0°C$. He also showed that storms with larger flash rates tended to have larger reflectivities above the $0°C$ isotherm in vertical profiles of maximum reflectivity with height (Fig. 7.58a). Rutledge and Petersen (1994) and Zipser and Lutz (1994) presented similar observations (Figs. 7.58b and c, respectively). Likewise, in a study of reflectivity parameters at fixed altitudes between roughly $-10°C$ and $-30°C$ in a Florida thunderstorm, Lhermitte and Krehbiel (1979) reported that the timing of a large increase in the flash rate agreed better with the timing of an increase in maximum reflectivity (Fig. 7.35) than with the timing of the increase in the integral of reflectivity mentioned above. Williams et al. (1992) found that storms with reflectivity of at least 35 dBZ extending through the mixed phase region always produced frequent flashes; storms with somewhat weaker reflectivities (25–35 dBZ) in the mixed phase region sometimes produced cloud flashes, but never produced ground flashes.

The hypothesis that precipitation is necessary to produce lightning has been disputed by Moore et al. (1958) and Vonnegut et al. (1959), as discussed in Section 7.7.1. In support of their argument against this hypothesis, they presented their observation that a weak, but increasing, foul weather electric field was consistently measured in clouds before radar echoes were detected (in agreement with other subsequent studies). Because the foul weather field continued to increase smoothly up to the time of lightning, they hypothesized that the same mechanism was responsible both for the weak initial electrification and for electrification sufficiently strong to cause lightning. However, they acknowledged the possibility that another mechanism might be responsible for strong electrification and noted that lightning was always preceded by radar detection of precipitation, although the reflectivity values that they observed in the vicinity of lightning typically were smaller than reported by other observers.

Their line of reasoning is controversial. If a second

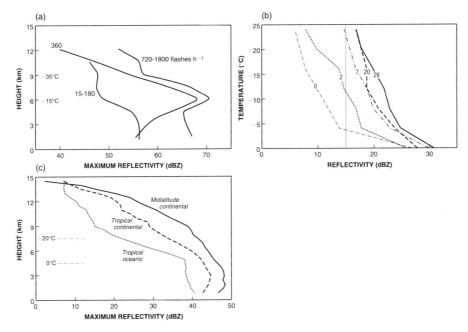

Fig. 7.58. Maximum reflectivity as a function of height or temperature for several storms. (a) Average profile for five to seven storms within each of three ranges of flash rates. (From Shackford 1960, with permission.) (b) Profiles for the stratiform regions of five storm systems, each labeled with the number of ground flashes produced in that region. (From Rutledge and Petersen 1994, with permission.) (c) Average profiles for 41–57 storms in each of three categories: tropical oceanic storms, tropical continental storms, and midlatitude continental storms. Flash rates tend to be larger from left to right, with tropical oceanic storms having little or no lightning and midlatitude continental storms having the largest rates. (From Zipser and Lutz 1994, with permission.)

mechanism begins to contribute to electrification, both its contribution and its rate of increase may well be initially small enough that there would be no discontinuity. The frequently recurring correlation between frozen precipitation and lightning production suggests that a second mechanism is important. On the other hand, lightning also has been documented in a few cases in tropical storms completely warmer than 0°C (see Sec. 8.4.8), and frozen precipitation obviously cannot be required in these cases. However, the existence of lightning in warm clouds does not by itself refute the hypothesis that ice is important for producing lightning in most thunderstorms.

7.18.5. Combinations of storm characteristics

Although no single storm parameter measured by conventional or Doppler radars is capable of identifying all thunderstorms, relatively little has been done to examine a combination of parameters. Grosh (1977) noted that storms were much more likely to become thunderstorms if two conditions were met: (1) their cloud bases were warm enough (at least approximately 10°C) that

updrafts could be substantial at temperatures colder than 0°C, and (2) their cloud tops were cold enough that there were significant amounts of both ice and liquid water. Buechler and Goodman's (1990) study used a combination of storm height and reflectivity threshold. For 15 storms in the southeastern United States, they found that lightning was imminent when reflectivity of 40 dBZ began to be observed at a height of −10°C and when the radar echo top exceeded 9 km.

Hondl and Eilts (1994) suggested that applications needing the maximum possible warning time might benefit from a combined approach that proceeds in stages having progressively greater reliability. To provide the earliest warning, they suggested that Doppler radar be used to detect the surface wind convergence boundaries that precede many storms, as had been suggested by Watson et al. (1987). However, using only wind convergence leads to many false alarms, so they suggested that additional radar-determined features could be used to give greater confidence and reliability. For example, detecting 10-dBZ reflectivity aloft would indicate increased probability of a thunderstorm developing, and then satisfying criteria such as proposed by Buechler and Goodman (1990) would provide the greatest certainty. Each step in the decision process would provide greater certainty, but reduced lead time of the forecast. However, we do not yet know what combination of radar parameters would give the greatest certainty, how reliable such a combination would be, or how the criteria would need to be modified for different regions.

New remote-sensing technology provides perhaps the greatest promise for thunderstorm detection using nonelectrical storm characteristics. If the noninductive graupel-ice mechanism is responsible for thunderstorm electrification, then formation of sufficient graupel and ice should precede electrification. Some parameters that can be measured by a multiparameter radar appear promising for detecting the formation of graupel and ice. Bringi et al. (1993a, b) presented case studies in which rapid electrification began when graupel and hail were detected and after cloud particles in the updraft froze to form small ice particles (a process called *glaciation*). Similarly, Goodman and Raghaven (1993) suggested that satellite microwave sounding systems such as the 58 MHZ and 52 MHZ channels on the SSC1 satellite are capable of detecting graupel and so can help detect thunderstorms. Mohr et al. (1995) and Mohr et al. (1996) compared ground flash rates and locations with output from a satellite-borne 85 GHz channel that responds to ice at middle levels of the troposphere. They noted that storm systems with colder brightness temperatures, indicative of more graupel, tended to have larger densities of negative ground flashes.

However, even techniques for detecting ice may be misleading when trying to identify thunderstorms. As will be discussed more fully in Chapter 9, Randell et al. (1994) and Solomon and Baker (1994) suggested on the basis of observations and modeling studies that electrification may remain too weak to produce lightning if graupel-ice interactions that deposit positive charge on graupel are roughly comparable in number to those that place negative charge on graupel. Thus, the presence of ice particles may not be sufficient by itself to make a thunderstorm. Another scenario in which this is true was reported by Black et al. (1993) and Zipser (1994), who studied hurricane and tropical storm electrification, respectively. Both found evidence that, typically, deep clouds that were not strongly electrified had many small ice particles, but little or no supercooled liquid cloud water particles or rimed graupel.

Note that many studies point out that more than one storm parameter must be within a favorable range of conditions for a storm to produce lightning. Reynolds and Brook (1956), for example, pointed out that storms with precipitation but little convective growth did not produce lightning. Dye et al. (1989), on the other hand, pointed out that there were many periods of cloud growth and convective activity that failed to produce an electric field >1 kV m^{-1}. Furthermore, Dye et al. noted that large reflectivity does not by itself imply frequent particle interactions. Dye et al. (1988) observed that space charge was offset from reflectivity maxima and tended to occur in a region having maximum particle collision rates. Using a dual-polarization radar to infer hydrometeor species, Maekawa et al. (1992) found that lightning was produced in winter storms after graupel (or large ice particles) and small ice crystals were detected in close proximity to each other and after graupel grew rapidly at the $-10°C$ level. Michimoto (1993b) and Kitagawa and Michimoto (1994) interpreted surface and radar reflectivity measurements similarly to indicate that graupel particles were present in winter storms that were strongly electrified and producing lightning, but were not present as storms dissipated, when the electric field was small and there was no lightning, even though the storm still was filled with snow and ice crystals. They suggested that the local space charge carried by graupel in winter storms, observed by Magono et al. (1983, 1984), was essential to produce strong enough electrification for lightning.

Additional evidence that ice particles and vigorous convection in the mixed phase region (typically $-10°C$ to $-40°C$) are both necessary ingredients for lightning production in most storms is provided by studies that have examined how flash rates depend on various storm properties. Some studies, for example, have noted a tendency for taller storms to have higher flash rates (Shackford 1960, Baughman and Fuquay 1970, Jacobson and Krider 1976, Livingston and Krider 1978, Williams 1985, Rutledge et al. 1992, Rutledge and Petersen 1994), but have noted that other storm parameters also affect flash rates, since some shorter storms have higher rates than some taller storms. Cherna and Stansbury (1986) found that: (1) Flash rates depended on both storm height and the area with 30-dBZ contours at a height of 7 km, (2) The relative impact of height and area on flash rates varied with different environmental conditions. Specifically, the dependence on height increased when the vertical shear in horizontal environmental winds increased. They suggested that since detrainment from the updraft was increased when vertical wind shear increased, greater detrainment of positively charged cloud particles from the updraft appeared to increase the effectiveness of large-scale charge separation.

Several studies suggested that the volume of some cloud region above the freezing level is more important than storm height in governing flash rates. Goodman et al. (1989) compared two storms whose tops grew at nearly identical rates and whose heights differed by approximately 1 km. The shorter storm produced roughly twice as many flashes and more than 50% larger peak flash rates. However, the shorter storm also was broader and had a much greater volume in both the mixed phase region and above the 0°C isotherm. It also had greater mass within echoes ≥30 dBZ above the freezing level. Similarly, in a multiparameter radar study of one Colorado storm, Carey and Rutledge (1996) found that cloud flash rates were proportional to the graupel/hail volume at temperatures colder than 0°C, while ground flash rates were proportional to the graupel/hail volume at temperatures warmer than 0°C. The dependence of ground flash rates on the graupel/hail volume in warmer regions was interpreted as providing support for the hypothesis that a lower region of positive charge is important in creating conditions energetically favorable for ground flashes.

López and Aubagnac (1997) used a multiparameter radar to examine the mass, instead of the volume, of various hydrometeor types in an Oklahoma hailstorm. They noted that the evolution of ground flash rates in this storm could be described as having a broad cycle lasting roughly 30 min, on which were superimposed shorter cycles lasting roughly 10 min. The broad cycle appeared similar to the variation in the mass of graupel at temperatures colder than 0°C. In contrast, the timing of the shorter cycles appeared similar to that of the variation in the mass of small hail at temperatures warmer than 0°C. The timing of variations in ground flashes and hail mass at warmer temperatures both lagged 3–4 min behind similar variations in the mass of small hail at 0°C to −10°C. Graupel mass at temperatures warmer than 0°C had no apparent relationship with ground flash rates. From these results, López and Aubagnac suggested that the growth of graupel was responsible for the overall level of electrification in the storm, but that the presence or descent of small hail appeared to enhance ground flash rates, possibly by providing positive charge below the main negative charge.

In conclusion, if the noninductive, graupel-ice mechanism is important to lightning production, as suggested by many of the studies in this section, then at least one factor that affects electrification and, hence, flash rates, is the number of graupel-ice interactions that occur under conditions favorable for electrification. We have seen that this depends on having sufficient concentrations of graupel, cloud ice, and supercooled cloud water particles simultaneously in the mixed phase region. It also depends on where the polarity of graupel charging is positive or negative, on the volume in which particle interactions occur, on the residence time of graupel in the mixed phase region, and on the subsequent trajectories of charged particles. These properties, in turn, are affected by the vertical and horizontal distribution of updraft speed, particularly above the lower boundary of the mixed phase region, by the structure of the wind field advecting charged particles, by the existence and prior history of particles that are injected into a cell from preexisting convection, and possibly by the way a storm's environment influences its microphysics (e.g., the concentration and types of cloud ice and condensation nuclei in the environment). In short, lightning is produced by a very complex system whose electrical state is difficult to determine from measurements of the nonelectrical properties of storms. However, some nonelectrical storm parameters do have enough consistency as predictors to be useful, as long as the required level of performance is not too demanding. Furthermore, it may be possible to improve performance by using combinations of these parameters or by incorporating new sensor technology to detect parameters that are more directly relevant to electrification.

7.19. LIGHTNING-RAINFALL RELATIONSHIPS

7.19.1. Observations of the rainfall yield per flash

We already have noted that storms with larger base-scan reflectivity often also have larger ground flash rates, and that trends in the rainfall produced by deep convective storms often are similar to trends in ground flash rates. Based on these observations, several studies attempted to quantify the relationship between cloud-to-ground lightning and surface rainfall. In one of the earliest attempts, Battan (1965) estimated from a rain gauge network and human observers of ground flashes that 52 storms in Arizona produced an average of roughly 3×10^7 kg of rainfall per ground flash; the range of values for individual storms extended two orders of magnitude (roughly 3×10^6 to 3×10^8 kg per flash) (Fig. 7.59a). An even larger range was found by Williams et al. (1992) in tropical storms (Fig. 7.59b). Table 7.6 shows values either found by or estimated from several studies. Of course, all values in the table omit storms that produced no ground flashes. Errors in the values of rainfall yield per ground flash are caused by errors in extrapolating point rainfall measurements to area totals, errors in estimating rainfall from base-scan radar data, and undercounting lightning flashes. Most studies indicated that their estimates of rainfall yield per flash probably were accurate only to within a factor of two to three. The range of estimates extends four orders of magnitude, much larger than can be explained by errors in the measurements.

There has been a similar range of values when estimates were based on time periods shorter than a storm's entire lifetime. For example, estimates of the rainfall per ground flash by Kinzer (1974) for 5-min periods in a single Oklahoma squall line varied from 4×10^6 kg to 2×10^8 kg. Estimates by López et al. (1991) for 1-h periods in all storms that they could observe one summer on the Florida coast varied from 5×10^6 kg to 1×10^{10} per ground flash (Fig. 7.59c). Estimates from hourly data presented by Holle et al. (1994) for a squall line in the central United States ranged from 3×10^8 kg to 2×10^9 kg per ground flash. Note, also, that ground flash counts and base-scan radar reflectivity analyzed on substorm distance scales at regular intervals during storms sometimes have been reported to be uncorrelated (e.g., Shackford 1960).

In many studies (Kinzer 1974, Williams et al. 1992, Maier et al. 1978, Buechler et al. 1990, and López et al. 1991), the median value of rainfall per ground flash tended to decrease as the number of ground flashes produced by a storm increased (Figs. 7.59b and c). It is unclear whether this was true in the study by Battan (1965) (Fig. 7.59a). However, in all studies the largest observed rainfall yields per flash occurred with the smallest number of ground flashes (typically less than 10–30 flashes).

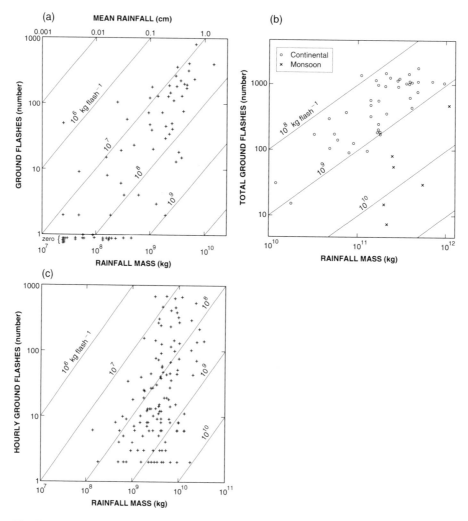

Fig. 7.59. Rainfall as a function of number of ground flashes. Diagonal lines indicate constant rainfall yield in kilograms per ground flash. (a) Arizona storms. (From Battan 1965, with permission.) (b) Daily totals for storms near Darwin, Australia. (From Williams et al. 1992, with permission.) (c) Hourly data from Florida storms. (From López et al. 1991, with permission.)

There have been few measurements of rainfall versus total lightning activity. In two storms, Piepgrass et al. (1982) determined that there was $0.7–0.9 \times 10^7$ kg of rainfall per flash. We are aware of no data on the range of rainfall yield per flash for total lightning activity across a broad spectrum of storms. However, simultaneous measurements of total flash rates and the area within contours of various reflectivity values at low levels in a few storms (e.g., MacGorman et al. 1989, Juvanon du Vachat and Cheze 1993) suggested that variations in rainfall yield per flash for different periods of a given storm would be similar to the variation in rainfall yield per ground flash found for different

storms and storm systems by Kinzer (1974), López et al. (1991), and Holle et al. (1994).

7.19.2. Reasons for variations in the rainfall yield per flash

Part of the reason for the wide variation in measured rainfall yield per flash is caused by variations in the yield per flash as a storm evolves and by storm-to-storm variations in the duration of different stages of storm evolution. For example, in Arizona Watson et al. (1994a) found that ground flashes sometimes began to occur before rain was detected at the surface, and ground flash

Table 7.6. Rainfall yield per ground flash for individual storms or storm days.

Study	Location	Cases	Yield (10^7 kg per ground flash)	Average Yield (10^7 kg per ground flash)
Battan (1965)	Arizona	52	0.3–30	3
Kinzer (1974)	Oklahoma	1[a]	2	2
Grosh (1978)	Illinois	1	3	3
Maier et al. (1978)	Florida	22	0.3–90	10[e]
Piepgrass et al. (1982)	Florida	2	2	2
Buechler et al. (1990)	Tennessee Valley	21	0.9–20	4
Buechler and Goodman (1991)	Florida	2	6–30	18
Williams et al. (1992)	Darwin, Australia	43[b]	9–100	50[e]
Williams et al. (1992)	Darwin, Australia	6[c]	200–3000	800[e]
Holle et al. (1994)	central United States	4[d]	30–80	50

[a]From the selected portion of a squall line; [b]Continental storms (as opposed to monsoon storms); [c]Monsoon storms; [d]Mesoscale convective systems; [e]A rough estimate from graphed data.

rates often decreased much more rapidly from peak values than rainfall. A similar interpretation can be made of data from a storm near St. Louis, Missouri, presented by Grosh (1978). Data on squall lines from Holle et al. (1994), Kane (1993b), Nielsen et al. (1994), and Rutledge and MacGorman (1988) indicate that widespread, light rainfall can continue for several hours after deep convection has dissipated, and rainfall decreases much less rapidly than ground flash rates of the squall line. However, variations in the duration of different stages of evolution can account for only part of the variation in rainfall yields; estimates of rainfall per flash typically have varied by only two orders of magnitude for any given storm during most of the period that lightning was produced.

The relationship between rainfall intensity and ground flash rates also can differ for different types of storms, and this would be expected to affect estimates of rainfall per ground flash. In isolated storms in Florida, for example, Piepgrass et al. (1982) found that peak rainfall occurred within 10 min of peak flash rates. However, the rainfall of squall lines often peaked 1–2 h from the time of the peak ground flash rate (e.g., Rutledge and MacGorman 1988, Holle et al. 1994, Kane 1993b).

Nonetheless, variations in the relationship between rainfall intensity and ground flash rates are not just a matter of different timing of rainfall and lightning for different storm types, but also are affected in other ways that are a function of the climatological region being studied. In a climatological study of lightning from April to September in Oklahoma and Kansas, Reap and MacGorman (1989) found that the number of ground flashes produced by a storm tended to increase as the peak base-scan reflectivity of a storm increased. Few ground flashes occurred in storms whose maximum base-scan reflectivity was small.

However, in mountainous regions of the western United States studied by Reap (1986), ground flashes frequently occurred in storms whose base-scan reflectivity was small or undetectable. The lower troposphere tends to be drier in this region than farther east, and this dryness often caused much of the precipitation to evaporate before reaching ground (e.g., Colson 1960). Furthermore, in another study of part of this region by Watson et al. (1994b), the number of ground flashes was largest on days in which 65–70% of surface stations reported rainfall and tended to be smaller on days when rainfall was more widespread.

In Florida—in contrast with the western United States—Maier et al. (1978) found that it was unusual for ground flashes to occur in storms with small base-scan reflectivities. The number of ground flashes produced by a storm tended to increase as rainfall increased up to a value of roughly 8×10^{10} kg of rainfall and then to decrease for larger values of rainfall. Thus, there are both similarities and differences in the relationship of ground flashes and rainfall in any two of these regions (Great Plains, western United States, and southeast coastal region).

The effect of atmospheric instability on rainfall yield per ground flash also can differ from one climatological region to another. Buechler et al. (1990), for example, found that the rainfall yield per ground flash appeared to be smaller on days with larger CAPE during five storm days in the Tennessee Valley. Williams et al. (1992) inferred that monsoon storms in northern Australia tended to have smaller values of CAPE and much larger rainfall yields per ground flash than continental storms; thus, the dependence on CAPE was similar to what was found in the Tennessee Valley. In Florida, however, Buechler and Goodman (1991) analyzed two storm days with similar values of CAPE that produced storms having much different rainfall yields per ground flash, and López et al. (1991) found only a weak correlation of rainfall yield per ground flash with CAPE for all storms observed during one summer.

López et al. pointed out that, during the warm season on the Florida coast, CAPE often is similar on days with and without thunderstorms and that other environmental characteristics are important in determining the character of convection. For example, they found that the direction of prevailing winds at heights between 1000 and 700 mb had a large effect because of interactions with the sea breeze along the coast. Storms that occurred on days when the prevailing wind blew from the northeast had a median rainfall yield per ground flash that was one to two orders of magnitude greater than the median for days when the wind blew from each of the other quadrants of the compass.

We have noted that the largest values of rainfall per ground flash occurred in storms that produced few ground flashes. The final extrapolation of this trend is that many storms produce rain at the surface, but no ground flashes. Buechler et al. (1990), for example, cited a storm system that produced a flood, but was not detected by a ground flash mapping system. The lack of lightning may include cloud flashes, too: Zipser and Lutz (1994) and Zipser (1994) documented cases in which tropical thunderstorms extending high above the $-20°C$ isotherm produced no lightning. Furthermore, rain showers and winter storms that produce no lightning are common at middle latitudes.

Therefore, a basic problem for using lightning to estimate precipitation is that precipitation can be produced by processes that contribute too little to the electrification of a storm to produce lightning. Because this is so we would expect that, even when storms do produce lightning, some precipitation growth would contribute little to the production of lightning and so would increase the precipitation yield per flash.

The existence of processes that produce precipitation, but little if any lightning, can interfere with climatological relationships between rainfall and lightning, which one might expect to be more robust than relationships for individual storms. For example, in a study of tropical storms, Zipser (1994) noted that large regions of the tropical oceans receive heavy rainfall, including rainfall from tall storm systems, with little if any lightning. In one case, a small atoll near the equator received over 500 mm of rain in 2 months, but no thunder or lightning was observed.

Precipitation unrelated to lightning also influences climatological relationships at middle latitudes. In a study of the annual rainfall and the annual number of ground flashes statewide during a 6-year period in Oklahoma, for example, MacGorman et al. (1997) found that both the sign and relative magnitude of year-to-year changes in rainfall and ground flash counts were sometimes different (Fig. 7.60). Also, unpublished data from Oklahoma showed that from the warm season to the cold season the percentage decrease in monthly ground flash counts was much greater than the percentage decrease in monthly rainfall.

The fact that precipitation can form without producing lightning does not depend on our knowledge of

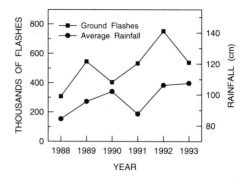

Fig. 7.60. Annual rainfall and annual number of ground flashes in Oklahoma. Annual rainfall is given as the average of the annual rainfall amounts received at every measuring site in the state. (From MacGorman et al. 1997, with permission.)

any charging mechanism; it is an empirical fact. However, to understand why particular storms do not produce lightning, we have to understand the mechanisms by which thunderstorms gain significant charge. In some respects this is discouraging because we know there are many processes by which hydrometeors can gain charge. Furthermore, there are gaps in our knowledge of almost every charging process, and we may be unaware of some processes that produce charge in thunderstorms.

However, if we assume that the noninductive graupel-ice mechanism is essential to producing lightning, the resulting predictions are consistent with a number of observations of storms that produce little or no lightning. (Note that the hypothesis is not that graupel-ice interactions are the only process relevant to electrification, only that they are essential for lightning. Other mechanisms still could contribute to electrification or may even produce charge that contributes to lightning production in some regions of a storm, as may be the case with screening-layer charge.) If the graupel-ice mechanism is essential in most storms, then precipitation growth that involves only liquid hydrometeors or involves cloud ice particles in the absence of either liquid water or larger ice particles could occur without causing lightning. Thus, precipitation could form without producing lightning in warm, shallow storms during the warm season or in some storms with ice, but no liquid water, during the cold season, as has been observed. Furthermore, it would explain how oceanic regions in the tropics in which annual rainfall from storm systems is large can have little lightning. Devlin (1995) noted that 85-GHz brightness temperatures measured by satellite indicated that oceanic storm systems in the tropics tended to have less graupel than continental storm systems. She suggested that the resulting impediment to the noninductive graupel-ice mechanism explained the lack of lightning in oceanic tropical regions.

7.19.3. Considerations for using lightning mapping systems to estimate rainfall

Radar systems detect rainfall more directly than lightning mapping systems, and so are preferable for rainfall measurement when radar data are available. However, even though lightning data have inherent limitations for rainfall detection, lighting data still could be useful in such applications when radar data are unavailable. For example, in mountainous regions susceptible to flash floods, radar coverage often is obstructed by the terrain. In these regions, lightning mapping systems readily indicate at least the presence of thunderstorms whenever lightning occurs. With further development testing and refining previous work and adjusting results for regional variations, it should be possible to make rough estimates of rainfall for storms that produce sufficient lightning. (Part of the development would be to determine the minimum number of flashes that could be considered sufficient.) A study of eight flash floods in Arizona by Holle and Bennett (1997) suggested that many flash floods in that region can be detected by using criteria based on the duration of ground flash activity. Lightning data also might be incorporated with satellite data to improve global rainfall estimates based on satellite imagery, by identifying which regions under cloud shields have convection deep and vigorous enough to be a thunderstorm.

We do not intend to imply that radar is superior to lightning mapping systems in all respects for detecting weather hazards. Warnings for activities susceptible to direct lightning damage can be provided more reliably and cheaply by lightning mapping systems. Lightning mapping systems also excel in identifying which storms are thunderstorms. Furthermore, the long range of VLF and ELF mapping systems discussed in Chapter 5 make them well suited for detecting thunderstorms over much longer ranges than can be achieved by a radar (and with more reliable thunderstorm identification than from cloud imaging systems carried by geostationary satellites). Lightning mapping systems have been used extensively by the British Meteorological Office to provide long-range, oceanic coverage (Lee 1986).

7.20. TYPICAL EVOLUTION OF LIGHTNING ACTIVITY RELATIVE TO STORM EVOLUTION

Although previous sections have examined how lightning is affected by various storm properties, we have not yet described how lightning activity typically evolves during the lifetime of a storm. Such a description is one of the principal tasks of this chapter. However, it and the conceptual models in the next section were deferred to the end of the chapter, because providing a reasonable treatment of each of the many diverse observations that must be incorporated in our description would have detracted considerably from the flow of the material. We hope that one benefit of deferring these topics is that now they can serve the additional purpose of integrating some of the observations of lightning that we have discussed separately.

Workman et al. (1942) suggested that the first flash typically occurred near or shortly after the time that the radar-detected storm height peaked, but the timing of the first flash relative to maximum storm height has varied considerably in other studies (Grosh 1978, Christian et al. 1980, Goodman et al. 1988a, b, Dye et al. 1989, Williams et al. 1989, Michimoto 1991, Kitagawa and Michimoto 1994). Christian et al. (1980) found that lightning activity ended when the radar echo reached its maximum height. More often, the first flash has occurred during vigorous convective growth, and the maximum flash rate has occurred approximately when the storm height peaked (Mazur et al. 1986, Goodman et al. 1988a, b, Williams et al. 1989, Weber et al. 1993) (Fig. 7.61).

Previous sections have discussed how lightning activity depends on complex interactions of many storm properties. Thus, it is likely that several properties of a given storm affect the timing of its lightning production relative to storm evolution. For example, if there is a typical threshold of storm height for lightning, such as the 9-km threshold suggested by Buechler and Goodman (1990), it may be that the maximum storm height in the New Mexico storms studied by Workman et al. (1942) was approximately that height, thereby making the beginning of lightning activity coincident with the maximum height. Also, the speed and structure of updrafts may affect the intensity and timing of electrification as suggested by Randell et al. (1994) and Solomon and Baker (1994). Takahashi (1984) modeled how grossly different microphysical particle development from one storm to another can affect the height and stage of storm development at which intense electrification first occurs. This is by no means an exhaustive list of storm characteristics that can affect the timing of electrification relative to storm evolution.

The first flash in an isolated thunderstorm usually is a cloud flash. In most storms, ground flashes tend to occur later in a storm's life cycle, after several cloud flashes have occurred (Workman et al. 1942, Livingston and Krider 1978, Krehbiel 1981, Goodman et al. 1988a, Williams et al. 1989, Maier and Krider 1986). In these referenced cases, the first ground flash typically occurred 5–15 min after the first cloud flash. However, during early stages of isolated storms that rapidly became severe on the Great Plains of the United States, there sometimes were few, if any, ground flashes for the first 30 min to 3 h of the storm's lifetime (e.g., Rust 1989, Shafer 1990, Keighton et al. 1991, MacGorman and Burgess 1994).

We can find only one storm mentioned in the scientific literature in which the first flash was a ground flash. In his overview, Krehbiel (1986) noted that the first flash was a ground flash in one small, mountain thunderstorm. He also has seen a few other cases in New Mexico and Florida (Paul Krehbiel, personal communication). Likewise, an analysis of field mill data at the Kennedy Space Center on the Florida coast showed that a ground flash was the first flash detected in several storms (Michael Maier, personal communication). However, no other information about characteristics of the storm and lightning activity have yet been presented in the published literature for any of these cases.

When ground flash rates have been examined relative to reflectivity structure, they have tended to peak as the reflectivity core descended to middle and lower levels of the storm, often after cloud flash rates peaked (Larson and Stansbury 1974, MacGorman et al. 1989, Goodman et al. 1988a, 1989, Williams et al. 1989). However, sometimes the descent of reflectivity cores occurred before ground flash rates increased, apparently because the descent was faster than could be explained by falling precipitation (MacGorman et al. 1989, Nisbet et al. 1990). Ground flash occurrence and the descent of reflectivity cores often are accompanied by the formation of downdrafts. Watson et al. (1987, 1991) noted that ground flash activity often occurred shortly after downdrafts were first detected by their divergence signature in surface winds during storms on the Florida coast, and Goodman et al. (1988a, b) and Williams et al. (1989) noted that ground flashes often occurred at roughly the time of microburst downdrafts in storms with heavy rain in Alabama (Fig. 7.61).

7.21. CONCEPTUAL MODELS OF INFLUENCES ON FLASH RATE AND TYPE

7.21.1. Flash rate

Our understanding of factors that govern flash rates and type is extremely limited. Of course, the rate at which large-scale charging occurs is a major factor in governing flash rates, but the geometry, distribution, and relative magnitudes of charge also should have an influence on flash rates and may have a dominant influence on flash type. Conceptual models have been offered to help explain how observed trends in lightning flash rates and type depend on the evolution of gross characteristics of thunderstorm charge distributions. For example, it appears reasonable that the height and separation of charge regions influence both flash rates and type.

Some insight can be gained about this issue by considering a vertical dipolar distribution of charge. As discussed in Chapter 3, if any mechanism relying on collisions of precipitation is responsible for electrification, then colliding pairs of hydrometeors do not contribute electric energy to a storm until they move apart, and whether they increase or decrease the storm's electric energy as they move apart depends on the preexisting electric field. As the oppositely charged particles begin to move apart early in a storm, the distance between the centers of regions of opposite net charge also grows. Initially, the preexisting field is weak everywhere, so the electric energy of the storm increases.

MacGorman et al. (1989) pointed out that, initially,

when the distance between charged regions is small, less charge is required in each region to create a large electric field, because the electric field is inversely proportional to the square of the distance from charge. As the charge centers move farther apart, more charge is required to create the same electric field magnitude. Thus, it should require a smaller charging rate to have large flash rates when charge regions are closer together. At the same time, only a relatively small volume of net charge can be unmasked when thunderstorm charge centers are close together, and the volume of net charge can increase as charge centers move farther apart.

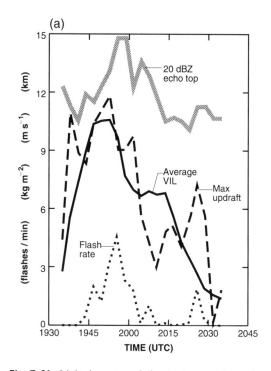

Fig. 7.61. Lightning rates relative to storm evolution in three storms. (a) Total flash rates relative to the maximum height of 20 dBZ, maximum updraft, and average Vertically Integrated Liquid water (VIL) in a Florida storm. (From Weber et al. 1993, with permission.) (b) Total and ground flash rates relative to storm parameters inferred from radar data for an Alabama storm. The top panel shows maximum reflectivity at each height as a function of time. In the middle panel, the unshaded histogram indicates total flash rates, while solid bars indicate the time of occurrence for six ground flashes. The bottom panel shows radar-inferred storm mass, rainfall, and the maximum difference in outflow velocities across the downdraft core at the surface. (From Goodman et al. 1988a: American Geophysical Union, and Goodman et al. 1989, with permission.) (c) Flash rates relative to other storm parameters for another Alabama storm. Storm height and the maximum difference in surface outflow velocities across the downdraft were inferred from radar. (From Williams et al. 1989, with permission.)

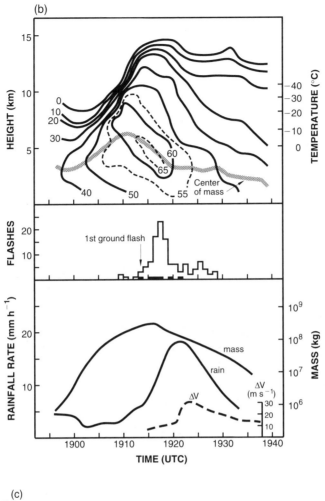

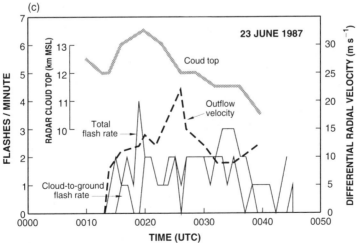

Fig. 7.61. (continued)

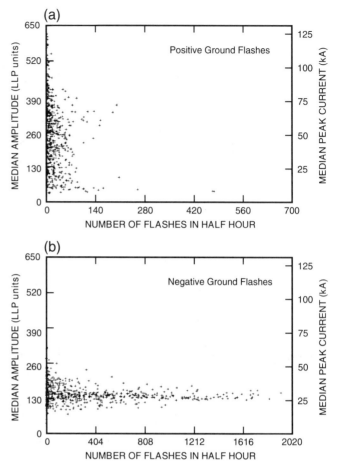

Fig. 7.62. Median amplitude (in LLP units) of the range-normalized signal from first return strokes as a function of number of flashes per 30-min period for 25 mesoscale convective systems occurring in the central United States in 1986. Data points indicate values for contiguous 30-min periods throughout the life cycle of each storm system. The peak current scale was calculated from the median amplitude scale by using the relationship found by Idone et al. (1993). (a) positive ground flashes; (b) negative ground flashes. (After Morgenstern 1991, with permission.)

These two trends (in the volume of charge and in the amount of charge required to reach a large electric field) lead us in this simple situation to expect increasing rates of flashes neutralizing relatively small charge magnitudes early in the electrically active stage of storm evolution, followed by decreasing rates of flashes neutralizing increasing charge magnitudes as the storm matures. Similarly, Williams (1985) suggested that the charge neutralized per flash in storms with high flash rates tends to be smaller than in storms with smaller flash rates. He cited storms studied by Livingston and Krider (1978), Lhermitte and Krehbiel (1979), and Williams (1981) as examples qualitatively demonstrating this behavior.

Although the relationship suggested by Williams (1985) may be true of cloud flashes (more data are needed for verification), it does not appear to be generally true for ground flashes. Morgenstern (1991) analyzed the distribution of peak signal strengths (which are approximately proportional to peak current) in an ensemble of distinct 30-min periods during the lifetime of 25 mesoscale convective systems. As shown in Fig. 7.62, the median peak current of negative ground flashes

was approximately 25–30 kA regardless of the number of flashes in the period, although there was greater scatter among periods with relatively few flashes, as would be expected statistically. The number of positive ground flashes, being smaller, made statistics less reliable, but at larger flash rates, positive ground flashes appeared to have median peak currents in one of two ranges: either 9–15 kA or 50–70 kA.

7.21.2. Flash type

The height and separation of thunderstorm charge centers also would be expected to have a relatively small influence on flash type. Since the electric field is largest between the main thunderstorm charges when they are close together, a thunderstorm with a vertical dipole would tend to produce predominantly cloud flashes at this time. As the distance between charges in the vertical dipole increases, the electric field magnitude below the negative charge becomes somewhat more comparable to the field magnitude between charges, and ground flashes would be expected to become slightly more probable. (Remember that image charges across the conducting surface of Earth contribute to the electric field.) An example of the effect of varying the height of charges is shown in the top two panels of Fig. 7.63. Note that the electric field magnitude below the negative charge would not equal the electric field magnitude between the positive and negative charges until the height of the negative charge was less than half the separation between the two charges.

Of course, a thunderstorm charge distribution is more complicated than a simple vertical dipole, so it would be unreasonable to expect a simple measure of the height and separation of charge centers to be useful for predicting flash type or rates. Even if it were possible to map the net charge everywhere in a thunderstorm, defining charge centers in an irregularly-shaped charge distribution in a way that would have meaning as a simple index would be difficult. For example, the geometry of charge regions can affect the maximum electric field magnitude produced by a given amount of charge (e.g., a sphere of charge produces a larger electric field at its lower boundary than a thin, horizontal disk having the same charge density and volume). Also, departures of both geometry and charge density from a symmetric, uniformly charged region may cause electric field variations over small distance scales that produce a much larger electric field than the larger scale average.

Furthermore, it is likely that other aspects of the charge distribution also would affect lightning activity. For example, several investigators (Malan and Schonland 1951, Clarence and Malan 1957, Workman 1967, Jacobson and Krider 1976, Marshall and Winn 1982, Jayaratne and Saunders 1984, Krehbiel 1986, Williams et al. 1989, Williams 1989) have noted that, if a small lower positive charge forms under the negative charge tapped by ground flashes, it can make the electric field below the negative charge larger and so can enhance initiation of ground flashes. (Note that adding a small positive charge below the negative charge in the lowest panel in Fig. 7.63 makes the electric field magnitude larger below the negative charge.) Therefore, several researchers have suggested that the delay between cloud flashes and ground flashes in a storm is caused by the time it takes for precipitation to descend to temperatures warmer than roughly -10 to $-15°C$, where the precipitation can begin to form a lower positive charge by the noninductive graupel-ice mechanism.

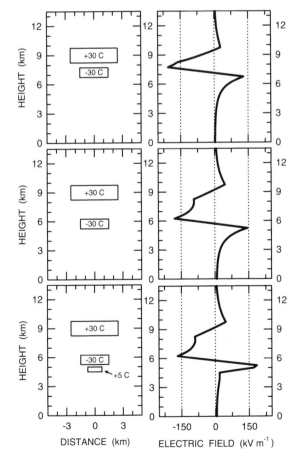

Fig. 7.63. Simple example of variations of the electric field profile due to changes in the relative heights of two oppositely charged regions and to the addition of a small, lower positive charge. Two disks, each containing approximately 30 C, are vertically separated in all panels. The upper positive disk is 1.5-km thick, 5 km in diameter, and centered at a height of 9 km in all panels. The negative disk is 1-km thick and 3 km in diameter. (a) The negative disk is centered at a height of 7 km. (b) The negative disk is centered at a height of 5.75 km. (c) Same as panel b, except that $+5$ C has been added in a disk 0.5-km thick, 1.5 km in diameter, and centered at a height of 4.75 km.

It also has been suggested that the probability of ground flashes is increased by a tendency for the storm to gain net negative charge with time. Solomon and Baker (1994) pointed out that the divergent outflow from the updraft near the top of the storm advects positive charge away from storm cells and so would contribute to the tendency for cells to gain negative charge with time. This is balanced to some extent by falling precipitation carrying negative charge out of the storm. Krehbiel (1986) suggested that a negative surplus would occur because air at the top of the storm is much more conductive than at the bottom of the storm. At the top of the storm, electric currents in the clear air above the storm deposit negative charge on the upper cloud boundary in response to the net positive charge in the upper part of the storm. Less positive charge is deposited by electric currents on the base of a storm, because the ohmic conductivity is much smaller at the base of a storm than at the top, so time scales to form a screening layer are much longer at the base (on the order of 10 min) and are comparable to time scales for significant changes in the air mass at cloud boundaries.

The mechanism by which currents to the top of a storm affect flash type may involve cloud flashes between the main upper positive charge and the negative screening layer. Taylor et al. (1984) observed frequent high flashes that he hypothesized were occurring between the upper positive charge and negative screening layer. MacGorman et al. (1996) suggested that such flashes tend to deplete positive charge in the upper part of the storm and consequently make the electric field between the main positive and negative charge smaller. While electrification mechanisms continue to operate, the relative amount of negative charge increases, as the high lightning flashes reduce the main positive charge, but not the main negative charge. Eventually, the electric field below the main negative charge and between the main negative and positive charges approach comparable magnitudes, thereby increasing the probability for initiation of ground flashes.

The last potential influence on flash type we consider is a consequence of the hypothesis, suggested by McCarthy and Parks (1992), that flashes are triggered by cosmic rays, which are ubiquitous, but sporadic. If this hypothesis is correct, then cosmic rays may introduce erratic variations in flash rate and type. In particular, if the breakeven electric field is attained in several regions of a storm, then the location of an energetic cosmic ray penetration may affect which region actually initiates a flash, and this may affect the type of flash produced.

7.21.3. Positive ground flashes

Little is known about why some ground flashes effectively lower positive instead of negative charge. Furthermore, relatively few reports have presented data on the charge structure of the storms that produce positive ground flashes or on the location in the storm of the charge neutralized by them. Takeuti et al. (1978), Brook et al. (1982), and Fuquay (1982) observed that the gross charge structure of thunderstorms with positive ground flashes still appeared to have positive charge over negative charge, as when negative ground flashes occurred. Pierce (1955b) suggested that positive ground flashes are initiated from the upper positive charge of thunderstorms after much of the lower negative charge has been removed by negative ground flashes. Takeuti et al. and Brook et al. suggested that positive ground flashes neutralized the upper positive charge because the upper positive charge extended horizontally beyond the lower negative charge and so was no longer shielded from the ground by the negative charge. Williams et al. (1994) suggested that positive ground flashes during the dissipating stage of storms are caused by the formation of increasing positive charge in the lower region of a storm, caused by microphysical particle interactions or the descent of particles during the end of storm oscillation observed in electric field records at the ground beneath storms (discussed in Sec. 7.2).

As discussed in Section 7.16.4, Krehbiel (1981) analyzed the charge neutralized by three positive ground flashes that occurred in the anvil or stratiform region of dissipating storms. Long, horizontal channels propagated from beyond the range of the measurements and removed positive charge from a layer in the radar bright band near the 0°C isotherm, where frozen precipitation was melting. These horizontal flashes produced a vertical channel to ground near heavier embedded precipitation and transported positive charge from a higher region than the horizontally propagating channels. Krehbiel also pointed out that horizontal channels from all three positive ground flashes propagated through the same general region of the same storm, so that positive charge appeared to be regenerated by some local charging process, which he suggested might involve the melting process. For ground flashes with long, horizontal channels propagating through layers of charge, one possibility is that the polarity of the ground flash is governed primarily by the polarity of charge in the layer along which it propagates. This could be tested in cases in which both positive and negative ground flashes occur from roughly stratified charge regions, as in some mesoscale convective systems. If true, the long, horizontal flashes that lower negative charge to ground would propagate along a negative layer.

Kitagawa (1992) suggested that one reason positive ground flashes in winter storms and stratiform regions tend to lower more charge to ground than is typical of negative ground flashes is that horizontally extensive charge regions store more charge than charge regions in deep convection. It requires more charge to produce a given magnitude of the electric field when charge is stored in two oppositely charged disks than when it is stored in two oppositely charged spheres of comparable volume and charge density to the disks. Furthermore, the amount of charge that can be stored increases with the horizontal area of the layer. Thus, more charge should be available to ground flashes that occur from large layers of charge.

8

Observations of the Electrical Characteristics of Thunderstorms

II. Severe, Winter, and Tropical Storms and Storm Systems

8.1. SEVERE AND SUPERCELL THUNDERSTORMS

8.1.1. Introduction

The term "severe storm" is used widely in the popular, and sometimes in the scientific, literature to mean, simply, a very strong storm, with broad latitude as to what is considered strong enough to be severe. Usually, the term is used in connection with some type of hazardous weather produced by the storm. A *severe thunderstorm* is defined in U.S. National Weather Service operations as one producing any of the following at the ground: a tornado, wind ≥ 26 m s^{-1} or hail having a diameter ≥ 1.9 cm (e.g., Johns and Doswell 1992). (We have converted the English units used by the National Weather Service to metric units.) This is the definition of a severe storm that we use in this book. Notice that lightning is not one of the hazards that makes a storm severe under this definition.

This section primarily will examine isolated severe storms. Storm systems that often produce severe weather will be discussed separately in Sections 8.2 and 8.5. Although many types of storms produce severe weather, including tornadoes and large hail, a class of storms called *supercell storms* (sometimes called simply *supercells*) produce most, if not all, of the more violent tornadoes and much of the large hail reported in the United States. Thus, supercell storms will be mentioned prominently in several parts of this section. Because supercell storms have some unique character-

istics, and their study has spawned specialized nomenclature, we will summarize some fundamental characteristics of the different types of supercell storms before examining the electrical properties of severe storms. For additional information about supercell storms, see reviews such as in Browning (1977) and Church et al. (1993).

Early studies based on radar reflectivity data (e.g., Browning 1965) identified several characteristics of storms categorized as supercell storms: Storms appeared to have a single cell that typically lasted 2–6 h and had strong reflectivities (≥ 55 dBZ) and rotating updrafts (vertical velocities often >40 m s^{-1}) for much of that time. The strong inflow and updrafts of a supercell storm often produced a pronounced notch of weak reflectivity intruding into regions of stronger reflectivities at low to middle altitudes. Larger reflectivities also extended over the top of the intrusion. Strong updrafts sometimes extended the weak-echo region far enough upward that the upper part was bounded horizontally on all sides by stronger reflectivity; now, this structure is called a *bounded weak echo region* (*BWER*). In the vicinity of the weak-echo region, a rotating cloud called the wall cloud, a manifestation of the rotating updraft, often was observed to protrude below cloud base. The overall structure of a supercell storm typically changed little throughout much of its lifetime, and so was said to be in a quasi-steady state. A sketch of a supercell is shown in Fig. 8.1.

Later studies found that the reflectivity structure of many long-lived severe storms with rotating updrafts differed significantly from the "classic" structure iden-

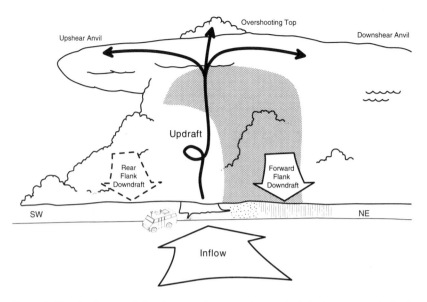

Fig. 8.1. Sketch of supercell thunderstorm. Important features include the strong updraft, elevated reflectivity core (top of shaded region), important downdrafts, and—below the main cloud base—the wall cloud, tornado, a region of large hail, and a wide region of rain. A mobile laboratory is shown in a preferred position for launching instrumented balloons into the updraft region.

tified by early studies. Other traits of these storms, such as lightning flash type and rates, hail size, tornado production, and rainfall rates, also varied considerably from storm to storm. Analyses of Doppler radar data (e.g., Lemon and Doswell 1978, Brandes 1984) and numerical cloud models (e.g., Weisman and Klemp 1982, Rotunno and Klemp 1985) suggested that the storm feature most characteristic of supercell storms is the *mesocyclone,* a region of persistent vertical vorticity (i.e., the horizontal component of wind rotates) that is a few kilometers in diameter, extends vertically over much of the storm depth, and forms in updrafts or at the interface of an updraft and downdraft. Consequently, Doswell and Burgess (1993) chose the mesocyclone as the single defining characteristic of supercell storms. Many tornadoes, especially the more violent ones, are produced by mesocyclones, so mesocyclone detection by Doppler radar is used to provide advance tornado warning.

In characterizing supercell storms, Doswell and Burgess (1993) found it useful to distinguish three subclasses, based on the amounts and spatial distribution of precipitation relative to airflow: (1) low-precipitation, (2) classic, and (3) heavy-precipitation supercell storms. Although this terminology has been the subject of some controversy, we will use it in this review, because each term is widely used in the scientific literature to indicate a particular part of the spectrum of supercell storms.

Low-precipitation supercell storms were so named by Bluestein and Parks (1983), because they produce relatively little rainfall, and yet display clear visual signs of mesocyclone rotation. Typically, a cloud tower with spiraling striations on its side encompasses the updraft and is topped by upshear and downshear anvil clouds. Absent is the large core of precipitation that normally extends to the ground downshear (typically to the north and northeast) of updrafts in other types of supercell storms. No precipitation cores are visible below middle levels. When large radar reflectivities are present at low to middle levels of the storm, rain gauge data and visual observations suggest that the reflectivity is caused by large but sparse hydrometeors (Bluestein and Woodall 1990). Although the striations and wall cloud are clearly visible indications of the mesocyclone in a low-precipitation storm, the mesocyclone sometimes occurs in weak reflectivities, and so is difficult to detect by radar. Low-precipitation supercell storms can produce tornadoes and large hail, but tornadoes are not common, and large hail usually falls outside the cloud tower.

As a group, classic supercell storms are closest to traditional conceptual models of supercell storms, such as presented by Browning (1965) and Lemon and Doswell (1978). These storms have moderate to heavy precipitation, indicated by large radar reflectivity (≥ 55 dBZ), extending to low levels downshear of updrafts and the mesocyclone. The mesocyclone sometimes produces an appendage of reflectivity called a hook echo at low heights, typically west or south of the weak-echo region but, if present, its reflectivity is generally less than that of the precipitation core. The mesocyclone typically occurs in weaker reflectivities and often displays a wall cloud beneath a precipitation-free cloud base upshear of precipitation cores. As the mesocyclone weakens and dissipates, however, it can fill with precipitation. If the mesocyclone has multiple sequential cores of circulation, old dissipating cores are replaced by new mesocyclone cores, which typically form in newer updrafts and weaker reflectivities (Burgess et al. 1982). Classic supercell storms can produce all types of severe weather, including violent tornadoes, but usually are not associated with flash floods.

Conceptual models of heavy-precipitation supercell storms have been formulated by Moller and Doswell (1988), Moller et al. (1990), Doswell and Burgess (1993), and Przybylinski et al. (1993). The primary characteristic of heavy-precipitation supercell storms is the mesocyclone's location in substantial precipitation for much of its lifetime (i.e., not just when mesocyclone cores are dissipating, as in some classic supercell storms). Because of the heavy precipitation, the mesocyclone is easily detected by radar; however, the heavy precipitation often obscures visible signs of rotation. On the right flank of the storm (i.e., typically south or southwest of the mesocyclone), there usually is extensive precipitation having reflectivity comparable to that of precipitation cores downshear of the mesocyclone. Heavy-precipitation supercell storms have a much greater variety of radar echo shapes and evolution than classic and low-precipitation supercell storms: Plan structures of entire storms include kidney bean, bowed, spiral, and S shapes. Furthermore, heavy-precipitation storms often have distinct, coexisting cells (i.e., they are multicellular). Like classic supercell storms, heavy-precipitation supercell storms produce all types of severe weather. However, they are more likely than classic supercell storms to produce torrential rainfall and flash floods.

An individual storm is not necessarily one type of supercell throughout its lifetime. Storms often exist for extended periods before or after they exhibit supercell characteristics. Furthermore, supercell storms can evolve from one type to another, usually from low-precipitation to classic supercell or from classic to heavy-precipitation supercell.

The following sections examine the electrical properties of severe storms, particularly of supercell storms. Most published data on this topic have described various properties of lightning, including flash rates, sferics rates, the polarity of current in ground flashes, and cloud flash percentages. The refereed literature presents a few measurements of E and particle charge in supercell storms. We know of no published measurements of particle charge or precipitation current made at the ground beneath severe storms.

One issue underlying much of the research on electrical properties of supercell storms is the extent to which electrical phenomena in supercell storms differ from phenomena observed in smaller isolated thunderstorms. Several of the investigators discussed in the following sections have suggested that the large flash rates observed in many supercell storms are caused by the unusually large updraft speeds in these storms. Vonnegut and Moore (1958), for example, compared "typical" and "giant" (probably supercell) thunderstorms and noted differences in electrical and other meteorological parameters between the two storm types. Besides having much larger flash rates, the "giant" storms were reported to have several luminous phenomena not usually observed in "typical" thunderstorms. Vonnegut and Moore suggested that the extraordinary electrification possibly was caused by the much larger size and updraft speeds of the "giant" storms. Using similar reasoning, Rust et al. (1981a) suggested that a simple scaling from small to large thunderstorms does not account for all the differences observed in the electrical phenomena of storms at the two extremes. For example, they noted possible links between severe storms and the occurrence of positive flashes to ground. Furthermore, MacGorman et al. (1989) hypothesized that the large cloud flash percentages produced by some supercell storms were caused by large updraft speeds increasing the height of the lowest region of substantial charge. This hypothesis will be discussed in more detail in Section 8.1.9.

8.1.2. Electric field at the ground

Since severe storms can be much larger in areal extent than airmass storms, we consider first what value a measurement of the electric field at the ground (E_{gnd}) has in determining anything about the electrification of the cloud. The measurement still indicates the dominant polarity of charge over the instrument, field changes from lightning, and perhaps the limiting value of E at the surface from corona-induced space charge. So E_{gnd} is as useful in analyzing big storms as small ones, with the possible exception that a particular field mill likely responds only to a small piece of a storm that becomes big. Noticeable in our (unpublished) records of E_{gnd} obtained on numerous severe storms is that the E_{gnd} often exhibits polarity reversals lasting a few minutes and that do not seem related to rain at the site. Severe storms often move rapidly across the terrain, and measurements of E_{gnd} beneath or near the precipitation core (or any other specific storm feature) are very difficult to obtain for extended periods of time in a constant storm-relative position. Our view is that we can tell little about the internal electrical nature of storms from the measurement of E_{gnd}; it does, however, remain a useful adjunct to other in situ measurements.

The above notwithstanding, we likely could learn something about the electrical characteristics (if any) of large tornado funnels if we could obtain close measurements. Not surprisingly, there are almost no measurements of E_{gnd} in the environment of, much less inside, tornadoes. The only known published records of E_{gnd} close to a tornado are those by Gunn (1956). He had eight sites with field mills in southern Kansas and reported that on May 25, 1955, the devastating Udall tornado probably came within 2 km of one of the sites. Our examination of the reproduced records indicates nothing remarkable: They show dominant negative charge overhead and numerous, frequent field changes from lightning. The records reveal a limiting value of $E_{gnd} \approx 5 \, kV \, m^{-1}$ at the site within 2 km of the tornado. Other sites in the area show that E_{gnd} may have had a limiting value of 10 kV m^{-1}. Also apparent in the records of two sites (the sixth and the eighth from the top of Gunn's Fig. 2) are reversals in the polarity of E_{gnd} lasting 2–3 min. Whether these are field excursions associated with precipitation is unknown.

We deviate slightly and refer you back to Chapter 6 on instrumentation and on the use of mobile laboratories in particular. Through the years of episodic field programs, we have made E_{gnd} measurements beneath severe storms, but they have never been analyzed extensively. In addition, the measurements often suffer from changing exposure to E of a mobile lab as it moves in and out of power lines, trees, overpasses, etc. There is at least one unusual recording of E_{gnd}. It was obtained beneath a mesocyclone with rotating wall cloud on 19 June 1980. It is shown in Fig. 8.2 with the lightning field changes removed. This record is also unusual in its large value of the maximum E_{gnd}. The large increase from 2–23 kV m^{-1} at 1823 CST was associated with the mobile lab being under the outer part of the wall cloud. The surface wind was flowing into the storm near the wall cloud and was consistently estimated to be $\geq 20 \, m \, s^{-1}$ with increases reported at 1830. Of course, the record is complicated by the relative motion of the storm and the mobile lab. However, parts of the record were made when the mobile lab was stationary; there is no obvious distinction between moving and stationary records. Regardless, the rotating wall cloud was close to the mobile lab at least from

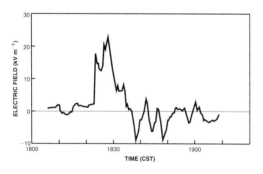

Fig. 8.2. First known record of E_{gnd} obtained near and beneath mesocyclone produced wall cloud. It was obtained with a mobile laboratory operated jointly by the National Severe Storms Laboratory and the University of Mississippi/Department of Physics on 19 June 1980. For simplicity, lightning transients have been removed from the record. The abrupt change in E at 1823 CST is associated with the rotating wall cloud moving over the mobile lab.

1820–1840 CST, with rotation observed overhead several times during and after the largest E_{gnd}. Rotation of the wall cloud was also overhead later when E_{gnd} had returned to typical values. From 1820 until at least 1840 there was frequent lightning. A very close ground strike occurred at 1829:15, but E_{gnd} was already decreasing then. There was no obvious connection between the lightning activity and E_{gnd}. It is perhaps worth noting that the only other time we see changes in E_{gnd} to high values is when we cross bridges over a large body of water whose surface limits the point-discharge ions available to screen the field. One speculation that might explain the increase and large E_{gnd} is that, since the field mill was about 2 m above ground, the strong winds swept any point-discharge ions away before they could screen the field to the normal extent (recall measurements such as those by Standler and Winn 1979). This argument is not convincing, because the wind stayed strong while E_{gnd} return to more typical values. Another possibility is that the negative charge aloft was lowering rapidly as the cloud top collapsed (the collapse was inferred from conventional radar).

Winn used a mobile field mill and several deployable, self-recording field mills to measure E_{gnd} in severe and tornadic storms during VORTEX-94 and -95, a project focused on the study of supercells and tornadoes (Rasmussen et al. 1994). Analyses of this study are underway, and we expect there will be interesting E_{gnd} and other accompanying meterological data from supercells.

8.1.3. E_{aloft} and total space charge density

We have tried to include a summary of all the published measurements of E_{aloft} that are clearly identified as being from severe thunderstorms. We begin with research conducted by scientists from Rice University who obtained a sounding though a severe storm in Oklahoma on June 10, 1983 (Byrne et al. 1987). This seems to have been the first published observations of the vertical profile of E inside a severe storm. The storm consisted of a line of cells trailed by about an 80-km wide area of stratiform precipitation. They described the storm as typical, since it fit the model of a multicell severe thunderstorm by Browning et al. (1976); the storm was in its mature stage during the sounding. Frequent lightning was observed. They used their coronasonde (see Sec. 6.2.3) to obtain the E sounding. The balloon initially went southeast in the outflowing gust front, came out the top of the outflowing air, and headed into the cloud in the inflow. Balloon tracking data did not begin until the balloon was at about 3.8 km. The remote tracking via a radio tracking antenna (GMD) introduced uncertainties in calculations of the balloon's position and ascent rate. The updraft was a few meters per second below 5.2 km MSL. The balloon flew in a downdraft of rapidly cooling, drier air between 5.2 and 6.2 km. Above that it returned to updraft whose peak

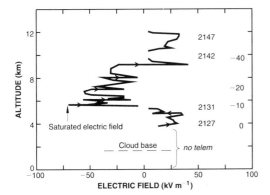

Fig. 8.3. Partial profile of vertical electric field, E_z, through multicell severe thunderstorm in central Oklahoma, 10 June 1983, 2116 CST. The temperatures (°C) were measured with the sonde used to track the balloon; E_z came from a calibrated coronasonde. Arrows on the E plot indicate lightning field changes of >25 kV m^{-1}; 17 smaller flashes were not shown. (After Byrne et al. 1987: American Geophysical Union, with permission.)

was 5 m s^{-1} at about 10 km. The balloon seemed to exit the storm at about 12 km, the radar top of the storm.

Byrne et al. stated their E sounding (Fig. 8.3) fit the dipole charge model, but they also noted that their data contained evidence for an upper screening layer. Using the criteria of Stolzenburg et al. (1994) for the use of Gauss's law to determine charges, we find that the gross features of the sounding indicate four regions of charge. Furthermore, we do not know what, if any, charge was in the lowest 2 km of the storm, because of a telemetry failure. They, as others since, have noted the presence of very shallow, high-density layers of charge embedded in a deeper but lower density layer. In a main negative charge region between about 5.2–5.6 km, $\rho_{tot} = -1.2$ to -2 nC m^{-3}. Near the top of that region, they found evidence of a layer 40–120 m thick with a $\rho_{tot} = -17$ nC m^{-3} at about -8°C. They noted clear evidence for a screening layer about 200 m thick at the top of the storm with a $\rho_{tot} = -1.5$ nC m^{-3}.

The same year as the above sounding, physicists from the University of Mississippi and the National Severe Storms Laboratory began their program of mobile ballooning (Rust 1989). Four soundings in severe thunderstorms were shown in Marshall and Rust (1991); no storm kinematic data were included. (The soundings are in their Fig. 1 and identified by year/day: 84156, 84157, 87148, and 83147.) Large peaks of both polarities were in the E profiles along with $|E_{max}| \leq$ 150 kV m^{-1}. From the E profiles in those four storms, Marshall and Rust inferred that the minimum number of charge layers was six. The largest total space charge density in significant charge regions that they found using Gauss's law was $+13$ nC m^{-3}.

Marshall et al. (1995a) reported on 11 E soundings in or near the convection of large storms that occurred

on the southern Great Plains. They used the 11 soundings to investigate the differences in electrical structure between small and large storms, to address the hypothesis of MacGorman et al. (1989), and to look for recurring patterns in the electrical structure of large storms. The storms were either multicells, supercells, or strong convection in a squall line. Marshall et al., as others before, used the balloon's ascent rate to estimate updraft. Since a balloon is buoyant, its ascent rate is greater than the vertical motion of the air in which it is rising. In still air, their balloons rose at 3–5 m s^{-1}, with the latter value more typical. Uncertainties in the estimate of true vertical air motion at the balloon came from possible unknown variations in the balloon's ascent rate caused by precipitation, as described by Davies-Jones (1974). To avoid the uncertainties in calculating the updraft speed, they plotted ascent rate instead. Where they estimated the updraft speed, they subtracted 5 m s^{-1} from the ascent rate.

Marshall et al. (1995a) divided the 11 soundings of E and accompanying thermodynamic variables into three groups. The soundings were named for a town or landmark near the launch site. For dates on which there were multiple launches from the same site, the sounding names include a flight number. In some soundings, they found a characteristic structure that they called "benchmark charge regions." Marshall et al. defined *benchmark charge regions* as two substantial charge regions that: had negative above positive charge, caused a significant horizontal V-shape in the E sounding, and were between 4–7 km MSL. Examples of their soundings and results follow.

(i) Soundings in weak updrafts. Soundings in this group had ascent rates mostly <10 m s^{-1}. These soundings were classified as being in weak updrafts. An example is the Wayne sounding from a multicellular storm that produced large hail (Fig. 8.4). As in all the E soundings they put in this group, the benchmark charge regions caused a horizontal V-shaped deflection in E; the tip of the V-shape is at 6.3 km in the Wayne sounding. The ascent rate of the balloon was mostly <10 m s^{-1}. The maximum E in the sounding was about $+115$ kV m^{-1}. As seen in the figure, Marshall et al. (1995a) identified 10 charge regions with densities ranging in magnitude between 0.1–2.4 nC m^{-3} and thicknesses ranging between 400–2600 m. These values are typical of those found in small, airmass thunderstorms.

(ii) Soundings in strong updrafts. Marshall et al. (1995a) defined this group as soundings with maximum ascent rates ≥ 15 m s^{-1} over a depth of ≥ 1 km. The $|E_{max}|$ ranged from 85–130 kV m^{-1}. These soundings had only three to five charge regions below 10 km. Also, the benchmark charges with their corresponding V-shape in the E profiles were not evident in these strong updraft soundings. They were typified instead by a large excursion in E and associated negative charge region between 6–10 km. Both the strong up-

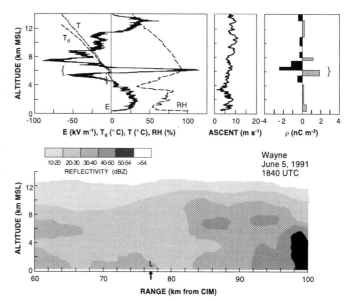

Fig. 8.4. Top: Sounding from behind the storm core near Wayne, Oklahoma, 5 June 1991, 1840 UTC. The left panel shows E, T_d (dewpoint), T, and RH (relative humidity with respect to water) versus altitude. The temperature data were faulty below ≈ 4 km. The braces denote benchmark charges. Ascent rate of the balloon and instruments is plotted in the middle panel. The right panel contains the main charge density regions. Bottom: Vertical cross section of radar reflectivity through the balloon launch site (L). The data are from the NSSL Cimarron (CIM) Doppler radar. (From Marshall et al. 1995a: American Geophysical Union, with permission.)

draft and elevated negative charge are obvious in the Prague sounding (Fig. 8.5). The sounding ended at 10 km because the instruments were destroyed by lightning. Only three charge regions were inferred up to the 10-km altitude. From the ascent rate of the balloon, they estimated 5–15 m s^{-1} updraft speeds, which increased with altitude. Tracking data placed the balloon in the reflectivity core aloft when lightning struck the instruments. This storm had produced a short-lived, very weak tornado about 20 min before the sounding was made.

Shown in Fig. 8.6 is another sounding in this group. The balloon was launched just ahead of a rotating wall cloud and was thought to have entered the mesocyclone. That electric field meter also was destroyed by lightning at about 10 km. The sounding appears similar to the Prague one in most respects. Marshall et al. (1995a) inferred a deep region of negative charge from 7–10 km in the updraft, whose maximum was ≈ 35 m s^{-1}. The lowest charge region was positive and in the rain below cloud base. The maximum value and shape of the ascent rate profile are nearly identical to the

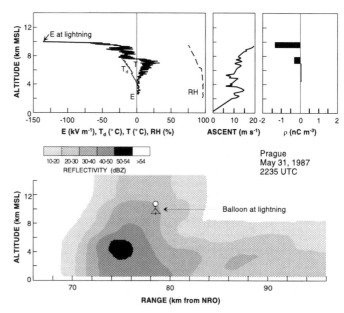

Fig. 8.5. Top: Sounding into tornadic storm near Prague, Oklahoma, 31 May 1987, 2220 UTC, about 20 min after the storm produced a small, weak tornado. The left panel shows E, T_d, T, and RH versus altitude. No data were obtained below 2 km. The arrow marks the E and balloon altitude when lightning destroyed it. Ascent rate of the balloon and instruments is plotted in the middle panel. The right panel contains the main charge density regions—to the greatest altitude at which the instruments acquired data. Bottom: Vertical cross section of radar reflectivity, with arrow showing balloon location when lightning struck. The data are from the NSSL Norman (NRO) Doppler radar. (From Marshall et al. 1995a: American Geophysical Union, with permission.)

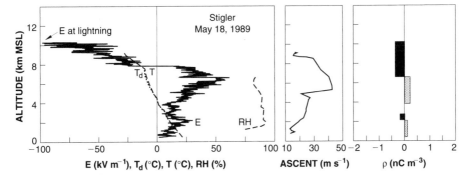

Fig. 8.6. Top: Sounding into tornadic storm near Stigler, Oklahoma, 18 May 1989, 2031 UTC. The left panel shows E, T_d, T, and RH versus altitude. The arrow marks the E and balloon altitude when lightning destroyed it. Ascent rate of the balloon and instruments is plotted in the middle panel. The right panel contains the main charge density regions—to the greatest altitude at which the instruments acquired data. There are no usable radar data. (From Marshall et al. 1995a: American Geophysical Union, with permission.)

Edmond profile in Davies-Jones (1974), which was obtained in squall line convection that had numerous reports of funnels.

(iii) Other soundings in convection. Of the two soundings through convection that did not fit well into the two groups above, one was very similar to the strong updraft group, but the other was very unusual. That sounding, Dalhart (Fig. 8.7), was made into a short line of storms that produced at least two tornadoes. (Marshall et al. (1995a) did not know if the storm cell that their balloon entered was tornadic.) Two of the most repeatable features in the E soundings of Marshall and Rust (1991) were a positive charge at upper levels capped by a negative charge at or near cloud top. In the Dalhart sounding, the polarities of these two charge regions were reversed.

By way of comparison, the E soundings through a few small thunderstorms (Marshall and Rust 1991;

Marshall and Marsh 1993; Marsh and Marshall 1993) often have revealed an especially simple charge structure with four charge regions that alternated in polarity with increasing altitude, beginning with positive charge at the bottom. This structure is similar to the tripole of Simpson and Robinson (1941) but with an additional charge at cloud top. Although also simple in appearance, the Dalhart storm charge structure had the opposite polarity to each of these four charge regions. Positive cloud-to-ground flashes were documented in this storm during and after the balloon flight, but there is no theory or observation that conclusively tell us that the thunderstorm charges must be inverted to change the polarity of ground flashes. The field changes of three lightning flashes seen in the E sounding at 6.2, 7.1, and 7.5 km were consistent with lowering positive charge to ground, although they could not verify that these flashes came to ground. They wondered if the opposite polarity of the Dalhart charge structure was as-

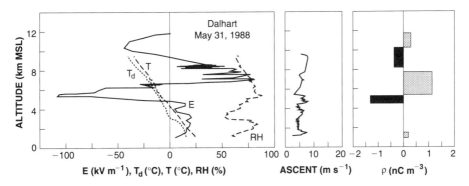

Fig. 8.7. Sounding into one storm in short line of storms near Dalhart in the Texas Panhandle, 31 May 1988, 0102 UTC. There are no usable radar data. (From Marshall et al. 1995a: American Geophysical Union, with permission.)

sociated with positive ground flashes. They had no other data to address this.

The findings showed that magnitudes of charge densities and thicknesses of charge regions in all 11 E soundings were similar to those in small, isolated thunderstorms. There was an apparent tendency for the maximum charge density magnitude in the weak updraft group to be greater than that in the strong updraft group. The number of charge regions in the weak updraft soundings fit with the 4–10 regions found by Marshall and Rust (1991) in some small isolated thunderstorms. In contrast, there were fewer charge regions in the strong updraft soundings below 10 km. In the strong updraft storms, three of the four balloons did not get above 10 km, and the fourth reached only 11.5 km, so Marshall et al. (1995a) could not determine the charge structure in the upper regions in the strong updraft storms. The Prague and Stigler tornadic storm soundings do not look like any E soundings from small, isolated thunderstorms because there were only two charge regions below 10 km in each. Their vertical velocity profiles in strong updrafts are similar in appearance to those in Davies-Jones (1974) and Bluestein et al. (1988, 1989). The link between strong updrafts and fewer charge regions below 10 km supports the hypothesis of elevated charge in strong updrafts (MacGorman et al. 1989), as the lowest significant charge was above 8 km and the updrafts were substantial.

The thermodynamic profiles in the strong updraft group were mostly close to saturation throughout, presumably because of the updraft. Several of the soundings in the other groups were also close to saturation throughout. The Dalhart sounding had a relative humidity mostly <75%, which was significantly drier than others. Prague, Wayne, and perhaps several other soundings showed a temperature inversion between 4–6 km. In at least three soundings, the temperature inversion was ≥5°C. Furthermore, the inversions were often coincident with a decrease in the ascent rate of the balloon, although most abrupt decreases in ascent rate were not accompanied by temperature inversions. The inversions could have been instrument errors (such as icing of the temperature sensor), but the decrease in balloon ascent rate at the inversion was also indicative of a decrease in buoyancy of the balloon because it moved into warmer air.

We expect additional information from soundings made in severe storms during VORTEX. Preliminary reports were given by Stolzenburg et al. (1994a) and Stolzenburg et al. (1996). Their preliminary interpretation that some charge regions were either elevated or absent in strong updrafts was similar to the interpretation of Marshall et al. (1995a). An analysis of all the soundings available in deep convection of supercell storms was given by Stolzenburg (1996). (Her analysis included 16 soundings—some partial—from VORTEX plus those reported by Marshall et al.) Stolzenburg inferred from the electric field profiles in the strong updrafts within convective regions of supercells that the charge structure consisted of four charges stacked vertically and alternating in polarity. Positive charge was lower most. Outside of the strong updraft, but still in the convective region, she inferred a more complex charge structure, with up to six charges stacked vertically. Note that the analysis used was the one-dimensional version of Gauss's law and so could conceivably overestimate the number of charges as discussed in Section 6.2.4. She also hypothesized a similarity among the electric field and charge structure of the convective regions in supercells, MCSs, and small thunderstorms. The center height of the main negative charge center increased with updraft speed, so the center of the main negative charge does indeed appear higher in supercell storms. Stolzenburg's results, which are based on the total collection of soundings in supercells, raise interesting questions about the model of storm charge structure, and they support the hypothesis that charge location is directly affected by updraft speed (also see Sec. 8.1.9).

(iv) Soundings in anvils of severe storms. The first known profile of E in the anvil region of a multicell severe storm was a partial sounding described in Marshall et al. (1989). The balloon's location, 50 km from the closest core edge as it entered the storm, is depicted in Fig. 8.8. The relative humidity with respect to ice went rapidly from 65% to 100% as the balloon entered the cloud, and E changed rapidly (Fig. 8.9). The balloon went through two charge regions, one of each polarity. The lower charge region was 300-m thick and likely was a screening layer in the base of the anvil. The electric field meter stopped rotating, so the upper half

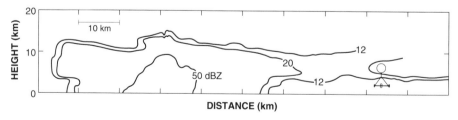

Fig. 8.8. Sketch of vertical cross section of radar reflectivity through a severe thunderstorm, 30 May 1982, in central Oklahoma. The balloon entered the anvil about 50 km from the core of the storm. (From Marshall et al. 1989: American Geophysical Union, with permission.)

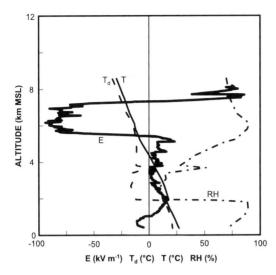

Fig. 8.9. Profile of E, T_d, T, and RH in anvil of a severe thunderstorm, 30 May 1982, in central Oklahoma. (After Marshall et al. 1989: American Geophysical Union, with permission.)

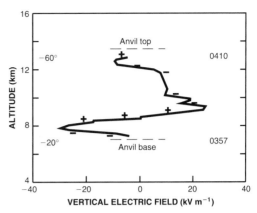

Fig. 8.10. Profile of the vertical component of E through severe storm anvil, 14 June 1983, at 0336 UTC in central Oklahoma. The sounding was 70 km upwind of the precipitation core, which had passed over the launch site earlier. (After Byrne et al. 1989: American Geophysical Union, with permission.)

of this anvil was unsampled. The $|E_{max}|$ was about 90 kV m^{-1}. A total space charge density of ≈ 2.5 nC m^{-3} of each polarity was calculated. This magnitude is twice that of the anvil in the airmass thunderstorm discussed in Section 7.8, but whether such differences typically are due to storm size is unknown.

An anvil from a supercell was probed by Byrne et al. (1989). The storm was in central Oklahoma and had a mesocyclone and a rotating wall cloud. The storm kept the radar characteristic of a mature supercell for about 4 h, and Byrne et al. made two flights during this period. They flew coronasondes and obtained one partial and one full sounding through the upwind anvil. The first balloon they launched was just on the back (west) side of the precipitation core. The first flight was about 20 km and the later one 70 km upwind of the precipitation core. The radar reflectivities through which the profiles were obtained in the anvils were <22 dBZ.

As the balloon entered the anvil cloud base, the relative humidity rose rapidly. Coincident with this was a rapid change in E, showing the presence of a lower negative screening layer of $\rho_{tot} \approx -0.4$ nC m^{-3}. Above that was positive charge of about 0.2 nC m^{-3}. Above the positive layer was a negative one. Data were lost at 11.5 km on the first flight. The second flight also showed a screening layer in the bottom 800 m of cloud. The sounding (Fig. 8.10) showed both a positive and a negative internal charge region, above which was a screening layer. Not shown in Fig. 8.10 is an E_h in the positive layer, which they interpreted as granularities in the charge structure. We estimated $\rho_{tot} = 2.8$ nC m^{-3} in the positive layer; they calculated a $\rho_{tot} = -0.4$ nC m^{-3} in the negative layer. As they noted, the unex-

pected finding supported by both flights was that the internal charge of the anvil was not simply a positive charge sandwiched between two negative screening layers. From the limited profiles available, we still do not know how universal this anvil structure is. To explain the extra region of charge inside the anvil, Byrne et al. proposed that particles, that had originally been charged as part of the upper negative screening layer, had fallen slowly from the cloud boundary into the anvil interior. The cloud then rescreened itself on top, but this time with positive charge. Their conceptual model, shown in Fig. 8.11, seems a reasonable hypothesis, but one awaiting more tests. Verification will require E profiles along with storm-relative airflow, particle trajectories, and particle charge and size distributions.

8.1.4. Particle charge aloft

Almost no measurements of particle charge in severe thunderstorms exist. Because of this limitation, we reproduce here unpublished data from inside a supercell (from visual appearance) that produced funnel clouds in southeast Kansas near El Dorado on June 15, 1990. As part of instrumentation development work, we intercepted the storm and launched a balloon carrying a sonde, an electric field meter, and a particle charge- and size-measuring instrument (Sec. 6.4.2). The launch was after dark, but our last glimpse of the storm in daylight revealed an apparent wall cloud, which was very close to the ground. It was raining heavily at the launch site, and the storm was producing frequent lightning. The sounding data are shown in Fig. 8.12. The ascent rates of <5 m s^{-1} show that the balloon's ascent was slowed down by precipitation loading of the balloon, a downdraft, or both. At 9.8 km, the balloon started to descend.

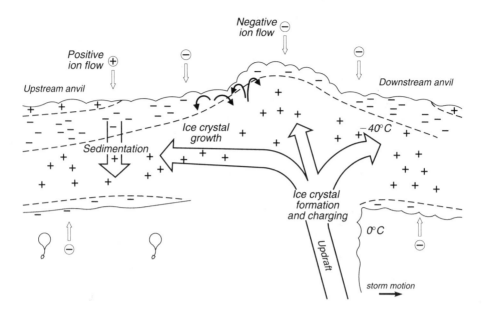

Fig. 8.11. Model of charge distribution in a severe storm anvil. The charge regions shown have all been observed. The circles with + and − represent small ions and their attraction to the cloud. The sedimentation of the upper negative screening layer to become an internal layer is new in this model. (From Byrne et al. 1989: American Geophysical Union, with permission.)

The particle charge and size distributions are shown in Fig. 8.13. The minimum detectable charge on that flight was 4 pC, so particles with $|q| < 4$ pC are plotted as zero charge. The particle charges are not different from what has been found in small and nonsevere thunderstorms. As with most distributions from small thunderstorms, there is no apparent charge-size correlation (Fig. 8.14). Though rare, little more can be done with these limited data, but we include them for future comparison if other particle measurements are ever made in a supercell.

8.1.5. Introduction to lightning observations

Most studies of lightning in severe storms have examined tornadic storms, and this preponderance of studies will be reflected in the remainder of our discussion of severe storms. Before discussing tornadic storms, however, we will examine lightning specifically relative to hail and to damaging wind other than tornadoes and tropical cyclones (e.g., hurricanes). Discussion of lightning relative to severe weather in large storm sys-

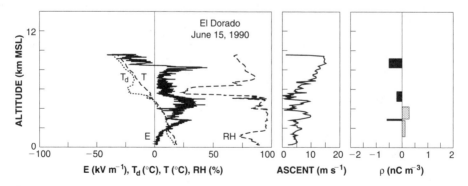

Fig. 8.12. Sounding into a severe thunderstorm near El Dorado, Kansas, 15 June 1990, 0211 UTC. The balloon only got to ≈10 km. There are no usable radar data. (From Marshall et al. 1995a: American Geophysical Union, with permission.)

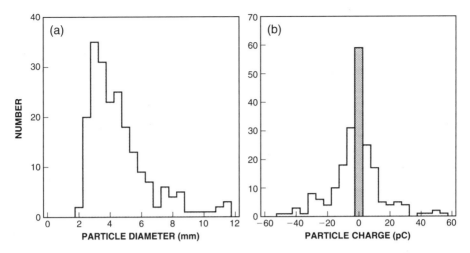

Fig. 8.13. Distributions of size and charge of precipitation particles inside severe thunderstorm near El Dorado, KS, 15 July 1990, as described by Bateman et al. (1990). The shaded bar in (b) denotes particles whose $|q| < 4$ pC, the minimum detectable. (Plot from Bateman, personal communication 1996.)

tems will be deferred to Section 8.2, and observations of lightning in tropical cyclones will be discussed in Section 8.5.

Prior to 1975, most quantitative measurements of lightning in severe storms examined the rates and characteristics of sferics, the electromagnetic noise radiated by lightning. These measurements were difficult to interpret because results depended partly on the characteristics of the sferics receiver, and because at best sferics instrumentation provided only the bearing to the lightning flash generating the sferics and often provided no location information at all. Furthermore, many of the sferics studies lacked radar data, and only one had any Doppler radar data. Equally important, most sferics studies occurred before there was much appreciation of the wide range of storm types that produce tornadoes and other severe weather. Thus, these studies made little systematic effort to relate the con-

siderable differences in the behavior of sferics rates among severe storms to the widely varying characteristics of the different storm types that produced them.

Better lightning data became available when systems were developed to map the location of lightning flashes, as described in Chapter 5. Beginning in the late 1970s, for example, various organizations began deploying newly developed systems for automatically mapping where lightning channels strike ground over ranges of a few hundred kilometers. Reap and MacGorman (1989) used ground strike data from a network covering much of the Great Plains of the United States to examine climatological relationships between ground flashes and severe weather. In Section 8.1.9, we will discuss the relationships that they found with positive ground flashes. However, the number of negative ground flashes produced by a storm had little relationship with severe weather: The probability of severe

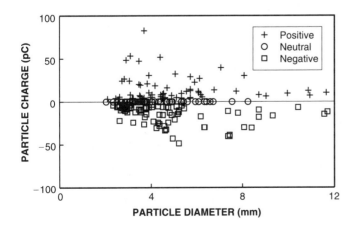

Fig. 8.14. Charge versus size for particles between 1–8 mm recorded up to $z \approx$ 10 km in a severe thunderstorm near El Dorado, KS, 15 July 1990, as described by Bateman et al. (1990). (Plot from Bateman, personal communication 1996.)

weather increased slightly for storms producing larger numbers of ground flashes up to 130 flashes per grid box (48×48 km^2) and then decreased as the number of ground flashes increased further.

The weakness of the trend found by Reap and MacGorman (1989) is at least partly a result of the considerable diversity of ground flash characteristics among severe storms. Although the rest of this section will review many studies of lightning in severe storms, readers should be aware that the number of studies with extensive data on both lightning and storm evolution is still small, so only a few early attempts have been made to develop hypotheses that relate the diversity of lightning evolution to the diverse characteristics and evolution of severe storms.

8.1.6. Lightning relative to damaging thunderstorm winds

Severe, straight line winds can be produced at the surface by a variety of storm types. Supercell storms frequently produce wind speeds at the surface large enough to be considered severe, but we are aware of no studies relating lightning flash rates or other lightning characteristics with supercell production of severe, straight line wind. Squall lines also can produce very strong winds at the surface, particularly when part of the squall line is bowed outward by downdrafts overtaking the line from the rear. In their study of a squall line with a bow echo, Schuur et al. (1991) found no obvious signature of the strong winds associated with a bow echo in observed lightning characteristics, except that there was a minimum in ground flash density and flash rates in the part of the squall line that was bowed. Once the bow feature began to appear, ground flash rates in that region decreased. A region of large ground flash densities initially just south of the convective region where the bow echo occurred shifted roughly 50 km farther south as the bow echo formed.

Most modern studies of lightning relative to damaging wind in isolated thunderstorms have examined microbursts, the surface winds caused by a transient strong downdraft diverging at the ground. Microbursts produce wind shears that are particularly hazardous to landing and departing aircraft. In the southeastern United States, microbursts often are accompanied by moderate to heavy rainfall, while in the drier environment of the western United States, they sometimes are accompanied only by light rainfall. As discussed in Sec. 7.20, Goodman et al. (1988a, b) and Williams et al. (1989a) found that microbursts in the southeastern United States tended to occur roughly 10 min after peaks in total lightning flash rates and were shortly after or roughly coincident with peaks in ground flash rates. Similar behavior of lightning flash rates relative to microbursts associated with moderate to heavy precipitation were found by Williams et al. (1989b) in the western United States and by Kane (1991) in the northeastern United States. However, Williams et al. (1989b) also found that storms in which microbursts were ac-

companied only by light precipitation produced much smaller lightning flash rates than storms with heavier precipitation. Thus, while lightning still preceded the microbursts associated with light precipitation, there was no obvious large peak in total lightning flash rates before the drier microbursts.

8.1.7. Lightning in hailstorms

There have been relatively few studies of relationships between lightning and large hail. Early studies examined lightning flash rates in storms that produced any hail, and many of the reported hail diameters in these studies were below the 1.9 cm threshold used by the U.S. National Weather Service to define severe hail. Shackford (1960) reported that no hail was observed at the surface from New England storms when flash rates were less than 10 min^{-1}, and that flash rates exceeded 100 h^{-1} in 60% of cases in which hail was observed. Blevins and Marwitz (1968) analyzed visually observed lightning stroke rates for storms in northeastern Colorado, although they noted that visual observations tend to undercount flashes and strokes. They noted that the probability of a storm producing hail ≥ 0.6 cm in diameter increased as stroke rates increased up to 70 min^{-1}, but that no hail was produced by storms with still larger stroke rates. However, individual lightning strokes in a flash occur so rapidly that it difficult to see how a visual count could separate them; thus, it seems likely that the reported stroke rates are closer to flash rates than stroke rates.

In the mountains of western Montana, Baughman and Fuquay (1970) compared hailstorms with storms that produced no hail. Flash rates and type were determined from recorded electric field changes combined with visual observations. The time and size of hail were noted by observers at five stations, and two 3-cm wavelength radars provided data on the structure of the storms. Only 3 out of 15 storms without hail produced flash rates comparable to most of the observed hailstorms, and the average number of flashes per storm without hail was less than half the average number for hailstorms. The storm that produced the largest number of flashes and largest flash rates (571 flashes at an average rate of 3.8 min^{-1}) also produced the largest hail (diameter of ≈ 5 cm), but in most of the Montana hailstorms the maximum diameter of hail recorded by the observer was <1 cm. Baughman and Fuquay found that hailstorms tended to be taller, to produce sustained lightning activity for a longer period, and to have larger flash rates. However, they also noted that flash rates in Montana storms tended to be smaller than flash rates reported in many other regions of the United States.

Pakiam and Maybank (1975) studied lightning in three hailstorms that occurred in Alberta, Canada. A 10-cm wavelength radar provided data on storm structure, and three electric field mills (two deployable and one mounted on a vehicle to make mobile measurements) were used to determine lightning flash rates. Hail sizes in all three storms were large enough to be considered severe. In two multicellular hailstorms, Pakiam and Maybank reported a slight tendency for hail sizes (diameters of 2–4 cm were measured) to increase with increasing total flash rates (ranging roughly from 25 min^{-1} to 42 min^{-1}), but there were few data points. In one storm that appears to have been a supercell, hail >5 cm in diameter was produced, but measured total lightning flash rates were fairly low (roughly 1 min^{-1}). Pakiam and Maybank suggested that low total flash rates might be characteristic of supercell storms, a hypothesis at odds with subsequent observations on the Great Plains of the United States, such as those made by Rust et al. (1981a) and Rust (1989). It appears possible from information given by Pakiam and Maybank that their electric field mills were too far from the supercell storm to detect many cloud flashes, which tend to produce smaller electric field changes at the surface than are produced by ground flashes. Thus, while ground flash rates probably were ≤ 1 min^{-1}, many cloud flashes may not have been detected.

As noted previously, Reap and MacGorman (1989) found that the probability that a storm on the Great Plains had produced severe weather increased slightly as the number of negative ground flashes it produced increased up to 130 in a 48 km $\times$ 48 km grid box and then decreased with still larger flash counts. This is similar to the relationship found by Blevins and Marwitz (1968). However, Reap and MacGorman also found that the probability of large hail increased considerably as the number of positive ground flashes produced by a storm increased. Several subsequent studies have found severe storms whose ground flash activity consists predominately of positive ground flashes, and some of these studies (e.g., MacGorman and Burgess 1994, Stolzenburg 1994) have suggested that these storms often produce large hail during the period when they are producing positive ground flashes. This relationship with the occurrence of positive ground flashes in severe storms will be discussed in considerably more detail in Sec. 8.1.10.

Changnon (1992) examined ground flash rates and strike locations relative to streaks of crop damage produced by hail, which he noted was not necessarily large hail. His analysis examined 48 hailstreaks that occurred in Illinois during 6 days in June–July 1989. All ground flashes associated with the hailstreaks were negative ground flashes. Of the hailstreaks for which there were radar data, 3 were associated with isolated cells, 11 were in groups of radar echoes, and 20 were in squall lines. None of the storms had a strong mesocyclone.

In agreement with Reap and MacGorman (1989), Changnon (1992) found that attempts to use ground flash rates to predict which cells would be associated with hail damage had essentially no measurable success; only 23% of centers of ground flash activity that occurred on the analyzed days were associated with hail damage. However, ground flashes and hailfall did evolve in characteristic patterns. Ground flashes associated with cells that produced hail typically began 8–10 min before the first hail damage, and their flash rates and area of coverage grew rapidly until damage began. Ground flash rates then decreased, but remained large, during hail damage and decreased more rapidly after hail damage ceased, with ground flash activity ending, typically, 8 min after the last hail damage. From estimates of the time required for hailstones to fall from the region in which they developed, Changnon suggested that hail was still developing aloft when the ground flash activity of the associated cell began.

Points where lightning struck ground were not colocated with hail damage, but tended to be distributed throughout the reflectivity core associated with the hail. The hailstreak typically was located on the edge of both the reflectivity core and the region of ground strikes. The location of lightning strikes around the periphery of hailfall varied considerably, especially for storms along stationary fronts. Hailstorms associated with cold fronts produced more ground flashes than those associated with stationary fronts.

A somewhat different evolution was found by Shafer (1990), who examined a storm system that produced large hail in Oklahoma. Storms that formed near a dryline (i.e., a surface boundary separating dry and moist air) produced large hail before producing ground flashes. However, once ground flashes were produced in these and other storms on the same day, the location of high densities of ground flashes agreed roughly with the location of large hail reports and with the location of large values of VIL (i.e., vertically integrated liquid water, which is calculated from radar reflectivity and tends to be well correlated with hail in storms on the Great Plains).

8.1.8. Visual observations of electrical phenomena associated with tornadoes

Reports of unusual electrical phenomena in tornadic storms date back at least to biblical time. Most reports are eyewitness descriptions, however, and it is sometimes difficult or impossible to determine reliably the nature of the observed phenomenon from these descriptions. Furthermore, typical eyewitness reports have almost no quantitative data on either the phenomenon or the storm producing it, so the reports are of little use in determining systematic relationships with tornadic storms.

Still, some reports are clear enough to provide at least a rudimentary description of the observed phenomena, and there are enough of these reports to indicate the range of visual electrical phenomena that can be produced by tornadic storms. Eyewitness reports of unusual electrical phenomena in tornadic storms have included scorching beneath tornado funnels, a steady or rapidly oscillating glow inside funnels, rapidly recurring small patches of light on the side of thunderstorms, and both unusually large or small flash rates (e.g., Church and Barnhart 1979, Vonnegut and Weyer 1966). The most common report of a luminous phenomenon other than lightning is of a discharge phenomenon that seems to be point discharge. As noted previously, most eyewitness reports, even modern reports, include almost no quantitative data, but a few of the more recent reports include still photographic images.

Bernard Vonnegut has been responsible for interviewing eyewitnesses and assembling many of the modern visual reports of luminous phenomena associated with tornadoes. One of the best documented cases is the photographic image of twin luminous pillars in tornadoes that occurred in Toledo, Ohio, on April 11, 1965 (Vonnegut and Weyer 1966). Tests showed that the pillars of light were not an artifact of the way in which the film was exposed or processed (Thompson and Johnson 1967). This image lends credence to the multitude of other similar eyewitness reports. However, some luminous events during tornadoes are from exploding electrical transformers or shorting electric power lines, so all reports need to be reviewed carefully before accepting them as descriptions of natural phenomena.

The reports of unusual luminous phenomena, particularly of luminosity in or below tornado funnels, led some investigators to consider whether tornadoes could be powered electrically. However, extrapolation of laboratory results showed that a maximum of 25% of the needed power could be generated electrically, so electrical generation of tornadoes appears unlikely (Watkins et al. 1978).

8.1.9. Lightning location and evolution in tornadic storms

(i) Sferics studies. Initial studies of sferics from tornadic storms examined radio bands from 10–500 kHz. (Sferics are classified normally by the frequency of the receiver used to detect them.) Dickson and McConahy (1956) found that sferics rates detected by a 10-kHz receiver often increased as storms rapidly grew taller, but peaked earlier in storms that had more violent severe weather. Rates peaked about 1.5 h before tornadoes and decreased during tornadoes to about 40% of peak value. At frequencies ≥ 150 kHz, sferics rates were found to increase until tornadoes occurred, and were exceptionally large during tornadoes (Jones 1951, 1958, Jones and Hess 1952, Kohl 1962, Kohl and Miller 1963). Kohl and Miller observed that sferics rates at 150 kHz usually peaked during severe weather and began decreasing prior to the end of severe weather. As might be expected from the above observations, Jones and Hess found that the ratio of the numbers of higher frequency to lower frequency sferics increased for sferics of large amplitude when storms were more severe; the ratio was 1:20 before and after tornadoes and in nonsevere thunderstorms versus 1:1 during tornadoes.

Later studies at frequency bands up to 150 MHz (Silberg 1965, Taylor 1973, Stanford et al. 1971, and Johnson et al. 1977) found that the increase in sferics rates during severe weather was greatest at frequencies >1 MHz. This is shown in Fig. 8.15, which presents a summary of Taylor's data concerning the dependence of sferics rates on receiver frequency. Silberg and

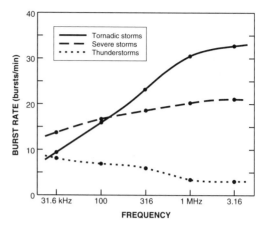

Fig. 8.15. Average sferics burst rate as a function of receiver frequency for various storm classifications. Data are from four tornadic storms, three nontornadic severe storms, and seven nonsevere thunderstorms. Sferics burst rates were measured at 10 receiver frequencies between 10 kHz and 137 MHz, indicated by dots. There was little change above 3 MHz, so higher frequencies are not shown. (From MacGorman et al. 1989, with permission.)

Taylor both found that the energy radiated as sferics increased at higher frequencies as storm severity increased. Radiated energy peaked at roughly 5 kHz for nonsevere storms (Taylor 1972), but the energy at higher frequencies gradually increased with increasing severity, until the energy became independent of frequency for tornadic storms.

There were exceptions to all of these trends. In several cases, high sferics rates did not occur in tornadic storms (Ward et al. 1965, Stanford et al. 1971, Lind et al. 1972, Taylor 1973, Greneker et al. 1976, Johnson et al. 1977). Taylor, who examined the largest number of storms, found that roughly 20% of tornadic storms did not have high sferics rates (MacGorman et al. 1989). It also was possible for high sferics rates to occur in storms without tornadoes (Ward et al. 1965, Stanford et al., MacGorman et al.). MacGorman et al. reported that 23% of nontornadic severe storms and 1% of nonsevere storms had sferics rates comparable to those of tornadic storms. However, 70% of the storms with high sferics rates were not tornadic, because the number of nonsevere storms was much larger than the number of tornadic storms. Weak tornadoes accounted for some tornadic storms without high sferics rates (e.g., Johnson et al.), and the combined effect of simultaneous storms accounted for some nonsevere storms with high sferics rates (e.g., MacGorman et al.). However, these situations did not account for all exceptions to the trend of having much higher 3-MHz sferics rates in tornadic storms than in nonsevere storms.

The most obvious inference from the increasing number of sferics at high frequencies and the decreasing or constant number at low frequencies was that cloud flash rates usually increased while ground flash rates decreased or remained fairly constant during the tornadic stage of storms. This interpretation was further supported by the theoretical work of Stanford (1971) and the airborne observations of Shanmugan and Pybus (1971). Both reported that horizontal polarization of sferics increased and vertical polarization decreased, indicating a greater preponderance of horizontal lightning channels, as the sferics rate and storm severity increased.

Other observations also supported this interpretation. Jones (1958) reported watching a storm when it was producing high sferics rates about 1 h before a tornado: No ground flashes could be seen, but relatively dim circular patches, about 0.5 km in diameter, were illuminated in rapid succession on the side of the storm. He suggested that these were an unusual form of intracloud lightning. Electric field records for this same storm indicated that the lightning flash rate was 12 min^{-1} during the tornado (Gunn 1956). Vonnegut and Moore (1957) inferred from recordings of the closest electric field sensor to the tornado that there were no ground flashes near the tornado. Unusually high lightning flash rates also were observed in two other tornadic storm systems by Orville and Vonnegut (1977) and Turman and Tettelbach (1980), who used satellite-borne optical detectors. Movie films of 18 tornadoes provided additional evidence that ground flash rates often were low near tornadoes: Davies-Jones and Golden (1975) found ground flashes near only 2 of the 18 filmed tornadoes, 1 with 12 ground flashes and 1 with a single ground flash. Note that movie cameras can miss flashes when the shutter is closed as the film advances, so actual flash counts may have been larger. Flash rates measured by mapping systems will be discussed in the next section.

It appears that the tornado itself is not likely to be the cause of unusual sferics rates. For example, Jones (1965) noted that sferics rates can be unusually high 60–90 min before tornadoes, although they may increase further when tornadoes occur. Scouten et al. (1972) further analyzed Jones's (1958) data at 150 kHz and concluded that touchdown of the tornado did not affect the high sferics rate. Although larger amplitude sferics clustered more closely about a core near the azimuth of the tornado, sferics as a whole were spread among azimuths spanning most of the storm. A similar lack of correlation with the time and location of tornado touchdown was reported for sferics in three bands between 300 kHz and 3 MHz (Taylor 1973) and for sferics at 1–50 kHz (Brown and Hughes 1978). Brown and Hughes also noted a lack of obvious relationships with the tornadic vortex signature, which is a column of reduced reflectivity co-located with the tornado (Lemon et al. 1982). In a study of sferics burst rates received in the 30–300 MHz band in two tornadic storms, Johnson (1980) found that the tornadoes began near the time of peaks in the sferics burst rate, but were not coincident with any peaks.

Several studies considered changes in tornadic storms that were correlated with the high sferics rates.

For example, several investigators found periodicities of 10–60 min in the sferics data and suggested that these resulted from the cyclical intensification and decay of thunderstorm cells (Kohl and Miller 1963, Stanford et al. 1971, Shanmugam and Pybus 1971, Lind et al. 1972, Trost and Nomikos 1975). Taylor (1973) reported that sferics at high frequencies originated along azimuths to low-level reflectivity cores and that sferics rates were larger from cores with higher reflectivity. Furthermore, Brown and Hughes (1978) found that the pattern in the number of very low frequency (VLF) sferics from the Union City tornadic storm was similar both in time and space to patterns in reflectivity at low levels of the storm. (VLF sferics are produced predominantly by return strokes.)

Although the sferics studies clearly demonstrated that lightning tended to evolve in a range of characteristic patterns in tornadic storms, there were a number of uncertainties: (1) There were uncertainties in extracting lightning flash type and rates from sferics data; (2) No sferics study examined the evolution of mesocyclones or considered differences in the types of storms that produced tornadoes to examine why sferics rates evolved differently; and (3) No sferics study examined the three-dimensional vector wind field for tornadic storms.

(ii) Lightning mapping studies.   Studies using lightning mapping systems have begun to address some of the shortcomings of sferics studies. However, most mapping studies thus far have examined only ground flashes, a basic drawback if the research is to improve understanding of lightning production by tornadic storms. All three of the mapping studies that examined either total or cloud flash rates (MacGorman et al. 1985, MacGorman et al. 1989, and Richard 1992) in tornadic supercell storms found the flash rates to be large during the tornadic stage, consistent with most of the sferics reports. In studies of two storms on different days that each produced a violent, long-lived tornado, MacGorman et al. (1985, 1989) found that cloud flash rates peaked during the violent tornado in each storm. Richard observed several tornadoes in a storm near Paris, France, during a period in which the storm also produced microbursts; thus, these tornadoes may have been associated with the formation of intense downdrafts more than with the mesocyclone itself. In the storm near Paris, cloud flash rates peaked strongly shortly before tornadoes began. Cloud flash rates increased to still larger values toward the end of the tornadic and downburst phase of the storm and reached their absolute maximum later, shortly before the storm began to produce large hail.

As discussed in Section 7.18.4, several investigators have observed that flash rates tend to increase in a storm as the updraft strengthens at altitudes colder than $-20°C$, so the large flash rates during the tornadic stage of many storms is not surprising. MacGorman et al. (1989) also observed that areas of large reflectivity ($>50–55$ dBZ) at heights of 8 km increased 5–10 min before periods of large cloud flash rates and at heights of 6 km increased at roughly the same time as cloud flash rates. Thus, they suggested that the increased flash rates probably were caused by increased particle interactions leading to reflectivity growth at 7–9 km in and near the strong updraft. Similar correlations of lightning rates with the horizontal area of large reflectivity at mid-levels of storms have been observed by Larson and Stansbury (1974), Cherna and Stansbury (1986), Lhermitte and Krehbiel (1979), and Keighton et al. (1991).

There has been more variability in the observed relationships of tornadoes with ground flash rates than with total flash rates. In three tornadic storms, ground flash rates were small before and during tornadoes (Orville et al. 1982, MacGorman et al. 1989, Richard 1992). In four, ground flash rates peaked during tornadoes, but were usually near a local minimum when the tornadoes began (Kane 1991, MacGorman and Nielsen 1991, Keighton et al. 1991, Seimon 1993). In two storms, tornadoes occurred after ground flash rates had begun to decrease from their peak, but before they had decreased to a local minimum (Kane 1991, MacGorman et al. 1985). Thirty-four of the 42 tornadoes produced on 4 days that were studied by MacGorman and Burgess (1994) began when ground flash rates were either <0.5 min^{-1} or near a local minimum value. During 7 of the 10 tornadoes that lasted $\geq$20 min, ground flash rates increased while the tornado was occurring.

Similar variability was observed by Knapp (1994), who analyzed ground flashes in 264 storms that produced tornadoes east of the Continental Divide of the United States on 35 days from 18 April until 1 July 1991. Of these storms, 60 produced fewer than 15 ground flashes during the 75-min period from 60 min before until 15 min after the tornado began. In 62 of the 204 storms with $\geq$15 ground flashes, at least 30% of ground flashes were positive; in 142, less than 30% of ground flashes were positive. Storms in the category with a larger positive ground flash percentage produced an average of 70 ground flashes (an average of 56% being positive) during the 75-min that was analyzed. These storms will be discussed further in Section 8.1.10. Storms in the category with a smaller positive ground flash percentage produced an average of 262 ground flashes (an average of 9% being positive).

Knapp (1994) also noted regional differences in the storms having a smaller positive ground flash percentage. He divided the analysis region into a southern coastal region (near either the Gulf of Mexico or the warm Atlantic Gulf Stream) and a more northern region. Tornadic storms with a smaller positive ground flash percentage in the southern coastal region produced an average of 401 ground flashes during the analyzed period; those in the more northern region produced an average of 187 ground flashes. In the southern coastal region, ground flash rates tended on average to be almost constant throughout the 75-min period. In the northern region, ground flash rates tended to increase

fairly steadily throughout the period. Note that the trends for both regions are averages over many storms, and so tend to mask the considerable variability normally observed from storm to storm within a region.

Results from MacGorman et al. (1989) and MacGorman and Nielsen (1991) suggest some reasons for the variability in ground flash rates of the two storms they analyzed. Before discussing the hypotheses offered by these two studies, however, we will review the observations. Both studies examined ground flash rates for flashes that struck within 10 km of the mesocyclone center. In the Binger tornadic storm studied by MacGorman et al. (Fig. 8.16), negative ground flash rates within this region of the storm were less than 1 min^{-1} until the violent Binger tornado began dissipating. Negative ground flash rates reached a relative maximum after the last tornado, as its mesocyclone core dissipated, and reached an absolute maximum (about 4 min^{-1}) as the last mesocyclone core in the storm dissipated. Cloud flash rates in this region reached an absolute maximum of approximately 14 min^{-1} during the most violent tornado and were well correlated with low-level cyclonic shear when the shear was greater than 1×10^{-2} s^{-1} during the tornado. (Cyclonic shear is proportional to vertical vorticity for ideal solid-body rotation of a cylinder. In calculating this shear, tornadic winds were ignored; shear was measured across the diameter of maximum tangential wind speed for the mesocyclone, typically a distance of a few kilometers.) After the last tornado, when there was a large decrease in low-level cyclonic shear but the mesocyclone was still strong at middle levels, cloud flash rates within 10 km of the mesocyclone center were only 0–2 min^{-1}.

In the Edmond storm studied by MacGorman and Nielsen (1991), there were no cloud flash data, but negative ground flash rates within 10 km of the mesocyclone center evolved differently from those in the Binger storm. Rates increased to a peak of 11 min^{-1} during tornadic activity and appeared to be correlated with cyclonic shear at the 5 km level (Fig. 8.17). Ground strike locations tended to cluster in the vicinity of a reflectivity core north of the mesocyclone during tornadic activity and to be more scattered before and after the tornadic stage of the storm (Fig. 8.18). Keighton et al. (1991) observed a similar increase in scattering of lightning strikes after tornadic activity in a storm on another day. During an extremely violent tornado, Seimon (1993) observed clustering of ground strike locations during a couple of 2-min periods, including the time when the tornado was extremely violent; Seimon observed no obvious clustering earlier or later during the tornado.

(iii) Hypothesized relationships in two tornadic storms.
One of the primary problems posed by the Binger and Edmond storms is how to explain why ground flash rates were so small near and during the time of the Binger tornado, but large during the Edmond tornado, particularly since total flash rates were large at

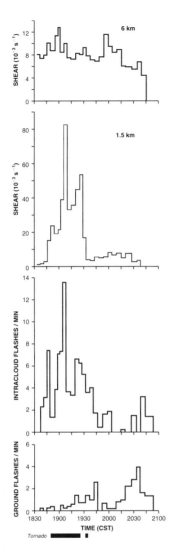

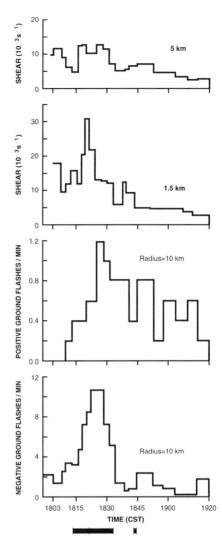

Fig. 8.16. Time-series plots of cyclonic shear at the 1.5- and 6-km levels and of ground and intracloud flash rates within 10 km of the mesocyclone center in the Binger storm of 22 May 1981. The bars on the bottom indicate when tornadoes occurred. (From MacGorman et al. 1989, with permission.)

Fig. 8.17. Time-series plots of cyclonic shear at the 1.5- and 5-km levels and of positive and negative ground flash rates within 10 km of the mesocyclone center in the Edmond, Oklahoma, storm of 8 May 1986. The bars on the bottom indicate the time of tornadoes. (From MacGorman and Nielsen 1991, with permission.)

this time in the Binger storm. MacGorman and Nielsen (1991) suggested that observed differences between the Binger and Edmond storms might account for the differences in the evolution of their ground flash rates. The Binger storm was a classic supercell storm, with a prominent weak-echo region, a strong, deep updraft core, and a long lasting mesocyclone having a family of mesocyclone cores (Burgess et al. 1982). Though the Edmond storm also was a supercell storm, it was weaker than the Binger storm by almost any measure: The duration of its supercell stage was only 30% that of the Binger storm; its mesocyclone was weaker, shal-

lower, and shorter lived; it produced fewer tornadoes; its most damaging tornado was smaller and less violent; and features such as the weak-echo vault and tornadic vortex signature were not as pronounced. MacGorman et al. (1989) suggested that the deeper and stronger updraft core and the pronounced weak-echo region of the Binger storm probably delayed ground flash activity more than in most storms and caused cloud flash activity to dominate because of effects on the thunderstorm charge distribution.

MacGorman et al. (1989) suggested that one primary effect of a strong, deep updraft on the charge dis-

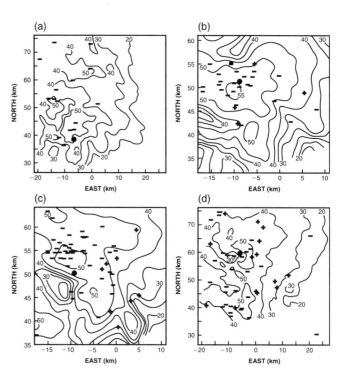

Fig. 8.18. Lightning ground strikes superimposed on radar reflectivity (in dBZ) for the 3-km level during four periods of the Edmond, Oklahoma, storm of 8 May 1986: (a) 1802–1806 CST, (b) 1814–1817 CST, and (c) 1824–1827 CST, and (d) 1834–1838 CST. The most violent tornado occurred during (b) and (c). Minuses indicate the strike point of flashes that effectively lowered negative charge to ground; pluses indicate the strike points of flashes that effectively lowered positive charge. The large dot marks the center of the mesocyclone core. The distance scale for (a) and (d) is different from that for (b) and (c). (From MacGorman and Nielsen, 1991: American Meteorological Society, with permission.)

tribution would be to keep negative charge higher than in most storms. This would increase the energy required for lightning to span the distance to ground and would decrease the electric field at most heights below the negative charge. Negative charge would be higher for three reasons: (1) The strong updraft core rapidly lifts all but the largest particles to upper levels of the storm; (2) Large updraft speeds do not allow hydrometeors to remain in a given layer as long as weaker updrafts. Therefore, hydrometeors have less time to grow and acquire charge, and positive and negative particles have less time to move apart after becoming charged, so there is less net negative or positive charge in a given layer of a strong updraft; and (3) Temperatures in a strong updraft core are warmer, and this causes the ice processes thought to be responsible for charging (see Sec. 3.5 and Sec. 7.18) to occur at a somewhat higher altitude.

A second important effect of a strong, deep updraft would be that regions of net positive and negative charge initially would be relatively close together near the top of the storm updraft. This occurs for two reasons: (1) Sedimentation and differing relative velocities of positively and negatively charged particles would not have time to separate the charges very far in a strong updraft core; and (2) The vertical shear of horizontal winds in a supercell storm causes precipitation to fall to the side of the updraft core, so much of the storm-scale separation of charge occurs outside the updraft.

When two charged regions are closer together, less net charge is needed in each region to create electric

field magnitudes sufficient to initiate lightning between them, as discussed in Section 7.21.1. Furthermore, the electric field required to initiate lightning decreases with height (see Ch. 5 and Sec. 7.13), so less charge would be required in charged regions near the top of the storm to initiate lightning there. Both of these effects would enhance cloud flash rates when charged regions are close together near the top of the storm. Note that, if less charge is required to initiate lightning between positive and negative regions, as under these conditions, then the net charge magnitude in each region is effectively limited to a smaller value by the lightning. This capping of the net negative charge, combined with the unusually large height of negative charge discussed above, makes ground flashes much less likely to occur near a very strong, deep updraft core. Ground flash rates increase later as the negative charge moves closer to ground and as the positive and negative charges move farther apart (thereby increasing the amount of net negative charge that can be separated before causing a cloud flash).

Having the main positive and negative charges close together near the top of the storm updraft could affect flash type in yet another way. If the magnitude of positive charge near the top of the storm is capped at smaller values, as we have just discussed, then the current that flows to the top of the cloud to form a negative screening layer also will be reduced. Krehbiel (1986) suggested that the flow of negative charge to the top of storms tends to make a storm more negative with time and so makes negative ground flashes increasingly likely as a storm ages. If this is so, then anything that reduces the current flow, such as a smaller net positive charge in upper regions of a storm, would tend to delay the onset of ground flashes.

Support for MacGorman et al.'s (1989) hypothesis has come from findings that: (1) The lowest charge determined from the first electric field sounding through a mesocyclone updraft was a region of negative charge starting at 9.5 km MSL, where the temperature was approximately $-37°C$ in the storm's environment and $-31°C$ in the updraft (Rust et al. 1990). This lower boundary was 3–5 km higher—and at environmental temperatures 15°–30°C colder—than the lower boundary normally reported for the main negative charge in continental thunderstorms; (2) Preliminary analysis of additional updraft soundings by Stolzenburg et al. (1996) found negative charge and larger values of $|E|$ occurring only above ≈ 8 km, with most positive charge occurring above 10 km; and (3) Ziegler and MacGorman (1994) used a kinematic retrieval that included electrification processes (Ziegler et al. 1991) to model the electric field and charge structure of the Binger storm. In agreement with the measured vertical soundings and the hypothesis, the lower boundary of negative charge near the updraft core in the retrieved charge distribution was elevated to a higher than normal altitude (Fig. 8.19).

MacGorman and Nielsen (1991) hypothesized that the charge distribution near the updraft of the Edmond storm was more like that of an ordinary thunderstorm and that this enabled ground flashes to occur frequently when the mesocyclone was strong. They suggested that because the Edmond updraft core was weaker, shallower, and shorter lived, and because 55-dBZ reflectivity cores extended to a height of only 5–6 km in the

Edmond storm versus >10 km in the Binger storm: (1) negative charge was lower; (2) a larger fraction of oppositely charged particles had time to separate; and (3) positive and negative charge were farther apart. Similarly, Keighton et al. (1991) observed another storm in which ground flash rates did not decrease during the tornadic stage and noted that it also had a weaker mesocyclone and updraft than the Binger storm.

8.1.10. Positive ground flashes in severe storms

Rust et al. (1981a, b) presented the first documentation that severe storms commonly produce positive ground flashes. Positive ground flashes were identified from recorded surface electric field changes that were coincident with either visual identification or a recorded video image of a cloud-to-ground channel. In supercell storms, Rust et al. observed positive ground flashes emerging from the downshear anvil, from high on the main cumulonimbus cloud tower, and from the rain-free cloud base near the wall cloud (Fig. 8.20). In severe squall lines, positive ground flashes were observed in the widespread region of light precipitation attached to the line of deep convective cells, as will be described further in Section 8.2.

Only a small percentage of ground flashes lowered positive charge to ground in the severe storms that Rust et al. (1981a, b) observed over the Great Plains of the United States, much like the overall percentage of total ground flash activity from all storms in this region. (The percentage of all ground flashes that lower positive charge was estimated to be 7% by Taylor (1963) in Oklahoma and Texas and to range from 4% to 8% for different years by MacGorman et al. (1997) in Oklahoma.) However, ground flashes occurring in some specific regions of supercell storms had a much higher probability of being positive. For example, most flashes emerging from the downshear anvil, tens of kilometers from the cloud tower, were positive ground flashes. (Rust et al. observed no negative ground flashes from this region, but some have since been observed.)

Ground strike mapping systems also have detected positive ground flashes in supercell storms. Typically, the majority of ground flashes in a given supercell storm has been negative ground flashes. We already have noted that Knapp (1994) found that only 62 of 264 tornadic storms had a positive ground flash percentage $\geq 30\%$ for at least 40 min. MacGorman and Nielsen (1991) reported that negative ground flash rates were larger than positive ground flash rates throughout a storm that produced a strong tornado and was classified as a supercell for an hour. Positive ground flashes began to occur a few minutes before the tornado touched down, and positive ground flash rates increased to their maximum value during the tornado, near the time when negative ground flash rates also peaked (Fig. 8.17). However, in a supercell storm that did not produce a tornado and had predominately negative ground flashes, Shafer (1990) noted that positive ground flash

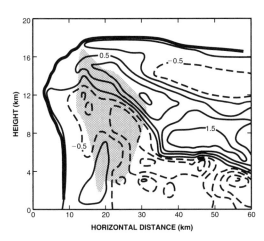

Fig. 8.19. Electric field from a numerical simulation of the Binger storm of 22 May 1981. Contours of electric field are labeled in kV m^{-1}. Note the more elevated region of high electric fields in the vicinity of the updraft. Shading indicates the updraft core. (After Ziegler and MacGorman 1994, with permission.)

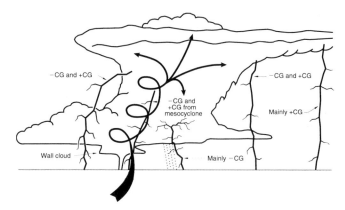

Fig. 8.20. Typical location of channels to ground for each polarity of ground flash relative to supercell storm structure when ground flash activity is not dominated by positive ground flashes. (After Rust et al. 1981a, with permission.)

rates were small during initial stages of the mesocyclone and increased to >1 min^{-1} as the mesocyclone began to dissipate.

Sometimes most of the ground flashes produced by a supercell storm for ≥30 min have been positive ground flashes, as was first reported by Rust et al. (1985a). Since that study, warm-season storms in which the majority of ground flashes lower positive charge to ground have been the topic of several studies. Takagi et al. (1986) observed a Japanese summertime storm in which 57% of ground flashes were positive, although no mention is made of whether the storm produced severe weather. Curran and Rust (1992) reported a case in which a supercell storm had predominately positive ground flashes while it was a low-precipitation supercell, and then had predominately negative ground flashes after it split and became a classic supercell. On a different day, Branick and Doswell (1992) observed that some storms had predominately positive ground flashes and others had predominately negative ground flashes. Storms dominated by positive ground flashes tended to have low-precipitation supercell characteristics, while storms dominated by negative ground flashes tended to be farther southeast and to have characteristics more like the classic or heavy-precipitation part of the spectrum of supercell storms.

Though many severe or tornadic storms do not have large numbers of positive ground flashes, evidence suggests that many storms that produce relatively high flash rates and densities of positive ground flashes also produce severe weather. In a climatological study of warm season storms occurring over the Great Plains of the United States, Reap and MacGorman (1989) reported that the probability of severe weather increased rapidly as the density of positive ground flashes produced by a storm increased (Fig. 8.21). Similarly, Rust et al. (1985a), Curran and Rust (1992), Branick and Doswell (1992), Seimon (1993), MacGorman and Burgess (1994), and Stolzenburg (1994) all noted that severe weather occurred in storms whose ground flash activity was dominated by positive ground flashes. However, remember that positive ground flashes also can occur in other types of warm season storms, in-

cluding shallow thunderstorms, the dissipating stage of small isolated thunderstorms, and the widespread, light precipitation associated with large storm systems. Usually these other types of storms produce smaller flash rates and densities of positive ground flashes than severe storms. No study has attempted to catalogue what fraction of storms that produce positive ground flashes can be attributed to each type of storm, but Knapp (1994) found that there were official reports of tornadoes or large hail in 36–43% of storms that produced ≥15 ground flashes and had a positive ground flash percentage ≥30% during 18 April–1 July 1991.

Furthermore, severe storms that produce relatively large positive ground flash percentages are not uncommon. In Knapp's (1994) study, 62 of 264 storms that produced tornadoes during 18 April–1 July 1991 also

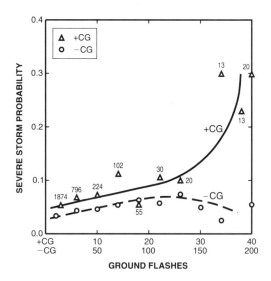

Fig. 8.21. Probability of severe weather versus the number of ground flashes in a 48 km × 48 km grid box. Numbers labeling the triangles indicate the number of grid boxes associated with the data point. (Adapted from Reap and MacGorman 1989, with permission.)

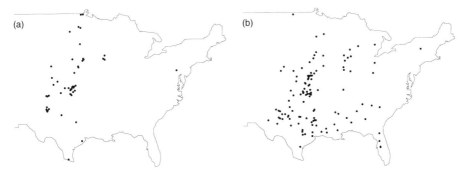

Fig. 8.22. Map of tornadic storms that occurred 18 April–1 July 1991 east of the Continental Divide and produced ≥15 ground flashes having (a) a positive ground flash percentage ≥30% for ≥40 min (average percentage = 58%), and (b) a smaller positive ground flash percentage (average percentage = 9%). (From Knapp 1994, with permission.)

produced a positive ground flash percentage ≥30% for at least 40 min of the 75 min analyzed for each storm. Tornadic storms with positive ground flash percentages ≥30% tended to occur in the central United States, in a swath stretching from the Texas panhandle and north-central Texas through western Minnesota (Fig. 8.22). Tornadic storms with a smaller positive ground flash percentage appeared to have no preferred location, being distributed over essentially the same region as the whole set of tornadic storms.

In studies of tornadic storms dominated by positive ground flashes on multiple days, MacGorman and Burgess (1994) and Stolzenburg (1994) found considerable variability also in the evolution of positive ground flash rates relative to storm evolution and severe weather occurrence. In some storms, positive ground flashes dominated ground flash activity throughout the lifetime of the storm. In others, as in the storm studied by Curran and Rust (1992), the dominant polarity of ground flash activity changed at least once as the storm evolved. We will consider an example of each of these two patterns of evolution.

The dominant ground flash polarity switched from positive to negative in a storm that occurred on 28 August 1990 and was studied by Seimon (1993) and MacGorman and Burgess (1994) (Fig. 8.23). (A similar transition was observed previously by Rust et al. 1985a and Curran and Rust 1992.) This storm produced large hail and six tornadoes, including an extremely violent tornado that struck Plainfield, Illinois. Before the Plainfield tornado began, most ground flashes in the storm were positive. During this period in which positive ground flashes dominated, the storm produced large hail and five tornadoes rated as high as F-2. (The Fujita F-scale varies from F-0 for weak tornadoes to F-5 for the most violent tornadoes.) Three of these tornadoes occurred during a period of very large positive ground flash rates.

As a hook echo began to form shortly before the mesocyclone produced the Plainfield tornado, ground flash rates decreased rapidly. The dominant polarity of ground flashes switched as the Plainfield tornado began, and most subsequent ground flashes were negative. The decrease in ground flash activity shortly before the Plainfield tornado is similar to that described earlier for negative ground flashes in some tornadic storms. Once negative ground flashes began to dominate, ground flash rates increased to a peak of approximately 40 per 5 min period during the Plainfield tornado, but increased to still larger values at the end of the time-series plot, when the storm was merging with other storms to form a squall line (similar to lightning behavior observed by Goodman and MacGorman (1986) when storms merged to form a mesoscale convective complex).

The Plainfield storm changed its character at approximately the time of the tornado and the switch in ground flash polarity. Before 1330 CST, there were insufficient data to determine when the storm was a supercell, but the storm displayed characteristics suggestive of a supercell for much of this period. After 1330 CST, the storm grew rapidly to become a classic supercell, and it remained so for roughly 30 min. During this period, the storm had a prominent, deep weak-echo region and produced large hail. At about the time that the Plainfield tornado began, heavy precipitation began to wrap around the mesocyclone, filling in the weak-echo region. Observers of the Plainfield tornado noted that it was obscured by heavy precipitation. Thus, during this period, the storm was classified as a heavy-precipitation supercell. The last report of large hail occurred shortly after both the beginning of the Plainfield tornado and the transition in ground flash polarity.

The second pattern, in which the dominant ground flash polarity is positive for almost the entire lifetime of a storm, was exhibited by a storm that produced weak tornadoes in south-central Kansas on 26 March 1991 (Fig. 8.24). This storm had low-precipitation supercell characteristics for much of its lifetime and produced large hail and two tornadoes. Both torna-

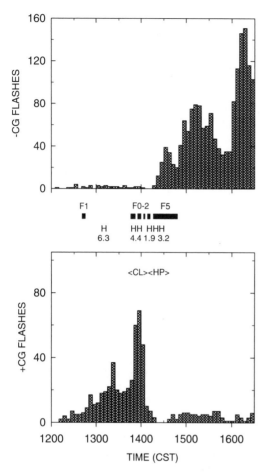

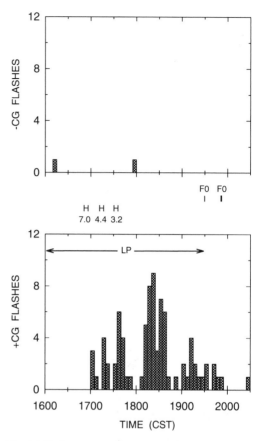

Fig. 8.23. Time-series plot of positive and negative ground flash rates in the Plainfield, Illinois, storm on 28 August 1990. Bars indicate the time of tornadoes; each is labeled with the maximum Fujita-scale damage rating produced by the tornado. The leading edge of Hs indicates the time of large hail reports, with the diameter given in centimeters. CL indicates when the storm was a classic supercell, and HP, when it was a heavy-precipitation supercell. (From MacGorman and Burgess 1994, with permission.)

Fig. 8.24. Time-series plot of positive and negative ground flash rates in a storm that produced tornadoes in south-central Kansas on 26 March 1991. The arrows labeled LP indicate when the storm was a low-precipitation supercell. (See the figure caption for Fig. 8.19.) (From MacGorman and Burgess 1994, with permission.)

does occurred after the peak in the positive ground flash rate.

MacGorman and Burgess (1994) found considerable variety in the details of the evolution of positive and negative ground flash rates and in the relationship of positive ground flashes to severe weather. For example, sometimes the dominant ground flash polarity was negative initially and switched more than once. However, MacGorman and Burgess noted tendencies toward some patterns:

1. On a given day when several supercell storms occurred, the region in which positive or negative ground flashes dominated appeared consistent, and the dominant polarity switched in roughly the same region for sequential storms following similar tracks. This did not appear to be related simply to a given stage of the storm, because on the same day the history of storms before their transition in dominant ground flash polarity sometimes varied considerably. Instead, MacGorman and Burgess (1994) suggested that the dominant ground flash polarity of a storm was influenced by environmental conditions that remained unchanged or changed slowly during storm activity on a given day.

2. Although all of the storms dominated by positive ground flashes in the study by MacGorman and Burgess (1994) produced severe weather, not all of them were supercell storms. Nonetheless, MacGorman and Burgess noted that the storms in their study tended to have low-precipitation or classic supercell characteristics when positive ground flashes dominated and to have classic or heavy-pre-

cipitation supercell characteristics when negative ground flashes dominated. A change in the dominant ground flash polarity often followed or was accompanied by changes in the character of the storm, such as a period of rapid growth or evolution from low-precipitation to more classic supercell characteristics or from classic to more heavy-precipitation supercell characteristics. However, it appears that no type of supercell storm is always dominated by a given polarity of ground flashes. Since publication of MacGorman and Burgess and Stolzenburg (1994), there have been unpublished observations of at least one low-precipitation supercell storm dominated by negative ground flashes and of a few heavy-precipitation supercell storms dominated by positive ground flashes.

3. In all tornadic storms in which positive ground flash rates had a prominent peak greater than 1.5 min^{-1}, the most damaging tornado produced by the storm began after the peak positive ground flash rate and (in storms in which the dominant ground flash polarity switched) before or near the time when negative ground flashes became dominant. Weaker tornadoes could be produced either earlier or later.

4. Sometimes ground flash rates decreased to near zero for 40–100 min after positive ground flash rates decreased from their peak and before negative ground flashes began to dominate ground flash activity. During this interval, storms generally appeared to remain strong and sometimes intensified, large hail continued to be reported, and often the most violent or longest-lived tornado began.

5. Usually when the predominant ground flash polarity was negative initially and then switched to positive, the initial negative ground flashes were infrequent and occurred while storms were weaker and less organized (i.e., did not yet have supercell organization). This was not true in an unpublished case in which a heavy-precipitation supercell storm was observed in Oklahoma on 6 May 1994. In that storm, the dominant polarity of ground flashes switched several times, and the early domination by negative ground flashes extended into the period when the storm was a supercell.

6. Although ground flash activity in hailstorms can be dominated by either polarity of lightning, hailstorms comprise a small percentage of storms dominated by negative ground flashes, but appear to comprise a large percentage of storms dominated by frequent positive ground flashes. All storms observed by MacGorman and Burgess (1994) with frequent positive strikes in regions where they were confident of hail verification had large hail during the period when positive ground flashes dominated ground flash activity. If the dominant ground flash polarity subsequently switched to negative, the diameter and frequency of large hail reports usually decreased. This relationship is consistent with the finding of Reap and MacGorman (1989) that severe weather, especially large hail, is more likely in storms that have a large density of positive ground flashes. MacGorman and Burgess cautioned, however, that many hailstorms are dominated by negative ground flashes, because the number of warm season storms dominated by negative ground flashes is much larger than the number dominated by positive ground flashes. Thus, domination by positive ground flashes may indicate that a storm is capable of producing large hail (more evidence is needed to establish this), but domination by negative ground flashes does not indicate that a storm is not a hailstorm.

In his study of the average evolution of positive and negative ground flash rates in 62 tornadic storms having a positive ground flash percentage ≥30%, Knapp (1994) observed evolution similar to the behavior noted above in item 3. On average, the ground flash rate reached a small local maximum 50 min before the tornado and then decreased until 20 min before the tornado. It began to increase rapidly again 15 min before the tornado and attained its largest value just before or at the time the tornado reached ground, and then the rate decreased. The positive ground flash percentage tended to remain >50% for the 45 min period preceding the tornado, but began decreasing steadily 20 min before the tornado, and dropped below 50% 5–10 min after the tornado began.

MacGorman and Nielsen (1991) and MacGorman and Burgess (1994) suggested two reasons why positive ground flashes might occur in severe storms. Note that more than one combination of factors may be responsible for producing positive ground flashes and that the factors may vary from storm to storm.

The first hypothesis suggested by MacGorman and Nielsen (1991) was similar to a hypothesis suggested by Brook et al. (1982) for winter storms. Positive charge in the upper part of winter storms that Brook et al. observed extended horizontally beyond the negative charge below it. Presumably, the positive charge extended far enough that it was effectively exposed to the ground (i.e., some electric field lines extended to the ground from a region of large field magnitudes near the upper positive charge). In supercell storms, precipitation at middle to low levels is displaced horizontally from the updraft core, and strong divergence at the top of the updraft creates pronounced upshear and downshear anvils. MacGorman and Nielsen suggested that positive charge on ice crystals around the top of the updraft core of supercell storms extends horizontally beyond the negative charge on precipitation at lower heights. Thus, the upper positive charge in a supercell storm might extend far enough that the lower negative charge no longer shields the positive charge from the ground.

Branick and Doswell (1992) suggested a variation of this mechanism applicable only to low-precipitation supercell storms: Not only does the horizontal displacement of the updraft and precipitation expose the upper positive charge to the ground, but there is little precipitation at low-to-middle levels in low-precipitation supercell storms, and this lack of precipitation may interfere with the formation of a large region of negative charge. To evaluate this variant, there are two outstanding basic issues: (1) Does the evaporation occur high enough in the storm to affect the negative charge on precipitation? and (2) If so, what happens to the negatively charged aerosols and ions left behind? These two issues have not yet been addressed.

The second mechanism discussed by MacGorman and Nielsen (1991) and MacGorman and Burgess (1994) was that there might be enough positive charge beneath the main negative charge of some severe thunderstorms that positive ground flashes are produced by this lower positively charged region. Lower positive charge might be formed in several ways, as discussed by Marshall and Winn (1982, 1985) and Jayaratne and Saunders (1984, 1985). One possibility is that it is formed by microphysical charging at temperatures and liquid water contents at which noninductive graupel-ice interactions put positive charge on graupel and negative charge on ice particles. MacGorman and Burgess suggested that positive charging of graupel might help explain the tendency for large hail to occur when positive ground flashes dominate ground flash activity. Rapid graupel and hail growth can occur in regions of large liquid water content or along lengthy particle trajectories across a broad updraft (e.g., Miller et al. 1990). Graupel tends to charge positively in regions of large liquid water content or along trajectories in a temperature range between 0°C and roughly −15°C, as discussed in Chapter 3.

Carey and Rutledge (1995) have presented the only multiparameter radar data thus far from a severe storm that produced a relatively high density and rate of positive ground flashes. The storm was not a supercell storm, but did produce large hail. The dominant polarity of ground flashes changed from negative (flash rates up to 2.8 min^{-1} for negative flashes and 0.2 min^{-1} for positive) to positive (up to 0.4 min^{-1} negative and 1.8 min^{-1} positive) and back to negative (0.5 min^{-1} negative and 0.1 min^{-1} positive). When most ground flashes were negative early in the storm, the region of precipitation was dominated by moderate rain, and no large hail was detected by the radar. Positive ground flashes began to dominate as a downdraft formed at middle levels of the storm and as graupel began descending to lower levels.

Positive ground flash rates increased to >0.5 min^{-1} 5–10 min before large hail was detected and remained high for 5–10 min after all detected large hail had fallen out of the storm. Before hail was detected, positive ground strike points were scattered in light rain, away from cores of large reflectivity. However, as large hail began to be detected, positive ground flashes began to strike in a tight cluster in light to moderate rain, adjacent to the hail shaft of a newer cell. Strike points of positive ground flashes continued to cluster near the hail shaft for 10 min after all large hail had fallen out of the storm.

Because positive ground flashes continued well after graupel and much of the rain had already fallen out of the storm, Carey and Rutledge (1995) discounted the possibility that domination by positive ground flashes was caused by a lower positive charge on descending graupel. Instead, they suggested that either a horizontal displacement or a lowering of the upper positive charge might have caused positive ground flashes. Although this is a reasonable interpretation of their data, it should be regarded as preliminary. There are too few data sets thus far to support or reject any hypothesis about the cause of predominately positive ground flash activity in severe storms.

8.2. MESOSCALE CONVECTIVE SYSTEMS

8.2.1. Definition and basic description

A *mesoscale convective system* (MCS) is a group of storms that interacts with and modifies the environment and subsequent storm evolution in such a way that it produces a long-lived storm system having dimensions much larger than individual storms. This range of dimensions and times is sometimes referred to as *meso-β* (20–200 km, <6 h) through *meso-α* (200–2000 km, >6 h) scales, but often is called simply *mesoscale*. An MCS involves synergisms among its individual storm cells and between itself and the environment. Zipser (1982) defined an MCS as having four broad attributes: (1) There must be a group or system of storms (i.e., dimensions are larger than a single cumulonimbus cloud) that have deep convection during most stages of the life cycle of the system; (2) The lifetime of the group must be several times the lifetime of individual cumulonimbus clouds (i.e., be roughly ≥2 h); (3) The anvils resulting from the upper level outflow of the individual storms in the group must eventually merge to form a single cloud shield; and (4) The individual downdrafts must combine to form a continuous zone of cool air in the lower troposphere.

The region containing the deep convection of all the individual storm cells of an MCS is called the *convective region*. Inside the cloud shield, but outside the convective region, there are often widespread, but weak, vertical motions (≤1 m s^{-1} in magnitude) called *mesoscale updrafts* and *mesoscale downdrafts*. The region of widespread, weak vertical motions within the cloud shield is called the *stratiform region*. Yet it is the precipitation, not the vertical motion, that is demonstratively stratiform. In many MCSs, there is a region between the stratiform region and the convective region called the *transition zone,* where the reflectivity at low altitudes is smaller than that in both the convective and stratiform regions. Although investigators have been aware for many years that some storm systems have stratiform regions (e.g., Newton 1950), early studies ignored the stratiform region to concentrate on the structure and dynamics of deep convection. Subsequent research (e.g., Zipser 1977, Sanders and Emanuel 1977, Ogura and Liou 1980, Smull and Houze 1985) has focused increasingly on mesoscale kinematics and structures to study the interdependence of stratiform and convective regions.

To describe the evolution of midlatitude MCSs, Zipser (1982) divided the life cycle into the formative, intensifying, mature, and dissipating stages, terms that had been defined to describe tropical MCSs (e.g., Leary and Houze 1979). A typical life cycle of an MCS begins with winds converging on the mesoscale under conditions that lead to organized convection. The production of individual storms marks the beginning of the formative stage of the MCS. Rainfall and downdrafts initially occur only in convective regions. As the individual storms grow and mature, the rainfall rate increases, and the mass flux in downdrafts increases.

The intensifying stage of the MCS begins with merger of the precipitation cores of individual storms into a line or cluster and merger of the individual cold downdrafts to form a pool of cold air having mesoscale dimensions. During this stage, upper level outflows from the updrafts of individual cells interact with each other and the environment over a few hours to create a widespread region in which stratiform precipitation forms and increases. Mesoscale updrafts tend to occur in the upper part of the stratiform region, and mesoscale downdrafts, to occur in the lower part, although mesoscale downdrafts may be less widespread than mesoscale updrafts. Stratiform precipitation forms as particles advected from the upper level outflow of the convective region continue to grow in the mesoscale updraft.

The mature stage begins when the stratiform region, with its associated precipitation and mesoscale updrafts and downdrafts, becomes well defined, so that stratiform precipitation and deep convection coexist in the MCS. During the mature stage, stratiform rainfall, mesoscale downdrafts, and the downward mass flux through low levels all increase in the stratiform region. The dissipating stage begins when the net vertical mass flux of the entire MCS through the 1-km level becomes negative, signifying the dominance of downdrafts, even though updrafts still can exist in the convective region and in the stratiform region.

Many of the advances in understanding interactions between mesoscale storm systems and their environments have come by concentrating on systems that are large enough to be sampled by operational atmospheric soundings and surface stations. Maddox (1980) defined and studied *mesoscale convective complexes* (MCCs), a class of large MCSs that generate much of the rainfall and severe weather in the central United States. Maddox's definition was based on satellite cloud top imagery: MCCs must have a roughly circular cloud shield (ellipticity of at least 0.7) with a horizontal dimension of ≥ 250 km and with at least 50,000 km^2 of the cloud top no warmer than $-52°C$ and at least 100,000 km^2 no warmer than $-32°C$ for at least 6 h.

Maddox found that the satellite-observed cloud shield of MCCs often evolved in the following characteristic pattern: (1) pre-MCC convective storm growth, (2) development of the roughly circular MCC cloud shield caused by formation of mesoscale updrafts, (3) rapid growth to maximum extent, and (4) dissipation. Maddox (1983) examined composite synoptic analyses for each stage of the MCC life cycle to develop a conceptual model of the characteristics of the different stages of MCC evolution. Cotton et al. (1989) extended Maddox's work by examining more cases and by improving temporal resolution.

Although MCC cloud shields tend to evolve in a characteristic pattern, it became apparent from case studies of individual MCCs that internal structure varies widely, from chaotically arranged cells of deep convection to various types of squall lines (Maddox et al. 1986, Smull and Houze 1987, Leary and Rappaport 1987, Watson et al. 1988). Because the size of MCCs makes it difficult, if not impossible, for a single radar to detect an entire MCC during most of its life cycle, progress in understanding MCC internal structural characteristics has been slow. Relatively few studies have examined the internal structure of multiple MCCs, and these have tended to be restricted to a relatively small region, to have reduced spatial or temporal resolution, or to include too few cases to reach definitive conclusions.

Kane et al. (1987) and McAnelly and Cotton (1989) used hourly precipitation data instead of radar, so spatial and temporal resolutions were reduced. Both studies examined warm season MCCs and found that those having a relatively simple life cycle (e.g., monotonic growth followed by monotonic shrinking of the cloud shield) had a repeatable pattern of rainfall development: The heaviest rainfall intensity occurred early in the MCC life cycle, the largest rainfall volume occurred near the time when cloud tops colder than $-54°C$ reached their largest horizontal extent, and the largest horizontal area of rainfall occurred about an hour later. McAnelly and Cotton (1986) noted that more complex patterns of evolution usually were associated with MCCs that consisted of several meso-β scale components having different lifetimes and phases.

As a group, the subset consisting of MCCs is similar in most respects to other MCSs. Kane et al. (1987) examined 74 MCCs and compared them with 32 other MCSs that failed to meet all the MCC criteria, such as size or duration. They found no significant differences between MCCs and the other MCSs, except that MCCs had a larger area of rainfall. In particular, the average and maximum rainfall rates at individual stations and the evolution and spatial pattern of rainfall were similar. Similarities between MCCs and some other MCSs have also been reported by Houze et al. (1990), who examined their internal reflectivity structure. Based on 63 radar-observed MCSs in Oklahoma, Houze et al. defined three categories of MCS structure, the first two of which are shown schematically in Fig. 8.25: (1) lead-

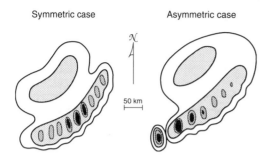

Symmetric case   Asymmetric case

50 km

Fig. 8.25. Schematic diagram of plan view of low-level radar reflectivity structure for two categories of MCSs: symmetric and asymmetric. (From Houze et al. 1990, with permission.)

ing line of deep convection with symmetrically placed trailing stratiform, (2) leading line of deep convection with asymmetrically placed trailing stratiform (usually displaced downstream, toward the north), and (3) chaotic, unclassifiable structure not bearing any resemblance to the leading-line/trailing stratiform archetype. The degree of symmetry also applies to the intensity of convection distributed along the line. The internal structure of those MCSs that were MCCs included all three categories of structure, no one of which was clearly favored. Houze et al. identified environmental characteristics conducive to each type.

The two types of leading line, trailing stratiform MCSs have been the topic of several electrification studies and so will be important in the rest of Section 8.2. Furthermore, several studies have examined their structure and evolution. Leary and Rappaport (1987) gave detailed documentation of the life cycle and internal structure of one such squall line that was an MCC. In agreement with previous investigators, they noted three prominent features in the reflectivity struc-

ture at low altitudes: (1) large reflectivity in the leading convective region, followed by (2) a region of lower reflectivity, and still farther back, (3) a secondary maximum of reflectivity in the heavier stratiform rain. In the stratiform region during the mature stage of the MCC, they identified regions of heavier precipitation that they called *rainbands,* which appeared to contain weak convective cells with high bases. Leary and Rappaport inferred that the bands were linked with a midlevel vortex of the mesoscale flow caused by the MCC. The development of the reflectivity structure went through four stages: (1) formation (lasting about 2 h), during which there was a broken line of isolated convective echoes, (2) intensification (lasting about 5 h), during which the isolated echoes merged to form intense leading-line convection, (3) maturation (lasting about 6 h), during which there was a broad area of light to moderate rain behind the convective line, and (4) dissipation, during which the leading convection diminished and the low-level reflectivity pattern disintegrated.

Houze et al. (1989) developed a conceptual model (Fig. 8.26) of the internal wind and reflectivity structure of a mature leading line, trailing stratiform squall line. Their model consists of a leading line of deep, convective cells with strong updrafts that periodically redevelop. New cells form at middle altitudes of the troposphere, along the leading edge of the convective line in the developing updrafts ahead of mature cells. Behind the mature, deep convective cells are older decaying cells. Farther back, there is often a gap or transition zone of lower radar reflectivity that is easily seen at low elevation angles on plan displays of radar reflectivity (Fig. 8.27).

Behind the transition zone, there is increased reflectivity over a broad area in the stratiform region, although peak reflectivities are not usually as great as in the convective line. Major kinematic features of the stratiform region are an ascending front-to-rear flow in

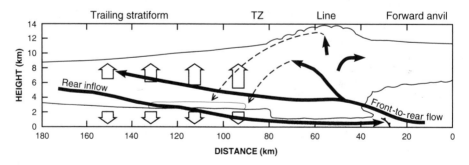

Fig. 8.26. Conceptual model of kinematic and precipitation structure in leading line, trailing stratiform MCS. The upper long arrow depicts ascending front-to-rear flow, and the lower arrow is the descending rear inflow, which is shown having merged with the gust front at the surface. The thin, vertically hatched area denotes the bright band with its secondary maximum of precipitation. The short, broad arrows denote the mesoscale updraft and downdraft. The dashed lines indicate particle trajectory paths. TZ is the transition region which separates the leading convection and the stratiform region. Storm motion is left to right. (After Houze et al. 1989, with permission.)

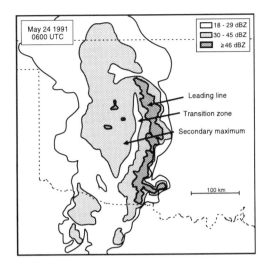

Fig. 8.27. Simplified base-scan radar reflectivity of a leading line/trailing stratiform MCS, 24 May 1991, 0600 UTC. (adapted from radar data from the WSI Corp.; after Marshall et al. 1996: American Geophysical Union.)

upper levels and a descending rear-inflow jet at lower levels. The widespread stratiform precipitation is caused by detrainment of ice particles from the deep convection back into the stratiform region at upper levels and by continued growth of precipitation through microphysical processes in the stratiform region. Both advection and growth of particles likely are important to the final electrical structure of the stratiform region of an MCS.

In the stratiform region, as ice particles fall and approach the 0°C isotherm, their stickiness increases and they grow faster. The number density of ice particles decreases, but the maximum size of aggregates increases with decreasing height. As a result, reflectivity begins to increase more rapidly with decreasing height near 0°C. Reflectivity increases even more when the aggregates begin to melt, because the reflectivity of water-coated ice particles is roughly five times larger than that of completely frozen ice particles of the same size and geometry. This enhanced reflectivity often shows as a bright band. As an aggregate falls farther, it eventually melts enough so that it collapses and its fall speed increases. The resulting decrease in both size and particle concentration causes reflectivity to decrease at lower heights. The reflectivity of the bright band is largest in a layer at temperatures slightly warmer than the 0°C isotherm.

The above overview provides only highlights of research results most needed for our discussion of electrification of MCSs. MCSs and MCCs remain the subjects of considerable research. For more information, the reader is referred to the above studies and to reviews by Maddox et al. (1986), Rutledge (1991), and Smull (1995).

8.2.2. Measurements of electric field and charge

Research on electrical aspects of MCSs is new, and as recently as 1990 almost nothing was known about the electrical structure of these important weather systems (Rutledge 1991). Just as with smaller storms, there is a need to complement electrical observations with an understanding of the other meteorological processes within storm systems. Although there are still considerable voids in our knowledge, much progress has been made. As we write this book, knowledge of MCS electrification is evolving rapidly.

(i) Measurements of E_{gnd}. The only published detailed record of E_{gnd} is that by Chauzy et al. (1985). They had a network of 11 field mills in an area of about 25×45 km in the northern Ivory Coast from which they showed a record of average E_{gnd} and rain rate for a 3.5-h period. They showed radar data from both the convection and the stratiform region and noted the presence of a radar bright band. They used five field mills that were approximately aligned with the direction of squall line propagation. These field mills recorded similar fluctuations in E, but the fluctuations were offset in time in a manner consistent with motion of the storm over the sites. Chauzy et al. converted their time-dependent data to a distance along the direction of propagation. Thus, they correlated features in their records of E_{gnd} with the vertical cross section of radar data that showed a typical leading line, trailing stratiform MCS (Fig. 8.28). They stated that the electrical and kinematic structures of the MCS, as evidenced in E_{gnd}, did not change appreciably over a 3.5-h period. We interpret their record as showing seven polarity reversals with horizontal distance across the stratiform region. They speculated that although intense, the main convective core was too shallow to produce a large electric field at the ground.

The analysis of the wind field indicated to them that ice particles were advected from the convection back into the stratiform. Chauzy et al. (1985) stated that the presence of a mesoscale updraft in the stratiform region would then increase the number of such particles. Shown in Fig. 8.28 are the dominant precipitation particle polarity at the ground during some periods. For example, the $N_-/N_+ = 1.7$ means that 1.7 times as many charged particles arriving at the ground were negative as positive. The dominant particle polarity was such that the falling precipitation could have accounted for, at least, the polarity of E_{gnd}; the mirror image effect was not observed. Chauzy et al. thought the magnitude of E_{gnd} was too low for significant ion-capture charging to have been occurring.

In looking at the time variation of the records for two squall lines from a network of corona points, which yielded the polarity of E_{gnd}, Engholm et al. (1990) calculated that the changes in the corona data matched the movement of features of the squall line over the corona points. They noted a dominance of positive charge

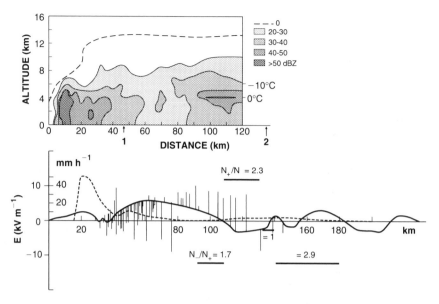

Fig. 8.28. Early observation of MCS electrification. The upper panel shows radar reflectivity from the convection ($\approx$5–40 km) through the transition zone ($\approx$40–70 km) and partially across the stratiform region ($\approx$70–120 km). E_{gnd} is the solid curve in the bottom plot: Note it is shown versus distance for comparison with the vertical structure. Spikes on E denote lightning. The ratios, e.g., N_+/N_-, are the number particles of the dominant polarity to the number of the other polarity observed at the ground; positive polarity dominates above the E = 0 axis, negative below. The horizontal line nearby comes from the time over which the ratio determined. The dashed curve is rain rate in mm h^{-1}. (After Chauzy et al. 1985: American Geophysical Union, with permission.)

overhead behind the convection. An interesting case was that of a stratiform region totally detached from the main convection (at the 0-dBZ level). The radar showed the stratiform region to be at least 40 km across and with a bright band that stretched for $\approx$25 km. As the stratiform region moved across the network, each corona point showed that E_{gnd} reversed polarity coincident with the disappearance of low-altitude radar reflectivity overhead. The sustained E_{gnd} for $\approx$1 h and the occurrence of a few cloud flashes in the stratiform region suggested that electrification mechanism(s) were active in the stratiform region independent of those active in the convective region, as has been proposed in other studies.

(ii) Early vertical profiles of E in MCS stratiform regions. We cannot be sure that some of the older profiles of E in the literature were not from MCSs, but the earliest appear to be those by Chauzy et al. (1980, in France; 1985, in the northern Ivory Coast). One of their E profiles was from a balloon-borne sounding through the transition zone of a leading line, trailing stratiform MCS in the northern Ivory Coast. They reported frequent lightning occurred in the transition zone. The first reported electric field sounding in the United States also went through a transition zone (Schuur et al. 1991). From that sounding, Schuur et al. attempted a

conceptual model of the electric charge structure of an MCS (Fig. 8.29). Since no data existed from the convective region, they showed a dipole (subsequent data cast substantial doubt on its validity). The structure of the forward anvil also was not derived directly from MCS observations, but from large severe storms. The charge structure in the model MCS stratiform region is at odds with a simple configuration such as a dipole or tripole. The observations of the bipolar ground strike pattern and the E soundings have led to attempts to determine the origin of the charge in the stratiform region.

Hunter et al. (1992) obtained an E profile near the rearward edge of the stratiform region of an MCS on 7 June 1989 that was not the leading line, trailing stratiform type. The low-level reflectivity structure (Fig. 8.30) appears similar to the structure of two storms in the "unclassifiable" category of Houze et al. (1990, Figs. 10a and f). Although the stratiform precipitation region was northeast of the convective zone, it was downwind of the convection relative to the prevailing upper level flow. The MCS moved toward the east-southeast, parallel to the convective line. The rear-inflow jet at the rearward flank of the stratiform region was parallel to the convective line, not perpendicular as it would tend to be in a leading line, trailing stratiform MCS. Another difference was the presence of migratory convective cells embedded in the stratiform re-

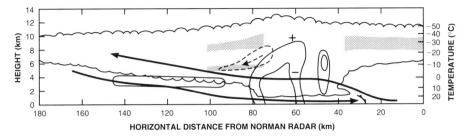

Fig. 8.29. First conceptual model of electrical structure in MCS based on existing kinematic and reflectivity model and balloon soundings of E. Movement of the MCS is left to right, corresponding to its tendency to move with a major west to east component. Solid contours in the core area indicate a mature cell in the core (echo to ground) and the reflectivity core centered aloft in front of a forming cell. (From Schuur et al. 1991, with permission.)

gion. Although they were not present in the vicinity of the sounding, convective cells of up to 50 dBZ were in the stratiform region during the MCS's mature stage.

The balloon ascended about 130 km from the primary convective zone. The region of the balloon flight was characterized by low-level subsidence (i.e., a mesoscale downdraft). The vertical and horizontal components of the electric field were large, and five main charge regions were within the cloud (Fig. 8.31). The structure of the E profile is similar to the one described by Schuur et al. (1991) for the transition zone of another MCS. Marshall and Rust (1993) found not only a similarity, but that the structures of these two E profiles were nearly identical when Hunter et al.'s (1992) profile was shifted up 1.5 km (Fig. 8.32). This

implies that the spacing and depth of corresponding charge regions were similar in the two MCSs. Most charge layers in the 7 June 1989 MCS were within the rear inflow. This was not true of the charge layers in the transition zone observed by Schuur et al. Hunter et al. speculated that collocation of charge in the rear inflow may have resulted from upper level, front-to-rear advection followed by some type of charged particle sedimentation into the rear inflow. They speculated that in situ charging seemed necessary as well.

Although there was almost no significant horizontal component of E (i.e., E_h) in the Schuur et al. (1991) sounding, soundings in several MCSs have had large E_h, such as indicated by the high-frequency oscillations in E in Fig. 8.31. Both the E_h and E_z profiles revealed a granularity in charge structure. One way to visualize granularity is that there are blobs with different total charge density embedded in the stratiform region; these blobs could create the observed E_h. The granularity may be similar to the electrical inhomogeneities described by Imyanitov et al. (1972) and the electrically active zones described by Brylev et al. (1989) for other types of stratiform clouds.

(iii) Soundings of E in MCS convective regions.
Marshall and Rust (1991) showed two soundings in convective lines. One had data only up to 8 km and very low E. The other (their Fig. 1, profile 87148) ascended into the convective line to a height of at least 3 km, above which position data were lacking. The second balloon may have been advected out of the line, as the ascent rate of the balloon suggested that it was not in an updraft. Stolzenburg et al. (1994b) compared this sounding with one from another stratiform region about 30 km from the convection of a different MCS and found the two soundings were quite similar. However in the convective region Stolzenburg et al. studied, E was greater, the charge was denser, and the bottom part of the profile was shifted upward to colder temperatures (by ≈2°C). Also in the convective region, there was no significant neutral layer between the densest positive and densest negative charge regions.

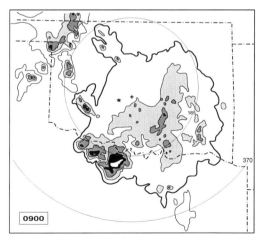

Fig. 8.30. Base-scan radar reflectivity, 7 June 1989 at 0900 UTC. Reflectivity contours starting with the heavy outer one are 18, 30, 41, 50, and 57 dBZ (the latter shown white inside black); range rings are kilometers from Norman, Oklahoma, denoted by the +. The star denotes location of balloon launch. (Radar data from the WSI Corp.; after Hunter et al. 1992, with permission.)

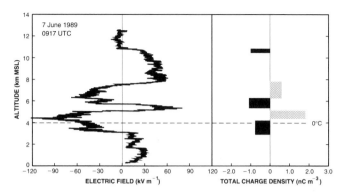

Fig. 8.31. Type A vertical profile of electric field and inferred charge distribution in stratiform region of an MCS, 7 June 1989. Sounding was made ≈130 km from the main convective zone. The total charge densities for major layers calculated from the E sounding are shown to the right. (From Marshall and Rust 1993, with permission.)

Stolzenburg et al. (1994b) showed a sounding (to 11.2 km) from 23 May 1991 when a balloon was launched directly into the apex of a bowed convective line, which often results from the rear inflow jet causing very high winds near or at the ground. They noted that the sounding had a mean updraft of ≈8 m s^{-1} between 1.7 and 7.8 km. The peak updraft was ≈20 m s^{-1} at 4 km. Stolzenburg et al. compared the E profile with that from the stratiform region in the 2–3 June 1991 MCS. The most notable differences were at 1.5–6 km; the convective sounding had deeper low-level positive charge (up to 3.7 km), elevated negative charge centered at 0°C (4.2 km), and much less positive charge at mid-levels. The pronounced dense positive charge at 0°C in the stratiform region was not seen in the convective region. Above 6 km, the convective and stratiform region soundings were similar. This observation is relevant to evaluating the plausibility of advective charging of the stratiform region, as discussed later.

(iv) Two types of vertical profiles in stratiform regions. The intriguing findings in the first soundings led to coordinated measurements of electrical and other meteorological parameters that culminated in a spring research project called COPS-91 in Oklahoma (Cooperative Oklahoma Profiler Studies, 1991, Jorgensen and Smull, 1993). Facilities for the project included ground-based Doppler radars; airborne Doppler radar on a NOAA P3 airplane; and balloon-borne measurements of E, atmospheric thermodynamics, wind and sometimes precipitation particle charge and size. Analyses of the soundings led Marshall and Rust (1993) to conclude that there are two distinct vertical structures of charge in stratiform regions of MCSs. Subsequent observations show there may be other types, but these two occur most of the time. A typical example of the structure called Type A by Marshall and Rust is the sounding in Fig. 8.31; another example is shown in Fig. 8.32. From all soundings available thus far, Type A stratiform regions have had four to six major vertically stacked charge layers in the cloud. These layers have the following properties: (1) The polarity of the layers alternates, with negative charge lowest; (2) The relative vertical spacings between three, and often four, of the charge layers are the same; (3) A fifth charge region, which appears to be a screening layer, is often observed near cloud top; (4) Sometimes there is a sixth charge below the screening layer; and (5) One of the layers is at or near 0°C and can be of either polarity.

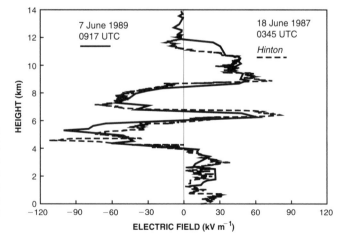

Fig. 8.32. Overlay of the two first MCS soundings in the U.S.; both are termed Type A. The nearly identical structure becomes obvious when the 7 June 1989 profile is moved upward 1.5 km. The 7 June data were low-pass filtered to simplify the plot. (From Marshall and Rust 1993, with permission.)

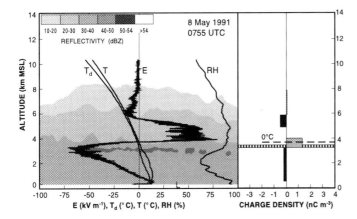

Fig. 8.33. Type B vertical profile of electric field along with thermodynamic variables and inferred charge distribution in trailing stratiform region of an MCS. L denotes the balloon launch location in the radar echo. The vertically hatched area denotes the quasi-isothermal layer near 0°C. (After Shepherd et al. 1996, with permission.)

The other structure of charge in stratiform regions, called Type B (Fig. 8.33), has been simpler in appearance and has had four to five major vertically stacked charge layers in the cloud. These layers have the following properties: (1) The polarity of the layers alternates, with negative charge lowest; (2) The relative vertical spacings between the centers of the lowest three are about the same; (3) The middle two charge layers seem tied to particular altitudes (or temperatures); (4) Sometimes there is a fifth charge layer near cloud top; and (5) One of the positive layers is always at or near 0°C.

Marshall and Rust (1993) found that within each the charge structure type, the E profiles and the charge distributions in MCS stratiform regions were remarkably similar from one location to another and from one MCS to another (e.g., Figs. 8.32 and 8.34). In both types, screening layers can occur at the upper and maybe lower cloud boundaries. The latter has not yet been conclusively shown owing to the difficulty in defining the lower cloud boundary. Zipser (1977) made aircraft observations beneath the trailing precipitation of a tropical squall line and noted that the cloud base was indefinite and composed of melting snow. Also, in both types there is often a dense (sometimes the most dense) region of charge near 0°C. Irrespective of polarity, this region often has the E_{max} for the entire profile. The polarity of charge at 0°C tended to be positive in Type B and negative in Type A electrical structures in the Marshall and Rust classification.

(v) Multiple soundings in same MCS stratiform region. Stolzenburg et al. (1994b) extended the MCS conceptual model by combining their spatially and temporally distributed electrical, kinematic, and reflectivity observations of an MCS. They analyzed five soundings from the same MCS, a Type B, which occurred during COPS-91. The MCS occurred on 2–3 June 1991, and one of the profiles is shown in Fig. 8.34. The conclusions from that extensive study are summarized here.

1. The basic reflectivity (Fig. 8.35) and kinematic structures of the stratiform region of this MCS were similar to the MCS conceptual model of Houze et

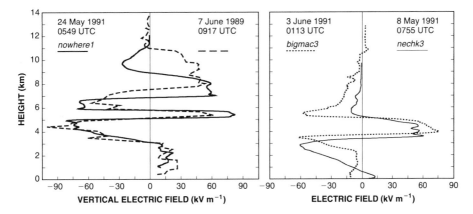

Fig. 8.34. Examples of similarities among different MCSs having Type A (left) and Type B (right) electrical structures. The Type A sounding from 24 May 1991 was shifted up ≈600 m; there was no shift of the Type B's. (After Marshall and Rust 1993, with permission.)

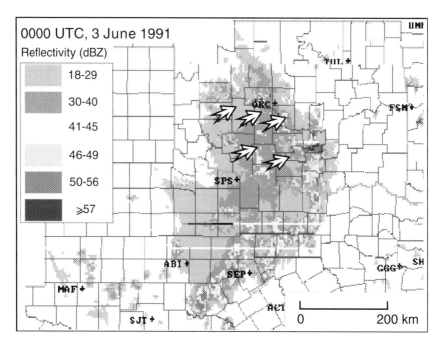

Fig. 8.35. Base-scan radar reflectivity on 3 June 1991, 0000 UTC. The stratiform-relative launch positions of the five balloons are indicated by arrowheads. These locations were chosen assuming a stratiform echo motion of 35 km h^{-1} from 285°. (Radar data from the WSI Corp.; after Stolzenburg et al. 1994b, with permission.)

al. (1989). The maximum areal extent of the low-level echo was 250 × 550 km. The radar bright band was the predominant feature in the reflectivity pattern with ≥45 dBZ through an area of at least 50 × 100 km. The reflectivity above the bright band was mostly horizontally homogeneous with gradual decreases in intensity and in the height of the echo top toward the rear of the system. The front-to-rear flow consisted of strong upper level southeasterlies, and the tendency toward rear inflow included the low-level westerlies that become weaker and shallower toward the front of the system. Separation between mesoscale ascent above and descent below occurred at an altitude of ≈5 km; vertical motions averaged over a 130 × 90 km domain were characteristically weak. Different analysis periods yielded similar values. For example, during one period the peak downdraft was −23 cm s^{-1} at a height (above ground) of 2.5 km, and weak updrafts of <10 cm s^{-1} were found above 5 km.

2. Electrical charge regions inside the stratiform cloud were horizontally extensive and layered. The E profiles were very similar over approximately 3 h and a storm-relative distance of 100 km. The horizontal homogeneity of the radar reflectivity pattern, especially above the bright band, supported the idea that the charges in MCS stratiform regions are layered.

3. The charge structure of the stratiform region was Type B with: negative charge near visual cloud base (i.e., the ceiling); positive charge mostly below the 0°C level and at least partially in a quasi-isothermal layer and in the radar bright band; the densest negative charge between 5–6 km; lower density positive charge 2–3 km deep between 6–10 km; and a fifth layer, with negative charge, near echo top in the part of the stratiform region closest to the convective region. In addition, there was low-density positive charge below cloud base. This conceptual model of charge distribution is significantly different from the one proposed before several soundings were available (Rutledge et al. 1993).

4. Variations in the charge structures of individual profiles may have been associated with variations in the intensity of precipitation, especially at low levels. More positive charge existed below 4 km in regions of heavy stratiform precipitation than in areas where precipitation was lighter.

5. No obvious correlations between ground flash activity and electric charge structure inside the stratiform cloud were detected.

6. A model of charge diffusion from a line source supported the hypothesis that preservation of advected charge originating in the MCS convective region is a viable mechanism for getting charge over the horizontal distances characteristic of trailing stratiform regions (this analysis is described in Sec. 8.2.4).

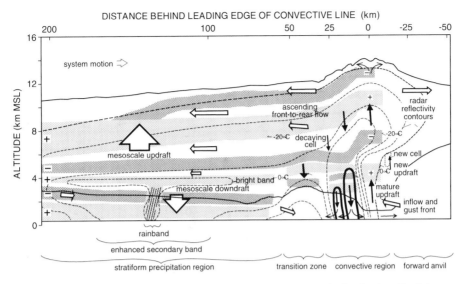

Fig. 8.36. Conceptual model of MCS with Type B electrical structure. Stylized radar reflectivity contours are at five levels of about 10–50 dBZ. Solid arrows depict convective region updrafts and downdrafts. Large open arrows indicate mesoscale updrafts and downdrafts. Small open arrows are system relative flow. No charge structure is shown in the forward anvil. Charge layers that seem related are connected between the regions labeled across the bottom. The radar bright band and a rainband beneath it are enclosed in dashed line. (From Stolzenburg 1996, with permission.)

7. The evidence suggested that charge advection from the convective region was important, especially at upper levels, and that one or more in situ charging mechanisms were important, especially at lower levels.

Note that the most recent update of the conceptual model of a Type B MCS is that developed in Stolzenburg (1996) and shown in Fig. 8.36.

(vi) Measurements of precipitation charge. There are only two sets of precipitation charge measurements that are clearly from MCSs: the measurements at the ground by Chauzy et al. (1985) and those aloft described by Bateman (1992) and Bateman et al. (1995). Chauzy et al. made their measurements beneath a squall line in southwestern France. Distributions of individual charges were not shown, but the ratios of particle charge (Fig. 8.28) suggested that the precipitation was driving the local field at the ground.

Bateman et al. (1995) described two soundings with balloons that carried a particle charge and size sensor, electric field meter, and rawinsonde. They showed precipitation data from one flight each into the stratiform region of a Type A MCS on 24 May 1991 and a Type B MCS on 8 May 1991. They found that both polarities of charge almost always coexisted. In the Type A MCS (Fig. 8.37), their measurements started in the back edge of the transition zone, but mostly were in the secondary reflectivity maximum. They had no way to tell whether

the charged precipitation particles were advected there or evolved in place. The particle size distributions were different between the flight in the Type A MCS and the Type B MCS (Fig. 8.38), but they observed particles up to 8-mm diameter in both. Bateman et al. concluded that there were two independent populations of precipitation sizes in the Type A MCS: The bimodal distribution had one peak at a diameter of 2.6 mm for particles at altitudes from about 5.2–6.6 km and had another peak at a diameter of 4.9 mm for particles at altitudes of 5.5–7.7 km. The bimodal distribution was not dependent on polarity (i.e., both polarities were in both distributions). The representativeness of the differences they found between precipitation in Types A and B is unknown since only these two cases exist.

As with measurements in smaller storms, no obvious relationship between the size and the charge of the precipitation was seen in the data. This included the particles whose charge was below the detection threshold; their diameters ranged across the detectable spectrum, especially in the Type A sounding.

In the stratiform region, Bateman et al. (1995) found that precipitation charge on individual particles ranged to >100 pC, but that the total charge densities calculated with Gauss's law and the E profiles were much greater than the net charge density calculated just from the precipitation. At some levels, the polarities of these two were opposite. An example from the secondary maximum in the stratiform region of a Type B

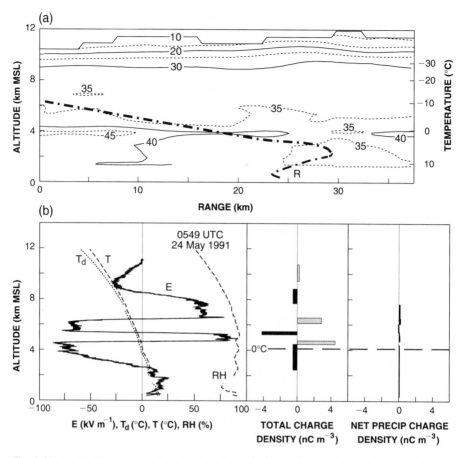

Fig. 8.37. (a) Vertical cross section of radar reflectivity (in dBZ) through secondary maximum in trailing stratiform region of a Type A MCS, 24 May 1991. The data were taken at 0549 UTC by the airborne Doppler radar on the NOAA P3; R is the release point of the balloon. Although smeared by the contouring of data, a well-defined bright band of ≥50 dBZ existed. The dash-dot line is the path along which precipitation data were acquired. (b) Profile of electric field, dew point, temperature, relative humidity, and calculated total and precipitation charge densities. (From Bateman et al. 1995: American Geo-physical Union, with permission.)

is shown in Fig. 8.38. The path of the balloon shows that measurements were obtained through a radar bright band. The implication from their charge density analyses is that significant charge was carried on smaller particles, (i.e., those below the detection thresholds of the particle instrument of ≈0.8 mm and 4 pC). This meant that small charged cloud particles were very important contributors to the electrical structure. The net precipitation charge density was small owing to the coexistence of both polarities. These results are different than those from small thunderstorms, in which the precipitation charge could account for most or all of the charge in the storm (see Sec. 7.12.2). This is a major difference; unpublished analysis of newer data obtained with the same instrument, but flown in small thunderstorms at Langmuir Laboratory, now suggest the difference is universal.

From their rudimentary analysis, Bateman et al. (1995) concluded that the charge separation process (at least up to the height data were obtained) was a steady-state process. Furthermore, they concluded that falling aggregates could have collected significant charge from cloud particles, as well as removing them. They further hypothesized that charge is separated locally in the bright band and is removed and recombined by precipitation—such a system could yield nearly constant regions of space charge for long periods.

Although the set of ground data and data from these two flights provided good information on precipitation charge, there are not nearly enough observations to have any confidence in the suggested generalities. As noted, the question of the relative importance of various sizes of charged hydrometeors to the net charge distribution remains.

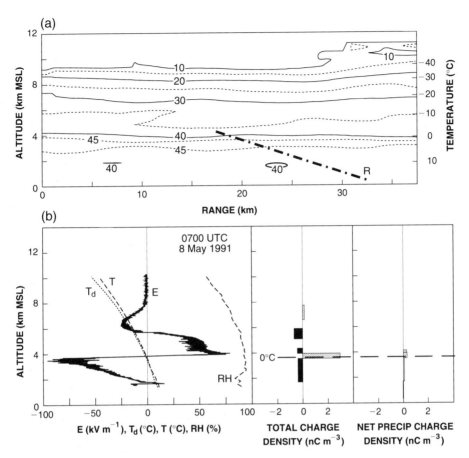

Fig. 8.38. (a) Vertical cross section of radar reflectivity (in dBZ) through secondary maximum in trailing stratiform region of a Type B MCS, 8 May 1991. The data were taken at 0653 UTC with the airborne Doppler radar on the NOAA P3; R is the release point of the balloon. Although smeared by the contouring, a well-defined bright band of ≥50 dBZ existed. The dash-dot line is the path along which precipitation data were acquired. (b) Profile of electric field, dew point, temperature, relative humidity, and calculated total and precipitation charge densities. The two transients at ≈1.7 km were due to lightning. (From Bateman et al. 1995: American Geophysical Union, with permission.)

(vii) E above MCS stratiform regions. The observation that red sprites and blue jets are prevalent over the stratiform region of large MCSs (Sec. 5.15) has produced considerable interest in the electric field above the tops of MCS stratiform regions. Though the initiation mechanism(s) for these phenomena is unknown, suggestions have included the above-cloud field change from positive ground flashes (e.g., Boccippio et al. 1995). Since sprites and jets have been associated closely with MCS stratiform regions, Marshall et al. (1996) examined their previous sounding data specifically for above-cloud E and field changes. Their preliminary conclusions, which contained no information about whether coincident sprites or jets occurred, follow. The data showed that lightning flashes in MCSs produce electric field changes that are detectable above the stratiform cloud, as other studies have shown for small thunderstorms (Sec. 7.7.3). Five of their eight soundings through and above trailing stratiform clouds of three different MCSs revealed one to four electric field changes. A total of 12 field changes, ranging from -1.1 to -4.0 kV m^{-1}, were observed at altitudes of 10–16 km. The electric field changes decayed exponentially, with a time constant proportional to the local conductivity. Nine of the 12 electric field changes observed above the electrical cloud top were coincident with positive flashes to ground. The coincident ground flashes included both single and multiple return stroke flashes detected by the National Lightning Detection Network, NLDN, with first-stroke peak currents ranging from 20–154 kA. Often two or three positive ground flashes were detected by the NLDN within 1 s of the electric field change measured above the cloud. No coincident ground flash was detected by the NLDN with two of the remaining three above-cloud field changes, but lightning was detected by a slow antenna at the ground on their mobile laboratory. Marshall et al. surmised that those flashes were either ground flashes missed by the national network or were cloud flashes. Not every positive ground flash caused a detectable field change above the cloud at the balloon's position. Only one negative ground flash was coincident with a field change above the cloud, but two positive ground flashes were also coincident (within 1 s) with that discontinuity. Because other negative ground flashes were in the vicinity of the electric field meters when they were above cloud top and because none of these negative flashes caused electric field changes detected by the balloon, Marshall et al. concluded that negative ground flashes do not usually cause electric field changes above MCS stratiform clouds, although such have been reported above other types of thunderstorms (see Sec. 7.7.3).

Marshall et al. (1996) used a simple model to show that discharging a horizontally extensive layer with charge densities of 1 or 3 nC m^{-3}, a thickness of a 400 m, and diameters of 20–200 km would produce field changes similar to those they observed. Since charge layers like those modeled are known to exist inside MCS trailing stratiform regions (Marshall and Rust 1993, Stolzenburg et al. 1994b) and since positive ground flashes that have extensive horizontal branches also occur in the same regions, Marshall et al. concluded it was possible that those ground flashes caused the electric field changes observed above cloud top. Further, they concluded that the observations and model results supported the suggestion of Boccippio et al. (1995) that sprites may be initiated by above-cloud field changes caused by positive ground flashes that discharge a horizontally extensive charge region in an MCS stratiform cloud. Although Marshall et al.'s simple model seemed to fit their observations, whether such a model is valid is unknown.

8.2.3. Lightning Observations

(i) Similarities to isolated storms. The location of lightning and the storm-relative evolution of flash rates in the convective regions of MCSs are similar in many respects to those in isolated storms. For example, in a squall line having a leading convective line and trailing stratiform region, Mazur et al. (1984) found that lightning occurred in two relatively compact regions, one near the top of the convective storm cells, just above 30-dBZ reflectivity contours, and the other at middle levels of the storm, inside and near 50-dBZ contours. Vertical dimensions of the regions were roughly 2-km thick, and horizontal dimensions were comparable to the width of individual cells. Very small flashes were restricted to the upper region; larger flashes could occur in either region. Either of the two regions could dominate lightning activity during a given period. The preference for one or the other region seemed to change systematically with the evolution of the storm. It appeared that the density of flashes was larger in the upper region as a cell was growing, and the density increased in the lower region as regions of high reflectivity descended from upper regions of the storm as the cell matured.

The few observations of total lightning in MCSs show that cloud flashes tend to dominate lightning activity: Mazur and Rust (1983) found that in one storm the number of cloud flashes was 40 times the number of ground flashes; Schuur et al. (1991) found 8 times more; Rutledge et al. (1992) found 3–15 times more (the factor increased with increasing total flash rate). As in isolated storms, ground flash rates for individual cells also have been reported to increase typically 10–15 min after the peak updraft, as reflectivity features descend from middle and upper levels (Billingsley 1994).

Goodman and MacGorman (1986), Stolzenburg (1990), and Morgenstern (1991) all have reported a tendency for the fraction of ground flashes having a single stroke to be largest and the average number of strokes per ground flash to be smallest during early stages of MCSs. Morgenstern noted that the average number of strokes also tended to have relative minima when the fraction of flashes lowering positive charge had relative maxima, because most positive ground flashes have only one stroke. (Nearly all positive ground flashes have only one stroke for all types and stages of storms.)

In MCSs on the Great Plains of the United States, ground flash rates often tend to increase as the area of the convective region increases (e.g., Kinzer 1974, Goodman and MacGorman 1986, Rutledge and MacGorman 1988). Similarly, Goodman and MacGorman found that contours of lightning ground flash density appeared similar to patterns in the distribution of rainfall (Fig. 8.39), and Rutledge and MacGorman determined that trends in negative ground flash rates were correlated with rainfall in the convective region of an MCS. However, ground flash rates may be sensitive to some storm parameters whose effect on rainfall at the ground varies from region to region and storm to storm; similarly, rainfall may be affected by parameters that have little effect on lightning production, as discussed in Sec. 7.19. Keighton et al. (1991) found that ground flash rates in one MCS were correlated much better with the area of large reflectivity at middle levels of the storm system than with the area at low levels. Billingsley (1994) found that the cells with stronger updrafts in one squall line produced fewer ground flashes, but considerable cloud flashes, much like some supercell storms.

(ii) Distinctive characteristics of lightning in MCSs. Lightning characteristics that are tied to MCSs involve patterns and evolution over spatial and temporal scales larger than in an isolated storm. Almost all studies of lightning in MCSs primarily have analyzed ground flashes. One exception was Mazur and Rust (1983), who observed total lightning in what was apparently a leading line, trailing stratiform MCS. Reflectivity struc-

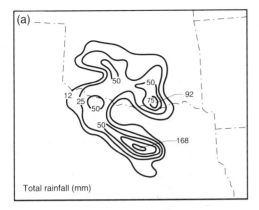

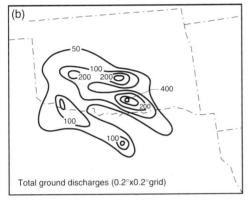

Fig. 8.39. Total precipitation (mm) and lightning ground flash density (km^{-2}) for 19.5 h between the time of the first $-32°$C cloud top on satellite imagery and the termination of the MCC on 13–14 June 1983. (From Goodman and MacGorman 1986, with permission.)

ture was from a 10-cm wavelength radar. Lightning was recorded with a 23-cm wavelength radar pointed roughly perpendicular to the leading convective region. Results are shown in Fig. 8.40. Flash rates in the stratiform region tended to be much smaller than those in the convective region. At almost all times, the maximum flash density detected in the squall line was toward the rear of the convective line, within or near regions of greatest reflectivity. However, there was a secondary maximum of flash density toward the leading edge of new cells. As a cell grew and its reflectivity increased, the flash density maximum moved toward the rear of the cell, eventually merging with lightning in older cells.

As cells aged, the radial length of the flashes they produced tended to become longer. Lightning having detected (i.e., radial) lengths <20 km occurred throughout the system, but longer flashes occurred only toward the rear of deep convection and in the stratiform region. Flashes with the longest horizontal extents tended to occur in the stratiform region. Several other investigators also have reported that lightning channels

frequently propagate tens of kilometers (sometimes more than 100 km) horizontally through the stratiform region (Rust et al. 1985b; Engholm et al. 1990; Schuur et al. 1991; Hunter et al. 1992). Observers located underneath the stratiform region frequently report seeing long horizontal flashes with an extensive network of channels that appear to develop along the lower cloud boundary. Rust et al. and Schuur et al. observed that these flashes propagated away from the convective line toward the rear of the stratiform region. This direction of development is the same as was seen by mapping systems (Ligda 1956, Krehbiel 1981) for long, horizontal flashes in multicell storms that had extensive regions of light precipitation adjoining heavier precipitation cores, but these storms may not have been mesoscale convective systems. Because of their appearance, flashes with complex channel structure beneath the stratiform cloud have been called *spider lightning* (Mazur et al. 1994). At least some of these long horizontal flashes are ground flashes, and our unpublished, anecdotal observations suggest that a single flash can strike ground in two or more locations several tens of kilometers apart.

The first study to examine ground flash evolution over entire system lifetimes was by Goodman and MacGorman (1986), who examined ten MCCs that occurred during the warm seasons of 1981–83 within range of a regional ground strike mapping network (Table 8.1). The network recorded only negative ground ($-$CG) flashes during most of the analyzed period, and part of the life cycle of several of the MCCs occurred outside the coverage of the network. Figure 8.41 shows hourly flash counts for the 5 most complete cases and a composite of all 10 cases relative to the time of the peak ground flash rate and to three of the stages of MCC evolution: (1) MCC initiation, when the MCC size criteria were first satisfied, (2) maximum extent of the cloud-shield, and (3) MCC termination, when the size criteria no longer were satisfied. Although there was considerable variability, there was a tendency for ground flash rates to follow a pattern: (1) From first storms until MCC initiation, hourly ground flash counts increased rapidly, with the rate of increase accelerating as individual storms intensified and merged together to form mesoscale structures. Peak strike densities also increased; (2) Usually the hourly flash count continued to increase after MCC initiation and peaked before the part of the cloud shield having temperatures $\leq -32°$C reached its maximum extent (an average of 2.6 h before). In only one case did hourly ground flash counts peak after the maximum extent of the cloud shield; (3) Hourly flash counts usually decreased rapidly from their peak. Typically the ground flash rate when the cloud shield reached its maximum extent was 65% of the maximum ground flash rate; and (4) Ground flash rates were almost always much smaller at MCC termination than at maximum extent and usually continued to decrease after MCC termination.

MCCs are often prolific producers of lightning. Goodman and MacGorman (1986) found peak ground

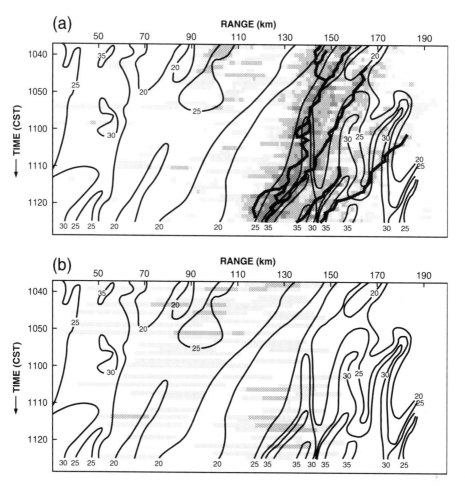

Fig. 8.40. Lightning echo flash density relative to storm reflectivity. Thin-line contours indicate reflectivity in dBZ. Shading denotes the number of lightning flashes along the radar beam per kilometer for each minute as follows: light = 1 km^{-1}, medium = 2–3 km^{-1}, and dark = 2–3 km^{-1}. Thick lines trace maxima in the lightning flash density. (a) Lightning flashes with echo lengths along the radar beam <20 km. (b) Lightning flashes with echo lengths along the radar beam ≥20 km. (Adapted from Mazur and Rust 1983: American Geophysical Union.)

flash rates of 1800–3300 h^{-1}. MCCs typically sustained rates of more than 1000 h^{-1} for nine consecutive hours. Neither the average flash rate nor the total number of flashes appeared to be a function of the maximum size of the cloud shield. In some regions, a single MCC can produce a significant fraction of the ground flashes that occur in a year, up to 25% of the annual total.

In an analysis of ground flashes in 25 MCSs, some of which were MCCs, Morgenstern (1991) showed that the evolution of ground flash rates of MCSs that did not satisfy MCC criteria was similar to the evolution for MCCs. In particular, relationships between ground flash rates and the evolution of the cloud shield for MCSs were similar to those found for MCCs by Goodman and MacGorman (1986), and Morgenstern found a similar lack of correlation between the total

number of ground flashes and the maximum size of the cloud shield. In fact, some MCCs analyzed by Morgenstern produced less than half as many ground flashes as the least prolific MCC studied by Goodman and MacGorman, and most smaller MCSs produced flash counts comparable to or larger than these smaller MCC flash counts. However, all cases in which >10,000 ground flashes were produced qualified as MCCs. Furthermore, even the smaller flash totals were comparable to those of strong thunderstorms; 22 of the 25 MCSs studied by Morgenstern and 3 of 4 MCSs studied by Holle et al. (1994) produced more than 3,000 ground flashes. Some characteristics of the 25 MCSs studied by Morgenstern are summarized in Table 8.2. The case with only 775 ground flashes had less than half as many ground flashes as any other MCS in the

Table 8.1. Minimum, average, and maximum values in ten MCCs studied by Goodman and MacGorman (1986).

Characteristic	Minimum	Average	Maximum
Duration of MCC[a] (h)	11.0	14.3	19.5
Max −52°C area[b] (km²)	74,000	160,000	242,000
Number of CG flashes	12,022	22,316	32,832
−CG flashes before MCC initiation[c] (%)	11.9	32.7	65.0
−CG flashes before max extent[d] (%)	53.7	76.9	97.1
Peak hourly −CG flash count[e]	1,830	2,679	3,299
Time of peak relative to max extent[f] (h)	−7.5	−2.6	3.0

[a]Time from first storm to termination of MCC; [b]Maximum area of cloud shield colder than −52°C; [c]Percentage of negative ground flashes before storm system satisfied criteria for MCC; [d]Percentage of negative ground flashes before maximum extent of −32°C cloud shield; [e]Maximum negative ground flash count during any hour of a given MCC; [f]Time of peak hourly flash count relative to time of maximum extent of −32°C cloud shield

study and also was unique in being the only MCS in which almost all detected ground flashes were positive.

Peak ground flash rates tend to occur later than reports of severe weather. All MCSs studied by Goodman and MacGorman (1986) and by Morgenstern (1991) produced severe weather. Most severe weather reports occurred when hourly ground flash counts were increasing or large. Goodman and MacGorman reported that 43% of severe weather reports had occurred by MCC initiation, at which time ground flash rates typically were increasing rapidly, and almost all severe weather occurred before the maximum extent of the MCC cloud shield, at which time ground flash rates typically were beginning to decrease from their peak. No severe weather was reported after ground flash rates dropped from their peak to below 1000 h⁻¹, although severe weather occurred with lower flash rates early in MCCs. Note that Maddox et al. (1986) reported similar timing of severe weather relative to MCC evolution: In 12 MCCs, the majority of severe weather reports occurred before MCC initiation, and almost all occurred before maximum extent of the cloud shield.

Lightning produced documented damage in 5 of the 10 MCCs studied by Goodman and MacGorman

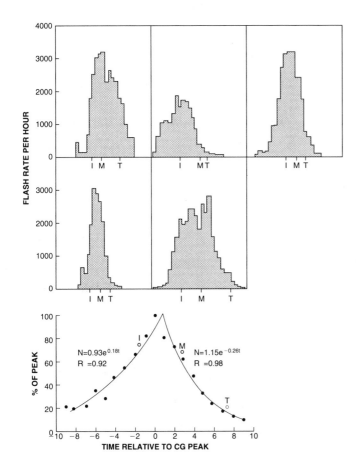

Fig. 8.41. Hourly ground flash rate for 5 MCCs (top) and a normalized composite of 10 MCCs (bottom). Phases of MCC evolution are indicated by letters on the horizontal axis: MCC initiation (I), maximum extent of the −32°C cloud shield (M), and MCC termination (T). The composite is plotted relative to the peak flash rate as the percentage of the peak in hours (±) from the peak. Regression equations are a fit to the fraction of flashes in a given hour, N, at time, t, for the composite. Time in the regression equations was measured from initiation or from peak flash rate for the respective equations. The correlation coefficient, R, of each curve is shown. (Adapted from Goodman and MacGorman 1986, with permission.)

Table 8.2. Minimum, quartile, and maximum values of cloud-to-ground (CG) lightning characteristics for 25 warm seasons MCSs studied by Morgenstern (1991). All cases occurred in 1986.

Characteristic	Minimum	25%	Median	75%	Maximum
Duration[a] (h)	7.2	9.7	11.9	16.7	22.7
Max $-52°C$ area[b] (km^2)	63,700	73,700	107,600	159,500	253,600
Number of CG flashes	775	4,000	6,800	12,900	40,888
Number of $-$CG flashes	11	3,650	6,150	12,400	39,063
Number of $+$CG flashes	137	225	465	800	2,762
Ratio of $+$CG to all CG (%)	1.5	3.5	6.0	13.5	98.6
Max 0.5-h $+$CG flash count[c]	20	35	70	115	645
Max 0.5-h $+$CG: all CG[d] (%)	4.8	6.5	19	46	100

[a]Time from the first to the last CG flash; [b]Maximum area of the $-52°C$ cloud shield; [c]Maximum number of positive ground flashes during any 0.5-h period of a given MCS; [d]Maximum ratio during any 0.5-h period of a given MCS.

(1986). Lightning damage tended to be more frequent when ground flash rates were large, and so tended to occur later than severe weather. Eighty percent of the incidents of lightning damage occurred between MCC initiation and maximum extent.

The only study of MCSs to examine the location of severe weather relative to lightning ground strikes was by Holle et al. (1994), who found severe weather in two of the four MCSs they studied. They analyzed one MCS in detail and found that severe weather tended to occur on the south and west sides of the convective region, near clusters of negative ground strikes. Consistent with previous reports, most of the severe weather occurred during the stage of rapid growth of both precipitation and lightning flash rates; the rest occurred during the mature stage, when mesoscale organization was most pronounced and lightning flash rates were large. Again, severe weather reports ended when ground flash rates dropped below 1000 h^{-1}.

One distinctive characteristic of many MCSs is a tendency for different polarities of ground flashes to occur in different regions. Figure 8.42 shows an example of lightning strikes superimposed on a satellite image of the cloud top temperature of an MCS. Negative strikes were clustered near or inside regions of deep convection. For negative strike points under the coldest cloud tops, this is obvious. Farther west, deep convection was either hidden by the cloud shield of older cells or was on the western edge of the cloud shield, where new cells were just starting to grow rapidly and become noticeable in the satellite imagery. Most of the lightning strikes northeast of the coldest cloud tops (i.e., in the stratiform region downshear of older cells) were positive ground flashes.

Typically, most ground flashes that strike in the convective region of an MCS are negative flashes. The fraction allocated to each ground flash polarity in the stratiform region has depended partly on how the stratiform

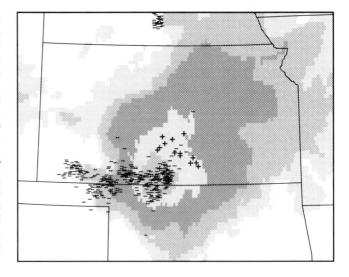

Fig. 8.42. Tracing of cloud-to-ground lightning strikes superimposed on satellite infrared imagery of cloud tops for an MCS on 14 August 1986 at 0000 CST. Lightning is from a 0.5-h period centered on the time of the satellite image. The location and polarity of ground strike points are indicated by $-$ for negative flashes and $+$ for positive flashes. Shading indicates cloud top temperature. The outermost shading indicates cloud top temperature $\leq -32°C$. Successive thresholds for changes of shading inward to the innermost shading are $\leq -46°C$, $\leq -52°C$, $\leq -64°C$, and $\leq -70°C$. The northern and southern boundaries of Kansas are at 40°N and 37°N, respectively, a separation of approximately 335 km. (From Morgenstern 1991, with permission.)

region was defined. In squall lines, there frequently were regions in which clusters of negative ground strikes extended into weaker reflectivity outside elongated reflectivity maxima at low levels in the convective line (e.g., Fig. 8.43). If the definition is based solely on radar reflectivity structure at low levels, these negative ground flashes are tabulated as occurring in the stratiform region. If extensions of convective lines are excluded from the stratiform region, the majority of ground flashes in the stratiform region usually are positive flashes. However, negative ground flashes do occur in the stratiform region, just as positive ground flashes occur in the convective region.

The appearance of spatial separation between positive and negative ground flashes gave rise to the term *bipole pattern,* or simply *bipole.* Orville et al. (1988) defined the bipole as a vector that connects a region of negative ground flashes to a region of positive ground flashes. Subsequently, characteristics of bipole patterns were studied in more cases by Engholm et al. (1990) and Stolzenburg (1990). However, it is difficult to determine how some of the results from these and other bipole studies apply to MCSs, because bipole patterns were found in a variety of storm situations that were grouped together with MCSs in statistical descriptions. Other situations in which bipoles have been documented include (1) two or more isolated storms or separate rainbands several tens of kilometers apart, each dominated by a different polarity of ground flash (e.g., Engholm et al.), (2) two parts of a single convective line dominated by opposite polarities (usually positive polarity was north of negative) (e.g., Engholm et al., Morgenstern 1991), and (3) the lightning distribution from the entire lifetime of a fast moving, single storm dominated by negative flashes while it was mature, but dominated by positive ground flashes as it dissipated (Brook et al. 1989).

The properties of bipoles that appear most relevant to MCSs are the following: (1) Ground flash rates in a bipole usually are much smaller in winter than in spring and summer; (2) Positive ground flashes often are a larger fraction of ground flash activity during the winter; (3) The orientation of the bipole is usually along the vertical shear vector of the horizontal wind between middle and upper levels (typically close to the direction of the geostrophic wind for winter storms), which governs the direction of storm-relative outflow from the upper region of deep updrafts; and (4) If the bipole vector is defined as pointing from negative ground flashes to positive, then it tends to point from deeper to shallower convection, whether the shallower convection is in another storm or in the stratiform region of the same storm system.

Positive and negative flash rates and strike locations usually evolve in different, but systematic, patterns in MCSs. We already have noted that negative ground flash rates tend to increase rapidly before and during MCS initiation and to peak before the maximum extent of the cloud shield. These relationships are an indication that negative ground flashes in MCSs are associated primarily with the deep convection that is characteristic of the convective region. Other observations provide further support that such an association exists. Morgenstern (1991) observed that negative ground flash rates tended to peak at approximately the time at which the cloud tops of MCSs attained their coldest temperatures and at which the area of the coldest cloud tops was largest, as shown in Fig. 8.44. Furthermore, in every study of ground flash location in spring and summer MCSs, most negative ground strikes have been located in the convective region throughout almost all of the MCS life cycle. For example, Holle et al. (1994) found that 86–90% of all ground strikes (mostly negative) occurred in the convective region (defined using only radar data—the percentage would have been even higher if lightning from extensions of the line were included in the convective region).

As the stratiform region forms, positive ground flashes frequently begin to occur within it (e.g., Rutledge and MacGorman 1988, Rutledge et al. 1990), although the density of strike points in the stratiform region is roughly an order of magnitude smaller than is typical of convective regions (Orville et al. 1988, Rutledge and MacGorman 1988, Stolzenburg 1990). Morgenstern (1991) noted that positive ground flashes occurred in the stratiform region in 17 of the 25 MCSs she analyzed. Many of the ground strike points in the stratiform region appear to cluster near regions of local reflectivity maxima ($\approx$50% of positive strikes were near reflectivity maxima in Rutledge and MacGorman)(Fig. 8.43). Rutledge and Petersen (1994) showed that low-level reflectivity maxima having ground flashes nearby in the stratiform region also tended to have larger reflectivities above the 0°C isotherm, and they hypothesized that noninductive charging was thereby enhanced.

As a stratiform region with positive ground flashes continues to grow in both area and reflectivity magnitude, positive ground flash rates usually increase (Fig. 8.45) (Rutledge and MacGorman 1988, Rutledge et al. 1990, Holle et al. 1994, Nielsen et al. 1994). Rutledge and MacGorman reported that positive ground flash rates increased as long as radar-inferred, stratiform rainfall also increased; the rate began decreasing once stratiform rainfall leveled off at a relatively constant value. Stratiform rainfall leveled off because regions of stronger stratiform precipitation began dissipating, although the area of stratiform precipitation continued to increase. This is consistent with the observation by Holle et al. that the number of positive flashes per unit rainfall volume peaked at about the time of transition from rapid growth to the mature stage. Furthermore, the analysis of satellite imagery by Morgenstern (1991) and our additional unpublished studies have suggested that positive ground flash rates decreased as the mesoscale updraft began to dissipate, which is consistent with a subsequent decrease in the area and precipitation rates associated with local reflectivity maxima in the stratiform region.

Morgenstern (1991) labeled the occurrence of positive ground flashes in the stratiform region the *strati-*

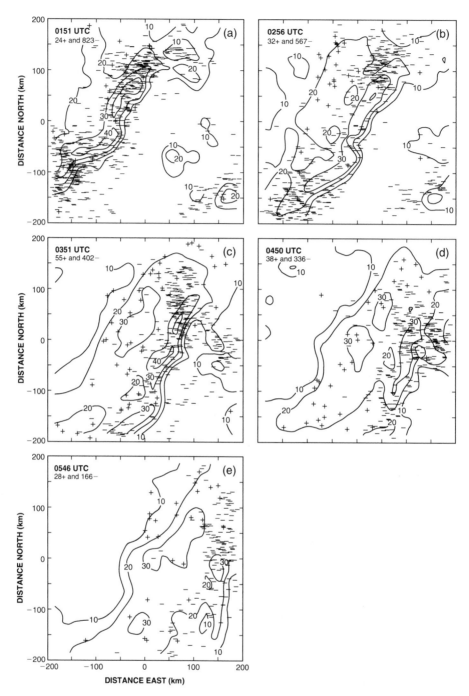

Fig. 8.43. Lightning ground strikes plotted relative to radar reflectivity contours (with ground clutter omitted) for an MCS on 10–11 June 1985. Contours indicate reflectivity (in dBZ) from a scan at an elevation angle of 1°. Ground strike points are plotted as − for negative and + for positive flashes for a 30-min period centered on the time of the radar scan. Strike locations were shifted to compensate for movement of the convective line. Scan time and the total number of ground flashes in the region are indicated in the upper left corner of each plot. The origin is the location of the radar. (From Rutledge and MacGorman 1988, with permission.)

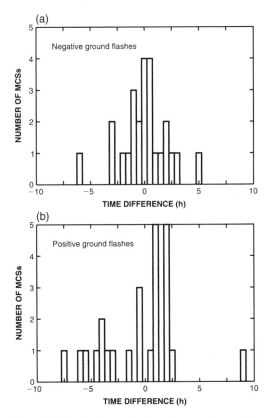

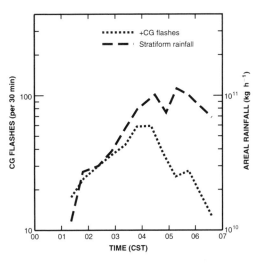

Fig. 8.45. Positive ground (+CG) flash counts per 30 min and hourly rate of precipitation mass integrated over the stratiform region of an MCS on 10–11 June 1985. Precipitation amounts were inferred from S-band radar data. (Adapted from Rutledge and MacGorman 1988, with permission.)

Fig. 8.44. The difference in time between the coldest cloud top and (a) the maximum number of −CG flashes per 0.5 h and (b) the maximum number of +CG flashes per 0.5 h for MCSs that occurred on the Great Plains during the warm season of 1986. Negative values indicate that flash rates peaked before cloud tops. (From Morgenstern 1991, with permission.)

form/dissipating mode to distinguish it from other storm situations in which positive ground flashes can occur (discussed in Sec. 7.15.2). It would have been simpler to label this mode the stratiform mode. However, it was possible that the simpler label would be misleading, because the local reflectivity maxima associated with positive ground flashes in stratiform regions can be produced by more than one process. In many MCSs, local reflectivity maxima form in the stratiform region, apparently because particles advected from the convective line along trajectories that reach the reflectivity maxima grow faster than particles that end up elsewhere in the stratiform region (e.g., Smull and Houze 1987, Rutledge and Houze 1987, Matejka and Schuur 1991, Braun and Houze 1994). Several studies have discussed possible effects of particle advection and growth on electrical characteristics of the stratiform region, as will be discussed further in Section 8.2.4. However in some MCSs, local reflectivity maxima in the stratiform region also have been

caused by cells that moved from one end of the convective line into the stratiform region as they dissipated (e.g., Hunter el al. 1992). Because dissipating storm cells often produce positive ground flashes, it is not clear that the reasons for the positive ground flashes when reflectivity maxima are caused by dissipating cells are the same as when reflectivity maxima are caused by particle advection and growth in the stratiform region. Because Morgenstern had no reflectivity data with which to identify local maxima in the stratiform region and to determine whether the maxima were due to dissipating storm cells or particle advection and growth in the stratiform region, she grouped both into a single category of positive ground flash occurrence, but she also incorporated both in the label as a cautionary measure in case of possible differences.

The stratiform region is not the only region of an MCS in which positive ground flashes can dominate ground flash activity. Several studies (Rutledge et al. 1990, Schuur et al. 1991, Morgenstern 1991, Holle et al. 1994) have noted that a substantial fraction (sometimes even a majority) of ground flashes in individual storms within the convective region of an MCS can be positive ground flashes (Fig. 8.46). Morgenstern called this mode of positive ground flash occurrence the *convective mode,* which for clarity we call the *+CG-convective mode.* She reported that it occurred in 9 of 25 MCSs. In this mode, positive ground flashes occurred in a dense cluster, much as negative ground flashes often do. During the +CG-convective mode, other cells in the convective region often were dominated by negative flashes, but positive ground flashes typically com-

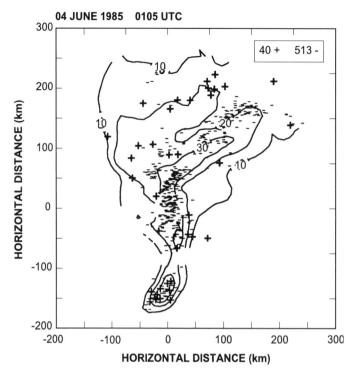

Fig. 8.46. Positive (+) and negative (−) lightning ground strikes for a 30-min period centered at 0105 UTC on 4 June 1985. Radar contours are labeled in dBZ. Note the concentration of +CG flashes in the southernmost storm cell. (From Rutledge et al. 1990, with permission.)

prised at least 20% of all ground flashes in the convective region. Usually the storm that was dominated by positive ground flashes occurred early in the lifetime of the MCS and on the southern or southwestern end of the convective region, where most severe weather occurred. This relationship between positive ground flashes and severe weather appeared similar to relationships in supercell storms dominated by positive ground flashes (Sec. 8.1.10). Most MCSs having the +CG-convective mode occurred in April–June in the study by Morgenstern, which focused on the Great Plains. Preliminary data suggested that the +CG-convective mode occurs somewhat later in the warm season in more northern locations, similar to the northern migration of severe storm occurrence (Kelly et al. 1985) and MCCs (Maddox et al. 1986).

Some MCSs exhibit both the stratiform/dissipating mode and the +CG-convective mode, and some MCSs do not appear to have characteristics of either mode. Three of the 25 MCSs studied by Morgenstern (1991) had both modes. In 2 of the 25, positive ground flashes were a small fraction of ground flash activity throughout the lifetime of each MCS and tended neither to occur in the stratiform region nor to occur frequently in any storm cell. Instead, a few positive ground flashes

were scattered among the numerous negative ground flashes in the convective region.

Morgenstern (1991) found that the evolution of positive ground flash rates relative to satellite cloud imagery tended to be quite different for the convective and stratiform/dissipating modes. In the +CG-convective mode, positive ground flash rates usually peaked at approximately the time of the coldest cloud top early in the lifetime of the MCS. (The absolute minimum temperature sometimes occurred later.) In the stratiform/dissipating mode, however, positive ground flash rates tended to peak much later. They usually peaked at approximately the time of the maximum area within an isotherm that encompassed much of the region of mesoscale updrafts in the stratiform region. The cloud top temperature giving the best correlation between cold cloud top area and positive flash rates in the stratiform/dissipating mode varied, but was typically $-50°$ to $-56°C$. The combination of the two modes caused the time of peak positive ground flash rates relative to the time of the coldest cloud tops to have the distribution shown in Fig. 8.44, with the maximum positive ground flash rate occurring before the coldest top in most convective cases and the coldest cloud top occurring first in most stratiform/dissipating cases.

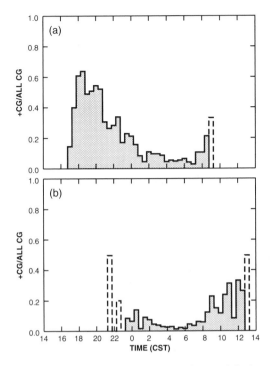

Fig. 8.47. Evolution of the fraction of ground flashes lowering positive charge per 30-min period for (a) an MCS in the +CG-convective mode on 31 March–1 April 1986 and (b) an MCS in the stratiform/dissipating mode on 8–9 August 1986. Dashed lines indicate that ≤5 ground flashes occurred during that 30-min period. (From Morgenstern 1991, with permission.)

Similarly, the two modes caused the fraction of ground flashes that lowered positive charge to evolve differently (Fig. 8.47). In the +CG-convective mode, the fraction was largest early in the lifetime of the MCS. In the stratiform/dissipating mode, the fraction was usually largest toward the end of the MCS, after both positive and negative ground flash rates had peaked. Although flash rates decreased for both polarities toward the end of an MCS's lifetime, the negative flash rate typically declined more rapidly than the positive flash rate, so the fraction of flashes that lowered positive charge increased. Similar evolution of the fraction has been reported by several other studies (Engholm et al. 1990, Stolzenburg 1990, Holle et al. 1994, Nielsen et al. 1994).

Another possibly distinctive characteristic of lightning in MCSs is the amplitude distribution of positive ground flashes. Recall that the range-normalized signal amplitude detected by lightning strike mapping systems is roughly proportional to the peak return stroke current (e.g., Orville 1991, Idone et al. 1993). When Morgenstern (1991) and MacGorman and Morgenstern (1997) examined median amplitudes every 30 min throughout the life cycle of MCSs, they found no

clearly systematic relationship between the evolution of median amplitudes and MCS evolution for either positive or negative ground flashes, although this conclusion is tentative and controversial. Furthermore, the distribution of amplitudes found by these investigators for negative ground flashes in MCSs was essentially the same as that found in larger data bases encompassing all types of warm season storms (e.g., Orville et al. 1987). However, the amplitude distribution of positive ground flashes in many MCSs appeared to be markedly different from that found in large, climatological data bases.

Morgenstern (1991) and MacGorman and Morgenstern (1997) divided the amplitude distributions of positive flashes produced by MCSs into small, medium-, and large-amplitude categories (Fig. 8.48). The medium-amplitude category is most similar to the distribution found for large, all-inclusive data bases. Because the number of flashes per amplitude interval decreased rapidly with increasing amplitude in the small-amplitude distribution, this distribution could be detected only within 150–250 km of mapping stations. However, the other categories were not caused simply by attenuation of the smallest amplitudes; middle- and large-amplitude distributions were identified when MCSs were close to lightning mapping stations, as well as when they were more distant. All but one of the nine MCSs having positive ground flashes in the +CG-convective mode had either the middle- or large-amplitude distribution. MCSs having positive flashes in the stratiform/dissipating mode were distributed almost equally among the three categories.

(iii) Source of lightning in the MCS stratiform region. A question that has received considerable discussion is, "What is the source of charge for ground flashes in the stratiform region?" Note that we include both positive and negative ground flashes, although in the stratiform region there has been more study of the positive strikes. If we adopt the view of Kasemir (1960, 1984) that the location of charges is not as important for lightning initiation as the distribution of the electric field, then the question might be rephrased as, "How do MCSs initiate flashes that strike ground in the stratiform region, and what configurations of the electric field are responsible for each polarity of ground flash?"

Published in situ data describe the electric field and inferred charge distributions in MCS stratiform regions. As yet, however, we have no data to determine where ground flashes in the stratiform region are initiated or how they propagate, so we have little with which to begin addressing these questions. If we assume that the charge that causes both the initiation and propagation of a ground flash is in the stratiform region in the vicinity of the strike point, then the stratiform charge structure is very relevant to understanding the production of these flashes. There is some evidence to support such an assumption: (1) Positive ground flashes in the stratiform region can continue for several hours after ground flashes have become infrequent in the con-

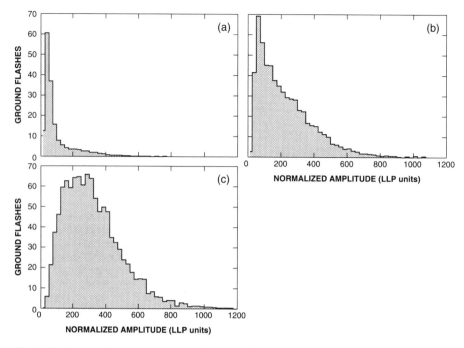

Fig. 8.48. The distribution of range-normalized amplitudes for the +CG flashes in (a) the 6 MCSs in the low-amplitude category, (b) 9 MCSs in the medium-amplitude category, and (c) 10 MCSs in the high-amplitude category. Peak currents can be estimated from measured signal amplitudes by using Eq. 5.22, a relationship determined by Idone et al. (1993). (From Morgenstern 1991, with permission.)

vective region, and they can outnumber the ground flashes in the convective region during late phases of MCS dissipation; and (2) Positive ground flashes sometimes occur in shallow storms isolated from deeper convection (i.e., no connecting cloud for several tens of kilometers) that has negative ground flashes.

However, the assumption that conditions in the stratiform region above a lightning strike point are related in any way to initiation of the flash may be unwarranted. Several investigators (e.g., Engholm et al. 1990, Marshall and Rust 1993, Stolzenburg et al. 1994b) have expressed their puzzlement as to how observed electric field structures in the stratiform region can cause positive ground flashes. Because long horizontal cloud and ground flashes are observed in the stratiform region, it could be that most ground flashes in that region are initiated far from the point at which they strike ground. Such was the case for positive ground flashes observed by Krehbiel (1981) in the stratiform cloud of a dissipating storm (probably not an MCS) in Florida.

If the strike point is far from the initiation point, then the observation that many positive ground strikes cluster around heavier stratiform precipitation may mean simply that flashes that were initiated elsewhere propagated horizontally until they approached the heavier precipitation, where they tended to produce a channel to ground. We expect that these long flashes propagate along a horizontal surface of small, relatively constant electrostatic field until they encounter large enough vertical perturbations in the approximately stratified charge distribution to drive a leader toward ground, as was discussed in Section 7.16.4. Furthermore, we have noted previously that the charge and electric field structure within regions of heavier stratiform precipitation often are somewhat different than observed in other parts of the same stratiform region. These inhomogeneities in the charge and electric field structure may be sufficient to change the direction of propagation of a previously horizontal flash. Note that even if ground strikes in the stratiform region are caused by flashes that begin far away, the ability of flashes to propagate through the stratiform region may still depend on the presence of highly charged layers; remember that Williams et al. (1985) hypothesized from their spark experiments that the distance lightning can propagate increases as the charge density through which it propagates increases. Data from two- or three-dimensional lightning mapping systems are needed to resolve the issue of where flashes in the stratiform region are initiated and how they propagate.

8.2.4. Source of charge in the MCS stratiform region

Whether or not charge in the stratiform region is responsible for producing electric fields that lead to initiation of ground flashes, the issue of how the stratiform region becomes charged still is of interest. Charge tends to occur in layers in the stratiform region, with estimates of the total charge contained in a single layer ranging up to 20,000 C (Marshall and Rust 1993). This is more than an order of magnitude larger than the charge that is thought to exist in the main charge regions of typical isolated thunderstorms, and so poses a considerable requirement on any hypothesized mechanisms for producing and maintaining the charge. We have discussed charging mechanisms previously, but we focus in this section on mechanisms that appear relevant to MCS stratiform regions. Recall that when we speak of charge generation, we mean both the initial separation of charge that occurs on a microphysical level and the systematic redistribution of charged particles (i.e., macroscopic separation) that leads to regions of net charge.

Several mechanisms have been proposed for producing the charge layers in the stratiform region. In this section, we examine two broad categories of mechanisms that encompass these hypotheses: (1) advection and (2) local generation. Although these terms can apply to charge anywhere in a thunderstorm, in this section we use them specifically to apply to the stratiform region. Thus, by *advection of charge,* we mean that charge separated in the convective region is transported into the stratiform region. By *local generation of charge,* we mean that particles gain charge by processes in the stratiform region; before gaining charge by local generation, particles can be either charged or uncharged, depending on their previous history. A combination of observations and simple numerical models have been used to evaluate the plausibility of each category of processes. Stolzenburg et al. (1994b) concluded that no single mechanism seems likely to account for all the charge layers repeatedly inferred from the vertical profiles of E. It is possible that both categories of mechanisms contribute to the charge structure of the stratiform region.

Some of the earlier studies we will discuss were done before the publication of measurements that clearly indicate there are four or more significant charge layers in the stratiform region. Thus, these earlier studies focused on explaining the monopolar or inverted dipolar structure that was assumed to be needed to produce positive ground flashes from horizontally extensive charge layers. Although we now know that the charge distribution in the stratiform region is not this simple, these studies still may be useful in explaining particular aspects of the charge distribution. We remind you that we do not know whether any particular charge region controls the polarity of ground flashes that occur in the stratiform region. Although many ground flashes obviously lower positive charge to ground in the stratiform region, it is unclear from vertical profiles of E why one polarity of ground flash dominates any particular area of the stratiform region.

(i) Advection into the stratiform region. If the stratiform region gains charge by advection of charged particles from the convective region, then obviously the particles became charged by some process or combination of processes in the convective region. We will not consider the charge generation mechanisms of the convective region in this section. Note also that winds in the stratiform region are not all directed away from convective cores. The flow in part of the stratiform region often is toward the convective cores (see Sec. 8.2.1). However, outflow from the upper part of the convective region is important in forming the stratiform region. Since charge definitely is separated in the convective cells, advection seems an obvious candidate for a charging mechanism.

Rutledge and MacGorman (1988) found that the maximum 30-min positive ground flash rate (mostly from flashes in the stratiform region) lagged the maximum negative ground flash rate (mostly from flashes in the convective region) by ≈ 2 h. Furthermore, they found that rainfall integrated over the stratiform region attained its maximum value ≈ 2 h after rainfall integrated over the convective region. They hypothesized, therefore, that positive charge was advected from the convective line into the stratiform region. To test this hypothesis, they calculated the time required for ice particles to travel from the upper levels of the convective line and descend to the melting level (from which they would fall rapidly to ground) in the stratiform region of the MCS they studied. They concluded that this time was comparable to the $\approx$2-h lag that they observed between positive and negative ground flashes and between stratiform and convective rainfall. Furthermore, the horizontal distance traversed by the particles was comparable to the distance of positive ground strikes and the region of heavier stratiform precipitation from the convective line. From these results, they suggested that advected particles could carry enough positive charge into the stratiform region to be responsible for the positive ground flashes there. They also noted that it was possible that particles would continue charging inside the stratiform region. They did not consider how long charged regions advected into the stratiform region would maintain sufficient charge density.

Several investigators have examined various factors affecting how fast charge densities decrease as parcels of charge are advected. Orville et al. (1988) noted that the charge relaxation time (i.e., the time in which ohmic leakage currents reduce the charge in a region by a factor of 2.7) measured in thunderstorms by Rust and Moore (1974) was 4400 s, and estimated that charge from the convective region could move a distance consistent with the length of observed bipoles in that amount of time. Hill (1988) investigated how rapidly charge density would decrease due to electrostatic repulsion between charged particles within a sphere of positive charge. He estimated that the radial velocity of expansion V_R would be

$$v_R = \left(\frac{8R_0^3 n_0}{3 \, C_D \rho_I} \right)^{\frac{1}{2}} \frac{q}{rR}, \qquad (8.1)$$

where R_0 is the initial radius of the sphere, R is the present radius, n_0 is the initial number density of charged particles within the sphere, C_D is the drag coefficient, and ρ_I, q, and r are the mass density, charge, and radius of the individual ice crystals, respectively. Although a spherical geometry overestimates the rate of expansion compared with an extensive horizontal slab, the rate of expansion still was very low, resulting in expansion of only 0.1 km in 4400 s and a doubling of the radius in 62,750 s. Rutledge et al. (1993) obtained a similar expansion rate from electrostatic repulsion using a different approach. However, Rutledge et al. pointed out that electrostatic diffusion would be much slower than diffusion by turbulent mixing of air parcels, and Bateman et al. (1995) concluded that gravity also would dissipate a charge region faster than mutual electrostatic repulsion of the charged particles.

Rutledge et al., therefore, used a simple model to estimate the effect of turbulent mixing of air parcels. They assumed that a spherical source 1 km in diameter continuously injected charge horizontally into a plume. While being advected downstream in the plume, the charge was dispersed by turbulent diffusion both vertically and horizontally perpendicular to the plume. In their expression for diffusion, taken from Turner's (1969) treatment of a point source, diffusion was a function of the advection velocity and of the initial charge density. With an advection velocity of 15 m s^{-1}, densities of 4 and 10 nC m^{-3} decreased by more than an order of magnitude within 10 km and were no more than 0.1 nC m^{-3} at a distance of 60 km. Rutledge et al. concluded that the contribution of charge advection to the stratiform region was of the order of 0.1 nC m^{-3}. Comparing this with estimates of charge from local charge generation, they concluded that local charge generation dominated advection of charge at distances $\geq$60 km. They suggested that advection could be a significant source of charge only closer to the convective line.

Stolzenburg et al. (1994b) pointed out that studies of squall lines established that they should be treated as horizontal line sources, instead of point sources, when considering advection of momentum, heat, and moisture into the stratiform region (LeMone et al. 1984, Lafore and Moncrieff 1989, Gallus and Johnson 1991), and so suggested that a line source also would be more appropriate for charge advection. Material blown downwind from a line source would diffuse only in the vertical direction. Therefore, the decrease in charge density from a line source would be slower than from a spherical source. Using the approach of LeMone et al., they estimated that the rate of change in charge density of a parcel would be given by

$$\frac{d\rho}{dt} = -\frac{\partial}{\partial z} \left(\kappa \, \frac{\partial \rho}{\partial z} \right) + Q, \qquad (8.2)$$

where κ is the eddy exchange coefficient and Q is a source of local charging in the stratiform region. To compare their results with those of Rutledge et al. (1993), they used the same advection velocity, an initial charge density of 10 nC m^{-3}, $\kappa = 10$–50 m^2 s^{-1}, and $Q = 0$ (i.e., no charging in the stratiform region). Initial charge density was assumed to decrease linearly with vertical distance above and below the middle of the layer to zero at its upper and lower boundaries. Using a layer depth of 1 km, which equaled the sphere radius Rutledge et al. had used, they estimated that the charge density in the middle of the layer decreased to 2.0–7.3 nC m^{-3} (depending on κ) after the charge was advected a distance of 60 km. When the depth of the layer was 2 km, which they thought was more reasonable for the squall line they studied, the initial 10 nC m^{-3} decreased only to 6.7–9.0 nC m^{-3}. Therefore, Stolzenburg et al. concluded that advection of charge was a viable mechanism over distances characteristic of trailing stratiform regions of midlatitude MCSs.

(ii) Local generation of charge in the stratiform region. Observations supporting the importance of local charge generation include the following: (1) The number of charge layers is larger than would be expected by advection of charge from a grossly dipolar thunderstorm charge distribution. (Of course, if the charge distribution of the convective cell has more than two charge regions, then there can be advection into additional charge layers. However, particle trajectories limit which charges in the convective region can be advected to a given layer of charge in the stratiform region, and so probably eliminate advection from consideration as a source of charge for some layers.); (2) The lower charge layers do not slant downward as fast as would be expected from sedimentation of particles carrying charge, and so they appear to be constantly modified, perhaps by local charge generation; (3) Conditions appear favorable for charge generation by the noninductive mechanism in at least some stratiform regions. Although charging rates are slow, time scales relevant to the stratiform region are long enough for large amounts of charge to be separated. Studies have found small amounts of supercooled liquid water with small ice particles and large aggregates (i.e., clusters of ice particles) in some stratiform regions, and there appear to be enough particle interactions for the noninductive mechanism to be significant. The primary unknown is whether charging of rimed aggregates is similar to the charging of graupel particles, which are practically nonexistent in stratiform regions; (4) Charge advection can be difficult to reconcile with the observed horizontal winds in some charge layers. For example, some charge layers have been observed at altitudes at which the horizontal wind was rear-to-front, not front-to-rear; (5) In some MCSs, the horizontal wind in the mesoscale updraft region has a significant component parallel to the line. In many of these cases, dissipating cells move from one end of the convective region into the stratiform region and cause some of the reflectivity maxima observed in the stratiform region. Although this might be considered advection, charge generation may well continue in the dissipating cell after it enters the stratiform region; and (6) The large charge found fairly consistently near the 0°C level indicates that charge separation via a melting process could be important in the lower charge layers.

Simple numerical cloud models have been used in a few limited tests of the plausibility of local charge separation in the stratiform region. All have considered only the noninductive mechanism discussed in Chapter 3. Rutledge et al. (1990) used a simple one-dimensional cloud model and vertical profiles of temperature and microphysical parameters, such as ice particle concentrations, taken from a combination of measurements and previous model results. The amount of charge separated per collision between an ice particle and a relatively low-density graupel particle (0.3 g cm^{-3}) was taken from laboratory studies by Jayaratne et al. (1983) and Saunders and Jayaratne (1986). Using parameterizations similar to those described in Chapter 9, they calculated the profile of charging rates shown in Fig. 8.49. The peak charging rate of 1100 fC m^{-3} s^{-1} at a height between 4–5 km gave 3.6 nC m^{-3} in an hour, which is comparable to observed values of charge density in the stratiform region. If either the amount of supercooled liquid water or the number concentration of ice particles was changed significantly, the charging rate changed in the same direction. Of course, the calculated particle charging rates do not result in regions of net charge unless oppositely charged particles move apart. Rutledge et al. estimated that particles would move apart almost 2 km vertically in 20 min, and so concluded that local charge generation could generate enough charge to be a significant source for the stratiform region.

Rutledge and Petersen (1994) attempted to determine whether there was likely to be supercooled liquid water in the mesoscale updraft of two MCSs. They used a simple one-dimensional model that was horizontally homogeneous. The condensation rate, C_{vl}, was estimated by

$$C_{vl} = w \frac{\partial q_s}{\partial z} - D_{vi}, \qquad (8.3)$$

where w is the updraft magnitude, q_s is the saturated specific humidity, and D_{vi} is the deposition rate of water vapor to ice crystals. The first term on the right represents the vertical advection of water vapor. The deposition rate was given by

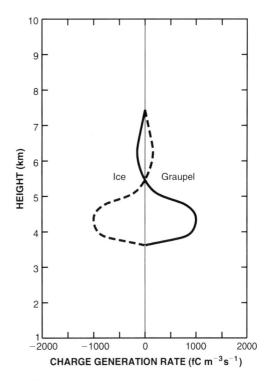

Fig. 8.49. Equilibrium charge generation rates produced by a simple one-dimensional model. The solid line indicates charge on graupel; the dashed line, charge on ice. The magnitudes of charge are equal, but of opposite polarity. (From Rutledge et al. 1990, with permission.)

resents the vertical advection of water vapor. The deposition rate was given by

$$D_{vi} = \sum_k n_k \frac{dm_k}{dt}$$

$$\frac{dm_k}{dt} = \frac{4\pi C_T \left(\dfrac{e_{sw}}{e_{si}} - 1 \right)}{\dfrac{R_V T}{e_{si} D_V} + \dfrac{L_S}{k_T T} \left(\dfrac{LS}{R_V T} - 1 \right)}, \qquad (8.4)$$

where n_k and m_k are the number density and mass, respectively, of ice crystals having a particular size and shape; C_T is the ice crystal capacitance (given by Rogers 1979); e_{si}, the ambient equilibrium vapor pressure over ice; e_{sw}, the ambient equilibrium vapor pressure over water; D_V, the diffusivity of water vapor; k_T, the thermal conductivity of air (units of joule per second per kelvin); L_S, the latent heat of sublimation; R_V, the ideal gas constant for water vapor; and T, environmental temperature (units of kelvin). For model input, ice particle concentrations and size were estimated from Doppler radar, vertical updrafts were taken from wind profiler measurements, and temperatures were taken from soundings through the stratiform region. For one tropical MCS that produced 15 positive ground flashes in the stratiform region, Rutledge and Petersen (1994) estimated that 0.2 g m^{-3} of liquid water was produced in mesoscale updrafts in a layer between $-5°$ and $-12°C$ and concluded that noninductive charging could have occurred. Another MCS, which occurred during a monsoon period, produced no ground flashes of either polarity in either the convective or stratiform regions. Rutledge and Petersen estimated that there was no more than 0.02 g m^{-3} of liquid water in the stratiform region of this second MCS, and so concluded that noninductive charging probably was too small to contribute significantly in this case.

As part of their investigation of the source of charge for the stratiform region, Rutledge and Petersen (1994) analyzed the vertical profile of reflectivity and the number of ground flashes in the stratiform region of each of five MCSs. The average number of positive ground flashes per 30 min was larger for MCSs with larger reflectivity above the 0°C level in the stratiform region. Although this is consistent with a relationship between particle growth and charge in the stratiform region—and so suggests that local, noninductive charging is important in the stratiform region—it does not provide conclusive evidence. For the MCSs in their sample, cases with greater reflectivity above 0°C probably also advected more particles into the stratiform region. Furthermore, we already have noted that flash initiation and, hence, flash rates may be governed by conditions far from the part of a stratiform region in which lightning strikes ground. Conditions near or in the convective line may have influenced ground flash rates in the stratiform region at least as much as conditions that caused the local reflectivity maxima and, for the five storms in the sample, also would plausibly explain the relative order of ground flash counts. Thus, the observational evidence for a linkage between number of flashes and local microphysics in the stratiform region still appears ambiguous, though the linkage is plausible.

Our conclusion from these modeling studies and measurements is that both advection and local generation appear to be feasible sources of the considerable charge found within the stratiform region. A combination of a series of electrical soundings along rearward moving parcel trajectories and more sophisticated MCS modeling studies would allow evaluation of the relative contributions of advection and local noninductive charging. It is possible, and seems likely to us, however, that mechanisms other than the noninductive mechanism contribute to electrification of the stratiform region. For example, Stolzenburg et al. (1994b) suggested that drop breakup, melting, and ion capture also might contribute. Testing these mechanisms will require experiments with numerical models and more definitive microphysical laboratory measurements under conditions approximating the stratiform region.

(iii) Charge in the bright-band regions of MCSs.
As discussed in Section 8.2.2, the observed electric

field and total space charge density at and near the 0°C level have been consistently large, often the largest in the sounding. Because we have not seen this feature consistently in the convective region of storms, we want to understand why it occurs in stratiform clouds. We first search for clues in previous studies of stratified, nonthunderstorm clouds, beginning with work carried out in the former Soviet Union (e.g., Imyanitov et al. 1972, Brylev et al. 1989). An example of large changes in the E profile near 0°C in a nimbostratus cloud is shown in Fig. 2.2. Also possibly related are the findings of Reiter (1965, 1988). He found in about 80% of cases of steady rain from nimbostratus clouds that the polarities of E and charged precipitation were negative at T $\leq$0°C; both E and precipitation charge had changed to positive by $+1°C$ (Fig. 2.1). He identified this phenomenon as the "melting zone effect." Whether this phenomenon in nimbostratus clouds has the same cause(s) as the electrification at and near 0°C in MCS stratiform regions is unknown. They are in very different weather regimes, and MCS stratiform regions seem to have a much larger maximum electric field.

Krehbiel (1981, 1986) described aspects of the charge structure and lightning associated with the stratiform cloud of a dissipating thunderstorm in Florida. Using slow antenna data, he found that horizontal lightning appeared to propagate just above the radar bright band. Among all the charges in the cloud, he inferred that the bright band contained negative charge, with positive charge just above it. Krehbiel hypothesized that charging of the stratiform cloud takes place at a lower rate than in convective thunderstorm cores. He associated the charge with the production and fall of snow, which begins melting just below the height of 0°C and can cause a radar bright band. The bright band during the end-of-storm-oscillation of thunderstorms seems to involve processes similar to those in the bright bands of MCSs, as discussed by Williams et al. (1994). Williams et al. postulated an intriguing link between mesoscale vertical velocities and E$_{gnd}$ for a Florida thunderstorm as presented in Section 7.2.1. The validity of their extension of this to processes in MCS stratiform regions is not yet clear.

Charging mechanisms proposed to explain the ubiquitous large E and ρ_{tot} at and near 0°C in MCS stratiform regions have included Drake's (1968) noninductive melting mechanism, Simpson's (1909) inductive melting mechanism, the noninductive ice-ice mechanism, screening layer formation, and embedded cellular convection. Stolzenburg et al. (1994b) suggested that the inductive and noninductive melting processes could explain their E data. In each process, previously uncharged precipitation particles would acquire charge as they melted by shedding charged cloud particles (either liquid or solid), though the mechanism for charging is different for each process. In the Drake mechanism, the melting precipitation particles shed negatively charged, cloud-sized particles, and the precipitation becomes positively charged. The result of the Simpson mechanism depends on the conditions under which small droplets break away from the melting particle. For example, in the presence of a vertical, upward-directed electric field, small droplets breaking off the top side of melting precipitation would become positively charged, and the melting precipitation would acquire negative charge. There are enough uncertainties because of the sparsity of in situ microphysical measurements and gaps in applicable laboratory results so that both of these mechanisms appear plausible as a source of charge in MCS stratiform regions. More hydrometeor observations, laboratory experiments, and tests using cloud models are needed.

Three key observables of a melting charging mechanism are: (1) A substantial charge density will grow at and near 0°C as more precipitation particles fall through the 0°C isotherm; (2) A smaller charge density of the opposite polarity will be carried toward the ground on the melting precipitation particles; and (3) The number of cloud-size particles at or near 0°C will increase with time. Shepherd et al. (1996) analyzed 12 soundings of E and thermodynamics, along with the associated radar data, from various types of electrified stratiform clouds, mostly in MCS stratiform regions, specifically to investigate the large E and ρ_{tot} at and near 0°C. An example of their data is in Fig. 8.33. At or near 0°C in the stratiform clouds, they observed $|\rho_{tot}|$ ≥ 1 nC m^{-3} in eight of the soundings and $|\rho_{tot}| \geq 2$ nC m^{-3} in four of those eight. Eleven of the stratiform soundings had an $|E| > 50$ kV m^{-1} near 0°C, and seven of those had an $|E| \geq 75$ kV m^{-1}.

When frozen precipitation melts, the solid-to-liquid phase change removes heat from, and thus cools, the surrounding air. Shepherd et al. (1996) showed that the temperature profile near 0°C had been affected in most of their MCS stratiform clouds: They contained quasi-isothermal layers just below the 0°C level. One sounding had a layer that approached being isothermal. They defined their quasi-isothermal layer as the region beginning at or near 0°C and extending downward with a temperature lapse rate ≤ 4°C km^{-1}; the lapse rate of the quasi-isothermal layer must have a differential of ≥ 3°C km^{-1} from the lapse rate immediately below the layer. (The dry adiabatic lapse rate is 9.8°C km^{-1}, and the moist adiabatic lapse rate is ≈ 6.5°C km^{-1}.) The tops of the quasi-isothermal layers were just above the most intense portion of the radar bright band. Shepherd et al. explained that although all particles begin melting just below the 0°C level and contribute to cooling that air, only the bigger melting particles that have become water coated contribute to the most intense region of the bright band. They speculated on two reasons why they did not find quasi-isothermal layers in a third of the cases. First, the radar reflectivity was lower in some soundings, which suggested fewer and smaller melting particles, and therefore less cooling. Second, the stratified cloud may have had insufficient time to develop the quasi-isothermal layer.

Shepherd et al. (1996) thought they had evidence in their data for the first two of the three key observables for melting charging. All soundings in stratiform regions showed a positive charge density at or near 0°C, and, with only one exception, they showed a less dense, oppositely charged layer that extended toward ground below the positive charge. They further used the particle concentrations found by Willis and Heymsfield (1989) and the precipitation charge and size data of Bateman et al. (1995) to show that only -0.1 to -0.2 pC on particles of diameter smaller than the 0.8-mm detection threshold of Bateman's instrument could account for the observed ρ_{tot}. Thus, Shepherd et al. concluded that the negative charge on precipitation falling toward ground could have been there, but could have been missed by direct measurements. We think that obtaining vertical soundings of precipitation particles and charges down to smaller magnitudes than previously obtained is absolutely critical to a conclusive answer.

The third key observable for melting charging seems to have been found in the measurements by Willis and Heymsfield (1989). They found that the number density of 10-μm cloud particles increased by two orders of magnitude between 0°C and +0.5°C. A possible explanation of this increase can be found in the results of Knight (1979), who showed that aggregates shed small particles during melting. Thus, the Drake mechanism appears to be a viable explanation. Further, it is plausible that if an E existed it would charge the shedding particle by induction. If a melting-charging process occurs in clouds, the precipitation types and their particular characteristics as they melt could control the amount of charging. Besides melting, the formation of ice particles that break away from larger particles after being weakened by evaporation, as described by Dong et al. (1994), may contribute to the charging process at and near 0°C. Dong et al. observed such evaporation at relative humidities, RH, of $\leq 90\%$, with very many particles produced quickly for RH $\leq 70\%$. Many soundings shown by Shepherd et al. (1996) had decreasing RH near and below the 0°C level. We think that at least a moderate number of observations exist to document to some extent the three key observables for melting charging.

Shepherd et al. (1996) noted that the charge density often extended a few hundred meters above the 0°C level. They found this observation to be the most troublesome regarding the hypothesized melting charging, since it was not obvious how the charged cloud particles, which could come from a melting mechanism, would move upward. They noted at least two mechanisms that could cause the charge to move higher. First, if the 0°C isotherm descended over time from cooling of the air, the charged cloud particles would not necessarily descend and would be found above the lowered 0°C level. Second, turbulent mixing or diffusion of small charged particles might spread the charge vertically over several hundred meters (Stolzenburg et al. 1994b). Shepherd et al. proposed that one mechanism for such mixing could be Kelvin-Helmholtz waves such as Houze et al. (1989) observed in an MCS stratiform region.

Shepherd et al. (1996) mentioned three alternatives to melting charging as possible explanations for the observations of E and ρ_{tot} at or near 0°C: (1) A screening layer charge could form at the lower cloud boundary. However, they noted that it was difficult to find the cloud base at which a screening layer would form and that even if a screening layer existed, it likely would be at altitudes too low to explain the observed charge structure near 0°C; (2) They noted that charge advection from the convective region had been proposed in several studies, but none of these studies advocated advection as the explanation of the charge density at and near 0°C; and (3) They concluded that the noninductive ice-ice mechanism could account for the positive charge density at or near 0°C if the positive charge was on precipitation. The negative charge layer often found above the positive layer supported this hypothesis, but they noted difficulties with the mechanism. The limited data on precipitation charge (Bateman et al. 1995) showed that the positive charge at or near 0°C was not on precipitation with equivalent diameters of 0.8–8.0 mm. Shepherd et al. stated that the product of $\rho\Delta z$ for the negative layers both above and below the positive layer at or near 0°C was consistently smaller than the $\rho\Delta z$ for the positive layer; they could reconcile this observation with a melting mechanism, but not with the ice-ice mechanism. Because many soundings had several charge layers above the one at 0°C, they decided that the ice-ice mechanism was more likely to be operating at temperatures colder than the reversal temperature, so this mechanism would produce negatively charged precipitation and positively charged ice crystals above the precipitation.

Shepherd et al. (1996) concluded that melting charging was the most viable explanation for the observations of the often dominant E and ρ_{tot} found at or near 0°C. A critical reason for their conclusion was that although melting of frozen precipitation must occur in all clouds that span the 0°C isotherm, the evidence clearly shows that a radar bright band, a quasi-isothermal layer, and perhaps a major electrical effect are observed only in stratiform clouds.

(iv) Effects of lightning on charge structure. While charge structure presumably affects lightning, the lightning also can affect charge structure. Thus, it is important to consider the possible effects of the observed long horizontal flashes on the charge distribution of the stratiform region. Krehbiel (1981) measured changes in the horizontal dipole moment of several thousand coulomb-kilometers from long (>40 km) flashes in a dissipating Florida thunderstorm. The laboratory simulations of Williams et al. (1985) produced long discharges that propagated through unipolar charge regions embedded in dielectric slabs, analogous to the charge layers in stratified clouds. As discussed in Section 8.2.3, long horizontal lightning flashes are prevalent in the stratiform region of MCSs, and observers frequently report seeing long horizontal flashes propagate along the visible cloud base. Such flashes also are common along the bottom of large anvils (e.g., Marshall et al. 1989, Mazur et al. 1994). Long hori-

zontal cloud flashes may also occur within upper layers or along the top of the stratiform region, as suggested by the horizontally extensive flashes astronauts have observed and filmed from space (Vonnegut et al. 1985). Whether the long horizontal flashes that occur in the stratiform region move through or affect more than one layer of charge is unknown. From the large currents and charges involved in many positive ground flashes, we might expect that at least some flashes moving horizontally within the stratiform region would modify its charge structure. However, the similarity of numerous vertical profiles of the electric field in MCSs (e.g., Marshall and Rust 1993) suggests that either the effect of lightning is small enough or the regeneration of the charge layers is fast enough that the major features of the charge structure in stratiform regions of MCSs remain relatively constant during periods of lightning activity. Krehbiel (1981) similarly concluded that charge on the stratiform precipitation of a dissipating storm also was being regenerated in the time between flashes.

8.3. WINTER THUNDERSTORMS

8.3.1. Introduction

Besides winter thunderstorms, winter clouds that become electrified, but do not produce lightning, are common. Work on winter storms in the United States has been limited. In many papers we have found on winter storms in various countries, the presence of lightning was often not specifically mentioned, making it difficult to determine whether the analyzed storms were thunderstorms or electrified, nonthunderstorm clouds. For example, the tendency in papers other than those dealing specifically with lightning is to use the term snowstorm to mean any snowing cloud, regardless of whether or not it produces lightning. Here, we will deal with electrified winter clouds, whether they are thunderstorms or not. Winter clouds have been studied extensively in Japan, Russia, and a few other countries.

8.3.2. Measurements of E_{gnd} and precipitation at the ground

Because there are so few measurements, we combine measurements of E and particle charge or precipitation current. The measurement of electrical parameters in snowstorms, with or without lightning, is fraught with difficulty, especially because of blowing snow in windy conditions. In such conditions, E_{gnd} can be dominated by the noisy signal from charged flakes blowing around. In early measurements, scientists were interested in determining the net charge on precipitation arriving at the ground and how it related to charge on the Earth. In some of the earliest work, Simpson (1909) measured the charge arriving at the ground during snowfall on nine winter days in Simla, India. He noted that lightning occurred on most of the days and that graupel was generally mixed with the snowflakes. He found that the proportion of large precipitation currents during snowfall was much greater than during rain; the snow was more highly charged per unit mass of water than rain was. More positive than negative charge arrived at the ground. Since the charge was not divided between that on snow particles and that on graupel, we do not know if one particle type dominated the total charge. Simpson did not report on any data of E_{gnd} to compare with the precipitation charge.

For several years, systematic observations were made at Kew Observatory in England. A part of those data contained information on winter thunderstorms. Simpson (1949) showed a small amount of ground-based data from Kew Observatory. In apparently non-thunderstorm clouds with maximum $|E_{gnd}| > 2$ kV m^{-1}, the mirror image effect often was evident on precipitation charge and was more pronounced with snow than with rain. He concluded there was no difference between rain and snow in that both collected point-discharge ions as they fell. He did state, however, that with the same point-discharge current snow had more charge than the same amount of rain.

Japanese scientists and engineers have carried out many studies of winter storms and the accompanying damage from lightning strikes. Because so many winter studies (at least percentage wise) have been done on these Japanese storms, we include a brief description of their nonelectrical aspects. For example, Goto and Narita (1991) described the meteorological conditions often associated with these storms. The storms form over the Sea of Japan, often on or to the rear of fronts. They are associated with troughs or cyclones moving across the Sea of Japan when the monsoon wind flows into the rear of cyclones. The temperature at 500 mb is $< -30°C$ and drops by more than 10°C per day as the system evolves. When the temperature of the sea surface is at least 42°C warmer than the air temperature at 500 mb, thunderstorms occur even if the conditions described in the previous sentence are not present; the energy generating these storms comes from the warm sea. Kitagawa (1992) reported that $\approx 60\%$ of the storms were caused by advection of Siberian air masses over the Sea of Japan, $\approx 20\%$ were formed by combined advection and frontal activity, and $\approx 20\%$ were formed by advection plus cyclonic scale forcing. There was not a preferential time of day for the formation of winter storms. The tropopause lowered from about 15 km in summer to 7–9 km in winter.

Simultaneous measurements of E and charge when graupel is present are scarce. For example, Isono et al. (1966) measured the charge, size, and habit of particles aloft, but made no measurements of E. They found different charge on particles depending on the location of the storms. For example, there was no significant charge on graupel falling over the sea (onto a peninsula jutting into the Japan Sea), but graupel falling inland was charged and carried both polarities. Isono et al. attributed the difference to the conditions in the cumuli-

form clouds over the sea versus inland. Specifically, they drew on laboratory results and hypothesized that the small numbers (or very small size) of snow crystals in the clouds over the sea meant that the graupel could not collide with crystals; hence, the graupel was uncharged. However, inland clouds had many snow crystals, which could collide with graupel and thereby generate charge.

Measurements of E_{gnd} by Magono et al. (1983a) in nonthunderstorm snow clouds on the Ishikari Coast, Hokkaido, often indicated $|E_{gnd}| > 1$ kV m^{-1} (1 kV m^{-1} was the saturation level of their instrument). They observed that the charge on snowflakes and graupel usually exhibited the mirror image effect. Their data showed that snowflakes occurred during negative E_{gnd} and graupel during positive E_{gnd}. Michimoto (1990) noted that the polarity of E often changed, and he attributed this to movement of the storm from the Sea of Japan across his measuring site on the west coastline of Japan. The surface wind was about 20 m s^{-1}. Michimoto showed a maximum $|E_{gnd}| = 15$ kV m^{-1}.

The behavior of E just above the ground is probably the same as for summer thunderstorms: The magnitude of E increases with height owing to space charge from point discharge at the ground. Using a helicopter to carry a dual field mill, Akiyama et al. (1984) reported that with a nearly constant $E_{gnd} = 8$ kV m^{-1}, they observed $E_{aloft} = 25$ kV m^{-1} by the time they ascended 70 m above the ground. They showed an exponential increase of E with height and a maximum space charge density just above the surface of about 10 nC m^{-3} from the point-discharge ions. Asuma et al. (1988) used a tethered balloon with a field mill and particle charge detector at a few levels 50 or 100 m apart. No lightning from the clouds was mentioned. Even in conditions when there was no point discharge, the mirror image relationship occurred for snow at the ground. Aloft this was not always so. During times of negative E large enough to cause negative charge release by point discharge, the snowflakes, which left the cloud base with positive charge, acquired negative charge as they fell. Unless they were highly charged positively at cloud base, they were negative when they arrived at the ground. (This was true in a general sense; the same snowflake could not be observed aloft and at the ground.) During snowfall with little wind, they observed $|E_{gnd}| \leq 8$ kV m^{-1}. The largest charges on the snowflakes exceeded 40 pC, but more typically were < 5 pC, with the polarity determined by the mirror image effect.

8.3.3. Measurements of E_{aloft} and particle charge aloft

As at the ground, measurements aloft of both E and particle charge are scarce. Isono et al. (1966) made eight soundings of particle charge with Takahashi's (1965) sonde in cumuliform winter clouds. They summarized two of the flights, both of which they believed were in snowflakes, since they did not observe graupel on the

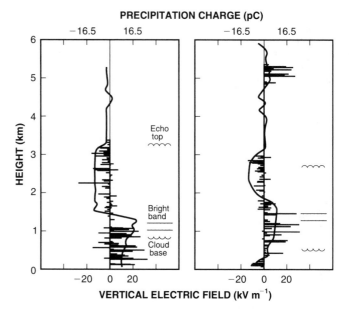

Fig. 8.50. The vertical distributions of E and particle charge on 10 Dec (left) and 15 Dec 1979 (right). We replaced the charge in electrostatic units on the particle charge axis with picocoulombs. They gave no explanation for the particle charges and E above the marked cloud top on 15 Dec.; perhaps it was another cloud layer. (After Magono et al. 1982, with permission.)

ground. The dominant polarity on particles was positive below 1300 m (T $\approx -5°C$) where Isono et al. thought the snow captured droplets in a dense cloud with supercooled water. Above 1300 m, the dominant polarity was negative, although both polarities were detected on particles. At other locations, graupel pellets were observed at the ground.

There are almost no soundings of E in winter thunderstorms. Magono (1980, p. 195) showed profiles made with balloon-borne alti-electrographs and with field-measuring devices. The graphs were plotted as the polarity of point discharge and did not have values of E. Magono et al. (1982) indicated in both their title and text that the two E soundings they reported were from thunderstorms. Their two profiles showed a maximum $|E_{aloft}| \leq 20$ kV m^{-1} and no lightning transients, which may have been removed before plotting, or may not have been detected during the sounding. They described their instrumentation as a balloon-borne field-measuring device and an electric charge meter (i.e., a particle charge sensor). Such data as these are rare, and their soundings are shown in Fig. 8.50. Both clouds had a radar bright band. Above the two bright bands was negative or mixed charge, while below was mostly positive charge. The mirror image effect was prevalent aloft.

With a sonde to measure particle charge and the vertical component of E, Magono et al. (1983b) made several soundings into active winter clouds. They did not describe or reference the instrumentation except to say that temperature came from standard, routine soundings about 50 km away. Magono et al. described the electrified clouds as stratocumulus. They did not mention, nor does any lightning show, on the soundings. The clouds typically had strong vertical shear in the

horizontal wind. The clouds were all electrified with a maximum $|E_{aloft}| \leq 30$ kV m^{-1}. The mirror image effect was not seen at the ground, but it was observed aloft. Magono et al. noted that the mirror image did not always show aloft. A single sounding of both E and particle charge was shown (Fig. 8.51). In their nine overlaid profiles of E, there is evidence of significant charge density at estimated temperatures of about 0° to $-5°C$: Three profiles had negative charge, and six had positive charge. Magono et al. identified the winter clouds as having the same polarity as in summer, but with the negative charge at lower heights and usually with very strong wind shear. They found charges on particles were mostly <10 pC. Similarly, Asuma et al. (1988) measured the maximum particle charge during each 5 s period and found values on snowflakes in the cloud base, which was about 200 m above ground, to be mostly <5 pC.

Magono and colleagues used an improved version of an electric charge sonde, which consisted of a shielded Faraday cup device to measure charge and a pressure sensor to determine the altitude, to measure the charge on precipitation during the winter of 1976 (Magono et al. 1984). There were no E sensors on the free balloons carrying the charge sonde. They described seven flights on four days from Sapporo, Japan. Temperatures were obtained from a standard radiosonde released simultaneously nearby; the temperature was typically $-2°C$ at the ground and $-20°C$ at about 2 km. E_{gnd} usually changed polarity for periods $\approx 1-10$ min during the flights. Ground measurements consistently showed the mirror image effect.

Magono et al.'s (1984) profile through a stratiform snow cloud, whose top was at 2 km, showed a unipolar precipitation charge density. Based on observations

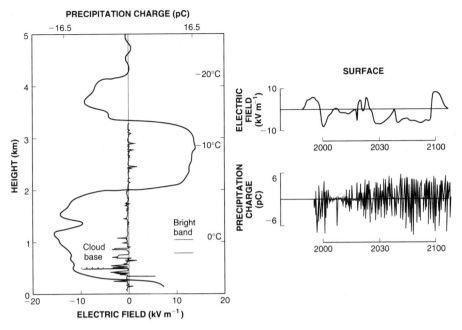

Fig. 8.51. Left: Sounding of E and particle q through a winter storm over the west coastline of Japan. Right: Simultaneous measurements at the surface (shown as sketches from their original. (After Magono et al. 1983b, with permission.)

at the ground, they determined the precipitation was snowflakes; all charges in the cloud were negative with a $\rho_P \lesssim 0.3$ nC m^{-3}. In a negative $E_{gnd} <1$ kV m^{-1}, charges on snowflakes at the ground were almost all negative with many particle charges >3 pC, the saturation level of their instrument. They concluded that the negative charge was the result of Wilson ion capture. In other profiles, they found that both polarities of charged precipitation coexisted at most levels in the clouds. Magono et al. also noted that graupel at temperatures warmer than $-10°C$ frequently carried positive charge, which they hypothesized was from collisions with unrimed snow crystals.

8.3.4. Lightning

There is a tendency in winter for a greater percentage of ground flashes to lower positive charge, as was reported as early as 1973 by Takeuti et al. (1973). Orville et al. (1987) found that along the east coast of the United States for June 1984 through May 1985, the monthly percentage of positive ground flashes was as high as 50% during the winter, compared with 5% during the summer. Most studies of lightning in winter storms have dealt with the physical characteristics of the lightning flashes rather than their relationship with the other coexisting meteorological conditions. Positive flashes often lower substantial amounts of charge as described elsewhere (e.g., Brook et al. 1982).

Brook et al. also found that the charges neutralized by positive flashes were systematically higher in the storm than those neutralized by negative ground flashes. A remarkable finding by Brook (1992) in winter storms in the United States was that the leaders were four times shorter in duration than for summer flashes, and this could not be accounted for by the lower height of the assumed region of charging during the winter. He hypothesized that the breakdown fields in winter storms must be greater than in summer. He noted that the difference could be from the type of precipitation particle from which breakdown occurs and suggested that research on the influence of precipitation on the breakdown electric field was needed. Please note that we have not included many studies concerned mostly with lightning physics; here we want to focus on available information concerning the possible links between lightning and winter storm meteorology.

Although positive ground flashes are more prevalent in winter, not all winter storms produce large positive ground flash percentages. For example, a winter study by Biswas and Hobbs (1990) using data collected over the Gulf Stream off the U.S. east coast (near North Carolina) during the field phase of the Genesis of Atlantic Lows Experiment (GALE) showed most ground flashes were negative and were associated with convective rainbands. In his analysis of the ground strike data from GALE, Orville (1990a) found that flash density contours appeared tied to the temperature contours for the Gulf Stream for one month. Thus, it seems we need to look to environmental or storm characteristics that are more common, but not always existent, in winter storms to explain the almost global tendency for a higher positive ground flash percentage in winter than in summer.

Lightning to ground in the winter has received more attention in the United States since the installation of a U.S. national lightning ground strike network. In their extensive analysis of a major snowstorm on February 11–12, 1983, Bosart and Sanders (1986) examined ground strike data and found only 69 flashes during a 12-hour period. More than half occurred in one of four clusters of flashes. The clusters seemed to be discrete episodes and tended to be along a line across which surface winds shifted. The lightning activity stayed in place geographically and did not move with the winds aloft or with the gravity wave in the system. A sounding indicated little instability and cloud-top temperatures of about −14°C. Bosart and Sanders noted that the geostrophic shear was above the threshold of wind shear that Brook et al. (1982) suggested was needed for production of positive ground flashes. There was no obvious link between the flashes and the measured accumulation of snowfall. The positive ground flashes were beneath relatively warm cloud tops. Bosart and Sanders speculated that cloud-top temperature, not updraft speed, determined the "electrical character," by which they meant the production of positive instead of negative flashes.

Engholm et al. (1990) documented a bipolar lightning ground strike pattern in the GALE data. In winter, they found the bipole aligned with the geostrophic wind. They inferred a systematic relationship between wind shear and the relative displacement of negative and positive lightning strikes. Engholm et al. noted that the shallow clouds still had radar reflectivity in the mixed phase region (i.e., 0 to −40°C). They linked the more moist and unstable air (i.e., higher potential equivalent temperature) to the negative end of the bipole and found that the positive ground flash percentage seemed to increase with decreasing cloud-top height. They concluded that both positive and negative ground flashes were associated with upright, not slantwise, convection.

Dodge and Burpee (1993) examined rainbands and found that with radar signals of 41–46 dBZ (VIP3) that there was an order of magnitude fewer ground flashes in their data than Reap and MacGorman (1989) found for a warm season in Oklahoma. The winter convection was much shallower than springtime convection having the same radar reflectivity (VIP level) in the southern Great Plains. About 99% of winter ground flashes was in rainbands and occurred during relatively strong midtropospheric cyclonic vorticity and weak, low-level, cold advection. Dodge and Burpee concluded that synoptic scale forcing was more important for the production of lightning strikes during winter. They did, however, question the universality of their conclusions since only two to three of the 20 data sets had more than

half of the storm cases with reflectivities ≥41 dBZ and had the majority of lightning.

Several studies of winter lightning in Japan have shown that ground strikes occur primarily to the sea surface and occur no more than 30 km inland from the coast (e.g., Hojo et al. 1989). Nakono (1979) showed conceptual diagrams of three types of winter storms, two of which had moderate or strong wind shear and a tilted dipole charge structure inferred from sounding data and positive ground flashes. In weak shear, the ground strikes were negative. To look for a cause of positive flashes, Brook et al. (1982) examined wind shear over the depth of the cloud (i.e., the difference in horizontal wind velocity between cloud base and cloud top). They used radar data to determine cloud top and measured cloud base visually or estimated the base from a sounding. The percentage of all ground flashes in a storm that were positive appeared to be linearly correlated with wind shear (a correlation coefficient of 0.95). They noted that the wind shear shifted the upper positive charge horizontally from the lower negative charge, and they hypothesized that this tilting facilitated production of positive ground flashes. Similar explanations have been proposed for warm season positive flash production.

However, several studies have failed to find a correlation between positive ground flash percentage and wind shear. For example, Kitagawa (1992) found no such correlation for a different sample of winter storms in Japan than that studied by Brook et al. (1982). Reap and MacGorman (1989) found no evidence of such a correlation in a climatological analysis of ground flashes during two warm seasons over the Great Plains. In an analysis of a severe storm in which ground flash activity was dominated for a while by positive ground flashes and then by negative ground flashes, Curran and Rust (1992) found that the environmental wind shear when positive ground flashes dominated was similar to the wind shear when negative ground flashes dominated. Reap and MacGorman and Curran and Rust both noted that wind shear often was above the threshold suggested by Brook et al., regardless of the dominant ground flash polarity during the storms they observed. Therefore, they suggested that strong wind shear might be necessary, but not sufficient, to cause positive ground flashes to dominate.

Fukao et al. (1991) used dual polarization radar to probe winter storms. They concluded that lightning was associated with the contact and mixing of graupel and ice crystals or snowflakes. This offered corroboration of Simpson's (1909) observations in India. They also believed that winter storms with their band structure are inherently different in many respects from isolated summer thunderstorms.

Others have examined lightning activity in the context of storm evolution. Radar observations in the Hokuriku region by Michimoto (1991) showed that the first lightning tended to occur when the top of the 30-dBZ radar echo was at a temperature colder than −20°C. Michimoto (1991, 1993) obtained radar data from several winter storms. He did not limit his study to storms with positive flashes or stratify storms by their positive ground flash percentage; wind shear did not appear in his conclusions. Michimoto (1991) observed the following in winter thunderstorms on the western coast of Japan: (1) The first flash occurred about 5 min after the 30-dBZ radar echo extended above the height of the −20°C isotherm; (2) Clouds produced no lightning when the −10°C isotherm was at a height of <1.4 km MSL; (3) Clouds produced little or no lightning when the −10°C isotherm was between 1.4–1.8 km MSL; and (4) Lightning activity peaked when echoes of ≥45 dBZ formed at the −10°C level and descended toward the 0°C level (found to be true in summer, as well). Michimoto (1993) stated that the following conditions were necessary for lightning to occur in midwinter (January) clouds: (1) The 30-dBZ echo top must extend above the height of the −20°C isotherm; (2) The −10°C level must be at a height above 1.4 km; and (3) The cloud must produce rapid development of the 40-dBZ echo region. These conditions are indicated by the scatter plot in Fig. 8.52. Near the demarcation level of 1.4 km, the clouds often produced only a single flash.

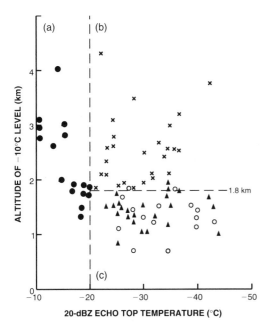

Fig. 8.52. Lightning activity relative to the altitude of the −10°C level and the temperature at the top the 20-dBZ contour. To the left (a) of the vertical dashed line at −20°C, there was no lightning, regardless of the altitude of the −10°C level (black circles). The x's are for convective clouds with more lightning (b). Open circles are for convective clouds with no lightning and whose −10°C level was <1.8 km; similarly the black triangles are for the same situation but with a little lightning (c). (From Michimoto 1993, with permission.)

Kitagawa (1992) and Michimoto (1993) suggested that the dependence on the height of the −10°C isotherm was related to the effect of this height on updraft strength. When the isotherm was at a height of ≤1.4 km MSL, updrafts at temperatures between −10°C and −40°C were too weak to generate charge by the non-inductive mechanism involving collisions of graupel and cloud ice. In Japan during the cold season, the monthly average altitude of the −10°C isotherm varies from 1.9 km MSL in January to 2.8 km MSL in March.

In their review of winter storms, Kitagawa and Michimoto (1994) summarized the characteristic features of winter thunderstorms as follows: (1) The electrically active period of a winter storm tended to be shorter than in a summer storm; (2) Japanese storms in midwinter had even less lightning than those in early winter; and (3) The percentage of ground flashes that were positive was about 33%, and positive ground flashes occasionally lowered extremely large amounts of charge (e.g., Brook et al. 1982). Kitagawa and Michimoto noted that there often was strong wind shear during winter thunderstorms, but they did not indicate any role for wind shear in their conceptual storm model. Two storm types with wind shear were associated with positive ground flashes; the nonsheared one with negative ground flashes. As discussed previously, however, other studies have shown no dependence on wind shear. Thus strong wind shear is at most a necessary, but not a sufficient, condition for the production of naturally occurring positive ground flashes.

With dual-polarization radar, Maekawa et al. (1992) examined storm conditions at the time of first lightning. They found the cloud tops reached 5–6 km, with the cloud sloping to the leeward from strong winds. The clouds had reflectivity cores ≥40 dBZ (Fig. 8.53). Maekawa et al. found that the developing cloud had ice crystals in the top portion and graupel in the middle. From a thermodynamic sounding about 80 km to the north, they surmised that the graupel-ice interface was at about the −10°C level. They observed the first lightning as the graupel began to descend. Kitagawa (1992) suggested that the falling graupel removed negative charge from the cloud and left a positively charged (monopole) cloud. No soundings of E through the clouds are available to help evaluate this hypothesis.

(i) Lightning and snowfall. Reports of lightning during snowfall, sometimes called thunder snowstorms, have been made for decades. Almost all are limited to anecdotal descriptions. To our knowledge, no detailed study has been conducted to determine the morphology of thunder snowstorms, the size or amount of snowfall relative to lightning rate, etc. The anecdotally reported flash rates have ranged from infrequent lightning to high flash rates. For example, Matthews (1964) observed a March thunderstorm from an observation room 25 m above ground at the Institute of Atmospheric Physics in the desert environment near Tucson, Arizona. The storm produced first rain and

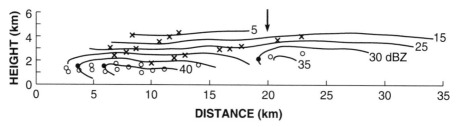

Fig. 8.53. RHI through winter thunderstorm near the west coastline of Japan, 18 Jan 1990, 1930–1935 LT. The x's denote ice crystals and the o's denote graupel as inferred from dual polarization radar. The arrow denotes the location in range of a lightning ground strike point. (From Maekawa et al. 1992: IEEE, with permission.)

then wet snow with snowflakes up to 5 cm in diameter. During the snowfall, he and others observed short duration lightning flashes at a rate of about 3 min^{-1}. He thought the lightning flashes were unusual in that they were not correlated with static on a radio and no thunder was heard. Distinct channels were not observed in the snow, but the illumination emanated from a single place in the cloud. Canovan (1972) described a few thunder snowstorms, including a towering cumulonimbus that produced a swath of snow outside and across London.

The equivalent of a rain gush has also been reported with snow—again these reports have been anecdotal. Tucker (1963) reported that a few hours after the onset of fine snowfall, a flash occurred within <1 km and was followed by a sudden increase in snowfall, characterized by heavy snowfall with big flakes. This was the only flash reported, but the heavy snowfall lasted for about 45 min and resulted in 4 cm of snow on the ground. In some other cases, the intensity of the snowfall and the size of the snowflakes were also reported to be unusually large after lightning occurred. An unusual winter storm occurred March 12–14, 1993 (often called the Blizzard of '93). It formed along the Texas Gulf Coast and moved eastward. As it moved, it set many records for snowfall and for low pressure at the surface. In this winter storm system, ground flash rates were high and both polarities of ground flashes occurred, with a peak flash rate of >5000 h^{-1}. Variations in the density and rates of positive ground flashes were reported (Orville 1993), but the reasons for these variations were not established.

A possible tie between significant winter weather and positive ground flashes was noted by Coleman (1990) who documented elevated thunderstorms above frontal surfaces in the absence of positive convective available potential energy, CAPE (see Eq. 7.2). Holle and Watson (1996) examined ground strike data during two outbreaks of weather events associated with major Arctic cold fronts. In both cases, the areas with lightning had winds from the southwest at speeds >20 m s^{-1} between the 900 and 800 mb levels; the air in this layer was saturated and colder than freezing. Lightning was associated with the areas and time periods in which pre-

cipitation developed most rapidly. Holle and Watson concluded that the combination of ground flash mapping data and radar echo tops would have aided an early diagnosis of low-level moist advection by forecasters and would have indicated regions where the heaviest ice accumulation at the ground was likely to occur.

In summary, we are aware of no thorough scientific investigation of causal relationships between the electrical state of winter storms and their snowfall. Extensive tests to evaluate the proposed hypotheses concerning possible links between lightning and the mesoscale and synoptic scale meteorology associated with winter storms have yet to be performed. However, the anecdotal evidence of a link between heavy snowfall and lightning is intriguing, and its study is scientifically important and may be of possible use in operational meteorology.

8.4. TROPICAL THUNDERSTORMS AND NONTHUNDERSTORM CLOUDS

8.4.1. Introduction

Both isolated thunderstorms and mesoscale convective systems occur in the tropics, which lie in a band of latitudes within ±23° of the equator. The only study of the electrical structure of a tropical MCS of which we are aware is a study by Chauzy et al. (1985). The vertical electrical structure they found fit the Type A of midlatitude MCSs discussed previously. Thus far, we know of no fundamental electrical difference between MCSs in the tropics and those at midlatitudes, but again little data are available on tropical MCSs.

Electrified warm clouds (i.e., the cloud top never goes above the 0°C level) are apparently more numerous in the tropics than at midlatitudes. A major contribution to the documentation of tropical clouds—both warm and cold—has come from the two decades of measurements by Tsutomu Takahashi. To study cloud electrification in the tropics, he combined laboratory studies, measurements using novel instrumentation at the ground and aloft in clouds, and numerical cloud

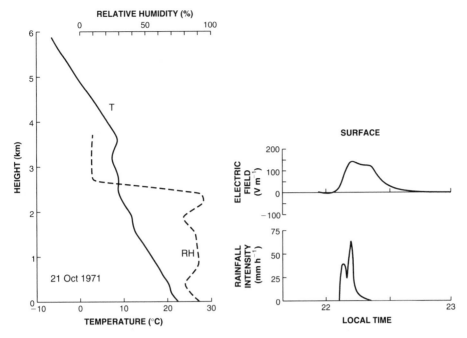

Fig. 8.54. Typical sounding of temperature and relative humidity in the environment of warm cumulus clouds in Hawaii. The simple E_{gnd} and accompanying rainfall rate for a warm rain cloud are shown. (After Takahashi 1975, with permission.)

models. He recognized that the entire spectrum of particle size and charge has to be measured to provide comprehensive and definitive data on the electrical structure of clouds, and he tried to cover this broad spectrum with his measurement schemes. We summarize aspects of his extensive work, along with that of others.

8.4.2. Electric field at the ground

Takahashi (1975) found that E_{gnd} always goes to foul weather polarity beneath raining, warm cumulus clouds in Hawaii (20°N latitude), even if the pattern of E_{gnd} ver-

sus time was simple. Examples of E_{gnd} and rain rate are shown in Fig. 8.54 along with vertical profiles of temperature and relative humidity through such a warm cloud. In Ponape, Micronesia (7°N latitude), Takahashi (1983) obtained E_{gnd} beneath nonthunderstorm clouds and thunderstorms. The recordings of E_{gnd} ranged from simple to complex in appearance. An example of a data set from the ground beneath a thunderstorm is shown in Fig. 8.55. The rainfall rates were often >100 mm h^{-1}. During the thunderstorm, he recorded a field excursion associated with precipitation (at ≈1220) just as can occur in midlatitude continental storms.

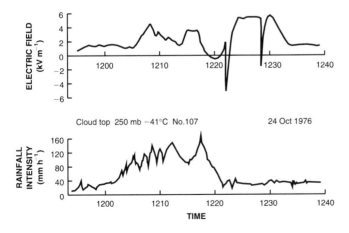

Fig. 8.55. Example of E_{gnd} and accompanying rain rate for a tropical thunderstorm on 24 Oct 1976. The broad peak at about 1220 was apparently a field excursion associated with precipitation. The two lightning flashes the storm produced are indicated by the spiked field changes between 1220–1230. (After Takahashi 1983, with permission.)

8.4.3. Particle charge at the ground beneath warm and cold tropical clouds

In examining very small particles having diameters, d, of 1–40 μm at the ground, Takahashi (1972) found that in fine weather with comparatively low humidity the particles were positive, but during high humidity and rain, they were predominantly negative. Takahashi and Craig (1973) extended the measurements to d = 120–700 μm and found that about 11% of all droplets had a charge magnitude of ≳3 fC, the minimum dectable. These measurements were made along Saddle Road at 1.6 km MSL on the island of Hawaii; the warm clouds extended from about 800–2600 m with tops well below the level of 0°C. Of the charged droplets, 90% were negatively charged. With increasing height in the cloud, the percentage of drops that carried charge increased. The mean negative and positive charges were roughly 1 fC. Measurements by Takahashi and Fullerton (1972) of warm cloud rain falling at rates <10 mm h^{-1} from lines of cumulus clouds on the windward side of the island of Hawaii showed that E_{gnd} was 100–500 V m^{-1} and increased with increasing rain rate. The percentage of drops having >3 fC of charge increased with drop size from 5% for d <1 mm to 40% for d >1.7 mm. Most raindrops of d = 0.7–1.4 mm carried negative charge. The number of positive drops and the number of negative drops having d = 1.4–1.7 mm were approximately equal. The net charge on particles of d ≳1.7 mm was positive.

The tendency was for mean drop charge for both polarities to increase in magnitude with increasing cloud top height. In follow-up measurements in Ponape, Takahashi (1983) found the same trend: Most drops were negative at d <1.2 mm. When d >2 mm, most were positive, and the E_{gnd} increased more in the foul weather sense. When the drop spectrum broadened more, drops of d = 2.5–3.5 mm were mostly negative, while the accompanying drops of d <2 mm were positive. Particles at the ground had about ten times more charge in Ponape than in Hawaii, and $|E_{gnd}|$ was greater. Takahashi concluded that the warm clouds at Ponape had the same electrical structure as those in Hawaii.

An interesting finding comes from Hawaii, where the warm clouds often rain over the ocean and then over the land as they move in the trade winds. Takahashi et al. (1969) found that raindrops arrived at the ocean surface with negative charge and that the charge was significantly greater than, and of opposite polarity to, the positive charge arriving at the ground once the cloud was over the island. Furthermore, the polarity E_{gnd} was reversed over the water: It was the typical foul weather positive E over land and enhanced fair weather negative E over the ocean. However, the two E profiles through warm clouds over the ocean did not show a clear difference from those over land. Considering the variability in the eight E profiles in warm clouds over land in Takahashi (1975), there does not appear to be a simple or even obvious difference between the over-

water and the over-land clouds, although some difference might have been expected from the opposite polarities of E_{gnd} in the two regimes. Furthermore, this appears to be yet another reinforcement of the concept that the internal structure of a cloud overhead is not necessarily revealed by a measurement of the electric field at the Earth's surface.

Electrical measurements also have been made beneath nimbostratus clouds in Poona (18°N latitude). During steady monsoon rain up to 30 mm h^{-1}, E_{gnd} was less than its fair weather value of 300 V m^{-1}, tending toward and usually being foul weather polarity, and little net charge was carried to ground by precipitation (Sivaramakrishnan 1962). In the same study, he found that when the cloud grew colder than 0°C, the rain had more net charge. In one case, a bright band whose top was at about 4 km existed for 7 h, during which the rainfall rate was 15 mm h^{-1} and E_{gnd} was ≤1000 V m^{-1}. Sivaramakrishnan reported the precipitation charge in terms of specific charge density (see Sec. 6.3). It was positive and ranged from 7 × 10^4 to 7 × 10^5 nC m^{-3}. (Such magnitudes of specific charge density on precipitation at the ground are common, e.g., Simpson 1949.) Furthermore, when E_{gnd} ranged from −100 to +600 V m^{-1} and no point discharge could occur, the rain still was positively charged. For a given E_{gnd}, the current carried by the precipitation was a function of rain rate. When long-lived stratified clouds did not produce rain, a noticeable perturbation of E_{gnd} was rare. This suggests either that the cloud overhead was not electrified or that the charge overhead in negative layers was approximately equal to that in positive layers. During continuous rain from tropical Ns clouds, the charge carried by precipitation and the perturbations of E_{gnd} were greater if the cloud top extended above the 0°C level.

Ette and Oladiran (1980) measured the precipitation current and other parameters for four rainy seasons at the ground beneath nonthunderstorm clouds and thunderstorms at Ibadan, Nigeria (≈7°N). They found an average precipitation current of 1 ± 0.1 × 10^{-10} A m^{-2}. The resulting average specific charge density was ρ_P = 4.3 ± 0.2 × 10^4 nC m^{-3}. Their calculated net current flowing beneath tropical thunderstorms from rain and point discharge was −0.9 ± 0.1 nA m^{-2}.

We now specifically consider tropical thunderstorms. Whether they exist with cloud tops below the 0°C level is key to the issue of warm-cloud lightning. We will return to that later, but it suffices here to say that most tropical thunderstorms extend above the level of 0°C, which has been interpreted as indicating that ice processes are likely important in charging. Beneath a thunderstorm, Takahashi and Craig (1973) found that 70% of the particles had q >30 fC, and positive charge dominated. For drops of d <0.4 mm, 60% were positive and for d >0.4 mm, 40% were positive. This trend of a decreasing percentage with increasing d is the opposite of the trend in the same measurements in warm clouds. No E_{gnd} was reported, so we cannot assess possible effects of point discharge. Takahashi (1983) found that when drop size distributions became larger: drops of d > 2.5 mm were predominantly negative; drops of d < 2 mm were mostly positive. When $E_{gnd} > 3000$ V m^{-1}, large drops had both polarities, and small drops were predominantly positive.

8.4.4. Electric field aloft

Fitzgerald and Byers (1959) reported measurements of trade wind cumulus clouds in the Caribbean ($\approx 21°$N) made with airplanes instrumented to measure one horizontal component of E. They made a total of 690 probes of such clouds whose tops ranged from 1.5–2.7 km MSL, which was below the altitude of 0°C. They concluded that early electrification was associated with the location of the most liquid water in the main updrafts and that it was typified by excess negative charge in the cloud. Since no E_z profiles were made, we cannot tell if those clouds had a unipolar or a more complex charge distribution. The values of the horizontal E were 100–1000 V m^{-1} in all but one additional cloud, which had a convective turret extending just above the 0°C level and a larger E of ≥ 8.8 kV m^{-1}, which was the instrument's full scale. From our analysis of Takahashi's (1975) E profile in a warm cloud (his Fig. 4; sounding 1), we found that the maximum $\rho_{tot} = -0.01$ nC m^{-3} was in the mature stage of a cloud with a top at 2.5 km and an $E_{max} = 250$ V m^{-1}. Furthermore, no repeatable, characteristic E profile and charge structure existed. Measurements of tropical clouds in Cuba were made by Mikhailovsky et al. (1992), who corroborated earlier findings that the charge in such clouds is tied to the location of maximum liquid water content and that, initially, the charge is negative. Furthermore, they found no difference in these tropical clouds and equivalent clouds in Russia.

8.4.5. Particle charge aloft

Takahashi (1975) developed and used different sondes to measure cloud particle charge, precipitation particle charge, space charge on particles of d <2 μm, and E_z. About 90 soundings were made in warm clouds near Hilo, Hawaii. A single sounding apparently did not have multiple sensors; e.g., no simultaneous precipitation charge density and E_{aloft} were shown. He developed composite electrical structures from multiple flights. In the cloud's developing stage, most cloud and rain particles were negative. The subsequent characteristics then depended on whether the cloud top, as inferred from the humidity measured by the sonde, was below or above 3 km. The 0°C level in the clouds was well above 3 km (at ≈ 5 km). The tendency for polarity change with altitude was the same for cloud and rain particles. In the developing stage, the predominant charge on droplets and drops was negative; the maximum rain drop charge density was 3×10^{-4} nC m^{-3}. In the mature stage with a cloud top >3 km, rain drops were predominantly positive, with large positive drops found near the cloud top. Both polarities of rain drop

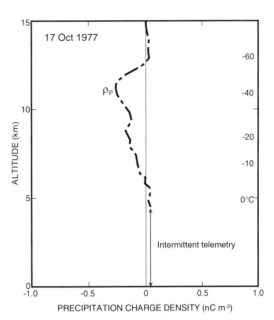

Fig. 8.56. Net precipitation charge density, ρ_P, derived from published profiles of negative and positive ρ_P in a tropical thunderstorm, 17 Oct. 1979. The first lightning was 1 h later. (Fig. 7a, Takahashi, 1978, with permission.)

charge density were up to 3×10^{-3} nC m^{-3}. During this mature stage, the cloud droplets were predominantly negative with a maximum of 3×10^{-3} nC m^{-3}. In dissipation, both polarities of rain drop occurred, but positive rain drops dominated as the cloud height increased, and maximum rain drop charge density was 2×10^{-3} nC m^{-3}.

Takahashi (1978) used a new sonde designed to measure particle charges of 1.6–3.0 pC and d of 0.8–10 mm. Individual charges were not shown, but he converted the 1-min sum of each polarity to a charge density versus altitude. One sounding into a cumulonimbus cloud was made about 1 h before lightning. Our plot of net charge density from his individual particle charge densities in the cloud gave a precipitation charge density magnitude of $|\rho_P| < 0.3$ nC m^{-3} (Fig. 8.56). He inferred the charge structure to be positive in the cloud top, negative in the middle, and positive in the lowest part (cloud base and top heights were not explicitly given). The summary given was that from 0° to $-40°$C, there was negatively charged graupel with individual charges of the order of -3 pC. At colder temperatures were positive particles having individual charges of about 0.3 pC.

8.4.6. Simultaneous measurements of E_{aloft} and particle charge aloft

We know of only about 10 published, simultaneous vertical profiles of E and precipitation charge density

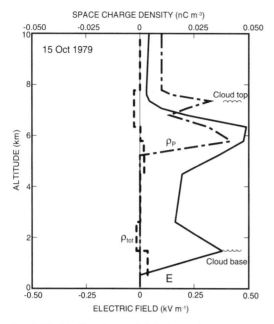

Fig. 8.57. Profiles from published sounding made in warm cloud, 15 Oct. 1979. These plots are not as smooth as the originals, but have all their essential features. Takahashi determined cloud top from humidity. (Derived from data shown in Fig. 4a in Takahashi 1983, with permission.)

in nonthunderstorm, warm clouds. Takahashi (1983) modified his precipitation charge and size measuring sonde to measure the potential gradient and to distinguish graupel. He stated that the minimum detectable diameter was 200 μm and the minimum charge was 2 fC. With his modified sonde, he obtained several soundings in nonthunderstorm clouds and thunderstorms at Ponape. All warm clouds had $|E_{max}| \leq 2$ kV m^{-1}. We took one of his warm cloud soundings and calculated a net ρ_P and a ρ_{tot} (Fig. 8.57). The precipitation charge did not match the total charge density, implying that charge on small cloud particles was significant. Another interesting result was that precipitation charge density was only positive, as opposed to that shown before with much negative precipitation. Of the charged precipitation particles, 75–90% were ice of diameter >1.2 mm or rain of diameter >1.5 mm, based on the microphone signals from the sonde. We infer from the E profile that there were three regions in the cloud.

Takahashi (1983) showed two E profiles in thunderstorms. One was a partial sounding in the storm's developing stage. This sounding had $E_{max} = 90$ kV m^{-1} at $-20°C$ (≈ 8 km MSL). From his plots for this storm, we calculated ρ_P and ρ_{tot} (Fig. 8.58). Again, the net precipitation charge density did not track with the total charge density, especially at $z \geq 8$ km. The maximum $|\rho_{tot}|$ was 1000 m thick, went upward from the

height of E_{max}, and was -0.9 nC m^{-3}. We infer at least four regions of charge from the sounding, even though data were missing in the lower and uppermost regions of the cloud. His other profile was in the decaying stage of a thunderstorm and had peak E of each polarity of 1 kV m^{-1}. The profile is similar to that found in a small, dying thunderstorm over Langmuir Laboratory (Fig. 7.3).

From his 11 soundings with the new sonde in 1979, Takahashi (1983) conceptualized five stages of tropical cloud development, as shown in Fig. 8.59. Stage 1 was smallest and had its top well below the level of 0°C. It consisted mostly of negatively charged drizzle. Stage 2 occurred as the cloud top approached the 0°C level; during this stage, the cloud developed positive rain drops whose individual charges were about 0.5 pC. They fell all the way to the ground. There were also negative drops in the cloud, and Takahashi stated that these caused the positive E_{gnd}. He thought, however, that the main charge carrier affecting E_{gnd} was negative ions, which had migrated into the cloud from the surrounding air. As the clouds grew above the $-10°C$ level, they developed into stage 3, with a secondary peak in their distribution of particle charges at about 5 pC from graupel. The cloud became stage 4 when it grew to about the $-20°C$ level. The polarity of precip-

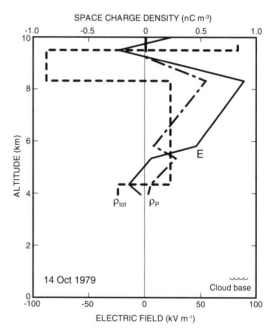

Fig. 8.58. Profiles created from published sounding made in tropical thunderstorm, 14 Oct. 1979. These plots are not as smooth as the originals, but show essential elements. A single ρ_{tot} from 4.4–8.4 km was used. The reason for no data below 4 km was not given; the sonde quit above 10 km. (Derived from data shown in Fig. 4b in Takahashi, 1983, with permission.)

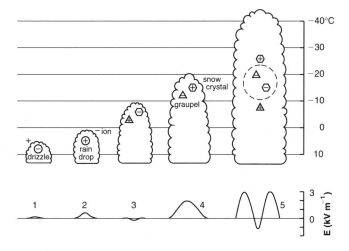

Fig. 8.59. Conceptual models of electrical properties of precipitation in tropical clouds and thunderstorms for five cloud stages. Stage 5 is a thunderstorm. Shown below each cloud is the typical E_{gnd}. The circles are liquid particles, triangles denote graupel, and hexagons denote snow crystals; the dominant charge polarity on those particles is indicated. The cloud stages are numbered and refer to summaries in the text. (From Takahashi 1983, with permission.)

itation aloft—apparently graupel—switched to negative. The E_{gnd} increased to >600 V m^{-1}. Such clouds contained large particles of the opposite polarity to the local E_{aloft}, suggesting to Takahashi that the large precipitation caused the field. Furthermore, Wilson ion capture seemed likely to be operating in the rain beneath the clouds. Stage 5 was the tropical thunderstorm, which was a taller combination of stages 3 and 4. Takahashi found graupel of both polarities of charge; the polarity was dependent on the temperature.

To summarize, we note that the |E| inside at least one tropical thunderstorm approached that in midlatitude thunderstorms. The total space charge density was similar, but the maximum density was less. In the warm clouds and thunderstorms, total and precipitation charge densities often did not agree. This implies that the total charge was not carried solely on precipitation-sized hydrometeors. Finally, we note that the combined cases studied by the investigators in this section included more than 700 warm clouds. When data were available for early electrification aloft, all showed that warm clouds begin with negative charge in the cloud. When the known cloud top was below the 0°C level, a cloud was never identified as a thunderstorm. This seems like compelling evidence against the possibility of lightning in warm clouds. However, we shall soon discuss equally compelling observations to the contrary.

8.4.7. Lightning in the tropics

Because of its frequency, lightning in the tropics may control the worldwide lightning flash rate, which historically has been estimated to be 100 s^{-1}. We will discuss both lightning in the tropics overall and examples of observations of lightning in individual storms. On this topic, as on many others, the available data are sparse, so our understanding should be considered embryonic. Lightning in tropical regions is an enigma in

that many very tall, heavily precipitating clouds produce no or very little observable lightning. The first question to ask might be whether this is caused by the sparsity of observation sites and the lack of automatic detection systems in much of the tropics. The answer seems to be no, as we shall see later. We will examine the observations of lightning on a large scale to ascertain whether lightning in the tropics is different from elsewhere.

We divide clouds and storms in tropical regions into two regimes: One is deep convection over land (and sometimes adjacent ocean areas) associated with highly unstable continental air; the other is clouds produced over the ocean and adjacent land in maritime air from monsoonal flow. The key to understanding the lightning enigma may be the identification of the type of airmass in which the storms develop.

Takeuti and Nagatani (1974) used a sferics detector on board a research vessel that sailed between ±20° latitude in the area of the Philippine Islands to Fiji. Their data covered 9 days and showed that oceanic storms had a broad diurnal peak, which started in the afternoon and went through the night and into early morning. This is much broader than the peak from continental non-MCS thunderstorms. They proposed that it was the heat capacity of the ocean that kept the solar influence on the "ground" minimized. They noted that lightning rates averaged about 0.5 min^{-1} for storm lifetimes of about 30 min. Takeuti and Nagatani concluded the lightning rate was less in the oceanic storms than in summer storms over land.

Takahashi (1978) defined the warm rain clouds at Ponape as those with tops at ≈5.5 km and with an estimated temperature of −3°C. (Although the cloud tops rose slightly above 0°C, he thought the microphysical processes would not involve the ice mechanism.) In the 77 clouds studied, he detected no lightning. This was in spite of rainfall rates of ≈200 mm h^{-1}. This result alone shows that significant total amounts and rates of rainfall are of themselves inadequate to generate an electric field capable of producing lightning. In the same paper, Takahashi reported on four thunderstorms whose tops reached 11 km and temperatures of $-41°C$. Data from one storm that moved across his site on 24 October 1977 showed a foul weather electric field at the ground for more than 40 min. For 20 min, the rainfall rates were >100 mm h^{-1}, but no lightning occurred. Two lightning flashes can be seen in his E_{gnd} record (see Fig. 8.55); both occurred after rainfall at the gauge had decreased to about 40 mm h^{-1}. There are insufficient data from these cases to learn the cause of the lightning; again, the mere presence of intense rainfall is inadequate. He concluded that the lightning-producing storms were formed within a convergence zone of a trough similar to those described for the Atlantic tropical areas by Riehl (1954). He also noted that the frequency of lightning was low for his observed storms. It seems to us that his observations of a dearth of lightning recorded near such storms illustrate one problem regarding thunderstorm identification. Without long-range detection and locating systems, it would be easy for such infrequent lightning to go unobserved. Takahashi hypothesized that only a few flashes occurred in the storms at Ponape owing to their very large liquid water content and the subsequent lack of significant electrification via graupel. Takahashi extended his observations with numerical cloud models of the electrical structures of these clouds (Ch. 9).

Using optical emissions recorded by satellites in the U.S. Defense Meteorological Satellite Program (DMSP), Orville and Spencer (1979) derived the latitudinal dependence of lightning. The data they used spanned a year and was limited to either local dusk or local midnight owing to the orbits of the satellites. The data are shown in Fig. 8.60 and clearly suggested that zonally integrated flash rates (i.e., integrated across all longitudes for a constant latitude) have a latitudinal dependence. Data for all times during the day were acquired by the Japanese Ionosphere Sounding Satellite, ISS-b, which received the radio frequency radiation from lightning. Those data spanned about 2 years, and their analysis by Kotaki et al. (1981) showed again that the zonally integrated lightning rate is a strong function of latitude (Fig. 7.25). That conclusion from these studies remains valid, although other proposed latitudinal dependencies have been refuted by observations.

As documented recently by Zipser (1994) with a relatively dense armada of research ships and 20,000 hours of observations during the GARP Atlantic Tropical Experiment (GATE), many tropical heavy rain events produced no lightning, even when their associated cloud tops reached 17 km. Zipser reiterated suggestions by others that the key to lightning production in tropical clouds is not cloud depth, but the vertical updraft velocity in the clouds: If updrafts do not exceed some threshold, lightning is absent or highly unlikely. Zipser and Lutz (1994) suggested that a threshold would be roughly a mean updraft speed of 6–7 m s^{-1} or a peak updraft speed of 10–12 m s^{-1}. Rutledge et al.

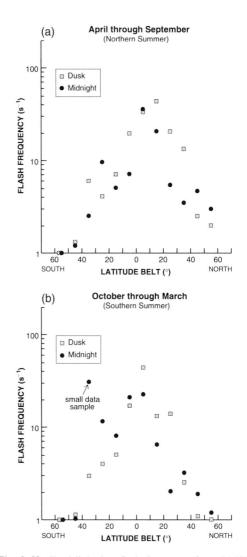

Fig. 8.60. Total-lightning flash frequency from DMSP satellites for 1-year period. The data were available at any given location at local dusk and midnight. Panel (a) is the northern summer and panel (b) the southern. (From Orville and Spencer 1979, with permission.)

(1992) proposed that even modest increases in CAPE in the tropics resulted in a large change in mass aloft in the cloud and thus was tied to electrification and lightning.

Examples from Darwin, Australia (12°S latitude), showed large variations in the cloud:ground flash ratio (often called the IC:CG ratio). Rutledge et al. (1992) used a combination of a field change sensor (slow antenna) and a ground strike location network and found that the IC:CG ratio increased with the total flash rate; in other words, the percentage of cloud discharges increased with increasing flash rate. They found the IC:CG ratio to be as high as 40, which is about that re-

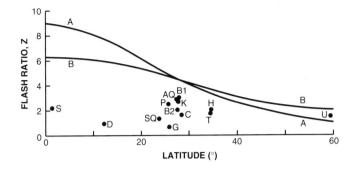

Fig. 8.61. Measured IC:CG ratio as a function of latitude (dots). There does not appear to be a latitudinal dependence. Lines A and B are published predictions. (From Baral and Mackerras 1992: American Geophysical Union, with permission. See references.)

ported in large midlatitude storms. Those large ratios and flash rates occurred during periods of large CAPE, leading them to propose that CAPE might be the controlling factor. The relationship that Rutledge et al. found for the ratio was

$$IC:CG = 2.7F^{\frac{1}{2}}, \qquad (8.5)$$

where F is the total flash rate in flashes per minute. The correlation coefficient was 0.93. We alert you that the literature contains relationships that use only ground flashes or that use only strokes instead of flashes, but what is used is not always clearly defined.

There has been a long-held notion that the IC:CG ratio depends on latitude. This idea was based on a sparse collection of data from a few latitudes; the curve fit through those data indicated that there should be a larger IC:CG ratio near the equator. Figure 8.61 shows evidence that began to refute such dependence (Baral and Mackerras, 1992); additional evidence was found by Mackerras and Darveniza (1994), which is shown in Fig. 8.62. Their recent measurements also support the idea that the flash density varies with latitude (Fig. 8.63). They found that the straight-line fit to zonally averaged variations in total flash density (N_{tot} in units of $km^{-2} y^{-1}$) from the tropics through 60° was

$$N_{tot} = e^{(3.7 - 0.07\phi)}, \qquad (8.6)$$

where ϕ is latitude.

Additional analyses of data from the Darwin area by Williams et al. (1992) documented some differences between two categories of lightning-producing storms:

storms embedded in monsoonal convection and storms during the break period typified by continental convection. They recorded total lightning within an area of $5 \times 10^3 \ km^2$ using a slow antenna (sec. 6.9.1) and recorded ground flash locations over about $10^5 \ km^2$. Precipitation measurements were from radar and from a rain gauge network that covered a circular area of $4 \times 10^4 \ km^2$ around the radar. They compared the ground flashes with precipitation because of the similar coverage. Williams et al. found that about 10 times more precipitation fell per ground flash from monsoon storms than from continental storms in the break periods (i.e., monsoon convection produced fewer ground flashes for the same rainfall). They attributed this increased electrification of continental storms compared with that of monsoonal storms to the influence of the convective available potential energy (CAPE) on the growth of ice. The continental convection occurred in an environment of larger CAPE and had much more ice aloft in the mixed-phase region. They found $5 \times 10^9 \ kg$ of precipitation per flash during monsoon convection and $5 \times 10^8 \ kg$ per flash during continental convection. The yields for continental convection at Darwin were greater by up to an order of magnitude than those reported in Piepgrass et al. (1982) for midlatitudes. Williams et al. concluded that the difference between the production of ground flashes over land and ocean was caused by differences in ice-phase condensate over land and over ocean. Their conclusion about why monsoon storms produced less lightning agreed with Takahashi's (1978)—lots of liquid precipitation with-

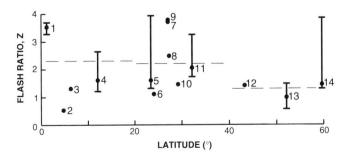

Fig. 8.62. IC:CG ratio as a function of latitude. The data are from 1987–91. The numbers denote the sites (see source reference for list), which ranged from 60°N to 27°S. The dashed lines show the mean Z, weighted by years of data for the range of latitude the line spans. Vertical lines show some yearly ranges of Z. (After Mackerras and Darveniza 1994: American Geo-physical Union, with permission. See references.)

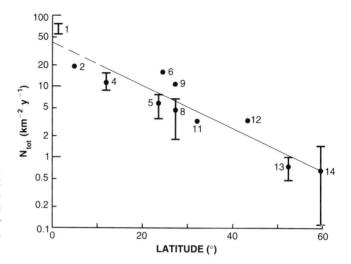

Fig. 8.63. Mean annual, total flash density, N_{tot}, versus latitude. Each site had at least 1 year of data. The numbers denote the sites (see source reference for list). The straight line is from Eq. 8.6. (After Mackerras and Darveniza 1994: American Geophysical Union, with permission.)

out adequate ice-phase precipitation results in weaker electrification and less lightning than in continental storms of comparable size. Williams et al. proposed that all oceanic activity, except in the western portions of oceans where surface wet-bulb temperatures are 1°–2°C warmer than elsewhere in the oceans, be classified as monsoonal convection.

There have been a few studies comparing the electromagnetic characteristics of flashes in the tropics with those in other places. All resulted in conclusions that cloud and ground flashes are the same in the tropics as elsewhere.

Research within the tropics has also included examinations of the continental squall lines that make up tropical mesoscale convective systems. An example was the COPT 81 study (Convection Profonde Tropicale 1981) during which scientists examined MCS data that included lightning, balloon soundings of the electric field, and electrical measurements at the ground (Sommeria and Testud 1984). Lightning was evident in the data of Chauzy et al. (1985), but lightning was not the main emphasis of the study and was not discussed much. Petersen and Rutledge (1992) compared examples of lightning production in eight tropical MCSs near Darwin with a midlatitude MCS. They found that peak currents from positive ground flashes were in the stratiform region in both locations. In both places, there was a large difference in the inferred peak positive currents between the convective regions, which had small currents, and stratiform regions, which had the larger peak currents. They discussed the possibility that the lightning peak current was affected by the volume of cloudy air or by intensification of the stratiform regions. Their results showed no real differences between lightning locations in tropical and midlatitude MCSs. They did concur with a modified version of Orville's (1990b) suggestion of peak current increasing with decreasing latitude, if the cloud top height were taken into account. Such a latitudinal dependence has yet to be proved conclusively.

8.4.8. Warm cloud lightning

As the name implies, we define *warm-cloud lightning* as lightning that occurs in a cloud whose top is not above the 0°C isotherm and has not extended above this level at any time in the cloud's life. The clouds are typified as small developing cumulus clouds that produce lightning. In the literature, some have called clouds warm if their tops did not exceed the height of −5° or −10°C, so the definition of warm cloud has been ambiguous. We will follow our stricter definition. In several studies, knowing if warm-cloud lightning occurred was muddled by uncertainties in the measurements of cloud top height or temperature. The phenomenon is important because, if it exists, it places a constraint on at least one type of cloud electrification mechanism; obviously frozen hydrometeors would not be a necessary condition for cloud electrification under at least some conditions. We include warm-cloud lightning here because most reports of lightning in warm clouds have been in tropical and subtropical regions. We also note that the existence of warm-cloud lightning is not universally accepted. Furthermore, mechanisms proposed to explain typical thunderstorm electrification usually place warm-cloud lightning in a separate category that is not treated. We think the observations are credible enough to warrant inclusion in any comprehensive discussion of the electrical nature of clouds, and we present a summary of the observations in the literature.

Foster (1950) reported observations of two narrow clouds flashing at a rate of once per 10 seconds near Guam (≈13.5°N). By the time their weather reconnaissance flight overflew the clouds, the tops were at a true radar altimeter reading of 2469 m. The cloud was

≤500 m in diameter and 1700 m thick. What was not stated in this report was whether the top ever exceeded the 0°C level; we interpret the article to indicate that the observers did not know.

The first compilation of reports of warm-cloud lightning probably was by Moore et al. (1960). Besides Foster's report, they noted several reports from pilots and weather observers. Moore et al. traveled to Grand Bahama Island to investigate the phenomenon. With an instrumented airplane, they measured electric field magnitudes up to 3 kV m^{-1} above and around warm clouds. They reported seeing lightning in clouds whose tops were below 0°C while they approached from 50 km away. By the time of the overflight, the clouds had risen to just colder than 0°C. There also was a 3-cm wavelength radar on the south shore of the island that was used in a vertically scanning mode. Sometimes they combined their visual and radar measurements to determine cloud top height. One example was a nighttime cloud that produced four flashes in 10 min. They photographed its luminosity, combined the photo with radar data, and concluded that the cloud top never was taller than the 0°C level. Their photographs and calculations showed that while it was producing lightning, the top of the cloud was 0.4–1.3 km below the height of 0°C, which they recorded as 4.6 km on that night.

Sartor (1964) recorded radio emissions from the early stages of cumulus cloud growth near Key West, Florida (≈24.5°N), which is just north of the demarcation of the tropics. His goal was to measure emissions that occurred before lightning. His instrumented aircraft measured temperature, vertical electric field, and radio emission at 30 and 50 MHz. He observed radiation not due to lightning from clouds whose tops were several degrees warmer than 0°C. The observational procedure indicated that the clouds had never attained the 0°C level, so there is little doubt that he observed nonlightning emissions from warm clouds. He ruled out corona from the airplane and mature thunderstorms as the source, because both were absent. He suggested that the likely source was emissions from charged particle interactions, which have been observed in laboratory experiments. Significant electrification of warm clouds was also reported by Moore and Vonnegut (1973) who measured magnitudes of E as they flew beside a cloud top that was growing at 5 m s^{-1} and that saturated their potential gradient sensor at its full scale of 3 kV m^{-1}. Although they recorded no lightning then, their data corroborated that significant electrification can occur in warm clouds.

Rossby (1966) flew an airborne, 600 MHz sferics detector near Key West, Florida, and found lightning in nighttime clouds. He reported that the character of the flashes was different from what he had observed in fully developed cumulonimbus clouds whose tops went well above 0°C. In the warm thunderclouds, the lightning rate was higher—about one every 8–15 s. They were exclusively cloud flashes, and their duration was very short. Rossby reported that on two nights the cloud tops were below their flight altitude of 3050 m.

Since the 0°C level was measured at 4780 m and 4480 m on those two nights, the clouds appeared to be well below the 0°C level. Left unstated was whether the clouds had ever been taller. One cloud continued to grow well above 0°C to a full cumulonimbus; Rossby reported this as unusual. Most warm-cloud lightning was followed by the demise of the cloud. Finally, the 600-MHz sferics from warm-cloud lightning was different from thunderstorms. The sferics from warm-cloud lightning were more "loosely packed" pulses. Interestingly, for the case that changed from a warm- to a cold-cloud thunderstorm, the sferics changed and became the typical, densely packed pulse trains such as were produced by lightning in other thunderstorms.

Moore et al. (1960) suggested that warm-cloud lightning was infrequently reported because the lightning could be seen easily only at twilight or in darkness, was mostly short-lived, and may have been linked preferentially to clouds over the ocean, where observations are less likely. These obstacles are still valid despite our present technologies and would be sufficient to make casual observation infrequent. Since there are only occasional reports of warm-cloud lightning, it appears to us that it will take a focused effort to provide reliable data to determine its existence more conclusively than has so far been done. If such proof were to be forthcoming, it would mean that at least under some condition, thunderstorm electrification does not require the presence of frozen or supercooled hydrometeors. There are enough credible reports of warm cloud lightning from different observers and scientists that the phenomenon seems to be one in legitimate need of additional observations.

8.5. TROPICAL CYCLONES

Observations of electrical effects in tropical cyclones (e.g., hurricanes or typhoons) are very limited. Overviews of these systems can be found in such texts as Cotton and Anthes (1989). For our purposes, it is worth noting that a major difference between tropical cyclones and midlatitude MCSs is that the updrafts are weaker in the tropical cyclones. For example, Jorgensen et al. (1985) found that maximum updrafts were <20 m s^{-1} for ≈99% of the data in tropical cyclones, with a median maximum vertical velocity of ≈3 m s^{-1}. Thus, updrafts (and downdrafts) are significantly less than those found in midlatitude convection. Also, less supercooled water is found in the tropical systems. Black and Hallett (1986) presented a conceptual model of hurricane microphysics applicable near 0°C. Features of the model included: maximum reflectivities in updraft regions with <10 L^{-1} raindrops and large graupel; lots of ice crystals in the downdraft on the outer edge of the reflectivity maximum; cloud liquid water content of >1 g m^{-3} only in updrafts warmer than −1°C or updrafts ≳5 m s^{-1}; supercooled water scarce except in eyewall updrafts; advection of much ice; and most water frozen by −10°C level. Tropical cyclones have extensive bright band regions outside the eyewall (Jorgensen 1984). These and other dynamic and microphysical features of hurricanes undoubtedly impact the electrical structure in these systems. As we will see, very few in situ electrical measurements have been made, and vertical soundings are nonexistent.

The earlier work on electrification of tropical cyclones was limited mostly to sferics detection of lightning, and served to confirm that it can occur in mature tropical cyclones. We give a brief report here on the few recent studies.

8.5.1. Electric field

We know of no systematic measurements of E at the ocean surface beneath tropical cyclones reported in the literature. Of particular interest would be in situ measurements of E_{aloft} in tropical cyclones. Attempts to gather such data have been undertaken, with preliminary results shown by Black et al. (1993) from four, rotating-vane field mills used on one of the hurricane hunting NOAA P3 airplanes. Although we report them here, we do know the reliability or accuracy of the E measurements inside these clouds, since the effect of charging of the P3 was not fully quantified; a preliminary report was given by Hallett et al. (1993). Black et al. measured E_{aloft} in Hurricane Tina with penetrations of the eye at temperature levels from $+5°$ to $-12°C$. They found little supercooled water, a main updraft of $\approx 5 \text{ m s}^{-1}$, and $|E_{aloft}| \lesssim 15 \text{ kV m}^{-1}$. The lack of a large E in convection near the center of the circulation may be related to the earlier findings by Black and Hallett (1986) that revealed little supercooled water existed, especially at temperatures colder than $-10°C$, in mature hurricanes. In contrast, Black and Hallett (1995) reported that Hurricane Claudette had a large convective cell which dominated its incomplete eyewall. In that eyewall region, they found characteristics more common of typical thunderstorm cells, including several grams per cubic meter of supercooled rain at $T > -5°C$. They showed E measurements whose magnitude exceed 20 kV m^{-1}, but they stated that there were uncertainties of a factor of two in the absolute calibration of the field mill system. Black and Hallett showed updrafts of $>10 \text{ m s}^{-1}$, and they postulated that it was likely that graupel carried the dominant portion of the negative charge in a temperature range of about $0°$ to $-5°C$. They concluded that excess negative charge existed in both updrafts and downdrafts at $T > -13°C$. They did not find much indication of positive charge except perhaps at the melting level.

These very preliminary results from in situ measurements provided a first insight into the electrical structure of hurricanes. Vertical profiles of E and a determination of whether the bright bands are associated with large E and a dominant space charge density near $0°C$ would be quite illuminating. Although challenging, this area is ripe for observational investigation to accompany existing modeling studies and in situ and radar observations of nonelectrical parameters.

8.5.2. Ground flashes in hurricanes

The lightning activity of a very few hurricanes has been observed. From the published figures in existing studies, we do not detect any bipole pattern of the two polarities of ground flash, similar to the pattern observed in midlatitude MCSs. Perhaps the most extensive study of lightning in hurricanes was by Molinari et al. (1994), who analyzed ground flashes in Hurricane Andrew. Most ground flashes consistently were in the outer bands, and almost none occurred within the eyewall. Molinari et al. noted that convection in the eyewall is typified by a lack of supercooled water and suggested that this lack of water was responsible for there being insufficient charge generation for lightning. Flashes did occur in the eyewall region when convection there grew rapidly. Because lightning did not occur in the eyewall region until just before hurricane intensification, they suggested that lightning might be used as an indicator of hurricane intensification, but they noted that more observations would be needed to test such an application.

Similar findings have also been reported by Black et al. (1986) and Lyons et al. (1989) and Lyons and Keen (1994). Lyons and Keen noted an increase in ground flash rate, which they called a "burst," near the eyewalls of Hurricanes Diana and Florence preceding hurricane intensification. They speculated that such bursts in ground flash rate are associated with supercells in the tropical cyclone and may have diagnostic value in identifying such cells. They proposed that in tropical storms: (1) Few ground flashes occur in the central region of mature hurricanes; (2) most flashes occur in the outer rainbands; (3) it may be possible to predict hurricane intensification up to 6–12 h in advance from an increase (i.e., a relative burst) in ground flash rates within 100–150 km of the circulation center.

Ground flash production was quite different in Hurricanes Hugo and Jerry, both studied by Samsury and Orville (1994). There were 33 detected ground flashes from Hugo and 691 from Jerry, although both were analyzed for an 18-h period. Most flashes struck in the right front and rear quadrants of the outer rainbands, although half the detected ground flashes in Hugo were near the eyewall.

Molinari et al. (1994) observed that the mean current of first negative strokes in the eyewall was 44% less than that for negative first strokes outside the eyewall. They found that this difference was statistically significant, but did not offer a physical reason for the difference. The distribution of ground flashes matched that of the plan convective structure of the mature hurricane. Ground flash density showed a minimum in the inner rainbands and a significant maximum in a region of the outer rainbands. Molinari et al. suggested that this arose from sedimentation of frozen particles ejected from the eyewall and from the subsequent removal of supercooled water from the vicinity of the frozen particles. The melting and evaporation below the 0°C level would then suppress convective updrafts. Molinari et al. stated that ground flashes in the eyewall were absent for hours, but that there were ground flashes in the eyewall for 3–6 h during intervals associated with storm intensification. This was noted particularly with a rapid deepening of convection just before hurricane landfall in Florida. No storm intensification occurred without the coexistence of increased ground flashes. While their findings add more credence to a connection between eyewall lightning and subsequent cyclone intensification, they noted that an inadequate number of cases have been studied to derive conclusive generalities. Nevertheless, they also noted that such potentially useful indicators are worth future exploration.

9

Numerical Models of Thunderstorm Electrification

9.1. INTRODUCTION

In its broadest sense, a numerical model of thunderstorm electrification can be considered to be any calculation used to examine some process involved in the electrification of clouds. This definition encompasses a broad range of sophistication. At the simplest end are short calculations for evaluating the plausibility of a given process. A simple model, for example, might assume "typical" number densities of a few classes of particles and calculate whether the resulting collision rates and vertical velocities of the particles can separate sufficient charge to cause lightning if a given interaction process operates. In the most sophisticated models, electrical processes are added to time varying, three-dimensional, numerical cloud models that stimulate a cloud from its genesis. These models integrate dynamic, microphysical, and electrodynamic processes to track what happens to several classes of particles as they move through the cloud and interact with each other and with the environment throughout the cloud's evolution.

Because of space limitations, we will concentrate in this chapter on numerical models that incorporate electrification processes in a cloud model. However, some thunderstorm electrification models include no microphysical or dynamic processes. Some of these numerical models treat electrification as interactions of electric circuit elements driven by a current, charge, or voltage source. We will mention an example of these models later, in a brief history of electrification modeling (Sec. 9.3).

When using models, it is prudent to keep their limitations in mind. No numerical thunderstorm model can reasonably be expected to replicate a thunderstorm exactly. To do so, it would have to (1) define a separate category of particles for every unique combination of particle characteristics and keep track of all relevant parameters of every category of particle in the storm, (2) incorporate an exact mathematical formulation of the physics of all significant interactions and behaviors for each particle category, and (3) handle simultaneous processes operating on temporal and spatial scales ranging from microseconds and microns to minutes and kilometers. The first and third tasks are prohibited by available computer memory and speed. The largest, fastest computer planned for the foreseeable future would be unable to treat the evolution of every unique class of particles in even a small three-dimensional thunderstorm or to track processes over the entire range of applicable scales in a reasonable amount of time. The second task requires more knowledge than we have. Although many processes are understood reasonably well, some are understood poorly, and others are understood primarily as ensemble-averaged behaviors of large populations.

We must, therefore, use simplified mathematical descriptions called *parameterizations* to deal with the spectrum of particle types, to estimate the effects of poorly understood physics, and to incorporate the effects of processes that occur on temporal and spatial scales too small to be included directly. The goal is for the parameterizations to incorporate enough physics on both the microphysical and storm scales and to retain enough resolution and detail in the hydrometeor, electric, wind, and thermodynamic fields that the model still can simulate accurately the aspect of the phenomenon we are studying. To some extent, success in addressing this goal is judged by examining how well the model simulates related storm properties that are observed, such as the location, amount, type, and evolution of precipitation and cloud particles; the location and magnitude of updrafts and downdrafts; the occurrence of lightning; and the electric field structure. However, it often is difficult to assess how well a particular model has succeeded in simulating a given phenomenon. Sometimes the difficulty is that relevant properties have not been or cannot be observed well enough to judge simulations. Sometimes the difficulty is in determining how good a match with observations is needed to establish that the model is adequate and that omitted processes are unimportant.

Although any model involves uncertainties that can be resolved only by additional measurements, models also have strengths that complement serious difficulties in interpreting measurements alone. Frequently, for example, a sensor (such as an airborne electric field mill) distorts what is being measured, or ambiguities arise when interpreting sensor output (such as back-scattered power received by a radar) in terms of desired storm parameters (such as hydrometeor size spectra or rainfall rate). An equally serious problem is that no combination of technologies in the foreseeable future is capable of observing simultaneously all of the thermodynamic, kinematic, electric, and hydrometeor fields in evolving clouds with enough temporal and spatial resolution to delineate all their significant behaviors and interactions. Modeling fills this lack by attempting to calculate all of the relevant fields in a physically consistent way from as close to first principles as is feasible. By providing a complete set of simulated observations of the very complex system that is a thunderstorm, modeling provides a useful vehicle for testing the plausibility and implications of ideas gained from theory and observations.

To do this, numerical models often are used in iterative cycles with observations: (1) Laboratory and field observations provide the knowledge needed to build a model; (2) Numerical modeling experiments provide insight into storm processes and help define predictions and model sensitivities that can be tested with observations; and (3) New laboratory and field observations are acquired under model guidance to test and refine the model. Then the cycle repeats itself with an improved model.

Our goal in this chapter is to give a basic understanding of the choices and some of the techniques that are necessary for modeling electrification, as well as to give an overview of some electrification research topics that have been pursued through modeling studies. However, the number of thunderstorm electrification models and modeling techniques that have been developed preclude a thorough treatment of each one in this book. Likewise, the amount of information that would be involved in describing the main features of even one cloud model in which an electrification model is imbedded is much greater than could fit in the space available here. Therefore, the rest of this chapter will include only a brief review of the microphysical and dynamic framework of cloud models and will present only selected examples of treatments of electrical processes. More information on modeling clouds is available in several books, including Pruppacher and Klett (1978), Cotton and Anthes (1989), and Houze (1993). More information about modeling cloud electrification is available in a review by Levin and Tzur (1986) and in many of the publications referenced in this chapter.

In this chapter, we have tried to use symbols typically used by modelers, when this can be done clearly and does not conflict with conventions commonly used in research on atmospheric electricity. However, because this chapter uses a large number of symbols, and different segments of the scientific literature commonly use the same symbols for different quantities, we sometimes differentiated symbols by subscripts. For example, ρ is used for both charge density and mass density, but has a lower case subscript when used for charge and an upper case subscript when used for mass density.

9.2. CLOUD MODEL PARAMETERIZATION

9.2.1. Cloud dynamics, thermodynamics, and microphysics

The behavior of a parcel of moist air in a cloud is described by an equation of motion, thermodynamic equations, and continuity equations. For clouds in general, the equation of motion given by Newton's second law of motion can be written as

$$\frac{d\vec{v}}{dt} = -\frac{1}{\rho_A}\vec{\nabla}P - g\hat{k} - 2\vec{\Omega} \times \vec{v} + \vec{a}_{\text{fr}}, \qquad (9.1)$$

where $\vec{v}$ is the vector velocity of the parcel, ρ_A is the mass density of air, P is pressure, $\hat{k}$ is the vertical unit vector, g is the acceleration of gravity, $\vec{\Omega}$ is Earth's angular velocity ($\vec{\Omega}$'s direction is upward along Earth's spin axis, and its magnitude is in units of radians s^{-1}), and $\vec{a}_{\text{fr}}$ is acceleration due to friction. The terms on the right are from pressure gradient, gravitational, Coriolis, and frictional forces, respectively. (See an elementary atmospheric dynamics text, such as Holton 1992, for an explanation of these forces.) Here and in the rest of this section, the time derivative $d\vec{v}/dt$ is the total derivative evaluated in a frame of reference moving with the parcel (called the *Lagrangian derivative*). The derivative with respect to a fixed frame of reference (called the *Eulerian derivative*) is given by $\partial v_i/\partial t = dv_i/dt - \Sigma v_j(\partial v_i/\partial x_j)$, where v_i and x_i are the ith velocity component and spatial coordinate, respectively, and the summation is over $j = 1$–3. (The Lagrangian and Eulerian derivatives of other parameters are related by analogous expressions.) Because the density of dry air, ρ_D, is much greater than the density of atmospheric water vapor, ρ_V, ρ_D can be used in place of ρ_A.

Numerical cloud models often use a slightly different form of the equation of motion based on perturbations from hydrostatic equilibrium. *Hydrostatic equilibrium* means that the weight per unit area of a column of atmosphere above an arbitrary point is equal to the pressure at that point, that is,

$$\frac{1}{\rho_A} \frac{\partial P}{\partial z} = -g. \tag{9.2}$$

Departures from hydrostatic equilibrium tend to be small in the troposphere, so the first two terms in Eq. 9.1 constitute a small difference of large quantities, a situation that makes accurate numerical solutions difficult to obtain. To derive a form of the equation that avoids this problem, we write Eq. 9.1 in terms of perturbations of pressure and density from a horizontally uniform, hydrostatically balanced state. The equation of motion then can be approximated closely by

$$\frac{d\vec{v}}{dt} = -\frac{1}{\rho_{A0}} \vec{\nabla}P^* + B\hat{k} - 2\vec{\Omega} \times \vec{v} + \vec{a}_{fr}, \tag{9.3}$$

where the subscript 0 refers to values of the hydrostatic state, the superscript * refers to deviations from the hydrostatic state, and B is *buoyancy*, defined by

$$B \equiv -g \frac{\rho_A^*}{\rho_{A0}}. \tag{9.4}$$

The thermodynamic equation of state for moist air is reasonably well approximated by the *Ideal Gas Law*

$$P = \rho_D R_D T + \rho_V R_V T, \tag{9.5}$$

where R is the gas constant, and T is temperature. Eq. 9.5 can be approximated by

$$P \approx \rho_D R_D T(1 + 0.61 q_V), \tag{9.6}$$

where q_V is the *mixing ratio* of water vapor in air (defined as the mass of water vapor per unit mass of dry air). Note that q with an upper case subscript denotes mixing ratio; q with a lower case subscript denotes charge. Changes in temperature are related to changes in density by the *First Law of Thermodynamics:*

$$c_V \frac{dT}{dt} + P \frac{d\alpha}{dt} = Q, \tag{9.7}$$

where c_V is the specific heat of dry air at constant volume, α is the *specific volume* ($\alpha = 1/\rho_D$), and Q is the net heating rate. The change in temperature also can be given in terms of changes in pressure by combining the equation of state with Eq. 9.7 to get

$$c_P \frac{dT}{dt} - \alpha \frac{dP}{dt} = Q, \tag{9.8}$$

where c_P is the specific heat of dry air at constant pressure.

Continuity equations for air and water constrain changes in the mass of each. For air, conservation of mass requires that

$$\frac{\partial \rho_A}{\partial t} = -\vec{\nabla} \cdot (\rho_A \vec{v}). \tag{9.9}$$

This form of the continuity equation is referred to as the *fully compressible continuity equation.* The same continuity equation can apply to dry air, with ρ_D in place of ρ_A. In some applications, the continuity equation for air is approximated as

$$\vec{\nabla} \cdot (\rho_A \vec{v}) = 0. \tag{9.10}$$

This approximation, called the *anelastic continuity equation,* retains variations of density with height, but eliminates time derivatives. The primary effect of eliminating time derivatives of density is to eliminate sound waves, and this often is considered a virtue in modeling clouds. If density is assumed to be approximately constant (as a scale analysis might show to be appropriate for shallow clouds), then Eq. 9.10 is approximated as

$$\vec{\nabla} \cdot \vec{v} = 0. \tag{9.11}$$

This form of the continuity equation is referred to as *incompressible.*

Continuity equations for water vapor and water substance are expressed in terms of mixing ratios. Like the mixing ratio for water vapor, defined earlier, the mixing ratio for the Nth type of hydrometeor is defined as the mass of the Nth hydrometeor per unit mass of dry air (often expressed as kilograms or grams of water per kilogram of dry air). The mixing ratio for total water substance, then, is the sum of the mixing ratios for all types of hydrometeors. For each type of hydrometeor, there is a continuity equation of the form

$$\frac{dq_N}{dt} = transport(q_N) + source(q_N) - sink(q_N), \tag{9.12}$$

where transport is the net transport of q_N into the volume by advection, turbulence, and diffusion; source is the sum of all sources of q_N in the volume; and sink is the sum of all loss mechanisms for q_N. Much of the challenge in parameterization is to develop physically realistic expressions for sources and sinks, once a partition of water into various categories is chosen.

The water substance in a cloud normally is partitioned into several categories. For example, the following categories are described by Houze (1993):

1. *Water vapor* (q_V) is water in the gaseous phase.
2. *Cloud liquid water* (q_{CLD}) consists of liquid droplets that are too small to have appreciable terminal fall speed (droplet radius less than roughly 100 μm).
3. *Precipitation liquid water* consists of liquid drops that are large enough to have an appreciable termi-

nal fall speed. This category sometimes is divided by terminal fall speed into *drizzle* (q_{Dr}) (radius roughly 0.1–0.25 mm) and *rain* (q_R) (radius >0.25 mm).

4. *Cloud ice* (q_I) indicates ice particles that are too small to have an appreciable terminal fall speed.
5. *Precipitation ice* is composed of ice particles that have a terminal fall speed of $\gtrsim 0.3$ m s^{-1}. This category often is subdivided by density and fall speed. For example, *snow* (q_S) has lower density and fall speeds of ≈ 0.3–1.5 m s^{-1}, *graupel* (q_G) is denser and falls at ≈ 1–3 m s^{-1}, and *hail* (q_H) is larger and even more dense, with fall speeds up to 50 m s^{-1}.

Models that omit all forms of ice are referred to as *warm cloud models*. Models that include equations for the ice phase are referred to as *cold cloud models* or *mixed-phase models*.

Each category of water substance interacts with the other categories to create sources and sinks. For example, cloud water droplets coalesce to form drizzle, a process that is a source for drizzle and a sink for cloud liquid water. There are several types of basic interactions:

1. *Condensation or deposition* of water vapor onto cloud nuclei (called *nucleation*) to form cloud liquid water droplets and cloud ice particles, respectively, having size spectra characteristic of the model cloud's environmental conditions
2. Hydrometeor growth through vapor *condensation or deposition*
3. *Collection* of particles to form larger particles (Collection of the various types of particles consists of collisions followed by sticking together. The ways in which two particles stick together are labeled by specific terms depending on the types of particles involved: *coalescence* involves two or more liquid water particles; *aggregation*, two or more ice particles; *riming*, cloud water droplets freezing to ice particles on contact. *Accretion* is used broadly to indicate collection of liquid particles by ice particles.)
4. *Breakup* of drops or *splintering* of ice (When this involves ice, it can make the number of ice particles larger than would be produced by nucleation alone. This increase is called *ice multiplication*.)
5. *Freezing* of liquid water (Liquid water exists as supercooled water at heights above the 0°C isotherm, but all liquid water usually is assumed to have frozen by the time a parcel reaches the −40°C isotherm.)
6. *Evaporation or sublimation* of water vapor from hydrometeors
7. *Melting* of ice
8. Precipitation reaching the ground

It is possible to expand considerably the number of categories and subcategories of hydrometeors. For example, a modeler might want to track different shapes (called *habits*) of ice particles separately. However, as more categories of water are used, the number of inter-

actions that must be considered increases rapidly. Figure 9.1 shows an example of the interactions that are needed in a model that has six categories of particles. (Here, drizzle and rain have been combined in the rain category.) The interactions obviously become very complex if more than a few categories are used. For that reason, modelers usually use only the categories and subcategories that are essential to simulating the particular phenomenon being studied.

Each category of water particle spans a range of sizes. For example, the size distribution of precipitating liquid often is described as an exponential function of the form

$$n(D) = n_0 \exp(-\Lambda D), \qquad (9.13)$$

where $n(D)$ is the number of particles per unit volume between D and $D + \delta D$, and n_0 and Λ are constants of the exponential function that have different values for different cloud conditions. Models can vary n_0 and Λ in space and time, as cloud conditions warrant. This form of the size distribution for rain often is referred to as a *Marshall-Palmer distribution* (Marshall and Palmer 1948). Other forms, such as gamma functions, also have been used to describe the size distribution of particles.

To parameterize microphysics, models handle the size distributions of the various categories in one of two ways, referred to as *bulk microphysics* and *explicit microphysics*. In *bulk microphysics*, the size distribution of particles is described by some simple function. The amount of water substance in a given category can be tracked at each grid point by a single parameter, such as the mixing ratio of water substance in that category, or by two or more parameters, such as the mixing ratio and number density of particles in a category. Distributions using two or more parameters provide a more versatile description than distributions using one parameter.

In bulk microphysics, each property of the category has a single representative value at a given grid point in the model domain that somehow averages the values for all of the category's particles within the corresponding volume. For example, consider the sedimentation of rain. Rain and cloud liquid water usually are parameterized by using concepts proposed by Kessler (1969) (sometimes called *Kessler warm cloud microphysics*). The terminal fall speed for raindrops is assumed to be related to drop diameter D, as discussed in Section 3.4.3. Then, since the mass of a raindrop is given by $(\pi/6)D^3 \rho_W$ (ρ_W is the density of liquid water, 1 g cm^{-3}), the mass-weighted terminal fall speed is

$$\overline{v_{RT}} = \frac{\int_0^\infty v_{RT}(D)\, D^3\, n_R(D)\, dD}{\int_0^\infty D^3\, n_R(D)\, dD}, \qquad (9.14)$$

where $v_{RT}(D)$ is the terminal fall speed for raindrops of diameter D and $n_R(D)$ is the raindrop size distribution. The assumption in bulk microphysics is that all raindrops in the volume centered on a grid point fall with the mass-weighted fall velocity. Then C_R, the net rate

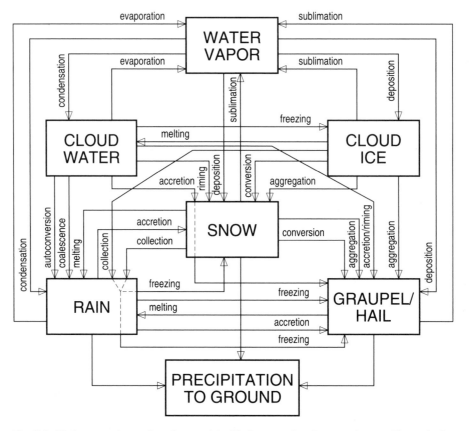

Fig. 9.1. Hydrometeor interactions for a model with six categories of water substance. If a particular interaction with hydrometeors in one of the categories does not add to the category, but results in conversion to another category, it is indicated by a dashed line continuing through that category box and connecting to the category to which the hydrometeor is converted. (Adapted from Rutledge and Hobbs 1984 and Helsdon and Farley 1987, with permission.)

of change in a grid point's rainwater mixing ratio, q_R, due to sedimentation of raindrops, is given by

$$C_R = \frac{\partial}{\partial z} (\overline{v_{RT}} \, q_R). \qquad (9.15)$$

When a model must retain the distinct microphysical behaviors of different sizes of hydrometeors, bulk microphysics is inadequate. Instead, the model must use a parameterization of microphysics called *explicit microphysics*, which subdivides one or more water substance categories according to size. For the Nth water substance type modeled by K size categories having a diameter increment of δD, the mixing ratio is approximated by

$$q_N = \sum_{j=1}^{K} q_{Nj}$$

$$\approx \frac{1}{\rho_D} \sum_{j=1}^{K} m_N(D_j) \, n_N(D_j), \qquad (9.16)$$

where q_{Nj} is the mixing ratio of the jth size category and $m_N(D_j)$ and $n_N(D_j)$ are the mass per particle and number density of particles, respectively, between D_j and $D_j + \delta D$. To represent the size distribution of particles accurately, roughly ten or more size categories must be included for each hydrometeor type being parameterized.

With explicit microphysics, it is unnecessary to assume that the size distribution has a particular form. The overall size distribution is allowed to evolve naturally as particles of different sizes gain and lose mass through their interactions. This is a more direct approach than bulk microphysics, but even with many size categories, an explicit parameterization is unlikely to be an exact treatment. There are gaps in our knowledge of particle interactions, particularly for some size ranges and particle types, and these gaps introduce uncertainties in explicit parameterizations. Furthermore, although particles within a size category are more uniform than in a category that spans all sizes in a bulk treatment, variations in properties, such as size, shape, or density may occur within categories being treated explicitly and can affect interactions involving these categories. When such variations are significant, explicit parameterizations need to avoid treating all particles in a size category identically, which would incorrectly cause all particles to transfer to a new size category at the same time. To avoid this, an explicit parameterization can treat particles in a given size category as having a distribution of sizes across the range of the category and can use statistical techniques to govern gradual changes in the population.

A disadvantage compared with bulk microphysics is that explicit microphysical treatments require considerably more computational resources to handle size categories: Each q_{N_j} must satisfy its own continuity equation (Eq. 9.12), and interactions must be included between size categories, as well as between water substance categories. Thus, explicit microphysics usually is reserved for addressing questions that cannot be addressed by bulk microphysics.

9.2.2. Other categorizations of cloud models

Besides the dynamic, thermodynamic, and microphysical considerations discussed in the previous section, other choices also influence the complexity and computational requirements of models. Most cloud models are *time dependent,* meaning that cloud properties are allowed to evolve, but some simpler models use *steady-state* dynamics. Another property by which models are classified is the number of spatial dimensions of the model grid. To study storms in which three-dimensional structure and circulations are important, it obviously is necessary to use a *three-dimensional* model. However, for simpler situations or more limited investigations, modelers can reduce the number of spatial dimensions they use. Because each spatial dimension added to a model typically increases computer memory and storage requirements by more than an order of magnitude, reducing the number of dimensions greatly reduces the computer resources required to run a model.

Some electrification models are essentially *zero-dimensional* cloud models because most parameters do not vary with either height or horizontal distance. These models normally assume that the upper and lower boundaries are infinite horizontal planes, so that the ambient electric field is constant between them. Microphysics and vertical winds also are kept uniform between the horizontal planes. On each boundary there is a source of hydrometeors, usually a source of small particles at the bottom and a source of large particles at the top. Particles and charge can vary with time, and charges are collected at the boundaries as particles reach them, thereby changing the ambient electric field. The model domain normally is considered to represent only part of a cloud, with the upper plate corresponding to the center of the upper positive charge, and the lower plate, the center of the lower negative charge in a thunderstorm charge distribution. Illingworth and Latham (1977) pointed out that a model with infinite planes will overestimate thunderstorm electric field magnitudes in most situations.

In a *one-dimensional, time-dependent* model, height is the only spatial coordinate that is retained. Cloud properties can vary with height, but not with horizontal position. Cloud properties also are functions of time, and so provide an indication of the cloud's evolution. The model domain normally is defined as a cylinder whose radius R(z) is specified. The use of a finite horizontal extent makes it possible to incorporate parameterizations of entrainment and turbulent eddy fluxes and makes electric field magnitudes more realistic.

In a *two-dimensional* model, height and one horizontal coordinate are retained, and variations in cloud properties over the remaining horizontal coordinate are set to zero. Two-dimensional models can be either slab-symmetric or axisymmetric. A *slab-symmetric model* uses Cartesian coordinates and sets variations along *x* or *y* to zero. This symmetry is applied sometimes to squall lines. An *axisymmetric model* uses cylindrical coordinates and sets azimuthal variations to zero. This symmetry is useful for small thunderstorms and some aspects of hurricanes. Unlike a one-dimensional model, two- and three-dimensional models do not need to specify cloud boundaries or entrainment. Instead, the model is provided the specified environmental thermodynamic and water vapor fields, and clouds form in the model wherever these fields interact to create them. (Slab-symmetric models also can use one horizontal component of environmental winds, but axisymmetric models can treat only radially converging or diverging horizontal winds.) Modeled air motions may be adequate to resolve entrainment, but eddy fluxes must be treated by parameterization of subgrid-scale turbulence.

Many one-dimensional models and some two- and three-dimensional models use the anelastic form of the continuity equation (Eq. 9.10). The more sophisticated two- and three-dimensional cloud models usually are fully compressible (Eq. 9.9). Although the anelastic approximation is reasonably accurate for clouds, the fully compressible form has computational advantages when used with some of the features often incorporated in newer models. However, the fully compressible form retains sound waves, which often are considered undesirable because their magnitude can be large and other processes are of more interest in modeling studies. Therefore, when the fully compressible form of the continuity equation is used, sound wave modes often are numerically split from the solutions.

A *full simulation model* produces clouds through appropriate initial conditions and simulates the subsequent evolution of all model fields. To simulate a cloud with a given one-dimensional, time-dependent model, a vertical profile of each relevant parameter of the ambient environment is specified, vertical forcing (e.g., a thermal perturbation or vertical velocity source) is applied at a low height, and the model then computes how deviations from environmental values evolve. Similarly, a two-dimensional, axisymmetric model generally applies forcing centered on $r = 0$. Usually, two-dimensional, slab-symmetric models and three-dimensional models are initiated by introducing some type of forcing with a more complicated geometry into the specified environmental conditions. For example, this forcing may be in the form of convergence in some region at low levels on which are superimposed thermal bubbles, whose size, location, and temperature excess above ambient conditions must be chosen (either given specific values or chosen randomly from some range). Storms that result from a given model experiment can be influenced by the form and magnitude of forcing that is applied, as well as by assumed environmental conditions; this is especially true of small thunderstorms driven primarily by local heating.

Besides full simulation models—which simulate the dynamics, thermodynamics, and microphysics of storms simultaneously, beginning with initial forcing in a prestorm atmospheric environment—another type of model, called a *kinematic model,* uses observed wind fields of storms to estimate the accompanying temperature and water vapor perturbations and microphysics in a particular atmospheric environment. In a kinematic model (e.g., Ziegler 1985, 1988), the thermodynamic equation of state and continuity equations for water substance are the same as in a full simulation model, but there is no equation for dynamics to govern the development of the wind field. Instead, the model ingests an observed cloud wind field from Doppler radars and ambient environmental conditions from an atmospheric sounding. Then the equation of state and continuity equations for water substance are solved to retrieve temperature, water vapor mixing ratio, and hydrometeor mixing ratios throughout the model domain. In a time-dependent, kinematic model, the various fields are allowed to evolve from their state at an early stage of the cloud. Changes in the wind field are calculated by interpolating between observed wind fields at each time step. Changes in the retrieved fields then are calculated from changes in the wind fields.

Kinematic models have the disadvantage that, because the modeled cloud does not begin with initial cloud formation, there may be significant errors in fields that are sensitive to the cloud's history of evolution. Furthermore, the wind field of the model is not influenced by microphysics and thermodynamics, and kinematic models can be used only for periods when Doppler wind fields are available. However, a kinematic model has the advantage that it produces a model storm consistent with both the observed storm wind field and the complete set of thermodynamic and continuity equations. It can be difficult to use full simulation models to investigate some storms, especially small storms, because of model sensitivities to the form of initial forcing, to boundary conditions, or to minor changes in the storm environment.

Most cloud models that have been used in electrification studies have tracked the properties of water substance categories at grid points fixed with respect to Earth (these are sometimes called *Eulerian models*). *Particle-tracing* or *Lagrangian* models (e.g., Kuettner et al. 1981), however, compute changes to the properties of individual particles or groups of identical particles along trajectories that follow the particles through the storm, instead of at fixed grid points. Model fields related to the particles are derived after each model time step by interpolating particle parameters from particle locations to the model grid. This type of treatment typically is used with simplifications of other aspects of the model to keep computations tractable. Sometimes an Eulerian model uses statistical techniques to mimic a Lagrangian treatment for a particular type of particle.

9.3. BRIEF HISTORY OF ELECTRIFICATION MODELING

In the 1950s and 1960s, there were several calculations of electrification rates from particle interactions that considered neither cloud dynamics nor the microphysics of the particles. The first attempt to include electrification in a numerical cloud model that considered these factors was by Pringle et al. (1973), who used a crude parameterization of charge separation that did not attempt to mimic any particular microphysical charging mechanism. Takahashi (1974) allowed hydrometeors to capture space charge in his one-dimensional, time-dependent cloud model, but did not include charge exchange between particles. Ziv and Levin (1974), Scott and Levin (1975), and Levin (1976) studied inductive charging by considering hydrometeors that moved vertically and interacted between the two plates of an infinite horizontal capacitor, but they did not include any cloud dynamics. Illingworth and Latham (1977) included explicit parameterizations of several microphysical charging mechanisms in a one-dimensional model with steady-state dynamics. Kuettner et al. (1981) incorporated parameterizations of inductive and noninductive charging in a two-dimensional Lagrangian model with steady-state dynamics. Although these last two studies were able to estimate the relative contributions to cloud electrification from the various mechanisms they modeled, the use of steady-state dynamics did not allow them to consider how electrification varied with evolving cloud dynamics. Relatively simple models continue to be used for tests of specific hypotheses (e.g., Mathpal and Varshneya 1982, Singh et al. 1986, Canosa et al. 1993).

The next step in the evolution of electrification modeling was to add electrification processes to cloud simulation models that coupled electrification with both microphysics and dynamics. Early attempts at this coupling included only warm rain processes (i.e., there was no freezing or ice). Takahashi (1979) used a two-dimensional, time-dependent axisymmetric model with explicit microphysics to study electrification of shallow, warm clouds. Chiu (1978) also developed a two-dimensional, time-dependent axisymmetric model, although with bulk microphysics, and was the first to include a parameterization of small ions and their interactions with particles in a cloud simulation model. By including ion-hydrometeor interactions, simulated clouds were able to form screening layers at cloud boundaries, and mechanisms such as the Wilson selective ion-capture mechanism (see Sec. 3.5.1) could affect the charge on particles in appropriate regions. Helsdon (1980) used a similar model that was slab-symmetric instead of axisymmetric to study whether cloud electrification could be modified by injecting metal-coated chaff fibers into a warm cloud.

Since several scientists already had suggested that ice-particle interactions were important for thunderstorm electrification by the noninductive mechanism, applications of warm cloud models were considered extremely limited, and there was considerable interest in developing electrification models that included ice processes. Rawlins (1982) included electrification in a three-dimensional cloud simulation model that parameterized ice processes. The model used bulk microphysics, a simple parameterization of the noninductive mechanism, and a traditional parameterization of the inductive mechanism. Electrification occurred only when graupel or hail interacted with snow or cloud ice, so no charge was generated by interactions involving liquid hydrometeors. Takahashi (1983, 1984) included ice processes in a two-dimensional, time-dependent axisymmetric model with explicit microphysics. Parameterized charging mechanisms included inductive charging, ion-hydrometeor interactions, and noninductive charging based on the laboratory data of Takahashi (1978). Takahashi's model had much better grid resolution than Rawlins's, but had a domain height of only 8 km, and so was limited to small storms.

Helsdon and Farley (1987) added a simple noninductive charging parameterization to Chiu's (1978) ion-capture and inductive charging parameterizations in a two-dimensional, time-dependent, slab-symmetric model with bulk ice microphysics. They used the model to simulate a storm that was observed to produce a single lightning flash during the Cooperative Convective Precipitation Experiment in Montana. Modeled space charge and electric field distributions were similar to those observed by two aircraft that penetrated the storm only in simulations in which both the noninductive and inductive mechanisms operated together. In that case, the time required for model electrification to increase to the point that lightning occurred was comparable to that required by the observed storm. After publication of Helsdon and Farley, an error was found in the formulation of the noninductive parameterization for experiments in which it was used alone. Wojcik (1994) repeated the investigation with the corrected formulation of the noninductive mechanism acting alone and found that it produced an electric field consistent with

the aircraft observations. Randell et al. (1994) used a noninductive parameterization similar to Takahashi's (1984) in a configuration of the Helsdon and Farley model that omitted the inductive mechanism (and used the corrected noninductive formulation). They simulated storms in three different environments to examine conditions under which the noninductive mechanism could produce a thunderstorm.

Electrification processes have been incorporated into another cloud simulation model based on a simpler geometry. Mitzeva and Saunders (1990) developed a one-dimensional model that included no inductive mechanism or ion capture, but used a sophisticated parameterization of noninductive charging based on laboratory studies of Jayaratne et al. (1983) and Keith and Saunders (1989), instead of Takahashi (1978). Their model employed bulk microphysics and was used primarily to examine the evolution of noninductive charging rates as a function of the intensity of precipitation produced by three storms.

The cloud models summarized above were simulation models, but some models have been kinematic models. Ziegler et al. (1986) used bulk microphysics in a one-dimensional, kinematic model whose domain consisted of a cylindrical cloud with a fixed radius. The only electrification mechanism was a parameterization of the noninductive mechanism suggested by Gardiner et al. (1985). Ziegler et al. (1991) used a two-dimensional, axisymmetric version of the model and modified the kinematic retrieval process to enable the model to assimilate radar-derived wind fields that were available every 3 min, from early in the storm's lifetime throughout much of its life. They also added parameterizations of the inductive charging mechanism and of screening-layer charge on the cloud boundary. Ziegler and MacGorman (1994) expanded the kinematic model to three dimensions to enable them to study a supercell storm (features that are not slab or axisymmetric are critical to an adequate treatment of supercell storms).

Norville et al. (1991) developed a kinematic model that used explicit microphysics. The only electrification mechanism that they included was a noninductive charging mechanism based on the laboratory work of Jayaratne et al. (1983), Baker et al. (1987), and Keith and Saunders (1989). The geometry of their model cloud consisted of two concentric cylinders to permit modeling of coexisting updrafts and downdrafts. Conditions were horizontally uniform inside the inner cylinder and different, but again uniform, between the inner and outer cylinders. This configuration has been called a one-and-one-half-dimensional model. Norville et al. simulated the same storm studied by Helsdon and Farley (1987) and Wojcik (1994). Like Wojcik's model, their model was able to produce electric field magnitudes comparable to observed values by using only the noninductive mechanism.

Most modeling studies, whether they used simulation or kinematic models, have examined electrification only in the absence of lightning. Without a lightning parameterization, models could simulate only initial electrification of thunderstorms, because lightning modifies the charge distribution and limits the maximum magnitude of E. Rawlins (1982) used a threshold electric field of 500 kV m^{-1} to initiate a lightning flash and a simple charge neutralization scheme. He found that the electric field regenerated quickly after a lightning flash. Helsdon and Farley (1987) and Helsdon et al. (1992) developed a detailed lightning flash parameterization that was used successfully to simulate a storm that produced a single lightning flash, but has not yet been used to study thunderstorms having more lightning. Takahashi (1987) incorporated a simple parameterization of a lightning flash in his model to examine factors influencing the height and location of lightning. Ziegler and MacGorman (1994) developed a simple parameterization of the net effect of several flashes per time step for their three-dimensional, kinematic model to study a supercell storm having high flash rates. MacGorman et al. (1996) recently developed a more sophisticated lightning parameterization that they incorporated into a three-dimensional simulation model developed by Straka (1989). Baker et al. (1995) added a simple parameterization of lightning to their one-and-one half-dimensional, kinematic model with explicit microphysics, and Solomon and Baker (1996) developed an analytical treatment of a vertical, one-dimensional lightning channel for the same model.

As we mentioned in this chapter's introduction, some models have treated thunderstorms and their environments as electric circuits and have included no hydrometeors, no microphysics, and no cloud dynamics. One such model was developed by Nisbet (1983) and applied by Nisbet (1985a) and Hager et al. (1989) to examine the role and location of current generators in coastal thunderstorms in Florida. Such models can be used with Maxwell current measurements at the ground to infer characteristics of thunderstorm current generation. They also interface naturally with models of global and magnetospheric electric circuits so that the response of these larger circuits to thunderstorms can be studied (Nisbet 1985b, Tzur and Roble 1985, Browning et al. 1987, Driscoll et al. 1992, Stansbery et al. 1993).

The rest of this chapter will describe some of the parameterizations of electrical processes used by the above cloud models and then will review some of the research themes that have been addressed by modeling experiments.

9.4. PARAMETERIZATION OF ELECTRICAL PROCESSES

To include electrification in numerical models, the various electrical processes of clouds must be parameterized and integrated with parameterizations of dynamic, thermodynamic, and microphysical processes. In this section, we discuss how numerical models produce and

transport charge on hydrometeors, how the electric field is calculated, and how lightning is parameterized. Our treatment of these topics will be more detailed than our treatments of cloud modeling in Section 9.2, because there are no other textbooks dealing with the broad issue of electrical parameterization for numerical models. However, as stated before, constraints on the length of our text do not allow treatment of all electrical parameterizations for every model that has been developed. In most cases, we will give only one or two examples of how a particular process has been parameterized.

9.4.1. Calculating the electric field

Once a modeled storm has produced charged hydrometeors, it is necessary to calculate the resulting electric field, both for analysis of model results and for use in the next time step to calculate the results of processes dependent on the electric field. Electric field calculations in zero- and one-dimensional models are much simpler than in two- and three-dimentional models. For zero-dimensional, infinite-layer models, the only nonzero component of the electric field is the vertical component, E_z, which can be calculated from Gauss's law. If the contribution of the image charge below the conducting Earth is included, E_z at z_0 due to charge at the kth grid level at height z_k is

$$E_z(z_0) = -\frac{\rho_k \Delta z}{\varepsilon} \quad \text{for } z_0 < z_k$$

$$= -\frac{\rho_k \Delta z}{2\varepsilon} \quad z_0 = z_k \quad (9.17)$$

$$= 0 \quad z_0 > z_k,$$

where Δz is the height increment between grid levels, and ρ_k is the charge density at the kth grid level. Because the electric field from an infinite layer is constant with height above and below the layer, this geometry is likely to give electric field profiles considerably different from profiles appropriate to most physically realistic thunderstorm geometries (although it might be suitable for extensive stratiform clouds).

More realistic electric field profiles usually can be obtained in a one-dimensional model in which the size of the cloud is limited to some radius. On the axis of a cloud with cylindrical symmetry, the only nonzero component of the electric field is again the vertical component. The contribution of an infinitesimal charge (including its image charge) to the vertical electric field on the axis at height, z_0, (Fig. 9.2) is given by

$$dE_z(z_0) = \frac{\rho_k}{4\pi\varepsilon}\left[\frac{z_k - z_0}{[r^2 + (z_k - z_0)^2]^{\frac{3}{2}}} + \frac{z_k + z_0}{[r^2 + (z_k + z_0)^2]^{\frac{3}{2}}}\right]$$
$$\cdot r\, dr\, d\phi\, dz. \quad (9.18)$$

The electric field on the axis due to a thin disk of charge of radius $R(z)$ (i.e., R can vary with height) at the kth grid level then is obtained by integrating over r and ϕ.

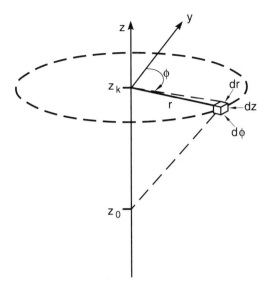

Fig. 9.2. Geometry for calculating $\vec{E}$ at height z_0 on the z axis from an infinitesimal element of a disk of charge at height z_k.

Some care is needed in determining the constants of integration to get

$$E_z(z_0) = \frac{\rho_k \Delta z}{2\varepsilon}\left[\frac{z_k - z_0}{[R^2 + (z_k - z_0)^2]^{\frac{1}{2}}} \right.$$
$$\left. + \frac{z_k + z_0}{[R^2 + (z_k + z_0)^2]^{\frac{1}{2}}} - C\right], \quad (9.19)$$

where $C = 0$ for $z_k < z_0$, $C = 1$ for $z_k = z_0$, and $C = 2$ for $z_k > z_0$.

For an arbitrary charge distribution in two or three dimensions, there are no simple expressions for the electric field. It would be possible to compute the electric potential Φ or the electric field at an arbitrary point by adding the contributions from the charge at every grid point (from the superposition principle), but computing the potential or field this way at every grid point would be too computationally intensive in many cases. Instead, what is done usually is to use standard numerical algorithms to solve for Φ at all grid points by inverting the Poisson equation,

$$\nabla^2 \Phi = -\frac{\rho}{\varepsilon}, \quad (9.20)$$

where ρ is the total space charge density at the point being evaluated. Then the electric field is computed from the potential by using the relationship $\vec{E} = -\vec{\nabla}\Phi$. Note that, in a two-dimensional model, the symmetry of the model requires the component of the electric field perpendicular to the plane of the model to be zero.

To use numerical Poisson solvers to determine Φ, it is necessary to specify boundary conditions for Φ on all

sides of the grid. Since the ground is a good conductor, hence an equipotential, it is obvious that the potential of the bottom layer of the grid at the ground should be set to a constant. The constant usually is set equal to zero, although any numerical value could be chosen without affecting the resulting electric fields. It is generally inappropriate to set the potential to a constant at the other grid boundaries, except possibly for the upper boundary if it is high enough to approximate the electrosphere. Choosing boundary conditions for the other sides is not as straightforward as for the ground, and choices can depend on the cloud size and geometry being modeled. At r = 0 in an axisymmetric, two-dimensional model, for example, E_z is the only nonzero component of $\vec{E}$, so the boundary condition there would be $\partial\Phi/\partial r = 0$. In three-dimensional or slab-symmetric, two-dimensional models, however, E_x might well be large if the storm is near the boundary at x = 0, so $\partial\Phi/\partial x = 0$ would not be an appropriate boundary condition on that side, and it might be better to use a boundary condition for the second derivative, setting $\partial^2\Phi/\partial x^2$ equal to a constant. It often is impossible to find simple boundary conditions that are exactly correct. However, if the cloud is completely contained by the grid, suitable boundary conditions for second or higher order derivatives of Φ can be used to produce reasonably accurate calculations of $\vec{E}$. Care and some ingenuity is needed if part of the storm (such as an electrified anvil) extends through a side of the grid.

9.4.2. Charge continuity

Continuity equations for charge are similar to the continuity equations for air and for water substance categories discussed earlier in this chapter. As for the hydrometeor mixing ratios, there must be a continuity equation for the charge on each water substance category. These equations govern how each water category gains and loses charge at a given grid point of a model. They express what in physics texts is called conservation of charge. The total charge at a grid point is then the sum of the charges on all the water substance categories that exist there, that is, if ρ is the total charge density at a grid point, and ρ_n is the charge density on the nth water substance category at that location, then

$$\rho = \sum_n \rho_n. \quad (9.21)$$

The charge density from small ions is included as a category if a model treats ion processes.

To derive expressions for the continuity equations, first consider particles with negligible terminal fall speed (i.e., either cloud-ice or cloud liquid-water particles), with charge density ρ_n in an updraft $w = w(z)$. The resulting electric current density at a given point is $\rho_n w$. Therefore, if we consider an infinitesimal box with dimensions δx, δy, δz, centered on an arbitrary point, as shown in Fig. 9.3, the net rate at which charge on the nth hydrometeor category flows vertically into the box is approximated by

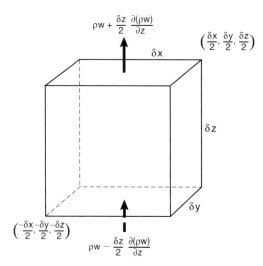

Fig. 9.3. Geometry for calculating divergence of current density carried by wind. At the center of the box, charge density, ρ, moves vertically in an updraft, w.

$$\frac{\partial Q_n}{\partial t} = \left[\rho_n w - \frac{\delta z}{2} \frac{\partial}{\partial z} (\rho_n w) \right] \delta x \, \delta y$$
$$- \left[\rho_n w + \frac{\delta z}{2} \frac{\partial}{\partial z} (\rho_n w) \right] \delta x \, \delta y, \quad (9.22)$$

where Q_n is the charge in the box. Simplifying the right-hand side and dividing both sides by the volume, $\delta x \, \delta y \, \delta z$, gives

$$\frac{\partial \rho_n}{\partial t} = - \frac{\partial}{\partial z} (\rho_n w). \quad (9.23)$$

Similar expressions can be found for wind in the x and y directions. Therefore, the net rate of change in charge per unit volume is

$$\frac{\partial \rho_n}{\partial t} = -\vec{\nabla} \cdot (\rho_n \vec{v}), \quad (9.24)$$

which is equivalent to

$$\frac{\partial \rho_n}{\partial t} = -\vec{v} \cdot \vec{\nabla} \rho_n - \rho_n \vec{\nabla} \cdot \vec{v}. \quad (9.25)$$

The first term on the right is due to advection of existing inhomogeneities in the charge distribution, while the second term is due to divergence of the wind. From the anelastic approximation of the continuity equation for air (Eq. 9.10) we get

$$\vec{\nabla} \cdot \vec{v} = - \frac{w}{\rho_D} \frac{\partial \rho_D}{\partial z}, \quad (9.26)$$

where ρ_D is the density of dry air. Substituting this expresstion in Eq. 9.25 gives

$$\frac{\partial \rho_n}{\partial t} = -\vec{v} \cdot \vec{\nabla} \rho_n + \frac{\rho_n w}{\rho_D} \frac{\partial \rho_D}{\partial z}. \quad (9.27)$$

Therefore, the second term on the right in Eq. 9.25 can be attributed to the compressibility of air.

Besides these terms, it is necessary to include terms for diffusion and for sources and sinks. The vertical component of the electric current density due to diffusion is $K_d(\partial \rho_n/\partial z)$, where K_d is the diffusion coefficient (Pruppacher and Klett 1978, p. 360); the expression for each of the horizontal components is analogous. From arguments similar to those used in deriving Eq. 9.24, it can be shown that the net change in charge density due to the flux of charge by diffusion is $\vec{\nabla} \cdot K_d \vec{\nabla} \rho_n$. If we denote the net contribution of sources and sinks as $S_{\rho n}$, the complete continuity equation for the charge density on cloud ice or cloud liquid water becomes

$$\frac{\partial \rho_n}{\partial t} = -\vec{v} \cdot \vec{\nabla} \rho_n + \frac{\rho_n w}{\rho_D} \frac{\partial \rho_D}{\partial z}$$
$$+ \vec{\nabla} \cdot (K_d \vec{\nabla} \rho_n) + S_{\rho n}. \quad (9.28)$$

Examples of sources and sinks for the nth category include ion capture, charge exchange during collisions with particles in another category, and mass loss or gain as particles are transferred from one category to another. When mass is lost from one category to another, the charge carried by the mass also must be transferred to the new category, thereby decreasing the magnitude of charge in the category losing mass. However, the magnitude of charge in the category gaining mass can either decrease or increase, because the polarity of the charge gained with the new mass can be either the same as the polarity of charge already on particles in the category or opposite to it. Because most processes can be either a source or a sink of charge for a given category of particle, it is not worthwhile to try to distinguish sources from sinks. Thus, we group all such processes together.

Note that the storm as a whole does not lose or gain charge unless charge is transferred to or from a region outside the storm. When the charge on particles in one category is altered by interactions with particles in other categories, a compensating change (equal in magnitude, opposite in sign) must occur in the charge of the other categories, so that the net charge summed over all categories remains the same. There are no compensating changes in the charge within the storm when the charge on particles is altered by a process involving charge from outside regions, such as ground flashes, cloud-to-air flashes, capture of ions emitted by point discharge from the ground, and screening-layer formation at cloud boundaries.

To derive the continuity equation for precipitation, the primary change that we must make to the derivation for cloud particles is to include the fall speed of precipitation. We can write the velocity of precipitation as $\vec{v} + \vec{v}_{nT}$, where $\vec{v}$ is the velocity of air and $\vec{v}_{nT}$ is the terminal fall speed of the nth type of precipitation, which has only a z-component. The air velocity contributes to the continuity equation in the same way as for cloud particles. The derivation of an expression for the effect of terminal velocity is analogous to the derivation of Eq. 9.23, if v_{nT} is substituted for w, and the expression appears similar to the right-hand side of that equation. The continuity equation for precipitation particles, therefore, is

$$\frac{\partial \rho_n}{\partial t} = -\vec{v} \cdot \vec{\nabla} \rho_n + \frac{\rho_n w}{\rho_D} \frac{\partial \rho_D}{\partial z} - \frac{\partial (\rho_n v_{nT})}{\partial z}$$
$$+ \vec{\nabla} \cdot (\mathrm{K_d} \vec{\nabla} \rho_n) + \mathrm{S}_{\rho n}. \qquad (9.29)$$

As discussed by Pruppacher and Klett (1978), v_{nT} can be generalized to include the effect of electrostatic force in addition to gravity. If this is done, the third term on the right of Eq. 9.29 should be replaced by $\vec{\nabla} \cdot (\rho_n \vec{v}_{nT})$, where $\vec{v}_{nT}$ is now a vector, not a scalar.

Many microphysical processes are capable of placing charge on particles, as discussed in Chapter 3. The main processes that have been included in cloud models are the noninductive mechanism, the inductive mechanism, and ion capture. The rest of Section 9.4 will consider treatments of each of these. Electrification mechanisms that have not yet been used in models will not be considered here, although other mechanisms may be significant in some situations (e.g., the possible role of melting processes in electrification of the stratiform precipitation region of storm systems, pointed out in Sec. 8.2.4.).

9.4.3. The noninductive charging mechanism

Parameterizing the noninductive mechanism requires that the charge per hydrometeor collision be determined and then included in an explicit or a bulk parameterization of all collisions involving a particular water substance category at a grid point during a time step of the model. Recall from Chapter 3 that (1) The noninductive mechanism appears to be most effective when rimed graupel collides with cloud-ice particles or snow in a region that also has liquid water (i.e., the mixed-phase region); and (2) The sign and magnitude of the charge that is transferred depends on ambient temperature, liquid water content, and impact speed. The simplest parameterization of the noninductive mechanism assumes that a constant value of charge is transferred per collision between particular hydrometeor types, as was done by Helsdon and Farley (1987), who also assumed that the sign of charge transfer reversed polarity at $-10°C$ (see Table 9.1).

Based on the work of Jayaratne et al. (1983), Gardiner et al. (1985) suggested that the charge transferred to a rimed graupel/hail particle when it collides with a cloud-ice/snow particle could be parameterized by

$$\delta q = \mathrm{k_q} \mathrm{D_i}^m (\Delta v_{gi})^n (LWC - LWC_{crit})$$
$$\cdot \mathrm{f}(\Delta T), \qquad (9.30)$$

where $\mathrm{k_q}$ is a constant of proportionality approximately equal to 73, $\mathrm{D_i}$ is the diameter of the cloud-ice particle

Table 9.1. Noninductive parameterization of Helsdon and Farley (1987)

Growth	Interacting Types	δq^a $(10^{-15}$ C)	Temperature
Dry	Graupel/cloud ice	2	$T > -10°C$
		-2	$T < -10°C$
	Graupel/snow	200	$T > -10°C$
		-200	$T < -10°C$
	Graupel/rain[b]	-100	All T
Wet	Graupel/cloud water[c]	0	All T
	Graupel/rain[c]	0	All T
Other	Rain/cloud water	0	All T

[a]Polarity is for charge transferred to graupel (to rain in growth mode Other); charge transferred to the particle colliding with graupel is of equal magnitude, but opposite polarity; [b]Splashing interactions; [c]Shedding or limited accretion

in meters, Δv_{gi} is the relative impact speed (in m s^{-1}) between the graupel particle and ice crystal given by the difference in their terminal velocities ($\Delta v_{gi} = |v_{gT} - v_{iT}|$), $m \approx 4$, $n \approx 3$, LWC is the liquid water content (in g m^{-3}), LWC_{crit} is the value of the liquid water content below which the sign of δq reverses (see plot of LWC_{crit} as a function of T by Jayaratne et al.), and ΔT is the number of degrees of supercooling ($\Delta T = T - 0°C$ for $T < 0°C$ and is 0 otherwise). The function f(ΔT) was a polynomial fit to the laboratory data of Jayaratne et al.:

$$\mathrm{f}(\Delta T) = \mathrm{a}\Delta T^3 + \mathrm{b}\Delta T^2 + \mathrm{c}\Delta T + \mathrm{d}. \qquad (9.31)$$

$\mathrm{a} = -1.7 \times 10^{-5}$, $\mathrm{b} = -0.003$, $\mathrm{c} = -0.05$, and $\mathrm{d} = 0.13$, when δq is in fC (10^{-15} C). Subsequent laboratory experiments showed that the increase in charge with ice-crystal diameter leveled off at large values of diameter, so the $\mathrm{D_i}^4$ dependence in Eq. 9.30 overestimated the charge transferred for large $\mathrm{D_i}$.

Saunders et al. (1991) suggested a new, more complicated parameterization for δq that was based on laboratory experiments over a broader range of cloud-ice size, liquid water content, and temperature. Their expression for charge (in fC) had a functional form similar to that used by Gardiner et al. (1985):

$$\delta q = \mathrm{k_q} \mathrm{D_i}^m (\Delta v_{gi})^n \mathrm{f}(T, EW), \qquad (9.32)$$

where EW is effective liquid water content, a parameter defined by Saunders et al. Unlike f(ΔT) in Eqs. 9.30 and 9.31, however, f(T, EW) had different functional forms in different regimes of temperature and effective liquid water content. Furthermore, k_q, m, and n depended on the size of the ice crystal and the polarity of charge transferred. These dependencies are shown in Table 9.2. Because data did not extend to temperatures warmer than $-7.4°C$, Wojcik (1994) linearly extrapolated f(EW, T) at a particular EW from the value given by the expression in Table 9.2 for $T = -7.4°C$ to zero at $T = 0°C$.

Table 9.2. Parameters for the noninductive charging parameterization of Saunders et al. (1991).

Valid For				& For			
T (°C)	EW (g m^{-3})	δq Polarity[a]	f(T, EW)[a] (fC)	D_i (μm)	k_q	m	n
<-20	<0.16	+	2042 EW − 129 (for $0.06 < EW < 0.12$)	<155	4.92×10^{13}	3.76	2.5
				155–452	4.04×10^{6}	1.9	
			−2900 EW + 463 (for $0.12 < EW < 0.16$)	>452	52.8	0.44	
-7.4 to -16	<0.22	−	−314EW + 7.9 (for $0.026 < EW < 0.14$)	<253	5.24×10^{8}	2.54	2.8
			419 EW − 92.6 (for $0.14 < EW < 0.22$)	>253	24	0.5	
-7.4 to T_r^{b}	>0.22	+	20.2 EW + 1.36 T + 10.1	<155	4.9×10^{13}	3.76	2.5
$\le -7.4^{c}$	$>1.1^{c}$			155–452	4×10^{6}	1.9	
				>452	52.8	0.44	
$<T_r^{b,d}$	$<1.1^{d}$	−	3.02 − 31.8 EW + 26.5 (EW)^2	<253	5.24×10^{8}	2.54	2.8
				>253	24	0.5	

[a]Polarity is for charge on graupel. Charge on cloud-ice has opposite polarity; [b]T_r is the temperature at which the polarity of charge gained by graupel reverses for a given value of liquid water content: $T_r = -15.06 \, EW - 7.38$, for EW in g m^{-3} and T_r in °C; [c]For regions where f(EW, T) is negative (at EW near 1.1 for $T \le -24$°C), use f(EW, T) = 0; [d]Must also be outside previous ranges in T-EW space. For regions where f(EW, T) is positive (at -16°C $< T < -20$°C and $EW \le 0.1$ g m^{-3}), use f(EW, T) = 0.

Saunders et al. (1991) used effective liquid water content instead of *LWC*, the liquid water content measured by in situ instruments and determined by most numerical cloud models, because their observations suggested that *EW* was more relevant to noninductive charging. *EW* is a modification of *LWC* that includes only the accreted fraction of liquid water content in the path of graupel. Therefore, it is given by the ambient liquid water content multiplied by the collection efficiency ($EW = \text{LWC} \, \mathscr{E}_{collect}$). The collection efficiency $\mathscr{E}_{collect}$ for graupel and water particles is equal to the product $\mathscr{E}_{colli} \, \mathscr{E}_{accrete}$, where $\mathscr{E}_{colli}$ is the collision efficiency, a factor ≤ 1 that reduces the geometric cross section of graupel to account for aerodynamic effects that sweep small particles around graupel (see Fig. 9.4), and $\mathscr{E}_{accrete}$ is the fraction of colliding particles accreted by graupel. Saunders et al. noted that in their experiments, *EW* typically was roughly equal to *LWC*/2.

The polarity of charge that graupel gains for each combination of temperature and effective liquid water content under the parameterization described in Table 9.2 is shown graphically in Fig. 9.5. The line dividing positive and negative charging for 0.22 g m^{-3} < *EW* < 1.1 g m^{-3} can be given in terms of critical values of either *EW* or *T*. Saunders et al. gave the following ex-

pression for EW_{crit} (in g m^{-3}), valid for -10.7°C $> T > -23.9$°C:

$$EW_{crit} = -0.49 - 6.64 \times 10^{-2}T. \quad (9.33)$$

The equation for T_r, given as a footnote in Table 9.2, was obtained by inverting this equation. Continued laboratory experiments (Brooks et al. 1997) have suggested that the riming rate also may affect the sign and magnitude of charge per collision and that this can be parameterized by substituting $EW \, \Delta v_{gi}/3$ in place of *EW* in Eq. 9.32 and Table 9.2.

Wojcik (1994) noted that the expressions given in Table 9.2 resulted in much more charge per collision than observed for some particle sizes and collision speeds at small values of *EW* (<0.22 g m^{-3} for positive charge and <0.16 g m^{-3} for negative charge). Furthermore, model simulations using these expressions tended to produce thunderstorm charge distributions in which the majority of negative charge was located above positive charge during most of the period being simulated. When Wojcik drastically reduced the amount of charge per collision to 10–20% of the value given by the expressions in Table 9.2 in regions of low *EW*, model simulations produced a charge distribution similar to the charge distribution inferred from in situ

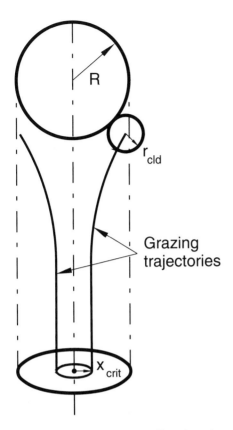

Fig. 9.4. Schematic diagram of the effect of aerodynamic forces on the interaction of a pair of spheres. In the volume swept out by the larger particle, small particles of radius r_{cld} in the outermost cylindrical shell are forced around the larger particle by aerodynamic forces, thereby reducing the cross section for collisions. The critical radius for a grazing trajectory of the small sphere is x_{crit}. (After Pruppacher and Klett 1978, with permission.)

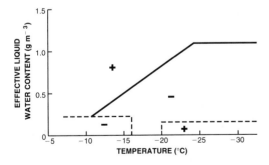

Fig. 9.5. Polarity of graupel charging as a function of temperature and effective liquid water content (From Saunders et al. 1991, with permission.)

measurements of the storm being analyzed. The resulting charge per collision is shown as a function of EW and T in Fig. 9.6.

In personal communications, C. Saunders noted that laboratory experiments at very low liquid water content were more difficult to perform. He suggested that the published magnitude of charging was consequently less reliable at very low liquid water contents, although he believed the polarity to be correct. A combination of more laboratory and modeling work is needed to address the issue of noninductive charging in regions of low liquid water content.

Instead of developing equations that fit laboratory data, Takahashi (1984) and Randell et al. (1994) used the laboratory results of Takahashi (1978) directly. Randell et al., for example, assigned δq from a table whose values were interpolated from Takahashi's (1978) laboratory measurements of δq versus T and LWC (see Fig. 3.12). Given the temperature and liquid water content at a grid point, δq was obtained from a two-dimensional array in which the table had been stored. No dependence on ice-particle diameter (D_i) and impact speed (Δv) was included by Randell et al. However, Takahashi (1984) parameterized the effect of particle size and impact speed described by Marshall et al. (1978) by multiplying laboratory values of δq by a factor of $5(D_i/D_0)^2|v_{gT} - v_{iT}|/v_0$, where $D_0 = 100\ \mu m$ and $v_0 = 8$ m s^{-1}. Takahashi restricted this factor to be no more than 10.

In models with explicit microphysics for graupel and cloud ice, the derivation of an expression for the rate at which graupel and cloud-ice charge density (ρ_g and ρ_i, respectively) build due to the noninductive mechanism is fairly simple. Per unit time, the volume in which a graupel particle of diameter D_g collides with cloud-ice particles of diameter D_i is just the product of the cross-sectional area $\pi(D_g + D_i)^2/4$ (the area of a circle in which graupel and ice particles just touch) times the vertical fall speed of graupel relative to cloud-ice particles, $\Delta v_{gi} = |v_{gT} - v_{iT}|$ (see Fig. 9.7). To compute a charging rate, the volume swept out by a graupel particle per unit time must be multiplied by the collision-separation efficiency for graupel and ice ($\mathcal{E}_{gi}$), which is the fraction of ice particles in this volume that collide with the graupel and separate from it. $\mathcal{E}_{gi}$ is equal to the product $\mathcal{E}_{colli}\mathcal{E}_{sep}$, where $\mathcal{E}_{colli}$ is a factor that accounts for aerodynamic effects, as discussed previously, and $\mathcal{E}_{sep}$ is the fraction of colliding particles that separate ($\mathcal{E}_{sep} = 1 - \mathcal{E}_{agg}$, where $\mathcal{E}_{agg}$ is the fraction aggregated by graupel). The modified volume per unit time is called the *collision kernel* K_{gi} and can be expressed as

$$K_{gi} = \frac{\pi}{4}(D_g + D_i)^2\ \Delta v_{gi}\ \mathcal{E}_{gi}. \qquad (9.34)$$

If n_N is the number density of the Nth particle type of diameter D_N, then the rate at which n_g graupel particles collide and separate from n_i cloud-ice particles is given by $K_{gi}n_g n_i$, and the rate at which this process charges graupel of diameter D_g and cloud ice of diameter D_i is

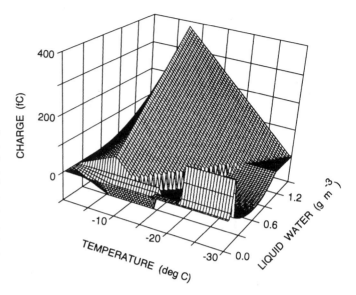

Fig. 9.6. Charge per collision from the noninductive, graupel-ice parameterization of Saunders et al. (1991), with the modifications of Wojcik (1994) at low liquid water content and at temperatures of 0°C to −7.4°C. Liquid water is the effective liquid water content (EW) defined by Saunders et al. The charge is for graupel colliding with ice crystals 0.3 mm in diameter at a relative vertical velocity of 3 m s⁻¹.

$$\frac{\partial \rho_g}{\partial t} = K_{gi} n_g n_i \, \delta q$$

$$= -\frac{\partial \rho_i}{\partial t}. \qquad (9.35)$$

The expression for noninductive charging between graupel and snow is the same as Eqs. 9.34 and 9.35, except that the parameters for cloud ice are replaced by those for snow (N_s, D_s, $\mathcal{E}_{gs}$, Δv_{gs}). Similarly for hail and snow, graupel parameters are changed to the val-

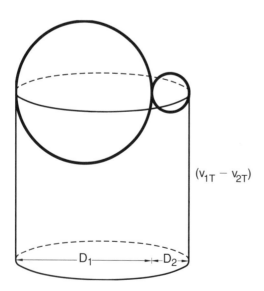

Fig. 9.7. Volume swept out in time, t, by graupel of diameter D_1 colliding with small ice particles of diameter D_2. $v_{1t} - v_{2t}$ is the relative vertical velocity between them.

ues for hail (N_h, D_h, $\mathcal{E}_{hs}$, Δv_{hs}), although δq has not yet been determined by laboratory experiments specifically for hail. To determine the rate of change in charge density for graupel in a particular size category of a model with explicit microphysics, it is necessary to add together the contributions given by Eq. 9.35 for every size category of snow and cloud ice that interacts with the graupel. Likewise, the rate of change in charge density on a particular size category of cloud ice or snow must include the contribution from all size categories of graupel and hail.

To determine the rate of change of charge density on a particular type of hydrometeor in a model with bulk microphysics, two approaches have been used. The first is to evaluate Eq. 9.35 at a grid point by determining mean values of δq and K_{gi} and the total concentrations across all sizes for N_g and N_i at that grid point. The second approach is to integrate the right-hand side of Eq. 9.35 across all sizes both of that hydrometeor category and of every other category with which that category interacts. For example, the change in charge density of graupel due to collisions with snow or cloud ice is

$$\frac{\partial \rho_g}{\partial t} = \frac{\pi}{4} \int \int (D_g + D_n)^2 |v_{gT} - v_{nT}|$$
$$\cdot \mathcal{E}_{gn} n_n(D_n) \, n_g(D_g) \delta q \, dD_n dD_g, \qquad (9.36)$$

where the subscript n refers to either snow or cloud ice. The size distributions $n_n(D_n)$ often are assumed to have the form of the Marshall-Palmer distribution (Eq. 9.13) with parameters n_{n0} and Λ_n. For cloud ice, because $D_i \ll D_g$ and $v_{iT} \ll v_{gT}$, the magnitude of sums and differences of these quantities can be approximated as D_g and v_{gT} in Eq. 9.36. To evaluate this integral for cloud ice, Ziegler et al. (1986) assumed further that, since cloud ice typically has a narrow size distribution, it can be approximated as a population with a single diameter D_i. If N_i is the total concentration of cloud-ice particles, this implies that

$$N_i = \frac{6\rho_A q_I}{\pi \rho_I D_i^3} = \frac{\rho_A q_I}{\rho_I V_i}, \qquad (9.37)$$

where ρ_A is the mass density of air, q_I is the cloud-ice mixing ratio, ρ_I is the mass density of cloud ice, and V_i is the volume of a cloud-ice particle. Both N_i and V_i can be allowed to vary as dictated by microphysical processes that affect the mass of cloud ice. The terminal fall speed of graupel, v_{gT}, can be approximated by

$$v_{gT} = \left(\frac{4g\rho_G}{3C_D\rho_A} \right)^{\frac{1}{2}} D_g^{\frac{1}{2}} \qquad (9.38)$$

(Wisner et al. 1972), where g is the acceleration of gravity, ρ_G is the mass density of graupel, and C_D is the drag coefficient of graupel (Ziegler et al. assumed $C_D \approx 1$). Because Ziegler et al. (1986) had already shown that the contribution of cloud ice to electrification was much less than the contribution of snow, Ziegler et al. (1991) assumed a constant charge per collision for cloud-ice particles and then integrated Eq. 9.36 with these simplifying assumptions to get

$$\frac{\partial \rho_g}{\partial t} = \frac{\pi}{4} \left(\frac{4g\rho_G}{3C_D\rho_A} \right)^{\frac{1}{2}} \mathcal{E}_{gi} N_i n_{g0} \Lambda_g^{-3.5} \Gamma(3.5) \delta q \qquad (9.39)$$

$$= 2.61 \ \left(\frac{4g\rho_G}{3C_D\rho_A} \right)^{\frac{1}{2}} \mathcal{E}_{gi} N_i n_{g0} \Lambda_g^{-3.5} \delta q,$$

where n_{g0} and Λ_g are the parameters of the Marshall-Palmer size distribution for graupel and $\Gamma(a)$ is the complete gamma function defined by

$$\Gamma(a) \equiv \int_0^\infty s^{a-1} e^{-s} ds, \qquad for \ a > 0. \quad (9.40)$$

The expression for noninductive interactions of graupel and snow is more complicated, because v_{sT} and D_s are not necessarily negligible compared with v_{gT} and D_g, the size distribution is broader, and δq varies too much to be approximated as a constant. Ziegler et al. (1991) used the parameterization for δq given by Gardiner et al. (1985), but assumed that LWC_{crit} is constant and modified the function f(ΔT) to be able to vary the temperature at which the charge gained by graupel reversed polarity. They also made the simplifying assumption that $|v_{gT} - v_{sT}|$ could be approximated by the following expression to facilitate the integration of Eq. 9.36:

$$|v_{gT} - v_{sT}| = \left[\left(\frac{4g\rho_G}{3C_D\rho_A} \right)^{\frac{1}{2}} - 4.83 \ \left(\frac{\rho_{Asl}}{\rho_A} \right)^{\frac{1}{2}} \right] \cdot \left(\frac{3.67}{\Lambda_s} \right)^{\frac{1}{4}} \left(\frac{3.67}{\Lambda_g} \right)^{-\frac{1}{2}} D_g^{\frac{1}{2}}, \qquad (9.41)$$

where ρ_{Asl} is the density of air at sea level. They then integrated Eq. 9.36 for snow to get

$$\frac{\partial \rho_g}{\partial t} = 5.73 \mathcal{E}_{gs}(LWC - LWC_{crit}) f(\Delta T) n_{s0} n_{g0}$$

$$\cdot \left[\left(\frac{4g\rho_G}{3C_D\rho_A} \right)^{\frac{1}{2}} - 4.83 \ \left(\frac{\rho_{Asl}}{\rho_A} \right)^{\frac{1}{2}} \right]$$

$$\cdot \left(\frac{3.67}{\Lambda_s} \right)^{\frac{1}{4}} \left(\frac{3.67}{\Lambda_g} \right)^{-\frac{1}{2}} \right]^4 \qquad (9.42)$$

$$\cdot \{ [\Gamma(5)]^2 \Lambda_s^{-5} \Lambda_g^{-5} + 2\Gamma(4)\Gamma(6)\Lambda_s^{-6}\Lambda_g^{-4} + \Gamma(3)\Gamma(7)\Lambda_s^{-7}\Lambda_g^{-3} \},$$

where $[\Gamma(5)]^2 = 576$, $2\Gamma(4) \Gamma(6) = 1440$, and $\Gamma(3) \Gamma(7) = 1440$.

9.4.4. The inductive charging mechanism

Recall from Chapter 3 that (1) The inductive mechanism occurs in the presence of an electric field; (2) The magnitude of charge transferred is a function of both the magnitude of the electric field and the angle between the electric field vector and the radial to the impact point on the particle; (3) Inductive charging during collisions of ice particles appears to be negligible because charge transfer is too slow in ice (Latham and Mason 1962); (4) Collisions of raindrops and cloud droplets do not contribute significant charge because, essentially, no cloud droplets separate after they collide with raindrops in an electric field; and (5) Only a small fraction of colliding raindrops or colliding graupel and cloud droplets subsequently separate, but enough charge is separated to be significant (Aufdermauer and Johnson 1972). As discussed in Section 3.5.2, the expression for the induced charge exchanged by a spherical graupel or hail particle colliding with a cloud droplet in an electric field is

$$\delta q = 4\pi\epsilon\gamma_1 |\vec{E}| r_{cld}^2 \cos\theta_{E,r} + AQ_g - BQ_{cld}, \qquad (9.43)$$

where r_{cld} is the radius of the cloud droplet; $\theta_{E,r}$ is the angle between the impact point and the electric field vector through the center of the graupel/hail particle (shown in Fig. 9.8); Q_g and Q_{cld} are the charge already on the graupel/hail particle and cloud droplet, respectively; γ_1, A, and B are dimensionless functions of r_{cld}/r_g, the ratio of the radii of the two particles. This is the same as Eq. 3.11, and a description of the dimensionless constants is given with that equation.

Parameterizing the inductive mechanism requires procedures similar to those used in parameterizing the noninductive mechanism, except it is necessary to take into account that the mechanism's effectiveness is a function of where on their surfaces two particles collide. The charge produced by the inductive mechanism is strongly dependent on the angle between the impact point and the electric field vector. As shown in Fig. 9.8 for a spherical graupel particle, the magnitude of the induced surface charge density is largest at the two ends of the diameter that parallels the electric field vector:

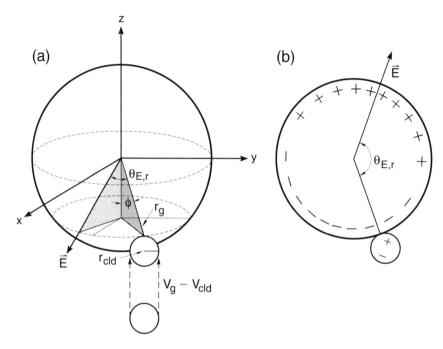

Fig. 9.8. (a) Geometry for the exchange of charge between large and small hydrometeors by the inductive mechanism. (Adapted from Chiu 1978, with permission.) (b) Schematic diagram of the variation of exchanged charge as a function of impact angle.

As the angle from the electric field increases to 90°, the surface charge density decreases to zero.

Besides this effect, the probability that colliding graupel and cloud droplets will separate can vary with the angle of the impact point from the vertical axis. Since colliding particles must separate for the inductive mechanism to work, the angular dependence of charge transfer and separation probability can interact in complicated ways. The mechanism will be most effective at the location on the graupel surface where the product of the induced surface charge density and the collision-separation probability is largest. Aufdermauer and Johnson (1972) and Moore (1975) suggested that the separation probability is much larger for glancing collisions than for head-on collisions. However, experiments by Sartor (1981), Mason (1988), and Brooks and Saunders (1994) demonstrated that cloud droplets can rebound from head-on, as well as glancing, collisions with rimed graupel.

Regardless of whether a model uses bulk or explicit microphysics, its inductive parameterization must handle the complication of these interacting factors. Chiu (1978) treated them by defining a mean separation probability $\langle S(\phi)\rangle$ and a mean impact cosine $\langle\cos\phi\rangle$

$$\langle S(\phi)\rangle = \frac{1}{\pi r_g^2}\int_0^{\frac{\pi}{2}} S(\phi)2\pi r_g^2\, 2\pi r_g^2 \sin\phi\,\cos\phi\, d\phi$$

$$\langle\cos\phi\rangle = \frac{1}{\langle S(\phi)\rangle}\int_0^{\frac{\pi}{2}} 2S(\phi)\sin\phi\,\cos^2\phi\, d\phi,$$

(9.44)

where the weighting factor under the first integral is an infinitesimal area in the horizontal cross section of the graupel particle. Then the mean charge exchanged by colliding graupel particles and cloud droplets is

$$\langle\delta q\rangle = \langle S(\phi)\rangle(4\pi\varepsilon\gamma_1|\vec{E}|r_{cld}^2\cos\theta_{E.z}\langle\cos\phi\rangle$$
$$+ A\langle Q_g\rangle - B\langle Q_{cld}\rangle),$$

(9.45)

where $\theta_{E.z}$ is the angle between the electric field vector and the lower vertical axis, and $\langle Q_g\rangle$ and $\langle Q_{cld}\rangle$ are the mean charge per particle on graupel and cloud droplets, respectively. Analogous to Eqs. 9.34 and 9.35, the rate at which this process charges graupel of radius r_g due to collisions with cloud droplets of radius r_{cld} is

$$\frac{\partial\rho_g(r_g, r_{cld})}{\partial t} = \pi r_g^2\Delta v_{g\,cld}\mathscr{E}_{colli}\, n_g(r_g)n_{cld}(r_{cld})\langle\delta q\rangle,$$

(9.46)

where $n_g(r_g)$ and $n_{cld}(r_{cld})$ are the number density of graupel and cloud liquid water particles of radius r_g and r_{cld}, respectively, and $\mathscr{E}_{colli}$ in this case is the collision efficiency for graupel and cloud particles (i.e., the fraction of cloud particles in the volume swept out by the graupel particle that actually collide with it). This expression neglects r_{cld} in the sum $r_{cld} + r_g$ for the cross sectional area.

The rate at which charge is generated by a parameterization of the inductive mechanism depends strongly on the values selected for $\langle S(\phi)\rangle$ and $\langle\cos\phi\rangle$. If the mean separation probability is 1 (i.e., all colliding particles separate), then $\langle cos\phi \rangle = 0.67$, and the mechanism's effectiveness is maximized. This is almost certainly an overestimate. Some studies, such as Aufdermauer and Johnson (1972), suggested that it is orders of magnitude too large. Helsdon and Farley (1987) used a mean separation probability of 0.015 and a mean cosine of 0.5 in their model experiment.

As for the noninductive mechanism, models with bulk microphysics must integrate the charge produced by the inductive mechanism across all sizes of interacting hydrometeors, so the total rate of change in the charge density of graupel would be given by integrating Eq. 9.46 over all values of cloud radius and graupel radius. Helsdon and Farley (1987) approximated the integration by replacing r_{cld}, r_g, $\Delta v_{g\,cld}$, and $\mathscr{E}_{colli}$ with their mean values at a grid point, thereby treating them as constants in the integral. The remaining integration of $n_g(r_g)n_{cld}(r_{cld})$ over all radii simply gave the total number density of graupel times the total number density of cloud droplets at the grid point being considered.

Ziegler et al. (1991) developed a bulk parameterization based on several simplifying assumptions. Consistent with their parameterization of the noninductive mechanism discussed above, they considered only $D_{cld} \ll D_g$ and $v_{cldT} \ll v_{gT}$, and so ignored the cloud droplet term in sums and differences of these quantities, and used Eq. 9.38 for v_{gT}. Furthermore, they assumed that the narrow droplet size spectrum could be approximated as a population having a single diameter D_{cld}, and they used values of γ_j, A, and B that were appropriate for $r_{cld}/r_g \approx 0$. In considering the collision process for inductive charging, they assumed that only a small fraction of graupel and cloud particles separate after they collide, as found by Aufdermauer and Johnson (1972), and assumed that rebounding occurs only during glancing collisions, as suggested by Moore (1975). Since the probability that any one cloud droplet will experience two rebounding collisions with graupel is much lower than the probability that it will experience one, the term involving preexisting droplet charge in Eq. 9.43 was omitted. Furthermore, they considered only a vertical electric field, because it was the vertical electric field that had been hypothesized to contribute to thunderstorm electrification by the inductive mechanism. The equation for the rate at which charge was gained by graupel from the inductive mechanism then became

$$\frac{\partial \rho_g}{\partial t} = \int \frac{\pi}{4} D_g^2 v_{gT} \mathscr{E}_{colli} \mathscr{E}_{sep} \, \alpha N_{cld} n_g (D_g)$$
$$\cdot \left[\frac{\pi^3}{2} \varepsilon E_z \cos \theta_{z,r} - \frac{\pi^2}{6} \frac{Q_g}{D_g^2} \right] D_{cld}^2 dD_g, \quad (9.47)$$

where α is the fraction of collisions that have glancing trajectories, N_{cld} is the total number concentration of cloud particles, $\mathscr{E}_{sep}$ is the fraction of glancing particles that separate, and E_z is the vertical electric field component. Ziegler et al. (1991) integrated this equation to get

$$\frac{\partial \rho_g}{\partial t} = \frac{\pi^3}{8} \mathscr{E}_{colli} \mathscr{E}_{sep} \, \alpha N_{cld} n_{g0} D_{cld}^2 \left(\frac{4g\rho_G}{3C_D \rho_A} \right)^{\frac{1}{2}}$$
$$\cdot \left[\pi \Gamma(3.5) \varepsilon E_z \Lambda_{gN}^{-3.5} \cos \theta_{z,r} - \frac{\rho_g \Lambda_{gN}^{-1.5}}{3N_g} \Gamma(1.5) \right], \quad (9.48)$$

where $\Gamma(1.5) = 0.886$ and $\Gamma(3.5) = 3.323$. They chose $\mathscr{E}_{colli} = 0.84$, $\mathscr{E}_{sep} = 0.1$, $\alpha = 0.022$, and $cos\theta = 0.1$, which gave a probability of rebounding collisions $\mathscr{E}_{colli} \mathscr{E}_{sep} \alpha$ near the lower end of the range found by Aufdermauer and Johnson (1972).

9.4.5. Ion attachment

Some models (e.g., Chiu 1978, Takahashi 1978, Helsdon and Farley 1987) explicitly treat space charge on free ions (as opposed to charge carried by hydrometeors) and incorporate ion capture by hydrometeors to examine how ion capture affects cloud electrification. Recall from Chapter 1: (1) Cosmic rays are the source of most ions in the fair weather troposphere, except near the ground; radioactive decay from the surface contributes up to half of the ions found near ground; (2) In fair weather, the number of positive ions is roughly equal to the number of negative ions, so that the net charge density results from small differences of large numbers of ions; (3) Ions move under the influence of E, and their average velocity in a given E is the mobility times the electric field $k_{\pm}E$, where the subscript gives the polarity of the ions; and (4) The $k_{\pm}$ increases with decreasing pressure and so increases with height.

Each polarity of ion must obey its own continuity equation, which is similar to Eq. 9.28 for cloud hydrometeors. For ions, however, a term must be added to account for charge motion under the influence of the electric field, since the resulting ion velocity can easily be much different than wind velocities. Also, the source/sink term often is split into sources and sinks occurring in fair weather and those requiring clouds or thunderstorms (e.g., Chiu 1978, Helsdon and Farley 1987). The continuity equation for free ions can be written

$$\frac{\partial n_{\pm}}{\partial t} = -\vec{v} \cdot \vec{\nabla} n_{\pm} + \frac{n_{\pm}w}{\rho_D} \frac{\partial \rho_D}{\partial z} + \vec{\nabla} \cdot (K_d \vec{\nabla} n_{\pm})$$
$$\mp \vec{\nabla} \cdot (n_{\pm} k_{\pm} \vec{E}) + G - \beta n_{+} n_{-} + S_{T\pm}, \quad (9.49)$$

where n_{+} is the number density of positive ions; n_{-}, the number density of negative ions; $G = G(z)$ is the nonthunderstorm ion generation rate (in $m^{-3}s^{-1}$) due primarily to cosmic rays; β is the ion recombination coefficient; and $S_{T\pm}$ is the net contribution of thunderstorm sources and sinks (e.g., lightning, evaporation of charged hydrometeors, ion capture by hydrometeors).

In the model developed by Chiu (1978), Helsdon (1980), and Helsdon and Farley (1987), the number densities of free ions initially are set to their fair weather values, which are assumed to be a function only of height. The model then allows the number densities to evolve in both the vertical and horizontal dimensions as a cloud forms and becomes electrified. Helsdon and Farley estimated the initial values of n_+ and n_- by using Gauss's law in the form

$$\frac{dE_z}{dz} = \frac{q_e}{\varepsilon}(n_+ - n_-), \qquad (9.50)$$

where q_e is the magnitude of charge on an electron, and by using the approximate relationship that the fair weather current density J_{FW} is constant with height,

$$J_{FW} = q_e(k_+ n_+ + k_- n_-)E_z. \qquad (9.51)$$

Solving Eqs. 9.50 and 9.51 for n_+ and n_-, they got

$$n_+ = \left(\frac{J_{FW}}{E_z} + k_- \varepsilon \frac{dE_z}{dz}\right)\frac{1}{q_e(k_+ + k_-)} \qquad (9.52)$$

$$n_- = n_+ - \frac{\varepsilon}{q_e}\frac{dE_z}{dz}.$$

Typical fair weather values for E_z, $k_\pm$, and J_{FW} are substituted in these two equations to obtain the initial height profiles of positive and negative ions.

The net fair weather ion generation rate $G(z)$ remains the same throughout a model run. To determine $G(z)$, Helsdon and Farley (1987) used the equilibrium condition for fair weather

$$\frac{\partial}{\partial z}(k_\pm n_\pm E_z) = \pm G(z) \mp \beta n_+ n_- \qquad (9.53)$$

and the expression for polar ionic mobilities from Shreve (1970)

$$k_\pm = c_\pm e^{az}, \qquad (9.54)$$

where $c_+ = 1.4 \times 10^{-4} \, m^2 \, V^{-1} \, s^{-1}$, $c_- = 1.9 \times 10^{-4}$ $m^2 \, V^{-1} \, s^{-1}$, and $a = 1.4 \times 10^{-4} \, m^{-1}$. By substituting the expressions for n_+ and k_+ in Eq. 9.53, Helsdon and Farley (1987) got

$$G(z) = \frac{k_+ k_- \varepsilon}{q_e(k_+ + k_-)}\left[aE_z\frac{\partial E_z}{\partial z} + \left(\frac{\partial E_z}{\partial z}\right)^2\right.$$
$$\left. + E_z\frac{\partial^2 E_z}{\partial z^2}\right] + \beta n_+ n_-. \qquad (9.55)$$

Besides fair weather sources of ions, it is necessary to add sources due to thunderstorms. Helsdon and Farley (1987) incorporated several thunderstorm sources, including lightning, corona from the ground, evaporation of charged cloud droplets, and sublimation of charged ice particles. Their treatment of lightning will be discussed later.

Other than recombination, the primary sink for free ions is capture of free ions by hydrometeors in clouds and thunderstorms. The rate at which ions are captured by hydrometeors depends on ion motion and on the preexisting charge on hydrometeors. Ion motion occurs by diffusion or is driven by the electric field. In his parameterization of ion capture, Chiu (1978) made the approximation that diffusion and electric fields could be treated independently, and the resulting capture

rates could then be added together. The rate of change in free charge density due to ion capture is then

$$\left(\frac{\partial \rho_f}{\partial t}\right)_{ion\ capture} = q_e\left[\left(\frac{\partial n_+}{\partial t}\right)_{dif} - \left(\frac{\partial n_-}{\partial t}\right)_{dif}\right.$$
$$\left. + \left(\frac{\partial n_+}{\partial t}\right)_E - \left(\frac{\partial n_-}{\partial t}\right)_E\right], \qquad (9.56)$$

where ρ_f is the net charge density on free ions, the subscript *dif* refers to diffusion-driven ion capture, and the subscript E refers to electric-field-driven capture. A derivation of expressions for ion attachment is beyond the scope of this text, but we will list the equations used by Chiu (1978). The rate at which free ions moving by diffusion attach themselves to hydrometeors in category j was obtained from Gunn (1954). The expression given by Chiu (1978), with a minor correction to convert from electrostatic units (esu) to MKS units, can be written as

$$\left(\frac{\partial n_\pm}{\partial t}\right)_{dif} = 4\pi r_j D_\pm n_\pm n_j(r_j)\left[\frac{\pm\dfrac{Q_j}{Q_D}}{\exp\left(\pm\dfrac{Q_j}{Q_D}\right) - 1}\right]$$
$$\cdot \left[1 + \left(\frac{r_j v_{jT}}{2\pi D_\pm}\right)^{\frac{1}{2}}\right], \qquad (9.57)$$

where D_+ and D_- are the diffusion coefficients of positive and negative ions, respectively; Q_j, $n_j(r_j)$, and v_{jT} are the charge, number density, and fall speed of the jth type of hydrometeor of radius r_j; and $Q_D = 4\pi\varepsilon r_j k_B T/q_e$ is the magnitude of charge on the hydrometeor for which the electric potential and thermal energies of an ion are comparable at the hydrometeor surface. (k_B is the Boltzmann constant; T is ambient air temperature in K; q_e is the charge of an electron.) The ion diffusion coefficient can be given in terms of ion mobility $k_\pm$ by $D_\pm = k_\pm k_B T/q_e$. The last bracketed factor of Eq. 9.57 accounts for the effect of hydrometeor ventilation on ion capture and approaches 1 as v_{jT} approaches 0 m s^{-1}. If Q/Q_D also approaches zero, the right side of Eq. 9.57 approaches $4\pi r_j D_\pm n_\pm n_j(r_j)$.

To determine the rate of attachment due to ion motion driven by an electric field, Chiu (1978) used an analysis of the Wilson ion-drop mechanism (Wilson 1929) from Whipple and Chalmers (1944). The expression for the rate of attachment depends on whether the hydrometeor is carrying enough charge that the electric field cannot polarize any part of the hydrometeor's surface, (i.e., the entire surface is positively or negatively charged). The magnitude of charge required to achieve this is

$$Q_M \equiv 12\pi\varepsilon|\vec{E}|r_j^2. \qquad (9.58)$$

The expression for the rate of attachment also depends on both the magnitude and direction of the mean ion velocity relative to the terminal fall velocity of the hydrometeor. Table 9.3 shows the expressions for various combinations of these variables.

Table 9.3. Expressions for ion capture driven by an electric field

Q_j	$\vec{E} \parallel \vec{v}_T$ or $\vec{E} \parallel -\vec{v}_T$	$k_+\|E\|$	$\partial n_+/\partial t^a$	$\partial n_-/\partial t^a$
$Q_j > Q_M{}^b$	Either	Any value	0	$n_- n_j k_- Q_j \varepsilon^{-1}$
$Q_j < -Q_M{}^c$	Either	Any value	$-n_+ n_j k_+ Q_j \varepsilon^{-1}$	0
$0 < Q_j < Q_M$ $-Q_M < Q_j < 0$	Parallel	$k_+\|E\| < \|v_T\|$	0 $-n_+ n_j k_+ Q_j \varepsilon^{-1}$	$n_- n_j k_- \|E\|(3\pi r_j^2) \cdot [1 + (Q_j/Q_M)]^2$
$-Q_M < Q_j < Q_M$		$k_+\|E\| > \|v_T\|$	$n_+ n_j k_+ \|E\|(3\pi r_j^2) \cdot [1 - (Q_j/Q_M)^2]$	
$0 < Q_j < Q_M$ $-Q_M < Q_j < 0$	Anti-parallel	$k_-\|E\| < \|v_T\|$	$n_+ n_j k_+ \|E\|(3\pi r_j^2) \cdot [1 - (Q_j/Q_M)]^2$	$n_- n_j k_- Q_j \varepsilon^{-1}$ 0
$-Q_M < Q_j < Q_M$		$k_-\|E\| > \|v_T\|$		$n_- n_j k_- \|E\|(3\pi r_j^2) \cdot [1 + (Q_j/Q_M)]^2$

$^a n_j = n_j(r_j)$; ^{b}Hydrometeor completely positive; ^{c}Hydrometeor completely negative; From Chiu 1978 based on Whipple and Chalmers 1994.

To evaluate the total rate of change in the number densities of ions from both diffusion-driven and electric-field-driven ion capture in the volume represented by a grid point, Chiu (1978) and Helsdon and Farley (1987) used mean values of parameters for each water substance category being considered to determine its contribution and then added the contributions of all categories together. For example, to calculate the contribution from raindrops, their model evaluated Eq. 9.57 and the applicable equations in Table 9.3 for a particular grid point by using the mean radius, terminal velocity, and charge of raindrops with appropriate values of other parameters for that location.

Note that even if ion-capture rates are large, the net change in hydrometeor charge can be small if capture rates are comparable for positive and negative ions. For electric-field-driven capture, the ratio of capture rates for positive and negative ions is $n_+ k_+ / n_- k_-$ when Q_j is small and the mean ion velocity is greater than the hydrometeor fall speed. For diffusion-driven ion capture, the ratio approaches $n_+ D_+ / n_- D_-$ as Q_j and v_{jT} approach zero. When these conditions are satisfied, the number densities, mobilities, and diffusion coefficients of small ions must be substantially different for the two polarities if the effect of ion capture on hydrometeor charge is to be significant. One situation in which the net charge from ion capture will be significant, for example, is when there is a predominance of one sign of ion, as when lightning injects ions into some region. Another situation is when hydrometeors already carry substantial charge; then the net effect of ions often is to reduce the magnitude of charge on the hydrometeors.

There are two numerical difficulties in treating ion budgets explicitly. One is that calculating either the net charge or the change in the net charge carried by free ions in a given region often involves taking relatively small differences of large quantities. Therefore, care is needed in the numerical techniques to preserve valid information by maintaining sufficient numerical precision. A second difficulty is that ion velocities, $k_+ E$, become large enough in regions of a thunderstorm with large magnitudes of E that variations in ion properties can move across more than one grid point in a single time step of a model (typically of the order of 1–10 s). The result can be numerical instability (e.g., see Haltiner and Williams 1980, page 170). This difficulty commonly is addressed by reducing the time step to treat ions. Randell et al. (1994), for example, used a dynamically adjusted time step that began at 3.75 s, but was reduced with increasing E until it was 0.5 s when $E \approx 200$ kV m^{-1}. In model experiments without lightning (Wojcik 1994), the time step became prohibitively small (on the order of 0.1 s) when E grew to roughly 700 kV/m, so the model experiment was terminated. However, lightning would be expected to occur at smaller values of E and would limit E to smaller values in simulated storms, so the size of the time step would not prevent the model from simulating storms with lightning.

Although an explicit treatment of ions creates difficulties, it also enables a model to deal with important phenomena including the early stages of electrification (when charging by other electrification processes is small or nonexistent), precipitation capture of ions emitted as corona beneath a storm, screening-layer charge at cloud boundaries, and the dispersal and capture of ions from a lightning channel. To avoid the numerical difficulties while still treating these phenomena, it may be desirable to use a hybrid approach that applies an explicit treatment of ions in situations only in which it is particularly needed to describe the evolving thunderstorm charge distribution.

9.4.6. Screening-layer charge

Instead of treating ion attachment to hydrometeors explicitly, Ziegler et al. (1991) included ion attachment only in a macroscopic treatment of screening-layer charge, one of the major consequences of ion capture by hydrometeors. In the detailed ion treatment described in the previous section, a screening layer arises because the change in mobility across the cloud boundary creates a divergence in polar currents ($\pm q_e n_\pm k_\pm E$), represented by the fourth term in Eq. 9.49), which causes a surplus of ions of one sign. The surplus then enhances the capture of that sign of charge by hydrometeors near the cloud boundary. Treating the screening layer macroscopically instead of treating the detailed microphysics of ion capture is considerably simpler and faster, but ignores the effects of ion capture in other circumstances.

To begin the macroscopic treatment, recall that the screening-layer charge is caused by the difference in electrical conductivity between cloudy and clear air. As discussed in Sec. 2.2, the equilibrium condition is that current be continuous across the discontinuity in conductivity, that is,

$$J_{n\,clr} = \sigma_{clr} E_{n\,clr} = \sigma_{cld} E_{n\,cld} = J_{n\,cld}, \quad (9.59)$$

where J is the current density, σ is conductivity, the subscript n means the component of $\vec{E}$ or $\vec{J}$ normal to the cloud boundary (with the convention that positive normal components point inward, as in Fig. 9.9), the subscript clr refers to clear air just outside the cloud boundary, and the subscript cld refers to values just inside the boundary. Near the cloud boundary, we can approximate the screening layer as an infinite plane, so the expression for the surface charge density, σ_q, of the screening layer, is derived in the same way as Eq. 2.3 and is

$$\sigma_q = 2\varepsilon \frac{\sigma_{clr} - \sigma_{cld}}{\sigma_{clr} + \sigma_{cld}} E_n, \quad (9.60)$$

where E is the field that would exist in the absence of a screening layer. (Note that the sign convention of E_n is opposite to that of E in Eq. 2.3 at cloud top.) If we approximate the screening-layer charge as a layer with thickness, L, then $\rho = \sigma_q/L$ is the approximate space charge density of the screening layer.

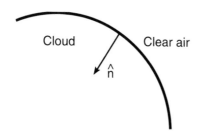

Fig. 9.9. Geometry of the unit vector normal to the cloud boundary for calculating screening-layer charge.

Unless the screening-layer charge approaches equilibrium within the time step of the model (typically <10 s), it is necessary to include the time dependence of screening-layer formation, so the model screening layer can adjust realistically to changes in the storm. For electric conductivities typical of the tops of storms, the electrical relaxation time constant is roughly 5–10 s, and for cloud base, it can be more than 10 min. Therefore, the screening-layer charge will require more than one time step to form in response to changes in the storm, and a parameterization that accounts for the time of formation is needed. For both computational convenience and physical reasons, the formulation for the increment of screening-layer charge to add at a given time step should be in terms of the electric field from all preexisting charges, including previously added screening-layer charges, which models do not track explicitly.

To derive an expression for the formation of screening-layer charge, we begin with an equation for the conservation of charge at the cloud boundary:

$$\frac{\partial \rho}{\partial t} = -\frac{\partial J_n}{\partial n} \quad (9.61)$$

$$\approx -\frac{\sigma_{cld} E_{n\,cld} - \sigma_{clr} E_{n\,clr}}{L},$$

where n is the unit vector normal to the cloud boundary. If $\delta\sigma_q \approx L\,\delta\rho$ is the increment of screening-layer surface charge density to be added in time step t, then Gauss's law may be applied to express the new electric field as the sum of the previous electric field and the field from the new screening charge increment:

$$E_{n\,clr}^t \approx E_{n\,clr}^{t-1} - \frac{L\delta\rho}{2\varepsilon}$$

$$E_{n\,cld}^t \approx E_{n\,cld}^{t-1} + \frac{L\delta\rho}{2\varepsilon}, \quad (9.62)$$

where the superscripts, t and $t-1$, only refer to the time step and are not exponents. The electric fields from previously formed screening-layer charges are included in E^{t-1}. Substituting these expressions for $E_{n\,clr}$ and $E_{n\,cld}$ in Eq. 9.61, gathering terms in ρ, and applying the boundary condition $\delta\rho(\Delta t) = 0$ at $\Delta t = 0$ gives the following expression for the screening-layer space charge formed during the time step Δt:

$$\delta\rho = \frac{2\varepsilon(\sigma_{clr} E_{n\,clr}^{t-1} - \sigma_{cld} E_{n\,cld}^{t-1})}{L(\sigma_{clr} + \sigma_{cld})}$$

$$\cdot \left\{ 1 - \exp\left[-\frac{(\sigma_{clr} + \sigma_{cld})\Delta t}{2\varepsilon} \right] \right\}. \quad (9.63)$$

Note that the screening-layer depth, L, has no effect on the amount of charge in the screening layer in this formulation; L affects only the volume through which the charge is distributed. Ziegler et al. (1991) set the screening layer depth L to 500 m (consistent with observations by Marshall et al. 1989) and assumed that the cloud boundary occurred at the last grid point in any direction before the cloud mixing ratio dropped below 0.01 g kg^{-1}. The normal component of the electric field at grid points adjacent to the cloud boundary was computed from the dot product of the electric field and the cloud water gradient. Screening-layer charge was divided between cloud ice and cloud water. At a given grid point just inside the cloud boundary, the fraction of the screening layer charge assigned to one of the two categories was the same as the fraction of the mass of total cloud water substance belonging to that category.

Screening-layer parameterizations are affected by some storm parameters that are not well understood. A key parameter is the conductivity of cloudy air. Although there have been several measurements of the conductivity of clear air, measurements of the conductivity of cloudy air are more difficult to make and have been the subject of controversy (see Sec. 7.6). Ziegler et al. (1991) used the result from Rust and Moore (1974) that the conductivity of cloudy air was typically 10% of the conductivity of clear air at the same altitude.

Screening-layer parameterizations also are affected by the location of the change in conductivity relative to the cloud particle field, but this would be expected to affect primarily the location of the screening layer, instead of the amount of charge in it. Rust (1973) noted that changes in conductivity do not necessarily coincide with the visual cloud boundary, but can begin near the cloud in what appears to be clear air. Similarly, Dolezalek (1963), Anderson and Trent (1966), Anderson (1977), and others have reported that the conductivity of air decreases before fog becomes visible, possibly due to a decrease in the number density of small ions. However, screening layers detected by balloon soundings (e.g., Marshall et al. 1989) have appeared to coincide with the visual cloud boundary, to the extent that this could be determined visually from the ground. Thus, we expect the main error in the location of screening layers from a model parameterization to be caused by uncertainties about the criteria used in the model to determine the location of cloud boundaries. (Note that the screening layer produced by an explicit treatment of ion capture also will be affected by how well the model locates the cloud boundary, simulates the location and magnitude of changes in ion mobility and small-ion number density in the vicinity of the cloud boundary, and simulates the number density and size of hydrometeors at the cloud boundary.)

The method by which a screening-layer parameterization distributes screening-layer charge among the various water substance categories can affect the subsequent history of the charge. It might be more accurate to distribute the charge by surface area instead of mass, because the surface area is proportional to a particle's cross section. Furthermore, although the preexisting charge on a particle retards the capture of additional charge, the effect of differences in the charge on different water substance categories has not yet been considered in apportioning the charge in a screening-layer parameterization.

9.4.7. Lightning

To model the electrification of storms once they have begun producing lightning, it is necessary to include a parameterization of lightning. A primary purpose of lightning parameterizations is to limit E magnitudes to observed values. Without a lightning parameterization, E would build to unrealistically large values, several times what is needed to cause electrical breakdown of air. A lightning parameterization also permits comparison with the observed timing and rates of lightning flashes, as well as permitting study of effects on cloud electrification, as we will discuss later.

There have been several treatments of lightning in models. Rawlins (1982), for example, included a simple parameterization that reduced the net charge density and the charge densities of individual water substance categories arbitrarily by 70% at each grid point when the discharge threshold $E = 500 \text{ kV m}^{-1}$ was attained. Takahashi (1987) also developed a simple parameterization of a lightning flash: A flash was initiated any time the electric field was at least 340 kV m^{-1} (the maximum field observed by Gunn 1948). The model then located the largest positive and negative charge densities, determined the smallest region about each maximum that encompassed 20 C (the mean charge neutralized by lightning flashes studied by Workman et al. 1942), apportioned the charge equally over all charged particles in the region, and subtracted the charge (positive charge from the positive region and negative from the negative region). A strict interpretation of Takahashi's use of an axisymmetric, two-dimensional model is that the geometry of the charge neutralized by a lightning flash (or by several flashes equivalent to the single simulated flash) was a torus formed by rotation of the neutralized two-dimensional charge about the axis. However, it also would be possible to interpret the lightning as occurring only in half of a vertical cross section of a simulated storm that otherwise had cylindrical symmetry, if flashes had a relatively minor effect on the thunderstorm charge distribution that produced subsequent flashes or if one was not concerned with subsequent evolution.

Ziegler and MacGorman (1994) used a similarly simple parameterization, but used a three-dimensional cloud model and treated the bulk effect of several lightning flashes, instead of treating an individual flash. Like Takahashi's (1987) parameterization, their parameterization initiated lightning when the electric field reached a threshold magnitude. However, they used a smaller threshold (200 kV m^{-1}) for two reasons: (1) The spatial resolution of the model for the large storm they studied was coarser (2 km horizontal grid spacing by 1 km vertical) than used in most modeling studies of isolated storms. If a large electric field existed only over small regions, as suggested by Winn and Byerley (1975), based on the rarity of observations of larger field magnitudes, it would not have been reproduced well by the coarser grid; and (2) Across substantial regions of the storm, the simulated electric field magnitude often was only slightly smaller than the lightning initiation threshold. As discussed in Chapter 7, there are few reliable observations of E larger than 200 kV m^{-1} in thunderstorms, so it was necessary to use a threshold no larger than 200 kV m^{-1} to limit the electric field of the model to commonly observed values.

Once the electric field threshold was exceeded, Ziegler and MacGorman (1994) neutralized some charge at all grid points where the magnitude of the charge density exceeded a threshold. Focusing the effect of lightning on charge regions was motivated by MacGorman et al.'s (1981) suggestion, based on observations of lightning structure inside clouds, that lightning channels often appeared to permeate regions where charge would be expected inside storms, with relatively few channels in regions of little charge. Additional support for this interpretation was provided by Williams et al. (1985), who examined self-propagating sparks in plastic blocks (polymethyl methacrylate) doped with charge in various regions. The parameterization did not restrict the effect of a given lightning event to charged regions in any specific part of the storm, in part because observed flash rates for the simulated storm were very large, so each lightning event was assumed to parameterize the effect of several flashes. Although this may have caused the parameterization to discharge a larger region than was actually involved in lightning activity in the observed storm, lightning events still occurred frequently in the simulation.

The magnitude of the charge neutralized by lightning was computed in three steps: (1) The model counted the number of grid points at which the magnitude of the net charge density exceeded a threshold. A preset fraction of the charge excess was summed over all grid points separately for each polarity to provide a first estimate of the positive and negative charge involved in the lightning event; (2) To insure that lightning neutralized equal amounts of positive and negative charge, the total magnitude of positive charge was compared with the total magnitude of negative charge, and a correction was computed for all grid points at which charge was to be neutralized. This was done because almost all flashes observed in the storm during the period being simulated were cloud flashes, which would not be expected to change the net charge of the storm; and (3) At each grid point where the magnitude of net charge exceeded the threshold, charge was neutralized by distributing the opposite charge over all hydrometeors at that location.

The total charge ($\delta\rho_{i,j,k}$) added at grid point (i, j, k) was given by

$$\delta\rho_{i,j,k} = 0 \qquad \text{for } |\rho_{i,j,k}| \leq \rho_{th}$$
$$\delta\rho_{i,j,k} = (|\rho_{i,j,k}| - \rho_{th})f_\rho - \delta\rho_{cor} \qquad \rho_{i,j,k} < -\rho_{th}$$
$$\delta\rho_{i,j,k} = -(\rho_{i,j,k} - \rho_{th})f_\rho - \delta\rho_{cor} \qquad \rho_{i,j,k} > \rho_{th}$$

$$\delta\rho_{cor} = \frac{1}{N_{dis}}\left\{\left[\sum_{i,j,k}(|\rho_{i,j,k}| - \rho_{th})f_\rho\right]_{neg} - \left[\sum_{i,j,k}(\rho_{i,j,k} - \rho_{th})f_\rho\right]_{pos}\right\}, \qquad (9.64)$$

where $\rho_{i,j,k}$ is the net charge density at (i, j, k) before the lightning event, ρ_{th} is the threshold charge density, f_ρ is the fraction of charge excess that is neutralized, N_{dis} is the number of grid points at which $|\rho_{i,j,k}| > \rho_{th}$, and the subscripts *pos* and *neg* on brackets indicate that the summations are over grid points with excess positive and negative charge density, respectively. Ziegler and MacGorman (1994) used $\rho_{th} = 0.5$ nC m^{-3} and $f_\rho = 0.33$ for their study, although other values were used in sensitivity analyses. Using $f_\rho = 0.1$ removed too little charge to prevent the electric field from exceeding values greater than needed to create a spark in clear air.

When Ziegler and MacGorman (1994) added charge of the opposite sign to all hydrometeors at a grid point, the charge was apportioned to each hydrometeor category according to its relative surface area. Thus, the charge density assigned to each category at a grid point was simply

$$\delta\rho_m = \frac{S_m}{\sum_m S_m}\delta\rho_{i,j,k} \qquad (9.65)$$

where $\delta\rho_m$ is the charge density deposited by lightning on hydrometeors in the *m*th category in the volume represented by grid point (i,j,k), S_m is the total surface area of hydrometeors in the same category and volume, and the summation is over all hydrometeor categories in that volume.

Helsdon et al. (1992) developed a lightning parameterization that estimated both the geometry and charge transport of an intracloud lightning flash. The basic theory for their parameterization was taken from Kasemir (1960, 1984), who hypothesized that lightning propagates bi-directionally from the region of initial breakdown and can be described as a conductor that is electrically neutral overall (i.e., the negative charge of one end is balanced by an equal positive charge at the opposite end). The parameterization used the local electrostatic field at the end of the channel as the parameter controlling the direction of propagation and the termination of a lightning channel, because Williams et al. (1985) identified the local field as the most important parameter in controlling laboratory discharges through plastic. The electric field threshold for lightning initiation was chosen to be 400 kV m^{-1}. Once initiated, one end of the lightning propagated parallel to the ambient electric field, and the other propagated antiparallel. Propagation direction was rounded off so that the channel developed in steps between adjoining grid points (either along a coordinate or diagonally). Lightning propagation was terminated when the ambi-

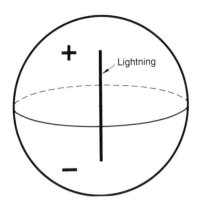

Fig. 9.10. Simplified model thunderstorm and lightning channel used by Helsdon et al. (1992) to develop a lightning parameterization. The upper hemisphere is filled with a uniform density of positive charge, and the lower hemisphere, with an equal density of negative charge.

ent electric field fell below 150 kV m^{-1}. The electric field from the channel itself was not included in computing either propagation direction or termination.

Once the lightning geometry was determined, Helsdon et al. (1992) computed the charge distribution along the lightning channel. The linear charge densities then were converted to ion number densities, which were added to preexisting ion densities. In subsequent time steps, ions were captured by hydrometeors, a process treated explicitly in their model.

To develop a parameterization for the charge distribution on a lightning channel, Helsdon et al. (1992) estimated the charge distribution induced on a long, thin spheroidal conductor when it is imbedded vertically in a spherical thunderstorm charge distribution (Fig. 9.10). For this estimate, they assumed that positive charge was in the upper hemisphere of the spherical charge distribution; negative charge, in the lower hemisphere; and lightning began at the center of the thunderstorm. They then solved for the linear charge density $\delta\lambda_q$ (in C m^{-1}) as a function of the location along the channel.

The solution for the lightning charge at a given location on the channel was a complicated function of the ambient electric potential $\Phi_{i,j,k}$ at that location, the potential of the initiation point Φ_0, the half length a and radius b of the channel, the radius R of the spherical charge distribution, and the distance, s, along the channel from the initiation point. However, according to Helsdon et al. (1992), if this complicated expression for the linear charge density at a grid point (i, j, k) on the lightning channel was evaluated at the end of the conductor, it became

$$\delta\lambda_{q(i,j,k)} = -k_C(\Phi_{i,j,k} - \Phi_0), \qquad (9.66)$$

where k_C is a constant that is the linear capacitance of the lightning channel. The value of k_C was insensitive to the radius R of the charge distribution as long as $R > a$. Helsdon et al. evaluated k_C for $R = 3$ km and several values of a and b. They then compared the charge density that would be computed along the channel by using the full function and by using Eq. 9.66. Table 9.4 shows the resulting values of k_C and the maximum error in the linear charge density along the channel caused by using k_C instead of the exact function. Because the error in $\delta\lambda_q$ from using the constant k_C was less than a factor of two and much less than other uncertainties, Helsdon et al. (1992) adopted Eq. 9.66 to parameterize the charge distribution on lightning channels. Since the range of values of k_C was relatively small, particularly for channel diameters of a few centimeters, their model did not compute k_C for each lightning flash in the model, but assumed a fixed value (approximately 5×10^{-12} F m^{-1}). The linear charge density along the channel then depended only on the potential difference between the point being evaluated and the initiation point, which tended to increase to a maximum and then decrease along the channel at increasing distances from the initiation point.

To ensure that the channel was electrically neutral overall, as hypothesized by Kasemir (1960), Helsdon et al. (1992) modified the channel ends to force the integral of lightning charge density for the whole channel to be zero. There were three steps:

Table 9.4. Values of k_C (in pF m^{-1}) and the percent error in $\delta\lambda_q$ as a function of channel half length a and radius b (in m) for $R = 3000$ m

	$a = 1000$ m		$a = 2000$ m		$a = 3000$ m	
b	k_C	Error[a]	k_C	Error[a]	k_C	Error[a]
0.001	4.11	0.4	3.88	1.9	3.66	5.7
0.01	4.96	0.5	4.63	2.3	4.33	6.8
0.1	6.23	0.7	5.72	2.8	5.26	8.4
1.0	8.40	0.9	7.49	3.8	6.7	11.1

[a]Maximum percentage error in $\delta\lambda_q$ at any point on the channel from using the constant value k_C everywhere along the channel instead of the full function at every point. (From Helsdon et al. 1992, with permission.)

1. The positive and negative channel segments were integrated separately to determine the total charge magnitude of each polarity;
2. The end of the channel with the polarity having the most charge was then extended four grid points, with the linear charge density at the four grid points being given by

$$\delta\lambda_{q\,(a\,+m)} = \delta\lambda_{q\,a}\exp[-\kappa_d(m\,\delta d)^2], \quad (9.67)$$

where $\delta\lambda_{qa}$ is the linear charge density at the termination point ($m = 0$), $m = 1$–4 for the extension, δd is the distance between grid points along the channel, and κ_d is chosen so that $\delta\lambda_{q(a+4)} = \delta\lambda_{qa}/1000$ ($\kappa_d = 1.094 \times 10^{-5}$ m^{-2} when $\delta d = 200$ m). The charge on these four points was added to the previous total for that polarity, and the difference in charge magnitude between the two polarities was recomputed.
3. The opposite end of the lightning channel, which had opposite polarity, was extended to adjoining grid points. The linear charge density placed on grid points in this extension decreased with distance from the termination point, as for the first end, but was chosen to make the net charge of the channel equal zero. Unlike the extension of the first end, the charge density added to the final grid point of the extension of the second end was not forced to be a factor of 1000 smaller than the density at the termination point. An example of the linear charge density along a lightning channel is shown in Fig. 9.11.

The next step in the lightning parameterization was to convert the linear charge densities along the lightning channel to volume charge densities in some region about the channel. The amount of charge in the region represented by a grid point was $\delta d \cdot \delta\lambda_{q(i,j,k)}$. Because the cloud model was two-dimensional and slab-symmetric, the charge was assumed to extend to infinity on both sides perpendicular to the plane of the model. The

charge density at a grid point represented a distance $\pm 0.5\ \delta d$ perpendicular to the plane and was constant over that distance. In the direction perpendicular to the channel, but in the plane of the model, charge was extended over four grid points in both directions away from the channel; the density was forced to decrease with distance from the channel by using an expression analogous to that for charge beyond the termination point (Eq. 9.67). If we consider a channel in the z direction and take x to be the orthogonal coordinate in the plane of the model and y to be perpendicular to the plane, then the charge distributed about grid point (i, j, k) on the lightning channel is

$$
\begin{aligned}
Q_{i,j,k} &= \delta d\,\delta\lambda_{q(i,j,k)} \\[4pt]
&= \delta\rho_{0(i,j,k)}\int_{-\frac{1}{2}\delta d}^{\frac{1}{2}\delta d}\int_{-\frac{1}{2}\delta d}^{\frac{1}{2}\delta d}\int_{-4\delta d}^{4\delta d} e^{-\kappa_d x^2}dz\,dy\,dx \\[4pt]
&= 2(\delta d)^2\delta\rho_{0(i,j,k)}\int_{0}^{4\delta d} e^{-\kappa_d x^2}dx,
\end{aligned}
$$

$$(9.68)$$

where $\delta\rho_{0(i,j,k)}$ is the average charge density in the grid box centered on (i, j, k) due to charge on the lightning channel. Helsdon et al. (1992) evaluated the integral numerically, used $\delta d = 200$ m, and rearranged it to get $\delta\rho_{0(i,j,k)} = \delta\lambda_{q(i,j,k)}/(1.079 \times 10^5\ \text{m}^2)$. The charge was then distributed in the perpendicular cross section according to an equation analogous to Eq. 9.67, with $\delta\lambda_{q(a+m)}$ replaced by $\delta\rho_{a+m}$ and $\delta\lambda_{q\,a}$ replaced by $\delta\rho_{0(i,j,k)}$. The parameter κ_d controlled how concentrated the charge was about the channel. If $\kappa_d = 1.094 \times 10^{-5}$ m^{-2}, as used by Helsdon et al., then 65% of the charge was within 400 m (2 grid points) of the channel, and 99% was within 600 m.

The last step in this lightning parameterization was to convert the lightning charge density at a given grid point to an ion number density. To do this, Helsdon et al. (1992) assumed that the ions were all singly ionized, and so divided the charge density, $\delta\rho_{i,j,k}$ by the magnitude of the electron charge. The resulting ion number density was then added to the preexisting number density. Once this was done for all grid points that participated in a lightning flash, the electric field was recalculated to take into account the effect of the modified charge distribution. This completed the lightning discharge process. In subsequent time steps, the free ions deposited by a lightning flash interacted with hydrometeors as described in Section 9.4.5.

The electric field threshold used by most models for lightning initiation has been constant with altitude. However, observations presented in Section 7.16.2 suggest that the large conductivity at upper altitudes of storms results in a smaller E threshold that allows lightning to be initiated between the negative screening layer and upper positive charge. Theory also suggests that the threshold for initiation decreases with altitude (Sec. 5.4). Thus, MacGorman et al. (1996) used the vertical profile of the breakeven electric field described

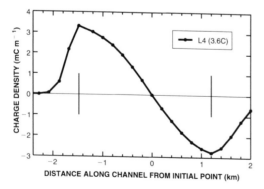

Fig. 9.11. Linear charge density along a modeled lightning channel. The vertical bars along the abscissa represent the extent of the channel before applying the charge balancing procedure. (From Helsdon et al. 1992, with permission.)

in Sections 5.4 and 7.13 for lightning initiation. Their parameterization of channel development was similar to the initial bi-directional propagation along electric field lines used by Helsdon et al. (1992). However, they traced electric field lines to a smaller E magnitude than that used by Helsdon et al., and if the grid point at which this threshold was reached was within a region of sufficient charge density, they assumed that the flash branched throughout the region of charge. To do this, the flash was extended to contiguous grid points having at least the threshold charge density within the volume bound by the equipotential surface passing through the grid point at which the threshold E was reached. This extension of flash development was motivated by the hypothesis, based on observations by MacGorman et al. (1981), Williams et al. (1985), and Shao and Krehbiel (1996), that lightning tends to propagate and branch through regions of charge, even if the ambient E is small. Procedures for determining how much charge was neutralized by the flash and for adding the opposite polarity of charge to hydrometeors at grid points involved in the flash to carry out the neutralization were similar to those developed by Ziegler and MacGorman (1994).

9.5. SOME THEMES OF ELECTRIFICATION MODEL RESEARCH

Issues that have been addressed by research using cloud electrification models include: (1) the viability of various electrification mechanisms, (2) the dependence of electrification on updraft characteristics and storm microphysics, (3) factors influencing lightning location and rates, and (4) lightning effects on thundercloud charge distributions. Modeling studies have improved understanding of a few electrification processes and have highlighted some gaps in knowledge where more field and laboratory experiments are needed. The rest of this section will examine aspects of the issues that have received the most attention in the literature.

9.5.1. Model evaluation of electrification mechanisms

By far the most frequent theme of modeling studies has been to examine the viability and relative contributions of various electrification mechanisms. Some studies, for example, have included ion capture by hydrometeors. To the extent that they parameterize cloud boundary microphysics, mixing, transport, and charge capture with enough fidelity and resolution, these could be used to evaluate some aspects of Vonnegut's (1953) hypothesis, which is frequently called the convective mechanism and which we call the Grenet-Vonnegut mechanism (see Ch. 3). Some studies (e.g., Ruhnke 1972, Chiu and Klett 1976) have examined this hypothesis explicitly. In these cases, the maximum electric field that could be generated by ion capture and subsequent transport as suggested by Vonnegut was

much weaker than observed in thunderstorms, but was comparable with observations of early electrification and of some nonthunderstorm clouds. Takahashi (1979) included both inductive charging and ion capture in a cloud model with warm cloud microphysics appropriate for some small maritime storms. In the model simulations, the most important electrification process was ion capture by hydrometeors, which was significantly enhanced during evaporation and condensation. The model reproduced observations of weakly electrified maritime clouds reasonably well, which suggests that ion capture can be important in these small storms. Sapkota and Varshneya (1988) used a simple one-dimensional model to evaluate a mechanism proposed by Wagner and Telford (1981) that involved ion capture at the boundaries of descending entrained air parcels of larger conductivity inside the storm. Again, the model was able to generate only weak electrification.

Although the results of these studies were affected by the parameterizations that were used, these results and results from other studies discussed previously, make it difficult to see how the Grenet-Vonnegut mechanism can be the primary charging mechanism for thunderstorm electrification, unless there are fundamental flaws in our understanding of at least part of the process. However, some modeling studies have found that ion capture can provide the initial electrification needed by the inductive mechanism to generate a thunderstorm electric field (Tzur and Levin 1981 used a one-dimensional model, and Chiu 1978 used a two-dimensional model). Other studies (Takahashi 1983, Ziegler and MacGorman 1994) found that transport and sedimentation of captured ions can contribute most of the charge in some regions of thunderstorms. Although there are significant doubts that the Grenet-Vonnegut mechanism can be the primary mechanism for electrifying most thunderstorms, it appears to play a role in producing at least some lightning flashes (suggested by the observations of Taylor et al. 1984 and Mazur et al. 1984) and may influence the relative timing and type of lightning flashes produced by a storm, as suggested by the modeling results of MacGorman et al. (1996).

Many modeling studies have concentrated on evaluating the inductive and noninductive mechanisms, either separately or together. Early model tests of the relative importance of the two mechanisms yielded mixed results. Some early studies using one-dimensional models that included both the inductive mechanism and a simple noninductive charging parameterization (Illingworth and Latham 1977, Tzur and Levin 1981, Mathpal et al. 1982) concluded that the most realistic electric field magnitudes, structure, and growth rates occurred when both inductive and noninductive mechanisms were active. Kuettner et al. (1981) used a simple two-dimensional model and reached the same conclusion. Noninductive mechanisms yielded fairly steady charging rates, but in most cases the maximum E generated by a noninductive mechanism alone was less than the threshold value thought necessary for lightning. Once moderate E was created by the noninductive mechanism, the inductive mechanism became important and was then able to generate E large enough to initiate lightning. Only Tzur and Levin and the three-dimensional model of Rawlins (1982) were able to generate large enough E for lightning with only a noninductive mechanism.

There are two reasons these early models failed to produce E large enough for lightning. First, data on the noninductive mechanism from the laboratory studies of Takahashi (1978), Jayaratne et al. (1983), Saunders et al. (1991), and others were not available at the time most of these studies were done. Most modeling studies used a constant charge transfer per hydrometeor collision, with little or no dependence on temperature, liquid water content, or particle size, and several of the models produced smaller charging rates than have been achieved with more recent parameterizations. Second, most of these studies assumed that the magnitude of E needed to produce lightning was substantially larger than is now thought necessary. (See discussions in Chs. 5 and 7 concerning maximum measured values of E, various mechanisms for reducing E needed for breakdown, and the breakeven electric field.) In some cases, the noninductive parameterization produced maximum values of E comparable to those produced in more recent simulations, but lightning was assumed to require still larger values.

Several modeling studies have used more complex parameterizations of the noninductive mechanism for collisions involving rimed graupel. Takahashi (1983, 1984, 1987) and Randell et al. (1994) included the dependence on liquid water content and temperature reported by Takahashi (1978). Takahashi (1983, 1984, 1987) also included dependence on particle size and impact velocity. Ziegler et al. (1986, 1991, 1994) used Gardiner et al.'s (1985) parameterization of results from Jayaratne et al. (1983), but modified it to allow the reversal temperature to be varied. Norville et al. (1991) and Solomon and Baker (1994) used measurements of noninductive charging by Jayaratne et al. (1983) and Baker et al. (1987). Wojcik (1994) tested each of three parameterizations of the noninductive mechanism: the parameterization developed by Helsdon and Farley (1987), one based on measurements by Takahashi (1978), and one based on measurements by Saunders et al. (1991). All of these studies using more complex parameterizations found that the noninductive mechanism produced realistic thunderstorm electric field magnitudes and distributions within the lifetime of a storm. None required the inductive mechanism to make either the magnitude or structure of the electric field more realistic. Although there are still uncertainties in our understanding of the noninductive graupel-ice mechanism, it seems likely that noninductive charging is adequate to produce thunderstorms in many situations. Warm cloud thunderstorms are an obvious exception.

Model evidence concerning the role of the inductive mechanism is even more confusing. Early models that used infinite horizontal plates for the upper and lower boundaries (Sartor 1967, Mason 1972, Ziv and Levin

1974, Scott and Levin 1975) generally found that the inductive mechanism could separate charge fast enough to create and maintain thunderstorm electric fields. However, Ziv and Levin pointed out that the contact time required for ice-ice collisions to separate charge increased with decreasing temperature, and they found that this behavior greatly diminished the effectiveness of the mechanism. Scott and Levin studied the effectiveness of the inductive mechanism as a function of the probability that charge would be transferred in a collision and placed limits on what the probability would need to be to create thunderstorm electric fields. Illingworth and Latham (1977) pointed out that models having infinite horizontal extent overestimate the magnitude of the electric field produced by a given charge density in the two charge regions. Furthermore, all of these infinite plane models simulated only part of a thunderstorm and assumed favorable conditions for the mechanism. The domain of the models was chosen such that the upper boundary was the center of positive charge, and the lower boundary, the center of negative charge. Particle and updraft velocities were assumed to allow precipitation to fall to the lower plate and cloud particles to be carried to the upper plate.

Results from two- and three-dimensional models and from one-dimensional models of finite extent usually have suggested a more limited role for the inductive mechanism. Initial electrification by a combination of ion capture at cloud boundaries to create a weak E and the inductive mechanism to intensify E to thunderstorm values usually was much slower than typically is observed for thunderstorms (Tzur and Levin 1981, Helsdon and Farley 1987, and Wojcik 1994) or required large rainfall rates (Chiu 1978, Rawlins 1982). Furthermore, Kuettner et al. (1981), Takahashi (1983), and Rawlins found that when the inductive mechanism was used without the noninductive mechanism, it produced charge distributions that often did not have even the correct overall polarities in the upper and lower regions of the storm. As discussed above, several studies (Illingworth and Latham 1977, Tzur and Levin 1981, Kuettner et al. 1981) found that the inductive mechanism was needed to supplement the noninductive mechanism by providing the final escalation of E needed to produce lightning. However, moderate to large precipitation rates were required for induction to work effectively, and models with more sophisticated treatments of the noninductive mechanism were able to produce large enough E to initiate lightning without the inductive mechanism, so even this role seems in question. In fact, Ziegler et al. (1991) found that the inductive mechanism reduced the magnitude of charge on hydrometeors throughout much of the cloud, and increased the charge density in only a relatively small region.

These modeling studies suggest that the inductive mechanism plays a secondary role in thunderstorms (although perhaps a larger role in weakly electrified clouds, as discussed by Takahashi 1979), but whether induction increases or decreases the magnitude of electric fields in thunderstorms is unknown. Some of the disagreement on this issue may be due to differences in one or more of the parameterizations. For example, all else in the model being equal, Ziegler et al.'s (1991) parameterization of the inductive mechanism produces charging rates more than 30 times smaller than those produced by Chiu's (1978), Helsdon and Farley's (1987), and Wojcik's (1994) parameterization, although both model parameterizations were within the range of laboratory results. It is not surprising that Chiu, Helsdon and Farley, and Wojcik observed a larger effect from induction in their studies. Nonetheless, it is difficult to tell whether an increase in the inductive mechanism's particle charging rate in Ziegler et al.'s model would cause even smaller E in many regions or would produce a more complex charge distribution with large regions of increased E. In the model, inductive charging is driven by E from preexisting charge in the thunderstorm, and particles that interact to separate charge by the inductive mechanism subsequently move under the influence of wind and gravity against a background of preexisting charge and simultaneous charging by other mechanisms. Weinheimer (1987) showed that changes in electrification depend on the relative motions of all charged particles relative to each other, and interactions often are complex.

Differences in the particle charging rates of the two inductive parameterizations underscore the need for laboratory studies of charging mechanisms to provide more definitive data. An example of the significant impact improved information can have is provided by recent laboratory studies of the inductive mechanism. The basic physics by which the electric field induces a dipolar charge distribution on hydrometeors is understood reasonably well, but there are still considerable uncertainties about the microphysics of hydrometeor interactions that govern contact time, separation probability, and separation mode for the particles. The classic inductive mechanism assumes that a small particle colliding with a large particle will either coalesce with or bounce off the large particle. McTaggart-Cowan and List (1975) and Low and List (1982) found that when two raindrops of different sizes collide, they do not rebound, as a small fraction of cloud droplets do when colliding with raindrops in a weak electric field. Instead, in a process called *disjection,* they coalesce, and then typically form a filament that breaks from the top of the larger particle to produce a mixture of small droplets and one or two larger drops. None of the resultant particles is necessarily the same size as either of the original particles. Disjection was found to be the only mode of separation for raindrop pairs down to diameters of 1.8 mm for the larger particle and 0.4 mm for the smaller.

Canosa and List (1993) studied the inductive mechanism for hydrometeor pairs in the size range where disjection occurred. They found charging consistent with the theory of Latham and Mason (1962) if the impact angle was replaced by the angle at which disjection occurred from the larger particle. A major consequence of this difference is that the small particle in a vertical E gets the sign of charge from the top of the large particle, not the bottom as during rebounds, so in-

ductive charging of raindrops is most likely to dissipate the ambient E, not enhance it. What is not clear is whether the effect of raindrop collisions with drizzle and cloud droplets is larger than the effect of collisions of raindrop pairs. Canosa et al. made an initial attempt to examine this issue by using an infinite horizontal model with 20 size categories of hydrometeors, 10 categories spanning the range 0.1–1mm, and 10 spanning 1–4mm. They found that, when the precipitation rate exceeded 2.4 mm h^{-1}, inductive charging dissipated E, and dissipation occurred faster with larger rainfall rates (see Fig. 9.12). Inductive charging enhanced E only during light rain (such as in cases studied by Takahashi 1979), but the rate of electric field growth was too slow to yield thunderstorm electrification during the lifetime of warm rain clouds. Additional laboratory and modeling experiments are needed to explore Canosa et al.'s observation further.

9.5.2. Effects of updraft strength and microphysics

Another frequent theme of modeling studies is the dependence of charge structure and electric field magnitudes on the characteristics of a storm's updraft and microphysics, both from storm to storm and during the evolution of a single storm. Many modeling studies (Rawlins 1982, Takahashi 1984, 1987, Ziegler et al. 1986, 1991, Helsdon and Farley 1987, Mitzeva and Saunders 1990, Norville et al. 1991, Randell et al. 1994, Solomon and Baker 1994; Ziegler and MacGorman 1994) have reproduced the observed behavior that E growth rates are larger in storms having larger updrafts in the mixed-phase region. All these studies have attributed the increased electrification to increased concentrations and sizes of snow and cloud ice particles as updraft magnitudes increase. For the noninductive mechanism, the increase in concentrations causes more collisions with graupel, and the increase in size produces more charge per collision.

Takahashi (1984), Ziegler et al. (1986, 1991), and Norville et al. (1991) pointed out that noninductive charging from graupel collisions with snow usually dominates the noninductive charging of thunderstorms because collisions with snow separate roughly one to two orders of magnitude more charge per collision than collisions with cloud ice. The most effective ice particle size for charging can be determined by the product $n(D)Q(D)$, where n is the concentration of particles with diameters in the category centered on D, and Q is the average charge per particle in that category. In model studies of thunderstorms, this product has always been largest for some diameter range of snow, although the diameter of the maximum has varied for different models and microphysical conditions (e.g., 0.6 and 1.3 mm in two simulations by Takahashi, 0.4 mm in Norville et al.). Norville et al. pointed out that snow crystals larger than 0.4 mm can become rimed and so can begin to play the role of graupel when they collide with smaller snow crystals. Such collisions usually

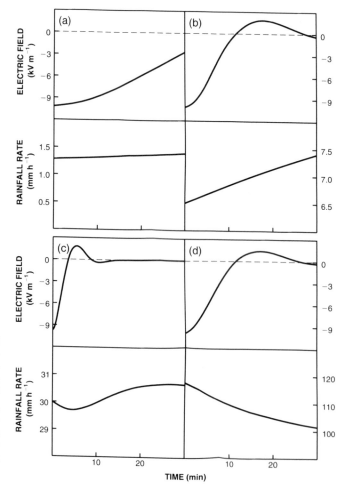

Fig. 9.12. Dissipation of the vertical electric field by inductive charging of colliding raindrops. The initial vertical electric field was -10 kV m^{-3}. Rainfall rates were (a) 1.3, (b) 6.5, (c) 30, and (d) 117 mm hr^{-1} for 30 min in all cases except (d), which lasted only 3 min. The time variation of rainfall rate is not caused by the electric field in the model, but is caused by the evolution of the drop spectra. (From Canosa et al. 1993, with permission.)

neutralized part of the charge gained by collisions with graupel, so Norville et al.'s inclusion of snow riming may account for some of the difference in the diameters of maximum charge between the two models.

As updrafts intensify at altitudes where there are ice particles, the size and number of snow crystals can increase rapidly. In fact, Norville et al. (1991) and Ziegler et al. (1991) showed that the frequently observed exponential growth of E during thunderstorm intensification, which had been cited as evidence for the feedback of the inductive mechanism, alternatively could be attributed to an exponential growth in the number of graupel-snow collisions, which creates a corresponding increase in charging by the noninductive mechanism (Fig. 9.13). To test this, Ziegler et al. suppressed snow and graupel growth by riming and deposition in their model, but used the same updraft as before. When growth was suppressed, the number of collisions did not increase (Fig. 9.13), and electrification of the storm was weak. These two studies, along with Randell et al. (1994), found that the growth of E became more

linear as the updraft began to dissipate.

The updraft also can affect the charge distribution in other ways. Ziegler et al. (1991) pointed out that the larger charge densities produced by a model thunderstorm cannot be explained by collisional charging rates alone, but require charge to be compressed by convergence of the vertical velocity of the charge carrier. As graupel or snow is carried upward in an updraft, it grows and its terminal velocity increases, while the updraft starts decreasing with height above mid-levels of the storm. As particles are carried upward, therefore, their ground-relative vertical velocity decreases with height until the terminal velocity balances the updraft. If the noninductive mechanism places negative charge on graupel and positive charge on snow, this convergence concentrates negative charge near the graupel balance level and positive charge near the snow balance level. The role of a particle balance level in storm electrification was discussed by Lhermitte and Williams (1985) and also was explored by Randell et al. (1994). Ziegler and MacGorman (1994) showed

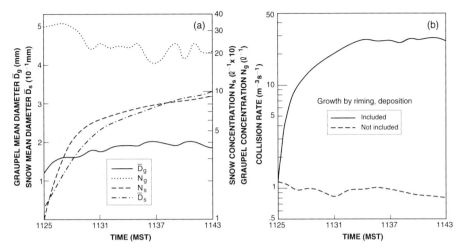

Fig. 9.13. Evolution of microphysical quantities at the location of the peak graupel-snow collision rate in a numerical model of a mountain thunderstorm: (a) Concentration and size of graupel and snow at the location of the peak collision rate. (b) Peak graupel-snow collision rates in two model cases, the one with complete model microphysics, the second without snow crystal and graupel growth by riming or deposition. (From Ziegler et al. 1991, with permission.)

that the very strong updrafts of a classic supercell tornadic storm caused both positive and negative charged regions to be shifted to significantly higher altitudes than in most storms. Note that particles will continue to grow when they are at their balance level and so eventually will start to descend. Particles also will start to descend if a horizontal component of wind carries them into weaker updrafts.

Takahashi (1984) examined the role of markedly different microphysics in controlling the evolution of the charge structure of thunderstorms. He pointed out that electrifying a thunderstorm by the noninductive mechanism required critical concentrations of graupel and snow to be present together. When there were high concentrations of ice nuclei in both continental (high concentrations of cloud water nuclei) and maritime (low concentrations of cloud nuclei) storms, this critical mixture of hydrometeors was able to form during storm intensification and caused the initial charge distribution to be at high altitudes in the storm. In some cases, there were enough graupel-snow interactions at high altitudes, where liquid water contents were low, that the polarity of charge separation reversed near the top of the storm. The graupel at upper levels of the storm then reinforced the positive charge on snow created by interactions at middle levels, and the snow at the top of the storm was negatively charged. Regardless of the sign of charge on graupel and snow, the charge distribution lowered as graupel and snow fell and the updraft weakened. When maritime storms had low concentrations of ice nuclei or continental storms had moderate concentrations, graupel formation required more time, so charging occurred later in the life cycle of the storm and consequently was lower in the storm. Takahashi also

found that under some conditions storms were unable to produce a lower positively charged region: Graupel gained enough negative charge that subsequent collisions at lower heights where it gained positive charge were not able to reverse the polarity of charge on graupel. However, the negative snow resulting from collisions at lower heights still helped increase charge concentrations in the main negative region of the storm.

To explain the infrequent occurrence of lightning in maritime storms, Takahashi suggested that maritime storms might have still lower concentrations of cloud nuclei than he modeled. If so, they would have rapid raindrop growth that might produce both graupel and snow through a large vertical depth of the storm. If this caused graupel particles to interact with snow in both positive and negative charging regimes, the space charge would be partially neutralized and electrification would be weak. However, none of the model experiments he presented produced this situation.

Randell et al. (1994) also compared maritime and continental storms, but they examined the effect of different values of convective available potential energy (CAPE) instead of different cloud nucleus concentrations. Analysis of data from northern Australia (Rutledge et al. 1992, Williams et al. 1992) had shown that the environment of maritime storms has much lower CAPE, typically $<<1000$ J kg^{-1}, than the environment of continental storms, often >2000 J kg^{-1}. Randell et al. modeled storms with three markedly different values of CAPE, one low, one moderate, and one high. Not surprisingly, the storm in the environment with largest CAPE had the strongest electrification. Although the degree of electrification was affected by the number of collisions between graupel and hail, it also was strongly affected by the location of these collisions relative to the height of the charge reversal level for the noninductive mechanism, as suggested by Takahashi (1984). If the charge reversal level was below most graupel-snow collisions, graupel was negatively charged, and snow was positively charged. If, however, the charge reversal level was in the middle of the region of graupel-snow collisions, graupel and snow were both charged positively and negatively, so a hydrometeor's charge was partially neutralized and storm electrification was weak. When this occurred, fluctuations in liquid water content sometimes caused graupel charging to reverse at essentially the same location, increasing the tendency toward neutralization.

The value of CAPE (and the magnitude of the updraft) had two effects on the relative heights of the charge reversal level and the graupel-snow interaction region. First, large CAPE values produced larger updrafts, which lifted graupel higher above the level of charge reversal and thereby made it more likely that the particle interaction zone where most charging occurs was well above the charge reversal level. Second, the larger updrafts produced larger mixing ratios of graupel, snow, cloud ice, and cloud water. Besides causing more particle growth and more graupel-snow collisions, the larger values of cloud water caused the temperature at which charge reversal occurred to be warmer, and therefore lower, in the cloud. Again, this increased the vertical distance between the charge reversal level and the particle interaction zone and made it less likely that opposing charging tendencies would weaken the electrification of a storm.

When storms have either coexisting or successive multiple cells, instead of a single isolated cell, interac-

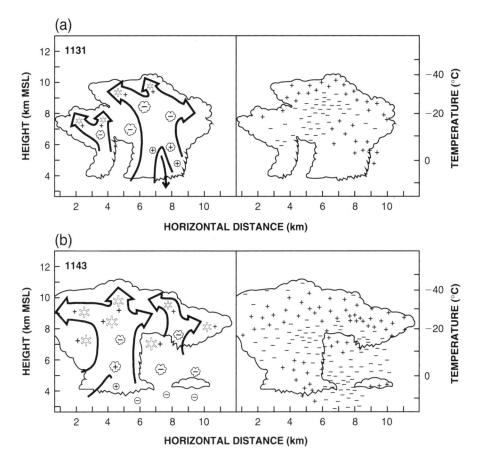

Fig. 9.14. Schematic depiction of the airflow, microphysics, and space charge in a model of a mountain thunderstorm. Left panels depict airflow (open arrow), cloud boundary (scalloped curve), precipitation habit (snow, graupel/hail, and rain), and sign of charge. Right panels depict the net space charge distribution relative to the cloud boundary. (a) Conditions during strong electrification of the first cell and early development of the second cell. (b) Conditions twelve min later, during decay of the first cell and strong electrification of the second. The lowest height plotted is the approximate height of the mountain top. (After Ziegler et al. 1991, with permission.)

tions can accelerate charging rates and complicate the charge structure of the storm. Ziegler et al. (1991) showed that graupel and ice particles that were advected from a first cell into a subsequent adjoining cell caused charging to begin earlier and charging rates to become larger in the second cell. Because the cells were at markedly different stages in their life cycles, the overall charge structure of the storm became more complicated when the second cell became electrified. Although the gross structure still could be recognized as having positive charge over negative charge, with small regions of lower positive charge, regions of positive charge were beside regions of negative charge at some heights (Fig. 9.14). Complicated multicell charge structures also were modeled by Takahashi (1987) and Randell et al. (1994). In fact, during the later stages of one of Randell et al.'s multicell simulations, the gross charge distribution no longer could be described as a simple dipole or tripole distribution (Fig. 9.15).

9.5.3. Effects of lightning

If the electric field threshold for lightning initiation varies with altitude, as would be expected, it may influence the ratio of cloud flashes to all flashes and the timing and location of the first flash of a storm. When MacGorman et al. (1996) used the vertical profile of the breakeven field for lightning initiation in simulations of tall supercell storms, they found that the small E threshold near the top of the storm caused the earliest lightning flashes (and frequent flashes thereafter) to occur between the negative screening layer at the top of the cloud and the upper positive charge. This depleted the upper positive charge compared with the lower negative charge. The relative magnitude of the negative charge eventually increased to the point that flashes were initiated in the lower part of the storm and propagated to ground. To test this effect, MacGorman et al. examined a second scenario in which the electric field required for lightning initiation was uniform with height. The resulting thunderstorm took much longer to produce ground flashes, and ground flash rates were much smaller than before. Furthermore, many of the ground flashes that did occur in this second scenario were positive ground flashes that began above negative charge and propagated along an electric field line that passed to the side of most negative charge and continued to the ground.

As mentioned previously, lightning serves the obvious purpose in models of limiting the maximum E that can exist in thunderstorms. When the electric field reaches the threshold for electric breakdown, lightning occurs and neutralizes some of the charge that created the electric field. Although it is possible that the reconfiguration of charge by a single lightning flash will create a large E that initiates a new flash in another region of the storm, the net effect of lightning activity is to limit the maximum E produced by models. If the threshold E for lightning initiation is assumed to vary with altitude, as discussed in Sections 7.13 and 7.16,

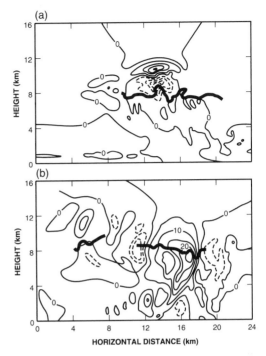

Fig. 9.15. Net charge density in a model of a vigorous continental thunderstorm. (a) Charge distribution at 44 min, the time of maximum electric field in the first cell of the model storm. The updraft of the first cell is weakening, and the new cell still is only weakly electrified. (b) Charge distribution at 52 min in the model run. The updraft of the first cell has dissipated, and the new cell has become strongly electrified. For both times, the contour interval is 0.6 nC m^{-3}. Solid contours indicate positive charge, and dashed contours, negative charge. The thick, black line indicates the level at which the polarity of noninductive charging reverses. (Adapted from Randell et al. 1994, with permission.)

the maxima in E produced by simulated thunderstorms also tend to vary with altitude, similar to the observed relationship reported by Marshall et al. (1995).

Modeling studies suggest that lightning may do more than simply neutralize charge and limit E. A lightning flash in a storm modeled by Helsdon et al. (1992), for example, created high enough charge densities of the opposite sign around the lightning channel that it reversed the net charge polarity in that region. This result appears plausible, but the magnitude of the effect obviously is sensitive to the amount of charge on the lightning channel and the volume over which the charge is distributed. Helsdon et al. estimated the line charge density on a lightning channel and distributed 65% of the charge within 400 m of the main channel and 99% within 600 m. If the charge were distributed over a larger volume (perhaps because of dendritic lightning branching) or if shielding by nearby branches in the lightning channel reduced the charge induced on

LIGHTNING CHARGE REDISTRIBUTION IN MODEL

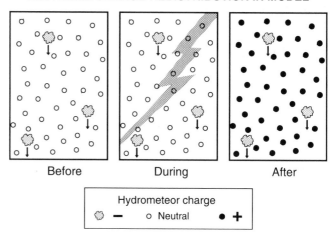

Fig. 9.16. Schematic illustration of the redistribution of space charge caused by parameterized lightning in a cloud model. Scalloped, closed curves denote graupel/hail particles; circles denote cloud droplets or cloud ice particles. Panels show conditions before, during, and after simulated lightning. (From Ziegler and MacGorman 1994, with permission.)

a given segment of a lightning channel, the magnitude of the effect at a grid point on the channel segment would be reduced. If charge were distributed over a smaller volume, however, the effect would be larger. There are no data on the volume over which charge is distributed, and no one has analyzed quantitatively how branching affects the line charge density of the channel or determined how to parameterize branched lightning structure realistically as a function of thunderstorm parameters.

Ziegler and MacGorman (1994) modeled another effect, which occurred when lightning neutralized a charged volume in which the dominant charge was carried by a relatively few large particles imbedded in a multitude of smaller particles. When lightning was assumed to neutralize charge by releasing oppositely charged ions that were captured by hydrometeors, most of the lightning charge was captured by the smaller particles, because smaller particles were much more numerous and so had a larger total cross section (Fig. 9.16). The net charge on the larger particles retained the same polarity. The resulting situation was similar to what occurs just after graupel-ice collisions in which particles become charged: Larger particles in the volume being neutralized had one sign of charge, while smaller particles had the opposite sign of charge. Shortly after a lightning flash, these charges did not contribute to the electric field, because they were mixed together and offset each other. However, subsequent differential sedimentation and advection of the two categories of particles separated the positive and negative charges, which could then contribute to electrification. Note that this process will not be effective if (1) the total cross section of the dominant charge carrier in a given volume is much larger than that of the background of uncharged or oppositely charged particles, or (2) the terminal fall speeds of the oppositely charged particles are essentially the same.

10

Electrical Effects on Cloud Microphysics

The connection between electricity and the formation of rain has long been a topic of interest. Based on his experiments and observations with drops in a vertical jet of water, Lord Rayleigh (1879) stated that "we have every reason to suppose that the results of an encounter will be different according to the electrical condition of the particles, and we may thus anticipate an explanation of the remarkable but hitherto mysterious connexion between rain and electrical manifestations." The evolution of study in this area has not been constant over time, and it has seen periods of reduced emphasis. For example, comparatively little new work has been done since the mid-1980s. This is puzzling because hydrometeor spectra, evolution, and so forth are a fundamental issue in predicting the evolution of clouds and precipitation (see related discussion in review by Beard 1987 and Cooper 1991). Numerical cloud models have become increasingly sophisticated and describe quite well many observed attributes of clouds. For the most part, however, electric forces have not been incorporated into this development. The focus of this chapter is on examining whether electric forces are important in the microphysical development of thunderstorms and in atmospheric chemistry.

10.1. RELEVANT CHARACTERISTICS AND TERMINOLOGY FOR HYDROMETEORS

As we are interested in the force of attraction (or repulsion) between two water particles, we need to know

that even distilled water has so short a relaxation time (10^{-4} s) that it can be treated as a conductor. A dielectric sphere behaves much like a pure conductor once the *dielectric constant*, κ, is > 80 (Grover 1976a). For liquid water, $\kappa \approx 88$, and for ice at 0°C κ is between 92 and 105, so both behave as conductors relative to electric forces between them. In theoretical calculations, water drops of diameter <1 mm behave as conducting spheres. Some theories use the Reynolds number and others use particle radius or diameter. The *Reynolds number, Re,* is the ratio of inertial to viscous forces and is defined by

$$Re \equiv \frac{v_T r}{\nu}, \qquad (10.1)$$

where v_T is the free stream (i.e., terminal) velocity of the particle, r is particle radius, and ν is the *kinematic viscosity* coefficient. It may be helpful to know that for $Re \lesssim 130$, which corresponds to a sphere of $r \leq 500$ μm, the flow can be considered as being past a rigid sphere; $Re > 1000$ corresponds to $r > 1000$ μm. Details of flow and Reynolds numbers may found in texts such as Pruppacher and Klett (1978).

In reading the literature, you will likely see inconsistent use of several terms relating to hydrometeors. We try to standardize here; we use drop for nonspecific statements describing liquid hydrometeors; we use droplet and drop when comparing different sizes of drops—droplet is for the smaller. To describe drop radii we will use the following: r is the undistorted drop radius if only one drop is being considered, and r is the droplet radius if more than one size is considered; R is

337

the radius of the larger of two drops; r_0 and R_0 are the initial radii before any shape deviation from a sphere or growth occurs. Often we find hydrometeor size expressed by its diameter, d, though the equations use the radius. For distorted drops, the axis ratio is important in theoretical treatments. Both a/b and b/a are used for the axis ratio, where a is the semimajor axis of an oblate spheroid, and b is the semiminor axis. (Which variable is applied to which axis is not consistent in the literature). As they fall, drops of diameter ≤ 280 μm are spheres, and larger drops of diameter ≤ 1 mm are approximately spherical. At d = 1 mm, the axis ratio b/a ≈ 0.98, as reviewed in Pruppacher and Klett (1978; p. 311).

The description of the force balance for a hydrometeor falling in a cloud is

$$m \frac{d\vec{v}}{dt} = m \left(\frac{\rho_W - \rho_A}{\rho_W} \right) \vec{g} + \vec{F}_D + q\vec{E}, \quad (10.2)$$

where m is the hydrometeor mass, v hydrometeor velocity, ρ_W the density of water, ρ_A the density of air, and $\vec{g}$ the acceleration of gravity. The first term on the right is the gravitational force. $\vec{F}_D$ is the drag force. The last term is the electric force, where $\vec{E}$ is the ambient electric field (not the field at the hydrometeor surface). The magnitude of the drag force can be written as

$$F_D = 6\pi r \eta v_T$$
$$= \frac{1}{2} \rho_A v_T^2 A C_D, \quad (10.3)$$

where A is the cross sectional area, η is the *dynamic viscosity* of air, and C_D is the *drag coefficient*. When a hydrometeor is at terminal velocity, all forces are balanced, and the term on the left side of Eq. 10.2 is zero. When drops are not severely deformed, A is the geometric area of the drop, which we use in subsequent equations.

Electrical effects on microphysics include those resulting in and from interactions between hydrometeors, which can in large part be addressed by examining collision, coalescence, and aggregation. We turn to the concept of collision efficiency; the often used symbol for this and other efficiency parameters is E, but we use $\mathcal{E}$ to avoid redundancy with the electric field, E. The *collision efficiency,* $\mathcal{E}_{colli}$, can be defined in terms of the drop and droplet radii by

$$\mathcal{E}_{colli} \equiv \left(\frac{x_{crit}}{R + r} \right)^2, \quad (10.4)$$

where the distance, x, is the separation between drop centers and is called the *impact parameter.* The critical value of the distance, x_{crit}, is the minimum value of the impact parameter for a grazing collision. Thus x_{crit} is the largest radius of a circle, through which the trajectories of the smaller droplet (of radius r) pass at right angles and result in a collision with the larger droplet (of radius R). The fraction $x_{crit}/(R + r)$ is the *linear col-*

lision efficiency, which is sometimes used. In some papers, the $\mathcal{E}_{colli}$ is defined just as

$$\mathcal{E}_{colli} \equiv \left(\frac{x_{crit}}{R} \right)^2, \quad (10.5)$$

Unless noted specifically, we will report results using Eq. 10.4. If charged drops and the electric field are involved, the expressions become more complicated, as we shall see.

The simplest definition of *coalescence efficiency* is that it is the fraction of all colliding drops that merge (coalesce) into one. Sartor (1954) defined coalescence efficiency as the percentage of smaller droplets that, on a collision trajectory with the larger drop at far distance, coalesce. *Far distance* is the minimum distance at which there was not any interaction between drops. Although we use it infrequently here, another parameter is the *collection efficiency.* It is defined as the fraction of drops initially on a collision course with another particle that collide and then stick together. For drops, it is the product of collision and coalescence efficiencies. *Aggregation efficiency* is the same as coalescence efficiency, but for ice particles.

Latham (1969a) reviewed the state of our understanding of electrical effects on the microphysical state of clouds. He broke the effects into two regimes: long-range forces and short-range or localized forces between drops. The long-range force creates pressure on the surface of drops and affects motion of particles. He suggested that the surface pressure force can cause drop disintegration, which in turn would modify the droplet population. Also in this category are the phenomena associated with droplets that become highly charged by nearby lightning breakdown. They possibly could move rapidly into the adjacent cloud and increase the growth of precipitation. All these phenomena require very strong E and likely are not operative in early cloud development. In the short-range regime, he included modification of collision efficiency of droplets and solid hydrometeors, alteration of coalescence efficiency of droplets, modification of ice particle aggregation, and the process of electrofreezing. In a subsequent review at an international conference in 1974, Latham (1977) concluded that a variety of electrical mechanisms exist to influence cloud microphysics in highly electrified situations, but there was no known process by which electric forces could exercise a crucial influence on the formation and growth of hydrometeors before a highly electrical state existed.

As we read the papers on numerical cloud modeling, we are struck by the lack of constraints from in situ measurements in assessing effects from large electric fields in thunderstorms. We now look at specific interactions. We will consider the evidence, almost entirely from theoretical and laboratory studies, for electrical effects on terminal fall speed, drop disruption, collisions, coalescence, electrofreezing, growth of ice, aggregation, and heavy precipitation.

10.2. MAXIMUM CHARGE POSSIBLE ON A DROP

There is a maximum charge that a drop can have before it disrupts from mutual repulsion caused by an electric surface stress that equals the surface tension stress. This condition is called the Rayleigh charge limit and comes from the experiments by Lord Rayleigh (1882). It can be derived from

$$E_d \sigma_q = E_d \left(\frac{q}{4\pi r^2} \right), \qquad (10.6)$$

Where E_d is the field at the drop surface, and σ_q is the surface charge density. From this, we get

$$E_d = \frac{2\zeta}{r}, \qquad (10.7)$$

where ζ is the surface tension, and the whole right-hand term is the surface tension force density (since $F = qE$, $E\sigma_q$ is force per unit area). The electric field tending to disrupt the drop is

$$E_d = \frac{1}{2} \left(\frac{q}{4\pi \varepsilon r^2} \right). \qquad (10.8)$$

When the repulsion just equals the surface tension, it follows that

$$q^2 = 64\pi^2 \varepsilon r^3 \zeta, \qquad (10.9)$$

and thus the *Rayleigh charge limit*, q_{Ra}, for a drop is

$$q_{Ra} = 8\pi(\varepsilon \zeta r^3)^{\frac{1}{2}}. \qquad (10.10)$$

10.3. CHANGES IN DROP SHAPE AND TERMINAL FALL VELOCITY

Drop shape is tied to terminal fall velocity because changes in shape will change some forces on the drop. With the arrival of multiparameter, dual-polarization radars, the interest in drop shape has increased owing to the different reflected signals from the same distorted drop. A few researchers have tried to develop theoretical models to include all the forces on drops to predict with greater accuracy the drop shapes and to improve rainfall estimates from dual-polarization radars. Because the drop size spectrum and thus rain rate are inferred from polarization radar measurements using drop shape constraints, an assessment of the electrical effects on drop shape is important. Reviews are found in Doviak and Zrnić (1993). Charged drops will have an electric, and also a gravitational, force acting to affect their terminal velocity. Under the approximation of spherical drops, this condition is one of equilibrium between the aerodynamic drag force and the combination of gravitational and electrostatic forces. When dv/dt = 0 (i.e., at terminal drop velocity), Eq. 10.2 becomes

$$\frac{1}{2} \rho_A v_T^2 C_D \pi r^2 = -m \left(\frac{\rho_W - \rho_A}{\rho_W} \right) g \pm qE. \qquad (10.11)$$

Whether the last term is positive or negative depends on the direction of the electric force. Now we examine studies of electrical effects on drop shape. Dawson and Warrender (1973) found the change in terminal velocity of uncharged drops of diameter 3.1–4.2 mm to be 0.1 m s^{-1} per 100 kV m^{-1} of applied vertical field. They concluded that the changes in terminal velocity in thunderstorms would be controlled by pressure (altitude) changes in the storm as opposed to changes in E. The effect on charged drops was not addressed.

Gay et al. (1974) calculated the effect on terminal velocity of E and q. To test their theory, they made measurements of the change in terminal velocity for two ranges of charged drops of radii 24–78 μm and 1.25–1.8 mm. The larger drops were examined in a wind tunnel in which they could be suspended in a vertical electric field. The change in terminal velocity depended on both the drop charge and E. For the smaller droplets of radius 24 μm, the change in v_T was about 0.1 m s^{-1} per 100 kV m^{-1}, in agreement with Dawson and Warrender. By a radius of 48 μm, there was slight deviation of experimental data from theory; by a radius of 78 μm, the deviation was substantial. The theoretically calculated terminal velocities were about 15–20% less than the measured ones, but otherwise the agreement was good. Their laboratory data showed that for a constant q ≈ −160 pC, a 1.5-mm radius drop changed its terminal velocity by ≈0.6 m s^{-1} per 100 kV m^{-1} for |E| = 0–300 kV m^{-1} (Fig. 10.1). For raindrops with large q, the effect of a high but reasonable value for a thunderstorm |E| = 200 kV m^{-1} was to alter v_T by ≤1 m s^{-1} (Fig. 10.2).

Beard (1980) developed an adjustment to the relationship for terminal velocity of drops used by Gay et al. (1974) that extended their relationship for any hydrometeor shape to different altitudes and electric fields. Beard's adjustment was a function of size, the density of air, charge, and electric field, and of the form $v = fv_0$, where the known velocity, v_0, was multiplied by a factor, f, to get an *adjusted velocity*. He found at E = 100 kV m^{-1}, a 32-μm diameter drop was levitated and a 300-μm diameter drop had its terminal velocity changed by 8%. Beard concluded that the velocity of smaller drops in thunderstorms would be significantly affected by the electric forces. His adjustment factor was divided into groups defined by Reynolds numbers of ≤0.2, 0.2–1,000, and >1,000, so the technique could also be applied to ice crystals, aggregates, graupel, and drops deformed in an E.

Beard et al. (1989) used a perturbation model to calculate terminal velocity. They compared their theory with the limited wind tunnel tests available, but noted that those tests could have been affected by turbulence. They used an equilibrium drop shape obtained from high-speed photographs of the slightly oscillating drops. Beard et al.'s theory and experiment agreed well for drops with diameters ≤3 mm. For larger drops, they concluded that their theory underestimated drop distortion. They examined the effects from drops in three

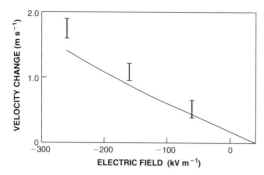

Fig. 10.1. Change in terminal fall velocity of a drop of r = 1.5 mm and q = −160 pC as a function of the magnitude of E. The solid line is a theoretically predicted value. (From Gay et al. 1974, with permission.)

situations: (1) no drop charge but in a vertical electric field, (2) charged drops but no field, and (3) charged drops in a vertical electric field. They found a negligible direct effect of drop charge, as the ratios of the drop charge to the Rayleigh limit, q/q_{Ra}, were always quite low. The reason the ratio was important was a result of their use of *internal drop pressure*, P_i, expressed as

$$P_i = \frac{2\zeta}{r_0} \left(1 - \frac{q^2}{q_{Ra}^2} \right). \qquad (10.12)$$

The effect of drop charge could thus be approximated by replacing surface tension, ζ, with $\zeta(1 - q^2/q_{Ra}^2)$. One shortcoming they noticed was that their model did not incorporate the electric force on a net charge embedded in E. Beard et al. found a tendency for large drops to elongate in a vertical electric field, but this tendency was partially offset by increased aerodynamic distortion from the greater fall velocity. They felt that

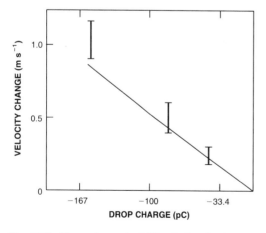

Fig. 10.2. Change in terminal fall velocity of a drop of r = 1.5 mm embedded in an |E| = 200 kV m⁻¹ versus drop charge. The solid line is a theoretically predicted value. (From Gay et al. 1974, with permission.)

the most likely condition for an electrical effect was just before lightning.

Beard et al. noticed that the combined effect of E and drop charge also resulted in shape changes, but only through the change in fall speed resulting from the force, qE. Shape changes would be more dramatic in areas of evaporation and in areas of large horizontal E. In the former, presumably the drop charge would become a larger fraction of the Rayleigh limit, and the latter would increase horizontal stretching of the drop and thus enhance the aerodynamic flattening. They also thought that the effects might be noticeable in radar data of drops in the large E prior to lightning. Implicit was that the effect on drops would exist only where the E was very large, and thus the effect would not be uniformly distributed in the storm.

Partially to get around the inadequacy of the perturbation model to describe larger diameters, Chuang and Beard (1990) developed a new model to predict drop shapes and thus improve rainfall estimates from polarization radar data. Their purpose was to determine drop shape with the inclusion of all relevant factors, which they listed as surface tension, hydrostatic pressure, dynamic pressure, and electrostatic stress. Even in the extreme, the maximum axis ratios for 1–6 mm diameter drops were 1.49–0.93 in vertical E of 1900–1170 kV m⁻¹, respectively. The increase in fall speed for a 5-mm diameter drop was >30% at 1200 kV m⁻¹. They found different shapes under upward and downward electric forces owing to altered fall speed. They found that an upward electric force decreased the aerodynamic distortion and caused an enhanced vertical stretch of the drop, which in turn meant that large raindrops could become unstable. The authors noted that for the much smaller, commonly observed magnitudes of E, the effect on shape was small. Because axis ratios contribute substantially to errors in polarization radar determination of rainfall rate, they suggested that an assessment of errors might be facilitated through use of model-predicted raindrop shapes.

Coquillat and Chauzy (1993) used a different approach with pressure balance on the drop and developed a simple model of water drops in an electric field. They gave their model's useful range as r = 0.5–2.5 mm. To account for the distorted drop shape, they used a cross sectional area A = πb² where they defined b as the distorted drop equatorial radius. Coquillat and Chauzy defined an axis ratio for a drop as the average of the two vertical elongations of the two half ellipsoids used to approximate the shape of a distorted drop. They concluded that vertical drop stretching caused an increase in the terminal velocity, as the distorted drop equatorial radius was reduced. When E tended to increase v_T, the aerodynamic pressure on the drop increased the distorted drop radius that in turn resulted in increased drag and decreased fall speed. Therefore, the increase in v_T from increasing E was self-limiting. They divided their tests into three scenarios of E and q: (1) When E = 0 and |q| > 0, there will be a negligible effect in thunderstorms, though the change in shape

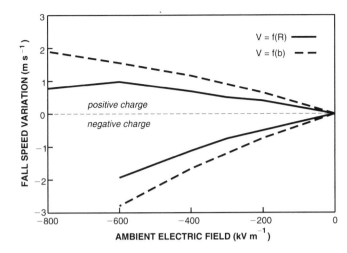

Fig. 10.3. Theoretical curves of changes in drop fall speed versus vertical E for drops of r = 2.5 mm and |q| = 640 pC. The dashed curves are from use of an undistorted drop radius, and the solid curves from use of the equatorial radius of a distorted drop. (After Coquillat and Chauzy 1993: American Geophysical Union, with permission.)

due to high q is not negligible; (2) when $|E| > 0$ and q = 0, the change in velocity $|\Delta v| \approx 0.1$ m s^{-1} per 100 kV m^{-1}, in agreement with Dawson and Warrender (1973). Coquillat and Chauzy stated that the Δv should not be neglected for accurate modeling of drops; and (3) when $|E| > 0$ and $|q| > 0$, the Δv was reduced by the distortion of the drop. For example (Fig. 10.3), when E = -200 kV m^{-1}, r = 2.5 mm, $\Delta v = -0.5$ m s^{-1} for a negative q and $\Delta v = +0.4$ m s^{-1} for q = $+640$ pC. Over the range of typically measured E in thunderstorms, the change in velocity would be $\lesssim 1$ m s^{-1}. The result of using the equatorial radius instead of the undisturbed one is to increase the change in v_T from E. They noted that in thunderstorm E = 200 kV m^{-1}, the fall velocities are quite different from the "standard" fall velocities for uncharged drops in zero field measured by Gunn and Kinzer (1949). They also concluded that the combined effect of E and drop q strongly affected the terminal fall velocity. Finally, we note that the values of drop charge used (e.g., 640 pC in Fig. 10.3) tended toward realistic, but were larger than the usual drop charge observed in storms. Georgis et al. (1995) modified the earlier model of Coquillat and Chauzy (1993) to include aerodynamic pressure, which led to oblate drops. They also showed that a small air gap between two drops is not necessarily equivalent to a large ambient E. They concluded that their advancement was only a step toward understanding corona initiation from a pair of drops in an E.

10.4. DISRUPTION OF DROPS

If drops change their shape enough, they can disrupt. Because of changing shape, v_T is affected. Furthermore, a *critical electric field*, E_{crit}, can cause the electric force to tear the drop apart by overcoming the forces holding the drop together. Just before disruption is the point of instability at which a spheroidal drop becomes conical. Latham and Roxburgh (1966) calculated that instabil-

ity for two closely spaced drops was different from a single drop. They found that very closely spaced drops in an E will distort to the point of disintegration even in weak external forces. This occurs from drop distortion due to polarization charges. The penetration of the air film between the droplets accounts for the disintegration and may increase the likelihood of a subsequent coalescence.

Richards and Dawson (1971) placed uncharged and charged drops in a vertical E in a wind tunnel. For charged drops, the polarity of the charge was chosen to lower the onset E threshold of drop instability and decrease fall speed. The minimum |E| for instability was 700 kV m^{-1}. That was for large drops in the air stream several minutes, and some have suggested that turbulence may have been involved in reducing the minimum E for instability. The instability was from the upper surface of the drop. The |E| for instability decreased with increasing drop charge. The charged drops first became unstable at the top surface (as did the uncharged drops), but then an instability occurred at the lower surface. The drops then elongated into a cone with the maximum point lasting <2.5 ms. Then the drop's upper surface collapsed, and the drop accelerated downward against the air flow. This resulted from their being highly charged after instability. Sometimes larger drops ejected droplets from their ends. Richards and Dawson thought that their laboratory results gave a lower bound for this instability field in the atmosphere. They reaffirmed the earlier work of Dawson and Richards (1970) that had shown large uncharged drops disrupt on their top.

Zrnić et al. (1984) conducted theoretical investigations regarding the change in drop shape due to the ambient electric field. To assess the importance of drop charge in changing oblateness, and thus the use of polarization diversity radars to determine precipitation type, they chose a very large drop charge of 900 pC, whatever the radius, to highlight any effect of charge on the differential reflectivity, Z_{DR}. They extended the formula describing drop shape in the absence of an electric field from Green (1975) to include an electric stress at the drop equator. They concluded that according to their simple theory, drop shapes would change insignificantly in commonly observed electric fields, although they noted that large charges could change the oblateness of the drop. Zrnić et al. restricted the applicability of their theory to drops of d < 2.3 mm and noted part of the observed change in Z_{DR} could be from electrical effects.

Rasmussen et al. (1985) investigated the effects of a vertical electric field on the shape of uncharged drops in a wind tunnel. They said that the drop shape at terminal velocity is the result of complicated interactions among aerodynamic, surface tension, hydrostatic, and electric forces. They thought that earlier studies had been limited by not having the interaction between electric forces from the external field and aerodynamic forces from flow around and inside the drop. For example, Zrnić et al. (1984) used Green's (1975) model that neglected hydrodynamic flow around and inside a falling drop. The predicted drop shapes would be incorrect when $r_0 > 1$ mm. They noted that Zrnić et al.'s prediction of drop shape strongly disagreed with those observed by Richards and Dawson (1971). The wind tunnel tests by Rasmussen et al. showed that the aerodynamic flow, and thus force around the drop, changed continuously as |E| increased. They felt their experiment was a realistic depiction of storms, as E acts on oblate deformed drops around which the aerodynamic flow changes constantly. They conducted their experiments at T = 15°C and P ≈ 1000 mb using equivalent radii of a_0 = 1–3 mm and a variable vertical E = 0 to −900 kV m^{-1}. For drops of equivalent radii, a_0 < 2.1 mm, and |E| < 400 kV m^{-1}, the aerodynamic and hydrostatic forces dominated the electric one. The electric forces began to offset and take over from aerodynamic and hydrostatic forces for a_0 > 2.1 mm. When

|E| > 400 kV m^{-1}, the electric forces dominated the other forces and the resultant drop shape for all drop sizes. However, when Rasmussen et al. extrapolated their results to smaller radii, they reported that for $a_0 \leq$ 1 mm, all |E| < 700 kV m^{-1} had no effect on their shapes. This was a result of the surface tension force dominating all other forces. The effects of E on drop shape denoted by its axis ratio are shown in Fig. 10.4. They found that the drop breakup in an E occurred from instabilities on top of the drop. Rasmussen et al. also noted the difference between this breakup of a single drop and that for a drop in E = 0, where the instability formed at the bottom of the drop. For a_0 = 2.5 mm, the drop deformation was nearly linear for |E| = 0–300 kV m^{-1}; it was only slightly less linear for |E| = 300–800 kV m^{-1} (Fig. 10.5).

Chuang and Beard (1990) extended their earlier model to include electric forces. They said that because there was no evidence of their model inducing more stability than is real, their model of E_{crit} = 1170–1350 kV m^{-1} for drop disruption should be considered a maximum—in contrast to the results of Richards and Dawson (1971), whose experiments yielded slightly lower limits. They also concluded that the difference in the experiments of Rasmussen et al. (1985) was likely due to drop oscillations and turbulence in their wind tunnel. Chuang and Beard's theory showed that electrically distorted drops had a highly nonlinear shape dependence on E. The instability of large drops was nearly independent of drop size and consistent with the experiment by Richards and Dawson. Chuang and Beard concluded that the E_{crit} for instability was about 1200 kV m^{-1}. Applicable to effects on dual-polarization radar measurements was the distortion of drops in strong vertical E tending toward a triangular shape in vertical cross section, again agreeing with laboratory experiments.

Wind tunnel experiments with an E = 0 by Kamra et al. (1991) showed that when a drop carried a charge

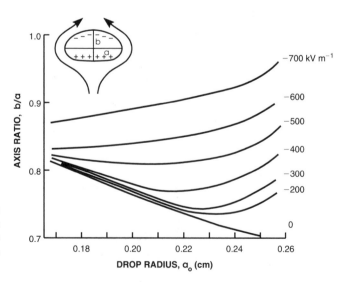

Fig. 10.4. Axis ratios of uncharged drop versus equivalent drop radius, a_0, as a function of ambient E in which the drop is embedded. (After Rasmussen et al. 1985, with permission.)

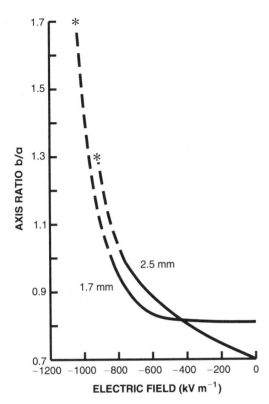

Fig. 10.5. Dependence of drop axis ratio, b/a, upon the ambient E for two uncharged drops of equivalent radius a_0 = 2.5 and 1.7 mm. The dashed portions are extrapolations of the present data to the critical breakup values (*) from Richards and Dawson (1971). (After Rasmussen et al. 1985, with permission.)

of 50 pC, its lifetime at terminal velocity was not significantly affected by an increase in size up to about 8 mm. (Drops this large are not common, but do occur, in nature.) If the drop charge was 500 pC, large drops of diameter 6.6–8 mm had reduced lifetimes before breakup as the drop size increased. For drops of diameter 8.2 mm and q = 0, there was little difference in lifetime from drops of 8 mm and q = 50 pC. They concluded that drop breakup is a random process. An important difference between small spherical drops and larger oblate spheroidal drops is that in the latter the charge is concentrated at the point of maximum curvature on its horizontal rim. The surface tension force is less able to counteract the aerodynamic force there. Furthermore, the curvature keeps changing from the drop oscillations as it falls. At the time the oscillation stretches the drop to its largest extent is also when the surface charge density is the highest at the horizontal extremity. Thus, the two forces may enhance each other. The net effect of drop charge opposing the surface tension that holds the drop together may be to elongate it in the horizontal direction and thus increase

its base. So a charged drop may break up sooner than an uncharged drop when initially they were the same diameter. The possible applicability of these conclusions to storms may be limited to those regions where E is very small.

Kamra et al. (1993) published the first experiments with initially uncharged drops falling at terminal velocity in a horizontal electric field, E_h. They used distilled water and a grounded conductor in the water near the place of drop formation to produce uncharged drops (no confirming charge measurements were made). For drops of diameter <6.6 mm, there were never any drop breakups if E_h = 0. If E_h > 0, such drops could break up. By suspending the drops in a wind tunnel, they observed that the drops are affected by the E_h differently as they rotate about their vertical axis. They reported that the drop deformation changed as it oscillated and rotated. Thus the E for the onset of corona depended on the mode of oscillation; this resulted in a lower minimum E_{crit} than previously reported, which Kamra et al. attributed to their experimental setup in which the drop was in the E_h for a long enough time for the different modes of oscillation to occur. They postulated that the drop rotation about the vertical axis in an E_h could result in different corona onset and breakup thresholds for the same initial drop size owing to changing deformation as it fell. Thus, measurements of E_{crit} will have a spread in values. They reasoned that in thunderstorms where both E_h is large and big drops are located, breakup could modify the drop size distribution of drops and droplets. Their lowest values of E_{crit} were for drops with d > 8 mm, where E_{crit} = 200 kV m^{-1}. However, they cautioned that the value in thunderstorms could be different. Errors in their E_{crit} may have arisen from turbulence in the airstream from a required back-pressure plate in the flow. This has recently been supported by Georgis (1996) who used freely falling drops at terminal velocity. His drops were only in the E_h for about 44 ms. He obtained higher values of the onset field for instability than those reported by Kamra et al., but lower than those in previously reported experiments in which the drops did not fall at terminal velocity. Georgis stated his concern that only the turbulence in the wind tunnel or the charge acquired because of its longer stay there could be the reason for their escape from the airstream and for the lower E for onset of instability noted by Kamra et al. (1991). Thus, the drop instability could have resulted from amplification of drop oscillations owing to the turbulence. However, Kamra et al. had referenced analysis by others that indicated the turbulence was not significant because it occurred at frequencies different from the drop oscillation frequencies (e.g., as reported in Kamra et al. 1991). Kombayasi et al. (1964) discussed the turbulence and drop oscillations in the free atmosphere versus a wind tunnel. For drops of d > 5.5 mm, drop breakup in a wind tunnel was the same as in the atmosphere. For drops of d < 5 mm, resonance in the tunnel was likely, and the drop lifetime was different from that in nature. Finally we note that Kamra et al. (1991) also showed that the probability of drop breakup increases with charge on the drop, so both the concerns expressed by Georgis have support in previous studies.

10.5. DROP NUCLEATION AND ELECTROFREEZING

As reviewed in Pruppacher and Klett (1978, p. 606), ions affect nucleation of drops in large values of supersaturation, but in the atmosphere such effects do not occur under ordinary conditions because the high values of supersaturation required do not exist. Drop nucleation thus has yet seen to be affected by electrical conditions.

Electrically induced changes in the nucleation of ice are not so easily ruled out. Pruppacher (1963a) performed laboratory experiments to see if the freezing of droplets could be influenced by an electrical effect—called *electrofreezing*. He decided that although it occurred, it was not an important process in the atmosphere. Smith et al. (1971) concluded from laboratory experiments that drop disruption in an E helped ice nucleation whether the drop was on a support or falling freely. The presence of a spark during disruption did not, however, affect the nucleation process. Cavitation seemed to them to be the most likely reason for the freezing. Thus a process that produced cavitation, whether from mechanical or electric forces, would facilitate freezing. Pruppacher (1973) reviewed the studies on electrofreezing; he noted that the necessary conditions were in thunderstorms where there already were large numbers of ice crystals. Thus, Pruppacher concluded that electrofreezing would have a negligible effect on overall precipitation, except possibly if it aids in breakup of drops into liquid and frozen fragments and results in a local rain gush.

Dawson and Cardell (1973) noted that the earlier work had left unanswered the question of whether freezing can be induced just by the electric field. They tried to avoid conditions that were unrealistic for the atmosphere and that had been covered in previous work. They used deionized drops of about 3-mm diameter in a wind tunnel at an altitude of 800 m MSL with temperatures of -8 to $-15°C$. Dawson and Cardell did not observe electrofreezing in $E \leq 400$ kV m^{-1}, even in situations of colliding uncharged or charged drops or in the presence of intentionally produced corona in their test chamber. They concluded that electrofreezing does not occur naturally in thunderstorms, with the possible exception of during large E transients from lightning, which they did not specifically study. In their brief review, Doolittle and Vali (1975) noted the importance of finding an answer to the question of whether the electric field can influence ice nucleation. They did experiments on a group of 121 drops with 1.3-mm radius placed in an $E = 600$ kV m^{-1}. They did not observe heterogeneous freezing nucleation. A difference from earlier experiments was that arcing of the drops did not occur, except inadvertently in one case, and that one

drop froze. They also concluded that electrically induced heterogeneous freezing nucleation is unlikely in natural clouds.

In contrast to these negative effects on nucleation, Tinsley and Deen (1991) offered a new possibility and outlined needed research—electrofreezing in which the charging of supercooled droplets high in the troposphere leads to ice crystal formation. They hypothesized the source of the charge as the flux into the atmosphere from above of high-energy, MeV-GeV (megaelectron volts-gigaelectron volts) particles. These particles are mostly from cosmic rays. If the hypothesized electrofreezing were to occur, it could be fundamental to the formation of seeder ice crystals, which in turn change precipitation, latent heat release, and so forth. The process would then be a tropospheric thermodynamic effect resulting from MeV-GeV particles. Their conceptual model proposed the initial change in the production rate of larger ice crystals, which in turn could gravitate and glaciate midlevel clouds. They proposed, and offered climatic evidence in support of, warm-core winter cyclones being intensified by the process. They suggested several specific experiments that could be conducted to test elements of their hypothesis.

10.6. SCAVENGING OF AEROSOLS

Scavenging of aerosols is the collection by larger particles of aerosols, which range from fractions to a few tens of microns in diameter. (Some references place the upper limit at about 1 μm, but there is no accepted single maximum.) The scavenging of aerosols by small droplets and crystals depends on forces related to Brownian diffusion, thermophoreses, diffusiophoreses, inertia, and electricity (e.g., see Young 1993, pp. 83–88). Aerosols are relevant in cloud processes because they often serve as nuclei and because precipitation scavenges them out of the atmosphere. A few studies have looked at the collection of aerosol particulates. First we consider two studies in which aerosol sizes ranged over radii from 0.001–10 μm. In these studies, the collision efficiency was defined by Eq. 10.4 and was assumed to equal the collection efficiency. The latter resulted from the assumption that if an aerosol collides with a droplet, it would always be captured. All studies summarized here have included the same relationship for charge on the particles of

$$|q| = 6.6 \times 10^{-7} r^2 \qquad (10.13)$$

for nonthunderstorm conditions. Here, r is in meters and q in coulombs, but often in the literature similar equations are in centimeters-grams-seconds (cgs) units of size and charge. Frequently, this equation is expressed as $|q| = 0.2 r^2$ with all variables in cgs units. For particle charges inside storms, these studies incorporated Takahashi's (1973) compilation of his and other's work:

$$|q| = 6.6 \times 10^{-6} r^2, \qquad (10.14)$$

which is an order of magnitude greater charge. It appears that for thunderstorms, this relationship approximately describes (it overestimates) the few data points on Takahashi's curve for cloud particles of $r \approx 5$ μm measured in a thunderstorm by Colgate and Romero (1970).

Grover et al. (1977) theoretically examined the effect of the electric field and charge over the range of aerosol radii of 0.5–10 μm being collected by water drops of radii 42–438 μm that fell at their terminal velocity. They noted the effects for $|E| > 0$ to 300 kV m^{-1}, but only for uncharged drops and aerosols. When charged, the drop and aerosol always had opposite polarities, with charges given by Eqs. 10.13 and 10.14 and were in an $E = 0$. Example results from their particle trajectory model are in Fig. 10.6. The decrease in collision efficiency (curves 2a–c) occurred even for the

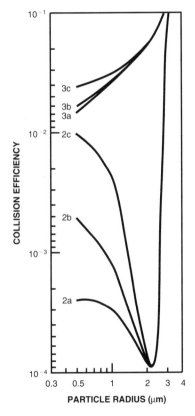

Fig. 10.6. Calculated collision efficiencies for drop of R = 42 μm at terminal velocity interacting with aerosols of bulk density 2×10^3 kg m^{-3} different radii in air of T = 10°C and P = 900 mb. Curves 2a–c are for equal, attractive charges each with surface charge density of 6.6×10^{-7} C m^{-2} in an $E = 0$. The RH = 100%, 95%, and 75% for curves 2a–c, respectively. All conditions for curves 3a–c are the same, except the surface charge densities were 6.6×10^{-6} C m^{-2}. (After Grover et al. 1977, with permission.)

smaller charge densities (derived from Eq. 10.13) in $E = 0$. The curves are very similar to the case of no electric forces, except collision efficiencies are slightly greater throughout. The minimum in the curves was from the interdependence of phoretic, hydrodynamic, and electric effects on aerosol trajectories. When the charge was increased to thunderstorm values using Eq. 10.14, the minimums tended to be eliminated in the curves (e.g., curves 3a–c in Fig. 10.6). Although the larger particle charges may be representative, an E = 0 is unlikely in most places in thunderstorms.

Overall, collision efficiency increased under any of these: greater ambient E, smaller collector drop for a given aerosol size, and the larger the charge on the drop and aerosol. Grover et al. concluded that there was no collection efficiency enhancement if the charge was only a tenth that calculated for particles in thunderstorms. However, for relative humidity of 50% and for thunderstorm charges, they calculated collection efficiencies up to a hundredfold over those considering only inertial impaction. At higher humidity, the increase was smaller. They found as R increased from 40 to 100 μm that collision efficiency decreased because hydrodynamic forces counteracted the phoretic and electric forces and swept the aerosol around the drop. For R > 100 μm, the efficiency increased with radius.

Wang et al. (1978) extended the work with a complementary model that treated aerosol radii of 0.001–0.5 μm and that was based on particle flux rather than trajectories. This model included Brownian diffusion, thermophoresis, diffusiophoresis, and the force due to charge (that from an electric field was not considered). They also considered particle charges to be in the two categories as defined by Eqs. 10.13–10.14. If the charge was only a tenth that expected in thunderstorms, they considered the effect on the collection efficiency as negligible. They combined the model with their earlier one (Grover et al. 1977) to calculate collision efficiencies for aerosols of r = 0.001–10 μm being collected by drops of R = 42–310 μm. They also assumed that all collisions would result in collection, so the model results gave, in effect, collection efficiencies. They noted the lack of comprehensive experimental data to compare with their theory, but such a comparison over the limited range of experimental data available showed good agreement. The modeled efficiencies had negligible change from charge on the interacting aerosol and drop both for r < 0.01 μm and for r > 1 μm. In between (in the Greenfield gap), however, the effect on efficiency was significant with $\mathscr{E}_{colli}$ increased by up to two orders of magnitude at an RH = 50% or if the charges on the aerosol and drop had the assumed thunderstorm values (Eq. 10.14). In contrast, the effect on $\mathscr{E}_{colli}$ was insignificant in the gap if the charges were the nonthunderstorm values.

Pitter (1977) studied the scavenging of aerosols by thin ice plates of semimajor axes 103–366 μm at T = −8°C and P = 400 mb. Attractive electric forces were important for aerosols of r ≲ 5 μm; at r > 5 μm, inertial forces dominated. If aerosols were either neutral or

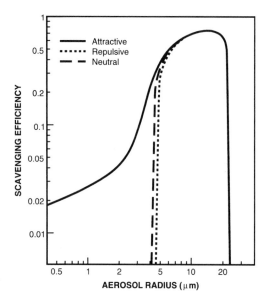

Fig. 10.7. Calculated scavenging efficiencies for ice crystal of R = 193 μm with aerosols of bulk density 2 × 10^3 kg m^{-3} but of different radii. The line codes indicate results for attractive charges, repulsive charges, and neutrals. (After Pitter 1977, with permission.)

the same charge as the crystal, the scavenging efficiency was greater than zero only for r = 2–5 μm, depending on the conditions. Attractive charges of thunderstorm magnitudes (Eq. 10.14) increased scavenging for all cases studied. Example results are in Fig. 10.7. The steep decrease at r = 23–24 μm was independent of charge. The decrease to cutoff at aerosol radii of 4.2 μm and 4.6 μm was for neutral particles and particles charged with the same polarity, respectively.

Laboratory experiments by Barlow and Latham (1983) revealed that the collection efficiency for drops of radii of 270–600 μm to collect aerosols of radii 0.2–1.0 μm increased from 0.01 to 0.1 as the charge on the drops ranged from 10 fC to 10 pC. They noted that the scatter in their data was about an order of magnitude and that the drops of R ≳ 0.6 mm were falling more slowly than their terminal velocity, which could have made their measured collection efficiency too large. Crude calculations of the scavenging of monodispersed aerosols by a charged rain at rates of 2 mm h^{-1} and 10 mm h^{-1} suggested that the electric forces were important.

10.7. CHANGES IN COLLISION EFFICIENCY

We now turn our attention from individual drops to interaction between drops. In his review, Latham (1969a) noted that the growth of droplets of radii >20 μm is

predominantly by coalescence and so the growth rate will be controlled by both collision and coalescence efficiencies. We now look at the evidence for changes in collision efficiencies caused by electric forces. Several investigators addressed the issue of whether the electric field, either external or solely from charge on particles, can affect the collision efficiency. For example, does the repulsive force of like-charged drops reduce collisions—and what happens when the drops are oppositely charged?

Sartor (1954) did an experiment to simulate electrical effects on droplets like those in natural clouds. To do this, he devised a laboratory chamber in which he tried to match the Reynolds numbers of drops in the chamber to those in the atmosphere. He described the basics of precipitation growth in warm clouds and noted that a diameter of about 30 μm is needed before growth by coalescence can occur (without electric forces). He also stated the often used assumption in calculations that the flow is undisturbed by the two drops is invalid when the drops are about the same size. To achieve the same Reynolds numbers, he used water drops in mineral oil. Parallel plate electrodes allowed a vertical electric field to be applied to a region through which the drops fell. In an E = 100 kV m^{-1}, a 7-μm radius droplet colliding with a 10-μm drop had a collision efficiency of 0.54, as compared to 0 based on theory in use at the time. Similarly, a 10-μm droplet colliding with a 15-μm one had a collision efficiency of 0.58, which was twice the theoretical value of 0.27. His observations indicated that drops are "selectively charged" during their movement about each other, and the resulting electric force acted in addition to the electric force from the electric field—both for top and bottom coalescence. In high E, he observed a discharge between the drops. When they were electrically connected via a discharge, they lost their electric attraction and moved apart to a distance as if they were in a field-free region. Sartor's results clearly indicated an electrical enhancement of collision efficiency, but Pruppacher (1963b) pointed out that the extrapolation of these results to natural clouds might not be correct, as the water-oil interface could behave differently from a water-air interface.

Sartor (1960) again examined collision efficiencies. He said that collision efficiency was modified by an electric field. He supported the work of Hocking (1959) that allowed for the interaction of the flows around two spheres that interact, but did not include electric forces. Sartor argued that if drop diameters were ≤60 μm, Hocking's analysis was appropriate for assessing the interaction forces and collision efficiencies. Sartor noted that for charged drops of the same polarity, the initial repulsive force would increase as they approached each other. They could reach a distance at which the force rapidly became an increasingly attractive force as the drops continued to approach each other. So electrostatic attraction can become important if the hydrodynamic forces are great enough to get the interacting drops past the point of the repulsion. For charged drops of different polarities, the force provided by an electric field is very weak at larger separations, but as the drops approach this force increases rapidly and becomes the dominant force at very small separations. Sartor concluded that in the early stage of cloud development, the electric field, not drop charge, was likely the dominant force in enhancing collision. He showed calculations for collisions between drops with radii of 15 µm and 12 µm, and drops of radii 19 µm and 15µm. Some of the calculated E required to increase the collision efficiency were larger than commonly found in the early stages of cumulus growth. However, it is important to remember that a cumulus congestus, which has typically undergone several vertical, grow-and-subside cycles, can grow to produce ice-phase precipitation and generate E adequate for lightning in a very few minutes. Thus, the E required to enhance collisions will certainly be present after the cloud has initially electrified.

Krasnogorskaya (1965) calculated collision efficiencies for charged and uncharged droplets. She noted the complexity of the process (i.e., the way that the electric field influenced the collisions depended on particle size, polarity and magnitude of their charge, and relative location of droplets). The electrical effects became stronger as the size decreased; this was due to the decrease in inertial forces with decreasing droplet mass. She found that the number of colliding droplets increased in the presence of E and charge. For example, small droplets with radii of 6 µm and 10 µm would normally not collide, but if they had a charge even as small as <1 fC, their collision efficiency became 0.3. The same droplets in the presence of $E_z = 120 \, kV \, m^{-1}$ had a fourfold increase in collision efficiency to 1.2. For R = 20 µm and r = 10 µm, the collision efficiency increased fortyfold from 0.13 when q = 0 on both to 5.05 for opposite $|q| = 3.3$ fC. In an $E_z = -60 \, kV \, m^{-1}$, uncharged drops of R = 20 µm and r = 6 µm had a collision efficiency of 0.49, as opposed to 0 in an E = 0. By an $E_z = -120 \, kV \, m^{-1}$, the efficiency was up to 1.95.

Plumlee and Semonin (1965) calculated the collision efficiencies of uncharged drops with and without an external electric field. They, like other researchers, set up their calculations in a coordinate system fixed with the larger drop, and the smaller droplet moved past the larger drop. This is physically equivalent to two drops falling in the atmosphere. The mutual, drop-droplet forces were not considered, so the trajectory of the droplet was governed by Eq. 10.2. Their calculations for droplets of radii 5–15 µm and for drops of radii of 30–50 µm showed that the angle of E and the line of approach of the drops affected the collision efficiency. A horizontal E produced a collision efficiency for the same sizes of drops that was 1.2–1.6 times greater than that in a vertical E; the variation of collision efficiency versus angle of E was nonlinear. The rate of increase in efficiency decreased with increasing size. The reason was that the relative flow velocity of the drops was faster, reducing the time for electric forces to help them

collide. Plumlee and Semonin concluded that for a given droplet size the collision efficiency decreased with increasing size of the collecting drop, despite whether or not an E was present. Similarly, for a given collecting drop size, the collision efficiency increased with increasing droplet size. The maximum effect of E was for a horizontal field, and the minimum effect occurred when E was at 42°. The collision efficiency varied up to 30% as a function of angle for 0–90° in an |E| = 210 kV m^{-1}. They used larger E also, but we limit our discussion to those magnitudes commonly measured in storms. In their calculations, the collision efficiency increased significantly and more rapidly for |E| > 100 kV m^{-1}.

Paluch (1970) developed a series of empirical equations for collision efficiencies of oppositely charged drops in vertical electric fields for three different drop-size regimes. When inertial forces were negligible and trajectories were not altered much by charge on the drops, she found that the collision efficiency, $\mathcal{E}_{colli}$, could be simply stated as

$$\mathcal{E}_{colli} = \frac{-2q_1q_2}{3\pi\eta r_1 r_2 (r_1 + r_2)\Delta v}, \quad (10.15)$$

where η is the dynamic viscosity of air ($\rho_A v$), and Δv is the difference in drop velocities. She noted that for collisions between many typical particle sizes, neither hydrodynamic nor inertial forces could be neglected, and other relationships had to be used. She found that the increase in collision efficiency from effects of charged drops were small if E = 0. For collisions of large, charged drops, she concluded that the calculated collision efficiency was too high in two situations: (1) in an E ≈ 0, and (2) in |E| > 0 when the free charge density of the drop dominated that induced by the external E. She provided an empirical adjustment that moved values to within about 5% of those calculated with the more complex numerical integration.

Abbott (1975) used Eq. 10.4 to measure the collision efficiency of droplets with radii ranging from 10 to 25 µm and a charge of about 30 fC. He compared the experimental results with the theoretical upper limit for collision efficiency for charged drops from Eq. 10.15. Abbott observed collision efficiencies about three orders of magnitude greater than that calculated by Davis and Sartor (1967) for uncharged drops. The deviation between the observations and theory at lower efficiencies was not due to simplifying assumptions. Abbott used Davis and Sartor's computer code and concluded that deviation of his observation from their theory at smaller collision efficiencies of <100 was not the result of the simplifying assumptions and could not be resolved by use of the more sophisticated expressions developed by Davis and Sartor.

Grover (1976b) did theoretical calculations of collision efficiencies for drop radii from 42 µm to 624 µm. He thought most of the experiments done previously did not determine or control charge on the particles well enough to compare with his work. As Krasnogorskaya

(1965) and Freire and List (1979) showed that fractions of femtocoulombs have an effect, Grover's concern seems well founded. He used Re = 1–400, T = 10°C, P = 900 mb, drop radii of 42–624 μm, and particles of density 2 g cm^{-3} and r = 0.5–10 μm. All E were downward pointing. The calculated collision efficiency for uncharged and various charged particles often ranged over two orders of magnitude for various q, E, and r (e.g., for an r = 0.5 μm, $\mathscr{E}_{colli}$ ranged from 0.001 at E = 0 to $\mathscr{E}_{colli}$ = 0.01 for E = 270 kV m^{-1}). Charge was important on the smaller collection drops, but not on the largest. Also, he found that as the larger drop increased in size, the effect of E or q on collision efficiency diminished—a result of dominance of the aerodynamic force. This led him to conclude that when R $\gtrsim$ 700 μm and r > 0.5 μm, the collision efficiency was nearly unaffected by electric forces.

Cohen and Gallily (1977) did theoretical calculations of collision efficiency with four drop sizes ranging from 20–75 μm for the larger and 2.5–10 μm for the smaller droplets. The drops were in external E = 0–1000 kV m^{-1}. They had only three cases in commonly observed thunderstorm fields (specifically E $\lesssim$ 100 kV m^{-1}). Example results included, for R = 20 μm, r = 10 μm, and both q = 0, a linear collision efficiency from 0.50–0.59 over the range of E = 0–100 kV m^{-1}. By E = 400 kV m^{-1}, the linear collision efficiency was 0.7. In general, they found an increase of the collision efficiency between oppositely charged drops and between one charged and one neutral drop. They concluded that for oppositely charged drops, the collision efficiency increased with increasing radius of the larger drop with no, or only small, charge. In contrast, the collision efficiency decreased for the same parameters, except for large charge on the larger drop. They also did the calculation for same-polarity drops and found efficiencies of approximately those of neutral droplets of the same size.

In a study restricted to small drops of radii $\leq$20 μm, Freire and List (1979) computed the collision efficiency between two weakly charged (0.03–0.3 fC) drops of opposite polarity when the electric force on the isolated, larger drop was in the same direction as gravity and when it was opposite to gravity. Among their findings were very large collision efficiencies for moderately to highly charged droplets. Collision efficiencies went from their viscous value of, for example, 0.004 to calculated maximum values of 19.0 for 10- and 9-μm radii drops and 11.2 for 10- and 2-μm drops. When the oppositely charged drops exerted force on the larger drop opposing the force of gravity, they concluded that the magnitude of the collision efficiencies left no doubt that effects from a vertical electric field provided an explanation for the rapid growth of cloud droplets beyond the size at which condensation is effective. They said that to neglect electrical effects in droplet growth was not justified. They suggested that tests of their conclusions required measurements and recommended incorporating electrical effects in models of drop spectrum evolution.

From all the studies, it seems clear that the fair weather field does not significantly affect the collision and, thus, the coalescence process. Once the electric field and/or drop charge becomes significant (which, as we have seen above, is defined differently depending on a variety of variables), the situation changes. Not so easily summarized are the various combinations of E, q, and r that can occur in clouds. Although not well defined for all aspects of cloud microphysics, it does seem clear that there are important electrical effects on the drop evolution and spectrum in electrified clouds.

10.8. COALESCENCE AND MASS GROWTH OF DROPS

From the collisions of drops, we move on to their coalescence, which sometimes results from those collisions. The film of air between colliding drops inhibits coalescence. Coalescence occurs only if the air film is removed between them and if the surface tension of each can be overridden to form a new drop. Coalescence does not have to be permanent; drops can coalesce and then separate into two or more drops. To try to explain rapid precipitation growth in warm clouds, Sartor (1954) ran experiments to find coalescence efficiencies. He stated that to understand coalescence, and not just collision, the viscous forces opposing the motion must be used. The smaller drops coalesced on the top side of the collector drop in weak E, but in strong E they coalesced on the bottom side. He found that coalescence was a linear function of increasing field for pairs of drops of radii 5–25 μm.

Goyer et al. (1960) used a Rayleigh water jet between horizontal plates and measured coalescence efficiencies for collisions between drops of R = 300–395 μm and a droplet of r $\approx$ 50 μm in vertical E. They did not have individual drop charge measurements, but calculated average charge per drop. Their measured coalescence efficiency increased from 29% for E = 0, q_R = 2.1 fC, and q_r = 2.4 fC to 95% for E_z = 3.8 kV m^{-1}, q_R = 4.2 pC, and q_r = 1.8 pC. Interestingly, for the first set of charges listed and at an E_z = 92 kV m^{-1}, the coalescence efficiency was zero. They attributed this to the repulsive forces of the drops of same polarity charge finally being large enough to prevent coalescence. They were unable to separate the effects of drop charge and polarization charge due to E_z on coalescence, although they preferred polarization in the electric field because it fit the observations better.

Telford and Thorndike (1961) studied drop coalescence when the electric field was absent, and they confirmed Hocking's (1959) calculations with their observation that colliding uncharged droplets of d $\approx$ 35 μm resulted in no coalescence. The collision of nearly equal diameter (45 μm) uncharged droplets resulted in occasional coalescence. (Uncharged in their case appears to have been |q| < 100 fC on the drop, below which they could not detect.) They found that E_h = 100–300 kV m^{-1} produced coalescence where there had been none before and suggested that in thunderstorms, there is a substantial contribution to rainfall from electric forces.

In her study of small drops, Krasnogorskaya (1965) concluded that colliding droplets of radii 5–25 μm with $|q| > 0.03$ fC in E > 10 kV m^{-1} would coalesce. She determined that there would have been no coalescence without the electric forces because these drop sizes were in the range where the probability of coalescence from gravity and hydrodynamic effects was zero. We digress to note that the ability to detect very small charges to bound tightly the category of drops supposedly with $|q| = 0$ was a function of the measuring capabilities in various experiments. For example, the minimal charge that Telford and Thorndike (1961) could detect apparently was much larger than Krasnogorskaya's. This alone could result in seemingly different conclusions about the effect of charge on the coalescence process.

Woods (1965) experimented with colliding droplets of the same radii, presumably in an E = 0. He varied the radii from 16 to 40 μm. With opposite polarity on drops of r < 16 μm, the charge had to read 17 fC before any increase in coalescence was noted. For 36-μm radii, the coalescence began at 26 fC and increased linearly with charge up to the maximum q = 70 fC used. For 40-μm radii and the same polarity of q = 66 fC on each, the coalescence decreased by a factor of five below that for uncharged droplets. For same polarity of droplets of r < μm, he observed no coalescence.

Latham (1969b) conducted experiments in Yellowstone National Park in the winter. He set up a laboratory apparatus outside in the cold, clean, water-vapor-laden air near geysers to study and experiment with the natural clouds that formed. In some experiments, he compared the simultaneous evolution of two particles: one grown in an $|E| > 0$, and another grown in E ≈ 0. The comparison of masses allowed him to estimate the influence of E, even though no absolute collection efficiency could be determined. He defined the *field effect on the growth rate in percent*, F_E, as

$$F_E = 100 \left(\frac{M_E - M_0}{M_0} \right), \qquad (10.16)$$

where M_E and M_0 are the mass growth for $|E| > 0$ and for E = 0, respectively. An example result for a collector drop of initial R = 130 μm and cloud droplets of median r = 7.5 μm is shown in Fig. 10.8. Changes in the collection of droplets had a threshold at E ≈ 150 kV m^{-1}, where coalescence increased about 20% over the no-field value. Of obvious interest in the figure is the rapid increase in growth beginning at 110 kV m^{-1}. By 160 kV m^{-1}, the mass growth was more than twice that at E = 0. He surmised that the reason was increased collision efficiency. The polarity of E did not affect growth, which he interpreted as showing no influence from any charge on the particles.

Sartor (1970) used a vertical E to study theoretically the collision and accretion of droplets, drops, and small

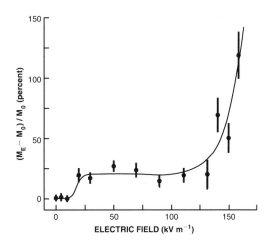

Fig. 10.8. Difference (%) in mass growth of water drop in presence of electric fields, M_E, and in zero field, M_0. Drop was of initial R = 130 μm and collected droplets of median r ≈ 7.5 μm in an airstream velocity of 0.8 m s^{-1} and a T = 2°C. (From Latham 1969b, with permission.)

hail. He pointed out that use of only a vertical E minimized the volume the larger drop swept out, thereby also minimizing collision efficiency and growth. Sartor also noted that if electric forces are to be important in modifying precipitation development, those forces must act under the condition of small drops with large charge or in high electric fields. The charge used in the calculations was proportional to the square of the droplet radius. The accumulated mass growth rate of uncharged, 10-μm radius drop was an order of magnitude greater when E = 30 kV m^{-1} than when E = 0. At an E = 300 kV m^{-1}, the increase was two orders of magnitude. If in addition to E = 300 kV m^{-1}, the drops were oppositely charged, initially only to 3 fC, the growth rate was three orders of magnitude greater than in the condition when E = 0 and q = 0. Sartor concluded that the increase in growth rate was most pronounced from the combination of charge and electric field. The rate of growth decreased rapidly as the collector drop radius, R, increased for each category of droplet radius, r. But even with an accreting drop of R = 500 μm, the combined effect of E and q was still a growth rate factor of two to three over that for the E = 0, q = 0 condition. In summary, the electric effect was most pronounced for cloud particles of R < 100 μm, but could still affect the growth of precipitation particles of R > 100 μm. For hail embryos of R = 5 mm and hail of R = 10 mm, he calculated little effect except for the collision with very small droplets. From his results, Sartor concluded that the electric forces cannot be ignored. This conclusion is strengthened owing to the general agreement (for similar conditions) with his and others' earlier, but less extensive, work.

To determine the coalescence between droplets as a function of charge, Brazier-Smith et al. (1972) used two needles to produce two streams of drops with radii of 150–750 μm. The relative velocity of the drops, Δv, ranged from 0.3–3.0 m s⁻¹. Under these conditions, they assumed all colliding drops would coalesce with droplets in their path. Brazier-Smith et al. wanted to examine whether the coalescence would be permanent or whether drops would separate after colliding. Note that, under their assumption, their equation for coalescence efficiency was the same as for collision efficiency (Eq. 10.4). They observed four collision behaviors: permanent coalescence, bounce (the drops separate intact after collision), separation into two drops following a temporary coalescence, and separation into two drops plus satellite drop(s) being produced. For many Δv, the coalescence efficiency was 0.1–0.4 for drops of equal radii. For drops with R = 2r, the coalescence efficiency ranged from 0.2–0.6. For drops of equal radii and equal, but opposite, charges, the coalescence efficiency increased with increasing drop charge from zero to about 3 pC. Above 3 pC, the coalescence efficiency remained constant with increasing charge (Fig. 10.9). As seen in the figure, the value of the coalescence efficiency was a function of the relative velocity between the interacting drops. (The Δv's were greater than the terminal velocity of 0.4 m s⁻¹ for such drops in an E = 0.) They also calculated coalescence efficiency as

$$\mathscr{E}_{coal} = 2.4 \left(\frac{\zeta}{r \rho_W \Delta v^2} \right) f \left(\frac{R}{r} \right). \quad (10.17)$$

The function f(R/r) ranged from 1.3 to 3.8 for R/r = 1–3, respectively. For this equation, the units for r are microns. They defined the term $r \rho_W \Delta v^2$ as a dimensionless parameter characteristic of the interaction process. Brazier-Smith et al. ruled out changes in collision efficiency as causing increased coalescence. Their estimates showed that drop charges were too small to alter drop trajectories. They concluded that increased coalescence efficiency was from huge electrical attraction between the charged drops once they were within a diameter of each other. This caused drops to move together, which reduced the distance between

the center of a drop and the trajectory of the other and increased the coalescence efficiency. When the drops had equal but opposite charges of ≳20 pC, bouncing stopped and coalescence always occurred.

Ochs and Czys (1987) cautioned against relying on applying previous laboratory experiments on cloud droplet coalescence to what happens in natural clouds. They noted large discrepancies between the coalescence of uncharged drops in drop-stream and suspended-drop experiments and the efficiencies of drops colliding while falling at terminal velocity. Their experiment was reported as the first to have both of the charged, colliding drops falling at terminal velocity. They used droplets of R = 340 μm and r = 190 μm with the same polarity of charge on each, and they examined more than 2500 collision events. The ambient electric field at the colliding drops apparently was assumed to be zero. The smaller drop had a $q_{avg} = 25 \pm 5$ fC, and they varied the charge on the larger drop from 2 fC to 2 pC. They defined an impact angle for the colliding pair as shown in Fig. 10.10. Whatever the charge on the drop, coalescence always occurred for impact angles less than the *critical impact angle* of 43 ± 1°. If the impact angle was >43°, what happened depended on charge, as follows. First, charge was treated in terms of the *mean relative charge,* $|q_R - q_r|$. The mean relative charge was >19 fC before coalescence was possible. When the mean relative charge was >300 fC, bounce was suppressed and temporary coalescence occurred in which a satellite droplet was produced. At 800 fC, coalescence occurred. If the charge difference was increased, the probability of a satellite droplet decreased. Finally, permanent coalescence was found for mean relative charges >2 pC. Ochs and Czys found that the lowest charge for permanent coalescence was at the largest impact angle (i.e., a grazing collision).

Ochs and Czys's (1987) study indicated that the result of drop collisions varies substantially over the range of the electrical environment found in clouds and thunderstorms. They decided that computations need to incorporate the more realistic coalescence assumption in which not all collisions of drops result in per-

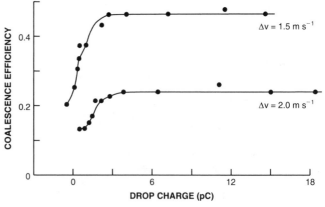

Fig. 10.9. Coalescence efficiency measured (dots) and theoretically calculated (solid lines) for drops of equal charge and $r_1 = r_2 = 500$ μm for differences in velocity, Δv = 1.5 and 2.0 m s⁻¹. (From Brazier-Smith et al. 1972, with permission.)

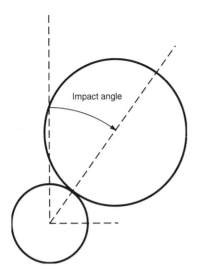

Fig. 10.10. Definition of impact angle for colliding drops used by Ochs and Czys (1987).

manent coalescence. In particular, they held that transition between bounce and coalescence induced by charge will significantly affect precipitation development. From their work, they concluded that the evolution of precipitation can be strongly influenced by its electrical environment.

Czys and Ochs (1988) reported more analysis from their study of 340- and 190-μm radii drops. They included more than 5500 events with +q always on the smaller droplet and varying polarity on the larger drop. Again we assume that there was no ambient E at the point of collision, with $|q_R - q_r|$ ranging from 9 fC to 2 pC. Their system had a minimum detectable q = 0.5 fC, and a minimum measurable q = 1 fC. Again they found the critical impact angle was 43°. The uncertainty of ±1° in the critical impact angle encompassed the outcomes of collisions that ranged from coalescence to noncoalescence. Collisions at ≤43° resulted in coalescence regardless of charge on the drops. For impact angles >43°, the minimum charge for coalescence was 20 fC. So if the mean relative charge $|q_R - q_r| <$ 20 fC, all collisions resulted in bounce. They stated that the results of collisions of oppositely charged drops were about the same as for collisions between the same drop polarities. They concluded that the effects of drop charge were more complicated than previously thought. A calculated charge of ≥0.27 pC required for electrical instability also was the observed charge that marked the transition between collisions resulting in a bounce and those leading to temporary coalescence. Czys and Ochs concluded that charge-induced instability could thus be responsible for coalescence. Furthermore, they pointed out that the previous lowest values of charge for coalescence in the literature had ranged from 30 fC to 40 pC, while here it was 7 fC for impact angles <43° (Fig. 10.11).

Using an improved camera system that provided an accurate measure of horizontal displacement between drops in their laboratory experiment, Ochs et al. (1991) restudied grazing coalescence, which occurs when the distance between the centers of the colliding drops is about the sum of their radii. Grazing coalescence thus

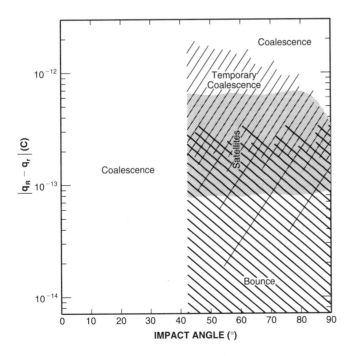

Fig. 10.11. Depiction of calculated results of collisions of drops as a function of impact angle and mean relative charge, $|q_R - q_r|$. The q_r was always 0.25 fC, while q_R varied from 1.5 fC to 2.1 pC in eight intervals. The areas of collision outcomes are labeled. The overlap of one style of hatching into another area is a qualitative indication of the probability of a particular outcome occurring in the region of another, more dominant outcome. (Reprinted with permission from *Nature*, Ochs and Czys 1987: Macmillan Magazines Ltd.)

occurs at the largest horizontal offsets between two colliding particles for which a coalescence can occur. The two drops fell at terminal velocity, had an R = 275 μm and an r = 200 μm, and were uncharged (i.e., q < 10 fC). This finding led the researchers to reexamine their work as reported in Ochs et al. (1986). They concluded that those earlier, experimentally determined coalescence efficiencies were about 20% too high, most likely because of "excess charge" on the drops (it had not been possible previously to control the drop charge as well). Including the contribution from grazing coalescence, their statistical best estimate of total coalescence efficiency within a 95% confidence interval was 46 ± 9%.

10.9. GROWTH OF ICE PARTICLES

In a search for electric field effects on aggregation of ice crystals, Latham and Saunders (1964) used a dual chamber in which they suspended identically shaped droplets (size not given), which were subsequently frozen. Ice crystals then flowed across each suspended ice sphere; one particle was in an adjustable, vertical electric field, and the other was in a field-free region. The latter served as a control; data were collected on mass growth as a function of E on the other. The mass of each particle was then determined by melting the particle and measuring the size of the resulting drop. Their results are shown in Fig. 10.12. The onset was an E ≈ 80 kV m^{-1}, above which the mass increased linearly. Their visual observations revealed that aggregation often occurred in long chains of crystals from the ice sphere. After reaching several hundred micrometers, the chains often folded inward to become an open, three-dimensional structure.

Latham (1969b) studied the effect of an electric field on the mass growth and growth rate of an ice particle in ice crystals and of ice spheres in supercooled

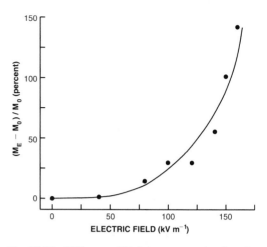

Fig. 10.13. Difference (%) in mass growth of an ice sphere by collecting ice crystals in the presence of an electric field, M_E, and in zero field, M_0. The collector had a R = 550 μm; the crystals r ≈ 60 μm. The collection velocity was 1.3 m s^{-1}, and T = −7°C. (From Latham 1969b, with permission.)

droplets. As he had done for the study of drops during the other experiments at Yellowstone National Park, Latham tried to make sure that there were no effects due to charge on the collector or colliding particles. He found an increase in the growth of ice particles by two processes: aggregating crystals and collecting supercooled droplets on frozen spheres. For ice crystals of r ≈ 60 μm colliding with an ice sphere of R = 550 μm, the field at which aggregation began to be enhanced was about 50 kV m^{-1}, with enhancement increasing with E (Fig. 10.13). The collection efficiency of ice crystals by an ice sphere diminished: For E = 0, the collection efficiency was insensitive to velocity, but at |E| = 150 kV m^{-1}, the efficiency decreased a factor of about 2.5 as velocity went from 0.5 to 4 m s^{-1}.

An ice sphere collected supercooled droplets at T = −8°C and with a relative velocity of 1.2 m s^{-1}. The sphere grew at a constant rate and independently of its size (R = 75–475 μm) when E = 0. However, when |E| = 150 kV m^{-1}, the growth rate was about 1.6 times as much for R ≈ 100 μm, and it decreased by about 15% by an R ≈ 475 μm. From Fig. 10.14, we see that the growth of the ice sphere began increasing at |E| = 90 kV m^{-1}, and it had doubled and was increasing rapidly by 160 kV m^{-1}. The growth rate of the sphere was independent of air temperature if it was colder than −4°C, but the rate decreased markedly as the temperature warmed toward 0°C. Latham concluded that growth enhancements were the result of an increased electrostatic adhesion force and that a strong E could produce significant growth of solid hydrometeors. We think that such a process might be important in certain stratiform cloud types, including the stratiform regions of mesoscale convective systems.

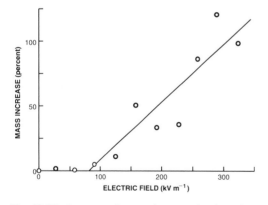

Fig. 10.12. Percentage increase in mass of an ice sphere in a stream of ice crystals at T = −9°C. (Reprinted with permission from *Nature,* Latham and Saunders 1964: Macmillan Magazines Ltd.)

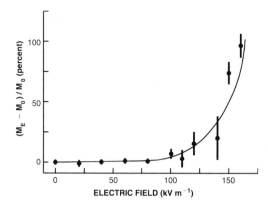

Fig. 10.14. Difference (%) in mass growth of an ice sphere by collecting supercooled droplets in the presence of an electric field, M_E, and in zero field, M_0. The collector had an R = 120 μm; the droplets median r ≈ 7.5 μm. The collection velocity was 1.2 m s^{-1}, and T = −8°C. (From Latham 1969b, with permission.)

Subsequently, uncharged and charged ice crystals in varying E also were investigated by Latham and Saunders (1970). They suspended an ice sphere of 2-mm diameter and collected 5-μm crystals in T = −7 to −27°C and in airflows of 0.2–8 m s^{-1}. Two experimental techniques were used: (1) analysis of movie camera films to measure the two-dimensional growth of an ice sphere, and (2) mass measurements with a microbalance. In both, there was simultaneous, side-by-side growth in a vertical |E| > 0 and an E = 0 to allow comparison under otherwise identical conditions.

The movie films of a suspended ice sphere of original diameter 1.5 mm showed that the growth increases with increasing E. Furthermore in higher E, there was both breaking off and folding in of the dendritic structure. Latham and Saunders noted the former was supported by visual observations of fragments breaking away and initially moving against the airstream. They

deduced that it was either from the mechanical stress alone or from the high electric field at the tips of the aggregates. When the rate of growth alone was considered, they found that it always increased with increasing E. There may be two competing processes in high E: aggregation and fragmentation. Both depend on the magnitude of E. Irrespective of the electric field, they found that aggregation occurred for temperatures as low as their coldest of −37°C. Furthermore, for |E| > 0, the collection efficiency increased at all temperatures as a function of E.

For quantitative measurements, Latham and Saunders did not rely on photographs of growing particles, but measured mass growth with a precision microbalance. They referenced the work of others, showing that such an ice sphere would have a v_T ≈ 3 m s^{-1}, so measurements were made at that airstream velocity. Each run of the experiment lasted 10 s, so they could compare the |E| > 0 and E = 0 cases without the collector sphere changing size so much that different collector sizes would add an additional variable. They defined the mass growth for aggregation as in Eq. 10.16. The aggregates grown in an |E| > 0 were less dense than the aggregates in E = 0, which they believed was due to loose packing of the crystals by their alignment along the lines of force of E. The aggregation of uncharged crystals increased with |E| up to about 150 kV m^{-1}, above which the growth rate declined due to fragmentation of the growing crystal. The field effect was independent of the temperature, with additional experiments showing this was so across the range of −7 to −37°C. Examples of their results are shown in Fig. 10.15.

From their mass measurements, Latham and Saunders determined the collection efficiency, which was defined as the fraction of ice crystals on a direct course with the collector that stuck to it. They determined collection efficiency by measuring the collector mass increase. The collection efficiency for |E| > 0 was found by multiplying F_E by the collection efficiency for E = 0. The absolute collection efficiencies were ≈0.3 with an uncertainty of 20% for T = −7 to −27°C at E = 0

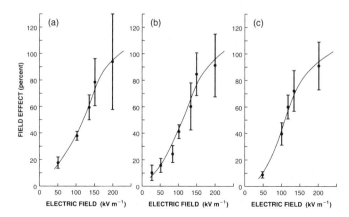

Fig. 10.15. Percentage increase in mass of aggregates created from an initial ice sphere of R = 1 mm at (a) T = −7°C, (b) T = −17°C, and (c) T = −27°C. The collection velocity was 3 m s^{-1}. (From Latham and Saunders 1970, with permission.)

for 5-μm particles at 3 m s^{-1}. They found it hard to tell if the increased collection efficiency was due to increased collision or adhesion of colliding crystals. Thus, the role of the electric field seemed to be to increase aggregation efficiency, not collision efficiency. This is a consequence of the very short range over which E would affect the two particles.

In an additional experiment, ice crystals charged to near maximum collided with a collector in E = 0. Again, the collector was a 2-mm diameter ice sphere embedded in an airstream of 3 m s^{-1}. The measured collection efficiencies were much less than they had found for crystals of q = 0 in E = 0. They surmised that this resulted from repulsion of crystals by the collector, which had acquired the same polarity as the crystals by collecting the crystals. Latham and Saunders interpreted this as further powerful evidence that E increases the aggregation efficiency, not the collision efficiency. Finally, they noted that natural crystals tumble as they fall and might behave differently from suspended particles.

Crowther and Saunders (1973) did laboratory experiments on the aggregation and fragmentation of ice crystals. A typical experimental situation was T = −19°C in which the crystal habits were about 65% hexagonal plates and 35% spatial dendrites in a radial (horizontal) electric field ranging from 0 to 150 kV m^{-1}. The crystals were allowed to grow as they fell in the chamber. The process of crystal growth and fall lasted about 40 s, after which the available vapor was depleted. Crystals were captured in formvar on the chamber bottom for observation. When E_h = 0, they were unbroken. When E_h > 0 was applied, the crystals showed increased aggregation, and many also showed evidence of fragmentation. As E_h increased up to 150 kV m^{-1}, the diameter of the unfragmented crystals decreased from about 32 μm down to 10 μm, indicating that the electric field can greatly affect the crystal population. As E_h increased, the aggregation of crystals approached 100%, but the same higher electric fields tended to cause fragmentation of the crystals it had just helped to aggregate. These trends are seen in Fig. 10.16 (the percentage aggregation is the percentage of aggregates in each sample, as opposed to a percent aggregated from the total crystal population). They interpreted the drop in percentages for E_h = 125 and 150 kV m^{-1} as showing that after aggregation, fragmentation occurred from the mechanical stresses due to high E_h acting on the induced charge of the aggregate. As the temperature was lowered, the aggregation and fragmentation both decreased. Even at E_h = 75–100 kV m^{-1}, aggregation was significantly greater than at E_h = 0. If their results are at all applicable to natural clouds, the electrical effects on microphysics in the large stratiform regions of MCSs (see Sec. 8.2), where large E and aggregates are thought typical could be immense.

Other experiments on the growth and fragmentation of ice crystals were conducted by Evans (1973). In the first experiment, he observed riming and crystal growth

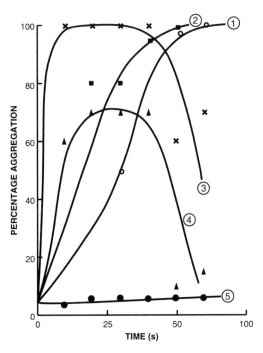

Fig. 10.16. Experimentally measured percentage of ice crystals that were aggregated versus time after nucleation of the test cloud of crystals for five values of horizontal electric field: (1) E_h = 75 kV m^{-1}; (2) E_h = 100 kV m^{-1}; (3) E_h = 125 kV m^{-1}; (4) E_h = 150 kV m^{-1}; (5) E_h = 0. The decrease in aggregation in curves 3 and 4 was due to fragmentation after initial aggregation. (From Crowther and Saunders 1973, with permission.)

in a radial E. Both processes were enhanced by the E, and dendrites tended to align with the E. He also grew crystals in a uniform vertical E to look for fragmentation; he reported that it was inconsequential. At temperatures warmer than −5°C, there were no dendrites. In the colder temperature range of −7 to −20°C, dendrites formed and fragmented, and the amount of fragmentation remained the same across the temperature range. The growth of dendrites was in the form of a chain of cloud droplets growing along the direction of E. Dendrites occurred only if the chain was >0.5 mm long, and chains extended up to 1 mm. On the longer chains, fragmentation of the dendrites occurred. Fragmentation of the dendrites on a chain required an E ≥ 200 kV m^{-1}. Evans decided that as well-developed needle crystals exist in thunderstorms, they could be fragmented and thus contribute to the formation of ice crystals, even though the process could not explain ice crystal multiplication.

Evans also examined and corroborated earlier theoretical work on electrogrowth of crystals from water vapor. He reviewed the electrodeposition of water vapor molecules on the highest radius of curvature of crystals to look for a role for E in the process. Based on his and others' calculations, Evans concluded that only for small, water ions was their migration sensitive to E. Such charged water molecules (essentially a small ion) would move quickly along the field lines toward the tips of crystals where the E was most enhanced. The availability of adequate charged water vapor to make the process important in thunderstorms was not known, but only enough vapor to grow an ice whisker at a rate of 50 μm s^{-1} was needed. He calculated this as corresponding to an ion current of 10^{-13} A, well within measured values. Once a whisker developed, it would grow laterally by diffusion. He concluded that the transport of charged water molecules could be the electrogrowth process in storms.

Finnegan and Pitter (1988) were interested in why aggregates of ice crystals often had the first two crystals shaped like a T (i.e., a tip-to-center aggregate). They considered the possible forces on aggregating ice crystals: aerodynamic, inertial, gravitational, and electrical. Finnegan and Pitter said that the combined effect of an electric field and charge on ice crystals was unknown. The also stated that neither a net charge on crystals nor a dipole charge structure in the crystal would lead to a T aggregate. They concluded that the simplest way to get a T shape was a linear quadrupole. They developed a laboratory experiment, which had its basis in the Workman-Reynolds effect. Finnegan and Pitter used a 7-m^3 chamber and conducted the experiment over a temperature range of -5 to $-30°C$. They tested 10–20-μm diameter droplets of deionized water and water solutions of sodium chloride, ammonium sulfate, and ammonium carbonate with about 10^{-4} normality (representative of the atmosphere) in LWC = 1.4–2.5 g m^{-3}. Different solutes could yield different polarity of the water with respect to the ice (e.g., sodium chloride yielded positive water and ammonium sulfate yielded negative water). At the suggestion of Bernard Vonnegut, they also made a mixed cloud by separately injecting droplets of both solutes into the chamber together. To test their hypothesis that an electric multipole (i.e., at least a quadrapole) was present in the growing crystals, Finnegan and Pitter looked for an effect of the solutes that were in the growing ice crystals on the orientation of those crystals in the aggregates. A run lasted 3–6 min, during which time the crystals were formed, grew, aggregated, and fell out of the observational volume to the chamber bottom. The researchers had an induction cylinder in the chamber to look for charges on the crystals; the minimum detectable charge was 1 fC. They also had a small horizontal grid of wires, which had alternating polarity voltage applied. These collected aggregates. They found aggregates of both columnar crystals and aggregates of plate-like crystals in five identifiable shapes: T, λ, point-to-point (i.e., end-to-end), V, and overlapping. They did not expect the high percentages of T and point-to-point junctions based solely on aerodynamic interactions. Shown in Table 10.1 are example results in which the large percentage of these two types of aggregates can be seen for each solute.

Table 10.1. Number of aggregates of different shapes grown from columnar crystals in different solutes at T = $-12°C$ by Finnegan and Pitter (1988).

	Shapes				
Solute	T	Point-to-Point	V	λ	Other
None	7	11	8	13	4
NaCl	9	6	8	7	5
$(NH_4)_2SO_4$	8	8	0	4	0
$(NH_4)_2CO_3$	45	10	10	21	11
$(NH_4)_2SO_4$ and NaCl	31	29	4	40	4

10.10. CHANGES IN ALIGNMENT OF HYDROMETEORS FROM E

We turn now to an electrical effect on the microphysics that could play a role in our ability to understand radar data from storms. As Hendry and Antar (1982) stated, the electric field can rotate the particles so that they give a reflection totally different from that if they were only under the influence of aerodynamic and gravitational forces. A few investigators have looked at the alignment of hydrometeors caused by the electric field. The consensus has been that the electric field can align particles to angles different from that in a field-free region and that lightning induces rapid realignment of ice crystals. For example, Hendry and Antar used a 1.8-cm wavelength radar to observe polarization and realignment of crystals from lightning. They concluded that some rapid reorientations high in thunderstorms and associated with lightning were caused by small, undetectable (for their radar) particles that were in the radar beam, rather than reorientation of detectable particles.

Krehbiel et al. (1992) used coherent correlations of the simultaneous co-polar and cross-polar returns from a circularly polarized, 3-cm wavelength radar to monitor the alignment of ice crystals above the 0°C level. They found that the effect came from propagation depolarization. They could discern the initial electrification before lightning in storms. Furthermore, from their real-time observations of the correlations, Krehbiel et al. developed an ability to predict the occurrence of lightning shortly before it occurs. They also found that when a storm appeared dead, it could still have aligned particles and fields high enough to create an occasional flash. Similarly, they could predict when the storm would produce no more lightning. In further studies, Krehbiel et al. (1996) concluded that the ice crystals have an electrical alignment that is predominately vertical, but their aerodynamic alignment is horizontal.

Metcalf (1993) used the cross-covariance amplitude ratio from an 11-cm wavelength polarimetric radar to examine changes in drop orientation from E. At low radar antenna elevation angles (looking into rain), there was no significant change in the cross-covariance amplitude ratio before and after lightning. At higher angles (looking into ice), there were changes. By 9 s after lightning, the propagation medium had realigned to the prelightning state. This led him to hypothesize that the cross correlation changes from observations by radars of wavelength <11 cm are all due to changes in the propagation medium and not due to aerodynamic reorientation in the backscatter medium. He noted that this hypothesis, if true, indicated important radar design considerations for both research and operational radars.

10.11. PRECIPITATION INTENSIFICATION ASSOCIATED WITH LIGHTNING

A fascinating aspect of storms is the phenomenon known as rain gush. When observed, the gush is typically a dramatic increase in precipitation arriving at the ground shortly after nearby lightning. It is most often called a rain gush, but the precipitation enhancement is not limited to rain: Gushes of small hail also occur. We discuss the observational and theoretical evidence for this phenomenon.

Although the anecdotal reports of rain gushes extend back centuries, we begin with the measurements made in the 1950s and early 1960s. Previous reports of the rain gush (or equivalent) existed in the scientific literature, but Moore et al. (1962) undertook the first concerted effort to document the phenomenon quantitatively. They used a vertically scanning, 3-cm wavelength radar to record cloud and precipitation echoes. The time to scan a full hemisphere was about 2 min, but near the zenith the time resolution was much higher. Their first experiment was conducted on Grand Bahama Island, and all the clouds reported had tops mostly below the 0°C level. Thus, the precipitation process was probably exclusively drop coalescence. They showed three data sets from lightning overhead followed by rapid growth aloft of precipitation, which subsequently fell to the ground where it was measured. The initial echo after the flash appeared within the warm sector of the cloud. The rain rate at the ground was about 100 mm h^{-1} with the gush. They also noted descent of the bottom edge of the radar echo at speeds up to 60 m s^{-1} (average was 40 m s^{-1}) and the coincidence of a radar echo intensification rate estimated at 10–40 dB min^{-1}. This was significantly higher than the 3 dB min^{-1} they referenced for typical coalescence growth. They examined the following possible sources of rapid precipitation development indicated in the radar reflectivity: advection of precipitation into the radar volume by wind, temporal resolution of radar scanning, and lightning induced. They produced data to rule out the first two. They preferred the latter for three reasons: (1) No intense rain echoes preceded the flash; (2) A new intense echo appeared and intensified within seconds after the lightning; and (3) The bottom of the new echo descended rapidly.

The initial studies were soon followed by two more seasons of similar data acquisition on Mt. Withington

in New Mexico. Again Moore et al. (1964) used a vertically scanning 3-cm wavelength radar, field change meter, and tipping-bucket rain gauges. They hypothesized two rain regimes: (1) the Bergeron ice mechanism, which was supported by the airplane observations showing snow aloft and then the presence of a radar bright band during such rain, and (2) convective storms, which had no bright band until later in the storm evolution. In the first regime, no gushes of precipitation were observed even if the cloud electrified and produced lightning. In the convective storms, they concluded that 52% of the total summer season rainfall occurred at high intensities immediately after lightning. Examples of radar data from 1962 were used to show rapid echo intensification after a nearby flash from radar data limited to new echoes occurring within 5° of zenith to allow the necessary, frequent recording of the evolving echo (the antenna scanned through zenith every 2.7 s). The restriction that winds aloft were low during the echo intensifications was also used to limit the data base. The initial echo had well-defined boundaries, but after 2 min were diffuse. Echo intensifications were up to a thousandfold, with rain increases to a hundredfold, which Moore et al. equated with a mass increase of larger drops of about two orders of magnitude. An example of a more modest, but easily identifiable, increase is shown in Fig. 10.17. They showed reflectivity that increased at an average rate of 120 dB min^{-1} in the first 10 s after the flash. The arrival at the ground of the lower edge of the recently formed echo coincided with the start of an identifiable gush. In addition, they noted that a large-scale downdraft was likely from the precipitation gush. (In the modern context, this raises the intriguing issue of whether lightning-induced rain gushes are the producers of at least some microbursts that are hazardous to aircraft flying through them as they approach landing.) In New Mexico, hail gushes were also observed. This process will be addressed later.

From these studies Moore et al. concluded that lighting caused the increased precipitation and that the increased rain must have been from coalescence that was in some way dramatically enhanced by the lightning. They noted that definitive data to assess the possible role of acoustic forces from thunder were lacking, but primarily on the grounds of the short duration of the force, they rejected it. The hypothesis they proposed, which is illustrated in Fig. 10.18, has as its focus the charges released by a lightning discharge and a resultant increase in particle coalescence. The following explanation of their hypothesis refers to the panels in the figure: (a) This example has positive charge on the lightning in a negatively charged cloud region; (b) The small ions created by the breakdown move out from the channels into the cloud under the influence of the local E; (c) The fast ions quickly attach to oppositely charged cloud particles, which neutralizes the original cloud particle charge and then charges it to the opposite polarity; (d) The particles become highly charged and also move under the force of E; (e) The electrified cloud

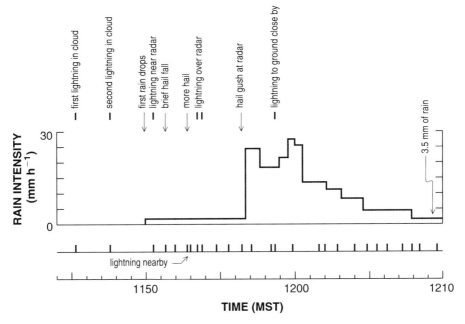

Fig. 10.17. Rain rate and observations of a stationary thunderstorm that produced a hail gush at Mt. Withington, New Mexico, 21 August 1962. The ticks on the line above the time are lightning flashes. (After Moore et al. 1964, with permission.)

particles produced by the lightning have many collisions, which cause them to gain mass but lose charge; and (f) The process produces larger particles and, as they fall faster, they continue to grow larger. These big particles make up the rain gush.

A decade later, Brazier-Smith et al. (1973) assessed the measurements and evidence for rain gush phenomena. They thought that much had been based on insufficient rain gauge data and with no vertical velocity, liquid water content, or other data aloft. In evaluating the work of Moore et al., they criticized the use of a monodisperse distribution of droplets in the calculations of drop growth. They further argued that the reflectivity required the high likelihood of 2-mm diameter drops before the nearby flash. Then growth to 3 mm, noted in Moore et al., could occur in the time observed without invoking electrically driven enhancements. In response, Moore and Vonnegut (1973) gave an updated summary of their earlier measurements and other clarifications. They revised upward the drop diameter at detection from the Bahamas to have been about 500 μm, but did not agree with the contention by Brazier-Smith et al. of 2-mm drops being present at flash time. Furthermore, Moore and Vonnegut concluded that more than a few drops made up the gush arriving at the ground. They reaffirmed the validity of their previous observational data.

Two decades after the radar observations of Moore and Vonnegut came the first direct observation of lightning and precipitation growth (but not a gush) in the same

cloud volume. At Langmuir Laboratory, Szymanski et al. (1980) used a field-change sensor, microphone arrays, and a steerable 11-cm wavelength radar to record lightning along with a vertically scanning 3-cm wavelength radar for cloud evolution. The lightning-detecting radar antenna was pointed at the storm of interest over Langmuir Laboratory, and the radar echo was recorded for several range gates along its beam. The radar reflectivity and the field change for a lightning flash are shown in Fig. 10.19. The occurrence of lightning is unmistakable, with a large radar signal occurring as the channel moves into the range gate at 5.5 km along the antenna beam. Evidence for precipitation enhancement from the lightning is shown in Fig. 10.20, which contains a sequence of radar reflectivity at ranges from 0–7 km. Each plot is the amplitude of radar reflectivity versus range at a particular time—called an A-scope display. Notice that before the lightning flash and its echo, there was no precipitation echo showing beyond about 4.5 km. The lightning echo at 5.5 km was followed in 3 s by a new precipitation echo, which grew in intensity and fell down. Szymanski et al. compared the radar location of the flash with an acoustic map of the entire flash. In turn, both were compared with the vertical and horizontal planes of radar reflectivity from the cloud obtained with the vertically scanning radar. The nearest precipitation of greater than −5 dBZ at or above the altitude of the lightning echo was 700 m away. With the measured cloud motion of about 10 m s^{-1}, it could not have advected into the volume in the time available.

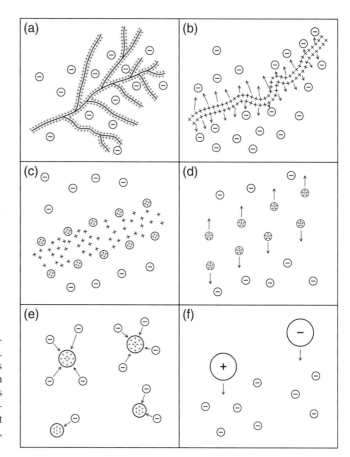

Fig. 10.18. Sketches illustrating hypothesis of how lightning can create a rain gush. The solid lines are channels; + and − signs without circles are ions; open circles with charges are drops of varying relative sizes with the number of + or − signs indicating relative amounts of charge. See the text for explanation. (From Moore et al. 1964, with permission.)

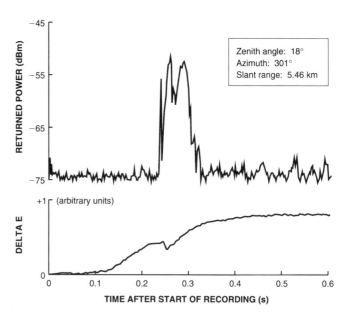

Fig. 10.19. Radar returned power (reflectivity) from a single range gate at 5.46 km from the radar in which an echo from a part of a lightning flash occurred, and the associated electric field change from the cloud flash that caused the echo. (From Szymanski et al. 1980: American Geophysical Union, witn permission.)

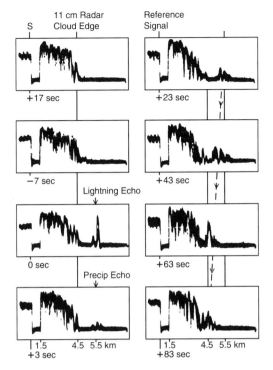

Fig. 10.20. Tracing of the 11-cm wavelength radar signal (called an A scope display) for precipitation and lightning echoes in thunderstorm over Langmuir Laboratory. The amplitude of the waveform is returned power (nonlinear and no units specified) and time moves left to right out from a reference signal, S. Time was converted (linearly) to range from the radar. The antenna was aimed 18° down from zenith. The precipitation echo first appeared at a range of about 5.5 km and then moved downward. (From Szymanski et al. 1980: American Geophysical Union, with permission.)

The estimated echo intensification rate of >24 dB min^{-1} in the volume with the lightning was several times that expected for growth by coalescence. Their data do not totally rule out the possibility that precipitation moved into the volume after the flash. However, it would have been quite a coincidence, with the added requirement that then the precipitation would have had to remain in the volume and fall out as shown in the sequence of reflectivity plots. They did not mention any detection of a gush at the ground. This is the only such dramatic case of precipitation enhancement tied so directly in space and time to a lightning flash of which we are aware.

Barker et al. (1983) did laboratory experiments to understand the physics driving the echo intensification process in the region where lightning occurred. Their experiments were limited to drops that interacted directly with corona streamers in an electric field; specif-

ically, they did not consider charging from the ions created by the channels as in the Moore-Vonnegut hypothesis (Fig. 10.18). Their crude calculations included an ambient E = 400 kV m^{-1}, which is higher than commonly observed. They concluded that the rapid precipitation descent observed by Szymanski et al (1980) could have been from increased velocity of directly charged drops of original radii 50–100 μm. Barker et al. further concluded from their rough calculations that the rate of echo intensification was just possible, but that a correct test required knowledge of the fraction of charged drops in the volume of interest.

Another theoretical assessment of the rain gush that was specifically tied to the idea of a lower positive charge center in a thunderstorm was done by Jayaratne and Saunders (1984). They proposed that falling, positively charged precipitation causes the lower positive charge that in turn causes lightning: The falling precipitation continues to fall and produces the gush. Marshall and Winn (1985) countered with a calculation showing that the proposed precipitation charging was a factor of 100 too small to produce the charge they obtained from their in situ measurements and that at best it was only one of several possible explanations. In their response, Jayaratne and Saunders (1985) pointed out that exceedingly high precipitation rates are not needed if a realistically possible graupel concentration of 10 l^{-1} were used. The link between the lower positive charge center and the rain gush remains an unknown one.

In studies with different emphases, several investigators looked at electrical effects on microphysics caused more indirectly by electricity (i.e., by nonelectrical aspects of lightning). First Goyer (1965) did experiments at Old Faithful geyser in Yellowstone National Park. He suspended line explosive (called Primacord) from a tethered balloon about 100 m high and at known distances from the geyser plume. The plume caused a precipitating supercooled cloud with a broad spectrum (not given) of water droplets. The explosive ranged from 2 to 40 m in length, and apparently was vertical, as opposed to tortuous. Immediately upon detonation, observers noted showers of small hail pellets. This indicated a hail gush from shock-induced freezing. Samples showed the hail was rimed ice pellets of diameter about 500 μm. Colder ambient air temperature yielded larger pellets. There was also evidence that some pellets collided and aggregated as they fell. The threshold for shock-induced freezing was an overpressure of 0.007 atm at T = −4°C and a detonation range of 60 m. The overpressure threshold depended on the temperature. Latham (1977) said that a remotely possible explanation of the hail gush was electrofreezing of rain drops into small hail.

Also observed after a single detonation was never-before-seen heavy precipitation about 600 m downwind from the geyser. Goyer explained this as possibly from the detonation shattering the large droplets into smaller ones that were then carried by the wind. In other cases, droplet spectra were recorded about 150 m downwind from the geyser. The number of droplets at that location increased by factors of 5 to 60 as compared with the absence of a detonation.

Rain gushes were also observed to result from detonations. They concluded that the rain gushes were the result of increased collisions and increased droplet coalescence rate in the radial wind from the detonation. No measurements of the wind were mentioned. All these results strongly suggested that the simulated thunder yielded several effects on the microphysics. Not addressed experimentally or theoretically was whether tortuosity in the form of bends in the Primacord would have mattered. In later laboratory studies and calculations, Goyer and Plooster (1968) noted that tortuosity was important, but hard to assess. Interestingly, they could not reproduce in the laboratory the nucleation of freezing that had occurred in the geyser, and they could not decide the importance of mechanical disturbances from lightning on nucleation. They noted that their laboratory cloud was not comparable to the geyser. They concluded that a cloud of micron-sized droplets was unaffected by a weak shock wave and that it seemed unlikely that rapid adiabatic expansion in a lightning-produced pressure wave would cause homogeneous nucleation of ice.

Foster and Pflaum (1988) used a numerical model to test the theory that freezing could occur near lightning channels for, as they noted, if it does the total volume of cloud involved from the lightning structure could be significant. They used a single frequency, ideal sound wave, which they cautioned does not simulate complex thunder acoustics. They suggested that acoustic waves could enhance collision and coalescence of drops by changing their relative motion and shift the drop spectrum toward larger drops. This, in turn, could influence processes tied to cloud droplet population such as riming and multiplication of ice. They concluded that the influence of thunder in a storm would be small, as the time of interaction is limited to about 1 s. The interaction of thunder with a droplet, however, is still ill-understood, and thus so are microphysical effects from thunder and shock waves.

The rain gush (and hail gush) continues to be largely unexplored and unexplained in any convincing manner. The phenomenon may be a complicated, multistage process, and whether electrical effects are irrelevant, dominant, or contribute to the rain gush remains to be learned. Furthermore, this chapter, and indeed the status of the understanding, leaves us with many questions as to the existence, extent, conditions, and so on under which electric forces can influence cloud microphysics, including precipitation. Based on the results summarized here, we feel compelled to caution against exclusion of electric forces and effects in attempts to model clouds, both for basic understanding and for forecasting purposes. Rather, we would encourage careful examination and tests of the phenomena likely to be involved in particular experiments or models.

10.12. EFFECTS ON ATMOSPHERIC CHEMISTRY

Chameides (1986) has suggested that corona currents from cloud particles may react with atmospheric gases to produce H_2O_2. The H_2O_2 dissolved in droplets, in turn, reacts with dissolved SO_2 to produce sulfuric acid, so Chameides pointed out that corona current may play a significant role in the formation of sulfuric acid in clouds. Other than this one possible effect of corona, most electrical effects on atmospheric chemistry are thought to be caused by lightning.

The intense pressure and high temperature inside lightning channels cause rapid chemical reactions that drastically change the composition of the air inside the channel. As we discussed in Section 5.11, the pressure in a lightning channel reaches several atmospheres, and the temperature can exceed 20,000 K. At these temperatures, the air quickly attains chemical equilibrium and is completely ionized.

As the channel cools through rapid expansion and turbulent mixing, the equilibrium composition changes to neutral atoms and then neutral molecules. At some temperature, called the *freeze-out temperature*, the time required for chemical reactions to bring the composition into the chemical equilibrium characteristic of that temperature becomes equal to the time for the channel to cool to ambient temperature. (For example, Borucki and Chameides (1984) noted that the time for NO to reach chemical equilibrium is less than 10^{-6} s at 5000 K and is 10^3 years at 1000 K.) The chemical composition at slightly cooler temperatures becomes fixed (or frozen) so that it persists after the channel dissipates. Although concentrations of the predominant atmospheric species (e.g., N_2 and O_2) return essentially to their ambient values, concentrations of some atmospheric trace gases (e.g., NO, CO, N_2O, and OH) are increased. Figure 10.21 shows how equilibrium volume mixing ratios (i.e., the fractional composition by volume) of selected atmospheric species vary as a function of temperature.

Of the trace gases produced by lightning, NO has received the most attention because of its affect on atmospheric ozone. Ozone plays a vital role in protecting Earth from ultraviolet radiation and in establishing the thermal balance of the troposphere. Direct production of ozone by lightning is thought to be insignificant compared with other sources. However, ozone is itself affected by other atmospheric trace gases. (For example, chlorofluorocarbons (CFCs) have received considerable attention because they deplete stratospheric ozone.) One such trace gas is NO_x (NO_x denotes NO plus NO_2), which acts as a catalyst in reactions with ozone and so survives to continue affecting ozone concentrations. NO_x has a long lifetime in the stratosphere, where it destroys ozone, and it is thought to have a major influence on stratospheric ozone concentrations. NO_x also is important in the upper troposphere, but there the reactions tend to increase ozone concentrations.

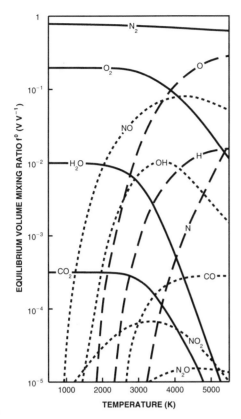

Fig. 10.21. Equilibrium volume mixing ratios (i.e., the fractional composition by volume) of selected atmospheric species as a function of temperature in heated tropospheric air. (From Chameides 1986, with permission.)

There is little question that lightning produces NO_x. A number of investigators have found enhanced NO_x concentrations in and near thunderstorms with moderate lightning flash rates (e.g., Noxon 1976, 1978; Drapcho et al. 1983; Chameides et al. 1987, Franzblau and Popp 1989, Ridley et al. 1996). However, formidable questions are unanswered: (1) How much NO_x is produced by thunderstorms locally and globally, relative to other sources? (2) Where is this NO_x advected? and (3) How significant are the effects of thunderstorm-generated NO_x on the chemistry and temperature of the upper troposphere and lower stratosphere?

Anthropogenic activities, such as industrial pollution and burning fossil fuels and biomass, produce NO_x at a rate of roughly 30×10^{12} g of N y^{-1} (Chameides 1986). Lightning appears to be a primary natural source for NO_x in the troposphere. (Other natural sources include stratospheric oxidation of N_2O, oxidation of NH_3, and NO emissions from soils.) Most estimates of lightning production of NO_x are between a factor of three and an order of magnitude smaller than production by anthropogenic sources (e.g., Tuck 1976, Borucki and Chameides 1984, Logan 1983). However, some investigators have suggested that previous estimates of NO_x production by lightning are systematically too low (Franzblau and Popp 1989, Liaw et al. 1990, Weinheimer 1989). Most estimates based on lightning channel length used values that were a factor of 3 to 8 smaller than typical lengths in most storms and were about a factor of 30 smaller than extreme lengths. A renormalization of previous estimates with more realistic lightning parameters gives new values ranging from 1 to 295 Tg N y^{-1} (Tg $= 10^{12}$ g) for theoretical estimates and from 70 to 220 Tg N y^{-1} for observational estimates (Liaw et al. 1990).

A major reason for this large range of estimates is the continuing uncertainty in our estimates of typical values of lightning parameters, primarily flash rates, channel length, and energy dissipation. However, lightning production of 100 Tg N y^{-1} would greatly exceed the total of all known natural loss mechanisms. Estimates of NO_x production that were made by balancing production with known loss mechanisms appear to place an upper bound of roughly 20 Tg N y^{-1} on lightning production, with a bound of 10 Tg N y^{-1} judged more likely (Logan 1983).

A simple comparison of lightning and anthropogenic sources of NO_x, such as those discussed in the previous paragraphs, understates the importance of lightning, because it ignores the distribution of the sources with height, which affects the lifetime of the injected NO_x. At lower altitudes, where most anthropogenic sources are located, NO_x has a relatively short lifetime before being lost to the atmosphere. In the summer boundary layer, for example, the lifetime of NO_x is only a few hours, and NO_x is not transported efficiently out of the boundary layer. Logan (1983) estimated that roughly 30% of the NO_x from combustion of fossil fuels is removed near its point of injection and that only about 40% survives long enough to be transported from the continent where it is generated. In spite of this reduction, anthropogenic sources probably play a major role in the lower troposphere. From measurements and transport models, for example, Singh et al. (1996) estimated that 65% of the tropospheric reservoir of reactive nitrogen (which includes other compounds besides NO_x) is from anthropogenic sources.

The lifetime of NO_x increases with altitude (it is as much as a week in the upper troposphere and about 2 months in the lower stratosphere). Therefore, the relative effectiveness of tropospheric sources will increase with the height of the source. Although thunderstorms lift NO_x from low levels to the upper troposphere (e.g., Dickerson 1987), thunderstorms are sporadic and lift NO_x from only their immediate vicinity. Singh et al. (1996) estimated that species produced by anthropogenic sources and lifted by convection contribute only 20% of NO_x in the upper troposphere, and measurements by several other investigators (e.g., Murphy et al. 1993, Smyth et al. 1996) have suggested a similar lack of NO_x in the upper troposphere (particularly in the tropics) from anthropogenic sources. Thus, a natural source is needed. One candidate besides lightning is transport of NO_x from the lower stratosphere. However, even with this source added to anthropogenic sources, there still must be another natural source, and lightning often has been suggested as a candidate (e.g., Ko et al. 1986, Murphy et al. 1993, Levy et al. 1996).

Because NO_x survives longer at higher altitudes, the elevated location of lightning would be expected to increase its relative effectiveness as a source, compared with sources near the ground. Most lightning channels are near or above the 0°C isotherm, and many channels are much higher in the troposphere. Furthermore, tall, vigorous storms tend to produce much more lightning than the small, isolated storms in which most NO_x measurements have been made. The longer lifetime of NO_x at higher altitudes and the much larger volume of the atmosphere that is remote from industrialized regions suggest that lightning can have a significant effect on the NO_x budget, and therefore, on the photochemistry of the atmosphere, even if it produces less NO_x than anthropogenic sources.

Possibly offsetting some of lightning's height advantage is a decrease with height in the ability of a lightning channel to produce NO_x. From the results of their numerical model, Goldenbaum and Dickerson (1993) suggested that a lightning channel will produce much less NO_x in the upper troposphere than in the lower troposphere. Although some reduction with height seems plausible, it is unclear whether the reduction would be as great as estimated by Goldenbaum and Dickerson. They noted, for example, that if the diameter of lightning channels tends to increase with height, their estimated reduction would be too large.

If the amount of NO_x that a channel can produce decreases with altitude, it does not appear to decrease fast enough to prevent lightning from being a major source of NO_x for the upper troposphere. The following are some of the observations substantiating that lightning probably is a major source:

1. Anthropogenic sources and transport from the stratosphere provide too little NO_x to account for observed concentrations in the upper troposphere, so there must be another natural source (Ko et al. 1986, Murphy et al. 1993, Smyth et al. 1996, Singh et al. 1996, Levy et al. 1996).
2. NO_x concentrations in the upper tropical troposphere are substantial, but highly variable, and this suggests that a transient source is important (e.g., Davis et al. 1987, Murphy et al. 1993).
3. Airborne sensors have found substantially enhanced NO_x concentrations in and near thunderstorm anvils, even when storms had moderate flash

rates (Chameides et al. 1987; Davis et al. 1987; Ridley et al. 1987, 1996; Murphy et al. 1993). From their measurements at anvil altitudes in two storms, Ridley et al. (1996) estimated that lightning produces 2–5 Tg N y^{-1} in the troposphere above 8 km. This would qualify lightning as a major source in the upper troposphere.

The effect of tropospheric sources of NO_x on the lower stratosphere is probably a function of the latitude, as well as the altitude, of the source. Although the transport of species between the troposphere and stratosphere is poorly understood (it is a subject of current climate research), the upwelling motion of the Hadley circulation in the tropics is generally expected to be much more favorable for transport into the stratosphere than the descending branch of the circulation at higher latitudes. At the equator, Ko et al. (1986) found that observed concentrations of NO_x below 30 km in the stratosphere were as much as an order of magnitude larger than would be expected without a tropospheric source. At 40°N, however, concentrations were much closer to those expected; observed values deviated from expected values over a smaller altitude range and with a smaller maximum excess. The expectation that transport from the troposphere occurs primarily in the tropics is also consistent with airborne NO_y (NO_x + all other reactive species of nitrogen) and ozone measurements made at various latitudes and at altitudes from the upper troposphere up to 21 km (Murphy et al. 1993). As we have discussed, lightning appears to be a major source of NO_x for the troposphere. Ko et al. (1986) noted that only a fraction of the NO_x produced by lightning globally needs to be transported to the stratosphere to account for observed abundances in the lower stratosphere.

Efforts to determine the role of NO_x produced by lightning are hampered by the limited data now available. There are large uncertainties in estimates of the amount and height distribution of NO_x produced per flash and in estimates of the global distribution of lightning flashes. Furthermore, a better understanding of NO_x transport is needed, including transport from low to high altitudes in the tropical troposphere, from lightning channels to the air surrounding thunderstorms, and from the upper troposphere to the lower stratosphere. Adequately quantifying lightning production of NO_x and the subsequent transport of NO_x by thunderstorms has been almost impossible, and it is still difficult. NO_x measurements in the vicinity of thunderstorms require care and ingenuity. Relating these measurements, which of necessity are usually at least a few kilometers from the lightning source, to lightning parameters and to storm winds compounds the problems.

Appendix A

Selected Symbols and Definitions

a	radius of droplet	q_R	rainwater mixing ratio; charge on larger drop of radius R
c	speed of light	q_{Ra}	Rayleigh charge limit on drop
c_s	speed of sound	q_s	saturated specific humidity
c_P	specific heat of dry air at constant pressure	q_V	mixing ratio of water vapor in air
c_V	specific heat of dry air at constant volume	r	radius of smaller drop or particle; radius
e_{si}	ambient equilibrium vapor pressure over ice	r_0	undistorted drop radius
e_{sw}	ambient equilibrium vapor pressure over water	r_{cld}	radius of cloud droplet
		t	time
f	frequency	t_{pk}	time of peak current or field
f_c	cloud flash fraction or percentage	u	surface wind; air speed through Gerdien cylinder; airplane speed
g	acceleration due to gravity; as a subscript, graupel	v	velocity
h	as a subscript, hail; height	v_T	terminal velocity
k	ion mobility	v_{TLM}	velocity of stroke from transmission-line model
k_B	Boltzmann's constant		
k_T	thermal conductivity of air	w	vertical velocity
m	molecular weight; mass	x_{crit}	minimum (critical) value of the impact parameter for a grazing collision
n	small-ion concentration; number density		
n_{crit}	number of ions with critical radius entering Gerdien cylinder	z	height or altitude
		z_-	height of 'main' negative charge center
n_0	constant of proportionality in Marshall-Palmer distribution	z_+	height of 'main' positive charge center
		z_{LPCC}	height of lower positive charge center
p	ion-pair production rate; dipole moment	A	"acceptance area" swept out by an induction ring; area
q	rate of ion pair production; photoionization rate; charge	B	magnetic field; buoyancy
		C	capacitance
q_e	charge of an electron, -1.6×10^{-19} C; also charge of an ion ($= 1.6 \times 10^{-19}$ C)	C_D	drag coefficient
		C_F	feedback capacitor
q_I	cloud ice mixing ratio	D	diameter; distance; electric displacement
q_r	recombination rate for ions; charge on smaller droplet of radius r	D_{cld}	diameter of cloud droplet
q_N	mixing ratio of the jth size category	D_i	diameter of the cloud-ice particle

D_+	diffusion coefficient of positive ions	N_N	total number density of *Nth* water substance category
D_-	diffusion coefficient of negative ions		
E	electric field; region of the ionosphere; as subscript, refers to electric field-driven capture	N_+	number of drops with positive charge
		N_-	number of drops with negative charge
		P	pressure; power
E_{aloft}	the electric field aloft	P_E	electrostatic pressure
E_{be}	breakeven electric field	Q	charge in a region; source of local charging; net heating rate
E_{cld}	electric field in the cloud		
E_{clr}	electric field in clear air	Q_-	charge in 'main' negative charge center
E_{crit}	electric field necessary for streamer propagation; electric field at which drop comes apart from electric forces	Q_+	charge in 'main' positive charge center
		Q_n	charge of nth water substance in box
		Q_{LPCC}	charge in lower positive charge center
E_{gnd}	electric field at the ground (Earth's surface)	R	range; resistance; return stroke; radius; radius of larger drop or particle; molar gas constant
E_h	horizontal component of electric field		
E_K	magnitude of vertical plus one horizontal component of electric field		
		R_D	gas constant for dry air
E_{on}	onset electric field for corona	R_F	feedback resistor
E_O	electric field outside a dielectric	R_V	ideal gas constant for water vapor
E_z	vertical component of E	Re	Reynolds number
$\mathscr{E}_{agg}$	aggregation efficiency	S	Poynting vector
$\mathscr{E}_{accrete}$	accretion efficiency	$S_{\rho n}$	net contribution of sources and sinks
$\mathscr{E}_{collect}$	collection efficiency	$S_{T\pm}$	net contribution of thunderstorm sources and sinks
$\mathscr{E}_{colli}$	collision efficiency		
$\mathscr{E}_{coal}$	coalescence efficiency	T	temperature
$\mathscr{E}_{gi}$	collision-separation efficiency for graupel and ice	V	voltage; volume; potential difference
		V_o	output voltage of circuit or instrument
$\mathscr{E}_{sep}$	separation efficiency	X	length of tube connected to Gerdien cylinder
EW	effective liquid water content		
F	force; total flash rate; ∇V	Z	flash ratio; radar reflectivity factor
F_D	drag force	Z_{DR}	differential reflectivity
F_E	field effect on mass growth	α	recombination coefficient between small ions; number of electrons created per unit length; specific volume
$\mathscr{F}$	form factor for elevated versus flush mounted instrument		
H	scale height	β	molecular cross section for ionization; attachment coefficient between small ions and aerosols; also $\beta = v/c$; ion recombination coefficient
H_E	height of the electrosphere		
I	current		
I_{PD}	point-discharge current		
J	volume current density	δq	charge exchanged by colliding particles
J_{CV}	unknown current density due to transport of charge by air motion such as convection	ϵ_0	permittivity for a vacuum
		ϵ	permittivity of air (8.86×10^{-12} F m^{-1})
		ϵ_{diel}	dielectric constant (a parameter of dielectrics)
J_D	displacement current density		
J_E	conduction current density due to the electric field	ζ	surface tension
		η	dynamic viscosity of air
J_{FW}	fair weather current density	θ	elevation angle; direction angle for wavefront
J_L	current density due to lightning transients		
J_M	Maxwell current density	κ	dielectric constant, relative permittivity; eddy exchange coefficient
J_P	precipitation current density		
J_{PD}	point-discharge current density	λ	wavelength
K_d	diffusion coefficient	λ_q	line charge density
L	length or thickness	μ	magnetic permeability
L_S	latent heat of sublimation	ν	kinematic viscosity coefficient
LWC	liquid water content	ρ	space charge density; mass density
M_E	mass growth for $|E| > 0$	ρ_P	space charge density carried on precipitation
M_0	mass growth for $E = 0$		
N	aerosol concentration	ρ_S	air density at standard temperature and pressure
N_e	electron number density		
N_i	number of particles with charge q_i	σ	small-ion conductivity; expected measurement error
N_{tot}	total flash density		

σ_{ab}	absorption cross section	$\Gamma(n)$	complete gamma function
σ_{clr}	conductivity in clear air	ΔE	electric field change
σ_{cld}	conductivity in cloud	ΔE_0	electric field change at the ground
σ_q	surface charge density	Δv	relative velocity of colliding hydrometeors
σ_+	positive ionic conductivity	Θ_e	equivalent potential temperature
σ_-	negative ionic conductivity	Λ	scale factor of exponential in Marshall-Palmer distribution
σ_{tot}	total ionic conductivity		
τ	relaxation time constant of atmosphere; retarded time; time delay	Φ	electric potential; volume flow rate in Gerdien cylinder
ϕ	latitude; azimuth	Φ_E	potential of the electrosphere
χ	chi-square statistic	Ψ	electric flux
ω	frequency of oscillation	Ω	Earth's angular rotation speed

Appendix B

Physical Constants and Conversion Factors

PHYSICAL CONSTANTS

Quantity	Symbol	Value	Units
Speed of light in vacuum	c	2.99792×10^8	$m\,s^{-1}$
Permeability of vacuum	μ_0	$4\pi \times 10^{-7}$ $= 12.5664 \times 10^{-7}$	$N\,A^{-2}$
Permittivity of vacuum	ε_0	$1/\mu_0 c^2$ $= 8.85419 \times 10^{-12}$	$F\,m^{-1}$
Newtonian constant of gravitation	G	6.67260×10^{-11}	$m^3\,kg^{-1}\,s^{-2}$
Planck constant	h	6.62608×10^{-34}	$J\,s$
Elementary charge	q_e	1.60218×10^{-19}	C
Electron mass	m_e	9.10939×10^{-31}	kg
Proton mass	m_p	1.67262×10^{-27}	kg
Molar gas constant	R	8.31451	$J\,mol^{-1}\,K^{-1}$
Boltzmann constant	k_B	1.38066×10^{-23}	$J\,K^{-1}$
Standard atmosphere	atm	1.01325×10^5	Pa
Standard acceleration of gravity	g	9.80665	$m\,s^{-2}$

UNITS

Quantity	Unit Name	Unit Symbol	Equivalent
Length	meter	m	—
Mass	kilogram	kg	—
Time[a]	second	s	—
Electric current	ampere	A	—
Thermodynamic temperature	kelvin	K	—
Amount of substance	mole	mol	—
Frequency	hertz	Hz	s^{-1}
Force	newton	N	$kg\,m\,s^{-2}$
Pressure	pascal	Pa	$N\,m^{-2}$
Work, energy, heat	joule	J	$N\,m,\ kg\,m^2\,s^{-2}$
Power	watt	W	$J\,s^{-1}$
Electric charge	coulomb[b]	C	$A\,s$
Electric potential	volt	V	$J\,C^{-1},\ W\,A^{-1}$
Electric resistance	ohm	Ω	$V\,A^{-1}$
Electric conductance	siemens	S	$A\,V^{-1},\ \Omega^{-1}$
Magnetic flux	weber	Wb	$V\,s$
Capacitance	farad	F	$C\,V^{-1}$
Magnetic field	tesla	T	$Wb\,m^{-2},\ N\,A^{-1}\,m^{-1}$
Celsius temperature	degree Celsius	°C	$= K - 273$

[a]Minute (min), hour (h), day (d), and year (y) are not SI units, but are accepted for use with SI units.
[b]The older literature on atmospheric electricity uses esu, where 1 esu $= 3.3 \times 10^{-10}$ C.
From Nelson, R. A., 1996: Guide for metric practice. *Phys. Today, August,* BG15–BG16.

UNIT PREFIXES

Factor	Prefix	Symbol
10^{21}	zetta	Z
10^{18}	exa	E
10^{15}	peta	P
10^{12}	tera	T
10^{9}	giga	G
10^{6}	mega	M
10^{3}	kilo	k
10^{2}	hecto	h
10^{1}	deka	da
10^{-1}	deci	d
10^{-2}	centi	c
10^{-3}	milli	m
10^{-6}	micro	µ
10^{-9}	nano	n
10^{-12}	pico	p
10^{-15}	femto	f
10^{-18}	atto	a
10^{-21}	zepto	z

From Cohen, E. R., and B. N. Taylor, 1987: The 1986 adjustment of the fundamental physical constants. *Rev. Mod. Phys.,* 57, 1121.

References

CHAPTER 1

Appleton, E.V., 1925: Replies to "Discussion on ionisation in the atmosphere." *Proc. Phys. Soc., Lond., 37,* 48D–50D.

Blakeslee, R.F., H.J. Christian, and B. Vonnegut, 1989: Electrical measurements over thunderstorms. *J. Geophys. Res., 94,* 13,135–40.

Brooks, C.E.P., 1925: The distribution of thunderstorms over the globe. *Geophys. Mem., Lond., 24,* 147–64.

Byrne, G.J., J.R. Benbrook, E.A. Bering, D. Oró, C.O. Seubert, and W.R. Sheldon, 1988: Observations of the stratospheric conductivity and its variation at three latitudes. *J. Geophys. Res., 93,* 3879–91.

Chalmers, J.A., 1967: *Atmospheric Electricity.* vol. 2, Pergamon, Oxford, p. 34.

Chamberlain, J.W., and D.M. Hunten, 1987: *Theory of Planetary Atmospheres,* Academic Press, San Diego, 481 pp.

Chapman, S., 1931a: The absorption and dissociative or ionizing effect of monochromatic radiation in an atmosphere on a rotating Earth, I. *Proc. Phys. Soc., 43,* 26–45.

Chapman, S., 1931b: The absorption and dissociative or ionizing effect of monochromatic radiation in an atmosphere on a rotating Earth, II. *Proc. Phys. Soc., 43,* 483–501.

Cleary, E.N., A.A. Few, and E.A.A. Bering, III, 1997: Statistics and trends of global atmospheric electricity measurements. *J. Geophys. Res.* Submitted.

Coulomb, C.A., 1795: *Mem. Acad. Sci.,* 616.

Crozier, W.D., 1965: Atmospheric electrical profiles below three meters. *J. Geophys. Res., 70,* 2785–92.

Dolezalek, H., 1972: Discussion of the fundamental problem of atmospheric electricity. *Pure and Appl. Geophys., 100,* 8–43.

Few, A.A., and A.J. Weinheimer, 1986: Factor of 2 error in balloon-borne atmospheric conduction current measurements. *J. Geophys. Res., 91,* 10,937–48.

Few, A.A., and A.J. Weinheimer, 1987: Reply. *J. Geophys. Res., 92,* 11,006–8.

Gerdien, H., 1905: Demonstration eines Apparates zur absoluten Messung der elektrischen Leitfähigkeit der Luft (Demonstration of a device for the absolute measurement of the electrical conductivity of air). *Phys., 6,* 647–66.

Gish, O.H., 1944: Evaluation and interpretation of the columnar resistance of the atmosphere. *Terr. Magn. Atmos. Elec., 49,* 159–68.

Gish, O.H., and K.L. Sherman, 1936: Electrical conductivity of air to an altitude of 23 km. *Nat. Geographic Soc. Stratosphere, Ser. 2,* 94–116.

Gish, O.H., and G.R. Wait, 1950: Thunderstorms and the Earth's general electrification. *J. Geophys. Res., 55,* 473–84.

Hanson, H.D., 1961: Structure of the ionosphere. In *Satellite Environment Handbook,* F.S. Johnson, ed., Stanford University Press, Stanford, Calif., pp. 27–46.

Heaps, M.G., 1978: Parametrization of the cosmic-ray ion-pair production rate above 18 km. *Planet. Space Sci., 26,* 513–17.

Heaviside, O., 1902: Theory of electric telegraphy. In *Electromagnetic Theory,* E. Weber, ed., Dover, New York, pp. 331–46.

Israël, H., 1970: *Atmospheric Electricity, Vol. 1, Fundamentals, Conductivity, Ions.* Israel Program for Scientific Translation, Jerusalem, pp. 114–16.

Israël, H., 1973: *Atmospheric Electricity, Vol. 2, Fields, Charges, Currents.* Israel Program for Scientific Translation, Jerusalem, p. 365.

Kasemir, H.W., 1977: Theoretical problems of the global atmospheric electric circuit. In *Electrical Processes in Atmospheres,* H. Dolezalek and R. Reiter, eds., Dr. Dietrich Steinkopff Verlag, Darmstadt, pp. 423–39.

Keese, R.G., and A.W. Castleman, Jr., 1985: Ions and cluster ions: Experimental studies and atmospheric observations. *J. Geophys. Res., 90,* 5885–90.

Kraakevik, J.H., 1961: Measurements of current density in the fair weather atmosphere. *J. Geophys. Res., 66,* 3735–48.

Markson, R., 1986: Tropical convection, ionospheric potentials and global circuit variation. *Nature, 320,* 588–94.

Markson, R., 1987: Comment on "Factor of 2 error in balloon-borne atmospheric conduction current measurements" by A.A. Few and A.J. Weinheimer, *J. Geophys. Res., 92,* 11,003–05.

Mason, B.J., 1971: *The Physics of Clouds,* Clarendon Press, Oxford, 671 pp.

Mitchell, J.D., and L.C. Hale, 1973: Observations of the lowest ionosphere *Space Res, XII,* 471–76.

Mohnen, V.A., 1977: Formation, nature, and mobility of ions of atmospheric importance. In *Electrical Processes in Atmospheres,* H. Dolezalek and R. Reiter, eds., Dr. Dietrich Steinkopff, Darmstadt, pp. 1–16.

Mühleisen, R., 1977: The global circuit and its parameters. In *Electrical Processes in Atmospheres,* H. Dolezalek and R. Reiter, eds., Dr. Dietrich Steinkopff, Darmstadt, pp. 467–76.

Ogawa, T., 1985: Fair-weather electricity. *J. Geophys. Res., 90,* 5951–60.

Peltier, A., 1842: Recherches sur la cause des phénomènes élec-triques de l'atmosphère et sur les moyen d'en recueiller la manifestation. *Ann. Chim. et de Phys., 4,* 385–433.

Reid, G.C., 1986: Electrical structure of the middle atmosphere. In *The Earth's Electrical Environment,* National Acad. Press, Washington, D.C., pp. 183–94.

Rosen, J.M., and D.J. Hofmann, 1981: Balloon-borne measurements of the small ion concentration. *J. Geophys. Res., 86,* 7399–7405.

Shen, J.S., W.E. Swartz, D.T. Farley, and R.M. Harper, 1976: Ionization layers in the nighttime *E* region valley above Arecibo. *J. Geophys. Res., 81,* 5517–26.

Stergis, C.G., G.C. Rein, and T. Kangas, 1957: Electric field measurements above thunderstorms, *J. Atmos. Terres. Phys. 11,* 83–91.

Thomson, W. (Lord Kelvin), 1860: On atmospheric electricity. *Roy. Inst. Lect: Papers on Electrostatics and Mag.* Also in *Physical Sciences The Royal Inst. Lib. of Sci., Vol. 1,* L. Bragg and G. Porter, ed., 1970, pp. 334–47.

Torreson, O.W., O.H. Gish, W.C. Parkinson, G.R. Wait, 1946: Scientific results of Cruise VII of the Carnegie during 1928–1929 under command of Captain J.P. Ault, Oceanography-III, Ocean atmospheric-electric results, Carnegie Inst. of Wash. Pub. 568, Washington, D.C.

Vonnegut, B., C.B. Moore, R.P. Espinola, and H.H. Blau, Jr., 1966: Electrical potential gradients above thunderstorms. *J. Atmos. Sci., 23,* 764–70.

Whipple, F.J.W., 1929: On the association of the diurnal variation of electric potential gradient in fine weather with the distribution of thunderstorms over the globe. *Q. J. Roy. Meteor. Soc., 55,* 1–17.

Whipple, F.J.W., and F.J. Scrase, 1936: Point-discharge in the electric field of the Earth. *Geophys. Mem. VII, No. 68,* 1–20.

Whitten, R.C., and I.G. Poppoff, 1971: *Fundamentals of Aeronomy,* John Wiley and Sons, New York, 446 pp.

Wilkening, M., 1985: Characteristics of atmospheric ions in contrasting environments. *J. Geophys. Res., 90,* 5933–35.

Willett, J., 1985: Atmospheric-electrical implications of ^{222}Rn daughter deposition on vegetated ground. *J. Geophys. Res., 90,* 5901–08.

Wilson, C.T.R., 1920: Investigations on lightning discharges and on the electric field of thunderstorms. *Phil. Trans. Roy. Soc. Lond., A, 221,* 73–115.

Woessner, R.H., W.E. Cobb, and R. Gunn, 1958: Simultaneous measurements of the positive and negative light-ion conductivities to 26 km. *J. Geophys. Res., 63,* 171–80.

CHAPTER 2

Allee, P.A., and B.B. Phillips, 1959: Measurements of cloud-droplet charge, electric field, and polar conductivities in supercooled clouds. *J. Appl. Meteor., 16,* 405–10.

Andreeva, S.I., and B.F. Evteev, 1974: The potential gradient of the electric field in nimbostratus clouds. In *Studies in Atmospheric Electricity,* V.P. Kolokilov and T.V. Lobodin, eds., trans. from Russian, Israel Prog. for Sci. Trans., Jerusalem, 1–5.

Brylev, G.B., S.B. Gashina, B.F. Yevteyev, and I.I. Kamaldyna, 1989: *Characteristics of Electrically Active Regions in Stratiform Clouds,* 303 pp. USAF translation, FTD-ID(RS)T-0698-89, of *Kharakteristiki Elektricheski Aktivnykh Zon v Sloistoobraznykh Oblakakh,* Gidrometeoizdat, Leningrad, 160 pp.

Burke, H.K., and A.A. Few, 1978: Direct measurements of the atmospheric conduction current. *J. Geophys. Res., 83,* 3093–98.

Chalmers, J.A., 1956: The vertical electrical currents during continuous rain and snow. *J. Atmos. Terres. Phys., 9,* 311–21.

Christian, H.C., V. Mazur, B.D. Fisher, L.H. Rhunke, K. Crouch, and R.P. Perala, 1989: The Atlas/Centaur lightning strike incident. *J. Geophys. Res., 94,* 13,169–77.

Chubarina, Ye. V., 1977: Large electric fields in the clouds of laminar form (in Russian). *Trans. GGO, Iss. 350,* 80–6.

Imyanitov, I.M., and Ye.V. Chubarina, 1967: *Electricity of the Free Atmosphere,* NASA Tech. Translation, NASA TT F-425, TT 67-51374 of *Elektrichestvo Svobodnoy Atmosfery,* Gidrometeoizdat, Leningrad, 1965, NTIS Accession no. N68–10079, 212 pp.

Imyanitov, I.M., Ye.V. Chubarina, and Ya.M. Shvarts, 1972: *Electricity of Clouds,* NASA Tech. Translation, NASA TT F-718, of *Elektrichestvo oblakov.* Hydrometeorological Press, Leningrad, 1971, 122 pp.

Phillips, B.B., and G.D. Kinzer, 1958: Measurements of the size and electrification of droplets in cumuliform clouds. *J. Meteor., 15,* 369–74.

Reiter, R., 1958: Observations on the electricity of nimbostratus clouds. In *Recent Advances in Atmospheric Electricity,* L.G. Smith, ed., Pergamon, New York, pp. 435–37.

Reiter, R., 1965: Precipitation and cloud electricity. *Q. J. Roy. Meteor. Soc., 91,* 60–72.

Rust, W.D., and C.B. Moore, 1974: Electrical conditions near the bases of thunderclouds over New Mexico, *Q. J. Roy. Meteor. Soc., 100,* 450–68.

Whitlock, W.S., and J.A. Chalmers, 1956: Short-period variations in the atmospheric electric potential gradient. *Q. J. Roy. Meteor. Soc. 16,* 325–36.

[WMO] World Meteorological Organization, 1969: *International Cloud Atlas (Abridged Atlas).* WMO, Geneva, 62 pp. 72 plates.

CHAPTER 3

Abbas, M.A., and J. Latham, 1967: The instability of evaporating charge drops. *J. Fluid Mech., 30,* 663–70.

Al-Saed, S.M., and C.P.R. Saunders, 1976: Electric charge transfer between colliding water drops. *J. Geophys. Res., 81,* 2650–2654.

Atlas, D., R.C., Srivastava, and R.S. Sekhon, 1973: Doppler radar characteristics of precipitation at vertical incidence. *Rev. Geophys. Space Phys., 2,* 1–35.

Aufdermaur, A.N., and D.A. Johnson, 1972: Charge separation due to riming in an electric field. *Q. J. Roy. Meteor. Soc., 98,* 369–82.

Avila, E.E., G.M. Caranti, and M.A. Lamfri, 1988: Charge reversal in individual ice-ice collisions. In *Proc. of the 8th Int'l Conf. on Atmos. Elec.,* Uppsala, Sweden, IAMAP, 245–50.

Avila, E.E., and G.M. Caranti, 1994: A laboratory study of static charging by fracture in ice growing by riming. *J. Geophys. Res., 99,* 10,611–20.

Avila, E.E., G.G.A. Varela, and G.M. Caranti. 1995: Temperature dependence of static charging in ice growing by riming. *J. Atmos. Sci., 52,* 4515–22.

Baker, B., M.B. Baker, E.R. Jayaratne, J. Latham, and C.P.R. Saunders, 1987: The influence of diffusional growth rate on the charge transfer accompanying rebounding collisions between ice crystals and hailstones. *Quart. J. Roy. Meteor. Soc., 113,* 1193–215.

Baker, M.B., and J.G. Dash, 1994: Mechanism of charge transfer between colliding ice particles in thunderstorms. *J. Geophys. Res., 99,* 10,621–26.

Baughman, B.G., and D.M. Fuquay, 1970: Hail and lightning occurrence in mountain thunderstorms. *J. Appl. Meteor., 9,* 657–60.

Beard, K.V.K., and H.T. Ochs, 1986: Charging mechanisms in clouds and thunderstorms. In *The Earth's Electrical Environment,* National Acad. Press, Washington, D.C., pp. 114–30.

Blakeslee, R.J., H.J. Christian, and B. Vonnegut, 1989: Electrical measurements over thunderstorms. *J. Geophys. Res., 94,* 13,135–40.

Blythe, A.M., W.A. Cooper, and J.B. Jensen, 1988: A study of the source of entrained air in Montana cumuli. *J. Atmos. Sci., 45,* 3944–64.

Bourdeau, C., and S. Chauzy, 1989: Maximum electric charge of a hydrometeor in the electric field of a thunderstorm. *J. Geophys. Res., 94,* 13,121–26.

Brooks, I.M., and C.P.R. Saunders, 1994: An experimental investigation of the inductive mechanism of thunderstorm electrification. *J. Geophys. Res., 99,* 10,627–32.

Brooks, I.M., and C.P.R. Saunders, 1995: Thunderstorm charging: Laboratory experiments clarified. *Atmos. Res., 39,* 263–73.

Brooks, I.M., C.P.R. Saunders, R.P. Mitzeva, and S.L. Peck, 1997: The effect on thunderstorm charging of the rate of rime accretion by graupel. *Atmos. Res. 43,* 277–95.

Brown, K.A., P.R. Krehbiel, C.B. Moore, and G.N. Sargent, 1971: Electrical screening layers around charged clouds. *J. Geophys. Res., 76,* 2825–35.

Buser, O., and A.N. Aufdermaur, 1977: Electrification by collisions of ice particles on ice or metal targets. In *Electrical Processes in Atmospheres,* H. Dolezalek and R. Reiter, eds., Dr. Dietrich Steinkopff, Darmstadt, Germany, pp. 294–300.

Byrne, G.J., A.A. Few, and M.E. Weber, 1983: Altitude, thickness, and charge concentrations of charge regions of four thunderstorms during TRIP 1981 based upon in situ balloon electric field measurements, *Geophys. Res. Lett., 10,* 39–42.

Canosa, E.R., and R. List, 1993: Measurements of inductive charges during drop breakup in horizontal electric fields. *J. Geophys. Res., 98,* 2619–26.

Caranti, J.M., and A.J. Illingworth, 1983: The contact potential of rimed ice. *J. Phys. Chem., 87,* 4125–30.

Caranti, J.M., A.J. Illingworth, and S.J. Marsh, 1985: The charging of ice by differences in contact potential. *J. Geophys. Res., 90,* 6041–46.

Caranti, J., E. Avila, and M. Re, 1991: The charge transfer during individual collisions in vapor growing ice. *J. Geophys. Res., 96,* 15,365–75.

Censor, D., and Z. Levin, 1973: Electrostatic interaction of axisymmetric liquid and solid aerosols. *Pub. ES 73-015,* Dept. of Environ. Sci., Tel Aviv Univ., Israel.

Chalmers, J.A., 1967: *Atmospheric Electricity.* Pergamon, Oxford, 515 pp.

Chauzy, S., and P. Raizonville, 1982: Space charge layers created by coronae at ground level below thunderclouds: Measurements and modelling. *J. Geophys. Res., 87,* 3143–48.

Chiu, C.S., 1978: Numerical study of cloud electrification in an axisymmetric, time-dependent cloud model. *J. Geophys. Res., 83,* 5025–49.

Chiu, C.S., and J.D. Klett, 1976: Convective electrification of clouds. *J. Geophys. Res., 81,* 1111–24.

Dash, J.G., 1989: Surface melting. *Contemp. Phys., 30,* 89–100.

Davis, M.H., 1964: Two charged spherical conductors in a uniform electric field: Forces and field strength. *Quart. J. Mech. Appl. Math., 17,* 499–511.

Dawson, G.A., 1973: Charge loss mechanism of highly charged water droplets in the atmosphere. *J. Geophys. Res., 78,* 6364–69.

Dinger, J.E., and R. Gunn, 1946: Electrical effects associated with a change of state of water. *Terr. Magn. Atmos. Electr., 51,* 477–94.

Dong, Y., and J.H. Hallett, 1992: Charge separation by ice and water drops during growth and evaporation. *J. Geophys. Res., 97,* 20,361–71.

Doyle, A., D.R. Moffet, and B. Vonnegut, 1964: Behavior of evaporating electrically charged droplets. *J. Coll. Sci., 19,* 136–43.

Dye, J.E., J.J. Jones, A.J. Weinheimer, and W.P. Winn, 1988: Observations within two regions of charge during initial thunderstorm electrification. *Q. J. Roy. Meteor. Soc., 114,* 1271–90.

Dye, J.E., W.P. Winn, J.J. Jones, and D.W. Breed, 1989: The electrification of New Mexico thunderstorms 1. Relationship between precipitation development and the onset of electrification. *J. Geophys. Res., 94,* 8643–56.

Elbaum, M., S. Lipson, and J.G. Dash, 1992: Optical study of surface melting of ice. *J. Cryst. Growth, 129,* 491.

Elster, J., and H., Geitel, 1888: Über eine Methode, die elektrische Natur der atmosphärischen Niederschläge zu bestimmen (About a method for determining the electric nature of atmospheric precipitation). *Meteor. Z., 5,* 95–100.

Faraday, M., 1860: Note on regelation. *Proc. Roy. Soc. London, 10,* 440.

Findeisen, W., 1940: Über die Entstehung der Gewitterelektrizität. *Meteorol. Z., 57,* 201–15.

Fletcher, N.H., 1962: Surface structure of water and ice. *Philos. Mag., 7,* 255–69.

Fletcher, N.H., 1968: Surface structure of water and ice, 2, A revised model. *Philos. Mag., 18,* 1287–1300.

Furukawa, Y., M. Yamamoto, and T. Kuroda, 1987: Ellipsometric study of the transition layer on the surface of an ice crystal. *J. Crystal Growth, 82,* 665–77.

Gaskell, W., 1981: A laboratory study of the inductive theory of thunderstorm electrification. *Quart. J. Roy. Meteor. Soc., 107,* 955–66.

Gaskell, W., A.J. Illingworth, J. Latham, C.B. Moore, 1978: Airborne studies of electric fields and the charge and size of precipitation elements in thunderstorms, *Q. J. Roy. Meteor. Soc., 104,* 447–60.

Gaskell, W., and A.J. Illingworth, 1980: Charge transfer accompanying individual collisions between ice particles and its role in thunderstorm electrification. *Quart. J. Roy. Meteor. Soc., 106,* 841–54.

Gish, O.H., and G.R. Wait, 1950: Thunderstorms and the Earth's general electrification. *J. Geophys. Res., 55,* 473–84.

Golecki, I., and C. Jaccard, 1978: Intrinsic surface disorder in ice near the melting point. *J. Phys. C, 11,* 4229.

Grenet, G., 1947: Essai d'explication de la charge électrique des nuages d'orages. *Ann. de Géophys., 3,* 306–07.

Gunn, R., 1957: The electrification of precipitation and thunderstorms. *Proc. IRE, 45,* 1331–1358.

Gunn, R., and G.D. Kinzer, 1949: The terminal velocity of fall for water droplets in stagnant air. *J. Meteor., 6,* 243–48.

Helsdon, J.H., Jr., and R.D. Farley, 1987: A numerical modeling study of a Montana thunderstorm: 2. Model results versus observations involving electrical aspects. *J. Geophys. Res., 92,* 5661–75.

Hoppel, W.A., and B.B. Phillips, 1971: The electrical shielding layer around charged clouds and its role in thunderstorm electricity. *J. Atmos. Sci., 28,* 1258–71.

Huzita, H., and T. Ogawa, 1950: Charge distribution in the average thunderstorm cloud. *J. Meteor. Jpn., 54,* 285–88.

Illingworth, A.J., 1985: Charge separation in thunderstorms: Small scale processes. *J. Geophys. Res., 90,* 6026–32.

Illingworth, A.J., and J.M. Caranti, 1985: Ice conductivity restraints on the inductive theory of thunderstorm electrification. *J. Geophys. Res., 95,* 6033–39.

Iribarne, J.V., 1972: The electrical double layer and electrification associated with water disruption processes. *J. Rech. Atmos., 6,* 265–81.

Jackson, J.D., 1962: *Classical Electrodynamics.* John Wiley and Sons, New York, 641 pp.

Jayaratne, E.R., 1993a: Temperature gradients in ice as a charge generation process in thunderstorms. *Atmos. Res., 29,* 247–60.

Jayaratne, E.R., 1993b: The heat balance of a riming graupel pellet and the charge separation during ice-ice collisions. *J. Atmos. Sci., 50,* 3185–93.

Jayaratne, E.R., C.P.R. Saunders, and J. Hallett, 1983: Laboratory studies of the charging of soft hail during ice crystal interactions. *Quart. J. Roy. Meteor. Soc., 109,* 609–30.

Jayaratne, E.R., and C.P.R. Saunders, 1985: Thunderstorm electrification: The effect of cloud droplets. *J. Geophys. Res., 90,* 13,063–66.

Jayaratne, E.R., and D.J. Griggs, 1991: Electric charge separation during the fragmentation of rime in an airflow. *J. Atmos. Sci., 48,* 2492–95.

Jennings, S.G., 1975: Charge separation due to water droplet and cloud droplet interactions in an electric field. *Quart. J. Roy. Meteor. Soc., 101,* 227–34.

Jensen, J.C., 1933: The branching of lightning and the polarity of thunderclouds. *J. Franklin Inst., 216,* 707–47.

Kasemir, H.W., 1965: The thundercloud. In *Problems of Atmospheric and Space Electricity,* S. Coroniti, ed., Elsevier, Amsterdam, pp. 215–31.

Keith, W.D., and C.P.R. Saunders, 1988: Light emission from colliding ice particles. *Nature, 336,* 362–64.

Keith, W.D., and C.P.R. Saunders, 1990: Further laboratory studies of the charging of graupel during ice crystal interactions. *Atmos. Res., 25,* 445–64.

Kikuchi, K., 1965: On the positive electrification of snow crystals in the process of their melting, III, IV. *J. Meteor. Soc. Jpn., 43,* 343–50, 351–58.

Klett, J.D., 1972: Charge screening layers around electrified clouds. *J. Geophys. Res., 77,* 3187–3195.

Krehbiel, P.R., 1981: An analysis of the electric field change produced by lightning. Ph.D. diss., U. Manchester Inst. Sci. & Tech. Published as Rpt. T-11, New Mex. Inst. Min. & Tech., Dec., 1981. Vol. 1 is 245 pp. and Vol. 2 is figures.

Krehbiel, P.R., 1986: The electrical structure of thunderstorms. In *The Earth's Electrical Environment,* National Acad. Press, Washington, D.C., pp. 90–113.

Krider, E.P., and J.A. Musser, 1982: Maxwell currents under thunderstorms. *J. Geophys. Res., 87,* 1982.

Krider, E.P., and R.J. Blakeslee, 1985: The electric currents produced by thunderclouds. *J. Electrostat., 16,* 369–78.

Kuettner, J., 1950: The electrical and meteorological conditions inside thunderstorms. *J. Meteor., 7,* 322–32.

Latham, J., 1981: The electrification of thunderstorms. *Quart. J. Roy. Meteor. Soc., 107,* 277–98.

Latham, J., and B.J. Mason, 1961: Electric charge transfer associated with temperature gradients in ice. *Proc. Roy. Soc. London, A260,* 523–36.

Latham, J., and Mason, B.J., 1962: Electric charging of hail pellets in a polarizing electric field. *Proc. Roy. Soc. London, A, 26,* 387–401.

Latham, J., and R. Warwicker, 1980: Charge transfer accompanying the splashing of supercooled raindrops on hailstones. *Quart. J. Roy. Meteor. Soc., 106,* 559–68.

Lenard, P., 1892: Über die Elektrizität der Wasserfälle (About the electricity of waterfalls). *Ann. Phys. (Leipzig), 46,* 584–636.

Livingston, J.M., and E.P. Krider, 1978: Electric fields produced by Florida thunderstorms, *J. Geophys. Res., 83,* 385–401.

Low, T.B., and R. List, 1982: Collision, coalescence and breakup of raindrops, I. Experimentally established coalescence efficiencies and fragment size distributions in breakup. *J. Atmos. Sci., 39,* 1591–1606.

MacGorman, D.R., C.L. Ziegler, and J.M. Straka, 1996: Considering the complexity of thunderstorms. In *Proceeding 10th International Conference on Atmospheric Electricity,* Osaka, Japan. pp. 128–131.

Magono, C., 1980: *Thunderstorms.* Elsevier Sci., Amsterdam, 261 pp.

Magono, C., and S. Koenuma, 1958: On the electrification of water drops by breaking due to the electrostatic induction under a moderate electric field. *J. Meteor. Soc. Jpn., 36,* 108–11.

Magono, C., and K. Kikuchi, 1965: On the positive electrification of snow crystals in the process of their melting. *J. Meteor. Soc. Jpn., 41,* 331–42.

Malan, D.J., 1952: Les décharges orageuses dans l'air et la charge inférieur positive d'un nuage orageux. *Ann. Géophys., 8,* 385–401.

Marshall, B.J.P., J. Latham, and C.P.R. Saunders, 1978: A laboratory study of charge transfer accompanying the collision of ice crystals with a simulated hailstone. *Quart. J. Roy. Meteor. Soc., 104,* 163–78.

Marshall, T.C., and W.P. Winn, 1982: Measurements of charged precipitation in a New Mexico thunderstorm: lower positive charge centers. *J. Geophys. Res., 87,* 7141–57.

Marshall, T.C., W.D. Rust, W.P. Winn, and K.E. Gilbert, 1989: Electrical structure in two thunderstorm anvil clouds. *J. Geophys. Res., 94,* 2171–81.

Marshall, T.C., and W.D. Rust, 1991: Electric field soundings through thunderstorms. *J. Geophys. Res., 96,* 22,297–306.

Mason, B.J., 1957: *The Physics of Clouds.* Oxford Univ. Press, London, pp. 364–419.

Mason, B.J., 1988: The generation of electric charges and fields in thunderstorms. *Proc. Roy. Soc. London, A327,* 303–15.

Mazur, V., J.C. Gerlach, and W.D. Rust, 1984: Lightning flash density versus altitude and storm structure from observations with UHF- and S-band radars. *Geophys. Res. Lett., 11,* 61–64.

McCappin, C.J., and W.C. Macklin, 1984: The crystalline structure of ice formed by droplet accretion. Part I: Fresh samples. *J. Atmos. Sci., 41,* 2437–45.

McKnight, C.V., and J. Hallett, 1978: X-ray topographic studies of dislocations in vapor-grown ice crystals. *J. Glaciol., 21,* 397–407.

McTaggart-Cowan, J.D., and R. List, 1975: Collision and breakup of water drops at terminal velocity. *J. Atmos. Sci., 32,* 1401–11.

Moore, C.B., 1975: Rebound limits on charge separation by falling precipitation. *J. Geophys. Res., 80,* 2658–62.

Moore, C.B., 1976: Reply, to "Further comments on Moore's criticisms of precipitation theories of thunderstorm electrification," by B.J. Mason. *Q. J. Roy. Meteor. Soc., 102,* 935–39.

Moore, C.B., and B. Vonnegut, 1977: The thundercloud. In *Lightning, Vol. 1: Physics of Lightning,* R.H. Golde, eds., Academic Press, San Diego, pp. 64–98.

Moore, C.B., B. Vonnegut, and D.N. Holden, 1989: Anomalous electric fields associated with clouds growing over a source of negative space charge. *J. Geophys. Res., 94,* 13,127–34.

Paluch, I.R., and J.D. Sartor, 1973a: Thunderstorm electrification by the inductive charging mechanism: I, Particle charges and electric fields. *J. Atmos. Sci., 30,* 1166–73.

Paluch, I.R., and J.D. Sartor, 1973b: Thunderstorm electrification by the inductive charging mechanism: II, Possible effects of updraft on the charge separation process. *J. Atmos. Sci., 30,* 1174–77.

Petrenko, V.F., and R.W. Whitworth, 1980: Charged dislocations and the plastic deformation of II-VI compounds. *Phil. Mag., 41A,* 681–99.

Pruppacher, H.R., and J.D. Klett, 1978: *Microphysics of Clouds and Precipitation.* D. Reidel, Dordrecht, pp. 580–624.

Rayleigh, Lord, 1879: The influence of electricity on colliding water drops. *Proc. Roy. Soc., XXVIII,* 406–09. Also in *Scientific Papers by Lord Rayleigh, Vol. II,* Dover Pub., New York, 1964, pp. 372–76.

Rayleigh, Lord, 1882a: On the equilibrium of liquid conducting masses charged with electricity. *Phil. Mag., XIV,* 184–86. Also in *Sci. Pap. by Lord Rayleigh, Vol. II* Dover Pub., New York, 1964, pp. 130–31.

Rayleigh, Lord, 1882b: Further observations upon liquid jets, in continuation of those recorded in the Royal Society's 'Proceedings' for March and May, 1879. *Proc. Roy. Soc., XXXIV,* 130–45. Also in *Scientific Papers by Lord Rayleigh, Vol. II,* Dover Pub., New York, 1964, pp. 103–17.

Reynolds, S.E., 1953: Thunderstorm-precipitation growth and electrical-charge generation. *Bull. Amer. Meteor. Soc., 34,* 117–23.

Reynolds, S.E., M. Brook, and M.F. Gourley, 1957: Thunderstorm charge separation. *J. Meteor., 14,* 426–36.

Roulleau, M., and M. Desbois, 1972: Study of evaporation and instability of charged water droplets. *J. Atmos. Sci., 29,* 565–69.

Rust, W.D., and C.B. Moore, 1974: Electrical conditions near the bases of thunderclouds over New Mexico, *Q. J. Roy. Meteor. Soc., 100,* 450–68.

Rust, W.D., and T.C. Marshall, 1996: On abandoning the thunderstorm tripole-charge paradigm. *J. Geophys. Res., 101,* 23,499–504.

Rydock, J., and E.R. Williams, 1991: Charge separation associated with frost growth. *Quart. J. Roy. Meteor. Soc., 117,* 409–20.

Sartor, D., 1954: A laboratory investigation of collision efficiencies, coalescence and electrical charging of simulated cloud droplets. *J. Meteor., 11,* 91–103.

Sartor, J.D., 1981: Induction charging of clouds. *J. Atmos. Sci., 38,* 218–20.

Saunders, C.P.R., 1993: A review of thunderstorm electrification processes. *J. Appl. Meteor. 32,* 642–55.

Saunders, C.P.R., 1994: Thunderstorm electrification laboratory experiments and charging mechanisms. *J. Geophys. Res., 99,* 10773–10779.

Saunders, C.P.R., W.D. Keith, and R.P. Mitzeva, 1991: The effect of liquid water on thunderstorm charging. *J. Geophys. Res., 96,* 11,007–17.

Saunders, C.P.R., and I.M. Brooks, 1992: The effects of high liquid water content on thunderstorm charging. *J. Geophys. Res., 97,* 14,671–76.

Saunders, C.P.R., H. Hickson, M.D. Malone, and J. von Richtofen, 1993: Charge separation during the fragmentation of rime and frost. *Atmos. Res., 29,* 261–70.

Simpson, G.C., 1909: On the electricity of rain and snow. *Proc. Roy. Soc. Lond., A, 83,* 394–404.

Simpson, Sir G., 1949: Atmospheric electricity during disturbed weather, *Geophys. Mem. No. 84, Lond.*

Simpson, Sir G., and F.J. Scrase, 1937: The distribution of electricity in thunderclouds, *Proc. Roy. Soc. Lond., A, 161,* 309–52.

Simpson, Sir G., and G.D. Robinson, 1941: The distribution of electricity in thunderclouds, II, *Proc. Roy. Soc. Lond., A, 177,* 281–328.

Soula, S., 1994: Transfer of electrical space charge from corona between ground and thundercloud: Measurements and modeling. *J. Geophys. Res., 99,* 10759–65.

Soula, S., and S. Chauzy, 1991: Multilevel measurement of the electric field underneath a thundercloud; 2. Dynamical evolution of a ground space charge layer. *J. Geophys. Res., 96,* 22327–36.

Sporn, P., and W.L. Lloyd, Jr., 1931: Lightning investigations on the transmission system of the American Gas and Electric Company. *Trans. AIEE, 49,* 1111–17.

Standler, R.B., 1980: Estimation of corona current beneath thunderclouds. *J. Geophys. Res., 85,* 4541–4544.

Standler, R.B., and W.P. Winn, 1979: Effects of coronae on electric fields beneath thunderstorms. *Q. J. Roy. Meteor. Soc., 105,* 285–302.

Stergis, C.G., G.C. Rein, and T. Kangas, 1957: Electric field measurements above thunderstorms, *J. Atmos. Terres. Phys., 11,* 83–91.

Stith, J.L., 1992: Observations of cloud-top entrainment in cumuli. *J. Atmos. Sci., 49,* 1334–47.

Stolzenburg, M., 1996: An observational study of electrical structure in convective regions of mesoscale convective systems. Ph.D. diss., Univ. of Oklahoma, Norman, 137 pp.

Stolzenburg, M., and T.C. Marshall, 1994: Testing models of thunderstorm charge distributions with Coulomb's law. *J. Geophys. Res., 99,* 25,921–32.

Stow, C.D., 1969: Atmospheric electricity. *Rep. Prog. Phys., 32,* 1–67.

Stromberg, I.M., 1971: Point discharge current measurements in a plantation of spruce trees using a new pulse technique, *J. Atmos. Terres. Phys., 33,* 485–95.

Takahashi, T., 1966: Thermoelectric effect in ice. *J. Atmos. Sci., 23,* 74–7.

Takahashi, T., 1969: Electric potential of liquid water on an ice surface. *J. Atmos. Sci., 26,* 1253–58.

Takahashi, T., 1978: Riming electrification as a charge generation mechanism in thunderstorms. *J. Atmos. Sci., 35,* 1536–48.

Takahashi, T., 1979: Warm cloud electricity in a shallow axisymmetric cloud model. *J. Atmos. Sci., 36,* 2236–58.

Tamura, Y., 1955: An analysis of electric field after lightning discharges. *J. Geomag. and Geoelec., 6,* 34–46.

Taylor, W.L., E.A. Brandes, W.D. Rust, and D.R. MacGorman, 1984: Lightning activity and severe storm structure. *Geophys. Res. Lett., 11,* 545–548.

Tzur, I., and R.G. Roble, 1985: The interaction of a dipolar thunderstorm with its global environment. *J. Geophys. Res., 90,* 5989–99.

Vonnegut, B., 1953: Possible mechanism for the formation of thunderstorm electricity. *Bull. Amer. Meteor. Soc., 34,* 378.

Vonnegut, B., 1955: Possible mechanism for the formation of thunderstorm electricity. *Geophys. Res. Pap. No. 42,* Proc. Conf. Atmos. Elec., May 19–21, 1954, 169–81.

Vonnegut, B., 1994: The atmospheric electricity paradigm. *Bull. Am. Meteor. Soc., 75,* 53–61.

Vonnegut, B., C.B. Moore, R.G. Semonin, J.W. Bullock, D.W. Staggs, and W.E. Bradley, 1962: Effect of atmospheric space charge on initial electrification of cumulus clouds. *J. Geophys. Res., 67,* 3909–21.

Wait, G.R., 1953: Aircraft measurements of electric charge carried to ground through thunderstorms. *Thunderstorm Electricity,* (ch. 10), H.R. Byers, eds., Univ. Chicago Press, Chicago, pp. 231–37.

Weinheimer, A.J., 1987: The electrostatic energy of a thunderstorm and its rate of change. *J. Geophys. Res., 92,* 9715–22.

Whipple, F.J.W., and F.J. Scrase, 1936: Point-discharge in the electric field of the Earth. *Geophys. Mem. VII, No. 68,* 1–20.

Whipple, F.J.W., and J.A. Chalmers, 1944: On Wilson's theory of the collection of charge by falling drops. *Quart. J. Roy. Met. Soc., 70,* 103.

Williams, E.R., 1985: Large-scale charge separation in thunderclouds. *J. Geophys. Res., 90,* 6013–25.

Williams, E.R., 1989: The tripole structure of thunderstorms. *J. Geophys. Res., 94,* 13,151–67.

Williams, E.R., R. Zhang, and R. Rydock, 1991: Mixed-phase microphysics and cloud electrification. *J. Atmos. Sci., 48,* 2195–2203.

Williams, E.R., R. Zhang, and D. Boccippio, 1994: Microphysical growth state of ice particles and large-scale electrical structure of clouds. *J. Geophys. Res., 99,* 10,787–92.

Wilson, C.T.R., 1916: On some determinations of the sign and magnitude of electric discharges in lightning flashes. *Proc. Roy. Soc. Lond., A, 92,* 555–74.

Wilson, C.T.R., 1920: Investigations on lightning discharges and on the electric field of thunderstorms. *Phil. Trans. Roy. Soc. Lond., A, 221,* 73–115.

Wilson, C.T.R., 1929: Some thundercloud problems. *J. Franklin Inst., 208,* 1–12.

Winn, W.P., 1992: Electrification of thunderclouds by the transport of bare charge: Implications of Grenet's and Vonnegut's hypothesis. *Eos, Trans. AGU, 73,* Fall meeting supplement, 110. Abstract.

Winn, W.P., C.B. Moore, and C.R. Holmes, 1981: Electric field structure in an active part of a small, isolated thundercloud. *J. Geophys. Res., 86,* 1187–93.

Winn, W.P., F. Han, J.J. Jones, D.J. Raymond, T.C. Marshall, and S.J. Marsh, 1988: Thunderstorm with anomalous charge. In *Proceedings 8th International Conference on Atmospheric Electricity,* Uppsala, Sweden, IAMAP, pp. 590–95.

Workman, E.J., and S.E. Reynolds, 1948: A suggested mechanism for the generation of thunderstorm electricity. *Phys. Rev., 74,* 709.

Workman, E.J., and S.E. Reynolds, 1950: Electrical phenomena occurring during the freezing of dilute aqueous solutions and their possible relationship to thunderstorm electricity. *Phys. Rev., 78,* 254–59.

Wormell, T.W., 1930: Vertical electric currents below thunderstorms and showers. *Proc. Roy. Soc. Lond., A, 127,* 567–90.

Wormell, T.W., 1939: The effect of thunderstorms and lightning discharges on the Earth's electric field. *Phil. Trans. Roy. Soc. Lond., A, 328,* 249–303.

Ziegler, C.L., D.R. MacGorman, P.S. Ray, and J.E. Dye, 1991: A model evaluation of noninductive graupel-ice charging in the early electrification of a mountain thunderstorm. *J. Geophys. Res., 96,* 12,833–55.

Ziegler, C.L., and D.R. MacGorman, 1994: Observed lightning morphology relative to modeled space charge and electric field distributions in a tornadic storm. *J. Atmos. Sci., 51,* 833–51.

CHAPTER 4

Chauzy, S., and P. Raizonville, 1982: Space charge layers created by coronae at ground level below thunderclouds: Measurements and modelling. *J. Geophys. Res., 87,* 3143–48.

Coquillat, S., and S. Chauzy, 1993: Behavior of precipitating water drops under the influence of electrical and aerodynamical forces. *J. Geophys. Res., 98,* 10,319–24.

Coquillat, S., and S. Chauzy, 1994: Computed conditions of corona emission from raindrops. *J. Geophys. Res., 99,* 16,897–905.

Crabb, J.A., and J. Latham 1974: Corona from colliding drops as a possible mechanism for the triggering of lightning. *Q. J. Roy. Meteor. Soc., 100,* 191–202.

Dawson, G.A., 1969. Pressure dependence of water-drop corona onset and its atmospheric importance. *J. Geophys. Res., 74,* 6859–68.

Ette, A.I.I., and E.U. Utah, 1973: Studies of point-discharge characteristics in the atmosphere. *J. Atmos. Terres. Phys., 35,* 1799–1809.

Griffiths, R.F., C.T. Phelps, and B. Vonnegut, 1973: Charge transfer from a highly electrically stressed water surface during drop impact. *J. Atmos. Terres. Phys., 35,* 1967–78.

Griffiths, R.F., and J. Latham, 1974: Electrical corona from ice hydrometeors. *Q. J. Roy. Meteor. Soc., 100,* 163–80.

Jhawar, D.S., 1968: The relation of point-discharge current to other parameters. *J. Atmos. Terres. Phys., 30,* 113–23.

Jhawar, D.S., and J.A. Chalmers, 1967: Point-discharge current through small trees in artificial fields. *J. Atmos. Terres. Phys., 29,* 1459–63.

Large, M.I., and E.T. Pierce, 1957: The dependence of point-discharge currents on wind, as examined by a new experimental approach. *J. Atmos. Terres. Phys., 10,* 251–57.

Latham, J., and I.M. Stromberg, 1977: Point-discharge. In *Lightning, Vol. 1: Physics of Lightning,* R.H. Golde, ed., Academic Press, San Diego, pp. 99–117.

Livingston, J.M., and E.P. Krider, 1978: Electric fields produced by Florida thunderstorms. *J. Geophys. Res., 83,* 385–401.

Loeb, L.B., 1965: *Electrical Coronas. Their Basic Physical Mechanisms.* Univ. Calif. Press, Berkeley, 694 pp.

McGinley, J., C. Ziegler, and D. Ziegler, 1982: Observations of steady glow and multicolored flashes associated with a thunderstorm. *Bull. Amer. Meteor. Soc., 63,* 189–90.

Standler, R.B., 1980: Estimation of corona current beneath thunderclouds. *J. Geophys. Res., 85,* 4541–44.

Standler, R.B., and W.P. Winn, 1979: Effects of coronae on electric fields beneath thunderstorms. *Q. J. Roy. Meteor. Soc., 105,* 285–302.

Stromberg, I.M., 1971: Point discharge current measurements in a plantation of spruce trees using a new pulse technique. *J. Atmos. Terres. Phys., 33,* 485–95.

Toland, R.B., and B. Vonnegut, 1977: Measurement of maximum electric field intensities over water during thunderstorms. *J. Geophys. Res., 82,* 438–40.

Vonnegut, B., 1974: Electric potential above ocean waves. *J. Geophys. Res., 79,* 3480–81.

Whipple, F.J.W., and F.J. Scrase, 1936: Point-discharge in the electric field of the Earth. *Geophys. Mem. No. 68,* 1–20.

Winn, W.P., and C.B. Moore, 1972: Reply. *J. Geophys. Res., 77,* 506–08.

Wormell, T.W., 1927: Currents carried by point-discharge beneath thunderclouds and showers. *Proc. Roy. Soc. Lond., A, 115,* 443–55.

CHAPTER 5

Ault, C., 1916: Thunder at sea. *Sci. Am., 218,* 525.

Balachandran, N.K., 1983: Acoustic and electrical signals from lightning. *J. Geophys. Res., 88,* 3879–84.

Barry, J.D., 1980: *Ball Lightning and Bead Lightning.* Plenum Press, New York, 298 pp.

Bass, H.E., and R.E. Losely, 1975: Effect of atmospheric absorption on the acoustic power spectrum of thunder. *J. Acous. Soc. Am., 57,* 822–23.

Beasley, W.H., 1995: Lightning research: 1991–1994. *Rev. Geophys., Supplement, U.S. Nat. Rpt. to Intl. Un. Geodesy and Geophys. 1991–1994,* Amer. Geophys. Un., 833–43.

Beasley, W.H., M.A. Uman, D.M. Jordan, and C. Ganesh, 1983a: Positive cloud-to-ground lightning return strokes. *J. Geophys. Res., 88,* 8478–82.

Beasley, W.H., M.A. Uman, D.M. Jordan, and C. Ganesh, 1983b: Simultaneous pulses in light and electric field from stepped leaders near ground level. *J. Geophys. Res., 88,* 8617–19.

Berger, K., 1967: Novel observations on lightning discharges: Results of research on Mount San Salvatore. *J. Franklin Inst., 283,* 478–525.

Berger, K., 1975: Development and properties of positive lightning flashes at Mount S. Salvatore with a short view to the problem of aviation protection. Presented at Conference on Lightning and Static Electricity, 14–17 April, Abingdon Berks, England.

Berger, K., 1977: The earth flash. In *Lightning, Vol. 1: Physics of Lightning,* R.H. Golde, ed., Academic Press, San Diego, pp. 119–90.

Berger, K., and E. Vogelsanger, 1966: Photographische Blitzuntersuchungen der Jahre 1955–1965 aug dem Monte San Salvatore, *Bull. Scheizerischer Electrotechnischer Verein., 57,* 1–22.

Berger, K., R.B. Anderson, and H. Kröninger, 1975: Parameters of lightning flashes. *Electra, 41,* 23–37.

Bils, J.R., E.M. Thomson, M.A. Uman, and D. Mackerras, 1988: Electric field pulses in close lightning cloud flashes. *J. Geophys. Res., 93,* 15,933–40.

Boccippio, D.J., E.R. Williams, S.J. Heckman, W.A. Lyons, I.T. Baker, and R. Boldi, 1995: Sprites, ELF transients, and positive ground strokes. *Science, 269,* 1088–91.

Boeck, W.L., O.H. Vaughan, Jr., R. Blakeslee, B. Vonnegut, and M. Brook, 1992: Lightning induced brightening in the airglow layer. *Geophys. Res. Lett., 19,* 99–102.

Bohannon, J.L., 1980: *Infrasonic Thunder: Explained.* Ph.D. diss. Rice University, Houston, TX.

Bohannon, J.L., A.A. Few, and A.J. Dessler, 1977: Detection of infrasonic pulses from thunderclouds. *Geophys. Res. Lett., 4,* 49–52.

Brantley, R.D., J.A. Tiller, M.A. Uman, 1975: Lightning properties in Florida thunderstorms from video tape records. *J. Geophys. Res., 80,* 3402–06.

Brook, M., 1969: Discussion on the Few-Dessler paper. In *Planetary Electrodynamics, Vol. I,* S.C. Coroniti and J. Hughes, eds., Gordon and Breach, New York, p. 580.

Brook, M., C. Rhodes, O.H. Vaughan, Jr., R.E. Orville, and B. Vonnegut, 1985: Nighttime observations of thunderstorm electrical activity from a high altitude airplane. *J. Geophys. Res., 90,* 6111–20.

Brown, E.H., and S.F. Clifford, 1976: On the attenuation of sound by turbulence. *J. Acous. Soc. Am., 60,* 788–94.

Burket, H.D., L.C. Walko, J. Reazer, and A. Serrano, 1988: In-flight lightning characterization program on a CV-580 aircraft. *AFWAL-TR-88-3024,* Flt. Dyn. Lab, Air Force Wright Aeron. Lab, Wright-Patterson AFB, Ohio, 172 pp.

Clayton, M., and C. Polk, 1977: Diurnal variation and absolute intensity of world-wide lightning activity, September 1970 to May 1971. In *Electrical Processes in Atmospheres,* H. Dolezalek and R. Reiter, eds., Dr. Dietrich Steinkopff, Darmstadt, pp. 440–49.

Dawson, G.A., and W.P. Winn, 1965: A model for streamer propagation. *Z. für Physik, 183,* 159–71.

Dawson, G.A., M.A. Uman, and R.E. Orville, 1968a: Discussion of paper by E.L. Hill and J.D. Robb, "Pressure Pulse from a Lightning Stroke." *J. Geophys. Res., 73,* 6595–97.

Dawson, G.A., C.N. Richards, E.P. Krider, and M.A. Uman, 1968b: The acoustic output of a long spark. *J. Geophys. Res., 73,* 815–16.

De L'Isle, J.N., 1783: Memoires pour servir à l'histoire et au progres del l'astronomie de la géographie et de la physique. *L'Imprimerie de l'Académie des Sciences,* St. Petersburg.

Dessler, A.J., 1973: Infrasonic thunder. *J. Geophys. Res., 78,* 1889–96.

Diendorfer, G., and M.A. Uman, 1990: An improved return stroke model with specified channel-base current. *J. Geophys. Res., 95,* 13,621–44.

Drellishak, K.S., 1964: *Partition Functions and Thermodynamic Properties of High Temperature Gases*. Defense Documentation Center, AD 429 210.

Eriksson, A.J., 1978: Lightning and tall structures. *Trans. S. Africa Inst. Elec. Eng., 69,* 238–52.

Evans, W.H., and R.L. Walker, 1963: High speed photographs of lightning at close range. *J. Geophys. Res., 68,* 4455.

Everett, W.H., 1903: Rocket lightning. *Nature, 68,* 599.

Few, A.A., 1968: Thunder. Ph.D. dissertation, Rice Univ., Houston, TX.

Few, A.A., 1969: Power spectrum of thunder. *J. Geophys. Res., 74,* 6926–34.

Few, A.A., 1974: Thunder signatures. *EOS, Trans. AGU, 55,* 508–14.

Few, A.A., 1975: Thunder. *Sci. Amer., 233,* 1 July, 80–90.

Few, A.A., 1982: Acoustic radiations from lightning. In *Handbook of Atmospherics,* Vol. II, H. Volland, ed., CRC Press, Boca Raton, FL, pp. 257–90.

Few, A.A., 1985: The production of lightning-associated infrasonic acoustic sources in thunderclouds. *J. Geophys. Res., 90,* 6175–80.

Few, A.A., and T.L. Teer, 1981: The accuracy of acoustic reconstruction of lightning channels. *J. Geophys. Res., 79,* 5007–11.

Fisher, R.J., G.H. Schnetzer, R. Thottappillil, V.A. Rakov, M.A. Uman, and J.D. Goldberg, 1993: Parameters of triggered-lightning flashes in Florida and Alabama. *J. Geophys. Res., 98,* 22,887–22,902.

Fleagle, R.G., 1949: The audibility of thunder. *J. Acous. Soc. Am., 21,* 411–12.

Franz, R.C., R.J. Nemzek, and J.R. Winckler, 1990: Television of a large upward electrical discharge above a thunderstorm system. *Science, 249,* 48–51.

Garbagnati, E., and G.B. Lo Piparo, 1982: Parameter von Blitzströmen. *Theoretische Elektrotechnik, 103,* 61–65.

Gurevich, A.V., G.M. Milikh, and R.A. Roussel-Dupre, 1992: Runaway electron mechanism for air breakdown and preconditioning during a thunderstorm. *Phys. Lett. A, 165,* 463–68.

Gurevich, A.V., G.M. Milikh, and R.A. Roussel-Dupre, 1994: Nonuniform runaway air breakdown. *Phys. Lett. A, 167,* 197–203.

Herodotou, N., W.A. Chishol, and W. Janischewskyj, 1992: Distribution of lightning peak stroke currents in Ontario using an LLP system. *IEEE/PPS 1992 Summer Mtg.,* IEEE Transmission and Distribution Comm., paper 92SM452-3, Seattle, WA, 12–16 July.

Hill, E.L., and J.D. Robb, 1968: Pressure pulse from a lightning stroke. *J. Geophys. Res., 73,* 1883–88.

Hill, R.D., 1968: Analysis of irregular paths of lightning channels. *J. Geophys. Res., 73,* 1897–1906.

Hill, R.D., 1971: Channel heating in return-stroke lightning. *J. Geophys. Res., 76,* 637–45.

Hill, R.D., 1977a: Thunder. In *Lightning, Vol. 1: Physics of Lightning,* R.H. Golde, ed., Academic Press, San Diego, 385–408.

Hill, R.D., 1977b: Energy dissipation in lightning. *J. Geophys. Res., 82,* 4967–68.

Hill, R.D., 1979: A survey of lightning energy measurements. *Rev. Geophys. Space Phys., 17,* 155–64.

Holmes, C.R., M. Brook, P. Krehbiel, and R. McCrory, 1971: On the power spectrum and mechanism of thunder. *J. Geophys. Res., 76,* 2106–15.

Holmes, C.R., E.W. Szymanski, S.J. Szymanski, and C.B. Moore, 1980: Radar and acoustic study of lightning. *J. Geophys. Res. 85,* 7517–32.

Holzworth, R., and H. Volland, 1986: Do we need a geoelectric index? *Eos. Trans. AGU, 67,* 545. Abstract.

Hubert, P., and G. Mouget, 1981: Return stroke velocity measurements in two triggered lightning flashes. *J. Geophys. Res., 86,* 5253–61.

Hubert, P., P. Laroche, A. Eybert-Berard, and L. Barret, 1984: Triggered lightning in New Mexico. *J. Geophys. Res., 89,* 2511–21.

Idone, V.P., 1990: Length bounds for connecting discharges in triggered lightning subsequent strokes. *J. Geophys. Res., 95,* 20,405–09.

Idone, V.P., 1992: The luminous development of Florida triggered lightning. *Res. Lett. Atmos. Elec., 12,* 23–28.

Idone, V.P., and R.E. Orville, 1982: Lightning return stroke velocities in the thunderstorm research international program (TRIP). *J. Geophys. Res., 87,* 4903–15.

Idone, V.P., and R.E. Orville, P. Hubert, L. Barret, and A. Eybert-Berard, 1984: Correlated observations of three triggered lightning flashes. *J. Geophys. Res., 89,* 1385–94.

Idone, V.P., and R.E. Orville, D.M. Mach, and W.D. Rust, 1987: The propagation speed of a positive lightning return stroke. *Geophys. Res. Lett., 14,* 1150–53.

Idone, V.P., and R.E. Orville, 1988: Channel tortuosity variation in Florida triggered lightning. *Geophys. Res. Lett., 15,* 645–48.

Idone, V.P., A.B. Saljoughy, R.W. Henderson, P.K. Moore, and R. Pyle, 1993: A reexamination of the peak current calibration of the national lightning detection network. *J. Geophys. Res., 98,* 18,323–32.

Jordan, D.M., V.P. Idone, V.A. Rakov, M.A. Uman, W.H. Beasley, and H. Jurenka, 1992: Observed dart leader speed in natural and triggered lightning. *J. Geophys. Res., 97,* 9951–57.

Kasemir, H.W., 1950: Qualitative Übersicht über Potential -, Feld -, und Ladungsverhaltnisse bei einer Blitzentlandung in der Gewitterwolke. In *Das Gewitter,* H. Israel, Akad. Verlags. Ges. Geest and Portig K.-G., Leipzig.

Kasemir, H.W., 1960: A contribution to the electrostatic theory of a lightning discharge. *J. Geophys. Res., 65,* 1873–78.

Kasemir, H.W., 1983: Static discharge and triggered lightning. Proc., 8th Intl. Aerosp. and Gnd. Conf. on Lightning and Static Elec., Ft. Worth, TX, June 21–23, (available as DOT/FAA/CT-83/25), 24.1–24.11

Kitagawa, N., and M. Brook, 1960: A comparison of intracloud and cloud-to-ground lightning discharges. *J. Geophys. Res., 65,* 1189–1201.

Kitagawa, N., M. Brook, and E.J. Workman, 1962: Continuing currents in cloud-to-ground lightning discharges. *J. Geophys. Res., 67,* 637–647.

Krehbiel, P.R., 1981: An analysis of the electric field change produced by lightning. Ph.D. Diss., Univ. Manchester Inst. Sci. & Tech. Published as Rpt. T-11, New Mex. Inst. Min. & Tech., Dec., 1981, Vol. 1 is 245 pp. and Vol. 2 is figures.

Krider, E.P., 1992: On the electromagnetic fields, Poynting vector, and peak power radiated by lightning return strokes. *J. Geophys. Res., 97,* 15,913–17.

Krider, E.P., and C. Guo, 1983: The peak electromagnetic power radiated by lightning return strokes. *J. Geophys. Res., 88,* 8471–74.

Kuhn, H.G., 1969: *Atomic Spectra*. Academic Press, New York, 472 pp.

Leteinturier, C., C. Weidman, and J. Hamelin, 1990: Current and electric field derivatives in triggered lightning return strokes. *J. Geophys. Res., 95,* 811–28.

Le Vine, D.M., and J.C. Willett, 1992: Comment on the transmission-line model for computing radiation from lightning. *J. Geophys. Res., 97,* 2601–10.

Lin, Y.T., M.A. Uman, and R.B. Standler, 1980: Lightning return stroke models. *J. Geophys. Res., 85,* 1571–83.

Loeb, L.B., 1965: *Electrical Coronas. Their Basic Physical Mechanisms.* Univ. Calif. Press, Berkeley, 694 pp.

Loeb, L.B., 1966: The mechanisms of stepped and dart leader in cloud-to-ground lightning strokes. *J. Geophys. Res., 71,* 4711–21.

Loeb, L.B., 1968: Confirmation and extension of a proposed mechanism of the stepped leader lightning stroke. *J. Geophys. Res., 73,* 5813–17.

Lyons, W.A., 1994: Low-light video observations of frequent luminous structures in the stratosphere above thunderstorms, *Mon. Wea. Rev., 122,* 1940–46.

Mach, D.M., and W.D. Rust, 1989a: A photoelectric technique for measuring lightning-channel propagation velocities from a mobile laboratory. *J. Atmos. Oceanic Tech., 6,* 439–45.

Mach, D.M., and W.D. Rust, 1989b: Photoelectric return-stroke velocity and peak current estimates in natural and triggered lightning. *J. Geophys. Res., 94,* 13,237–47.

Mach, D.M., and W.D. Rust, 1993: Two-dimensional velocity, optical risetime, and peak current estimates for natural positive lightning return strokes. *J. Geophys. Res., 98,* 2635–38.

MacGorman, D.R., A.A. Few, and T.L. Teer, 1981: Layered lightning activity. *J. Geophys. Res., 86,* 9900–10.

Malan, D.J., 1963: *Physics of Lightning.* English Univ. Press, London, 176 pp. (see p. 47).

Marshall, T.C., M.P. McCarthy, and W.D. Rust, 1995: Electric field magnitudes and lightning initiation in thunderstorms. *J. Geophys. Res., 100,* 7097–7103.

Master, M.J., M.A. Uman, Y.T. Lin, and R.B. Standler, 1981: Calculations of lightning return stroke electric and magnetic fields above ground. *J. Geophys. Res., 86,* 12,127–32.

Mazur, V., 1986: Rapidly occurring short duration discharges in thunderstorms as indicators of a lightning-triggering mechanism. *Geophys. Res. Lett., 13,* 355–58.

Mazur, V., 1989a: A physical model of lightning initiation on aircraft in thunderstorms. *J. Geophys. Res., 94,* 3326–40.

Mazur, V., 1989b: Triggered lightning strikes to aircraft and natural intracloud discharges. *J. Geophys. Res., 94,* 3311–25.

Mazur, V., and W.D. Rust, 1983: Lightning propagation and flash density in squall lines as determined with radar. *J. Geophys. Res., 88,* 1495–1502.

Mazur, V., B.D. Fisher, and J.C. Gerlach, 1984: Lightning strikes to an airplane in a thunderstorm. *J. Aircraft, 21,* 607–11.

Mazur, V., and B.D. Fisher, 1990: Cloud-to-ground strikes to the NASA F-106 airplane. *J. Aircraft, 27,* 466–68.

Mazur, V., and J.-P. Moreau, 1992: Aircraft-triggered lightning: processes following strike initiation that affect aircraft. *J. Aircraft, 29,* 575–80.

Mazur, V., and L. Ruhnke, 1993: Common physical processes in natural and artificially triggered lightning. *J. Geophys. Res., 98,* 12,913–30.

Mazur, V., X-M. Shao, and P.R. Krehbiel, 1994: "Spider" intracloud and positive cloud-to-ground lightning in the late stage of Florida storms. Proc. of 1994 Intl. Aerosp. and Gnd. Conf. on Lightning and Static Elec., May 24–27, Mannheim, Ger., pp. 33–41.

Mazur, V., P.R. Krehbiel, and X-M. Shao, 1995: Correlated high-speed video and radio interferometric observations of a cloud-to-ground lightning flash. *J. Geophys. Res., 100,* 25,731–53.

McCarthy, M.P., and G.K. Parks, 1992: On the modulation of X ray fluxes in thunderstorms. *J. Geophys. Res., 97,* 5857–64.

Nakono, M., T. Takeuti, Z-I Kawasaki, and N. Takagi, 1983: Leader and return stroke velocity measurements in lightning from a tall chimney. *J. Meteor. Soc. Jpn., 61,* 339–45.

Nucci, C.A., G. Diendorfer, M.A. Uman, F. Rachidi, M. Ianoz, and C. Mazzetti, 1990: Lightning return stroke current models with specified channel-base current: a review and comparison. *J. Geophys. Res., 95,* 20,395–408.

Ogawa, T., and M. Brook, 1964: The mechanism of the intracloud lightning discharge. *J. Geophys. Res., 69,* 5141–50.

Orville, R.E., 1968a: A high-speed time-resolved spectroscopic study of the lightning return stroke: Part I. A quantitative analysis. *J. Atmos. Sci., 25,* 827–38.

Orville, R.E., 1968b: A high-speed time-resolved spectroscopic study of the lightning return stroke: Part II. A quantitative analysis. *J. Atmos. Sci., 25,* 839–51.

Orville, R.E., 1968c: A high-speed time-resolved spectroscopic study of the lightning return stroke: Part III. A time-dependent model. *J. Atmos. Sci., 25,* 852–56.

Orville, R.E., 1968d: Spectrum of the dart leaders. *J. Geophys. Res., 73,* 6999–7008.

Orville, R.E., 1977a: Wind profile in the sub-cloud layer of a thunderstorm. *J. Geophys. Res., 82,* 3453–56.

Orville, R.E., 1977b: Lightning spectroscopy. In *Lightning, Vol. 1: Physics of Lightning,* R.H. Golde, eds., Academic Press, San Diego, pp. 281–306.

Orville, R.E., 1991: Calibration of a magnetic direction finding network using measured triggered lightning return stroke peak currents. *J. Geophys. Res., 96,* 17,135–42.

Orville, R.E., M.A. Uman, and A.M. Sletten, 1967: Temperature and electron density in long air sparks. *J. Appl. Phys., 38,* 895–96.

Orville, R.E., G.G. Lala, and V.P. Idone, 1978: Daylight time-resolved photographs of lightning. *Science, 201,* 59–61.

Orville, R.E., and V.P. Idone, 1982: Lightning leader characteristics in the Thunderstorm Research International Program TRIP. *J. Geophys Res., 87,* 11,177–92.

Phelps, C.T., 1971: Field-enhanced propagation of corona streamers. *J. Geophys. Res., 76,* 5799–5806.

Phelps, C.T., 1974: Positive streamer system intensification and its possible role in lightning initiation. *J. Atmos. Terres. Phys., 36,* 103–11.

Phelps, C.T., and R.F. Griffiths, 1976: Dependence of positive corona streamer propagation on air pressure and water vapor content *J. Appl. Phys., 47,* 2929–34.

Pierce, E.T., 1963: Excitation of earth-ionosphere cavity resonances by lightning flashes. *J. Geophys. Res., 68,* 4125–27.

Plooster, M.N., 1971a: Numerical simulation of spark discharges in air. *Phys. Fluids, 14,* 2111–23.

Plooster, M.N., 1971b: Numerical model of the return stroke of the lightning discharge. *Phys. Fluids, 14,* 2124–33.

Polk, C., 1969: Relation of ELF noise and Schumann resonances to thunderstorm activity. In *Planetary Electrodynamics, Vol. 2,* S.C. Coroniti and J. Hughes, eds., Gordon and Breach, New York, pp. 55–82.

Polk, C., 1982: Schumann resonances. In *CRC Handbook of Atmospherics, Vol. I,* H. Volland, ed., CRC Press, Boca Raton, FL, pp. 111–78.

Proctor, D.E., 1981: VHF radio pictures of cloud flashes. *J. Geophys. Res., 86,* 4041–71.

Proctor, D.E., 1983: Lightning and precipitation in a small multicellular thunderstorm. *J. Geophys. Res., 88,* 5421–40.

Proctor, D.E., 1991: Regions where lightning flashes began. *J. Geophys. Res., 96,* 5099–5112.

Proctor, D.E., R. Uytenbogaardt, B.M. Meredith, 1988: VHF radio pictures of lightning flashes to ground. *J. Geophys. Res., 93,* 12,683–727.

Richard, P., J. Appel, and F. Broutet, 1985: A three-dimensional interferometric imaging system for the spatial characterization of lightning discharges. 10th Intl. Aerosp. and Gnd. Conf. on Ltng. and Static Elec., Paris, June 10–13.

Ribner, H.S., and D. Roy, 1982: Acoustics of thunder: A quasilinear model for tortuous lightning. *J. Acous. Soc. Am., 72b,* 1911–25.

Roussel-Dupre, R.A., A.V. Gurevich, T. Tunnel, and G.M. Milikh, 1992: Kinetic theory of runaway air breakdown. *Phys. Rev. E, 49,* 2257–71.

Rust, W.D., 1986: Positive cloud-to-ground lightning. In *The Earth's Electrical Environment,* National Acad. Press, Washington, D.C., pp. 41–45.

Rust, W.D., D.R. MacGorman, and R.T. Arnold, 1981: Positive cloud-to-ground lightning flashes in severe storms. *Geophys. Res. Lett., 8,* 791–94.

Rust, W.D., W.L. Taylor, D.R. MacGorman, E. Brandes, V. Mazur, R., Arnold, T. Marshall, H. Christian, and S. J. Goodman, 1985: Lightning and related phenomena in isolated thunderstorms and squall line systems. *J. Aircraft, 22,* 449–54.

Salanave, L.E., 1980: *Lightning and Its Spectrum.* Univ. of Arizona Press, Tucson, 136 pp.

Schonland, B.F.J., 1938: Progressive lightning IV. *Proc. Roy. Soc. Lond., A, 164,* 132–50.

Schonland, Sir B., 1964: *The Fight of Thunderbolts.* Clarendon Press, Oxford, 182 pp.

Schonland, B.F.J., and H. Collens, 1934: Progressive Lightning, I. *Proc. Roy. Soc. Lond., A, 143,* 654–74.

Schumann, W.O., 1952: Uber die Dampfung der elecktromagnetischen Eigenschwingungen des Systems Erde—Lufte—Ionosphare. *Z. Naturforsch., 7a,* 250–52.

Sentman, D.D., and B.J. Fraser, 1991: Simultaneous observations of Schumann Resonances in California and Australia: Evidence for intensity modulation by the local height of the D region. *J. Geophys. Res. (Space Res.), 96,* 15,973–84.

Sentman, D.D., and E.M. Wescott, 1993: Observations of upper atmospheric optical flashes recorded from an aircraft. *Geophys. Res. Lett., 20,* 2857–60.

Sentman, D.D., E.M. Wescott, D.L. Osborne, D.L. Hampton, and M.J. Heavner, 1995: Preliminary results from the Sprites94 aircraft campaign: 1. Red sprites. *Geophys. Res. Lett., 22,* 1205–08.

Taylor, W.L., E.A. Brandes, W.D. Rust, and D.R. MacGorman, 1984: Lightning activity and severe storm structure. *Geophys. Res. Lett., 11,* 545–48.

Teer, T.L., 1973: *Lightning Channel Structure Inside an Arizona Thunderstorm.* Ph.D. diss., Rice Univ., Houston, TX.

Thottappillil, R., D.K. McLain, M.A. Uman, and G. Diendorfer, 1991: Extension of the Diendorfer-Uman lightning return stroke model to the case of a variable upward return stroke speed and a variable downward discharge current speed. *J. Geophys. Res., 96,* 17,143–50.

Uman, M.A., 1969: *Lightning.* McGraw-Hill, New York, 264 pp. (Also published by Dover, 1984).

Uman, M.A., 1971: *Understanding Lightning.* Bek Technical Pub. Inc., Carnegie, PA, 166 pp.

Uman, M., 1987: *The Lightning Discharge.* Academic Press, 377 pp.

Uman, M.A., and R.E. Voshall, 1968: Time interval between lightning strokes and the initiation of dart leaders. *J. Geophys. Res., 73,* 497–506.

Uman, M.A., D.K. McLain, and F. Myers, 1968: Sound from Line Sources with Application to Thunder. Westinghouse Research Laboratory Rep. 68-9E4-HIVOL-R1.

Uman, M.A., A.H. Cookson, and J.B. Moreland, 1970: Shock wave from a four-meter spark. *J. Appl. Phys. 41,* 3148–55.

Uman, M.A., D.K. McLain, and E.P. Krider, 1975: The electromagnetic radiation from a finite antenna. *Amer. J. Phys, 43,* 33–38.

Uman, M.A., and E.P. Krider, 1982: A review of natural lightning: experimental data and modeling. *IEEE Trans. Elecmag. Compatibility, EMC-24,* 79–112.

Uman, M.A., M.J. Master, and E.P. Krider, 1982: A comparison of lightning electromagnetic fields with the nuclear electromagnetic pulse in the frequency range 10^4–10^7 Hz. *IEEE Trans. Elecmag. Compatibility, EMC-24,* 410–16.

Vaughan, O.H., Jr., and B. Vonnegut, 1989: Recent observations of lightning discharges from the top of a thundercloud into the clear air above. *J. Geophys. Res., 94,* 13,179–13,182.

Vaughan, O.H., Jr., R. Blakeslee, W.I. Boeck, B. Vonnegut, M. Brook, and J. McKune, Jr., 1992: A cloud-to-space lightning as recorded by the space shuttle payload-bay TV cameras. *Mon. Wea. Rev., 120,* 1459–60.

Veenema, L.C., 1920: The audibility of thunder. *Mon. Wea. Rev., 48,* 162.

Villanueva, Y., V.A. Rakov, M.A. Uman, and M. Brook, 1994: Microsecond-scale electric field pulses in cloud lightning discharges. *J. Geophys. Res., 99,* 14,353–60.

Vonnegut, B., and C.B. Moore, 1965: Nucleation of ice formation in supercooled clouds as the result of lightning. *J. Appl. Meteor., 4,* 640–42.

Wagner, K.H., 1966: Die Entwicklung der Elektronenlawine in en Plasmakanal, untersucht mit Bildverstärker und Wischverschluss. *Z. Physik, 189,* 495–515.

Weidman, C., and P. Krider, 1984: Variations à l' échelle submicroseconde des champs électromagnétiques rayonnés par la foudre. *Ann. des Télécomun. FRA., 39,* 163–74.

Wescott, E.M., D. Sentman, D. Osborne, D. Hampton, and M. Heavner, 1995: Preliminary results from the Sprites94 aircraft campaign: 2. Blue jets. *Geophys. Res. Lett., 22,* 1209–12.

Whipple, F.J.W., and F.J. Scrase, 1936: Point-discharge in the electric field of the Earth. *Geophys. Mem. VII, No. 68,* 1–20.

Willett, J.C., V.P. Idone, R.E. Orville, C. Leteinturier, A. Eybert-Berard, L. Barret, and E.P. Krider, 1988: An experimental test of the "transmission-line model" of electromagnetic radiation from triggered lightning return strokes. *J. Geophys. Res., 93,* 3867–78.

Willett, J.C., J.C. Bailey, V.P. Idone, A. Eybert-Berard, and L. Barret, 1989: Submicrosecond intercomparison of radiation fields and current, in triggered lighting return strokes based on the transmission-line model. *J. Geophys. Res., 94,* 13,275–86.

Williams, E., 1992: The Schumann resonance: A global tropical thermometer. *Science, 256,* 1184–87.

Wilson, C.T.R., 1920: Investigations on lightning discharges and on the electric field of thunderstorms, *Phil. Trans. Roy. Soc. Lond., A, 221,* 73–115.

Wilson, C.T.R., 1925: The electric field of a thundercloud and some of its effects. *Proc. Roy. Soc. Lond., 37,* 32D–37D.

Wilson, C.T.R., 1956: A theory of thundercloud electricity. *Proc. Roy. Soc. Lond., A, 236,* 297–317.

Wright, W.M., 1964: Experimental study of acoustical n waves. *J. Acous. Soc. Am., 36,* 1032.

Wright, W.M., and N.W. Medendorp, 1967: Acoustic radiation from a finite line source with n-wave excitation. *J. Acous. Soc. Am., 43,* 966–71.

CHAPTER 6

Anderson, R.V., and J.C. Bailey, 1991: Errors in the Gerdien measurement of atmospheric electric conductivity. *Meteor. Atmos. Phys., 46,* 101–12.

Bateman, M.G., W.D. Rust, and T.C. Marshall, 1994: A balloon-borne instrument for measuring the charge and size of precipitation particles inside thunderstorms. *J. Atmos. Oceanic Tech., 11,* 161–69.

Bendat, J.S., and A.G. Piersol, 1971: *Random Data: Analysis and Measurement Procedures.* John Wiley and Sons, New York, 407 pp.

Bevington, P.R., 1969: Data reduction and error analysis for the physical sciences. McGraw Hill, New York, 334 pp.

Blakeslee, R.J., and E.P. Krider, 1985: The electric currents under thunderstorms at the NASA Kennedy Space Center. *Preprints,* 7th Intl. Conf. on Atmospheric Electricity, Amer. Meteor. Soc, 265–68.

Blakeslee, R.J., H.J. Christian, and B. Vonnegut, 1989: Electrical measurements over thunderstorms. *J. Geophys. Res., 94,* 13,135–40.

Bohannon, J.L., 1978: Infrasonic pulses from thunderstorms. Master's thesis, Rice Univ., Houston.

Boys, C.V., 1926: Progressive lightning. *Nature, 118,* 749–50.

Brook, M., R. Tennis, C. Rhodes, P. Krehbiel, B. Vonnegut, and O.H. Vaughan, Jr., 1980: Simultaneous observations of lightning radiations from above and below clouds. *Geophys. Res. Lett., 7,* 267–70.

Brook, M., R.W. Henderson, and R. B. Pyle, 1989: Positive lightning strokes to ground. *J. Geophys. Res., 94,* 13,295–303.

Byrne, G.J., A.A. Few, and M.F. Stewart, 1986: The effects of atmospheric parameters on a corona probe used in measuring thunderstorm electric fields. *J. Geophys. Res., 91,* 9911–20.

Chalmers, J.A., 1957: Point-discharge current, potential gradient and wind speed. *J. Atmos. Terres. Phys., 11,* 301–02.

Chalmers, J.A., 1967: *Atmospheric Electricity.* Pergamon Press, Oxford, 515 pp.

Chauzy, S., and P. Raizonville, 1982: Space charge layers created by coronae at ground level below thunderclouds: Measurements and modelling. *J. Geophys. Res., 87,* 3143–48.

Chauzy, S., J-C Médale, S. Prieur, and S. Soula, 1991: Multilevel measurement of the electric field underneath a thundercloud 1. A new system and the associated data processing. *J. Geophys. Res., 96,* 22,319–26.

Christian, H.J., R.J. Blakeslee, and S.J. Goodman, 1989: The detection of lightning from geostationary orbit. *J. Geophys. Res., 94,* 13,329–37.

Christian, H.J., R.J. Blakeslee, and S.J. Goodman, 1992: Lightning imaging sensor (LIS) for the Earth observing system. *NASA* Technical Memorandum 4350, 36 pp.

Christian, H.J., K.T. Driscoll, S.J. Goodman, R.J. Blakeslee, D.M. Mach, and D.E. Buechler, 1996: Seasonal variation and distribution of lightning activity. *Eos, Trans. Amer. Geophys. Un., 77,* F80. Abstract.

Colgate, S.A., and J.M. Romero, 1970. Charge versus drop size in an electrified cloud. *J. Geophys. Res., 75,* 5873–81.

Colladon, D., 1826: Ablenkung der Mangetnadel durch den Strom einer gewöhnlichen Elektrisier-Maschine und der atmosphärischen Elektrizität. (Deflection of the magnetic needle by the current of a conventional electrification machine and atmospheric electricity.) *Ann. Phys. Chem., 8,*336–52.

Cummins, K.L., E.A. Bardo, W.L. Hiscox, R. Bent, and A.E. Pifer, 1995: NLDN'95: A combined TOA/MDF technology upgrade of the U.S. National Lightning Detection Network. *Preprints,* 1995 International Aerospace and Ground Conference on Lightning and Static Electricity, Sept. 26–28, Williamsburg, VA.

Davis, M.H., M. Brook, H. Christian, B.G. Heikes, R.E. Orville, C.G. Park, R.A. Roble, and B. Vonnegut 1983: Some scientific objectives of a satellite-borne lightning mapper. *Bull. Amer. Meteor. Soc., 64,* 114–19.

Elster, J. and H. Geitel, 1888. Über eine Methode, die elektrische Natur der atmosphärischen Niederschläge zu bestimmen (About a method for determining the electric nature of atmospheric precipitation). *Meteor. Z., 5,* 95–100.

Few, A.A., 1970: Lightning channel reconstruction from thunder measurements. *J. Geophys. Res., 75,* 7517–23.

Few, A.A., and T.L. Teer, 1974: The accuracy of acoustic reconstructions of lightning channels *J. Geophys. Res., 79,* 5007–11.

Fieux, R.P., C. Gary, and P. Hubert, 1975: Artificially triggered lightning above land. *Nature (London), 257,* 212–14.

Fieux, R.P., C.H. Gary, B.P. Hutzler, A.R. Eybert-Berard, P.L. Hubert, A.C. Meesters, P.H. Perroud, J.H. Hamelin, and J.M. Person, 1978: Research on artificially triggered lightning in France. *IEEE Trans. Power Appar. Sys., PAS-97,* 725–33.

Fisher, R.J., G.H. Schnetzer, R. Thottappillil, V.A. Rakov, M.A. Uman, and J.D. Goldberg, 1993: Parameters of triggered-lightning flashes in Florida and Alabama. *J. Geophys. Res., 98,* 22,887–902.

Frier, G.D., 1972: Comments on "Electric field measurements in thunderclouds by instrumented rockets," by W.P. Winn and C.B. Moore. *J. Geophys. Res., 77,* 505.

Gerdien, H., 1905: Demonstration eines Apparates zur absoluten Messung der elektrischen Leitfähigkeit der Luft, (Demonstration of a device for the absolute measurement of the electrical conductivity of air). *Phys. Z. 6,* 647–66.

Gondot, P., 1987: Definition and use of sensors for measuring electrical conductivity in thunderclouds. European Space Agency, *ESA-TT-1047.* Translation of *Definition et Exploitation de Capteurs Adaptes a la Mesure de la Conductivite Electrique dans les Nuages d'Orages,* ESA, Paris, 142 pp.

Goodman, S.J., H.J. Christian, and W.D. Rust, 1988: A comparison of the optical pulse characteristics of intracloud and cloud-to-ground lightning as observed above clouds. *J. Appl. Meteor, 27,* 1369–81.

Gschwend, P.O., 1922. Beobachtungen über die elektrischen Ladungen einzelner Regentropfen und Schneeflocken (Observations about the electric charge of individual raindrops and snowflakes). *Jahrbuch der Radioakt. und Elektronik, 17,* 62–79.

Gunn, R., 1947: The electrical charge on precipitation at various altitudes and its relation to thunderstorms. *Phys. Rev., 71,* 181–86.

Gunn, R., and G.D. Kinzer, 1949. The terminal velocity of fall for water droplets in stagnant air. *J. Meteor., 6,* 243–48.

Guo, C., and E.P. Krider, 1982: The optical and radiation field signatures produced by lightning return strokes. *J. Geophys. Res., 87,* 8913–22.

Hayenga, C.O., and J.W. Warwick, 1981: Two-dimensional interferometric positions of VHF lightning sources. *J. Geophys. Res., 81,* 7451–62.

Hojo, J., M. Ishii, T. Kawamura, F. Suzuki, H. Komuro, and M. Shiogama, 1989: Seasonal variation of cloud-to-ground lightning flash characteristics in the coastal area of the Sea of Japan, *J. Geophys. Res., 94,* 13,207–13,212.

Holmes, C.R., E.W. Szymanski, S.J. Szymanski, and C.B. Moore, 1980: Radar and acoustic study of lightning. *J. Geophys. Res., 85,* 7517–32.

Horii, K., K. Nakamura, H. Sakurano, S. Sumi, M. Yoda, Z. Kawasaki, N. Tahara, M. Nakano, K. Wakamatsu, K. Yamamoto, K. Kumazaki, T. Inaba, and H. Sakai, 1996: Triggered lightning experiments in 1994–1995 for winter thunderstorms. In *Proceedings, 10th International Conference on Atmospheric Electricity,* Osaka, Japan, pp. 648–51.

Horner, F., 1954: The accuracy of the location of sources of atmospherics by radio direction-finding. *Proc. IEEE, Rad. Sec., 101,* 383–90.

Hubert, P., and G. Mouget, 1981: Return stroke velocity measurements in two triggered lightning flashes. *J. Geophys. Res. 86,* 5253–61.

Idone, V.P., and R.E. Orville, 1982: Lightning return stroke velocities in the thunderstorm research international program (TRIP). *J. Geophys. Res., 87,* 4903–15.

Idone, V.P., and R.E. Orville, D.M. Mach, and W.D. Rust, 1987: The propagation speed of a positive lightning return stroke. *Geophys. Res. Lett., 14,* 1150–53.

Israël, H., 1970: *Atmospheric Electricity, Vol. 1, Fundamentals, Conductivity, Ions.* Israel Program for Scientific Translation, Jerusalem, pp. 216–20.

Jacobson, E.A., and E.P. Krider, 1976: Electrostatic field changes produced by Florida lightning. *J. Atmos. Sci., 33,* 103–17.

Jones, J.J., 1990: Electric charge acquired by airplanes penetrating thunderstorms. *J. Geophys. Res., 95,* 16,589–600.

Jones, J.J., W.P. Winn, and F. Han, 1993: Electric field measurements with an airplane: problems caused by emitted charge. *J. Geophys. Res., 98,* 5235–44.

Jonsson, H.H., 1990: Possible sources of errors in electrical measurements made in thunderstorms with balloon-borne instrumentation. *J. Geophys. Res., 95,* 22,539–45.

Jonsson, H.H., and B. Vonnegut, 1995: Comment on "Negatively charged precipitation in a New Mexico thunderstorm" by Thomas C. Marshall and Stephen J. Marsh and "Charged precipitation measurements before the first lightning flash in a thunderstorm" by Stephen J. Marsh and Thomas C. Marshall. *J. Geophys. Res., 100,* 16,867–16,868.

Kasemir, H.W., 1951: Die Feldkomponentenmühle Ein Gerät zur Messung der drei Komponenten des luftelektrischen Feldes und der Flugzeugeigenladung bei Flugzeugaufstiegen (The field component meter—an apparatus for the measurement of the three components of the atmospheric electric field and the aircraft self charge when airborne). (In German) *Tellus, 3,* 240–47.

Kasemir, H.W., 1972. The cylindrical field mill. *Meteor. Rundshau, 25,* 33–38.

Kawasaki, Z-I, T. Kanao, K. Matsuura, and K-I Nakamura, 1991: The electric field changes and UHF radiations caused by the triggered lightning in Japan. *Geophys. Res. Lett., 18,* 1711–14.

Kawasaki, Z-I., K. Yamamoto, K. Matsuura, P. Richard, T. Matsui, Y. Sonoi, and N. Shimokura, 1994: SAFIR operation and evaluation of its performance. *Geophys. Res. Lett., 21,* 1133–36.

Keithley Instruments, 1972: *Electrometer Measurements.* Keithley Instruments, Cleveland, 80 pp.

Koshak, W.J., and E.P. Krider, 1989: Analysis of lighting field changes during active Florida thunderstorms. *J. Geophys. Res., 94,* 1165–86.

Koshak, W.J., and E.P. Krider, 1994: A linear method for analyzing lightning field changes. *J. Atmos. Sci., 51,* 473–88.

Kraakevik, J.H., 1958: The airborne measurement of atmospheric conductivity. *J. Geophys. Res., 63,* 161–69.

Krehbiel, P.R., 1981: An analysis of the electric field change produced by lightning. Ph.D. diss., U. Manchester Inst. Sci. & Tech. Published as Rpt. T-11, New Mex. Inst. Min. & Tech., Dec., 1981. Vol. 1 is 245 pp. and Vol. 2 is figures.

Krehbiel, P.R., M. Brook, R.A. McCrory, 1979: An analysis of the charge structure of lightning discharges to ground. *J. Geophys. Res., 84,* 2432–56.

Krehbiel, P., T. Chen, S. McCrary, W. Rison, G. Gray, and M. Brook, 1996: The use of dual channel circular-polarization radar observations for remotely sensing storm electrification. *Meteor. and Atmos. Phys., 59,* 65–82.

Krider, E.P., 1989: Electric field changes and cloud electrical structure. *J. Geophys. Res., 94,* 13,145–49.

Krider, E.P., R.C. Noggle, and M.A. Uman, 1976: A gated, wideband magnetic direction finder for lightning return strokes. *J. Appl. Meteor., 15,* 301–06.

Krider, E.P., A.E. Pifer, and D.L. Vance. 1980: Lightning direction-finding systems for forest fire detection. *Bull. Amer. Meteor. Soc., 61,* 980–86.

Krider, E.P., and R.J. Blakeslee, 1985: The electric currents produced by thunderclouds. *J. Electrostat., 16,* 369–78.

Landweber, L., 1951: An iteration formula for Fredholm integral equations of the first kind. *Amer. J. Math., 73,* 615–24.

Latham, J., and C.D. Stow, (1969): Airborne studies of the electrical properties of large convective clouds. *Q. J. Roy. Meteor. Soc., 95,* 486–500.

Lee, A.C.L., 1986a: An experimental study of the remote location of lightning flashes using a VLF arrival time difference technique. *Quart. J. Roy. Meteor. Soc., 112,* 203–29.

Lee, A.C.L., 1986b: An operational system for the remote location of lightning flashes using a VLF arrival time difference technique. *J. Atmos. Ocean. Tech., 3,* 630–42.

Lee, A.C.L., 1989a: The limiting accuracy of long wavelength lightning flash location. *J. Atmos. Ocean. Tech., 6,* 43–49.

Lee, A.C.L., 1989b: Ground truth confirmation and theoretical limits of an experimental VLF arrival time difference lightning flash locating system. *Quart. J. Roy. Meteor. Soc., 115,* 1147–66.

Lee, A.C.L., 1990: Bias elimination and scatter in lightning location by the VLF arrival time difference technique. *J. Atmos. Ocean. Tech, 7,* 719–33.

Lenard, P., 1892. Über die Elektrizität der Wasserfälle (About the electricity of waterfalls). *Ann. Phys. 46,* 584–636.

Lennon, C., and L. Maier, 1991: Lightning mapping system. Proceedings of the International Aerospace and Ground Conference on Lightning and Static Electricity, Cocoa Beach, FL, NASA Conference Publication 3106, Vol. II, 89.1–89.10.

Lhermitte, R., and P.R. Krehbiel, 1979: Doppler radar and radio observations of thunderstorms, *IEEE Trans. Geosci. Electron., GE-17,* 162–71.

Lhermitte, R., and E. Williams, 1985: Thunderstorm electrification: A case study. *J. Geophys. Res., 90,* 6071–78.

MacGorman, D.R., 1977: Lightning in a Colorado thunderstorm. Master's thesis. Rice Univ. Houston, 133 pp.

MacGorman, D.R., 1978: Lightning in a storm with strong wind shear. Ph.D. diss., Rice Univ. Houston.

MacGorman, D.R., A.A. Few, and T.L. Teer, 1981: Layered lightning activity. *J. Geophys. Res., 86,* 9900–9910.

MacGorman, D.R., M.W. Maier, and W.D. Rust, 1984: *Lightning Strike Density for the Contiguous United States from Thunderstorm Duration Records.* Report to the U.S. Nuclear Regulatory Commission. Available as NUREG/CR-3759, 51 pp.

MacGorman, D.R., and W.D. Rust, 1989: An evaluation of the LLP and LPATS lightning ground strike mapping systems. *Preprints,* 5th Int'l Conf. on Interactive Information and Processing Systems, Amer. Meteor. Soc., 249–54.

Mach, D.M., and W.D. Rust, 1989: A photoelectric technique for measuring lightning-channel propagation velocities from a mobile laboratory. *J. Atmos. Oceanic Tech., 6,* 439–45.

Mach, D., D. Buechler, K. Driscoll, R. Raghavan, W. Boeck, R. Blakeslee, H. Christian, S. Goodman, and W. Koshak, 1996: A flash discrimination algorithm based on the temporal and spatial characteristics of lightning optical pulses observed from space. *Eos, Trans. Amer. Geophys. Un., 77,* F92. Abstract.

Maier, L.M., C. Lennon, T. Britt, and S. Schaefer, 1995a: Lightning detection and ranging (LDAR) system performance analysis. *Preprints,* 6th Conf. Aviation Weather Systems, Amer. Meteor. Soc.

Maier, L.M., C. Lennon, P. Krehbiel, M. Stanley, and M. Robison, 1995b: Comparison of lightning and radar observations from the KSC LDAR and NEXRAD radar systems. *Preprints,* 27th Conf. Radar Meteorology, Amer. Meteor. Soc., 648–50.

Marsh, S.J., and T.C. Marshall, 1993: Charged precipitation measurements before the first lightning flash in a thunderstorm. *J. Geophys. Res., 98,* 16,605–11.

Marshall, T.C., and W.P. Winn, 1982: Measurements of charged precipitation in a New Mexico thunderstorm: lower positive charge centers. *J. Geophys. Res., 87,* 7141–57.

Marshall, T.C., and W.D. Rust, 1991: Electric field soundings through thunderstorms. *J. Geophys. Res., 96,* 2297–306.

Marshall, T.C., and B. Lin, 1992: Electricity in dying thunderstorms. *J. Geophys. Res., 97,* 9913–9918.

Marshall, T.C., and S.J. Marsh, 1993: Negatively charged precipitation in a New Mexico thunderstorm. *J. Geophys. Res., 98,* 14,909–16.

Marshall, T.C., and S.J. Marsh, 1995: Reply. *J. Geophys. Res., 100,* 16,869–71.

Marshall, T.C., W.D. Rust, and M. Stolzenburg, 1995a: Electrical structure and updraft speeds in thunderstorms over the southern Great Plains. *J. Geophys. Res., 100,* 1001–15.

Marshall, T.C., W. Rison, W.D. Rust, M. Stolzenburg, J.C. Willett, and W.P. Winn, 1995b: Rocket and balloon observations of electric field in two thunderstorms. *J. Geophys. Res., 100,* 20,815–28.

Mazur, V., D.S. Zrnić, and W.D. Rust, 1985: Lightning channel properties determined with a vertically pointing Doppler radar. *J. Geophys. Res., 90,* 6165–74.

Mazur, V., L.H. Ruhnke, and T. Rudolph, 1987: Effect of E-field mill location on accuracy of electric field measurements with instrumented airplane. *J. Geophys. Res., 92,* 12,013–19.

Mazur, V., E. Williams, R. Boldi, L. Maier, and D.E. Proctor. 1995: Comparison of lightning mapping with operational time-of-arrival and interferometric systems. Proc. of the 1995 Int'l. Aerospace and Ground Conf. on Lightning and Static Electricity, Williamsburg, 60.1–60.20.

Mazur, V., E. Williams, R. Boldi, L. Maier, D.E. Proctor, 1997: Initial comparison of lightning mapping with operational time-of-arrival and interferometric systems. *J. Geophys. Res. 102,* 11,071–85.

McDonald, J.E., 1958: The physics of cloud modification. In *Advances in Geophysics, 5,* H.E. Landsberg and J. Van Mieghem, eds., Academic Press, New York, pp. 223–313 (see p. 244).

Milner, J.W., and J.A. Chalmers, 1961: Point discharge from natural and artificial points (Pt. II). *Q. J. Roy. Meteor. Soc., 87,* 592–96.

Moore, C.B., B. Vonnegut, and A.T. Botka, 1958: Results of an experiment to determine initial precedence of organized electrification and precipitation in thunderstorms. In *Recent Advances in Atmospheric Electricity,* L.G. Smith, eds., Pergamon, New York, pp. 333–60.

Moore, C.B., B. Vonnegut, and F.J. Mallahan, 1961: Airborne filters for the measurement of atmospheric space charge. *J. Geophys. Res., 66,* 3219–26.

Nakamura, K., K. Horii, M. Nakano, and S. Sumi, 1992: Experiments on rocket triggered lightning. *Res. Lett. Atmos. Elec., 12,* 29–35.

Nakono, M., T. Takeuti, Z-I Kawasaki, and N. Takagi, 1983: Leader and return stroke velocity measurements in lightning from a tall chimney. *J. Meteor. Soc. Jpn., 61,* 339–45.

Newman, M.M., 1958: Lightning discharge channel characteristics and related atmospherics. In *Recent Advances in Atmospheric Electricity,* L.G. Smith, ed., Pergamon, New York, pp. 475–84.

Newman, M.M., 1965: Use of triggered lightning to study the discharge process in the channel and application to V.L.F. propagation studies. In *Problems of Atmospheric and Space Electricity,* S. Coroniti, eds., Elsevier, Amsterdam, pp. 482–90.

Nisbet, J.S., T.A. Barnard, G.S. Forbes, E.P. Krider, R. Lhermitte, and C.L. Lennon, 1990: A case study of the Thunderstorm Research International Project Storm of July 11, 1978. Part 1: Analysis of the data base. *J. Geophys. Res., 95,* 5417–33.

Oetzel, G.N., and E.T. Pierce, 1969: VHF technique for locating lightning. *Radio Sci., 4,* 199–202.

Orville, R.E., 1994: Cloud-to-ground lightning flash characteristics in the contiguous United States: 1989–1991. *J. Geophys. Res., 99,* 10,833–41.

Orville, R.E., R.A. Weisman, R.B. Pyle, R.W. Henderson, and R.E. Orville, Jr., 1987: Cloud-to-ground lightning flash characteristics from June 1984 through May 1985. *J. Geophys. Res., 92,* 5640–44.

Poehler, H.A., and C.L. Lennon, 1979: Lightning Detection and Ranging System, LDAR, System Description and Performance Objectives. NASA Technical Memorandum 741005.

Proctor, D.E., 1971: A hyperbolic system for obtaining VHF radio pictures of lightning. *J. Geophys. Res., 76,* 1478–89.

Proctor, D.E., 1981: VHF radio pictures of cloud flashes. *J. Geophys. Res., 86,* 4041–71.

Proctor, D.E., 1983: Lightning and precipitation in a small multicellular thunderstorm. *J. Geophys. Res., 88,* 5421–40.

Rhodes, C.T., X.M. Shao, P.R. Krehbiel, R.J. Thomas, and C.O. Hayenga, 1994: Observations of lightning phenomena using radio interferometry. *J. Geophys. Res., 99,* 13,059–82.

Richard, P., 1990: SAFIR system: An application of real-time VHF lightning localization to thunderstorm monitoring. *Preprints,* 16th Conf. Severe Local Storms and Conf. Atmos. Elec., Amer. Meteor. Soc., J21–J26.

Richard, P., 1992: Monitoring of total lightning activity of thunderstorms. Applications of atmospheric electricity to severe weather forecasting. *Proc. 9th Intl. Conf. on Atmos. Elec., Vol. III,* 15–19 June, St. Petersburg, Russia, pp. 925–28.

Richard, P., A. Delannoy, G. Labaune, and P. Laroche, 1986: Results of spatial and temporal characterization of the VHF-UHF radiation of lighting. *J. Geophys. Res., 91,* 1248–60.

Richard, P., A. Soulage, P. Laroche, and J. Appel, 1988: The SAFIR lightning monitoring and warning system, application to aerospace activities. Proc. Int'l Aerospace and Ground Conf. Lightning and Static Elect., Oklahoma City, NOAA Special Report, pp. 383–90.

Robertson, F.R., G.S. Wilson, H.J. Christian, Jr., S.J. Goodman, G.H. Fichtl, and W.W. Vaughan, 1984: Atmospheric science experiments applicable to space shuttle spacelab missions. *Bull. Amer. Meteor. Soc., 65,* 692–700.

Rosen, J.M., and D.J. Hofmann, 1981a: Balloon-borne measurements of the small ion concentration. *J. Geophys. Res., 86,* 7399–7405.

Rosen, J.M., and D.J. Hofmann, 1981b: Balloon-borne measurements of electrical conductivity, mobility, and the recombination coefficient. *J. Geophys. Res., 86,* 7405–10.

Rust, W.D., 1989: Utilization of a mobile laboratory for storm electricity measurements. *J. Geophys. Res., 94,* 13,305–11.

Rust, W.D., and C.B. Moore, 1974: Electrical conditions near the bases of thunderclouds over New Mexico. *Q. J. Roy. Meteor. Soc., 100,* 450–68.

Rust, W.D., W.L. Taylor, D.R. MacGorman, and R.T. Arnold, 1981: Research on electrical properties of severe thunderstorms in the Great Plains. *Bull. Amer. Meteor. Soc., 62,* 1286–93.

Rust, W.D., and D.R. MacGorman, 1988: Techniques for measuring electrical parameters of thunderstorms. In *Thunderstorms: A Social, Scientific, and Technological Documentary, Vol. 3, Instruments and Techniques for Thunderstorm Observation and Analysis,* E. Kessler, ed., Univ. of Oklahoma Press, Norman, pp. 91–118.

Rust, W.D., and T.C. Marshall, 1989: Mobile, high-wind, balloon-launching apparatus. *J. Atmos. Oceanic Tech., 6,* 215–17.

Rust, W.D., R. Davies-Jones, D.W. Burgess, R.A. Maddox, L.C. Showell, T.C. Marshall, and D.K. Lauritsen, 1990: Testing a mobile version of a cross-chain loran atmospheric sounding system (M-CLASS). *Bull. Amer. Meteor. Soc., 71,* 173–80.

Saunders, C.P.R., 1978: Electrification experiments on Elk Mountain. *J. Geophys. Res., 83,* 5050–56.

Saunders, C.P.R., R.L. Pitter, B.A. Gardiner, and J. Hallett, 1988: An airborne precipitation cloud particle charge measurement device and analysis system. *J. Atmos. Oceanic Tech., 5,* 149–53.

Schuur, T.J., B.F. Smull, W.D. Rust, and T.C. Marshall, 1991: Electrical and kinematic structure of the stratiform precipitation region trailing an Oklahoma squall line. *J. Atmos. Sci., 48,* 825–41.

Scrase, F.J., 1938: Electricity on rain. *Geophys Mem., Lon., 9,* No. 75, 3–20.

Shao, X.M., P.R. Krehbiel, R.J. Thomas, and W. Rison, 1995: Radio interferometric observations of cloud-to-ground lightning phenomena in Florida. *J. Geophys. Res., 100,* 2749–83.

Shao, X.M., and P.R. Krehbiel, 1996: The spatial and temporal development of intracloud lightning. *J. Geophys. Res. 101,* 26,641–668.

Simpson, G.C., 1909: On the electricity of rain and snow. *Proc. Roy. Soc. Lond., A, 83,* 394–404.

Simpson, G.C., 1949: Atmospheric electricity during disturbed weather. *Geophys. Mem. No. 84, Lon.* (see plates XV–XVII).

Simpson, Sir G., and F.J. Scrase, 1937: The distribution of electricity in thunderclouds, *Proc. Roy. Soc. Lond., A, 161,* 309–52.

Simpson, Sir G., and G.D. Robinson, 1941: The distribution of electricity in thunderclouds, II, *Proc. Roy. Soc. Lond. A, 177,* 281–328.

Stolzenburg, M., 1993: Electrical and meteorological conditions in the stratiform cloud region of a mesoscale convective system. Master's thesis, Oklahoma Univ., Norman, 153 pp.

Stolzenburg, M., and T.C. Marshall, 1994: Testing models of thunderstorm charge distributions with Coulomb's law. *J. Geophys. Res., 99,* 25,921–32.

Stolzenburg, M., T.C. Marshall, W.D. Rust, B.F. Smull, 1994: Horizontal distribution of electrical and meteorological conditions across the stratiform region of a mesoscale convective system. *Mon. Weath. Rev., 122,* 1777–97.

Szymanski, E.W., S.J. Szymanski, C.R. Holmes, and C.B. Moore, 1980: An observation of a precipitation echo intensification associated with lightning. *J. Geophys. Res., 85,* 1951–53.

Takahashi, T., 1975: Electric charge life cycle in warm clouds. *J. Atmos. Sci., 32,* 123–42.

Takahashi, T., 1983: Electric structure of oceanic tropical clouds and charge separation processes. *J. Meteor. Soc. Jpn., 61,* 656–69.

Takahashi, T., and T. Craig, 1973. Charge-size measurement of drizzle drops in warm cloud. *J. Meteor. Soc. Jpn., 51,* 191–96.

Taylor, W.L., 1978: A VHF technique for space-time mapping of lightning discharge processes. *J. Geophys. Res. 83,* 3575–83.

Thomson, E.M., P.J. Medelius, and S. Davis, 1994: System for locating the sources of wideband dE/dt from lightning. *J. Geophys. Res., 99,* 22,793–802.

Toland, R.B., and B. Vonnegut, 1977: Measurement of maximum electric field intensities over water during thunderstorms. *J. Geophys. Res., 82,* 438–40.

Uman, M.A., 1987: *The Lightning Discharge.* Academic Press, Orlando, pp. 215–28.

Uman, M.A., W.H. Beasley, J.A. Tiller, Y. Lin, E.P. Krider, C.D. Weidmann, P.R. Krehbiel, M. Brook, A.A. Few, Jr., J.L. Bohannon, C.L. Lennon, H.A. Poehler, W. Jafferis, J.R. Gulick, and J.R. Nicholson, 1978: An unusual lightning flash at Kennedy Space Center. *Science, 201,* 9–16.

Uman, M.A., V.A. Rakov, K.J. Rambo, T.W. Vaught, M.I. Fernandez, J.A. Bach, Y. Su, A. Eybert-Berard, J.P. Berlandis, B. Bador, P. Lalande, A. Bonamy, F. Audran, R. Morillon, P. Laroche, A. Bondiou-Clergerie, S. Chauzy, S. Soula, C.D. Weidman, F. Rachidi, M. Rubinstein, C.A. Nucci, S. Guerrieri, H.K.H. HØyidalen, and V. Cooray, 1996: 1995 Triggered lightning experiment in Florida. In *Proceedings, 10th International Conference on Atmospheric Electricity,* Osaka, Japan, pp. 644–47.

Vonnegut, B., C.B. Moore, and F.J. Mallahan, 1961: Adjustable potential-gradient-measuring apparatus for airplane use. *J. Geophys. Res., 66,* 2393–97.

Vonnegut, B., O.H. Vaughan, Jr., M. Brook, and P. Krehbiel, 1985: Mesoscale observations of lightning from space shuttle. *Bull. Amer. Meteor. Soc., 66,* 20–29.

Warwick, J.W., C.O. Hayenga, and J.W. Broshahan, 1979: Interferometric directions of lightning sources at 34 MHz. *J. Geophys. Res., 84,* 2457–63.

Weinheimer, A.J., 1988. The charge induced on a conducting cylinder by a point charge and its application to the measurement of charge on precipitation. *J. Atmos. Oceanic Tech., 5,* 298–304.

Weinheimer, A.J., J.E. Dye, D.W. Breed, M.P. Spowart, J.L. Parrish, T.L. Hoglin, and T.C. Marshall, 1991: Simultaneous measurements of the charge, size, and shape of hydrometeors in an electrified cloud. *J. Geophys. Res., 96,* 20,809–29.

Willett, J.C., D.C. Curtis, A.R. Driesman, R.K. Longstreth, W. Rison, W.P. Winn, and J.J. Jones, 1992: The rocket electric field sounding (REFS) program: Prototype design and successful first launch. Phillips Lab. Tech. Rpt.-92-2015. Available from NTIS, 126 pp.

Williams, E.R., S.G. Geotis, and A.B. Bhattacharya, 1989: A radar study of the plasma and geometry of lightning. *J. Atmos. Sci., 46,* 1173–85.

Williams, E.R., V. Mazur, and S.G. Geotis, 1990: Lightning investigation with radar. In *Radar in Meteorology,* D. Atlas, ed., Amer. Meteor. Soc., Boston, pp. 143–150.

Winn, W.P., 1968: An electrostatic theory for instruments which measure the radii of water drops by detecting a change in capacity due to the presence of a drop. *J. Appl. Meteor., 7,* 929–37.

Winn, W.P., 1993: Aircraft measurement of electric field: Self-calibration. *J. Geophys. Res., 98,* 7351–365.

Winn, W.P., and C.B. Moore, 1971: Electric field measurements in thunderclouds using instrumented rockets, *J. Geophys. Res., 76,* 5003–17.

Winn, W.P., and C.B. Moore, 1972: Reply. *J. Geophys. Res., 77,* 506–08.

Winn, W.P., and LG. Byerley, III, 1975: Electric field growth in thunderclouds. *Q. J. Roy. Meteor. Soc., 101,* 979–94.

Winn, W.P., C.B. Moore, C.R. Holmes, and L.G. Byerley, III, 1978: Thunderstorm on July 16, 1975, over Langmuir Laboratory: a case study. *J. Geophys. Res., 83,* 3079–92.

Ziegler, C.L., and D.R. MacGorman, 1994: Observed lightning morphology relative to modeled space charge and electric field distributions in a tornadic storm. *J. Atmos. Sci., 51,* 833–51.

CHAPTER 7

Allee, P.A., and B.B. Phillips, 1959: Measurements of cloud-droplet charge, electric field, and polar conductivities in supercooled clouds. *J. Meteor., 16,* 405–10.

Asuma, Y., and K Kikuchi, 1987: Numerical experiments of the charging mechanism of precipitation particles by the ion-capture process below the cloud base. *J. Meteor. Soc. Jpn., 65,* 973–89.

Atlas, D., 1958: Radar lightning echoes and atmospherics in vertical cross section. In *Recent Advances in Atmospheric Electricity,* L.G. Smith, ed., Pergamon, New York, pp. 441–59.

Baral, K.N., and D. Mackerras, 1992: The cloud-to-ground flash ratio and other lightning occurrence characteristics in Kathmandu thunderstorms. *J. Geophys. Res., 97,* 931–38.

Baral, K.N., and D. Mackerras, 1993: Positive cloud-to-ground lightning discharges in Kathmandu thunderstorms. *J. Geophys. Res., 98,* 10,331–40.

Battan, L.J., 1965: Some factors governing precipitation and lightning from convective clouds. *J. Atmos. Sci., 22,* 79–84.

Battan, L.J., 1973: *Radar Observations of the Atmosphere.* University of Chicago Press, Chicago, 324 pp.

Baughman, R.G., and D.M. Fuquay, 1970: Hail and lightning occurrence in mountain thunderstorms. *J. Appl. Meteor., 9,* 657–60.

Black, R.A., J. Hallett, and C.P.R. Saunders, 1993: Aircraft studies of precipitation and electrification in hurricanes. *Proceedings, 17th Conf. on Severe Local Storms and Conf. on Atmos. Elec.,* St. Louis, MO, J20–J25, 4–8 October.

Blakeslee, R.J., and E.P. Krider, 1985: The electric currents under thunderstorms at the NASA Kennedy Space Center. *Preprints,* 7th Intl. Conf. on Atmospheric Electricity, Amer. Meteor. Soc, 265–68.

Blakeslee, R.J., H.J. Christian, and B. Vonnegut, 1989: Electrical measurements over thunderstorms. *J. Geophys. Res., 94,* 13,135–40.

Bluestein, H.B., 1992: *Synoptic-Dynamic Meteorology. Vols. 1 and 2.* Oxford Univ. Press, New York; Vol. 1, 431 pp.; Vol. 2 593 pp.

Bly, R.T., Jr., and J.E. Nanevicz, 1977: Aerial measurements of the electric field in the vicinity of Florida thunderstorms: Analysis and results. Final Rpt., Stanford Research Institute Proj. 5537, Menlo Park, CA, Contract NAS10-9013.

Bosart, L.F., and F. Sanders, 1986: Mesoscale structure in the megalopolitan snowstorm of 11–12 February 1983. Part III: A large-amplitude gravity wave. *J. Atmos. Sci., 43,* 924–39.

Branick, M.L., and C.A. Doswell III, 1992: An observation of the relationship between supercell structure and lightning ground strike polarity. *Weath. Forecast., 7,* 143–49.

Breed, D.W., and J.E. Dye, 1989: The electrification of New Mexico thunderstorms 2. Electric field growth during initial electrification. *J. Geophys. Res., 94,* 14,841–54.

Bringi, V.N., I.J. Caylor, J. Turk, and L. Liu, 1993a: Microphysical and electrical evolution of a convective storm using multiparameter radar and aircraft data during CaPE. *Preprints,* 26th Intl. Conf. Radar Meteorology, Amer. Meteorol. Soc., 312–14.

Bringi, V.N., A. Detwiler, V. Chandrasekar, P.L. Smith, L. Liu, I.J. Caylor, and D. Musil, 1993b: Multiparameter radar and aircraft study of the transition from early to mature storm during CaPE: the case of 9 August 1991. *Preprints, 26th Intl. Conf. Radar Meteorology, Amer. Meteorol. Soc.,* 318–20.

Brook, M., and N. Kitagawa, 1960: Some aspects of lightning activity and related meteorological conditions *J. Geophys. Res., 65,* 1203–10.

Brook, M., M. Nakano, P. Krehbiel and T. Takeuti, 1982: The electrical structure of the Hokuriku winter thunderstorms, *J. Geophys. Res., 87,* 1207–15.

Brooks, C.E.P., 1925: The distribution of thunderstorms over the globe. *Geophys. Mem., Lond., 3,* No. 24, 147–64.

Browning, K.A., J.C. Fankhauser, J.P. Chalon, P.J. Eccles, R.C. Strauch, F.H. Merrem, D.J. Musil, E.L. May, and W.R. Sand, 1976: Structure of an evolving hailstorms. Part V: Synthesis and implications for hail growth and hail suppression. *Mon. Wea. Rev., 104,* 603–10.

Buechler, D.E., and S.J. Goodman, 1990: Echo size and asymmetry: Impact on NEXRAD storm identification. *J. Appl. Meteor. 29,* 962–69.

Buechler, D.E., P.D. Wright, and S.J. Goodman, 1990: Lightning/rainfall relationships during COHMEX. *Preprints, 16th Conf. on Severe Local Storms, Amer. Meteor. Soc.,* 710–14.

Buechler, D.E., and S.J. Goodman, 1991: Radar characteristics of cloud-to-ground lightning producing storms in Florida. *Preprints, 25th International Conf. on Radar Meteorology, Amer. Meteor. Soc.,* 897–900.

Byers, H.R., and R.R. Braham, 1949: *The Thunderstorm. Report of the Thunderstorm Project.* U.S. Weather Bureau, U.S. Govt. Printing Office, 287 pp.

Byrne, G.J., A.A. Few, and M.E. Weber, 1983: Altitude, thickness, and charge concentrations of charged regions of four thunderstorms during TRIP 1981 based upon in situ balloon electric field measurements, *Geophys. Res. Lett., 10,* 39–42.

Carey, L.D., and S.A. Rutledge, 1996: A multiparameter radar case study of the microphysical and kinematic evolution of a lightning producing storm. *J. Meteor. and Atmos. Phys., 59,* 33–64.

Chapman, S., 1953: Thundercloud electrification in relation to rain and snow particles. In *Thunderstorm Electricity,* H.R. Byers, ed., Univ. Chicago Press, Chicago, pp. 207–30.

Cherna, E.V., and E.J. Stansbury, 1986: Sferics rate in relation to thunderstorm dimensions. *J. Geophys. Res., 91,* 8701–07.

Christian, H.J., C.R. Holmes, J.W. Bullock, W. Gaskell, A.J. Illingworth, and J. Latham, 1980: Airborne and ground-based studies of thunderstorms in the vicinity of Langmuir Laboratory. *Q. J. Roy. Meteor. Soc., 106,* 159–74.

Christian, H.J., K.T. Driscoll, S.J. Goodman, R.J. Blakeslee, D.M. Mach, and D.E. Buechler, 1996: Seasonal variation and distribution of lightning activity. *Eos, Trans. Amer. Geophys. Un., 77,* F80. Abstract.

Clarence, N.D., and D.J. Malan, 1957: Preliminary discharge processes in lightning flashes to ground. *Quart. J. Roy. Meteor. Soc., 83,* 161–72.

Clodman, S., and W. Chisholm, 1993: Storms with very high lightning density in the southern Great Lakes area. *Preprints, 17th Conf. on Severe Local Storms, Amer. Meteorol. Soc.,* 803–7.

Cobb, W.E., 1975: Electric fields in Florida cumuli. *Eos, Trans. Amer. Geophys. Un., 56,* 990.

Colson, D., 1960: High level thunderstorms of July 31–August 1, 1959. *Mon. Wea. Rev., 88,* 279–85.

Curran, E.B., and W.D. Rust, 1992: Positive ground flashes produced by low-precipitation thunderstorms in Oklahoma on 26 April 1984, *Mon. Weath. Rev., 120,* 544–53.

Deaver, L.E., and E.P. Krider, 1991: Electric fields and current densities under small Florida thunderstorms. *J. Geophys. Res., 96,* 22,273–81.

Devlin, K.I., 1995: *Application of the 85 GHz Ice Scattering Signature to a Global Study of Mesoscale Convective Systems.* Master's thesis, Texas A&M Univ., College Station, 100 pp.

Doviak, R.J., and D.S. Zrnić , 1993: *Doppler Radar and Weather Observations,* 2d ed., Academic Press, San Diego, 562 pp.

Driscoll, K.T., R.J. Blakeslee, and W.J. Koshak, 1994: Time-averaged current analysis of a thunderstorm using ground-based measurements. *J. Geophys. Res., 99,* 10,653–61.

Driscoll, K.T., Christian, H.J., S.J. Goodman, R.J. Blakeslee, D.J. Boccippio, 1996: Diurnal global lightning distribution as observed by the optical transient detector. *Eos, Trans. Amer. Geophys. Un., 77,* F92. Abstract.

Dye, J.E., J.J. Jones, W.P. Winn, T.A. Cerni, B. Gardiner, D. Lamb, R.L. Pitter, J. Hallet, and C.P.R. Saunders, 1986: Early electrification and precipitation development in a small isolated Montana cumulonimbus, *J. Geophys. Res., 91,* 1231–47.

Dye, J.E., J.J. Jones, A.J. Weinheimer, and W.P. Winn, 1988: Observations within two regions of charge during initial thunderstorm electrification *Q. J. Roy. Meteor. Soc., 114* 1271–90.

Dye, J.E., W.P. Winn, J.J. Jones, and D.W. Breed, 1989: The electrification of New Mexico thunderstorms 1. Relationship between precipitation development and the onset of electrification. *J. Geophys. Res., 94,* 8643–56.

Engholm, C.D., E.R. Williams, and R.M. Dole, 1990: Meteorological and electrical conditions associated with positive cloud-to-ground lightning. *Mon. Weath. Rev., 118,* 470–87.

Evans, W.H., 1969a: Electric fields and conductivity in thunderclouds. *J. Geophys. Res., 74,* 939–48.

Evans, W.H., 1969b: Reply. *J. Geophys. Res., 74,* 7056–57.

Fitzgerald, D.R., and H.R. Byers, 1958: Aircraft observations of convective cloud electrification. *J. Atmos. Terres. Phys., 15,* 254–61.

Freeman, W.B., 1977: *The Distribution of Thunderstorm Days, Lightning Discharges, and the Incidence of Lightning Discharge Derived from VLF Sferics Data,* Report 77-0112, Air Force Geophys. Lab., Hanscom AFB, Mass.

Funaki, K., N. Kitagawa, M. Nakano, and T. Takeuit, 1981a: Multistation measurement of lightning discharges to ground in Hokuriku winter thunderstorms. *Res. Lett. Atmos. Elect., 1,* 19–26.

Funaki, K., K. Sakamoto, R. Tanaka, and N. Kitagawa, 1981b: A comparison of cloud and ground lightning discharges observed in south-Kanto summer thunderstorms, 1980. *Res. Lett. Atmos. Electricity, 1,* 99–103.

Fuquay, D.M., 1982: Positive cloud-to-ground lightning in summer thunderstorms. *J. Geophys. Res., 87,* 7131–40.

Gallimberti, I., 1979: The mechanism of the long spark. *J. de Physique, Colloque C7,* 193–250.

Gardiner, B., D. Lamb, R.L. Pitter, J. Hallett, and C.P.R. Saunders, 1985: Measurements of initial potential gradient and particle charges in a Montana summer thunderstorm. *J. Geophys. Res., 90,* 6079–86.

Gaskell, W., A.J. Illingworth, J. Latham, C.B. Moore, 1978:

Airborne studies of electric fields and the charge and size of precipitation elements in thunderstorms, *Q. J. Roy. Meteor. Soc., 104,* 447–60.

Gish, O. H., and G.R. Wait, 1950: Thunderstorms and the Earth's general electrification. *J. Geophys. Res., 55,* 473–84.

Goodman, S.J., and D.R. MacGorman, 1986: Cloud-to-ground lightning activity in mesoscale convective complexes. *Mon. Wea. Rev., 114,* 2320–28.

Goodman, S.J., D.E. Buechler, P.D. Wright, and W.D. Rust, 1988a: Lightning and precipitation history of a microburst-producing storm. *Geophys. Res. Lett., 15,* 1185–88.

Goodman, S.J., D.E. Buechler, and P.J. Meyer, 1988b: Convective tendency images derived from a combination of lightning and satellite data. *Wea. Forecast., 3,* 173–88.

Goodman, S.J., D.E. Buechler, P.D. Wright, W.D. Rust, and K.E. Nielsen, 1989: Polarization radar and electrical observations of miroburst producing storms during COHMEX. *Preprints,* 24th Conf. on Radar Meteorology, Amer. Meteorol. Soc., Boston, 109–12.

Goodman, S.J., and R. Raghaven, 1993: Investigating the relation between precipitation and lightning using polarimetric radar observations. *Preprints,* 26th Conf. on Radar Meteorology, Amer. Meteorol. Soc., 793–95.

Griffiths, R.F., and J. Latham, 1974: Electrical corona from ice hydrometeors. *Q. J. Roy. Meteor. Soc., 100,* 163–80.

Griffiths, R.F., J. Latham, and V. Myers, 1974: The ionic conductivity of electrified clouds. *Q. J. Roy. Meteor. Soc., 100,* 181–90.

Griffiths, R.F., and C.T. Phelps, 1976a: The effects of air pressure and water vapour content on the propagation of positive corona streamers and their implications to lightning initiation. *Q. J. Roy. Meteorol. Soc., 102,* 419–26.

Griffiths, R.F., and C.T. Phelps, 1976b: A model for lightning initiation arising from positive corona streamer development. *J. Geophys. Res., 81,*3671–76.

Grosh, R.C., 1977: Relationships between severe weather and echo tops in central Illinois. *Preprints,* 10th Conf. on Severe Local Storms, Amer. Meteorol. Soc., 231–38.

Grosh, R.C., 1978: Lightning and precipitation—the life history of isolated thunderstorms. *Preprints,* Conf. on Cloud Physics and Atmospheric Electricity, Amer. Meteorol. Soc., 617–24.

Gunn, R., 1935: The electricity of rain and thunderstorms. *J. Terr. Magn., 40,* 79–106.

Gunn, R., 1950: The free electrical charge on precipitation inside an active thunderstorm. *J. Geophys. Res., 55,* 171–78.

Gurevich, A.V., G.M. Milikh, and R.A. Roussel-Dupre, 1992: Runaway electron mechanism for air breakdown and preconditioning during a thunderstorm. *Phys. Lett. A, 165,* 463–68.

Gurevich, A.V., G.M. Milikh, and R.A. Roussel-Dupre, 1994: Nonuniform runaway air breakdown. *Phys. Lett. A, 167,* 197–203.

Hatakeyama, H., 1958: The distribution of the sudden change of electric field on the earth's surface due to lightning discharge. In *Recent Advances in Atmospheric Electricity,* L.G. Smith, eds., Pergamon, New York, pp. 289–98.

Hayenga, C.O., and J.W. Warwick, 1981: Two-dimensional interferometric positions of VHF lightning sources. *J. Geophys. Res., 81,* 7451–62.

Holle, R.L., and M.W. Maier, 1982: Radar echo height related to cloud-ground lightning in south Florida. *Preprints,*

12th Conf. on Severe Local Storms, American Meteorological Society, 330–33.

Holle, R.L., A.I. Watson, R.E. López, D.R. MacGorman, R. Ortiz, and W.D. Otto, 1994: The life cycle of lightning and severe weather in a 3–4 June 1985 PRE-STORM mesoscale convective system. *Mon. Wea. Rev., 122,* 1798–1808.

Holle, R.L., and S.P. Bennett, 1997: Lightning ground flashes associated with summer 1990 flash floods and streamflow in Tucson, Arizona: An exploratory study. *Mon. Wea. Rev. 125,* 1526–36.

Holmes, C.R., E.W. Szymanski, S.J. Szymanski, and C.B. Moore, 1980: Radar and acoustic study of lightning. *J. Geophys. Res., 85,* 7517–32.

Hondl, K.D., and M.D. Eilts, 1994: Doppler radar signatures of developing thunderstorms and their potential to indicate the onset of cloud-to-ground lightning. *Mon. Wea. Rev., 122,* 1818–36.

Hoppel, W.A., and B.B. Phillips, 1971: The electrical shielding layer around charged clouds and its role in thunderstorm electricity. *J. Atmos. Sci., 28,* 1258–71.

Houze, Jr., R.A., 1993: *Cloud Dynamics.* Academic Press, San Diego, 573 pp.

Idone, V.P., A.B. Saljoughy, R.W. Henderson, P.K. Moore, and R. Pyle, 1993: A reexamination of the peak current calibration of the national lightning detection network. *J. Geophys. Res. 98,* 18,323–32.

Imyanitov, I.M., B.F. Evteev, and I.I. Kamaldina, 1969: A thunderstorm cloud. In *Planetary Electrodynamics,* Vol. 1, S.C. Coroniti and J. Hughes, eds., Gordon and Breach, New York, pp. 401–25.

Imyanitov, I.M., Ye.V. Chubarina, and Ya.M. Shvarts, 1972: *Electricity of Clouds.* 122 pp, NASA Tech. Translation, NASA TT F-718, of *Elektrichestvo oblakov.* Hydrometeorological Press, Leningrad, 1971.

International Radio Consultative Committee, 1964: *World Distribution and Characteristics of Atmospheric Radio Noise,* Report 332, Intl. Telecomm. Union, Geneva, 78 pp.

Ishihara, M., H. Sakakibara, Z. Yanagisawa, K. Matsuura, and J. Aoyagi, 1987: Internal structure of thunderstorms in Kanto observed by two Doppler radars. *Tenki, 34,* 321–32. (In Japanese)

Ishii, M., T. Kawamura, J. Hojyo, and T. Iwaizumi, 1981: Ground flash density in winter thunderstorm. *Res. Lett. Atmos. Elec. 1,* 105–8.

Ishikawa, H., 1961: Nature of lightning discharges as sources of atmospherics. *Proc. Res. Inst. of Atmospherics Nagoya Univ., 8A,* 1–274.

Jacobson, E.A., and E.P. Krider, 1976: Electrostatic field changes produced by Florida lightning. *J. Atmos. Sci., 33,* 103–17.

Jayaratne, E.R., and C.P.R. Saunders, 1984: The "rain-gush," lightning, and the lower positive charge center in thunderstorms. *J. Geophys. Res., 89,* 11,816–18.

Jones, R.F., 1950: The temperature at the tops of radar echoes associated with various cloud systems. *Quart. J. Roy. Meteorol. Soc., 76,* 312–30.

Juvanon du Vachat, R., and J. Cheze, 1993: The ASPIC project, Presentation of a short-range forecasting system for storms and precipitations, preliminary evaluation of the SAFIR lightning system. *Preprints,* 17th Conf. on Severe Local Storms, Amer. Meteor. Soc., 756–60.

Kamra, A.K., 1979: Contributions of cloud and precipitation particles to the electrical conductivity and the relaxation time of air in thunderstorms. *J. Geophys. Res., 84,* 5034–38.

Kamra, A.K., 1981: Reply. *J. Geophys. Res., 86,* 4302–04.

Kane, R.J., 1991: Correlating lightning to severe local storms in the northeastern United States, *Weath. Forecast., 6,* 3–12.

Kane, R.J., 1993a: Lightning-rainfall relationships in an isolated thunderstorm over the mid-Atlantic states. *Natl. Wea. Digest, 18,* 2–14.

Kane, R.J., 1993b: A case study of lightning-radar characteristics in a mesoscale convective complex. *Preprints,* 17th Conf. on Severe Local Storms, Amer. Meteor. Soc., 816–22.

Kawasaki, Z-I., K. Yamamoto, K. Matsuura, P. Richard, T. Matsui, Y. Sonoi, and N. Shimokura, 1994: SAFIR operation and evaluation of its performance. *Geophys. Res. Lett., 21,* 1133–36.

Keener, J.E., C.W. Ulbrich, M.F. Larsen, and P.B. Chilson, 1993: A study of tropical thunderstorm lightning with the use of a dual wavelength Doppler radar. *Preprints,* 26th International Conf. Radar Meteorology, Amer. Meteorol. Soc., 324–26.

Keener, J.E., and C.W. Ulbrich, 1995: Spatial locations of lightning channel echoes at VHF related to the dynamics and microphysics in a tropical thunderstorm. *Preprints,* 27th International Conf. Radar Meteorology, Amer. Meteor. Soc., 639–41.

Keighton, S.J., H.B. Bluestein, and D.R. MacGorman, 1991: The evolution of a severe mesoscale convective system: Cloud-to-ground lightning location and storm structure. *Mon. Wea. Rev., 119,* 1533–56.

Kessler, E. (ed.), 1986: *Thunderstorm Morphology and Dynamics.* 2d ed., Univ. Oklahoma Press, Norman, 411 pp.

Kinzer, G.D., 1974: Cloud-to-ground lightning versus radar reflectivity in Oklahoma thunderstorms. *J. Atmos. Sci., 31,* 787–99.

Kitagawa, N., 1992: Charge distribution in winter thunderclouds. *Res. Lett. Atmos. Elec., 12,* 143–53.

Kitagawa, N., and M. Brook, 1960: A comparison of intracloud and cloud-to-ground lightning discharges. *J. Geophys. Res., 65,* 1189–1201.

Kitagawa, N., and K. Michimoto, 1994: Meteorological and electrical aspects of winter thunderclouds. *J. Geophys. Res., 99,* 10,713–21.

Koshak, W.J., and E.P. Krider, 1989: Analysis of lightning field changes during active Florida thunderstorms. *J. Geophys. Res., 94,* 1165–86.

Kotaki, M., I. Kuriki, C. Katoh, and H. Sugiuchi, 1981: Global distribution of thunderstorm activity observed with ISS-b. *J. Radio Res. Lab., 28,* 49–71.

Kraakevik, J.H., 1958: The airborne measurement of atmospheric conductivity. *J. Geophys. Res., 63,* 161–69.

Krehbiel, P.R., 1969: Conductivity in clouds in the presence of electric fields. *Eos, Trans., Amer. Geophys. Un., 50,* 618. Abstract.

Krehbiel, P.R., 1981: An analysis of the electric field change produced by lightning, Ph.D. diss. Univ. Manchester Inst. Sci. & Tech. Published as Rpt. T-11, New Mex. Inst. Min. & Tech., Dec., 1981, Vol. 1 is 245 pp. and Vol. 2 is figures.

Krehbiel, P.R., 1986: The electrical structure of thunderstorms. In *The Earth's Electrical Environment,* National Acad. Press, Washington, D.C. pp. 90–113.

Krehbiel, P.R., M. Brook, R.A. McCrory, 1979: An analysis of the charge structure of lightning discharges to ground. *J. Geophys. Res. 84,* 2432–56.

Krider, E.P., 1988: Spatial distribution of lightning strikes to ground during small thunderstorms in Florida. *Proceedings,* 1988 International Aerospace and Ground Conf. on Lightning and Static Electricity, NOAA Special Report, 318–23.

Krider, E.P., 1989: Electric field changes and cloud electrical structure. *J. Geophys. Res., 94,* 13,145–49.

Krider, E.P., and J.A. Musser, 1982: Maxwell currents under thunderstorms. *J. Geophys. Res., 87,* 1982.

Krider, E.P., and R.J. Blakeslee, 1985: The electric currents produced by thunderclouds. *J. Electrostat. 16,* 369–78.

Kuettner, J., 1950: The electrical and meteorological conditions inside thunderstorms. *J. Meteor., 7,* 322–32.

Larson, H.R., and E.J. Stansbury, 1974: Association of lightning flashes with precipitation cores extending to height 7 km. *J. Atmos. Terrestrial Phys., 36,* 1547–53.

Latham, J. and C.D. Stow, 1969: Airborne studies of the electrical properties of large convective clouds. *Q. J. Roy. Meteor. Soc., 95,* 486–500.

Lee, A.C.L., 1986: An operational system for the remote location of lightning flashes using a VLF arrival time difference technique. *J. Atmos. Ocean. Tech., 3,* 630–42.

Lhermitte, R., and P.R. Krehbiel, 1979: Doppler radar and radio observations of thunderstorms, *IEEE Trans. on Geoscience Electron., GE-17,* 162–71.

Lhermitte, R., and E. Williams, 1985: Thunderstorm electrification: A case study. *J. Geophys. Res., 90,* 6071–78.

Ligda, M.H., 1956: The radar observation of lightning. *J. Atmos. Terr. Phys., 3,* 329–46.

Liu, X., and P.R. Krehbiel, 1985: The initial streamer of intracloud lightning flashes. *J. Geophys. Res., 90,* 6211–18.

Livingston, J.M., and E.P. Krider, 1978: Electric fields produced by Florida thunderstorms, *J. Geophys. Res., 83,* 385–401.

López, R.E., W.D. Otto, R. Ortíz, and R.L. Holle, 1990: The lightning characteristics of convective cloud systems in northeastern Colorado. *Preprints,* 16th Conf. on Severe Local Storms, Amer. Meteorol. Soc., 727–31.

López, R.E., R. Ortíz, W.D. Otto, and R.L. Holle, 1991: The lightning activity and precipitation yield of convective cloud systems in central Florida. *Preprints,* 25th International Conf. on Radar Meteorology, Amer. Meteor. Soc., 907–10.

López, R.E., and J.-P. Aubagnac, 1997: The lightning activity of a hailstorm as a function of changes in its microphysical characteristics inferred from polarimetric radar observations. *J. Geophys. Res.,* in press.

Lucas, C., E.J. Zipser, and B.S. Ferrier, 1995: Warm-pool cumulonimbus and the ice phase. *Preprints,* Conf. on Cloud Physics, Amer. Meteor. Soc., 318–20.

Ludlam, F.H., 1980: *Clouds and Storms.* Pennsylvania State Univ. Press, Univ. Park, 405 pp.

MacGorman, D.R., 1978: Lightning in a storm with strong wind shear. Ph.D. diss., Rice Univ., Houston.

MacGorman, D.R., A.A. Few, and T.L. Teer, 1981: Layered lightning activity. *J. Geophys. Res., 86,* 9900–10.

MacGorman, D.R., W.L. Taylor, and A.A. Few, 1983: Lightning location from acoustic and VHF techniques relative to storm structure from 10-cm radar. In *Proceedings in Atmospheric Electricity,* L.H. Ruhnke and J. Latham, ed., A. Deepak Publishing, Hampton, pp. 377–380.

MacGorman, D.R., M.W. Maier, and W.D. Rust, 1984: *Lightning Strike Density for the Contiguous United States from Thunderstorm Duration Records.* Report to the U.S. Nuclear Regulatory Commission, available as NUREG/CR-3759, 51 pp.

MacGorman, D.R., D.W. Burgess, V. Mazur, W.D. Rust, W.L. Taylor, and B.C. Johnson, 1989: Lightning rates relative to tornadic storm evolution on 22 May 1981. *J. Atmos. Sci., 46,* 221–50.

MacGorman, D.R., and K.E. Nielsen, 1991: Cloud-to-ground lightning in a tornadic storm on 8 May 1986, *Mon. Wea. Rev., 119,* 1557–74.

MacGorman, D.R., and D.W. Burgess, 1994: Positive cloud-to-ground lightning in tornadic storms and hailstorms. *Mon. Wea. Rev., 122* 1671–97.

MacGorman, D.R., C.L. Ziegler, and J.M. Straka, 1996: Considering the complexity of thunderstorms. *Proceedings, 10th International Conference on Atmospheric Electricity,* Osaka, Japan, pp. 128–31.

MacGorman, D.R., K. Crawford, H. Xia, 1997: A lightning strike climatology of Oklahoma: Part I—Geographic and interannual variability. *J. Climate.* Submitted.

Mach, D.M., and K.R. Knupp, 1993: Radar reflectivity as a tool to remotely infer the presence of significant fields aloft. *Preprints,* 26th Intl. Conf. on Radar Meteorology, Amer. Meteorol, Soc., 315–17.

Mackerras, D. 1985: Automatic short-range measurement of the cloud flash to ground flash ratio in thunderstorms. *J. Geophys. Res., 90,* 6195–6201.

Mackerras, D. and M. Darveniza, 1994: Latitudinal variation of lightning occurrence characteristics. *J. Geophys. Res., 99,* 10813–21.

Maekawa, Y., S. Fukao, Y. Sonoi, F. Yoshino, 1992: Dual polarization radar observations of anomalous wintertime thunderclouds in Japan. *IEEE Trans. Geosci. Rem. Sens., 30,* 838–844.

Magono, C., 1980: *Thunderstorms.* Elsevier Science, Amsterdam, 261 pp.

Magono, C., T. Endoh, and T. Shigeno, 1983: The electrical structure of snow clouds Part I: Relationship to their precipitation-physical properties. *J. Meteor. Soc. Jpn., 61,* 325–38.

Magono, C., H. Sakamoto, T. Endoh, and T. Taniguchi, 1984: The electrical structure of snow clouds, Part 2: Vertical distribution of precipitation charge. *J. Meteor. Soc. Japan, 62,* 323–34.

Maier, L.M., E.P. Krider, and M.W. Maier, 1984: Average diurnal variation of summer lightning over the Florida peninsula. *Mon. Wea. Rev., 112,* 1134–40.

Maier, L.M., and E.P. Krider, 1986: The charges that are deposited by cloud-to-ground lightning in Florida. *J. Geophys. Res., 91,* 13275–89.

Maier, L.M., C. Lennon, P. Krehbiel, M. Stanley, and M. Robison, 1995: Comparison of lightning and radar observations from the KSC LDAR and NEXRAD radar systems. *Preprints,* 17th Conf. Radar Meteorology, Amer. Meteorol. Soc., 648–50.

Maier, M.W., A.G. Boulanger, and J. Sarlat, 1978: Cloud-to-ground lightning frequency over south Florida. *Preprints,* Conf. on Cloud Physics and Atmospheric Electricity, Amer. Meteor. Soc., 605–10.

Malan, D.J., and Schonland, B.F.J., 1951: The distribution of electricity in thunderclouds. *Proc. Roy. Soc. London A, 209,* 158–77.

Marsh, S.J., and T.C. Marshall, 1993: Charged precipitation measurements before the first lightning flash in a thunderstorm. *J. Geophys. Res., 98,* 16,605–11.

Marshall, J.S., and W.M. Palmer, 1948: The distribution of raindrops with size. *J. Meteor. 5,* 165–66.

Marshall, J.S., and S. Radhakant, 1978: Radar precipitation maps as lightning indicators. *J. Appl. Meteor., 17,* 206–12.

Marshall, T.C., 1993: A review of charges on thunderstorm precipitation particles. *Eos. Trans. Amer. Geophys. Un., 74,* 153.

Marshall, T.C., and W.P. Winn, 1982: Measurements of charged precipitation in a New Mexico thunderstorm: lower positive charge centers. *J. Geophys. Res., 87,* 7141–57.

Marshall, T.C., W.D. Rust, W.P. Winn, and K.E. Gilbert, 1989: Electrical structure in two thunderstorm anvil clouds. *J. Geophys. Res., 94,* 2171–81.

Marshall, T.C., and W.D. Rust, 1991: Electric field soundings through thunderstorms. *J. Geophys. Res., 96,* 22,297–306.

Marshall, T.C., and B. Lin, 1992: Electricity in dying thunderstorms. *J. Geophys. Res., 97,* 9913–18.

Marshall, T.C., and S.J. Marsh, 1993: Negatively charged precipitation in a New Mexico thunderstorm. *J. Geophys. Res., 98,* 14,909–16.

Marshall, T.C., W. Rison, W.D. Rust, M. Stolzenburg, J.C. Willett, and W.P. Winn, 1995a: Rocket and balloon observations of electric field in two thunderstorms. *J. Geophys. Res., 100,* 20,815–28.

Marshall, T.C., M.P. McCarthy, and W.D. Rust, 1995b: Electric field magnitudes and lightning initiation in thunderstorms. *J. Geophys. Res., 100,* 7097–7103.

Marwitz, J.D., 1972: The structure and motion of severe hailstorms. Part II: Multicell storms. *J. Appl. Meteor., 11,* 180–88.

Mazur, V., 1989: Triggered lightning strikes to aircraft and natural intracloud discharges. *J. Geophys. Res., 94,* 3311–25.

Mazur, V., and W.D. Rust, 1983: Lightning propagation and flash density in squall lines as determined with radar. *J. Geophys. Res., 88,* 1495–1502.

Mazur, V., J.C. Gerlach, and W.D. Rust, 1984: Lightning flash density versus altitude and storm structure from observations with UHF- and S-band radars. *Geophys. Res. Lett., 11,* 61–4.

Mazur, V., J.C. Gerlach, and W.D. Rust, 1986: Evolution of lightning flash density and reflectivity structure in a multicell thunderstorm. *J. Geophys. Res., 91,* 8690–8700.

McCarthy, M.P., and G.K. Parks, 1992: On the modulation of X ray fluxes in thunderstorms. *J. Geophys. Res., 97,* 5857–64.

Michimoto, K., 1991: A study of radar echoes and their relation to lightning discharge of thunderclouds in the Hokuriku District. Part I: Observation and analysis of thunderclouds in summer and winter. *J. Meteor. Soc. Japan, 69,* 327–35.

Michimoto, K., 1993a: A study of radar echoes and their relation to lightning discharge of thunderclouds in the Hokuriku District. Part II: Observation and analysis of "single-flash" thunderclouds in midwinter. *J. Meteor. Soc. Japan, 71,* 195–204.

Michimoto, K., 1993b: A study of the charge distribution in winter thunderclouds by means of network recording of surface electric fields and radar observation of clouds structure in the Hokuriku District. *Res. Lett. Atmos. Elect., 13,* 33–46.

Miyazaki, T., Y. Nishida, S. Kokubu, and S. Kawamata, 1982: On the echo-top height of thunderclouds in each month. *Res. Lett. Atmos. Electricity, 2,* 45–8. (Japanese with English abstract)

Mohr, K.I., E.R. Toracinta, R.E. Orville, and E.J. Zipser, 1995: A comparison of WSR-88D reflectivities, SSM/I brightness temperatures and lightning for mesoscale convective systems in Texas. *Preprints,* Conf. on Cloud Physics, Amer. Meteor. Soc., 547–52.

Mohr, K.I., R.E. Toracinta, E.J. Zipser, and R.E. Orville, 1996: A comparison of WSR-88D reflectivities, SSM/I brightness temperatures, and lightning for mesoscale convective systems in Texas. Part II: SSM/I brightness temperatures and lightning. *J. Appl. Meteor., 35,* 919–31.

Moore, C.B., B. Vonnegut, and A.T. Botka, 1958: Results of an experiment to determine initial precedence of organized electrification and precipitation in thunderstorms. In *Recent Advances in Atmospheric Electricity,* L.G. Smith, eds., Pergamon, New York, pp. 333–60.

Moore, C.B., and B. Vonnegut, 1977: The thundercloud. In *Lightning, Vol. 1: Physics of Lightning,* R.H. Golde, eds., Academic Press, San Diego, 64–98.

Moore, C.B., B. Vonnegut, T.D. Rolan, J.W. Cobb, D.N. Holden, R.T. Hignight, S.M. McWilliams, and G.W. Cadwell, 1986: Abnormal polarity of thunderclouds grown from negatively charged air. *Science, 233,* 1413–16.

Moore, C.B., B. Vonnegut, and D.N. Holden, 1989: Anomalous electric fields associated with clouds growing over a source of negative space charge. *J. Geophys. Res., 94,* 13,127–34.

Moore, C.B., B. Vonnegut, and D.N. Holden, 1991: Correction to "Anomalous electric fields associated with clouds growing over a source of negative space charge" by C.B. Moore, B. Vonnegut, and D.N. Holden. *J. Geophys. Res., 96,* 7589.

Morgenstern, C.D., 1991: *Cloud-to-Ground Lightning Characteristics in Mesoscale Convective Systems, April–September 1986,* master's thesis, Univ. Oklahoma, Norman, 109 pp.

Nakano, M., 1973: Lightning channel determined by thunder. *Proc Res. Inst. Atmospherics, Nagoya Univ., 20,* 1–7.

Nakano, M., 1976: Characteristics of lightning channel in thunderclouds determined by thunder. *J. Meteor. Soc. Japan, 54,* 441–47.

Nakano, M., 1979a: The cloud discharge in winter thunderstorms of the Hokuriku coast. *J. Meteor. Soc. Japan, 57,* 444–51.

Nakano, M., 1979b: Initial streamer of the cloud discharge in winter thunderstorms, of the Hokuriku coast. *J. Meteor. Soc. Japan, 57,* 452–57.

Nakano, M., T. Takeuti, K. Funaki, N. Kitagawa, and C. Takahashi, 1984: Oceanic tropical lightning at Ponape, Micronesia. *Res. Lett. Atmos. Electric. 4,* 29–33.

Nielsen, K.E., R.A. Maddox, and S.V. Vasiloff, 1994: The evolution of cloud-to-ground lightning within a portion of the 10–11 June 1985 squall line. *Mon. Wea. Rev., 122,* 1809–17.

Nisbet, J.S., T.A. Barnard, G.S. Forbes, E.P. Krider, R. Lhermitte, and C.L. Lennon, 1990: A case study of the Thunderstorm Research International Project Storm of July 11, 1978. Part 1: Analysis of the data base. *J. Geophys. Res., 95,* 5417–33.

Norinder, H., and E. Knudsen, 1961: Some features of thunderstorm activity. *Ark. Geofys., 3,* 367–74.

Ogawa, T., and M. Brook, 1969: Charge distribution in thunderstorm clouds. *Q. J. R. Meteorol. Soc., 95,* 513–25.

Orville, R.E., 1994: Cloud-to-ground lightning flash characteristics in the contiguous United States: 1989–91. *J. Geophys. Res., 99,* 10,833–41.

Orville, R.E., and D.W. Spencer, 1979: Global lightning flash frequency. *Mon. Weath. Rev., 107,* 934–43.

Orville, R.E., and R.W. Henderson, 1986: Global distribution of midnight lightning: September 1977 to August 1978. *Mon. Wea. Rev., 114,* 2640–53.

Orville, R.E., R.A. Weisman, R.B. Pyle, R.W. Henderson, and R.E. Orville, Jr., 1987: Cloud-to-ground lightning flash characteristics from June 1984 through May 1985. *J. Geophys. Res., 92,* 5640–44.

Peckham, D.W., M.A. Uman, and C.E. Wilcox, Jr., 1984: Lightning phenomenology in the Tampa Bay Area. *J. Geophys. Res., 89,* 11,789–805.

Petersen, W.A., R. Cifelli, S.A. Rutledge, and B.F. Smull, 1995: Cloud-to-ground lightning and the related kinematic structures of two tropical oceanic MCS's: Contrasting cases. *Preprints,* Conf. on Cloud Phys., Amer. Meteorol. Soc., 553–58.

Petersen, W.A., S.A. Rutledge, and R.E. Orville, 1996: Cloud-to-ground lightning observations to TOGA COARE: Selected results and lightning location algorithms. *Mon. Wea. Rev., 124.* 602–620.

Phelps, C.T., 1974: Positive streamer system intensification and its possible role in lightning initiation. *J. Atmos. Terres. Phys., 36,* 103–11.

Phelps, C.T., and R.F. Griffiths, 1976: Dependence of positive corona streamer propagation on air pressure and water vapor content. *J. Appl. Phys., 47,* 2929–34.

Phillips, B.B., 1967: Ionic equilibrium and the electrical conductivity in thunderclouds. *Mon. Weath. Rev., 95,* 854–62.

Piepgrass, M.V., E.P. Krider, and C.B. Moore, 1982: Lightning and surface rainfall during Florida thunderstorms. *J. Geophys. Res., 87,* 11,193–201.

Pierce, E.T., 1955a: Electrostatic field-changes due to lightning discharges *Q. J. R. Meteorol. Soc., 81,* 211–28.

Pierce, E.T., 1955b: The development of lightning discharges. *Q. J. R. Meteorol. Soc., 81,* 229–40.

Pierce, E.T., 1970: Latitudinal variation of lightning parameters. *J. Appl. Meteorol., 9,* 194–95.

Pierce, E.T., and T.W. Wormell, 1953: Field changes due to lightning discharge. In *Thunderstorm Electricity,* H.R. Byers, ed., Univ. Chicago Press, Chicago, pp. 251–66.

Prentice, S.A., 1960: Thunderstorms in the Brisbane area. *J. Inst. Eng. Aust., 32,* 33–45.

Prentice, S.A., and D. Mackerras, 1977: The ratio of cloud-to-ground lightning flashes in thunderstorms. *J. Appl. Meteorol., 16,* 545–50.

Price, C., and D. Rind, 1992: A simple lightning parameterization for calculating global lightning distributions. *J. Geophys. Res., 97,* 9919–33.

Price, C., and D. Rind, 1993: What determines the cloud-to-ground lightning fraction in thunderstorms? *Geophys. Res. Lett., 20,* 463–66.

Proctor, D.E., 1981: VHF radio pictures of cloud flashes. *J. Geophys. Res., 86,* 4041–71.

Proctor, D.E., 1983: Lightning and precipitation in a small multicellular thunderstorm. *J. Geophys. Res., 88,* 5421–40.

Proctor, D.E., 1991: Regions where lightning flashes began. *J. Geophys. Res., 96,* 5099–5112.

Proctor, D.E., R. Uytenbogaardt, and B.M. Meredith, 1988: VHF radio pictures of lightning flashes to ground. *J. Geophys. Res., 93,* 12,683–727.

Pruppacher, H.R., and J.D. Klett, 1978: *Microphysics of Clouds and Precipitation.* D. Reidel, Dordrecht, pp. 585–93.

Ramsay, M.W. and J.A. Chalmers, 1960: Measurement on the electricity of precipitation. *Q. J. Roy. Meteor. Soc., 86,* 530–39.

Randell, S.C., S.A. Rutledge, R.D. Farley, and J.H. Helsdon, Jr., 1994: A modeling study on the early electrical development of tropical convection: Continental and oceanic (monsoon) storms. *Mon. Wea. Rev., 122,* 1852–77.

Ray, P.S., D.R. MacGorman, W.D. Rust, W.L. Taylor, and L.W. Rasmussen, 1987: Lightning location relative to storm structure in a supercell storm and a multicell storm. *J. Geophys. Res., 92,* 5713–24.

Raymond, D.J., R. Solomon, and A.M. Blyth, 1991: Mass fluxes in New Mexico mountain thunderstorms from radar and aircraft measurements. *Q. J. Roy. Meteor. Soc., 117,* 587–621.

Reap, R.M., 1986: Evaluation of cloud-to-ground lightning data from the western United States for the 1983–84 summer seasons. *J. Clim. Appl. Meteorol., 25,* 785–99.

Reap, R.M., 1991: Climatological characteristics and objective prediction of thunderstorms over Alaska. *Wea. Forecast., 6,* 309–19.

Reap, R.M., and D.R. MacGorman, 1989: Cloud-to-ground lightning: Climatological characteristics and relationships to model fields, radar observations, and severe local storms, *Mon. Wea. Rev., 117,* 518–35.

Reynolds, S.E., and H.W. Neill, 1955: The distribution and discharge of thunderstorm charge-centers. *J. Meteor., 12,* 1–12.

Reynolds, S.E., and M. Brook, 1956: Correlation of the initial electric field and the radar echo in thunderstorms. *J. Meteor., 13,* 376–80.

Rhodes, C.T., X.M. Shao, P.R. Krehbiel, R.J. Thomas, and C.O. Hayenga, 1994: Observations of lightning phenomena using radio interferometry. *J. Geophys. Res., 99,* 13,059–82.

Richard, P., 1991: Localization of atmospheric discharges, a new way for severe weather nowcasting. *Preprints,* 25th International Conference on Radar Meteorology, Amer. Meteor. Soc., Boston, 911–15.

Roohr, P.B., and T.H. Vonder Harr, 1994: A comparative analysis of the temporal variability of lightning observations and GOES imagery. *J. Appl. Meteor., 33,* 1271–90.

Roussel-Dupre, R.A., A.V. Gurevich, T. Tunnel, and G.M. Milikh, 1992: Kinetic theory of runaway air breakdown. *Phys. Rev. E, 49,* 2257–71.

Rust, W.D., 1973: Electrical conditions near the bases of thunderclouds. Ph.D. diss. New Mexico Inst. of Min. and Tech, Socorro, (Figs. 27–29).

Rust, W.D., 1989: Utilization of a mobile laboratory for storm electricity measurements. *J. Geophys. Res., 94,* 13,305–11.

Rust, W.D., and C.B. Moore, 1974: Electrical conditions near the bases of thunderclouds over New Mexico, *Q. J. Roy. Meteor. Soc., 100,* 450–68.

Rust, W.D., W.L. Taylor, D.R. MacGorman, and R.T. Arnold, 1981a: Research on electrical properties of severe thunderstorms in the Great Plains. *Bull. Amer. Meteor. Soc., 62,* 1286–93.

Rust, W.D., W.L. Taylor, and D. MacGorman, 1981b: Preliminary result of the study of lightning relative to storm structure and dynamics. Paper AIAA-81-0080, American Institute of Aeronautics and Astronautics 19th Aerospace Sciences Meeting, 7 pp.

Rust, W.D., D.R. MacGorman, and R.T. Arnold, 1981c: Positive cloud-to-ground lightning flashes in severe storms. *Geophys. Res. Lett., 8,* 791–94.

Rust, W.D., W.L. Taylor, D.R. MacGorman, E. Brandes, V. Mazur, R. Arnold, T. Marshall, H. Christian, and S.J. Goodman, 1985: Lightning and related phenomena in isolated thunderstorms and squall line systems. *J. Aircraft, 22,* 449–54.

Rutledge, S.A., and D.R. MacGorman, 1988: Cloud-to-ground lightning activity in the 10–11 June 1985 mesoscale convective system observed during the Oklahoma-Kansas PRE-STORM project. *Mon. Wea. Rev., 116,* 1393–1408.

Rutledge, S.A., C. Lu, and D.R. MacGorman, 1990: Positive cloud-to-ground lightning flashes in mesoscale convective systems. *J. Atmos. Sci., 47,* 2085–2100.

Rutledge, S.A., E.R. Williams, and T.D. Kennan, 1992: The down under Doppler and electricity experiment (DUNDEE): Overview and preliminary results. *Bull. Amer. Meteor. Soc., 73,* 3–16.

Rutledge, S.A., and W.A. Petersen, 1994: Vertical radar reflectivity structure and cloud-to-ground lightning in the stratiform region of MCS's: Further evidence for in-situ charging in the stratiform region. *Mon. Wea. Rev., 122,* 1760–76.

Schonland, B.F.J., 1928: The polarity of thunderclouds. *Proc. Roy. Sc. London, A118,* 233–51.

Scott, J.P., and W.H. Evans, 1969: The electrical conductivity of clouds. *Pure and Appl. Geophys., 75,* 219–32.

Scott, R., P. Krehbiel, M. Stanley, and S. McCarty, 1995: Relation of lightning channels to storm structure from interferometer and dual-polarization radar observations. *Preprints,* 27th Conf. Radar Meteorology, Amer. Meteorol. Soc., 645–47.

Seimon, A., 1993: Anomalous cloud-to-ground lightning in an F5-tornado-producing supercell thunderstorm on 28 August 1990. *Bull. Amer. Meteorol. Soc., 74,* 189–203.

Shackford, C.R., 1960: Radar indications of a precipitation-lightning relationship in New England thunderstorms. *J. Meteorology, 17,* 15–19.

Shafer, M.A., 1990: *Cloud-to-ground Lightning in Relation to Digitized Radar Data in Severe Storms.* master's thesis, Univ. of Oklahoma, 93 pp.

Shao, X.M., 1993: *The Development and Structure of Lighting Discharges Observed by VHF Radio Interferometer.* Ph.D. diss. New Mexico Inst. Mining and Tech., Socorro, NM, 324 pp.

Shao, X.M., P.R. Krehbiel, R.J. Thomas, and W. Rison, 1995: Radio interferometric observations of cloud-to-ground lightning phenomena in Florida. *J. Geophys. Res., 100,* 2749–83.

Shao, X.M., and P.R. Krehbiel, 1996: The spatial and temporal development of intracloud lightning. *J. Geophys. Res., 100,* 26, 641–68.

Simpson, G.C., 1909: On the electricity of rain and snow. *Proc. Roy. Soc. Lond., A, 83,* 394–404.

Simpson, G.C., 1949: Atmospheric Electricity during Disturbed Weather. *Geophys. Memoirs No. 84.* 51, pp. plus plates.

Simpson, Sir G., and F.J. Scrase, 1937: The distribution of electricity in thunderclouds. *Proc. Roy. Soc. Lond., A, 161,* 309–52.

Simpson, Sir G., and G.D. Robinson, 1941: The distribution of electricity in thunderclouds, II, *Proc. Roy. Soc. Lond., A, 177,* 281–328.

Solomon, R., and M. Baker, 1994: Electrification of New Mexico thunderstorms. *Mon. Wea. Rev., 122,* 1878–88.

Soula, S., and S. Chauzy, 1991: Multilevel measurement of the electric field underneath a thundercloud 2. Dynamical evolution of a ground space charge layer. *J. Geophys. Res., 96,* 22,327–36.

Sparrow, J.G., and E.P. Ney, 1971: Lightning observations by satellite. *Nature, 232,* 540–41.

Standler, R.B., and W.P. Winn, 1979: Effects of coronae on electric fields beneath thunderstorms. *Q. J. Roy. Meteor. Soc., 105,* 285–302.

Stergis, C.G., G.C. Rein, and T. Kangas, 1957: Electric field measurements above thunderstorms, *J. Atmos. Terres. Phys., 11,* 83–91.

Stolzenburg, M., 1990: Characteristics of the bipolar pattern of lighting locations observed in 1988 thunderstorms. *Bull. Amer. Meteor. Soc., 71,* 1331–38.

Stolzenburg, M., 1994: Observations of high ground flash densities of positive lightning in summertime thunderstorms. *Mon. Wea. Rev., 122,* 1740–50.

Stow, C.D, 1980: Large hydrometeor charges below thunderstorms. *J. Meteor. Soc., Jpn., 58,* 217–24.

Tabata, A., S. Nakazawa, Y. Yasutomi, H. Sakakibara, M. Ishihara, and K. Akaeda, 1989: The structure of a long-lasting single cell convective cloud. *Tenki, 36,* 499–507. (In Japanese)

Takagi, N., T. Watanabe, I. Arima, T. Takeuti, M. Nakano, and H. Kinosita, 1986: An unusual summer thunderstorm in Japan. *Res. Lett. Atmos. Elec., 6,* 43–48.

Takahashi, T., 1970: Vertical distribution of electric charge on precipitation elements in the cloud obtained by radiosonde. *J. Meteor. Soc. Jpn., 48,* 85–90.

Takahashi, T., 1978: Oceanic tropical thunderstorm electricity at Ponape, Micronesia. *Preprints,* Conf. on Cloud Physics and Atmospheric Electricity, Amer. Meteor. Soc., 641–48.

Takahashi, T., 1984: Thunderstorm electrification—a numerical study. *J. Atmos. Sci., 41,* 2541–58.

Takeuti, T., M. Nakano, H. Ishikawa, and S. Israelsson, 1977: On the two types of thunderstorms deduced from cloud-to-ground discharges observed in Sweden and Japan. *J. Meteor. Soc. Jpn, 55,* 613–16.

Takeuti, T., M. Nakano, M. Brook, D.J. Raymond, and P. Krehbiel, 1978: The anomalous winter thunderstorms of the Hokuriku coast. *J. Geophys. Res., 83,* 2385–94.

Tamura, Y., 1958: Investigations on the electrical structure of thunderstorms. In *Recent Advances in Atmospheric Electricity,* L.G. Smith, ed., Pergamon, New York, pp. 269–76.

Taylor, W.L., 1983: Lightning location and progression using VHF space-time mapping technique. In *Proceedings in Atmospheric Electricity,* L.H. Ruhnke and J. Latham, eds., A. Deepak Publishing, Hampton, pp. 381–84.

Taylor, W.L., E.A. Brandes, W.D. Rust, and D.R. MacGorman, 1984: Lightning activity and severe storm structure. *Geophys. Res. Lett., 11,* 545–48.

Teer, T.L., 1973: *Lightning Channel Structure Inside an Arizona Thunderstorm.* Ph.D. diss., Rice Univ., Houston.

Teer, T.L., and A.A. Few, 1974: Horizontal lightning. *J. Geophys. Res., 79,* 3436–41.

Tohsha, M., and I. Ichimura, 1961: Studies on shower and thunderstorm by radar. *Pap. Meteor. Geophys., 12,* 18–29.

Toland, R.B., and B. Vonnegut, 1977: Measurement of maximum electric field intensities over water during thunderstorms. *J. Geophys. Res., 82,* 438–40.

Turman, B.N., and B.C. Edgar, 1982: Global lightning distributions at dawn and dusk. *J. Geophys. Res., 87,* 1191–1206.

Uman, M.A., 1987: *The Lightning Discharge.* Academic Press, Orlando, FL, 377 pp.

Uman, M.A., A.H. Cookson, and J.B. Moreland, 1970: Shock wave from a four-meter spark. *J. Appl. Phys., 41,* 3148–55.

Uman, M.A., W.H. Beasley, J.A. Tiller, Y. Lin, E.P. Krider, C.D. Weidmann, P.R. Krehbiel, M. Brook, A.A. Few, Jr., J.L. Bohannon, C.L. Lennon, H.A. Poehler, W. Jafferis, J.R. Gulick, and J.R. Nicholson, 1978: An unusual lightning flash at Kennedy Space Center. *Science, 201,* 9–16.

Vonnegut, B., 1963: Some facts and speculations concerning the origin and role of thunderstorm electricity. *Meteorolog. Monogr. 5,* 224–41.

Vonnegut, B., 1969: Discussion of paper by Evans, 'Electric fields and conductivity in thunderclouds.' *J. Geophys. Res., 74,* 7053–55.

Vonnegut, B., C.B. Moore, and A.T. Botka, 1959: Preliminary results of an experiment to determine initial precedence of organized electrification and precipitation in thunderstorms. *J. Geophys. Res., 64,* 347–57.

Vonnegut, B., C.B. Moore, R.G. Semonin, J.W. Bullock, D.W. Staggs, and W.E. Bradley, 1962: Effect of atmospheric space charge on initial electrification of cumulus clouds. *J. Geophys. Res., 67,* 3909–21.

Vonnegut, B., C.B. Moore, R.P. Espinola, and H.H. Blau, Jr., 1966: Electrical potential gradients above thunderstorms. *J. Atmos. Sci., 23,* 764–70.

Vorpahl, J.A., J.G. Sparrow, and E.P. Ney, 1970: Satellite observations of lightning. *Science, 169,* 860–62.

Watson, A.I., R.E. López, R.L. Holle, and J.R. Daugherty, 1987: The relationship of lightning to surface convergence at Kennedy Space Center: A preliminary study. *Wea. Forecast., 2,* 140–57.

Watson, A.I., R.L. Holle, R.E. López, and R. Ortiz, 1991: Surface wind convergence as a short-term predictor of cloud-to-ground lightning at Kennedy Space Center. *Wea. Forecast., 6,* 49–64.

Watson, A.I., R.E. López, and R.L. Holle, 1994a: Diurnal cloud-to-ground lightning patterns in Arizona during the southwest monsoon. *Mon. Wea. Rev., 122,* 1716–25.

Watson, A.I., R.L. Holle, and A.I. Watson, 1994b: Cloud-to-ground lightning and upper-air patterns during bursts and breaks in the southwest monsoon. *Mon. Wea. Rev., 122,* 1726–39.

Watson, A.I., R.L. Holle, and R.E. López, 1995: Lightning from two national detection networks related to vertically integrated liquid and echo-top information from WSR-88D radar. *Wea. Forecast. 10,* 592–605.

Weber, M.E., H.J. Christian, A.A. Few, and M.F. Stewart, 1982: A thundercloud electric field sounding: charge distribution and lightning. *J. Geophys. Res., 87,* 7158–69.

Weber, M., R. Boldi, P. Laroche, P. Krehbiel, and X. Shao, 1993: Use of high resolution lightning detection and localization sensors for hazardous aviation weather nowcasting. *Preprints,* 17th Conf. on Severe Local Storms, Amer. Meteor. Soc., 739–44.

Weinheimer, A.J., 1987: The electrostatic energy of a thunderstorm and its rate of change. *J. Geophys. Res., 92,* 9715–22.

Weinheimer, A.J., and A.A. Few, Jr., 1981: Comment on 'Contributions of cloud and precipitation particles to the electrical conductivity and the relaxation time of the air in thunderstorms' by A.K. Kamra. *J. Geophys. Res., 86,* 4302–04.

Weinheimer, A.J., J.E. Dye, D.W. Breed, M.P. Spowart, J.L. Parrish, T.L. Hoglin, and T.C. Marshall, 1991: Simultaneous measurements of the charge, size, and shape of hydrometeors in an electrified cloud. *J. Geophys. Res., 96,* 20,809–29.

Wilkening, M., 1970. Radon 222 concentrations in the convective patterns of a mountain environment. *J. Geophys. Res., 75,* 1733–40.

Williams, E.R., 1981: *Thunderstorm Electrification: Precipitation versus convection.* Ph.D. diss., Massachusetts Inst. Technology, Cambridge.

Williams, E.R., 1985: Large-scale charge separation in thunderclouds. *J. Geophys. Res., 90,* 6013–25.

Williams, E.R., 1989: The tripole structure of thunderstorms. *J. Geophys. Res., 94,* 13,151–67.

Williams, E.R., C.M. Cooke, and K.A. Wright, 1985: Electrical discharge propagation in and around space charge clouds. *J. Geophys. Res., 90,* 6059–70.

Williams, E.R., M.E. Weber, and R.E. Orville, 1989: The relationship between lightning type and convective state of thunderclouds. *J. Geophys. Res., 94,* 13,213–20.

Williams, E.R., S.A. Rutledge, S.G. Geotis, N. Renno, E. Rasmussen, and T. Rickenbach, 1992: A radar and electrical study of tropical "hot towers." *J. Atmos. Sci., 49,* 1386–95.

Williams, E.R., and N. Renno, 1993: An analysis of the conditional instability of the tropical atmosphere. *Mon. Wea. Rev., 121,* 21–36.

Williams, E.R., R. Zhang, and D. Boccippio, 1994: Microphysical growth state of ice particles and large-scale electrical structure of clouds. *J. Geophys. Res., 99,* 10,787–92.

Wilson, C.T.R., 1929: Some thundercloud problems, *J. Franklin Inst., 208,* 1–12.

Winn, W.P., G.W. Schwade, and C.B. Moore, 1974: Measurements of electric fields in thunderclouds. *J. Geophys. Res., 79,* 1761–67.

Winn, W.P., and L.G. Byerley, III, 1975: Electric field growth in thunderclouds. *Q. J. Roy. Meteor. Soc., 101,* 979–94.

Winn, W.P., C.B. Moore, C.R. Holmes, and L.G. Byerley, III, 1978: Thunderstorm on July 16, 1975, over Langmuir Laboratory: a case study. *J. Geophys. Res., 83,* 3079–92.

Winn, W.P., C.B. Moore, and C.R. Holmes, 1981: Electric field structure in an active part of a small, isolated thundercloud. *J. Geophys. Res., 86,* 1187–93.

Woessner, R.H., W.E. Cobb, and R. Gunn, 1958: Simultaneous measurements of the positive and negative light-ion conductivities to 26 km. *J. Geophys. Res., 63,* 171–80.

Workman, E.J., 1967: The production of thunderstorm electricity. *J. Franklin Inst., 283,* 540–57.

Workman, E.J., R.E. Holzer, and G.T. Pelsor, 1942: *The Electrical Structure of Thunderstorms,* Tech. Report No. 864, National Advisory Committee for Aeronautics, Washington, D.C., 66 pp.

Workman, E.J., and S.E. Reynolds, 1949: Electrical activity as related to thunderstorm cell growth. *Bull. Amer. Meteorol. Soc., 30,* 142–44.

World Meteorological Organization, 1956: *World Distribution of Thunderstorm Days, Part 2:* Tables of Marine Data and World Maps, Report WMO/OMM–No. 21.

Zipser, E.J., 1994: Deep cumulonimbus cloud systems in the tropics with and without lightning. *Mon. Wea. Rev., 122,* 1837–51.

Zipser, E.J., and K.R. Lutz, 1994: The vertical profile of radar reflectivity of convective cells: A strong indicator of storm intensity and lightning probability? *Mon. Wea. Rev., 122,* 1751–59.

CHAPTER 8

Akiyama, H., H. Yamashita, and K. Horii, 1984: Electric field and space charge below the winter thunderclouds in Hokuriku. *Trans. IEE of Japan, 104,* 68.

Asuma, Y., K. Kikuchi, T. Taniguchi, S. Jujui, 1988: Vertical structures of the atmospheric electric potential gradient and the behavior of precipitation charges during snow-

falls near the ground surface. *J. Meteor. Soc. Jpn. 66,* 473–87.

Baral, K.N., and D. Mackerras, 1992: The cloud-to-ground flash ratio and other lightning occurrence characteristics in Kathmandu thunderstorms. *J. Geophys. Res., 97,* 931–38.

Bateman, M.G., 1992: The charge and size of precipitation particles inside thunderstorms and mesoscale convective systems on the Great Plains. Ph.D. diss., Univ. Okla., 137 pp.

Bateman, M.G., W.D. Rust, and T.C. Marshall, 1990: Precipitation particle measurements from Great Plains thunderstorms. *Eos, Trans. Amer. Geophys, Un., 71,* 1238. Abstract.

Bateman, M.G., W.D. Rust, T.C. Marshall, and B.F. Smull, 1995: Precipitation charge and size measurements in the stratiform region of two mesoscale convective systems. *J. Geophys. Res., 100,* 16,341–56.

Baughman, R.G., and D.M. Fuquay, 1970: Hail and lightning occurrence in mountain thunderstorms. *J. Appl. Meteor., 9,* 657–60.

Bent, R.B., and W.A. Lyons, 1984: Theoretical evaluations and initial operational experiences of LPATS (Lightning Position and Tracking System) to monitor lightning ground strikes using a time-of-arrival (TOA) technique. *Preprints,* 7th Intl. Conf. Atmospheric Electricity, Amer. Meteor. Soc, 317–24.

Billingsley, D.B., 1994: Evolution of cloud-to-ground lightning characteristics within the convective region of a midlatitude squall line, Master's thesis, Texas A&M Univ., College Station, 192 pp.

Biswas, K.R., and P.V. Hobbs, 1990: Lightning over the Gulf Stream. *Geophys. Res. Lett., 17,* 941–43.

Black, P.G., R.A. Black, J. Hallett, and W.A. Lyons, 1986: Electrical activity of the hurricane. *Preprints,* 23rd Conf. Radar Meteor. and Conf. Cloud Phys., Amer. Meteor. Soc., J277–J280.

Black, R.A., and J. Hallett, 1986: Observations of the distribution of ice in hurricanes. *J. Atmos. Sci., 43,* 802–22.

Black, R.A., J. Hallett, and C.P.R. Saunders, 1993: Aircraft studies of precipitation and electrification in hurricanes. *Preprints,* 17th Conf. Severe Local Storms and Conf. Atmos. Elec., Amer., Meteor. Soc., J20–J25.

Black, R.A., and J. Hallett, 1995: The relationship between the evolution of the ice phase hydrometeors with altitude and the establishment of strong electric fields in Hurricane Claudette. *Preprints,* Conf. Cloud Phys., Amer. Meteor. Soc., 535–40.

Blevins, L.L., and J.D. Marwitz, 1968: Visual observations of lightning in some Great Plains hailstorms. *Weather, 23,* 192–94.

Bluestein, H.B., and C.R. Parks, 1983: A synoptic and photographic climatology of low precipitation thunderstorms in the Southern Plains. *Mon. Wea. Rev., 111,*2034–46.

Bluestein, H.B., E.W. McCaul, Jr., G.P. Byrd, G.R. Woodall, 1988: Mobile sounding observations of a thunderstorm near the dryline: the Canadian, Texas storm of 7 May 1986. *Mon. Wea. Rev., 116,* 1790–1804.

Bluestein, H.B., E.W. McCaul, Jr., G.P. Byrd, G.R. Woodall, G. Martin, S. Keighton, and L.C. Showell, 1989: Mobile sounding observations of a thunderstorm near the dryline: the Grover, Texas storm complex of 25 May 1987. *Mon. Wea. Rev., 117,* 244–50.

Bluestein, H.B., and G.R. Woodall, 1990: Doppler-radar analysis of a low-precipitation severe storm. *Mon. Wea. Rev., 118,* 1640–64.

Boccippio, D.J., E.R. Williams, S.J. Heckman, W.A. Lyons, I.T. Baker, and R. Boldi, 1995: Sprites, ELF transients, and positive ground strokes. *Science, 269,* 1088–91.

Bosart, L.F., and F. Sanders, 1986: Mesoscale structure in the megalopolitan snowstorm of 11–12 February 1983. Part III: A large-amplitude gravity wave. *J. Atmos. Sci., 43,* 924–39.

Brandes, E.A., 1984: Vertical vorticity generation and meso-cyclone sustenance in tornadic thunderstorms: The observational evidence. *Mon. Wea. Rev., 112,* 2253–69.

Branick, M.L., and C.A. Doswell III, 1992: An observation of the relationship between supercell structure and lightning ground strike polarity. *Wea. Forecast., 7,* 143–49.

Braun, S.A., and R.A. Houze, Jr., 1994: The transition zone and secondary maximum of radar reflectivity behind a midlatitude squall line: Results retrieved from Doppler radar data. *J. Atmos. Sci., 51,* 2733–55.

Brook, M., 1992: Breakdown electric fields in winter storms. *Res. Lett. Atmos. Elec., 12,* 47–52.

Brook, M., M. Nakano, P. Krehbiel and T. Takeuti, 1982: The electrical structure of the Hokuriku winter thunderstorms. *J. Geophys. Res., 87,* 1207–15.

Brook, M., R.W. Henderson, and R.B. Pyle, 1989: Positive lightning strokes to ground. *J. Geophys. Res., 94,* 13,295–303.

Brown, R.A., and H.G. Hughes, 1978: Directional VLF sferics from the Union City, Oklahoma, tornadic storm. *J. Geophys. Res., 83,* 3571–74.

Browning, K.A., 1965: The evolution of tornadic storms. *J. Atmos. Sci., 22,* 664–68.

Browning, K.A., 1977: The structure and mechanisms of hailstorms. In *Hail: A Review of Hail Science and Hail Suppression,* G.B. Foote and C.A. Knight, eds., Meteor. Monogr., 16, Amer. Meteor. Soc., pp. 1–43.

Browning, K.A., J.C. Frankauser, J.P. Chalon, P.C. Eccles, R.G. Strauch, F.H. Merrem, D.J. Musil, E.L. May, and W.R. Sand, 1976: Structure of an evolving hailstorm, part V: Synthesis and implication for hail growth and hail suppression. *Mon. Wea. Rev., 104,* 603–10.

Brylev, G.B., S.B. Gashina, B.F. Yevteyev, and I.I. Kamaldyna, 1989. *Characteristics of Electrically Active Regions in Stratiform Clouds,* 303 pp. USAF translation, FTD-ID(RS)T-0698-89, of *Kharakteristiki Elektricheski Aktivnykh Zon v Sloistoobraznykh Oblakakh.* Gidrometeoizdat, Leningrad, 1989, 160 pp.

Burgess, D.W., V.T. Wood, and R.A. Brown, 1982: Mesocyclone evolution statistics. *Preprints,* 12th Conf. Severe Local Storms, Amer. Meteor. Soc., 422–24.

Byrne, G.J., A.A. Few, and M.E. Weber, 1983: Altitude, thickness, and charge concentrations of charged regions of four thunderstorms during TRIP 1981 based upon in situ balloon electric field measurements. *Geophys. Res. Lett., 10,* 39–42.

Byrne, G.J., A.A. Few, M.F. Stewart, A.C. Conrad, and R.L. Torczon, 1987: In situ measurements and radar observations of a severe storm: Electricity, kinematics, and precipitation. *J. Geophys. Res., 92,* 1017–31.

Byrne, G.J., A.A. Few, and M.F. Stewart, 1989: Electric field measurements within a severe thunderstorm anvil. *J. Geophys. Res., 94,* 6297–6307.

Canovan, R.A., 1972: Snow accompanied by thunder. *Weather, 27,* 257–58.

Carey, L.D., and S.A. Rutledge, 1995: Positive cloud-to-ground lightning in severe hailstorms: A multiparameter radar study. *Preprints,* 27th Conference Radar Meteorol. Amer. Meteor. Soc., 629–32.

Changnon, S.A., 1992: Temporal and spatial relations between hail and lightning. *J. Appl. Meteor., 31,* 587–604.

Chauzy, S., P. Raizonville, D. Hauser, and F. Roux, 1980: Electrical and dynamical description of a frontal storm deduced from LANDES 79 experiment. *J. Rech. Atmos., 14,* 457–67.

Chauzy, S., M. Chong, A. Delannoy, and S. Despiau, 1985: The June 22 tropical squall line observed during COPT 81 experiment: Electrical signature associated with dynamical structure and precipitation. *J. Geophys. Res., 90,* 6091–98.

Cherna, E.V., and E.J. Stansbury, 1986: Sferics rate in relation to thunderstorm dimensions. *J. Geophys. Res., 91,* 8701–07.

Church, C.R., and B.J. Barnhart, 1979: A review of electrical phenomena associated with tornadoes. *Preprints,* 11th Conf. Severe Local Storms, Amer. Meteor. Soc., 337–42.

Church, C., D. Burgess, R. Davies-Jones, and C. Doswell, eds., 1993: *The Tornado: Its Structure, Dynamics, Prediction, and Hazards,* Geophys. Monogr. 79, Amer. Geophys. Union, 637 pp.

Coleman, R.B., 1990: Thunderstorms above frontal surfaces in environments without positive CAPE. Part I: A climatology. *Mon. Wea. Rev., 118,* 1123–44.

Cotton, W.R., and R.A. Anthes, 1989: *Storm and Cloud Dynamics,* Academic Press, San Diego, 883 pp.

Cotton, W.R., M. Lin, R.L. McAnelly, and C.J. Tremback, 1989: A composite model of mesoscale convective complexes. *Mon. Wea. Rev., 117,* 765–83.

Curran, E.B., and W.D. Rust, 1992: Positive ground flashes produced by low-precipitation thunderstorms in Oklahoma on 26 April 1984. *Mon. Wea. Rev., 120,* 544–53.

Davies-Jones, R.P., 1974: Discussion of measurements inside high-speed thunderstorm updrafts. *J. Appl. Meteor., 13,* 710–13.

Davies-Jones, R.P., and J.H. Golden, 1975: On the relation of electrical activity to tornadoes. *J. Geophys. Res.,* 1614–16.

Dickson, E.B., and R.J. McConahy, 1956: Sferics readings on windstorms and tornadoes. *Bull. Amer. Meteor. Soc., 37,* 410–12.

Dodge, P.V., and R.W. Burpee, 1993: Characteristics of rainbands, radar echoes, and lightning near the North Carolina coast during GALE. *Mon. Wea. Rev., 121,* 1936–55.

Dong, Y., R.G. Oraltay, and J. Hallett, 1994: Ice particle generation during evaporation. *Atmos. Res., 32,* 45–53.

Doswell, C.A., III, and D.W. Burgess, 1993: Tornadoes and tornadic storms: A review of conceptual models. In *The Tornado: Its Structure, Dynamics, Prediction, and Hazards,* Geophys. Monogr. 79, Amer. Geophys. Union, Washington, D.C., pp. 161–72.

Drake, J.C., 1968: Electrification accompanying the melting of ice particles. *Q. J. Roy. Meteor. Soc., 94,* 176–91.

Dye, J.E., J.J. Jones, W.P. Winn, T.A. Cerni, B. Gardiner, D. Lamb, R.L. Pitter, J. Hallet, and C.P.R. Saunders, 1986: Early electrification and precipitation development in a small isolated Montana cumulonimbus. *J. Geophys. Res., 91,* 1231–47.

Engholm, C.D., E.R. Williams, and R.M. Dole, 1990: Meteorological and electrical conditions associated with positive cloud-to-ground lightning. *Mon. Wea. Rev., 118,* 470–87.

Ette, A.I., and E.O. Oladiran, 1980: The characteristics of rain electricity in Nigeria. I—Magnitude and variations. *Pageoph, 118,* 753–64.

Fitzgerald, D.R., and H.R. Byers, 1959: Aircraft observations of convective cloud electrification. *J. Atmos. Terres. Phys., 15,* 254–61.

Foster, H., 1950: An unusual observation of lightning. *Bull. Amer. Meteor. Soc., 31,* 140–41.

Fukao, S., Y. Maekawa, Y. Sonoi, and F. Yoshino, 1991: Dual polarization radar observation of thunderclouds on the coast of the Sea of Japan in the winter season. *Geophys. Res. Lett., 18,* 179–82.

Gallus, W.A., Jr., and R.H. Johnson, 1991: Heat and moisture budgets of an intense midlatitude squall line. *J. Atmos. Sci., 48,* 122–46.

Goodman, S.J., and D.R. MacGorman, 1986: Cloud-to-ground lightning activity in mesoscale convective complexes. *Mon. Wea. Rev., 114,* 2320–28.

Goodman, S.J., D.E. Buechler, P.D. Wright, and W.D. Rust, 1988a: Lightning and precipitation history of a microburst-producing storm. *Geophys Res. Lett., 15,* 1185–88.

Goodman, S.J., D.E. Buechler, and P.J. Meyer, 1988b: Convective tendency images derived from a combination of lightning and satellite data. *Wea. Forecast., 3,* 173–88.

Goto, Y., and K.I. Narita, 1991: The meteorological conditions for the occurrence of winter thunderstorms. *Res. Lett. Atmos. Elec., 11,* 61–91.

Greneker, E.F., C.S. Wilson, and J.I. Metcalf, 1976: The Atlanta tornado of 1975. *Mon. Wea. Rev., 104,* 1052–57.

Gunn, R., 1956: Electric field intensity at the ground under active thunderstorms and tornadoes. *J. Meteor., 13,* 269–73.

Hallett, J., W. Hendricks, R.A. Black, C.P.R. Saunders, and I.M. Brooks, 1993: Aircraft observations of precipitation development and hydrometeor charge in Florida cumuli. *Preprints,* 17th Conf. Severe Local Storms and Conf. Atmos. Elec., Amer. Meteor. Soc., 785–90.

Hill, R.D., 1988: Interpretation of bipole pattern in a mesoscale storm. *Geophys. Res. Lett., 15,* 643–44.

Hojo, J., M. Ishii, T. Kawamura, F. Suzuki, H. Komuro, and M. Shiogama, 1989: Seasonal variation of cloud-to-ground lightning flash characteristics in the coastal area of the Sea of Japan. *J. Geophys. Res., 94,* 13,207–12.

Holle, R.L., A.I. Watson, R.E. Lopez, and D.R. MacGorman, 1994: The life cycle of lightning and severe weather in a 3–4 June 1985 PRE-STORM mesoscale convective system. *Mon. Wea. Rev., 122,* 1798–1808.

Holle, R.L., and A.I. Watson, 1996: Lightning during two central U.S. winter precipitation events. *Wea. Forecast. 11,* 599–614.

Houze, R.A., Jr., S.A. Rutledge, M.I. Biggerstaff, and B.F. Smull, 1989: Interpretation of Doppler weather radar displays of midlatitude mesoscale convective systems. *Bull. Amer. Meteor. Soc., 70,* 608–19.

Houze, R.A., Jr., B.F. Smull, and P. Dodge, 1990: Mesoscale organization of springtime rainstorms in Oklahoma. *Mon. Wea. Rev., 118,* 613–54.

Hunter, S.M., T.J. Schuur, T.C. Marshall, and W.D. Rust, 1992: Electrical and kinematic structure of the Oklahoma mesoscale convective system of 7 June 1989. *Mon. Wea. Rev., 120,* 2226–39.

Idone, V.P., A.B. Saljoughy, R.W. Henderson, P.K. Moore, and R. Pyle, 1993: A reexamination of the peak current calibration of the national lightning detection network. *J. Geophys. Res., 98,* 18,323–32.

Imyanitov, I.M., Ye.V. Chubarina, and Ya.M. Shvarts, 1972: *Electricity of Clouds,* 122 pp. NASA Tech. Translation, NASA TT F-718, of *Elektrichestvo oblakov,* Hydrometeorological Press, Leningrad, 1971.

Isono, K., M. Komabayasi, and T. Takahashi, 1966: A physical study of solid precipitation from convective clouds over the sea: Part III, Measurement of electric charge of snow crystals. *J. Meteor. Soc. Jpn., 44,* 227–33.

Jayaratne, E.R., C.P.R. Saunders, and J. Hallett, 1983: Laboratory studies of the charging of soft hail during ice crystal interactions. *Quart. J. Roy. Meteor. Soc., 109,* 609–30.

Jayaratne, E.R., and C.P.R. Saunders, 1984: The "rain-gush", lightning, and the lower positive charge center in thunderstorms. *J. Geophys. Res., 89,* 11,816–18.

Jayaratne, E.R., and C.P.R. Saunders, 1985: Reply. *J. Geophys. Res., 90,* 10,755.

Johns, R.H., and C.A. Doswell, III, 1992: Severe local storms forecasting. *Wea. Forecast., 7,* 588–612.

Johnson, H.L., R.D. Hart, M.A. Lind, R.E. Powell, and J.L. Stanford, 1977: Measurements of radio frequency noise from severe and nonsevere thunderstorms. *Mon. Wea. Rev., 105,* 734–47.

Johnson, R.L., 1980: Bimodal distribution of atmospherics associated with tornadic events. *J. Geophys. Res., 85,* 5519–22.

Jones, H.L., 1951: A sferic method of tornado identification and tracking. *Bull. Amer. Meteor. Soc., 32,* 380–85.

Jones, H.L., 1958: The identification of lightning discharges by sferic characteristics. In *Recent Advances in Atmospheric Electricity,* L.G. Smith, eds., Pergamon, New York, pp. 543–56.

Jones, H.L., 1965: The tornado pulse generator. *Weatherwise, 18,* 78–79, 85.

Jones, H.L., and P.N. Hess, 1952: Identification of tornadoes by observation of waveform atmospherics. *Proc. I.R.E., 40,* 1049–52.

Jorgensen, D.P., 1984: Mesoscale and convective-scale characteristics of mature hurricanes. Part I: General observations by research aircraft. *J. Atmos. Sci., 41,* 1268–85.

Jorgensen, D.P., E.J. Zipser, and M.A. LeMone, 1985: Vertical motions in intense hurricanes. *J. Atmos. Sci., 42,* 839–56.

Jorgensen, D.P., and B.F. Smull, 1993: Mesovortex circulations seen by airborne Doppler radar within a bow-echo mesoscale convective system. *Bull. Amer. Meteor. Soc., 74,* 2146–57.

Kane, R.J., 1991: Correlating lightning to severe local storms in the northeastern United States. *Wea. Forecasting, 6,* 3–12.

Kane, R.J., C.R. Chelius, and J.M. Fritsch, 1987: Precipitation characteristics of mesoscale convective weather systems. *J. Clim. Appl. Meteor. 26,* 1345–57.

Kasemir, H.W., 1960: A contribution to the electrostatic theory of a lightning discharge. *J. Geophys. Res., 65,* 1873–78.

Kasemir, H.W., 1984: Theoretical and experimental determination of field, charge and current on an aircraft hit by natural and triggered lightning. *Preprints,* International Aerospace and Ground Conference on Lightning and Static Electricity, National Interagency Coordinating Group, Orlando, Fla.

Keighton, S.J., H.B. Bluestein, and D.R. MacGorman, 1991: The evolution of a severe mesoscale convective system: Cloud-to-ground lightning location and storm structure. *Mon. Wea. Rev., 119,* 1533–56.

Kelly, D.L., J.T. Schaefer, and C.A. Doswell, III, 1985: Climatology of nontornadic severe thunderstorm events in the United States. *Mon. Wea. Rev., 113,* 1997–2014.

Kinzer, G.D., 1974: Cloud-to-ground lightning versus radar

reflectivity in Oklahoma thunderstorms. *J. Atmos. Sci., 31,* 787–99.

Kitagawa, N., 1992: Charge distribution in winter thunderclouds. *Res. Lett. Atmos. Elec., 12,* 143–53.

Kitagawa, N., and K. Michimoto, 1994: Meteorological and electrical aspects of winter thunderclouds. *J. Geophys. Res., 99,* 10,713–21.

Knapp, D.I., 1994: Using cloud-to-ground lightning data to identify tornadic thunderstorm signatures and nowcast severe weather. *Natl. Wea. Assoc. Digest, 19,* 35–42.

Knight, C.A., 1979: Observations of the morphology of melting snow. *J. Atmos. Sci., 36,* 1123–30.

Kohl, D.A., 1962: Sferics amplitude distribution jump identification of a tornado event. *Mon. Wea. Rev., 90,* 451–56.

Kohl, D.A., and J.E. Miller, 1963: 500 kc/sec sferics analysis of severe weather events. *Mon. Wea. Rev., 91,* 207–14.

Kotaki, M., I. Kuriki, C. Katoh, and H. Sugiuchi, 1981: Global distribution of thunderstorm activity observed with ISS-b. *J. Radio. Res. Labs., 28,* 49–71.

Krehbiel, P.R., 1981: *An Analysis of the Electric Field Change Produced by Lightning.* Ph.D. diss., Univ. Manchester Inst. Sci. & Tech., England. Published as Rpt. T-11, New Mex. Inst. Min. & Tech., 1981, Vol. 1 is 245 pp. and Vol. 2 is figures.

Krehbiel, P.R., 1986: The electrical structure of thunderstorms. In *The Earth's Electrical Environment,* National Acad. Press, Washington, D.C., pp. 90–113.

Krider, E.P., A.E. Pifer, and D.L. Vance, 1980: Lightning direction-finding systems for forest fire detection. *Bull. Amer. Meteor. Soc., 61,* 980–86.

Lafore, J.P., and M.W. Moncrief, 1989: A numerical investigation of the organization and interaction of the convective and stratiform region of tropical squall lines. *J. Atmos. Sci., 46,* 521–44.

Larson, H.R., and E.J. Stansbury, 1974: Association of lightning flashes with precipitation cores extending to height 7 km. *J. Atmos. Terres. Phys., 36,* 1547–53.

Leary, C.A., and Houze, R.A., Jr., 1979: The structure and evolution of convection in a tropical cloud cluster. *J. Atmos. Sci., 36,* 437–57.

Leary, C.A., and E.N. Rappaport, 1987: The life cycle and internal structure of a mesoscale convective complex. *Mon. Wea. Rev., 115,* 1503–27.

LeMone, M.A., G.M. Barnes, and E.J. Zipser, 1984: Momentum flux by lines of cumulonimbus over tropical oceans. *J. Atmos. Sci., 41,* 1914–32.

Lemon, L.R., and C.A. Doswell III, 1978: Tornadic storm airflow and morphology derived from single-Doppler radar measurements. *Mon. Wea. Rev., 106,* 48–61.

Lemon, L.R., D.W. Burgess, and R.A. Brown, 1979: Severe thunderstorm evolution and mesocyclone structure as related to tornadogenesis. *Mon. Wea. Rev., 107,* 1184–97.

Lemon, L.R., D.W. Burgess, and L.D. Hennington, 1982: A tornado extending to extreme heights as revealed by Doppler radar. *Preprints,* 12th Conf. Severe Local Storms, Amer. Meteor. Soc., 430–32.

Lhermitte, R., and P.R. Krehbiel, 1979: Doppler radar and radio observations of thunderstorms. *IEEE Trans. Geosci. Electron., GE-17,* 162–71.

Ligda, M.H., 1956: The radar observation of lightning. *J. Atmos. Terr. Phys., 3,* 329–46.

Lind, M.A., J.S. Hartman, E.S. Takle, and J.L. Stanford, 1972: Radio noise studies of several severe weather events in Iowa in 1971. *J. Atmos. Sci, 29,* 1220–23.

Lyons, W.A., M.G. Venne, P.G. Black, and R.C. Gentry, 1989: Hurricane lightning: A new diagnostic tool for tropical storm forecasting? *Preprints,* 18th Conf. Hurr. and Trop. Meteor., Amer. Meteor. Soc., 113–14.

Lyons, W.A., and C.S. Keen, 1994: Observations of lightning in convective supercells within tropical storms and hurricanes. *Mon. Wea. Rev., 122* 1897–1916.

MacGorman, D.R., 1993: Lightning in tornadic storms: A review. In *The Tornado: Its Structure, Dynamics, Prediction, and Hazards.* C. Church, D. Burgess, R. Davies-Jones, and C. Doswell, eds., Geophys. Monogr. 79, Amer. Geophys. Union, pp. 173–82.

MacGorman, D.R., M.W. Maier, and W.D. Rust, 1984: *Lightning Strike Density for the Contiguous United States from Thunderstorm Duration Records.* Report to the U.S. Nuclear Regulatory Commission, available as NUREG/CR-3759, 51 pp.

MacGorman, D.R., W.D. Rust, and V. Mazur, 1985: Lightning activity and mesocyclone evolution, 17 May 1981. *Preprints,* 14th Conf. Severe Local Storms, Amer. Meteor. Soc., 355–58.

MacGorman, D.R., and W.D. Rust, 1989: An evaluation of the LLP and LPATS lightning ground strike mapping systems. *Preprints,* 5th Intl. Conf. Interactive Information Processing Systems, Amer. Meteor. Soc., 249–54.

MacGorman, D.R., and W.L. Taylor, 1989: Positive cloud-to-ground lightning detection by a direction-finder network. *J. Geophys. Res., 94,* 13,313–18.

MacGorman, D.R., D.W. Burgess, V. Mazur, W.D. Rust, W.L. Taylor, and B.C. Johnson, 1989: Lightning rates relative to tornadic storm evolution on 22 May 1981. *J. Atmos. Sci., 46,* 221–50.

MacGorman, D.R., and K.E. Nielsen, 1991: Cloud-to-ground lightning in a tornadic storm on 8 May 1986. *Mon. Wea. Rev., 119,* 1557–74.

MacGorman, D.R., and D.W. Burgess, 1994: Positive cloud-to-ground lightning in tornadic storms and hailstorms. *Mon. Wea. Rev., 122,* 1671–97.

MacGorman, D.R., K. Crawford, H. Xia, 1997: A lightning strike climatology of Oklahoma: Part I—Geographic and interannual variability. *J. Climate.* in press.

MacGorman, D.R., and C.D. Morgenstern, 1997: Some characteristics of cloud-to-ground lightning in mesoscale convective systems. *J. Geophys. Res., 102,* in press.

Mackerras, D., and M. Darveniza, 1994: Latitudinal variation of lightning occurrence characteristics. *J. Geophys. Res., 99,* 10813–21.

Maddox, R.A., 1980: Mesoscale convective complexes. *Bull. Amer. Meteor. Soc., 61,* 1374–87.

Maddox, R.A., 1983: Large-scale meteorological conditions associated with midlatitude, mesoscale convective complexes. *Mon. Wea. Rev., 111,* 1475–93.

Maddox, R.A., K.W. Howard, D.L. Bartels, and D.M. Rodgers, 1986: Mesoscale convective complexes in the middle latitudes. In *Mesoscale Meteorology and Forecasting,* P.S. Ray, ed., Amer. Meteor. Soc., Boston, pp. 390–413.

Maekawa, Y., S. Fukao, Y. Sonoi, F. Yoshino, 1992: Dual polarization radar observations of anomalous wintertime thunderclouds in Japan. *IEEE Trans. Geosci. Rem. Sens., 30,* 838–44.

Magono, C., 1980: *Thunderstorms.* Elsevier Sci. Amsterdam, 261 pp.

Magono, C., T. Endoh, and T. Taniguchi, 1982: The mirror image relation in the vertical distributions of electric field and precipitation charge in winter thunderclouds. *J. Meteor. Soc., Jpn., 61,* 1188–93.

Magono, C., T. Endoh, and T. Shigeno, 1983a: The electrical structure of snow clouds Part I: Relationship to their precipitation-physical properties. *J. Meteor. Soc. Jpn., 61,* 325–38.

Magono, C., T. Endoh, and T. Taniguchi, 1983b: Charge distribution in active winter clouds. In *Proceedings in Atmospheric Electricity,* L.H. Rhunke and J. Latham, eds., A. Deepak Pub., Hampton, pp. 242–45.

Magono, C., H. Sakamoto, T. Endoh, and T. Taniguchi, 1984: The electrical structure of snow clouds. Part II, vertical distribution of precipitation charge. *J. Meteor. Soc. Jpn., 62,* 323–34.

Marsh, S.J., and T.C. Marshall, 1993: Charged precipitation measurements before the first lightning flash in a thunderstorm. *J. Geophys. Res., 98,* 16,605–11.

Marshall, T.C., and W.P. Winn, 1982: Measurements of charged precipitation in a New Mexico thunderstorm: Lower positive charge centers. *J. Geophys. Res., 87,* 7141–57.

Marshall, T.C., and W.P. Winn, 1985: Comments on "The 'rain gush,' lightning, and the lower positive center in thunderstorms. *J. Geophys. Res., 90,* 10753–54.

Marshall, T.C., W.D. Rust, W.P. Winn,. and K.E. Gilbert, 1989: Electrical structure in two thunderstorm anvil clouds. *J. Geophys. Res., 94,* 2171–81.

Marshall, T.C., and W.D. Rust, 1991: Electric field soundings through thunderstorms. *J. Geophys. Res., 96,* 22,297–306.

Marshall, T.C., and S.J. Marsh, 1993: Negatively charged precipitation in a New Mexico thunderstorm. *J. Geophys. Res., 98,* 14,909–16.

Marshall, T.C. and W.D. Rust, 1993: Two types of vertical electrical structures in stratiform precipitation regions of mesoscale convective systems. *Bull. Amer. Meteor. Soc., 74,* 2159–70.

Marshall, T.C., W.D. Rust, and M. Stolzenburg, 1995: Electrical structure and updraft speeds in thunderstorms over the southern Great Plains. *J. Geophys. Res., 100,* 1001–15.

Marshall, T.C., M. Stolzenburg, and W.D. Rust, 1996: Electric field measurements above mesoscale convective systems. *J. Geophys. Res., 101,* 6979–96.

Matejka, T., and T.J. Schuur, 1991: The relation between vertical air motions and the precipitation band in the stratiform region of a squall line. *Preprints,* 25th Conf. Radar Meteor., Amer. Meteor. Soc., 501–4.

Matthews, J.B., 1964: An unusual type of lightning. *Weather, 19,* 291–92.

Mazur, V., and W.D. Rust, 1983: Lightning propagation and flash density in squall lines as determined with radar. *J. Geophys. Res., 88,* 1495–1502.

Mazur, V., J.C. Gerlach, and W.D. Rust, 1984: Lightning flash density versus altitude and storm structure from observations with UHF- and S-band radars. *Geophys. Res. Lett., 11,* 61–64.

Mazur, V., X-M. Shao, and P.R. Krehbiel, 1994: "Spider" intracloud and positive cloud-to-ground lightning in the late stage of Florida storms. In *Proc. 1994 Intl. Aerosp. Gnd. Conf. Lightning Static Elec.,* Mannheim, Germany, pp. 33–41.

McAnelly, R.L., and W.R. Cotton, 1986: Meso-β-scale characteristics of an episode of meso-α-scale convective complexes. *Mon. Wea. Rev., 114,* 1740–70.

McAnelly, R.L., and W.R. Cotton, 1989: The precipitation life cycle of mesoscale convective complexes over the central United States. *Mon. Wea. Rev., 117,* 784–808.

Michimoto, K., 1990: Relation between surface electric field and radar echo due to winter thunderclouds over the Hokuriku district. *Res. Lett. Atmos. Elec., 10,* 17–23.

Michimoto, K., 1991: A study of radar echoes and their relation to lightning discharge of thunderclouds in the Hokuriku district. Part I: Observations and analysis of thunderclouds in summer and winter. *J. Meteor. Soc. Jpn., 69,* 327–66.

Michimoto, K., 1993: A study of radar echoes and their relation to lightning discharges of thunderclouds in the Hokuriku district. Part II: Observation and analysis of "single-flash" thunderclouds in midwinter. *J. Meteor. Soc. Jpn., 71,* 195–204.

Mikhailovsky, Yu, Yu Agapov, and B. Koloskov, 1992: Electrification of the tropical clouds. In *Proc. 9th Intl. Conf. Atmos. Elec.,* vol. 1, St. Petersburg, Russia, pp. 176–78.

Miller, L.J., J.D. Tuttle, and G.B. Foote, 1990: Precipitation production in a large Montana hailstorm: Airflow and particle growth trajectories. *J. Atmos. Sci., 46,* 1619–46.

Molinari, J., P.K. Moore, V.P. Idone, R.W. Henderson, and A.B. Saljoughy, 1994: Cloud-to-ground lightning in Hurricane Andrew. *J. Geophys. Res., 99,* 16,665–76.

Moller, A.R., and C.A. Doswell III, 1988: A proposed advanced storm spotter's training program. *Preprints,* 15th Conf. Severe Local Storms, Amer. Meteor. Soc., 173–77.

Moller, A.R., C.A. Doswell III, and R.W. Przybylinski, 1990: High precipitation supercells: A conceptual model and documentation. *Preprints,* 16th Conf. Severe Local Storms, Amer. Meteor. Soc., 52–57.

Moore, C.B., B. Vonnegut, B.A. Stein, and H.J. Survilas, 1960: Observations of electrification and lightning in warm clouds. *J. Geophys. Res., 65,* 1907–10.

Moore, C.B., and B. Vonnegut, 1973: Reply (to Brazier-Smith et al.). *Q. J. Roy. Meteor. Soc., 99,* 779–86.

Morgenstern, C.D., 1991: *Cloud-to-Ground Lightning Characteristics in Mesoscale Convective Systems, April–September 1986.* Master's thesis, Univ. Oklahoma, Norman, 109 pp.

Nakono, M., 1979: The cloud discharge in winter thunderstorms of the Hokuriku Coast. *J. Meteor. Soc. Jpn, 57,* 444–51.

Newton, C.W., 1950: Structure and mechanisms of the prefrontal squall line. *J. Meteor., 7,* 210–22.

Nielsen, K.E., R.A. Maddox, and S.V. Vasiloff, 1994: The evolution of cloud-to-ground lightning within a portion of the 10–11 June 1985 squall line. *Mon. Wea. Rev., 122,* 1809–17.

Ogura, Y., and M.T. Liou, 1980: The structure of a midlatitude squall line: A case study. *J. Atmos. Sci., 37,* 553–67.

Orville, R.E., 1990a: Winter lightning along the East Coast. *Geophys. Res. Lett., 17,* 713–15.

Orville, R.E., 1990b: Peak-current variations of lightning return strokes as a function of latitude. *Nature, 343* 149–51.

Orville, R.E., 1991: Calibration of a magnetic direction finding network using measured triggered lightning return stroke peak currents. *J. Geophys. Res., 96,* 17,135–42.

Orville, R.E., 1993: Cloud-to-ground lightning in the blizzard of '93. *Geophys. Res. Lett., 20,* 1367–70.

Orville, R.E., and B. Vonnegut, 1977: Lightning detection from satellites. In *Electrical Processes in Atmospheres,* H. Dolezalek and R. Reiter, eds., Dr. Dietrich Steinkopff, Verlag, Darmstadt, pp. 750–53.

Orville, R.E., and D.W. Spencer, 1979: Global lightning flash frequency. *Mon. Wea. Rev., 107,* 934–43.

Orville, R.E., M.W. Maier, F.R. Mosher, D.P. Wylie, and W.D. Rust, 1982: The simultaneous display in a severe storm of lightning ground strike locations onto satellite images and radar reflectivity patterns. *Preprints,* 12th Conf. Severe Local Storms, Amer. Meteor. Soc., 448–51.

Orville, R.E., R.W. Henderson, and L.F. Bosart, 1983: An east coast lightning detection network. *Bull. Amer. Meteor. Soc., 64,* 1029–37.

Orville, R.E., R.A. Weisman, R.B. Pyle, R.W. Henderson, and R.E. Orville, Jr., 1987: Cloud-to-ground lightning flash characteristics from June 1984 through May 1985. *J. Geophys. Res., 92,* 5640–44.

Orville, R.E., R.W. Henderson, and L.F. Bosart, 1988: Bipole patterns revealed by lightning locations in mesoscale storm systems. *Geophys. Res. Lett., 15,* 129–32.

Pakiam, J.E., and J. Maybank, 1975: The electrical characteristics of some severe hailstorms in Alberta, Canada. *J. Meteor. Soc. Jpn. 53,* 363–83.

Petersen, W.A., and S.A. Rutledge, 1992: Some characteristics of cloud-to-ground lightning in tropical northern Australia. *J. Geophys. Res., 97,* 11,553–60.

Piepgrass, M.V., E.P. Krider, and C.B. Moore, 1982: Lightning and surface rainfall during Florida thunderstorms. *J. Geophys. Res., 87,* 11,193–201.

Przybylinski, R.W., J.T. Snow, E.M. Agee, and J.T. Curran, 1993: The use of volumetric radar data to identify supercells: A case study of June 2, 1990. In *The Tornado: Its Structure, Dynamics, Prediction, and Hazards,* Geophys. Monogr. 79, Amer. Geophys. Union, Washington, D.C. pp. 241–50.

Rasmussen, E., J.M. Straka, R., Davies-Jones, C.A. Doswell III, F.H. Carr, M.D. Eilts, and D.R. MacGorman, 1994: Verification of the origins of rotation in tornadoes experiment: VORTEX. *Bull. Amer. Meteor. Soc., 75,* 995–1006.

Reap, R.M., and D.R. MacGorman, 1989: Cloud-to-ground lightning: Climatological characteristics and relationships to model fields, radar observations, and severe local storms. *Mon. Wea. Rev., 117,* 518–35.

Reiter, R., 1965: Precipitation and cloud electricity. *Q. J. Roy. Meteor. Soc., 91,* 60–72.

Reiter, R., 1988. Precipitation and cloud electricity—New observations and results. *Proc., 8th Intl. Conf. Atmos. Elec.,* Uppsala, Sweden, 705–10.

Richard, P., 1992: Application of atmospheric discharges localization to thunderstorm nowcasting. *Proc. WMO Tech. Conf. Instruments and Methods of Observation,* World Meteor. Org.

Riehl, H., 1954: *Tropical Meteorology.* McGraw-Hill, New York, 392 pp.

Rogers, R.R., 1979: A Short Course in Cloud Physics, Pergamon, Oxford, 233 pp.

Rossby, S.A., 1966: Sferics from lightning in warm clouds. *J. Geophys. Res., 71,* 3807–09.

Rotunno, R., and J.B. Klemp, 1985: On the rotation and propagation of simulated supercell thunderstorms. *J. Atmos. Sci., 42,* 271–92.

Rust, W.D., 1989: Utilization of a mobile laboratory for storm electricity measurements. *J. Geophys. Res., 94,* 13305–11.

Rust, W.D., and C.B. Moore, 1974: Electrical conditions near the bases of thunderclouds over New Mexico. *Q. J. Roy. Meteor. Soc., 100,* 450–68.

Rust, W.D., W.L. Taylor, D.R. MacGorman, and R.T. Arnold, 1981a: Research on electrical properties of severe thunderstorms in the Great Plains. *Bull. Amer. Meteor. Soc., 62,* 1286–93.

Rust, W.D., D.R. MacGorman, and R.T. Arnold, 1981b: Positive cloud-to-ground lightning flashes in severe storms. *Geophys. Res. Lett., 8,* 791–94.

Rust, W.D., D.R. MacGorman, and W.J. Goodman, 1985a: Unusual positive cloud-to-ground lightning in Oklahoma storms on 13 May 1983. *Preprints,* 14th Conf. Severe Local Storms, Amer. Meteor. Soc., 372–75.

Rust, W.D., W.L. Taylor, D.R. MacGorman, E. Brandes, V. Mazur, R. Arnold, T. Marshall, H. Christian, and S.J. Goodman, 1985b: Lightning and related phenomena in isolated thunderstorms and squall lines. *J. Aircraft, 22,* 449–54.

Rust, W.D., Davies-Jones, R., D.W. Burgess, R.A. Maddox, L.C. Showell, T.C. Marshall, and D.K. Lauritsen, 1990: Testing a mobile version of a cross-chain Loran atmospheric sounding system (M-CLASS). *Bull. Amer. Meteor. Soc., 71,* 173–80.

Rutledge, S.A., 1991: Middle latitude and tropical mesoscale convective systems. *U.S. Natl. Rpt. Intl. Un. Geodesy. Geophys. 1987–1990,* Amer. Geophys. Union, 88–97.

Rutledge, S.A., and R.A. Houze, Jr., 1987: A diagnostic modeling study of the trailing stratiform region of a midlatitude squall line. *J. Atmos. Sci., 44,* 2640–56.

Rutledge, S.A., and D.R. MacGorman, 1988: Cloud-to-ground lightning activity in the 10–11 June 1985 mesoscale convective system observed during the Oklahoma-Kansas PRE-STORM project. *Mon. Wea. Rev., 116,* 1393–1408.

Rutledge, S.A., C. Lu, and D.R. MacGorman, 1990: Positive cloud-to-ground lightning flashes in mesoscale convective systems. *J. Atmos. Sci., 47,* 2085–2100.

Rutledge, S.A., E.R. Williams, and T.D. Kennan, 1992: The down under Doppler and electricity experiment (DUNDEE): Overview and preliminary results. *Bull. Amer. Meteor. Soc., 73,* 3–16.

Rutledge, S.A., E.R. Williams, and W.A. Petersen, 1993: Lightning and electrical structure of mesoscale convective systems. *Atmos. Res., 29,* 27–53.

Rutledge, S.A., and W.A. Petersen, 1994: Vertical radar reflectivity structure an cloud-to-ground lightning in the stratiform region of MCS's: Further evidence for in-situ charging in the stratiform region. *Mon. Wea. Rev., 122,* 1760–76.

Samsury, C.E., and R.E. Orville, 1994: Cloud-to-ground lightning in tropical cyclones: A study of Hurricanes Hugo (1989) and Jerry (1989). *Mon. Wea. Rev., 122,* 1887–96.

Sanders, F., and K.A. Emmanuel, 1977: The momentum budget and temporal evolution of a mesoscale convective system. *J. Atmos. Sci., 34,* 322–30.

Sartor, D., 1964: Radio observations of the electromagnetic emission from warm clouds. *Science, 143,* 948–50.

Saunders, C.P.R., and E.R. Jayaratne, 1986: Thunderstorm charge transfer values. *Preprints,* 23rd Conf. Radar Meteorology and Conf. Cloud Physics, Amer. Meteor. Soc., Boston, 260–63.

Schuur, T.J., B.F. Smull, W.D. Rust, and T.C. Marshall, 1991: Electrical and kinematic structure of the stratiform precipitation region trailing an Oklahoma squall line. *J. Atmos. Sci., 48,* 825–41.

Scouten, D.C., D.T. Stephenson, and W.G. Biggs, 1972: A sferic rate azimuth-profile of the 1955 Blackwell, Oklahoma, tornado. *J. Atmos. Sci., 29,* 929–36.

Seimon, A., 1993: Anomalous cloud-to-ground lightning in an F5-tornado-producing supercell thunderstorm on 28 August 1990. *Bull. Amer. Meteor. Soc., 74,* 189–203.

Shackford, C.R., 1960: Radar indications of a precipitation-lightning relationship in New England thunderstorms. *J. Meteor., 17,* 15–19.

Shafer, M.A., 1990: *Cloud-to-ground Lightning in Relation to*

Digitized Radar Data in Severe Storms. Master's thesis, Univ. Oklahoma, 93 pp.

Shanmugam, K., and E.J. Pybus, 1971: A note on the electrical characteristics of locally severe storms. *Preprints*, 7th Conf. Severe Local Storms, Amer. Meteor. Soc., 86–90.

Shepherd, T.R., W.D. Rust, and T.C. Marshall, 1996: Electric fields and charges near 0°C in stratiform clouds. *Mon. Wea. Rev., 124*, 919–38.

Silberg, P.A., 1965: Passive electrical measurements from three Oklahoma tornados. *Proc. IEEE, 53*, 1197–1204.

Simpson, G.C., 1909: On the electricity of rain and snow. *Proc. Roy. Soc. Lond., A, 83*, 394–404.

Simpson, G.C., 1949: Atmospheric Electricity during Disturbed Weather, *Geophys. Memoirs No. 84*.

Simpson, Sir G., and G.D. Robinson, 1941: The distribution of electricity in thunderclouds, II, *Proc. Roy. Soc. Lond., A, 177*, 281–328.

Sivaramakrishnan, M.V., 1962: The origin of electricity carried by continuous, quiet type of rain in the tropics. *Ind. J. Meteor. Geophys., 13*, 196–208.

Smull, B.F., 1995: Convectively induced mesoscale weather systems in the tropical and warm-season midlatitude atmosphere. *Rev. Geoph., 33 Supplement, U.S. Nat. Rpt. to Intl. Union Geodesy. Geophys. 1987–1990*, Amer. Geophys. Union, 897–906.

Smull, B.F., and R.A. Houze, Jr., 1985: A midlatitude squall line with a trailing region of stratiform rain: radar and satellite observations. *Mon. Wea. Rev., 113*, 117–33.

Smull, B.F., and R.A. Houze, Jr., 1987: Dual-Doppler radar analysis of a midlatitude squall line with a trailing region of stratiform rain. *J. Atmos. Sci., 44*, 2128–48.

Sommeria, G., and J. Testud, 1984: COPT81: A field experiment designed for the study of dynamics and electrical activity of deep convection in continental tropical regions. *Bull. Amer. Meteor. Soc., 65*, 4–10.

Standler, R.B., and W.P. Winn, 1979: Effects of coronae on electric fields beneath thunderstorms. *Q. J. Roy. Meteor. Soc., 105*, 285–302.

Stanford, J.L., 1971: Polarization of 500 kHz electromagnetic noise from thunderstorms: A new interpretation of existing data. *J. Atmos. Sci., 28*, 116–19.

Stanford, J.L., M.A. Lind, and G.S. Takle, 1971: Electromagnetic noise studies of severe convective storms in Iowa: The 1970 storm season. *J. Atmos. Sci., 28*, 436–48.

Stolzenburg, M., 1990: Characteristics of the bipolar pattern of lightning locations observed in 1988 thunderstorms. *Bull. Amer. Meteor. Soc. 71*, 1331–38.

Stolzenburg, M., 1994: Observation of high ground flash densities of positive lightning in summertime thunderstorms. *Mon. Wea. Rev., 122*, 1740–50.

Stolzenburg, M., 1996: An observational study of electrical structure in convective regions of mesoscale convective systems. Ph.D. diss., Univ. Oklahoma, Norman, 137 pp.

Stolzenburg, M., W.D. Rust, and T.C. Marshall, 1994a: Electric field soundings through strong updrafts of severe, supercellular thunderstorms. *Eos, Trans. Amer. Geophys. Un., 75*, 100. Abstract.

Stolzenburg, M., T.C. Marshall, W.D. Rust, and B.F. Smull, 1994b: Horizontal distribution of electrical and meteorological conditions across the stratiform region of a mesoscale convective system. *Mon. Wea. Rev., 122*, 1777–97.

Stolzenburg, M., W.D. Rust, and T.C. Marshall, 1996: Electric field soundings through strong updrafts of supercell thunderstorms. *Preprints*, 18th Conf. Severe Local Storms, Amer. Meteor. Soc., 10.4, 437–41.

Takagi, N., T. Watanabe, I. Arima, T. Takeuti, M. Nakano, and H. Kinosita, 1986: An unusual summer thunderstorm in Japan. *Res. Lett. Atmos. Electr., 6*, 43–48.

Takahashi, T., 1965: Measurement of electric charge in the thundercloud by means of radiosonde. *J. Meteor. Soc. Jpn., 43*, 206–17.

Takahashi, T., 1972: Electric charge of small particles (1–40μ). *J. Atmos. Sci., 29*, 921–28.

Takahashi, T., 1975: Electric charge life cycle in warm clouds. *J. Atmos. Sci., 32*, 123–42.

Takahashi, T., 1978: Electrical properties of oceanic tropical clouds at Ponape, Micronesia. *Mon. Wea. Rev., 106*, 1598–1612.

Takahashi, T., 1983: Electric structure of oceanic tropical clouds and charge separation processes. *J. Meteor. Soc.Jpn., 61*, 656–69.

Takahashi, T., R. Uchida, and C.M. Fullerton, 1969: Surface measurements of the electrical properties of warm clouds over the sea. *Nature, 224*, 1013–14.

Takahashi, T., and C.M. Fullerton, 1972: Raindrop charge-size measurements in warm rain. *J. Geophys. Res., 77*, 1630–36.

Takahashi, T., and T. Craig, 1973. Charge-size measurement of drizzle drops in warm cloud. *J. Meteor. Soc., Jpn., 51*, 191–96.

Takeuti, T., M. Nakano, M. Nagatani, and H. Nakada, 1973: On lightning discharges in winter thunderstorms. *J. Meteor. Soc. Jpn., 51*, 494–96.

Takeuti, T., and M. Nagatani, 1974: Oceanic thunderstorms in the tropical and subtropical Pacific. *J. Meteor. Soc. Jpn., 52*, 509–11.

Taylor, W.L., 1963: Radiation field characteristics of lightning discharges in the band 1 kc/s to 100 kc/s. *J. Res. NBS, 670*, 539–50.

Taylor, W.L., 1972: Atmospherics and severe storms. In *Remote Sensing of the Troposphere*, V.E. Derr, ed., available from the Superintendent of Documents, U.S. Gov't Printing Office, Washington, D.C. 20402.

Taylor, W.L., 1973: Electromagnetic radiation from severe storms in Oklahoma during April 29–30, 1970. *J. Geophys. Res., 78*, 8761–77.

Thompson, B.J., and R.H. Johnson, 1967: Tornadoes: Puzzling phenomena and photographs. *Science, 155*, 29–32. Letter.

Trost, T.F., and C.E. Nomikos, 1975: VHF radio emissions associated with tornadoes. *J. Geophys. Res., 80*, 4117–18.

Tucker, J.M., 1963: Unusual lightning. *Weather, 18*, 286.

Turman, B.N., and R.J. Tettelbach, 1980: Synoptic-scale satellite lightning observations in conjunction with tornadoes. *Mon. Wea. Rev., 108*, 1878–82.

Turner, D.B., 1969: *Workbook of Atmospheric Dispersion Estimates*, U.S. Dept. Health, Education, and Welfare, Consumer Protection and Environmental Health Service, National Air Pollution Control Admin., Cincinnati, Pub. No. 999-AP-26, 84 pp.

Vonnegut, B., and C.B. Moore, 1957: Electrical activity associated with the Blackwell-Udall tornado. *J. Meteor., 14*, 284–85.

Vonnegut, B., and C.B. Moore, 1958: Giant Electrical Storms. In *Recent Advances in Atmospheric Electricity*, L.G. Smith, ed., Pergamon, New York, pp. 399–410.

Vonnegut, B., and J.R. Weyer, 1966: Luminous phenomena in nocturnal tornadoes. *Science, 153*, 1213–20.

Vonnegut, B., O.H. Vaughan, Jr., M. Brook, and P. Krehbiel, 1985: Mesoscale observations of lightning from space shuttle. *Bull. Amer. Meteor. Soc., 66*, 20–29.

Ward, N.B., C.H. Meeks, and E. Kessler, 1965: Sferics reception at 500 kc/sec, radar echoes, and severe weather. In *Papers on Weather Radar, Atmospheric Turbulence, Sferics, and Data Processing,* U.S. Dept. of Commerce, Technical Note 3-NSSL-24, available from National Technical Information Service, Springfield, VA, pp. 39–71.

Watkins, D.C., J.D. Cobine, and B. Vonnegut, 1978: Electric discharges inside tornadoes. *Science, 199,* 171–74.

Watson, A.I., J.G. Meitin, and J.B. Cunning, 1988: Evolution of the kinetic structure and precipitation characteristics of a mesoscale convective system on 20 May 1979. *Mon. Wea. Rev., 116,* 1555–67.

Weisman, M.L., and J.B. Klemp, 1982: The dependence of numerically simulated convective storms on vertical wind shear and buoyancy. *Mon. Wea. Rev., 110,* 504–20.

Williams, E.R., C.M. Cooke, and K.A. Wright, 1985: Electrical discharge propagation in and around space charge clouds. *J. Geophys. Res., 90,* 6059–70.

Williams, E.R., M.E. Weber, and R.E. Orville, 1989a: The relationship between lightning type and convective state of thunderclouds. *J. Geophys. Res., 94,* 13,213–20.

Williams, E.R., M.E. Weber, and C.D. Engholm, 1989b: Electrical characteristics of microburst-producing storms in Denver. *Preprints,* 24th Conference Radar Meteorology, Amer. Meteor. Soc., 89–92.

Williams, E.R., S.A. Rutledge, S.G. Geotis, N. Renno, E. Rasmussen, and T. Rickenbach, 1992: A radar and electrical study of tropical hot towers. *J. Atmos. Sci., 49,*1386–95.

Williams, E.R., R. Zhang, and D. Boccippio, 1994: Microphysical growth state of ice particles and large-scale electrical structure of clouds. *J. Geophys. Res., 99,* 10,787–92.

Willis, P.T., and A.J. Heymsfield, 1989: Structure of the melting layer in mesoscale convective system stratiform precipitation. *J. Atmos. Sci., 46,* 2008–25.

Ziegler, C.L., D.R. MacGorman, P.S. Ray, and J.E. Dye, 1991: A model evaluation of non-inductive graupel-ice charging in the early electrification of a mountain thunderstorm. *J. Geophys. Res., 96,* 12,833–55.

Ziegler, C.L., and D.R. MacGorman, 1994: Observed lightning morphology relative to modeled space charge and electric field distributions in a tornadic storm. *J. Atmos. Sci., 51,* 833–51.

Zipser, E.J., 1977: Mesoscale and convective-scale downdrafts as distinct components of squall-line circulation. *Mon. Wea. Rev., 105,* 1568–89.

Zipser, E.J., 1982: Use of a conceptual model of the life-cycle of mesoscale convective systems to improve very-short-range forecasts. In *Nowcasting,* K.A. Browning, ed., Academic Press, London, pp. 191–204.

Zipser, E.J., 1994: Deep cumulus cloud systems in the tropics with and without lightning. *Mon. Wea. Rev., 122,* 1837–51.

Zipser, E.J., and K.R. Lutz, 1994: The vertical profile of radar reflectivity of convective cells: A strong indicator of storm intensity and lightning probability? *Mon. Wea. Rev., 122,* 1751–59.

CHAPTER 9

Anderson, R.V., 1977: Atmospheric electricity in the real world. In *Electrical Processes in Atmospheres,* H. Dolezalek and R. Reiter, eds., Dr. Dietrich Steinkopff, Darmstadt, pp. 87–99.

Anderson, R.V., and E.M. Trent, 1966: Evaluation of the Use of Atmospheric Electricity Recordings in Fog Forecasting. *Naval Res. Lab. Report No. 6426.*

Aufdermauer, A.N., and D.A. Johnson, 1972: Charge separation due to riming in an electric field. *Quart. J. Roy. Meteor. Soc., 98,* 369–82.

Baker, B., M.B. Baker, E.R. Jayaratne, J. Latham, and C.P.R. Saunders, 1987: The influence of diffusional growth rates on the charge transfer accompanying rebounding collisions between ice crystals and hailstones. *Q. J. Roy. Meteor. Soc., 113,*1193–1215.

Baker, M.B., H.J. Christian, and J. Latham, 1995: A computational study of the relationships linking lightning frequency and other thundercloud parameters. *Quart. J. Roy. Meteor. Soc., 121,* 1525–48.

Brooks, I.M., and C.P.R. Saunders, 1994: An experimental investigation of the inductive mechanism of thunderstorm electrification. *J. Geophys. Res., 99,* 10627–32.

Brooks, I.M., C.P.R. Saunders, R.P. Mitzeva, and S.L. Peck, 1997: The effect on thunderstorm charging of the rate of rime accretion by graupel. *Atmos Res., 43,* 277–95.

Browning, G.L., I. Tzur, and R.G. Roble, 1987: A global time-dependent model of thunderstorm electricity, I. Mathematical properties of the physical and numerical models. *J. Atmos. Sci., 44,* 2166–77.

Canosa, E.R., and R. List, 1993: Measurements of inductive charges during drop breakup in horizontal electric fields. *J. Geophys. Res., 98,* 2619–26.

Canosa, E.F., R. List, and R.E. Stewart, 1993: Modeling of inductive charge separation in rainshafts with variable vertical electric fields. *J. Geophys. Res., 98,* 2627–33.

Chiu, C.S., 1978: Numerical study of cloud electrification in an axisymmetric time-dependent cloud model. *J. Geophys. Res., 83,* 5025–49.

Chiu, C.S., and J.N. Klett, 1976: Convective electrification of clouds. *J. Geophys. Res, 81,* 1111–24.

Cotton, W.R., and R.A. Anthes, 1989: *Storm and Cloud Dynamics.* Academic Press, San Diego, 883 pp.

Dolezalek, H., 1963: The atmospheric electric fog effect. *Rev. of Geophys., 1,* 231–82.

Driscoll, K.T., R.J. Blakeslee, and M.E. Baginski, 1992: A modeling study of the time-averaged electric current in the vicinity of isolated thunderstorms. *J. Geophys. Res., 97,* 11535–51.

Gardiner, B., D. Lamb, R.L. Pitter, J. Hallett, and C.P.R. Saunders, 1985: Measurements of initial potential gradient and particle charges in a Montana summer thunderstorm. *J. Geophys. Res., 90,* 6079–86.

Gunn, R., 1948: Electric field intensity inside of natural clouds. *J. Appl. Phys., 19,* 481–84.

Gunn, R., 1954: Diffusion charging of atmospheric droplets by ions and the resulting combination coefficients. *J. Meteor., 11,* 339–47.

Hager, W.W., J.S. Nisbet, J.R. Kasha, and W. Shann, 1989: Simulations of electric fields within a thunderstorm. *J. Atmos. Sci., 46,* 3542–58.

Haltiner, G.J., and R.T. Williams, 1980: *Numerical Prediction and Dynamic Meteorology.* John Wiley and Sons, New York, 477 pp.

Helsdon, J.H., Jr., 1980: Chaff seeding effects in a dynamical electrical cloud model. *J. Appl. Meteor., 19,* 1101–25.

Helsdon, J.H., Jr., and R.D. Farley, 1987: A numerical modeling study of a Montana thunderstorm: 2. Model results versus observations involving electrical aspects. *J. Geophys. Res., 92,* 5661–75.

Helsdon, J.H., Jr., G. Wu, and R.D. Farley, 1992: An intracloud lightning parameterization scheme for a storm electrification model. *J. Geophys. Res., 97,* 5865–84.

Holton, J.R., 1992: *An Introduction to Dynamic Meteorology.* Academic Press, San Diego, 511 pp.

Houze, Jr., R.A., 1993: *Cloud Dynamics.* Academic Press, San Diego, 573 pp.

Illingworth, A.J., and J. Latham, 1977: Calculations of electric field growth, field structure, and charge distributions in thunderstorms. *Q. J. Roy. Meteor. Soc., 103,* 277–98.

Jayaratne, E.R., C.P.R. Saunders, and J. Hallett, 1983: Laboratory studies of the charging of soft hail during ice crystal interactions. *Quart. J. Roy. Meteor. Soc., 109,* 609–30.

Kasemir, H. W., 1960: A contribution to the electrostatic theory of a lightning discharge. *J. Geophys. Res., 65,* 1873–78.

Kasemir, H.W., 1984: Theoretical and experimental determination of field, charge and current on an aircraft hit by natural and triggered lightning. *Preprints,* Intl. Aerosp. Ground Conf. Lightn. Static Electric., National Interagency Coordinating Group, Orlando, Fla.

Keith, W.D., and C.P.R. Saunders, 1989: Charge transfer during multiple large ice-crystal interactions with a riming target. *J. Geophys. Res., 94,* 13103–106.

Kessler, E., 1969: On the distribution and continuity of water substance in atmospheric circulations. *Meteor. Monogr., 10,* No. 32, 84 pp.

Keuttner, J.P., Z. Levin, and J.D. Sartor, 1981: Thunderstorm electrification—Inductive or non-inductive? *J. Atmos. Sci., 38,* 2470–84.

Latham, J., and B.J. Mason, 1962: Electrical charging of hail pellets in a polarizing electric field. *Proc. Roy. Soc. London, Ser. A, 266,* 387–401.

Levin, Z. 1976: A refined charge distribution in a stochastic electrical model of an infinite cloud. *J. Atmos. Sci., 33,* 1756–62.

Levin, Z., and I. Tzur, 1986: Models of the development of the electrical structure of clouds. In *The Earth's Electrical Environment,* National Acad. Press, Washington, D.C., pp. 131–45.

Lhermitte, R., and E. Williams, 1985: Thunderstorm electrification: A case study. *J. Geophys. Res., 90,* 6071–78.

Low, T.B., and R. List, 1982: Collision, coalescence and breakup of raindrops, I. Experimentally established coalescence efficiencies and fragment size distributions in breakup. *J. Atmos. Sci., 39,* 1591–1606.

MacGorman, D.R., A.A. Few, and T.L. Teer, 1981: Layered lightning activity. *J. Geophys. Res., 86,* 9900–10.

MacGorman, D.R., C.L. Ziegler, and J.M. Straka, 1996: Considering the complexity of thunderstorm electrification. *Preprints,* 10th Intl. Conf. on Atmos. Elec., Osaka, Jpn., 128–31.

Marshall, B.J.P., J. Latham, and C.P.R. Saunders, 1978: A laboratory study of charge transfer accompanying collision of ice crystals with a simulated hailstone. *Quart. J. Roy. Meteor. Soc., 104,* 163–78.

Marshall, J.S., and W.M. Palmer, 1948: The distribution of raindrops with size. *J. Meteor., 5,* 165–66.

Marshall, T.C., W.D. Rust, W.P. Winn, and K.E. Gilbert, 1989: Electrical structure in two thunderstorm anvil clouds. *J. Geophys. Res., 94,* 2171–81.

Marshall, T.C., M.P. McCarthy, and W.D. Rust, 1995: Electric field magnitudes and lightning initiation in thunderstorms. *J. Geophys. Res., 100,* 7097–7103.

Mason, B.J., 1972: The physics of thunderstorms. *Proc. R. Soc. London, A327,* 433–66.

Mason, B.J., 1988: The generation of electric charges and fields in thunderstorms. *Proc. Roy. Soc. London, A415,* 303–315.

Mathpal, K.C., and N.C. Varshneya, 1982: Riming electrification mechanism for charge generation within a thundercloud of finite dimensions. *Ann. de Géophys., 38,* 167–75.

Mazur, V., J.C. Gerlach, and W.D. Rust, 1984: Lightning flash density versus altitude and storm structure from observations with UHF- and S-band radars. *Geophys. Res. Lett., 11,* 61–64.

McTaggart-Cowan, J.D., and R. List, 1975: Collision and breakup of water drops at terminal velocity. *J. Atmos. Sci., 32,* 1401–11.

Mitzeva, R., and C.P.R. Saunders, 1990: Thunderstorm charging: Calculations of the effect of ice crystal size and graupel velocity. *J. Atmos. Terrest. Phys., 52,* 241–45.

Moore, C.B., 1975: Rebound limits on charge separation by falling precipitation. *J. Geophys. Res., 80,* 2658–62.

Nisbet, J.S., 1983: A dynamical model of thundercloud electric fields. *J. Atmos. Sci., 40,* 2855–73.

Nisbet, J.S., 1985a: Thundercloud current determination from measurements at the earth's surface. *J. Geophys. Res., 90,* 5840–56.

Nisbet, J.S., 1985b: Currents to the ionosphere from thunderstorm generators: A model study. *J. Geophys. Res., 90,* 9831–44.

Norville, K., M. Baker, and J. Latham, 1991: A numerical study of thunderstorm electrification: Model development and case study. *J. Geophys. Res., 96,* 7463–81.

Pringle, J.E., H.D. Orville, and T.D. Stechmann, 1973: Numerical simulation of atmospheric electricity effects in a cloud model. *J. Geophys. Res., 78,* 4508–14.

Pruppacher, H.R., and J.D. Klett, 1978: *Microphysics of Clouds and Precipitation.* D. Reidel Publishing, Dordrecht, 714 pp.

Randall, S.C., S.A. Rutledge, R.D. Farley, and J.H. Helsdon, Jr., 1994: A modeling study on the early electrical development of tropical convection: Continental and oceanic (monsoon) storms. *Mon. Wea. Rev., 122,* 1852–77.

Rawlins, F., 1982: A numerical study of thunderstorm electrification using a three dimensional model incorporating the ice phase. *Q. J. Roy. Meteor. Soc., 108,* 779–800.

Ruhnke, L.H., 1972: Atmospheric electron cloud modeling. *Meteor. Res., 25,* 38–41.

Rust, W.D., 1973: Electrical conditions near the bases of thunderclouds. Ph.D. diss., New Mexico Inst. of Min. and Tech, Socorro.

Rust, W.D., and C.B. Moore, 1974: Electrical conditions near the bases of thunderclouds over New Mexico, *Q. J. Roy. Meteor. Soc., 100,* 450–68.

Rutledge, S.A., and P.V. Hobbs, 1984: The mesoscale and microscale structure and organization of clouds and precipitation in midlatitude cyclones. XII: A diagnostic modeling study of precipitation development in narrow cold-frontal rainbands. *J. Atmos. Sci., 41,* 2949–72.

Rutledge, S.A., E.R. Williams, and T.D. Keenan, 1992: The Down Under Doppler and Electricity Experiment (DUNDEE): Overview and preliminary results. *Bull. Amer. Meteor. Soc., 73,* 3–16.

Sapkota, B.K., and N.C. Varshneya, 1988: Electrification of thundercloud by an entrainment mechanism. *Meteor. Atmos. Phys., 39,* 213–22.

Sartor, J.D., 1967: The role of particle interactions in the distribution of electricity in thunderstorms. *J. Atmos. Sci., 24,* 601–15.

Sartor, J.D., 1981: Induction charging of clouds. *J. Atmos. Sci., 38,* 218–20.

Saunders, C.P.R., W.D. Keith, and R.P. Mitzeva, 1991: The effect of liquid water on thunderstorm charging. *J. Geophys. Res., 96,* 11007–17.

Scott, W.D., and Z. Levin, 1975: A stochastic electrical model of an infinite cloud: Charge generation and precipitation development. *J. Atmos. Sci., 32,* 1814–28.

Shao, X.M., and P.R. Krehbiel, 1996: The spatial and temporal development of intracloud lightning. *J. Geophys. Res., 101,* 26,641–668.

Shreve, E.L., 1970: Theoretical derivation of atmospheric ion concentrations, conductivity, space charge density, electric field, and generation rate from 0 to 60 km. *J. Atmos. Sci., 27,* 1186–94.

Singh, P., T.S. Verma, and N.C. Varshneya, 1986: Some theoretical aspects of electric field and precipitation growth in a finite thundercloud. *Proc. Indian Acad. Sci, 95,* 293–98.

Solomon, R., and M. Baker, 1994: Electrification of New Mexico thunderstorms. *Mon. Wea. Rev., 122,* 1878–88.

Solomon, R., and M. Baker, 1996: A one-dimensional lightning parameterization. *J. Geophys. Res., 101,* 14,983–90.

Stansbery, E.K., A.A. Few, and P.B. Geis, 1993: A global model of thunderstorm electricity. *J. Geophys. Res., 98,* 16591–16603.

Straka, J.M., 1989: *Hail Growth in a Highly Glaciated Central High Plains Multi-cellular Hailstorm.* Ph.D. diss., Univ. Wisconsin, Madison, 413 pp.

Takahashi, T., 1974: Numerical simulation of warm cloud electricity. *J. Atmos. Sci., 31,* 2160–81.

Takahashi, T., 1978: Riming electrification as a charging generation mechanism in thunderstorms. *J. Atmos. Sci., 35,* 1536–48.

Takahashi, T., 1979: Warm cloud electricity in a shallow axisymmetric cloud model. *J. Atmos. Sci., 36,* 2236–58.

Takahashi, T., 1983: A numerical simulation of winter cumulus electrification. Part 1: Shallow cloud. *J. Atmos. Sci., 40,* 1257–80.

Takahashi, T., 1984: Thunderstorm electrification—A numerical study. *J. Atmos. Sci., 41,* 2541–58.

Takahashi, T., 1987: Determination of lightning origins in a thunderstorm model. *J. Meteor. Soc. Jpn, 65,* 777–94.

Taylor, W.L., E.A. Brandes, W.D. Rust, and D.R. MacGorman, 1984: Lightning activity and severe storm structure. *Geophys. Res. Lett, 11,* 545–48.

Tzur, I., and Z. Levin, 1981: Ions and precipitation charging in warm and cold clouds as simulated in a one-dimensional, time-dependent model. *J. Atmos. Sci., 38,* 2444–61.

Tzur, I., and R.G. Roble, 1985: The interaction of a dipolar thunderstorm with its global electrical environment. *J. Geophys. Res., 90,* 5989–99.

Vonnegut, B., 1953: Possible mechanism for the formation of thunderstorm electricity. *Bull. Amer. Meteor. Soc., 34,* 378–81.

Wagner, P.G., and J.W. Telford, 1981: Cloud dynamics and an electric charge separation mechanism in convective clouds. *J. Rech. Atmos., 15,* 97–120.

Weinheimer, A.J., 1987: The electrostatic energy of a thunderstorm and its rate of change. *J. Geophys. Res., 92,* 9715–22.

Whipple, F.J.W., and J.A. Chalmers, 1944: On Wilson's theory of the collection of charge by falling drops. *Quart. J. Roy. Meteor. Soc., 70,* 103–19.

Williams, E.R., C.M. Cooke, and K.A. Wright, 1985: Electrical discharge propagation in and around space charge clouds. *J. Geophys. Res., 90,* 6054–70.

Williams, E.R., S.A. Rutledge, S.G. Geotis, N. Renno, E. Rasmussen, and T. Rickenbach, 1992: A radar and electrical study of tropical hot towers. *J. Atmos. Sci., 49,* 1386–95.

Wilson, C.T.R., 1929: Some thundercloud problems. *J. Franklin Inst. 208,* 1–12.

Winn, W.P., and L.G. Byerley, III, 1975: Electric field growth in thunderclouds, *Quart. J. Roy. Meteor. Soc., 101,* 979–94.

Wisner, C., H.D. Orville, and C. Meyers, 1972: A numerical model of a hail-bearing cloud. *J. Atmos. Sci., 29,* 1160–81.

Wojcik, W.A., 1994: *An Examination of Thunderstorm Charging Mechanisms Using the IAS 2D Storm Electrification Model.* Master's thesis, So. Dakota Schl. Mines Technol., Rapid City, 113 pp.

Workman, E.J., R.E. Holzer, and G.T. Pelsor, 1942: *The Electrical Structure of Thunderstorms.* Technical Notes of Natural Advisory Committee for Aeronautics, No. 0864, 47 pp.

Ziegler, C.L., 1985: Retrieval of thermal and microphysical variables in observed convective storms: Part 1. Model development and preliminary testing. *J. Atmos. Sci., 42,* 1487–1509.

Ziegler, C.L., 1988: Retrieval of thermal and microphysical variables in observed convective storms: Part II. Sensitivity of cloud processes to variation of the microphysical parameterization. *J. Atmos. Sci., 45,* 1072–90.

Ziegler, C.L., P.S. Ray, and D.R. MacGorman, 1986: Relations of kinematics, microphysics and electrification in an isolated mountain thunderstorm. *J. Atmos. Sci., 43,* 2098–2114.

Ziegler, C.L., D.R. MacGorman, J.E. Dye, and P.S. Ray, 1991: A model evaluation of non-inductive graupel-ice charging in the early electrification of a mountain thunderstorm. *J. Geophys. Res., 96,* 12,833–55.

Ziegler, C.L., and D.R. MacGorman, 1994: Observed lightning morphology relative to modeled space charge and electric field distributions in a tornadic storm. *J. Atmos. Sci., 51,* 833–51.

Ziv, A., and Z. Levin, 1974: Thundercloud electrification: Cloud growth and electrical development. *J. Atmos. Sci., 31,* 1652–61.

CHAPTER 10

Abbott, C.E., 1975: Charged droplet collision efficiency measurements. *J. Appl. Meteor., 14,* 87–90.

Barker, E., J.A. Bicknell, R.F. Griffiths, J. Latham, and T.S. Verma, 1983: The scavenging of particles by electrified drops: Radar and echo intensification following lightning. *Q. J. Roy. Meteor. Soc., 109,* 631–44.

Barlow, A.K., and J. Latham, 1983: A laboratory study of the scavenging of sub-micron aerosol by charged raindrops. *Q. J. Roy. Meteor. Soc., 109,* 763–70.

Beard, K.V., 1980: The effects of altitude and electrical forces on terminal velocity of hydrometeors. *J. Atmos. Sci., 37,* 1363–74.

Beard, K.V., 1987: Cloud and precipitation physics research 1983–86. In *Review of Geophysics 25, U.S. Natl. Rpt. to Intl. Un. Geod. Geophys., 1983–1986.* pp. 357–70.

Beard, K.V., J.Q. Feng, C. Chung, 1989: A simple perturbation model for the electrostatic shape of falling drops. *J. Atmos. Sci., 46,* 2404–18.

Borucki, W.J., and W.L. Chameides, 1984: Lightning: Estimates of the rates of energy dissipation and nitrogen fixation. *Rev. Geophys. 22,* 363–72.

Brazier-Smith, P.R., S.G. Jennings, and J. Latham, 1972: The interaction of falling water drops: coalescence. *Proc. Roy. Soc. Lond., A, 326,* 393–408.

Brazier-Smith, P.R., S.G. Jennings, and J. Latham, 1973: Increased rates of rainfall production in electrified clouds. *Q. J. Roy. Meteor. Soc., 99,* 776–79.

Chameides, W.L., 1986: The role of lightning in the chemistry of the atmosphere. In *The Earth's Electrical Environment,* National Acad. Press, Washington, D.C. pp. 70–77.

Chameides, W.L., D.D. Davis, J. Bradshaw, M. Rodgers, S. Sandholm, and D.B. Bai, 1987: An estimate of the NO_x production rate in electrified clouds based on NO observations from the GTE/CITE 1 Fall 1983 field operation. *J. Geophys. Res., 92,* 2153–56.

Chuang, C.C., and K.B. Beard, 1990: A numerical model for the equilibrium shape of electrified raindrops. *J. Atmos. Sci., 47,* 1374–89.

Cohen, A-H., and I. Gallily, 1977: On the collision efficiency and the coalescence of water droplets under the influence of electric forces II: Calculations, small Reynolds numbers. *J. Atmos. Sci., 34,* 827–42.

Colgate, S.A., and J.M. Romero, 1970: Charge versus drop size in an electrified cloud. *J. Geophys. Res., 75,* 5873–81.

Cooper, W.A., 1991: Research in cloud and precipitation physics: Review of U.S. theoretical and observational studies, 1987–1990. In *Review of Geophysics 25, U.S. Natl. Rpt. to Intl. Un. Geod. Geophys., 1987–1990* pp. 69–79.

Coquillat, S., and S. Chauzy, 1993: Behavior of precipitating water drops under the influence of electrical and aerodynamical forces. *J. Geophys. Res., 98,* 10,319–24.

Crowther, A.G., and C.P.R. Saunders, 1973: On the aggregation and fragmentation of freely-falling ice crystals in an electric field. *J. Meteor. Soc. Jpn., 51,* 490–93.

Czys, R.R., and H.T. Ochs, III, 1988: The influence of charge on the coalescence of water drops in free fall. *J. Atmos. Sci., 45,* 3161–68.

Davis, M.H., and J.D. Sartor, 1967: Fast growth rates for cloud droplets due to electrical forces. In *Planetary Electrodynamics,* S.C. Coronity and J. Hughes, eds., Gordon and Breach, New York, pp. 339–44.

Dawson, G.A., and C.N. Richards, 1970: Discussion of paper by J. Latham and V. Myers, 'Loss of charge and mass from raindrops falling in intense electric fields.' *J. Geophys. Res., 75,* 4589–92.

Dawson, G.A., and G.R. Cardell, 1973: Electrofreezing of supercooled water drops. *J. Geophys. Res., 78,* 8864–66.

Dawson, G.A., and R.A. Warrender, 1973: The terminal velocity of raindrops under vertical electric stress. *J. Geophys. Res., 78,* 3619–20.

Dickerson, R.R., 1987: Thunderstorms: An important mechanism in the transport of air pollutants. *Science, 235,* 460–465.

Doolittle, J.B., and G. Vali, 1975: Heterogeneous freezing nucleation in electric fields. *J. Atmos. Sci., 32,* 375–79.

Doviak, R.J., and D.S. Zrnić, 1993: *Doppler Radar and Weather Observations,* 2d ed., Academic Press, San Diego, 562 pp.

Drapcho, D.I., D. Sisterson, and R. Kumar, 1983: Nitrogen fixation by lightning activity in a thunderstorm. *Atmos. Environ., 17,* 729–34.

Evans, L.F., 1973: The growth and fragmentation of ice crystals in an electric field. *J. Atmos. Sci., 30,* 1657–64.

Finnegan, W.G., and R.L. Pitter, 1988: A postulate of electric multipoles in growing ice crystals: Their role in the formation of ice crystal aggregates. *Atmos. Res., 22,* 235–50.

Foster, M.P., and J.C. Pflaum, 1988: The behavior of cloud droplets in an acoustic field: a numerical investigation. *J. Geophys. Res. 93,* 747–58.

Franzblau, E., and C.J. Popp, 1989: Nitrogen oxides produced from lightning. *J. Geophys. Res., 94,* 11089–104.

Freire, E., and R. List, 1979: Collision enhancement for droplet pairs with electrically reduced approach speed. *J. Atmos. Sci., 36,* 1777–86.

Gay, M.J., R.F. Griffiths, J. Latham, C.P.R. Saunders, 1974: The terminal velocity of charged raindrops and cloud droplets falling in strong electric fields. *Q. J. Roy. Meteor. Soc., 100,* 682–87.

Georgis, J-F., 1996: Comportement et interactions électriques des gouttes précipitantes dans un nuage d'orage. (Behavior and electrical interactions of precipitating water drops in a thundercloud.) Ph.D. diss., Doctorat de l'Université Paul Sabatier, Toulouse, France.

Georgis, J-F., S. Coquillat, and S. Chauzy, 1995: Modelling of interaction processes between two raindrops in an electrical environment. *Q. J. Roy. Meteor. Soc., 121,* 745–61.

Goldenbaum, G.C., and R.R. Dickerson, 1993: Nitric oxide production by lightning discharges. *J. Geophys. Res., 98,* 18333–38.

Goyer, G.G., 1965: Mechanical effects of simulated lightning discharge on the water droplets of 'Old Faithful' geyser. *Nature, 206,* 1302–04.

Goyer, G.I., I McDonald, F. Baer, and R. Braham, Jr., 1960: Effects of electric field on water droplet coalescence. *J. Meteor., 17,* 442–45.

Goyer, G.G., and M.N. Plooster, 1968: On the role of shock waves and adiabatic cooling in the nucleation of ice crystals by the lightning discharge. *J. Atmos. Sci., 25,* 857–62.

Green, A.W., 1975: An approximation for the shapes of large raindrops. *J. Appl. Meteor. 14,* 1578–83.

Grover, S.N., 1976a: The electrostatic force between a charged conducting sphere and a charged dielectric sphere in an arbitrarily oriented external field. *Pageoph, 114,* 521–39.

Grover, S.N., 1976b: A numerical determination of the effect of electric fields and charges on the efficiency with which cloud drops and small raindrops collide with aerosol particles. *Pure Appl. Geophys., 114,* 509–20.

Grover, S.N., H.R. Pruppacher, and A. E. Hamielec, 1977: A numerical determination of the efficiency with which spherical aerosol particles collide with spherical water drops due to inertial impaction and phoretic and electric forces. *J. Atmos. Sci., 34,* 1655–63.

Gunn, R., and G.D. Kinzer, 1949: The terminal velocity of fall for water droplets in stagnant air. *J. Meteor., 6,* 243–48.

Hendry, A., and Y.M.M. Antar, 1982: Radar observations of polarization characteristics and lightning induced realignment of atmospheric crystals. *Rad. Sci., 17,* 1243–50.

Hocking, L.M., 1959: The collision efficiency of small drops. *Q. J. Roy. Meteor. Soc., 85,* 44–50.

Jayaratne, E.R., and C.P.R. Saunders, 1984: The "rain-gush," lightning, and the lower positive charge center in thunderstorms. *J. Geophys. Res., 89,* 11,816–18.

Jayaratne, E.R., and C.P.R. Saunders, 1985: Reply. *J. Geophys. Res., 90,* 10,755.

Kamra, AK, R.V. Bhalwankar, and A.B. Sathe, 1991: Spontaneous breakup of charged and uncharged water drops freely suspended in a wind tunnel. *J. Geophys, Res., 96,* 17,159–68.

Kamra, A.K., R.V. Bhalwankar, and A.B. Sathe, 1993: The onset of disintegration and corona in water drops falling at terminal velocity in horizontal electric fields. *J. Geophys. Res., 98,* 12,901–12.

Ko, M.K.W., M.B. McElroy, D.K. Weisenstein, and N.D. Sze, 1986: Lightning: A possible source of stratospheric odd nitrogen. *J. Geophys. Res., 91,* 5395–5404.

Komabayasi, M., T. Gonda, and K. Isono, 1964: Life time of water drops before breaking and size distribution of fragment drops. *J. Meteor. Soc. Jpn., 42,* 330–40.

Krasnogorskaya, N.V., 1965: Effect of electrical forces on the coalescence of particles of comparable sizes. *Izv., Acad. Sci., USSR, Atmos. Oceanic Phys., 1,* 339–35. (In Russian, translated by J.S. Sweet).

Krehbiel, P.R., T. Chen, S. McCrary, W. Rison, T. Blackman, and M. Brook, 1992: Dual-polarization radar signatures of the potential for lightning in electrified storms. *Proc. 9th Intl. Conf. Atmos. Elec.,* June 15–19, St. Petersburg, Russia, pp. 166–69.

Krehbiel, P., T. Chen, S. McCrary, W. Rison, G. Gray, and M. Brook 1996: The use of dual channel circular-polarization radar observations for remotely sensing storm electrification. *Meteor. Atmos. Phys., 59,* 65–82.

Latham, J., 1969a: Influence of electrical forces on cloud physical phenomenon. In *Planetary Electrodynamics, Vol., 1,* S.C. Coroniti and J. Hughes, eds., Gordon and Breach, New York, pp. 359–72.

Latham, J., 1969b: Experimental studies of the effect of electric fields on the growth of cloud particles. *Q. J. Roy. Meteor. Soc., 95,* 349–61.

Latham, J., 1977: Some electrical effects in clouds. In *Electrical Processes in Atmospheres,* H. Dolezalek and R. Reiter, eds., Dr. Dietrich Steinkopff Verlag, Darmstadt, pp. 263–71.

Latham, J., and C.P.R. Saunders, 1964: Aggregation of ice crystals in strong electric fields. *Nature, 244,* 1293–94.

Latham, J., and I.W. Roxburgh, 1966: Disintegration of pairs of water drops in an electric field. *Proc. Roy. Soc. Lond., A295,* 84–97.

Latham, J., and C.P.R. Saunders, 1970: Experimental measurements of the collection efficiencies of ice crystals in electric fields. *Q. J. Roy. Meteor. Soc., 96,* 257–265.

Levy, II, H., W.J. Moxim, and P.S. Kasibhatla, 1996: A global three-dimensional time-dependent lightning source of tropospheric NO_x. *J. Geophys. Res., 101,* 22911–22.

Liaw, Y.P., D.L. Sisterson, and N.L. Miller, 1990: Comparison of field, laboratory, and theoretical estimates of global nitrogen fixation by lightning. *J. Geophys. Res., 95,* 22489–94.

Logan, J.A. 1983: Nitrogen oxides in the troposphere: Global and regional budgets. *J. Geophys. Res., 88,* 10785–807.

Marshall, T.C., and W.P. Winn, 1985: Comments on "The 'rain-gush', lightning, and the lower positive center in thunderstorms" by E.R. Jayaratne and C.P.R. Saunders. *J. Geophys. Res., 90,* 10,753–754.

Metcalf, J.I., 1993: Observation of the effects of changing electric fields on the orientation of hydrometeors in a thunderstorm. *Bull. Amer. Meteor. Soc., 74,* 1080–83.

Moore, C.B., B. Vonnegut, J.A. Machado, and H.J. Survilas, 1962: Radar observations of rain gushes following overhead lightning stokes. *J. Geophys. Res., 67,* 207–20.

Moore, C.B., B. Vonnegut, E.A. Vrablik, and D.A. McCaig, 1964: Gushes of rain and hail after lightning. *J. Atmos. Sci., 21,* 646–1565.

Moore, C.B., and B. Vonnegut, 1973: Reply (to Brazier-Smith et al.). *Q. J. Roy. Meteor. Soc., 99,* 779–86.

Murphy, D.M., D.W. Fahey, M.H. Proffitt, S.C. Liu, K.R. Chan, C.S. Eubank, S.R. Kawa, and K.K. Kelly, 1993:

Reactive nitrogen and its correlation with ozone in the lower stratosphere and upper troposphere. *J. Geophys. Res., 98,* 8751–73.

Noxon, J.G., 1976: Atmospheric nitrogen fixation by lightning. *Geophys. Res. Lett., 3,* 463–65.

Noxon, J.F., 1978: Tropospheric NO_2. *J. Geophys. Res., 83,* 3051–57.

Ochs III, H.T., R.R. Czys, and K.V. Beard, 1986: Laboratory measurements of coalescence efficiencies for small precipitation drops. *J. Atmos. Sci., 43,* 225–32.

Ochs III, H.T., and R.R. Czys, 1987: Charge effects on the coalescence of water drops in free fall. *Nature, 327,* 606–08.

Ochs, III, H.T., D.E. Schaufelberger, and J.Q. Feng, 1991: Improved coalescence efficiency measurements for small precipitation drops. *J. Atmos. Sci., 48,* 946–51.

Paluch, I.R., 1970: Theoretical collision efficiencies of charged cloud droplets. *J. Geophys. Res., 75,* 1633–40.

Pitter, R.L., 1977: Scavenging efficiency of electrostatically charged thin ice plates and spherical aerosol particles. *J. Atmos. Sci., 34,* 1797–1800.

Plumlee, H.R., and R.G. Semonin, 1965: Cloud droplet collision efficiency in electric fields. *Tellus, XVII,* 356–64.

Pruppacher, H.R., 1963a: The effect of an external electric field on the supercooling of water drops. *J. Geophys. Res., 68,* 4463–74.

Pruppacher, H.R., 1963b: The effects of electric fields on cloud physical processes. *ZAMP, 14,* 590–98.

Pruppacher, H.R., 1973: Electrofreezing of supercooled water. *Pageoph, 104,* 623–34.

Pruppacher, H.R., and J.D. Klett, 1978: *Microphysics of Clouds and Precipitation.* D. Reidel Pub. Dordrecht, 714 pp.

Rasmussen, R., C. Walcek, H.R. Pruppacher, S.K. Mitra, J. Lew, V. Levinzzani, P.K. Wang, and U. Barth, 1985: A wind tunnel investigation of the effect of an external vertical electric field on the shape of electrically uncharged rain drops. *J. Atmos. Sci., 42,* 1647–52.

Rayleigh, Lord, 1879: The influence of electricity on colliding water drops. *Proc. Roy. Soc., XXVIII,* 406–09. (also in *Scientific Papers by Lord Rayleigh, Vol., II,* Dover, New York, 1964, pp. 372–76.)

Rayleigh, Lord, 1882: On the equilibrium of liquid conducting masses charged with electricity. *Phil. Mag., XIV,* 184–186. (also in *Scientific Papers by Lord Rayleigh, Vol. II,* Dover, New York, 1964, pp. 130–31.

Richards, C.N., and G.A. Dawson, 1971: The hydrodynamic instability of water drops falling at terminal velocity in vertical electric fields. *J. Geophys. Res., 76,* 3445–55.

Ridley, B.A., J.E. Dye, J.G. Walega, J. Zheng, F.E. Grahek, and W. Rison, 1996: On the production of active nitrogen by thunderstorms over New Mexico. *J. Geophys. Res., 101,* 20985–21005.

Sartor, D., 1954: A laboratory investigation of collision efficiencies, coalescence and electrical charging of simulated cloud droplets. *J. Meteor., 11,* 91–103.

Sartor, D., 1960: Some electrostatic cloud-droplet collision efficiencies. *J. Geophys. Res., 65,* 1953–57.

Sartor, D., 1970: Accretion rates of cloud drops, raindrops, and small hail in mature thunderstorms. *J. Geophys. Res., 75,* 7547–58.

Singh, H.B., D. Herlth, R. Kolyer, L. Salas, J.D. Bradshaw, S.T. Sandholm, D.D. Davis, J. Crawford, Y. Kondo, M. Koike, R. Talbot, G.L. Gregory, G.W. Sachse, E. Browell, D.R. Blake, F.S. Rowland, R. Newell, J. Merril, B. Heikes, S.C. Liu, P.J. Crutzen, M. Kanakidou,

1996: Reactive nitrogen and ozone over the western Pacific: Distribution, partitioning, and sources. *J. Geophys. Res., 101,* 1793–1808.

Smith, M.M., R.F. Griffiths, and J. Latham, 1971: The freezing of raindrops falling through strong electric fields. *Q. J. Roy. Meteor. Soc., 97,* 495–505.

Smyth, S., J. Bradshaw, S. Sandholm, S. Liu, S. McKeen, G. Gregory, B. Anderson, R. Talbot, D. Blake, S. Rowland, E. Browell, M. Fenn, J. Merrill, S. Bachmeier, G. Sachse, J. Collins, D. Thornton, D. Davis, and H. Singh, 1996: Comparison of free tropospheric western Pacific air mass classification schemes for the PEM-West A experiment. *J. Geophys. Res., 101,* 1743–62.

Szymanski, E.W., S.J. Szymanski, C.R. Holmes, and C.B. Moore, 1980: An observation of a precipitation echo intensification associated with lightning. *J. Geophys. Res., 85,* 1951–53.

Takahashi, T., 1973: Measurement of electric charge of cloud droplets, drizzle, and raindrops. *Rev. Geophys. Sp. Phys., 11,* 903–24.

Telford, J.W. and N.S.C. Thorndike, 1961: Observations of small drop collisions. *J. Meteor., 18,* 382–87.

Tinsley, B.A., and G.W. Deen, 1991: Apparent tropospheric response to MeV-GeV particle flux variations: A connection via electrofreezing of supercooled water in high-level clouds? *J. Geophys. Res., 96,* 22,283–96.

Tuck, A.F., 1976: Production of nitrogen oxides by lightning discharges *Quart. J. Roy. Meteor. Soc., 102,* 749–55.

Wang, P.K., S.N. Grover, and H.R. Pruppacher, 1978: On the effect of electric charges on the scavenging of aerosol particles by clouds and small raindrops. *J. Atmos. Sci., 35,* 1735–43.

Weinheimer, A.J., 1989: Chemical production by atmospheric electrical discharges: A review of field and laboratory observations. *Eos, Trans. AGU, 70,* 1009. Abstract.

Woods, J.D., 1965: The effect of electric charges upon collisions between equal-size water drops in air. *Q. J. Roy. Meteor. Soc., 91,* 353–55.

Young, K.C., 1993: *Microphysical Processes in Clouds.* Oxford Univ. Press, New York, 427 pp.

Zrnić, D., R.J. Doviak, and P.R. Mahapatra, 1984: The effect of charge and electric field on the shape of raindrops. *Radio Sci., 19,* 75–80.

Index